The biological chemistry of the elements

The inorganic chemistry of life

The biological chemistry of the elements

The inorganic chemistry of life

J. J. R. Fraústo da Silva
Professor of Analytical Chemistry, Instituto Superior Técnico, Universidade Técnico de Lisboa

and

R. J. P. Williams
Royal Society Napier Research Professor, University of Oxford

OXFORD
UNIVERSITY PRESS

OXFORD
UNIVERSITY PRESS

Oxford University Press is a department of the University of Oxford.
It furthers the University's objective of excellence in research, scholarship,
and education by publishing worldwide. Oxford is a registered trademark of
Oxford University Press in the UK and in certain other countries

Published in India by
Oxford University Press
22 Workspace, 2nd Floor, 1/22 Asaf Ali Road, New Delhi 110 002, India

First edition 1991
Second edition 2001

ISBN-13: 978-0-19-886258-1

Reprinted 2020
Printed in India by Replika Press Pvt. Ltd.

Preface to the Second Edition

Ten years have elapsed since we wrote this book, during which the subject of biological inorganic chemistry has developed very considerably. In the first edition, after introducing the principles of biological inorganic chemistry, we largely described the chemical properties and the biological functions of individual elements, referring mostly to their interactions with selected proteins. Much of the more recent work has been devoted to (1) the further structural and spectroscopic study of particular proteins and DNA involving several new specific metal ion or non-metal compounds, (2) the ways in which each metal ion is selected, carried and transferred within an organism, and (3) the detailed mechanisms of their action. The manner in which, after selective uptake, some elements equilibrate between many binding sites while others are inserted kinetically in few sites is now rather well appreciated. All this new knowledge has been readily incorporated within the existing chapters from 8-19 in Part II. Four aspects of progress concerning principles are more difficult to incorporate in a simple way and some larger changes have had to be made in all the chapters of Part I and Chapter 20 of the book. The changes are necessary since the new knowledge of genetic structure, even of whole genomes, throws new light upon the involvement of the chemical elements in organisms and especially upon their controlled use and the evolution of that use. To introduce these genetic aspects, much of Chapter 6 (previously Chapter 7) has been reorganized and sections on genes have been inserted there and into various chapters. A second change derives from the increasing knowledge concerning the mechanisms of uptake since kinetic controls have been shown to be of greater consequence than we considered earlier. Several elements are now known to be transferred by pumps and carriers, and then inserted into their final sites rather than being in equilibrated exchange. We have put much of this new material into Chapters 2 to 5, as well as into the chapters on selected elements. The third increase in knowledge concerns the networking of reactions of elements in cells, linked to control functions stemming from the coded information in DNA. This has required changes in Chapter 7 (previously Chapter 6) and Chapter 20. In our opinion it is not possible to understand a cellular *system* just from the study of isolated molecules and the relationship of these chemicals to the genetic structure. In addition we need to show why a given system, represented only in part by a particular genome, has greater survival value than others in given environmental conditions, and the explanation of this fact has to be given in terms of the *functional cooperative activities* of the chemical elements in the system. This requires knowledge of the quantities of elements and compounds in a cell, their location, their uptake and rejection and so on, none of which can be deduced from DNA. We shall refer therefore not just to the genome and the proteome as characteristics of cells or cell compartments, but also to the free metallome, which gives the free element concentrations in compartments and which is linked to the environmental availability of elements and energy. An introduction to this discussion is included in the new Chapter 7, after being mentioned in Chapter 1, but again it affects attitudes to the operational roles of all elements so that in each of the chapters concerning specific elements, Chapters 9-19, we have added a section on networks of interaction relevant to the particular element. Nearly all this new material together with that on genes adds to the sense of the total organization of activity in cells and the way this has evolved. The fourth addition relates to these evolutionary considerations, which are incorporated in short sections in Chapters 8–19 as well as in Chapter 20. We have also attempted to give a brief explanation of whole systems and their organization in this chapter, but for a more extensive discussion we refer the reader to our two more recent books in which we have looked at the problems concerning organization of element chemistry in more detail [R.J.P. Williams and J.J.R. Fraústo da Silva (1996). *The natural selection of the chemical elements-the environment and life's chemistry.* Oxford University Press, and R.J.P. Williams and J.J.R. Fraústo da Silva (1999). *Bringing chemistry to life-from matter to man.* Oxford University Press].

The amount of new material, and this includes an up-dated selection of references including especially recent books and reviews, forced us to make some sacrifices to keep the length of the book within reasonable limits. Thus, excluding minor cuts, we have simplified the contents of (the new) Chapters 6 and 7, omitted the discussion of models in Chapter 17, and left out entirely the former Chapter 20 (on biominerals) and Chapter 22 (man's use of chemical elements, especially in medicine). In both cases there are now recent books and extended monographies to which the reader is referred. Note that little detailed discussion of individual structures is included. There are two *Handbooks of Metalloproteins* to be published in 2001. One is edited by R. Huber, K. Wieghardt, T. Poulos, and A. Messerschmidt (Wiley, Chichester UK)) and the other is edited by I. Bertini, A. Sigel, and H. Sigel (Marcel Dekker, New York). Together they contain a vast assembly of structural data on more than 100 metalloproteins.

Lisbon and Oxford J.J.R.F. da S
2000 R.J.P.W.

Preface to the First Edition

The majority of present-day living organisms require rather strict environmental conditions: temperature from 0 to about 40°C, pressure of the order of 1 atm., salinity up to about 4 per cent, pH in the range 4 to 9, redox potential in the region of stability of water (-0.4 to + 0.8 V at pH = 7 versus the standard H^+/H_2 electrode), water activity between 0.7 and 1.0 mole fraction. In a few cases these limits can be greatly exceeded and it is known that, in the earth's past, conditions existed which were quite different from the present ones but did not hinder the development of life, although they did impose constraints upon the *forms* of life that could exist. Indeed, this is an essential feature of life itself: the continuous adaptation of living organisms to changing environmental conditions through the development of defence devices and mechanisms against those changes, while sometimes turning new (adverse) factors into favourable ones. The result was, and still is, the survival and predominant proliferation of the fittest, and the history of evolution, which started as development under constraints of fixed conditions, became the history of the continuous adaptation of living organisms to changing conditions. In fact, and as we shall see, organisms themselves have caused much of the chemical environmental change.

Although ideas concerning the evolution of species and forms are well accepted, they are less recognized when transferred to the small molecule chemistry scale, even though it has become customary to call 'chemical evolution' the series of events which is thought to have taken place between the beginnings of prebiotic synthesis of large organic molecules and the appearance of the first vesicles ('protocells') to which one might assign some 'vital' properties. Throughout the (unfinished) biological evolution there has been, in addition, a continual selection toward the best chemical components for the necessary functions and a continuous improvement in the conditions of their utilization, leading to the formation of separate compartments in cells, to the functional separation of cells, to the specialization of different multicellular organisms and to their diversification and development. As the species became more complex, so did their chemistry, as evidenced by the increased complexity of the materials used and the manipulation of dynamic factors. The development of materials is again well recognized in the evolution of biological polymers but it can only be thoroughly appreciated, we believe, in terms of the structures and patterns of the flow of elements and chemicals, of ionic and electronic charge, and of energy directed in biological space. The strict environmental conditions that are required today by living organisms also point to the necessity of strict internal control and regulation of these processes, and this again means time and space organization of the biological substances, both large and small, and of their reactions. Much of this control is included in the now well-known classical treatment of the organic chemistry of life but, diverging from the classical tradition, we shall regard it in this book as the interplay of not just the great number of small and very large organic molecules but also of a group of about 25 chemical elements, free as ions, combined as complexes, or as precipitates, organized in biological space and time. Most of these elements are classified as 'inorganic'. We shall go on to insist that a proper appreciation of living systems and their evolution cannot be obtained without reference to these elements.

Central to this group of elements is a series of metals, each one of them used in many ways within a given biological system. The position of each metal element in space and the variation of this position in time, as well as the exact chemical partners with which it interacts, have been selected and perfected by the demands of evolution seemingly in such a way that the element's function has become virtually optimal and its role separate and distinct. Hence, several such elements are essential for life. In order to fully appreciate the subtleties of the function that Nature makes of these *ca.* 15 metal elements, it is necessary to understand how they are selected, in terms of uptake, and how they are combined, transported, stored, and tuned to their specific functions. Evolution refined function by acting on the interaction of the metals with their partners (usually proteins) or by abandoning some in favour of others more appropriate to conditions. Much of this chemistry involves ionic interactions.

Of course, in the description of the way in which the chemical elements are involved in life we cannot concentrate on metal ions alone. Some non-metals actually function in ways not too different from those of the metals, but as anions not cations, e.g. Cl^-. Others, the major elements of life, C, N, O, and S, to which we refer to rather a small degree since their biological chemistry is well described elsewhere, behave very differently, much as their chemistry is also very different, being loosely called covalent rather than ionic. Between the extremes lies the chemistry of phosphorus, as phosphate in biology, and hydrogen, which have both ionic and covalent chemistry. Our point will be that all the elements we find in biology are used in some optimal manner developed from their chemical possibilities. Biological chemistry is concerned, therefore, with this intricate, interwoven, 'designed', and sometimes symbiotic use of organic (non-metal) molecules and inorganic (metal) chemical elements in cells and tissues and cannot be restricted or related to so-called organic chemistry rather than to so-called inorganic chemistry. This book is then about biological chemistry seen from a particular inorganic perspective where a certain inorganic *bias* is inevitable to compensate for the traditional approach which leans heavily on organic chemistry. Accordingly, we have named it *The biological chemistry of the elements*.

Before entering the main body of the text, the reader is warned that every biological device is unlikely to be the best of all possible devices. Not only has biology general chemical limitations, but it is also evolving in changing conditions, many of which are created by life itself. A consequence of this slow progression is that now and then a less than perfect feature will survive, which has evolved from a committed route earlier in evolution and then found a new semi-effective use. It also follows that, should environmental conditions, the air, the sea, or the land, change very rapidly, complex life forms are at risk. There is a very general warning here, as man's activities saturate the ability of the environment to buffer the chemical changes he makes.

The book is organized in the following manner. In Chapter 1 we analyse the occurrence of the chemical elements in biology showing that of the 92 elements in the periodic table some 25-30 are especially important in biology. Much of the reason behind this selection lies in the limitations imposed by the abundance of the elements (even in the universe) and their availability on earth. Availability is further constrained by the concentration in waters and in the air and depends on local conditions. In the absence of organic chemicals most elements are bound in simple oxygen- or sulphur-containing compounds, as hydrates or precipitates. On entering biology the elements undergo special combinations with organic molecules. In the next five chapters we shall be considering this chemical speciation in a great variety of compounds in biology. We shall describe in the first instance in Chapter 2 how the biological uptake of the elements from natural sources can be controlled by selective (energized) chemistry of the elements and leads on to selective chemical combinations, e.g. of metals, M, with organic ligands, L. In Chapter 3 we extend the discussion to physical transfer and separation of M, L, or ML across membranes into compartments. Only at this stage will it be possible to start to explore function. Function is dependent upon the ability of elements to undergo chemical change (within compounds). The kinetics of activity will therefore be analysed in Chapter 4. Chapter 5 will illustrate how energy is put into compounds within biology. This is an obvious necessity both in synthesis and in transfer. We shall then turn in Chapter 6 to one major concern of this book-the functionally necessary requirement for life of a variety of elements. Chapter 7 will be a rather lengthy but necessary interjection concerning the biopolymers of life. The second part of the book, Chapters 8 to 19, will consider the essential functions of individual inorganic elements in life's processes in detail and especially in relation to the principles we outline in the first part, before we attempt to pull all the (inorganic) chemistry of life together, in the last part of the book.

We begin the difficult final analysis with an enquiry into the nature of minerals in biology (Chapter 20), since this allows us to look at a peculiar biological problem, the relationship between chemical activity and morphology. Shape is so clearly an essential feature of living things. Having established some simple principles, we move in Chapter 21 to a consideration of the simultaneous homeostasis of shape and chemistry. It is here that the interactive feedback relationships between the functions of elements become so intriguing. It almost goes without saying at this stage that the balance of elements in life is related to the elements available and it follows that changing the availability must affect life. We look at problems of man's interference with the natural supply of elements in Chapter 22.

Each chapter has only a limited number of references and most of them are to reviews. We realize that this approach may offend some but we found it impossible to be sure that we could give proper priority amongst the vast number of primary sources we have used.

Lisbon and Oxford — J.J.R.F.d.S
January 1991 — R.J.P.W

Acknowledgements

Our special thanks go to Mrs S. Compton of the Inorganic Chemistry Laboratory, Oxford. We also thank Profs H.A.O. Hill and J. Reedijk for many corrections and suggestions relative to the first edition. Many scientists throughout the world have also made specific comments which we have incorporated. They are too numerous to thank individually. We have done our best, but some deficiencies certainly remain, for which we are entirely responsible. As to some spelling aspects which were criticized, use of *ph* instead of the now adopted *f*, for example in sulphur, corresponded to an editorial practice at the time and it was not viable now to alter the entire book. So the practice was maintained.

J.J.R. Fraústo da Silva acknowledges the Fundação para a Ciência e Tecnologia (Ministério da Ciência e Tecnologia), Portugal, for general support of research activities and thanks especially Prof. J. Armando L. da Silva for introducing him to the art of Internet searching (without which this second edition might have proved to be an impossible task) and Mrs Teresa Maria Carreiras da Silva and Ms Cristina Sequeira da Silva for invaluable secretarial assistance.

R.J.P. Williams acknowledges the continued support of the Inorganic Chemistry Laboratory, Oxford University, and many exchanges with its members.

Contents

General introduction

This book is not a traditional bio-inorganic chemistry text for undergraduates, centred on the description at an elementary level of the structures, physical properties and functions of *isolated* biological molecules, such as proteins and enzymes which contain or interact with one or more metal elements. Neither is it concerned with the physical methods used in structural studies nor with *in vitro* modelling studies of important biological reactions using small non-biological molecules. There are now several good introductory textbooks and monographs which cover these topics. Our main aim is to describe the functional value of the chemical elements in living organisms, free or combined with the *actual* biological molecules, that is, *in vivo*, the reasons for the selection of such elements, the processes of their uptake, transport and final localization in cells, the regulation of these processes and the control of their reactions in a very complex holistic system which includes the interactions with the environment. The book is therefore addressed mainly to graduate students, researchers, and teachers who want to have a general overview of what biological chemistry is about today, providing them with the basic information, discussing the main issues, and giving references to more detailed texts (mostly books and reviews given the obvious impossibility of referring to the thousands of contributions of individual researchers).

In effect, we believe that the understanding of the chemistry of life has been seriously hampered by the ways in which chemists have looked upon living organisms. At first, this study was conducted very largely by (bio-)organic chemists who made remarkable discoveries concerning the small molecules which go to make biological substrates (sugars, amino-acids, fatty substances and so on, and then the polymers) proteins, polysaccharides, and finally reaching the genetic material, DNA and RNA. Simultaneously the mechanisms by which they are produced were analysed. This study of organic molecules, their pathways of synthesis and degradation and their activities, the traditional biochemistry, left to one side certain functional features of organisms such as the ways in which materials and energy are captured, the manner of nerve conduction, the process of muscle contraction, etc., which were studied by physiologists concerned more with currents and potentials than with the chemicals involved. Here we know now, of course, that inorganic elements such as Na, K, Cl, Mg, and Ca dominate. Left to one side also were other mineral elements of life, including Fe, Mn, Co, Ni, Cu, and Zn, present in many catalysts. In effect, until about 1950 little attention was paid to this inorganic content of cells and its function. Only since that time has it been increasingly realized that several 'inorganic elements' have an essential role in the *activities* of all organisms, much though some are present in quite small amounts.

A useful 'Glossary of terms used in bioinorganic chemistry' is given by De Bolster, M.W.G. (1997). *Pure and Applied Chemistry*, **69**, 1251–1303.

Given the realization that some twenty elements are vital for function in selected roles and a few others are also absolutely required by some organisms, it is necessary to appreciate biological chemistry in a manner different from the approach based just on organic molecular chemistry but including it. In effect, to understand life it is required to adopt an integrated organic/inorganic chemical approach. This is the approach we use in this book, which is largely concerned with the molecular chemistry associated with the essential elements naturally selected by organisms. This, however, is a first requisite since we know that underlying this molecular chemistry is the holistic systemic activity of a complex network of reactions which gives rise to life, so we develop our enquiry in this direction. Of course, any chemical approach to such systems has to be based, in the first instance, on elemental analysis, so that we know the variables of composition which have been used to create life. This is our first objective in Chapter 1. The analysis then becomes more complicated as the examination proceeds to an uncovering of the molecular combinations of the elements and then their interactions. The study of combination, structure and property is part of the analysis of the local combinational chemistry as opposed to just the *compositional* variable of chemistry. The way in which elements come together is largely decided by the energy of atomic element interactions. This topic is not so well analysed in inorganic texts especially when it applies to reversible association in aqueous solvents though there are specialist books available. Hence we give in Chapter 2 an examination of the stability of reversible association of elements in complexes, particularly those involving an organic molecule, a ligand, and an element in aqueous solution relevant to cellular chemistry. We observe that stability is an interesting function of composition, different for different elements and in different structures. In inorganic chemistry textbooks all of this description of stability against composition as well as its variation with *temperature* and *pressure* is dealt with under the heading of thermodynamics of *chemical systems*. In living organisms temperature and pressure are fixed within a small range, so that we can omit these variables and concentrate upon variation in composition. It is curious and instructive then to appreciate the resultant selections observed in living organisms against the background knowledge of observed speciation driven by thermodynamics (equilibrium) in coordination chemistry, while being aware of the limitations to this exercise.

Once we have introduced the thermodynamics of inorganic complexes in Chapter 2, and noted that this is only a useful approximation and not a final approach to all speciation and energetics of biological systems, since we also observe that many elements are irreversibly inserted into sites, we can turn in Chapter 3 to the biologically relevant problem of how ions and molecules are distributed in compartments. This is part of the variable distribution in space of molecules, much as partition coefficients relate to molecular distributions in organic solvent systems. It obviously depends, in part, on energy input.

The final variable which we must introduce is *time*, treated first under kinetics of molecular reactions, Chapter 4, and then under flow patterns (Chapter 7). Several of the resultant combinations we observe in cells do not follow from thermodynamic considerations and we shall have to consider combinations brought about as in inorganic and organic chemistry by kinetic control, e.g. by pumping. Once again we must warn the reader that chemical kinetics of reactions of molecules moving randomly in free solution is not the only problem of the influence of the variable time in a biological cell. Here time has an extra irreversible character due to controlled flow in directions, both on physical and chemical pathways, stimulated by energy input. The input of energy is described in Chapter 5, both to the placement of compounds in cells, to the flow and to the energy of compound formation. As stated above, apart from the considerations of the properties of biological molecules, many containing inorganic elements, we have to appreciate the nature of the external environment, its element content and its energy gradients, without which reproductive life cannot exist.

In Chapter 6 we examine the polymers which for the most part bind the elements, exert control over their concentrations and are largely responsible for producing organized activity. We refer also to the genes in an organism and their active products, the proteins and further downstream the lipids and polysaccharides. All these organic molecules, synthezised and positioned in cells and with the added essential 'inorganic' elements, make up the cellular composition, *in toto* and locally in compartments, but we must stress that they all arise from the environment. Furthermore, they are all part of one larger holistic flow system for it is easily shown that conversion of environmental chemicals such as CO_2, N_2, and H_2O into useful organic chemicals requires these inorganic elements as does the energy transduction necessary for their synthesis. Thus, it is absolutely essential that we recognize that all these organic and inorganic chemicals in cells must be tightly controlled relative to and in harmony with a changing environment.

It is now apparent that only by seeing the biological chemistry of the elements in three distinct ways, namely as (1) isolated molecules with given properties, (2) positioned and/or moving in a cell, and (3) active as separate species in certain time periods within an holistic system, shall we be able to appreciate biological chemistry (Chapter 7). A major part of our further understanding of biological systems must be the undercovering of the reason organisms are broken down into species and are not part of a continuum although they may have a common origin. Separate problems are the origin and the evolution of these species. In this book we shall not consider the origin of life in detail, but in Chapter 20 we will examine survival of given open *organizations* as functions of compositional and structural variables opposite a changing inorganic environment in time. The problem is not totally unlike that of understanding thermodynamic stability of structures against similar variables, but the introduction of compartments, time, flow, and coded information (DNA) complicate the subject enormously. We will start our analysis in Chapter 20 considering first the *composition* variable of systems and then the basic inorganic chemical variable in organisms. This corresponds to a *quantitative* examination of the content of organisms. Later, we attempt to relate this knowledge to the *linear* coded controlling molecule, DNA (the genome), which by itself will not reveal the composition except of protein types (but not quantities of each) and possible modes of regulation and of their synthesis. To define proteins quantitatively is the purpose, separate from analysis of the genome, of proteome studies which include incorporated metal ions. However, to define the cell in relation to its environment we also need the analysis of the levels of free elements in each cellular compartment. We shall call this extra characteristic the free metallome. While the genome and the proteome are not readily related to the environment the free metallome is in direct contact with it. We shall also refer to the total metallome, which includes free and bound concentrations of elements. We trust that our approach provides a novel view of the nature of living systems and their evolution.

PART I

The chemical and physical factors controlling the elements of life

CHAPTER 1

The chemical elements in biology

1.1 The element content of living systems

1.1.1 The biological elements

The analytical element content of cells as we know them today is in some way related to the limitations imposed by the genome, DNA, of the cell. However, the chemistry of a cell including the functioning of DNA and its synthesis only operates in a particular ionic medium and using a particular set of some accumulated twenty elements. Each cell also needs to collect energy. Hence, survival of a cell line, an organism, depends not just on reproducing DNA but on the uptake and use of selected elements and energy, all three being in a feed-back relationship. In fact, historically, DNA, or any other coded set of instructions such as RNA, could only have been an extra way of supporting and reproducing patterns of chemical reactions which predated it. The feed-back connection between DNA and the element content is brought about by proteins. It is then possible to describe a cell in terms of composition in at least three related and dependent ways: by its *genome* (DNA), by its proteins, a cell's *proteome*, and by its element content, which we may refer to as the cells' *metallome*. Other possibilities will be described later. In this book the approach to life starts from the simplest of the three, the element content, and then progresses to the more chemically complicated components and processes while stress remains on those molecular units which can be grouped together by their selected inorganic element content. The relationship to the genome and consequently to biological speciation will become clear later.

There are now known to be almost two million different 'species' of living organisms on earth (see Aside) and besides their phenotypic and phylogenetical characteristics we need to know their chemical similarities and differences to be able to understand what life is and how it works. An obvious start is the identification of the 'biological' elements. Analyses of many bacteria, a few hundred plant species, among the known 0.4 million, and of about two hundred among the catalogued 1.1 million animal species as well as of organs, tissues, and other substances of biological origin, have enabled us to establish the number and identity of the chemical elements present in biological systems and to recognize those that are 'essential' for bacterial, plant, and animal life. It became clear in these and related studies that, amongst the chemical elements, all biological systems concentrate certain elements while rejecting others and that some of the processes involved require energy. One could, therefore, speak of a natural selection of the chemical elements by biological systems, which involves a readjustment of the element distribution on the earth's local scale by utilizing energy ultimately provided by the sun.

Number of known living organisms	
Animals	1 032 000
Higher plants	248 400
Fungi	69 000
Protozoa	30 800
Algae	26 900
Bacteria	4 800
Viruses	1 000
TOTAL	1 412 900

See E.O. Wilson, in the references, for more details of known species. Note that *existing* species may well be over 10 millions, and these may be some 10 per cent or less of those that ever existed.

Table 1.1 Percentage of atoms in the human body

Element	Atom percentage (wet wt)
Hydrogen	62.8
Oxygen	25.4
Carbon	9.4
Nitrogen	1.4
Others	1.0

Analysis has shown that *eleven* elements appear to be approximately constant and predominant in all biological systems. In the human body these constitute 99.9 per cent of the total number of atoms present, but just *four* of them—carbon, oxygen, hydrogen, and nitrogen—correspond to 99 per cent of that total (see Table 1.1). The very large percentages of oxygen and hydrogen arise from the high water content of all living systems. Carbon and nitrogen, next in importance in Table 1.1, together with oxygen and hydrogen, are the basic elements of the organic structures and metabolites of living systems.

The other *seven* elements, which together represent about 0.9 per cent of the total number of atoms in the human body, are sodium, potassium, calcium, magnesium, phosphorus, sulphur, and chlorine. Besides these eleven elements, which are absolutely essential, some *ten* others—metals and non-metals italicized in the list that follows—are required by most biological systems, but not necessarily by *all* biological systems. There are still a further eight elements, some of which may be generally required by plants and animals, whereas others may be required by just plants or just animals or sometimes by a relatively few species of plants or animals. These last 18 elements are: vanadium, chromium, *manganese, iron, cobalt, nickel, copper, zinc, molybdenum (tungsten), boron, silicon, selenium*, fluorine, iodine, arsenic(?), bromine(?), and tin(?). The levels of requirement of these 18 elements are generally very different and we note particularly the relatively large amounts of iron and zinc present in most if not all species. Finally, we note that elements such as cadmium, strontium, and barium are known to be important in the chemistry of one or two particular species.

Table 1.2 Unusual accumulation of elements by plants or animals

Element	Plant	Animal
Al	Club moss, hydrangea, tea plant	
As	Brown algae, Ferns, *P. vittata*	Coelenterates
B	Brown algae, *Plumbaginaceae*	Sponges
Br	Brown algae	
Ba	Rhizopods, Brazil nuts, *Loxedes*	Coelenterates, molluscs
Cl	Numerous marine and salt desert plants	Soft coelenterates
Cu	*Caryophyllaceae*	Annelids, molluscs, etc.
F	*Dichapetalum cymosum*	Vertebrates
I	Brown algae, diatoms	Marine annelids
Li	*Thalictrum* sp., *Cirsium* sp.	
Mn	Ferns, *Digitalis purpurea*	Marine crustacea
Mo	*Papilonaceae*	
Ni	*Alyssium* sp., *Hybanthus floribundus*	
Se	*Micrococcus selenifera, Cruciferae, Astragalus racemosus*	
Si	Horse tails, *Radiolaria*	Protozoa. sponges
Sr	Brown algae	Alcyonaria, octocorallia Acantharia
V	Some *Amanita*, fungi, brown algae	Ascidians, some worms
Zn	*Thlaspi calaminare*	Coelenterates

Sources: Bowen. H. J. M. (1966). *Trace elements in biochemistry,* Academic Press. London. Hewit, E. J. and Smith, T. A. (1975). *Plant mineral nutrition.* The English Universities Press, London.

The limited number of species examined, the difficulties of the analytical work (particularly for elements at the trace level) and the lack of detailed knowledge concerning the role of each element make it necessary to check their hypothetical 'essentiality' by delicate tests, usually by following the developmental growth of species, animal or plant, while giving them carefully prepared 'diets' deficient in the particular element considered.

In some cases these studies have indeed enabled not only the classification of a certain element as 'essential' or 'not essential' for a particular species but also the evaluation of the effects of its deficiency. In other cases the results are ambiguous and no definite conclusion has been reached. At present, elements such as arsenic, bromine, and tin are considered as 'possibly essential' while others, like aluminum (see Table 1.2), are considered to have no definite general biological role, perhaps because the effects of their deficiency have not been sufficiently tested or because the requirements by the species used in the tests are so low that the needs are satisfied with the extremely low amounts present even in carefully purified diets.

Uptake of a certain element by a biological system, taken as isolated evidence, is no proof or guarantee of essentiality and even less is the detection of a certain (trace) element in a particular biological system. The analytical difficulties, the possibility of contaminations at the trace and ultratrace level, and the likelihood of competition of different elements for the same chemical sites leave room for considerable doubts as to the conclusions to be extracted from these experiments. Moreover, some biological systems accumulate elements even though they may not need them (Table 1.2). It follows that the classification of a certain element as essential or not essential implies a prescriptive definition of 'essentiality'. In general terms we will consider as essential an element consistently present in a certain biological species such that its deficiency in the nutritive sources of that species leads to disease, metabolic anomalies, or perturbations in its development. More precise criteria have been put forward and are commonly adopted by nutritionists.

The results obtained in the studies of essentiality are summarized in Fig. 1.1—the periodic table of the elements—where the elements that are today considered to be essential for life are conveniently identified. A first conclusion to be derived from this figure is that the 'chemistry of life' is, predominantly, the chemistry of the *lighter elements*, metals and non-metals of

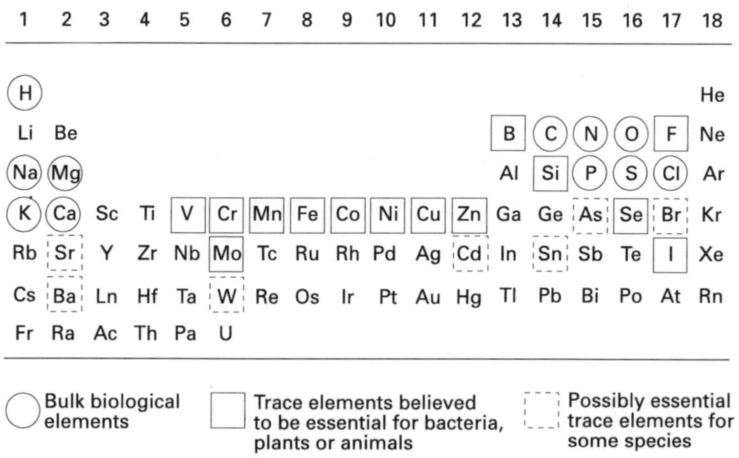

Fig. 1.1 The distribution of elements essential for life in the periodic table

atomic number less than 36. Of confirmed required elements only molybdenum, iodine, and tungsten have higher atomic numbers and all of these do not seem to be generally required by *all* living species. On the other hand, the biological elements seem to have been selected from practically all groups and subgroups of the periodic table (the only exceptions are groups 3 and 4, besides that of the inert gases) and this means that practically all kinds of chemical properties are associated with life processes within the limits imposed by environmental constraints. A little thought will show that such a selection is to be expected from an optimalized evolving chemical system of an organism. We conclude by stating that *all organisms require about twenty elements* though the precise requirement differs somewhat within different species.

1.1.2 Chemical speciation: limitations of life chemistry and physics

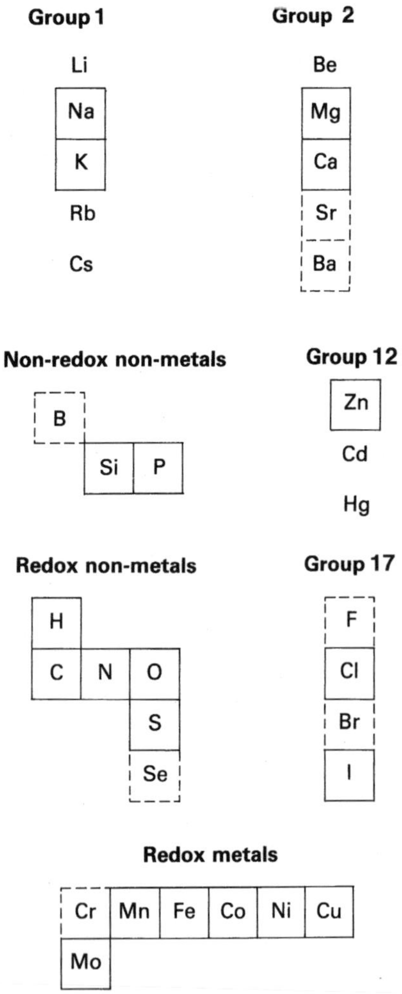

To describe the limitations which the elements (in compounds) face in life one must recall that oxygen and hydrogen are, to a large extent, combined with each other in the form of water and that most life chemistry takes place in aqueous media. This fact implies that, generally, only processes compatible with the presence of this solvent are possible, restricting the utilizable range of redox potentials, pH. and temperature. Bulk metals or alloys as such, for instance, cannot be used by biological systems since they are not preparable in water and usually are unstable in it. Processes depending on metallic electronic conduction which might be valuable, e.g. to send messages, must be carried out by other chemical agents or systems. The same can be said for temperature constraints. Thus, today's industry uses operations requiring high energy input to produce products that biological systems also need, e.g. NH_3 obtained from N_2; in biological systems these operations must be carried out at normal physiological temperatures which is only possible by using extremely sophisticated catalytic processes developed over millions of years.

To explore these aspects in somewhat more detail we note that the elements in the periodic table which concern us are not just those that constitute the majority of the organic biomolecules, but also include some closed shell metals, especially of groups 1 and 2, most of the first series transition metals, groups 6 to 11, Zn in group 12 and also a more extensive group of non-metals of groups 13 to 17 including (B), F, Si, P, S, Cl, Se, Br, and I, besides molybdenum, which can be considered, in certain respects, as a non-metal. Given that water is the solvent, these later non-metal elements tend to be found as anions, in contrast with the elements Na, K, Mg, Ca, Zn and several transition metals which are found as cations.

Although the distinctive properties *in aqueous solution* of each of these conventionally separated four classes of elements, i.e. groups 1 and 2, transition metals, group 12 metals and non-metals, do not always provide sharp divisions, it is useful to distinguish some general chemical properties associated with these classes. Hence, groups 1 and 2 elements are associated with equilibrium ionic-model chemistry. The transition metals are also generally associated with aqueous ionic chemistry, apart from organo-metallic chemistry, which is unusual in organisms and additionally with one-electron redox reactions, but across the series they exhibit increasing covalent chemistry and increasing Lewis acid strength, not necessarily at equilibrium with their surroundings. This means that in biology they become largely bound to organic biological molecules in which equilibrium is harder to achieve. Group 12 metal ions, e.g. zinc, display a compromise in their chemistry having strong Lewis acid properties while maintaining faster ionic equilibration but with limited or no redox activity. They too usually become bound in biological systems. There is next a class of non-metals (H, C, N, O, S, P) in which covalence introduces gross kinetic control in aqueous solution. It is this that allows the development of the biopolymers as well as kinetic control over two- and (a few) one-electron oxidation state changes and over many acid–base reactions, with Lewis base properties dominating over Lewis acid properties. The extreme non-metals appear overwhelmingly as simple anions (we specially note the presence of chlorine as chloride), when their chemistry is classed with the group 1 and 2 elements. The aqueous conditions which allow the classification are maintained in biological systems.

Biological systems took immediate advantage of these differences of chemical properties, but we shall see that the distinctions between the various individual elements have been heightened during evolution in such a way that each of them has been selected to perform one or more particular functions with high specificity, much more so than in ordinary bench chemistry. This is not just a matter of selecting partners but normally involves an energized state of the element brought about by separation across membranes, for example. These general points are so important that it is worth digressing to give a simple example.

1.2 An aside: some biological chemistry of hydrogen

There is one very special element in chemistry and biology—hydrogen. It is special in that it has three very distinct states, H^+ (a cation), H- (a covalently bound state), and H^- (an anion). As H^+, the proton, it is an extremely strong acid and extremely mobile in protic media such as water and the water/polymer interfaces of biology. In other words, as an acid it is unique. As H- it is a special element in that it forms very stable non-metal bonds in the form of C–H and N–H which are terminal blocking units limiting reactivity at carbon and nitrogen. These bonds are kinetically stable even in the presence of dioxygen and provide great stability to organic molecules in biology. Third, under the influence of catalysts hydrogen can be transferred from a non-metal such as carbon not just as H^+ or H- but as H^-. Which of these forms is transferred depends upon the group which the hydrogen is leaving. Thus, hydrogen atoms can be involved in one- or two-electron processes. Great use of this redox chemistry is made in biological redox reactions. Even more importantly, the internal redox energy of organic molecules, C–H bonds, versus dioxygen, both derived from sunlight (Fig. 1.2), can be converted within membrane systems, using oxidation/reduction, to different forms of more usable energy. Here hydrogen is transferred first as H^-, then effectively as $H^\bullet$, and finally as H^+, to produce eventually proton gradients across membranes in water (Fig. 1.2). The reactions require the separate movement of electrons, of course. The strong acid, H^+, which is freely mobile, is seen to be readily connected to covalent, kinetically-controlled, redox chemistry, as well as to ionic acid/base chemistry. The last step in Fig. 1.2 is to make ATP, a form of pyrophosphate useful in synthesis, from the proton gradient. The gradient can also be used to pump other materials into a cell. *No other element has these different capacities to the same degree* in water and, as a consequence, hydrogen is at the heart of energy transduction mechanisms in biology (see

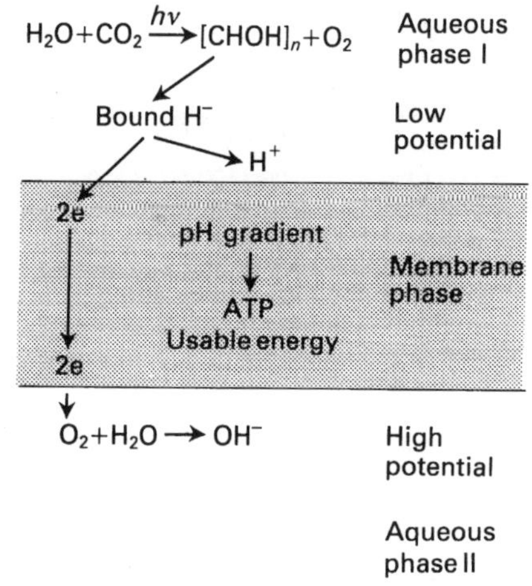

Fig. 1.2 Some chemical functional values of the element hydrogen which have been realized in biology. Hydrogen atoms are initially incorporated into carbon compounds by the action of light. They can be transferred in different redox states e.g. H^+, $H^\bullet$ or H^-. If the transfer operations are energized and diffusion limited, say, by a membrane, while electrons are moved separately as shown, then H^+/OH^- gradients can be built. Hydrogen is the only element which can be used in this manner to generate electrolytic cell energy from light. The H^+ gradient is subsequently used to aid the transfer of other ions and molecules across membranes or, as shown, to make adenosine triphosphate, ATP, the first form of chemically stored energy (see Chapter 5 for a more detailed explanation). Notice that a redox potential difference is established across the membrane.

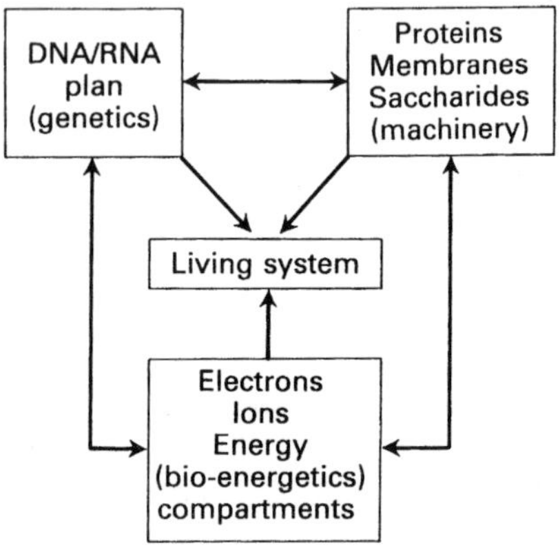

Fig. 1.3 The scheme shows the interrelationship between the three interlocking features of a biological system. The DNA/RNA unit is that of inheritable information. The protein unit covers structural aspects, including the lipid membranes, the proteins, and the polysaccharides, within which there is the mechanical and catalytic machinery of the cell. The third unit is a source of energy, which is necessary for the activity of cells including all syntheses and the uptake of chemicals. The role of inorganic elements is dominant in the last unit, is generally very important within the machinery, and has a place in the DNA/RNA unit.

Chapter 5). A further major feature of hydrogen chemistry is its ability to form H-bond structures which in part give rise to the properties of water, proteins, and DNA.

We give this example to stress the way in which life's chemistry is built around elements which have essential functional potential. In the case of other elements this is the case for the transmission of information (Na^+, K^+, Ca^{2+}), for catalysis (Fe, Cu, Zn, Mo), for the transfer of electrons (Fe), and for the building of solid structures (Ca, P, Si). All of these features are essential in the same sense as it is true to say that C, N, H, and O are essential for the synthesis of biological macromolecules (Fig. 1.3).

Before delving further into these matters, it is appropriate to give some justification for the various trends observed in the distribution in biology of the elements of the periodic table.

1.3 The economical use of resources: abundance and availability

In general, an element that is to be used by biological systems must have a useful function but it must also be abundant in the environment and must be in an easily extractable form, i.e. it must be 'biologically available' from the environment. A cursory examination of Fig. 1.1 immediately shows that the elements essential for life are the lighter elements which are also the most abundant in the universe (Fig. 1.4). They are also the elements which are most abundant on earth. It is true that the abundance in the planets, particularly in the earth, does not exactly correspond to that of the solar system and deeper considerations of this aspect must go down to the level of the detailed composition of the earth's crust, oceans, atmosphere, etc. Even then, however, and considering the amounts of the relevant elements locally, it is not necessary to modify the discussion greatly: generally, biological elements are still those which are the most universally abundant.

Some elements, e.g. lithium, caesium, beryllium, and strontium, are either of low abundance or are less abundant than others which have analogous properties and can play a similar biological role. Presumably one reason why the bulk biological metals, sodium, potassium, magnesium, and calcium, were selected in preference to lithium, rubidium, caesium, beryllium, strontium, and barium is that they are more abundant and can perform in large measure the tasks that the other elements of the same groups of the periodic table can perform. The periodic table (Fig. 1.1) is, to some degree, a table of redundancy. This, like many other generalizations, has interesting exceptions, which will be considered later in this section.

Besides the problem of abundance there is that of availability in water solutions. Availability is concerned with chemical combination and solubility of compounds as well as abundance. Many of the compounds of the elements of groups 1, 2, and 17 as well as of O and S in group 16 are soluble and found in the sea in quantity (see Fig. 1.5). The parallel between the elemental composition of animals or plants and that of sea water in some respects is quite striking, but, since life has modified the composition of the sea itself, no simple deductions are possible. Group 3 and 4 elements are of relatively low availability, i.e., although they form amphoteric oxides of high abundance, they are insoluble and it requires great energy to obtain measurable amounts of such elements at pH = 7.0 (Fig. 1.6). In fact, these elements are notable for their absence from biology.

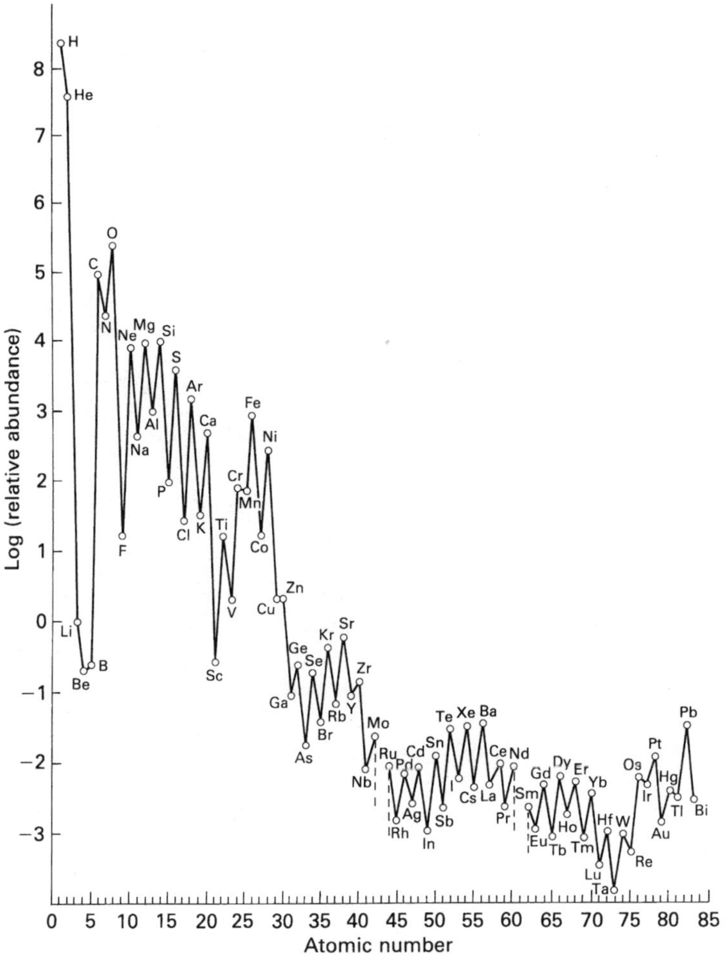

Fig. 1.4 Abundance of the elements in the universe relative to Si = 10^4 as a function of the atomic number.

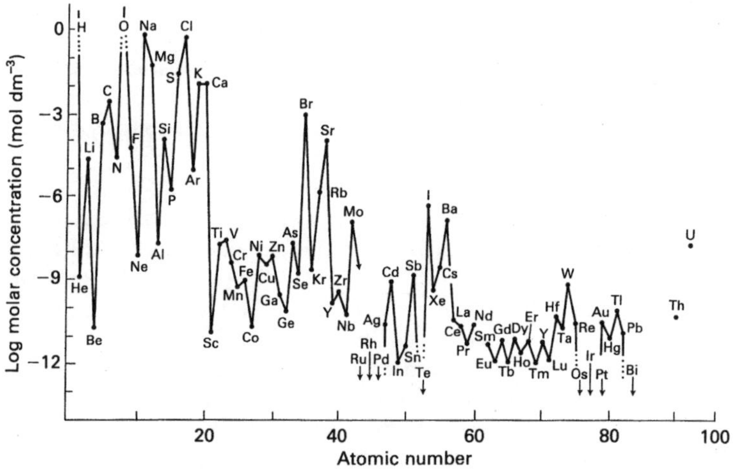

Fig. 1.5 The concentrations of elements in the sea (After P. A. Cox, see 'Further Reading'.)

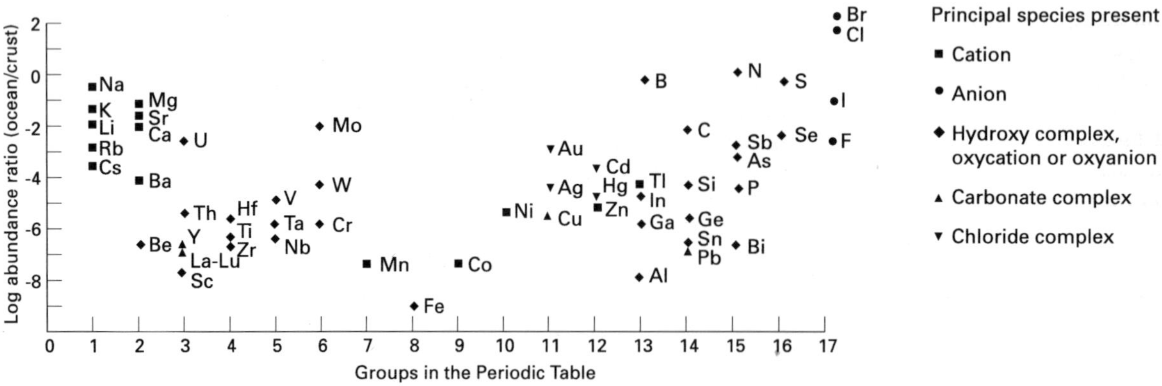

Fig. 1.6 The figure shows the ease of extraction of elements from the earth to the sea and is a guide to the ease with which biological systems will have access to elements, no matter what the absolute abundance. The right side of the figure indicates the speciation.

If we extend this analysis to other groups, one or two elements turn out to be common in biology but not really available. A striking problem, revealed in Fig. 1.6, is the difficulty biological systems face today with the lack of availability of iron. The special lengths to which biology has been forced in order to obtain this precious element will be seen later to have a controlling influence on the whole of living systems, yet biology created this difficulty for itself by generating dioxygen (see Chapter 20). Inspection of Figs 1.5 and 1.6 and the data given earlier for the element content of biological systems indicates how absolute is the demand for some specific elements. Why?

One may therefore say, in a general way, that in its selection of chemical elements Nature responded to abundance and availability following a principle of 'economical utilization of resources', i.e. choosing those elements less costly in terms of energy required for uptake, *given the function for which they are required*.

It is important not to consider this principle as other than dynamic, since biological systems are adaptable and can to some extent manipulate their chemistry and their position in the environment. While it would appear that the solvent for life has always been water, other conditions, such as those of temperature and the dominant chemical species in the atmosphere, have changed with time. Life, while itself producing many of these changes, has evolved with the changes. This will always involve compromises, such as the greater ease of access to one element at the cost of lesser access to another. Again, geological conditions today provide and always have provided a variety of niches, which are indeed special environments into which species may 'guide' themselves (see Chapter 20). It is also the case that different elements are accumulated in biological systems to take advantage of special conditions. A simple example is the accumulation of large quantities of dense $SrSO_4$ and $BaSO_4$ in some protozoa and algae (Table 1.2), which allows these organisms to remain in relatively deep (anaerobic) waters of relatively high density. We must note too that the changes in the atmosphere and the consequent changes in the sea and in the earth's surface over millenia have had a considerable effect especially on rare element distribution and uptake. Some of the elements found in biology were selected a long time ago in very different circumstances; iron is a striking example. Even so, the remarks made above concerning the economy of uptake have a general quality across many species.

Against the background of these figures we turn to the detailed reasons today for the observed availability of the elements and their speciation in water at pH = 7, given exposure to molecular dioxygen of the air. At this stage of discussion we are not concerned with organic chemicals at all but only with the elements and their simple oxides and hydrides which will be used later to form all compounds *in vivo*.

1.4 **Biological environment and element availability**

1.4.1 **pH and redox potential considerations**

If we are to appreciate the skill displayed by the living systems in the uptake and utilization of the chemical elements, both those they need as well as those they reject because of their potential toxicity, we must study the chemical forms, their speciation, in which all these elements occur in the natural surroundings of life (Fig. 1.6). While doing this we must also take into account the limitations imposed not only by the solvent water and its pH but also the prevailing (present-day) redox conditions.

Recalling that at pH 7 the redox potentials in water are largely restricted to the range –0.4 to +0.8 V relative to the standard hydrogen electrode, which corresponds to the systems H^+/H_2 and O_2/OH^-, i.e. to the values below which and above which water is reduced to H_2 or oxidized to O_2, respectively, then a transition metal as an ion has only a small range of stable oxidation states, e.g. iron (+2, +3), copper (+1, +2), etc., but a non-metal may have a very wide range, e.g. sulphur (–2 to +6), carbon (–4 to +4)—see the oxidation state diagrams in Figs 1.7 and 1.8. This difference arises from the different electronic energies involved in the two cases. In the first transition series it is the 3d electron core that is ionized and this process, which is of rapidly increasing energy with increasing positive charge, cannot be easily compensated by chemical binding to water, but in the non-metals the ionization is of sp electrons and compensation by chemical bond formation is largely possible. Note that we shall use both arabic and roman numerals alternatively for oxidation state, *n*: M^{n+} stresses free ion character while M(III) stresses bound character, e.g. compare Fe^{3+} and Fe(III) [see Phillips and Williams (1966) in the 'Further reading' section].

One element of biological interest for which these generalizations fail is molybdenum, which is a second-row transition *metal* element but behaves more like a non-metal. It has a wide variety of oxidation states in water within the referred restricted range of potentials (Fig. 1.7). Note that here it is the 4d core which is ionized. The case of vanadium is also interesting in that it has an unusual set of available high-valence states. Although its biological role is not so extensive, it will be seen later to afford another example of unexpected solutions provided by Nature to face specific necessities in particular cases (see Chapter 17). Molybdenum and vanadium can perform similar functions, e.g. N_2-fixation. Tungsten has properties similar to molybdenum and is similarly used in a few primitive organisms.

Another conclusion that can be derived from the oxidation state diagrams is that there are metals and non-metals with only one very stable state. This is the case of the alkali and alkaline-earth metals restricted, respectively, to the +1 and +2 oxidation states; of zinc and cadmium, restricted to the +2 oxidation

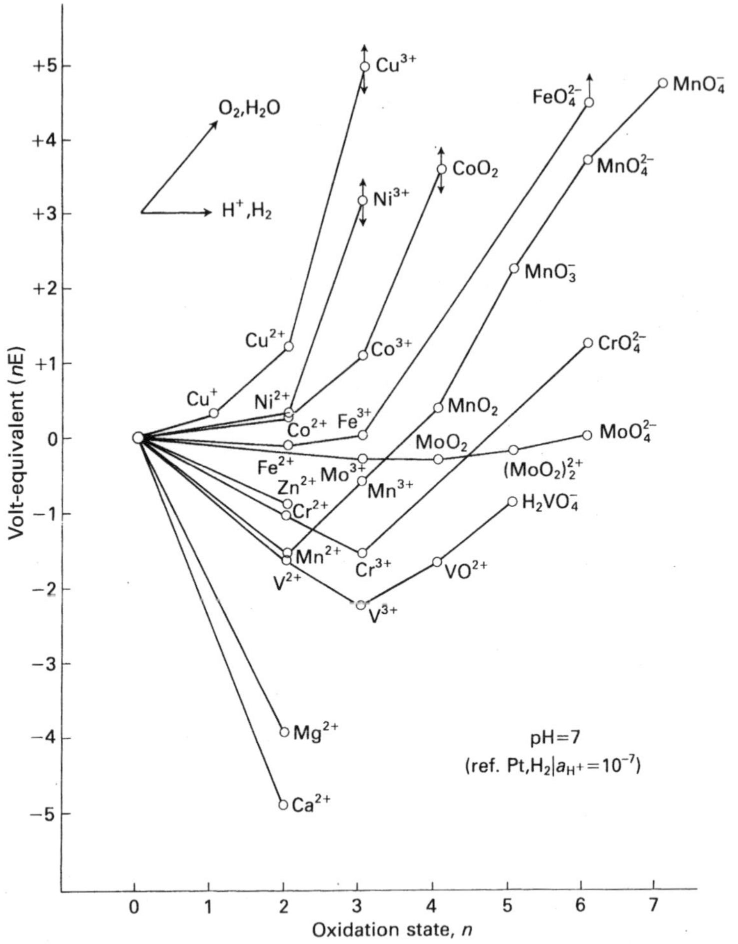

Fig. 1.7 Oxidation state diagram for some metals at pH = 7 (*n E* versus *n*) taking into account the formation of hydroxides, where *n* is the oxidation state and *E* the electrode potential from the element to that state.

state; and of several non-metals, notably boron, silicon, and phosphorus, which appear only in the highest oxidation states of +3, +4 and +5, respectively, as borate, silica, and phosphate. The predominant form of chlorine is as a chloride ion and the few rare compounds of biological origin containing covalently combined chlorine are often antibiotics. (This observation should be carefully examined in its implications for the rather careless use that man still makes of organo-chlorine compounds, insecticides, and other industrial products, which are being banned by progressively more restrictive legislation due to their detrimental biological effects.) As examples of the opposite behaviour, a few other non-metals are unstable at the extremes of their available redox states, for example nitrogen as nitrate, which tends to be converted to NH compounds, and selenate, which is incorporated as covalently combined seleno-thiol and analogous species.

This shows that only a very limited number of non-metals can be used by life as *reversible* redox switches. Hydrogen (see Section 1.2) and carbon occupy overwhelmingly dominant positions, with available oxidation states from –1 to +1 and –4 to +4, respectively, and all are interconvertible at redox potentials close to zero. Sulphur comes close to carbon and selenium comes

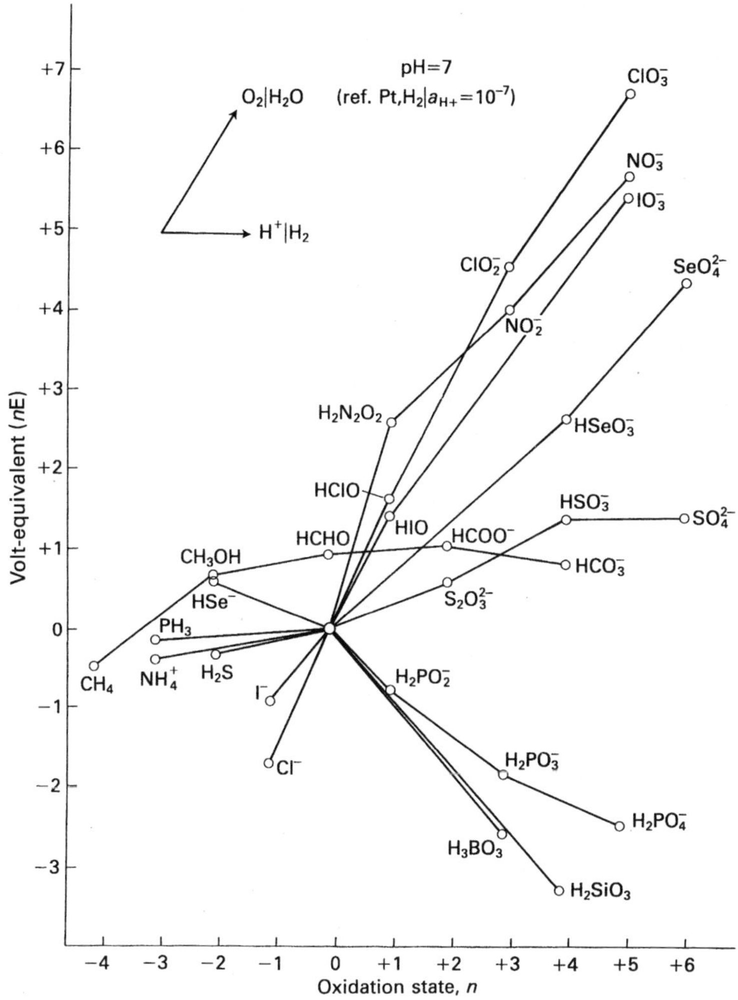

Fig. 1.8 Oxidation state diagram for non-metals at pH = 7 (*n E* versus *n*) taking into account protonation reactions.

next, but the number of sulphur and selenium states in biologically relevant chemical species is much more limited than that of carbon. Against the background of this non-metal chemistry it is interesting to note again the restricted valence state range of the transition metals (Fig. 1.9) as is also illustrated by the steepness of the curves in Fig. 1.7 as opposed to those in the centre of Fig. 1.8. Note that the data apply strictly to pH = 7.0 and that some compartments in cells are at a very different pH.

1.4.2 Element concentrations in aerobic environments and in some organisms

From what has been said, it can be concluded that the biological elements are either available from the atmosphere, i.e. as gases, in the cases of C(CO_2), of O (O_2), of N (N_2), and of H (H_2O), or from water solutions in which they are likely to be present in simple or not very complex species, as shown in Tables 1.3 and 1.4, due to the oxidizing conditions of present-day waters.

Some of the peculiarities of the influences of redox potentials of the ele-

Table 1.3 Forms in which main biological elements occur in aerated soil, water, rivers, lakes, and sea (or in blood plasma)—simple species

Cations	Anions	Neutral species
NH_4^+, H_3O^+	HCO_3^-, CO_3^{2-} NO_3^-	H_2O, $B(OH)_3$
Na^+, K^+	$H_2PO_4^-$ HPO_4^{2-}	CO_2, SiO_2, (nH_2O)
Mg^{2+}, Ca^{2+}	OH^-, F^-, Cl^-, Br^-, I^-, SO_4^{2-}	N_2, NH_3, O_2

Table 1.4 Inorganic speciation of some trace metals (essential, not essential, and toxic) in marine systems at 25°C, 1 atm. pressure, 3.5 per cent salinity, and pH = 8 (computer model calculation from known total concentrations and available stability constant data)

Metal	Main species
Al^{3+}	$Al(OH)_3$ 100%
Cr^{3+}	$Cr(OH)_3$ 100%
Mn^{2+}	Mn^{2+} 58%, $MnCl^+$ 37%, $MnSO_4$ 4%, $MnCO_3$ 1%
Fe^{2+}	Fe^{2+} 69%, $FeCl^+$ 20%, $FeCO_3$ 5%, $FeSO_4$ 4%, $Fe(OH)^+$ 2%
Fe^{3+}	$Fe(OH)_3$ 100%
Co^{2+}	Co^{2+} 58%, $CoCl^+$ 30%, $CoCO_3$ 6%, $CoSO_4$ 5%, $Co(OH)^+$ 1%
Ni^{2+}	Ni^{2+} 47%, $NiCl^+$ 34%, $NiCO_3$ 14%, $NiSO_4$ 4%, $Ni(OH)^+$ 1%
Cu^{2+}	Cu^{2+} 9%, $CuCO_3$ 79%, $Cu(OH)^+$ 8%, $CuCl^+$ 3%, $CuSO_4$ 1%
Zn^{2+}	Zn^{2+} 46%, $ZnCl^+$ 35%, $Zn(OH)^+$ 12%, $ZnSO_4$ 4%, $ZnCO_3$ 3%
Mo(VI)	MoO_4^{2-}(100%)
Cd^{2+}	Cd^{2+} 3%, $CdCl^+$ 97%
Pb^{2+}	Pb^{2+} 3%, $PbCl^+$ 47%, $PbCO_3$ 41%, $Pb(OH)^+$ 9%, $Pb(SiO_4^-)$ 1%
Si(IV)	H_4SiO_4 92%, $H_3SiO_4^-$ 5%

Data from: Turner, D., Whitfield, M., and Dickson, A.G. (1981). *Geochimica et Cosmochimica Acta*. 45, 855

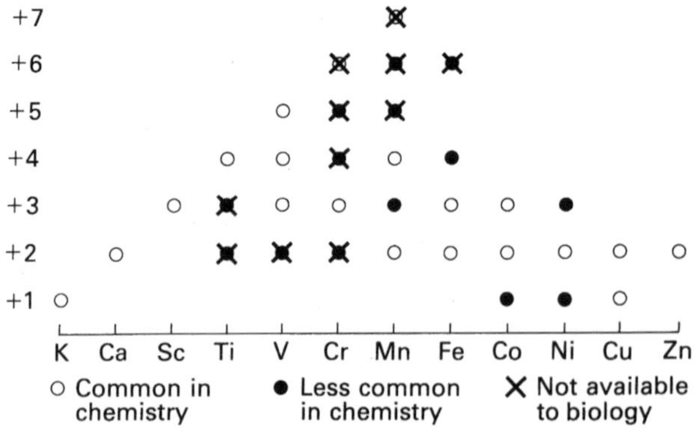

Fig. 1.9 Oxidation state patterns for the elements from potassium to zinc, compiled from a survey of aqueous monatomic ions and aqueous oxyions, stoichiometric oxides, halides, and other simple salts.

ments at pH = 7 appear in these tables, but, in addition, it is the hydrolytic reactions of the elements with water in the presence of dioxygen which have determined the availability of elements today. Thus, in the transition series, iron is only present as a minute trace in water as the redox potential restricts this element to the state Fe^{3+}, which precipitates as insoluble $Fe(OH)_3$, so that

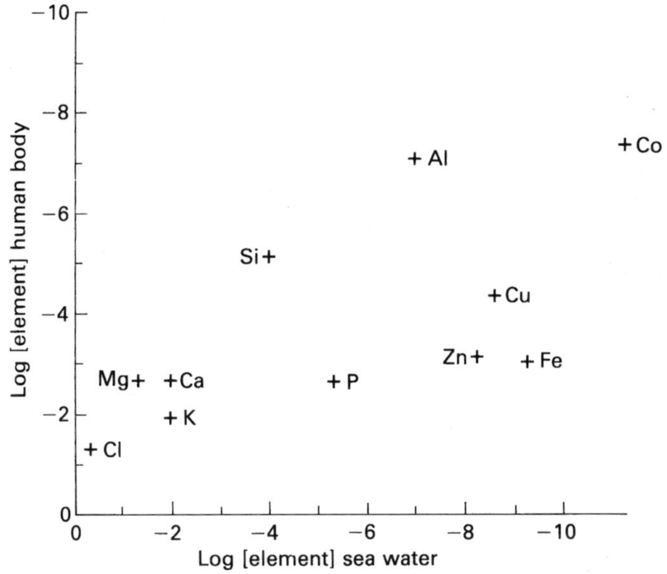

Fig. 1.10 A comparison between the molar concentrations of elements in the human body and in sea water. Against a general trend there are strong variations indicative of some of the problems faced in maintaining an element balance in man.

it cannot be obtained as a soluble cation, e.g. Fe^{2+}, or a soluble anion, such as FeO_4^{2-}. The above environmental availability restricts to some extent the possibilities of development of life, but, as we have seen, Nature has found ways of circumventing this type of limitation when there is a strong requirement for some function as a result of evolutionary pressure, and can extract the elements it needs, sometimes to a quite remarkable extent, as can be appreciated from Table 1.5, and Fig. 1.10.

It is worth stressing that Table 1.5 includes elements that are not considered

Table 1.5 Abundance of chemical elements in sea water and in some marine organisms (dry weight)

Chemical element	Total concentration in sea water (p.p.b.)		Maximum concentration (p.p.b.) in marine organisms
	*	†	‡
Ti	<1	0.5	10 000
V	1.5	1–2	280 000
Cr	0.3	0.2–0.25	1400
Mn	0.03	0.04–0.1	4100
Fe	0.05	0.01–0.4	86 000
Cu	0.3	0.08–0.3	7500
Mo	10	10	6400
Nb	0.01	0.001	98 000
Tl	0.01	0.01	700

p.p.b., parts per billion

*Bruland, K.W. (1983). In *Chemical oceanography*, Vol. 8. (ed. J.P. Riley and R. Chester), pp. 157–220. Academic Press, London.

†min–max values: Emsley, J. (1991). *The elements*. Clarendon Press, Oxford.

‡Goldberg, E.D. (1957). *Memoirs of the Geological Society of America*, 67, 345.

as 'essential', e.g. Ti, Nb, and Tl. Their extraction may be determined by competition with the 'naturally' required element, by substitutive capabilities (as is the case for Nb, which replaces V in some marine species of *Ascidiae*) or because the element extracted is indeed essential for the species concerned but essentiality has not yet been established. This same reasoning applies to many other elements which occur irregularly in living tissues and can be accumulated by certain species of plants and animals (cf. Table 1.2) but do not meet the requirements for essentiality as far as such studies have gone. Elements such as aluminium, bismuth, germanium, lithium, lead, rubidium, and titanium are likely candidates, but the prevalent view is still that these or other (non-essential) elements (about 30 of them) may have been acquired through contact with the environment, either by physical processes (adsorption, ingestion of particles, contamination of skin, etc.) or by chemical processes, in competition with the 'normal' essential elements for their binding sites in biological systems. In this case it is possible, if the competition becomes permanent, that some of these biological systems will adapt to the 'abnormal' conditions and change the (in principle) adverse effect of a non-essential element into a beneficial one. They will then have started to 'use' the new element; this is actually one of the general principles of evolution whereby, by adapting to new conditions, living species become more fit and increase their possibilities of survival. The cases of dioxygen, zinc, and copper are clear examples of this trend.

Once again we stress too that the above description of hydrolytic equilibria applies to the majority of environments of life today. It does not apply to strictly anaerobic environments, to high-temperature reducing waters such as are found in deep ocean vents, nor did it apply in the early history of life. Table 1.6 shows some changes which occur on changing from anaerobic (early history of earth) to aerobic (recent history of earth) conditions. In fact life underwent a dramatic transition with the coming of dioxygen (Fig. 1.11). We return to this problem in Section 1.5.

Figures 1.4 and 1.5 point to another problem for living systems. The avail-

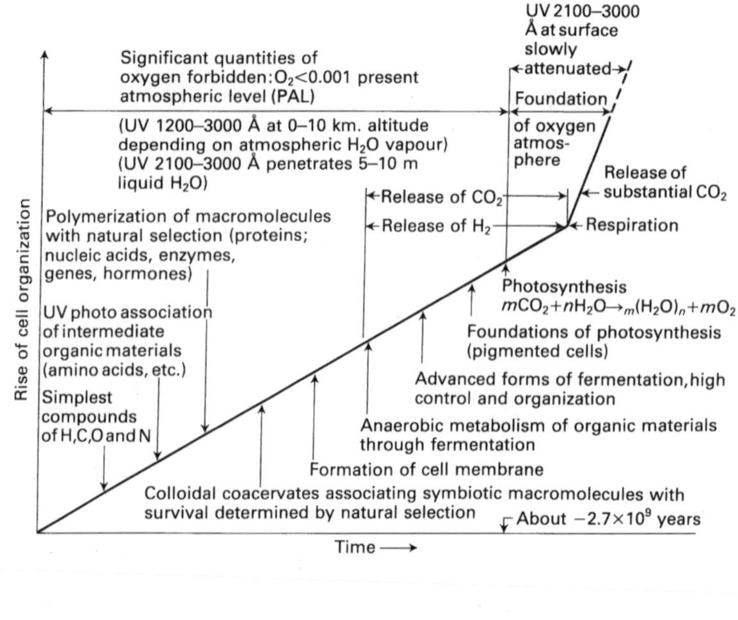

Fig. 1.11 The involvement of the elements in life has to be seen against the progression of the development of organization over historical time. We suppose earth has existed for over 4×10^9 years. We guess that organic molecules were formed haphazardly for a long period both through the action of light and on catalytic surfaces. Organization of chemistry in space and time evolved and is called living once it could reproduce. For a long period only simple cells, prokaryotes, developed. They had no internal boundaries. At some stage from about 2.7×10^9 years an evolution of dioxygen in the atmosphere developed due to prokaryote metabolism. A vast change took place by some 570×10^6 years ago bringing internal cellular structures of great diversity and multicellular organisms. [This figure is after Berkner, J.V. and Marshall, L.C. (1965).

Table 1.6 Forms of elements and (availability)

Element	Reducing environment	Oxidizing environment
Iron	Fe (II) (high)	Fe (III) (low)
Copper	As sulphide (low)	Cu (II) (moderate)
Sulphur	HS^- (high)	SO_4^{2-} (high)
Molybdenum	$[MoO_nS_{4-n}]^{2-}$, MoS_2 (low)	MoO_4^{2-} (moderate)
Vanadium	V^{3+}, V(IV) sulphides (moderate)	VO_4^{3-} (moderate)

Note the switch in availability of iron relative to copper that may have has a profound effect on evolution as we will show later. Vanadium sulphides are quite soluble.

ability of the elements is a bio-geological problem on an immense time scale. In the past it has adjusted slowly allowing adaptation of species and development, but man is now having an effect on element availability over a short period of time. He has learned to apply enormous energy sources in an intense fashion to mining and agriculture (note the effect on the atmosphere). The benefits to human kind today are very substantial, but a glance at the figures in this chapter gives cause for concern. Biological systems are acting in adverse circumstances keeping out some common elements and taking in others which are not very available. What will happen if the availability of any element, even a rare one, changes by a factor of, say, 10 within, say, a period of 500 years? One particular concern today is with the acidity of water and release of aluminium. While concentrating upon this problem, constant vigilance will be needed in the immediate future to avoid the adjustment of availability of many other elements, e.g. Sn, Pb, Hg, Ni, and Be. Added to these problems are those of radioactive nuclides, which are not of concern in this book. A further problem is the changing amount of O_3, CO_2, and so on in the atmosphere, but these last are problems of speciation, i.e. of (compounds), not of elements.

1.4.3 Solubility products

In the oxidation state diagram for metals at pH = 7.0 (Fig. 1.7) certain of the elements are in simple hydrated ionic states or are present as simple anions. As we have stated, it is this that makes them highly available. Other elements are present as insoluble precipitates, e.g. TiO_2, and in such states these elements are virtually not available. The magnitude of solubility products of the hydroxides (oxides) of monovalent and divalent ions (Fig. 1.12) does not cause serious precipitation of any of the related elements of major biological interest, although there is an increase in insolubility following the Irving-Williams series. The situation for the trivalent ions is very different and all trivalent hydroxides are very insoluble at pH = 7; for example, $[Fe^{3+}]$ cannot exceed 10^{-18} M. It is this that removes elements such as Sc, Ti, Cr, Mn, Fe, and Ga from aqueous oxygenated waters. (Note that Mn^{2+}, Co^{2+}, Ni^{2+}, and Cu^{2+} do not give precipitates from 10^{-4} M solutions at pH = 7 while Ti (TiO_2), Cr^{3+}, Fe^{3+}, Co^{3+}, and Ga^{3+} ($M(OH)_3$) all do). While these precipitations had and have a profound effect on life in aerobic conditions, much as does the insolubility of calcium, strontium, and barium (large cations) with oxyanions (such as phosphate and sulphate), there must have been an even more important influence on early life through the insolubility of sulphides (Fig. 1.12). We shall see in Chapter 20 how the concentrations of free ions follow approximately the inverse of Irving–Williams stability series here and inside all cells so that this series is a characteristic of life's chemistry. Like the genome in

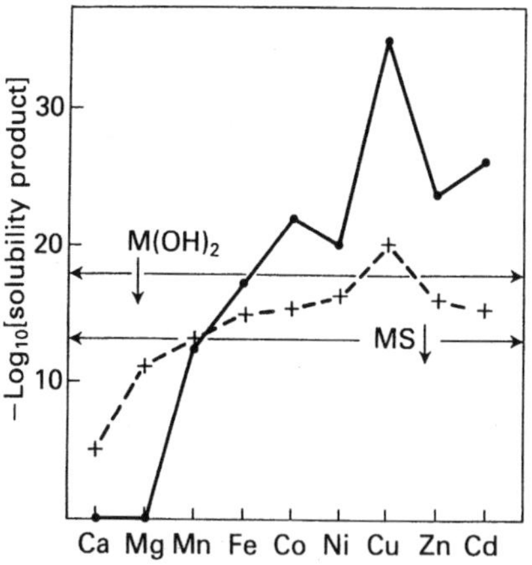

Fig. 1.12 The solubility products of hydroxides (- - - -) and sulphides (———) which have been of extreme importance in evolution. The horizontal line indicates that solubility product which gives a precipitate at pH = 7.0 when [M] = 10^{-4} M

which there is but one code there is one basic metallome in the cytoplasm, which is related to the environment.

1.5 Sulphide chemistry in water (anaerobes)

On good grounds it is believed that the early atmosphere and the early sea did not have a redox potential range between the H_2 and the O_2 potentials at pH = 7.0. These are the conditions we have described for aqueous solution chemistry so far. It is much more probable that the early reaction systems in water including life were limited by the H_2S/S_n and H_2/H^+ potentials which cover the range 0.0 to –0.4 V at pH = 7. We should therefore reconsider the oxidation state diagrams, Figs 1.7 and 1.8, so that we include the reactions of H_2S/S_n as being more important than those of H_2O/O_2 and ignore the possibility of the establishment of redox potentials outside the anaerobic range. Since the redox potential of $(CHOH)_n$ type organic compounds is also around 0.0 V both in oxidizing and reducing directions this allows the full range of organic compounds from CO_2 to CH_4. Nitrogen may have been reduced to NH_3 or HCN when it would have been available. However, these early redox conditions had a dramatic effect on the availability of other elements, especially metals, and their reactions (Table 1.6). We attempt a semi-quantitative analysis, and to do so a value has to be put on the free H_2S concentration. We shall assume that it was around 10^{-3} M in water.

We shall start the discussion of metal partition between oxides and sulphides with one of the most important reactions for biology or indeed mineralogy under these conditions which can be written schematically as a competition between sulphide and oxide for hydrogen and metals

$$MO + H_2S \rightleftarrows MS + H_2O.$$

Here we include under MO and MS the variety of hydrolysed forms of oxides and sulphides from OH^- and SH^- compounds to the oxy- and thiolate anions such as MO_4^{2-} and MS_4^{2-}. Early in the periodic table in both of the first two rows, i.e. from hydrogen to chlorine, and in the first four groups, sulphide chemistry is of little consequence. However, as soon as we consider transition metal or groups 12 to 16 chemistry we find remarkably stable sulphur compounds in water given even this relatively low H_2S presence. The easiest way in which to see this chemistry quickly is to look at the old-fashioned (unfortunately) qualitative analysis tables for students. The separation procedures used in those tables depend on the above equilibria and allow first quantitative separation in acid solution of the most insoluble sulphides, i.e. of the elements Mo, Cu, Ag, Cd, Sn, Pb, As, Sb, Bi, and many later second and third transition series metals which are never seriously involved in biological systems except as drugs. Notice immediately that, while copper is precipitated as Cu(I), many of the other elements are precipitated in high oxidation states, e.g. Mo, Sb, Sn, and As. Of these metals, Mo, Sn, As, and Sb sulphides dissolve in an alkaline H_2S solution to give thio-anions. V and W sulphides are soluble in both acid and alkali, forming soluble anions, but they precipitate when saturated H_2S solutions of the alkaline thiosalts are acidified somewhat below pH 7. Thus sulphide removed Cu, Ag, Cd, Hg, and Pb very effectively from participation in early biology but it left some V, (Mo), W, Sn, As, and Sb as available elements at pH = 7. In turn the removal of the first group of heavy metals left selenium available as H_2Se. Reduction of Mo to give MoS_2, contrast W, greatly lowered its possible use.

Notice too that the chemical switch from oxide to sulphide chemistry going from P to As (and Sb), from Ge to Sn, from Cr to Mo is very dramatic and allows strict separation of these pairs of elements to be made easily. The distinction occurs in both the higher and the lower valence states of these elements and, whatever the theoretical explanation given, it was critical for the development of the geochemistry of the earth and its biological systems as we shall see in Chapter 20.

Now, at pH = 7 (and in more alkaline solutions) the qualitative analysis tables show that a further groups of elements precipitate as sulphides, including Fe, Co, Ni, Zn, and less markedly, Mn (see Fig. 1.12). In the presence of 10^{-3} M H_2S, i.e. the presumed condition of early life, the concentration of all these metals was therefore limited, but we must remember that at a pH close to the solubility product of any precipitate there will always exist small clusters in solution. We may well believe that biology took advantage of these sulphide clusters in the cases of Fe, Co, and Ni, while Cu was virtually excluded since it is so insoluble (Fig. 1.12). There is little or no copper in archaebacteria to this day and Zn tends to be removed with copper and does not readily give clusters.

Returning to the equilibrium

$$MO + H_2S \rightleftarrows MS + H_2O,$$

Fig. 1.12 shows that for two elements, Mn and Fe, the equation is balanced in a very intriguing and different manner. The situation is complicated by the variety of oxidation states and the geometry demands (electronic states) of the two elements. However, the following points can be established: firstly, in the presence of H_2S at pH = 7 we need only consider oxidation states of these elements from II to IV, since higher oxidation states are very powerful oxidizing agents which are not stable with either oxide or sulphide as ligands in water but generate dioxygen and sulphur respectively. Considering the lower oxidation states we find that the tendency to form sulphide rather than oxide links is

$$\text{Fe(II)} > \text{Fe(III)} \gg \text{Mn(II)} > \text{Mn(III)} \gg \text{Mn(IV)} \gg \text{Fe(IV)}.$$

(It is probable that this order is a consequence of the exclusion of sulphide chemistry by redox reactions of Fe(IV) and Mn(III) and by the geometric demands (of Mn(IV).) The consequences of these stabilities allowed Fe/S cluster chemistry, Fe(II) and Fe(III), to dominate early biology since iron must have been the most abundant metal and the Fe(II) sulphide is quite soluble below pH = 7. The link between availability of metal sulphide complexes and the evolution of biological systems will be stressed later in the cases of nickel, molybdenum, vanadium, and tungsten, as these elements with iron still dominate the simple chemistry of N_2, H_2, O_2, CO, CH_4, etc., in biology and within sulphide compounds.

Turning to manganese we see that its chemistry is more closely connected to groups 1 and 2 in that the binding to sulphide is never dominant over binding to oxide (Fig. 1.12). The fact that it also has, apart from Mn(II), two just accessible oxidation states Mn(III) and Mn(IV) (see Fig. 1.7) made it possible for the fluctuating fortunes of evolution to uncover the photochemical generation of dioxygen unimpeded by sulphide reactions on manganese centres. The whole of advanced life depended and depends on the fact that the chemistry of manganese is dominated by oxide and not by sulphide under biological conditions, while the reverse is true for Fe, Ni, Zn, Mo, and Cu.

Amongst the non-metals we have already mentioned that oxide chemistry dominates them all except for hydrogen itself (H_2S) and sulphur (and seleni-

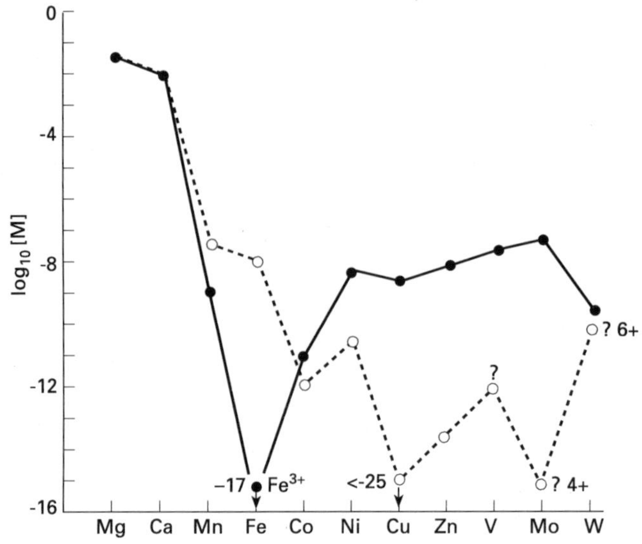

Fig. 1.13 The concentrations of free elements [M] in the sea; ○ in the primitive sulphide sea, ● in the aerated sea of today.

um). This is not due to a lower absolute affinity for oxygen but a depleted availability of this element on earth after combining with other elements for which it has relatively higher affinity. In the reducing conditions *inside* cells to this day H_2S has a limited presence while sulphate is not stable. Outside cells in aerobic conditions sulphide is not stable whilst sulphate is.

It can be seen from this section that the forms of life are dependent on the state (and availability) of chemical elements, while the state of the chemical elements in the environment can slowly become dependent on the nature of life (Fig. 1.13). The chemistry of the surface of any planet dictates living forms and, as growth intensifies, this surface chemistry is adjusted by life and the system, planet and life, evolves interactively until they could become almost one. Lovelock and others proposed that earth has reached this last condition (the so-called Gaia hypothesis) an idea which was received with deep doubt by some but has now gained some acceptance (see Lovelock and Volk in the references, and Chapter 20). In fact, the sea is not in equilibrium with the land and the atmosphere and the ability of organisms to respond to certain changes may well be very limited.

Against this background we introduce a note about the care with which biological systems maintain the element and small compound concentrations within themselves, so-called homeostasis, which in a sense is opposed to evolution. There is a very strong conservative character in biology which makes its forms very slowly adaptable. The importance of homeostasis will become apparent in Chapter 20, and here we give but a brief note.

1.6 A note on homeostasis

The famous French physiologist Claude Bernard called attention to the fact that all living species have two different 'environments'; one, the external surroundings, the 'habitat', and the other, the internal 'milieu', made up of the biological fluids, e.g. the blood plasma. The external environment in its physical condition or chemical composition can change to a large extent and even abruptly, e.g. in tidal waters, but the internal medium must be kept approximately constant or else the metabolic reactions cannot be co-ordinated, the mechanisms of adaptation cannot function, and the species will die. (Bernard was not to know that the environment is largely biologically generated as far as, especially, its inorganic chemistry is concerned and that availability of elements has changed during evolution). To see this clearly it is enough to observe some simple physical facts as to what happens in our everyday life: the body temperature may not be below 35° nor above 42°C and, when the external conditions are such that there is a risk of these limits being exceeded, several corrective mechanisms start operating to restore the normal conditions, e.g. muscle shivering to increase the temperature and sweating to lower it. For other animals there are similar mechanisms and protective devices, such as insulating fat or fur in cold environments, increased surface to ease losses by convection (e.g. an elephant's ears) or to promote evaporation (e.g. a dog's tongue) in the cases of excess heat. The same can be said for the chemical composition of biological fluids, e.g. even small deviations from the normal salt (NaCl) composition of the blood reflect a state of disease and prolonged abnormal situations will endanger life. This means that only those species with adequate regulating internal mechanisms can survive in external environments subject to changes. Many plant and animal species do not have very well regulated mechanisms and they must remain in biological niches where conditions are stable. This is the case, for example, for some primitive organisms, e.g. many archaebacteria, which, lacking the adequate adaptative possibilities, are forced to live in anoxic media. Similarly, we can only live in aerobic media.

The same argument has been invoked to support the sea or, perhaps better, some sheltered closed ponds, as the birthplace or at least the major development environment for life. The earth's crust is too much exposed to sudden changes of temperature, pressure, or humidity and to natural phenomena, e.g. winds, volcanoes, freezing, etc., to be suitable for the staging of the first phases of the processes leading to life. The large water masses afford far better conditions since the high specific heat and viscosity of water ensure fewer thermic or mechanical disturbances and aquatic media are clearly suitable for chemical reactions, enable easier transport of nutrients, and increase their availability. As mentioned in Section 1.5, even CO_2, H_2S, and O_2 are reasonably soluble in water. Of course, there are other possibilities here aiding the origin of life, amongst which is the interface between water and a solid surface which will be discussed later.

This conservation of conditions in a living system, homeostasis, see Chapter 20, does not correspond to a state of equilibrium in thermodynamic terms (which would be equivalent to death) but to a series of related and controlled steady states in a dynamic process of continuous material, charge, and energy flows through the cells, with forced and fixed directions. Inside the cells there will then be a vast array of chemical transformations, leading to hundreds of compounds, development and reproduction, and the overall process requires energy to effect reactions (Fig. 1.3). It is necessary to pump in reagents, to pump out the products, and to maintain the physical stress needed for the structural organization of these cells, as well as the osmotic balance and material or electric potential gradients. Due to these gradients the cell can respond to the outside environment and its internal metabolic reactions can be compensated for changes of external conditions, given that the system as a whole has feedback regulation. We shall have to examine these points very carefully. We shall see that they are very dependent on inorganic chemicals.

We can therefore say that the steady homeostatic states are definable in terms of energy supply and available concentrations of simple compounds *plus* energy utilization and concentration gradients. This addition to the more traditional concept of homeostasis is so important that it may be at the origin of the first manifestation of 'life' properties. We come again to this subject in Chapter 20

1.7 **Survival and evolution**

Coincidental with the general development of living organisms has been the division into a multiplicity of species. This separation of species within the larger divisions, such as those of the plants and animals, is known to be related to the inability to cross-fertilize, i.e. to combine genes (see Sections 2.21 and 6.14). However, failure of genetic crosses cannot explain why only certain kinds of species exist. For example, there are no photosynthesizing animals and no plants with nervous systems. The implication is that certain systems of chemical processes and hence of elements have a higher survival value than others (see Aside). Clearly, we have to show not just the relationship between genetic structure and the system, in the organism, important though this is (see Section 6.13) but why the whole living system has arisen utilizing a selected group of chemical elements, and why small variations of this composition leads to speciation (see our more recent books and Chapters 7 and 20). We have to recognize that a given free element content of a cell or organism, its 'metallome', reflects its speciation in a given state just as does the DNA, its genome, which is then 'expressed' (in the proteome). There must be a link between the two associated with protein production, the proteome, and feedback from element and protein content to DNA expression. Clearly, however, the uptake of elements is partly dependent upon their environmental availability and external energy. In effect as the environment changes it must force evolution and changes in DNA are then necessary. The manner of this 'forcing' is at present uncertain. In Sections 2.21 and 6.14, we therefore give a very short introduction to genes and their transcription. Later in sections of Part II we show how the system operates for individual elements separately as far as we know it as yet. We also give a brief introduction to complete cells, organized systems of all twenty essential elements in Chapter 7, and a fuller version in Chapter 20. The integrated systems are analysed in detail in our two more recent books listed in the Preface to the Second Edition where the issue of survival of dynamic systems is treated at greater length.

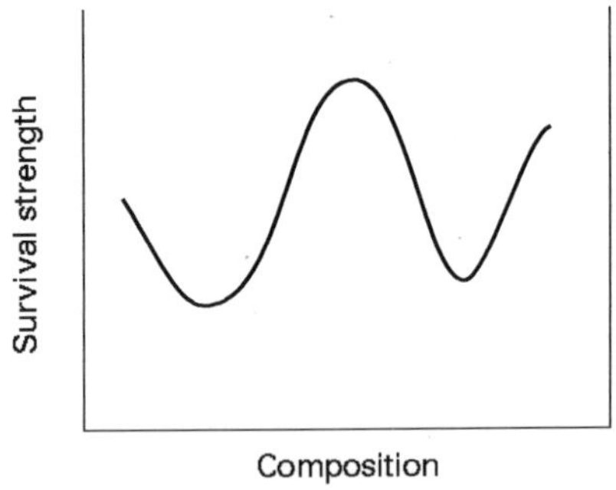

The variation of survival value with composition.

1.8 **Conclusion**

In this chapter we have described the abundance and availability of the elements and their overall distribution in living systems. The only chemical combinations of elements with other elements which have been considered in any detail were those formed between all the elements and hydrogen, oxygen, and sulphur. These are the basis of the redox–hydrolytic equilibria of elements in water, which dominate availability through inorganic reactions in the environment (see Fig. 1.13). The nature of living systems is different. Through the use of energy they make a great variety of 'organic' compounds, often based on just C, N, H, O, sometimes with P and S, from a selected set of small molecules and often in controlled reducing conditions, to which other 'inorganic' elements can gain controlled access and with which they may combine. We have seen that there is a common basic set of these elements and their reactions (organic/inorganic) observed in biological chemistry, in all life-forms.

In effect, the variety of living organisms must not be allowed to hide their common physical and chemical features which have existed for almost 4×10^9 years at a closely fixed temperature and pressure. Apart from the obvious organic chemical combinations between C, H, N, O, P, and S which give both physical cellular constructs and take part in chemical activities, *all* cells, from all periods of time, have essential quantitative demands for Na, K, Mg, Ca, Fe and/or (some of) Mn, Co, Zn, Mo, Se, Cl. Together they generate the chemical system which we call life, whether or not DNA or RNA is essential. In different forms of life there are limited quantitative variations of these elements and some additional elements have been incorporated including Sr, Ba, B, Si, As, Br, I, V, (Cr), Ni, Cu, Cd, and W, but the essential system is fixed. We can, therefore, view evolution in terms of the developing quantitative use of the absolutely required elements together with the innovative employment of this secondary set, no matter how their uses became coded, (see Fig. 1.3). Thus, we can analyse quite separately from the consideration of genes much though they allow reproduction and qualitative indication of protein content.

Finally, we note that in this book we shall have very little need to refer to organo-metallic chemistry with the exception of a few cobalt, nickel, and iron compounds described in the corresponding chapters. Biological chemistry is overwhelmingly chemistry in water—all cells are 80–90 per cent water. Again, we shall not refer to the analytical methods through the use of which much of the data were collected. We stress that we wish to appreciate the functional value of elements *in vivo* and their complex network of interactions which is often hidden by all forms of reductionist examination, based on extraction of molecules, much though we need this approach to help our formulation of the whole system.

Further reading

Berkner, I.V. and Marshall, L.C. (1965). On the origin and rise of oxygen concentration in the Earth's atmosphere. *Journal of Atmospheric Sciences*, **22**, 225–61

Bowen, H.J.M. (1966). *Trace elements in biochemistry*. Academic Press, London.

Cox, P.A. (a) (1989). *The elements*. Oxford University Press, Oxford. (b) (1995). *The elements on earth*. Oxford University Press, Oxford.

Emsley, J. (1991). *The elements*. Clarendon Press, Oxford.

Gerstein M. and Levitt, M. (1998). Simulating water and the molecules of life. *Scientific American*, **279**, 75–9.

Goldberg, E.D. (1957). The biogeochemistry of trace elements. *Memoirs of the Geological Society of America*, **67**, 345.

Hewitt, E.J. and Smith, T.A. (1985). *Plant mineral nutrition*. English Universities Press, London.

Ledin, M. (2000). Accumulation of metals by microorganisms. *Earth-Science Reviews*. **51**, 1–31.

Lovelock, J. (a) (1979). *Gaia—a new look at life on earth*. Oxford University Press, Oxford. (b) (1988). *The ages of Gaia*. Oxford University Press, Oxford.

Margulis, L. and Sagan, D. (1995). *What is life*? Simon & Schuster, New York.

McDowell, L. and Russell (1992). *Minerals in animal and human nutrition*. Academic Press, New York.

Phillips, C.S.G. and Williams, R.J.P. (1966). *Inorganic chemistry*. Oxford University Press, Oxford.

Turner, D., Whitfield, M. and Dickson, A.G. (1981). Inorganic speciation of trace metals. *Geochimica Cosmochimica Acta*, **45**, 855.

Volk, T. (1997). *Gaia's body*. Copernicus, New York.

Wilson, E. O. (1992). *The diversity of life*. Allen Lane, The Penguin Press, London.

Woese, C.R., Kandler, O. and Wheelis, (1993). Towards a natural system of organisms: proposal for the domains Archaea, Bacteria and Eukaria. *Proceedings of the National Academy of Sciences USA*, **87**, 4576.

See also our two recent books:

Williams R.J.P. and Fraústo da Silva, J.J.R. (1996). *The natural selection of the chemical elements*. Oxford University Press, Oxford

Williams R.J.P. and Fraústo da Silva, J.J.R. (1999). *Bringing chemistry to life*. Oxford University Press, Oxford

CHAPTER 2

The principles of the uptake and chemical speciation of the elements in biology

2.1 Introduction to chemical speciation

The essential features of the study of the chemistry of a compound or system containing any number of elements involve (1) isolation of the compound, (2) analysis of its composition, (3) description of its structure, and (4) examination of properties, including stability, spectral features, and kinetic activities. These reductive approaches provide one way to approach an understanding of integrated dynamic systems, here living organisms, hoping that the knowledge of the isolated compounds will help in a description of a cell. As stated in Chapter 1, we know the elementary compositions of many living organisms and even those of many of their molecular units. We treat these specific molecular units and their structures in Chapters 8–19, handling each element separately. A more difficult problem is to see how the elements, in compounds or not, could become incorporated in the organisms, before we ask how they operate together to give a cell's organization or system. One obvious way is to examine their selected combinations in cells to see if each one has its strongest equilibrium thermodynamic binding under the cellular conditions. Since we cannot be sure of the range of compounds which cells can produce we need to study *designed model systems* to uncover the strengths of the selective binding constants of the available elements to possible different model organic molecule partners using the full range of O, N, S potential donors of cells. Another possible mode of incorporation is a kinetic pathway which starts from an initial complex formation followed by passage through cell membranes and solutions, while exchanging partners but not the free element, and ends with the final insertion of the element into an organic site. We shall look for the general ways in which this kinetic pathway incorporation can be managed in this chapter, but we describe the rate controlling steps in Chapter 4. We shall find that different elements are bound or incorporated in different ways in cells (see Fig. 2.1).

We start this chapter, therefore, with a more detailed look at general features of modes of uptake based on molecular combinations before describing binding strengths of model complexes and kinetic modes of incorporation. We stress again that cells are limited by the fact that their combinations arise in aqueous solution, in a narrow temperature range and at a pressure close to one atmosphere. Life can exist under much more extreme conditions, but overwhelmingly it is confined as described, so we restrict model studies to the same conditions. At the end of the chapter we shall show that, in essence, the kinds of observed element distribution in cells, not the intimate details, can be appreciated in fair measure on the basis of the different stabilities and insertion reactions of model complexes.

2.1.1 General aspects of uptake and rejection

We shall now examine how biological systems can take up and distribute differentially the elements they need for their functions. This will require us to discuss in detail the general mechanism used to achieve physicochemical movement and capture and the basis of the competition of the elements for biological organic molecular or solid structures, most frequently in aqueous solutions. Since the plants are the primary sources of chemical elements for animal life they will be considered first, but even plants depend on microorganisms for several elements they need, e.g. nitrogen. We will have to be concerned with the mechanisms operating at the cell level in different types of cells. General principles are discernible, but not all biological systems behave alike with regard to all elements.

Let us briefly examine how a plant appropriates elements present in its natural surroundings, e.g. the soil. The uptake of a certain element, M, involves a series of reactions which are interrelated as

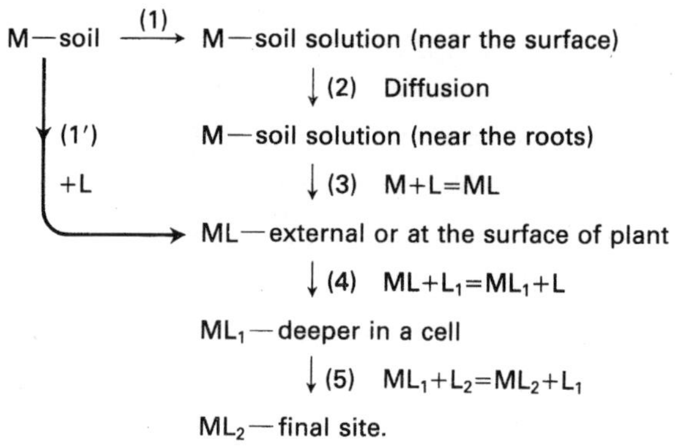

Many of these reactions will not come to rapid equilibrium and we shall consider both thermodynamic and kinetic controls in the overall uptake. Step 1 is determined, essentially, by the occurrence and then the solubility in water of the mineral containing the element M in the soil (see Chapter 1). In some cases step 1′ is operative, when the plant (or micro-organism) sends out a complexing agent L which solubilizes and captures the necessary element. Note that some plants acidify the soil around their roots to aid these steps. Step 2 is a diffusional process and depends on the nature of the chemical species containing M, on the distance to the plant roots, and on the nature (physical and chemical) of the soils. Plants break up soils to aid diffusion. Steps 1′, 3, 4, and 5 are complexing reactions under thermodynamic or kinetic biological control; in particular, step 5 can be a kinetic device to ensure irreversible uptake of an element which may take the form of a simple barrier to

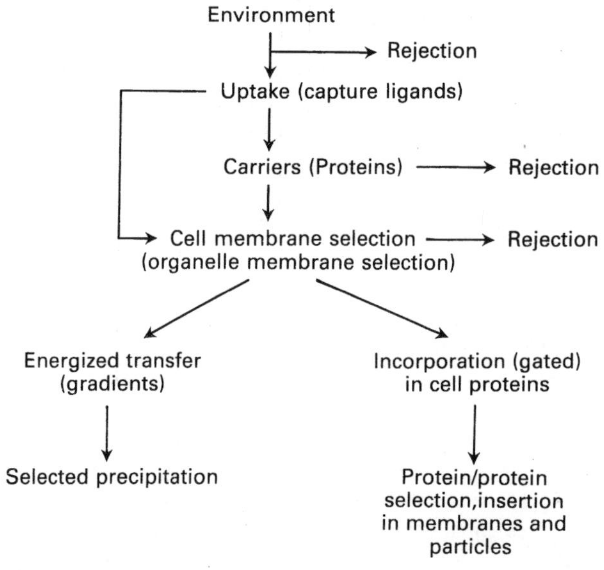

Fig. 2.1 Some pathways for the uptake and incorporation of elements.

Table 2.1 Affinity of metal ions for some plant surface tissues

Ion	Species	Organ or tissue	log *K*
K^+	*Hordeum vulgare* (barley)	Excised roots	1.68
K^+	*Zea mays* (maize)	Leaf tissue	1.42
K^+	*Elodea densa* (waterweed)	Leaf (whole)	1.00
Rb^+	*Hordeum vulgare* (barley)	Excised roots	1.77
Rb^+	*Zea mays* (maize)	Leaf tissue	1.82
Sr^{2+}	*C. alpestris* (lichen)	Powdered plant	1.95
Ni^{2+}	*U. muhlenbergii*	Powdered plant (pK_a = 5.40)	3.30
Ni^{2+}	*U. muhlenbergii*	Powdered plant (pK_a = 2.85)	3.05
Cu^{2+}	Sugar cane	Leaf	1.87
Zn^{2+}	Sugar cane	Leaf	1.95
Mn^{2+}	Sugar cane	Leaf	1.79

K (for acid centres) = $[MA^+]/[M^{2+}]\,[A^-]$ where A is a surface group. The pK_a, where given, shows the proton affinity for A.

NB There is evidence that lichens have special binding enolates for higher valent metals which makes lichens peculiarly useful in studies of heavy metal pollution. See (for lichens): Tuominen, Y. (1964). *Annales Botanici Fennici*, **4**, 1, Nieboer, E., Puckett, K. J., and Grace, B. (1976). *Canadian Journal of Botany*, **54**, 724. Data for other plants adapted from: Clarkson, D. (1974). *Ion transport and cell structure in plants*, pp. 282–3. McGraw Hill, Maidenhead.

free diffusion or an irreversible reaction (Fig. 2.1). Note that energy is used in two ways: (1) in synthesis of ligands; and (2) in pumping of M, possibly as ML, across boundaries. As one might expect, in some of these steps oxidation reduction reactions may occur, as is the case for the metals Fe^{3+} and MoO_4^{2-} and the non-metals NO_3^-, SO_4^{2-}, O_2, and N_2, and many of these redox steps are also effectively irreversible.

Many of the uptake mechanisms are not truly specific, at least in the primary steps, and, given the possibility of competition, initial selection is based on the concentration and binding of the various elements near the plant root (or the absorbing cells). Since the affinity of the elements for common surface binding centres (organic carboxylates, phosphates, sulphates, etc.) does not vary much, selectivity is poor there (Table 2.1). For instance, the affinity of the alkaline-earth and first-series transition-metal ions for such ligands, represented by the stability constants of their complexes, shows little selectivity (see Table 2.2). Hence, many plants and especially lichens can concentrate elements on their surfaces, some of which are not considered as 'biological' (cf. Table 1.2 and Fig. 1.2) and, in some cases, can be toxic. *All* elements, even the essential ones, can be toxic if the uptake is excessive.

The overall process described so far readily explains why, when non-essential elements are relatively abundant and become available to plants (micro-organisms or animals) for natural reasons or because of accidental contamination or environmental pollution, they may compete successfully with the essential elements and may even be preferentially captured. Over the long term this change in availability is part of evolution. We shall need to see how uptake of essential elements and protection from poisonous ones are managed also by genetic controls (Chapter 6).

Table 2.2 Stability constants (log K_{ML}) of alkaline-earth and first-series transition metal ion 1:1 complexes with some carboxylate, phosphate, and sulphate ligands (T = 25°C)

	log K_{ML}									
	Mg^{2+}	Ca^{2+}	Sr^{2+}	Ba^{2+}	Mn^{2+}	Fe^{2+}	Co^{2+}	Ni^{2+}	Cu^{2+}	Zn^{2+}
Acetate	1.25	1.24	1.19	1.15	1.40	—	1.46	1.43	2.23	1.57
Malonate	1.95	1.85	(1.25)	1.34	(2.3)	(2.5)	2.98	3.30	5.55	2.97
Phosphate	1.60	1.33	1.0	—	2.58	—	2.18	2.08	3.2	2.4
Pyrophosphate	5.4	4.9	(4.7)	(4.6)	—	—	6.1	6.98	7.3	(5.1)
Methylphosphate	1.52	1.49	—	—	2.19	—	2.00	1.91	—	2.16
Sulphate	2.20	2.31	2.3	2.3	2.0	2.3	2.47	2.40	2.4	2.3
Thiosulphate	1.84	1.98	2.04	2.21	1.95	2.17	2.05	2.06	—	2.30

Source: Chemical Society (1964). *Stability constants.* The Chemical Society Special Publications, Nos 17 and 25. The Chemical Society, London.

2.2 The separations that biology can achieve

A simple (prokaryotic) cell has two main features on top of the above external surface as traps for chemical elements: an external membrane system which does not allow diffusion of ions or large molecules; and an internal (cytoplasmic) concentration of trapped synthesized organic molecules, especially polymers, that can bind to many elements. Some such cells have two membranes within the outer wall and the space between them is called the periplasmic space. More evolved cells also have internal compartments (organelles and vesicles) to which the same considerations apply. A cell can, therefore, utilize five types of aqueous traps for elements (see Fig. 2.2):

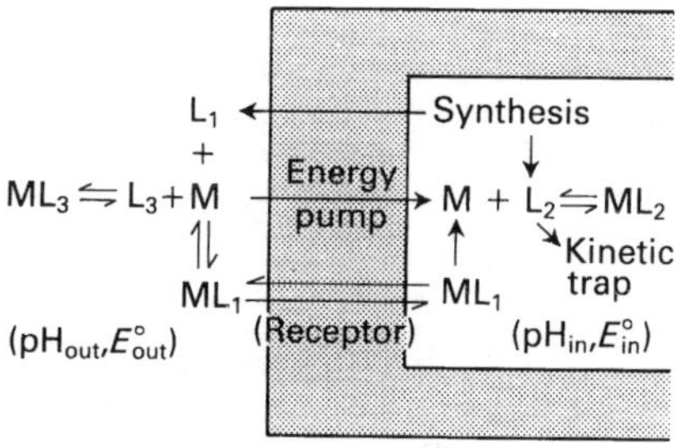

Fig. 2.2 A scheme for steps in the transfer of elements, M, through membranes into cells using ligands. L. Note the energy-dependent control of pH, E^0, synthesis, and pumping.

1. The cell membrane can be used as a physical barrier to diffusion of a free or bound ion, or it can contain a selective pump, in or out.
2. An element can be stored, free or bound, within an internal vesicle or organelle after passing a second membrane.
3. Internal polymers can bind to an element at equilibrium.
4. Internal polymers can incorporate an element forming thermodynamically unstable but kinetically stable bonds, e.g. covalent bonds in organic compounds (insertion reactions may be required).
5. Two trapped elements can combine to form a precipitate at equilibrium.

(Note in passing that many of these steps can be used to reject elements.)

Traps 1, 2, and 4 are always kinetically controlled; 1, 2, and 5 hinder free diffusion, and 4 eliminates the possibility of exchange, though the polymer may or may not diffuse. The non-metals H, C, N, O, P, S and Se are very largely trapped according to trap 4, i.e. by forming kinetically stable but not thermodynamically stable molecules and polymers, such as proteins and polynucleotides. In a few cases this is also true for the halogens Cl, Br and I, but their incorporation requires unusual oxidizing conditions and organic molecules with covalently bound chlorine and bromine rarely occur in higher plants or animals.

This incorporation of the more common non-metals is usually carried out in three steps: reduction of simple molecules or ions, e.g. CO_2, N_2, O_2, SO_4^{2-}, and SeO_4^{2-}, then condensation of the products into monomeric small molecules and finally into polymers. The processes require energy and, although the subsequent degradation is usually a downhill reaction, it has a high activation energy; hence the incorporation of these elements is associated with strong kinetic barriers. (Notice that there is always a steady state concentration of small-molecule non-metal ionic forms too, e.g. HCO_3^-, HPO_4^{2-}, OH^-, H^+, amino acids, and a great variety of small 'substrates'.)

The metals cannot be readily trapped in this way. Sodium and potassium, particularly, form relatively weak complexes; hence they exchange and diffuse rapidly as simple ions. The major possible ways, steps 1 and 2, in which to retain them are by physical barriers to ionic diffusion, i.e. by membranes enclosing aqueous domains into which these ions must be driven using the energy of 'pumps'. This is not to deny that even these ions can be retained in special cyclic chelates, but this is an unusual mode of retention. The small anions containing non-metals can be retained in the same way.

For magnesium and calcium, as well as for the transition and group 11 and 12 metal ions, it becomes increasingly possible to retain them by thermodynamic traps (step 3) in polymers (proteins, polysaccharides), but their final site often is required to be in a state from which exchange is very slow, i.e. for which the thermodynamic binding strength and/or the kinetic activation energy for escape is made very large. It is immediately obvious that carrier and transport complexes (L_1) in Fig. 2.2 must differ chemically from final sites (L_2) since a required property for the first is easy release of the metal.

In some systems a cation and an anion are pumped into a cell or one of its compartments until the solubility product of the salt ML is reached, i.e. mode 5. The solid is sometimes expelled out of the cells. Frequently, it grows within an organic matrix to produce the common biominerals: $SiO_2(H_2O)_n$, $CaCO_3$, $Ca_2(OH)PO_4$, $Fe(OH)_3$, or $Ca(C_2O_4)$. Once again these minerals must be correctly placed in space and the control of shape in the process of formation of the solid phases is of utmost importance. The minerals must remain in some steady state relationship with their free ions.

The way in which a certain element is trapped immediately imposes constraints upon its functional role. Very mobile elements, like Na^+, K^+, or Cl^-, are of little use as reaction centres, for instance in enzymes, but they can transfer ionic charge, a function that fixed ions cannot accomplish. But before we turn to the uses of the chemical elements in their selected sites we must see in detail how many of them can be selectively captured by more specific chemistry and sometimes even forced into particular sites, and here we shall consider passage across a single membrane only. This will be the purpose of the next sections. Note that we shall be interested at first in the principles of *equilibrium chemical* selectivity, Chapter 3 describes the several *physically* separated compartments into which different chemicals are distributed. When we put the two together we shall see how selectively elements are handled in biology (Chapters 8 to 19).

2.3 The selective uptake of metal ions

To a first approximation we can consider that when any organic surface is immersed in an aqueous media, e.g. roots in soil or plants in the sea, the uptake is based upon equilibrium considerations, i.e.

Amount bound varies as K_{ML} · [concentration of free ion]

where K_{ML} is the (effective) stability (affinity) constant (see Section 2.4) of the 1:1 complex of each metal ion with the ligand(s) at the surface. Since the concentration of the free metal ions in natural aerated aqueous media falls dramatically (see Chapter 1), say, roughly as

K^+, Na^+	Mg^{2+}, Ca^{2+}	Zn^{2+}	Cu^{2+}	Fe^{3+}
10^{-1} M	$\le 10^{-2}$ M	$<10^{-8}$M	$<10^{-9}$ M	10^{-17} M,

it is clearly difficult to incorporate large quantities of any other metal ions except K^+, Na^+, Mg^{2+}, and Ca^{2+}. Some special chemicals are clearly necessary to make K_{ML} values large and different from metal ion to metal ion, e.g. selective ligands based on different donor atoms (Table 2.3), to retain most metals at the levels found in cells. We know that for the alkali metals these are neutral oxygen atom, O, donors, e.g. crown ethers (see Fig. 2.6), but for the transition metals they are N/S donors (see Table 2.4). This means that deeper behind cell membranes there must be much stronger binding sites than those at the surface described in Tables 2.1 and 2.2, and, indeed, several lines of evidence show that these metal binding sites exist, undergo slower exchange, may become saturated, and may be coupled to energized steps which transfer the element inward (or outward). In order that the various metal ions can be selectively captured different types of selective ligands must therefore be used, obeying conditions such as those shown in Table 2.4 where some simple ligands are suggested merely as examples.

The separation of elements into compounds by group, such as the three types in Table 2.4, may therefore be easily achieved by chemical synthesis of reagents followed by controlled reactions. A parallel description at this level will be given in Sections 2.14 and 2.15 for the selective uptake of classes of anions. It is still not clear at this stage how the individual ions in each of the three groups in the table can be separated during their uptake into cells.

This complicated problem can be approached by separating all the individual possible steps as in Fig. 2.2 (neglecting, for the moment, competition

Table 2.3 Selective ligands based on different donor atoms

O-donor	$^-O_2C-CH_2-CO-CH_2-CO_2^-$
N,O-donor	$^-O_2C-CH_2-NH-CH_2-CO_2^-$
N-donor	$H_2N-CH_2-CH_2-NH-CH_2-CH_2-NH_2$
O-donor (phenolate)	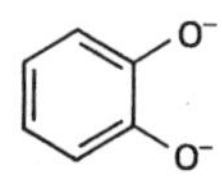
N-donor (aromatic)	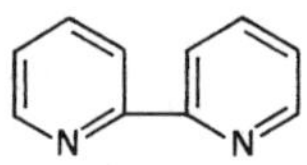
O,S-donor	$^-O_2C-CH_2-S^-$
N,S-donor	$H_2N-CH_2-CH_2-S^-$
S,S-donor	$^-S-CH_2-CH_2-S^-$
O-donor (hydroxamate)	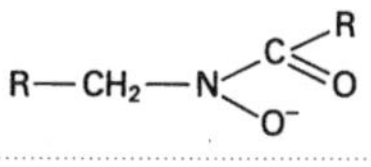
EDTA	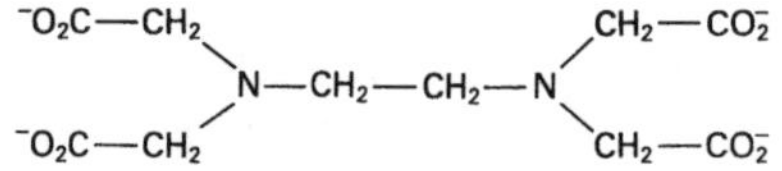

Table 2.4 Differentiating ligands for some cations: stability constant (K_{ML}) and uptake factor (K_{ML}•[M])

	Na^+, K^+	Mg^{2+}, Ca^{2+}	Zn^{2+}, Cu^{2+}	Class of ligand
Type I				
K_{ML}	>10	<10^2	<10^6	O-Macrocycles (large hole)
Uptake factor	>1	<0.1	<0.1	(Fig. 2.6)
Type II				
K_{ML}	1.0	>10^3	<10^6	Di-or-tri carboxylates
Uptake factor	<0.1	>1.0	<0.1	(O-donors)
Type III				
K_{ML}	0.1	<10^2	>10^9	Nitrogen (or sulphur)
Uptake factor	<0.1	<0.1	>1.0	ligands

The concentration of M, [M], is given in Section 2.3.

between different elements, including the proton, and different ligands) and considering the following:

1. The ability of different ligands L_1 to compete with an external ligand, L_3, for M. This gives the thermodynamic ability to capture metals, M, and is in part a problem of relative stability constants, but ML_1 must be soluble in the organic membrane phase, hence we must consider, besides the stability constants K_{ML} in aqueous media, as above, the distribution coefficient between water and the liquid membrane, D, for K_{ML_1} (which may be governed by a receptor). The ability of M to enter a membrane is then proportional to the product $K_{ML_1} D$ at equilibrium.
2. The rates of formation and of dissociation of the complexes ML_3 and ML_1 and any rate steps involved in crossing the interfaces.
3. The further formation of ML_2 when similar considerations apply.
4. The input of synthetic and degradative energy in the binding of M to L_1 and L_2 in the cell, taking into account the possibility of variations of pH and any redox changes of M due to the metabolic activity of the cell, as well as the stabilization of L_2 by M in ML_2.
5. An additional energy-coupled transport of ML_1 through the membrane (in or out) for which the necessary energy is provided by special energy-giving chemical processes coupled to the receptor in Fig. 2.2.
6. The movement of ions M through selective channels, energized or not, that bypass the need for the formation of ML_1 (see the energy pump in Fig. 2.2).
7. Finally, the synthesis of scavenging and carrier ligands can be coupled to the concentration levels of the metal so that there is a feedback control of uptake even back to gene expression (see Section 2.20 and Chapters 6 and 20).

The possibilities considered under (5) and (in some cases) (6) are what we call active or energized transport. The processes (1) to (4) can be treated as equilibration within a steady state of the cell. Over relatively short periods this second assumption is valid since any further passage of M from ML_2 to, say,

ML_4 can be considered as a slow irreversible step. A full rigorous treatment of all these processes is obviously impossible for lack of the necessary information, but some simplifying assumptions can be introduced and consequences derived on a semiquantative basis.

We will illustrate first (Sections 2.4 to 2.10) an application of the thermodynamic principles for the case of a single cell with an outer bathing solution, a porous wall (outside the cell), an ion impermeable membrane, an inner cytoplasm (inside the cell), and no other divisions or compartments. This is very like a prokaryote cell, e.g. a bacterium. We shall assume that the wall contains, overwhelmingly, (charged) O-donor ligands, but that the membrane and the cytoplasm have a range of N/S-donors as well as O-donors (mostly proteins). Taking the outside solution to be high in Na, K, Mg, and Ca, such that these elements are in excess over the ligands, and relatively low in other metals (see beginning of this section), then the outer face of the cell will probably be a kind of Ca/Mg loaded ion-exchange resin surrounded by an aqueous phase rich in these free cations with which the ions in the wall can exchange. (This is, in fact, the case, but note peculiarities in some cells, e.g. of lichens; see Table 2.1.) Other ions, ML_1, and ML_3 species will pass through this wall even by simple diffusion. These trace elements can be complexed outside the cells by special ligands L_1 or L_3, e.g. iron by the siderophores, or exchanged from the wall ligands to other ligands L_1 at the membrane. Generally, the alkali (and alkaline earth) metals do not form sufficiently stable species to be dealt with by these mechanisms, and their uptake (or their rejection) is usually by selective gated channels and pumping devices which pass them through membranes. In fact it is rather their loss from cells which is mediated by ionophore complexing drugs which allow transport of them out through the membrane.

Inside cells (and this includes the membranes), besides other O-donor ligands, there are a variety of N/S-donor sites designed to capture particular *trace* elements using energy and to protect against damage from unwanted elements (Table 2.4). The unwanted elements may well be rejected.

From this general outline of possible events we can go to the finer details of the uptake and separation of the individual metal ions, but two important aspects must be stressed: *first*, there is a fundamental difference between the conditions outside the cells and inside the cells (or in the membrane), including differences in ion concentration and redox potential as well as pH. Outside cells the concentration of ions can be much higher than that of the (weak) ligands at the surface, whereas inside cells and in the membrane the opposite is true, i.e. the concentration of stronger ligands is under metabolic control and probably is (just) slightly higher than that of the essential trace ions transported. This suggests immediately that the supply of ligands and control of transport is under feedback control by [M] or [ML] through their genetic apparatus for the production of L (see Section 2.20 and Chapter 6). This point will be shown using iron as an example in Chapter 12 and in more detail in Chapter 20. *Second*, at equilibrium, the stability constants that should be used to describe the binding of metal ions to the ligands are not usual stoichiometric stability constants K_{ML} determined in aqueous solutions for simple metal–ligand equilibria, but the so-called *conditional* or *effective* binding constants K_{ML}^{eff}, which take into account the competition of the proton and of other metal ions for the sites of M and also the competition of other ligands present for M. This competition has very important effects on selectivity.

2.4 Effective stability constants

The effective (conditional) stability constants are related to the conventional stability constants by the expressions (considering competitive complex formation for only two metals M_1 and M_2 and two ligands L_1 and L_2 and when only M_1L_1 and M_2L_1 complexes are formed)

$$\log K^{\text{eff}}_{M_1L_1} = \log K_{M_1L_1} - \log \alpha_L - \log \alpha_M$$

where

$$\alpha_M = 1 + \sum_{i=1}^{i=N} \beta_i \cdot [L_2]^i$$

for M_1 $(L_2)_i$ complexes (and the simplest case is when there is only of these complexes, M_1L_2) and

$$\alpha_L = 1 + \sum_{i=1}^{i=N} \beta_i^{\,H}[H^+]^i + K_{M_2L_1}[M_2].$$

The β_i^H are the overall formation constants of the species L_1H_i, $K_{M_2L_1}$ is the stability constant of the complex formed between M_2 and L_1, and the β_i are the overall stability constants of any M_1 $(L_2)_i$ complexes formed. In effect, α_L is the reduction in L_1 available to M_1 due to the presence of H^+ and M_2, while α_M is the reduction of M_1 available to L_1 due to the removal of M_1 by binding to L_2. The values of $K^{\text{eff}}_{M_1L_1}$ may be much smaller than $K_{M_2L_1}$. The various equilibria locked in K^{eff}_{ML} generally are frequently so complex that only with the aid of computers can we assess the likelihood of formation of any ML in a determined set of conditions (see Da Silva and Williams) (1976) in the 'Further reading' section and the diagram in the margin).

To illustrate the considerations affecting effective constants one may recall some simple examples, e.g. that of nickel ions that bind to ammonia molecules with a 1:1 binding constant for $[Ni(NH_3)(H_2O)_5]^{2+}$ of $\log K = 3.0$. However, protons bind to NH_3 with a $\log K = 9.0$; hence, at $pH < 6$, nickel ions do not bind NH_3 appreciably, i.e. the ability of ammonia to bind to nickel is limited. For plants growing in nickel-rich soils, the acidity of the storage vacuoles, for example, does not allow simple amines to bind to nickel and storage is associated with ligands of low acid dissociation constants (pK_a), e.g. organic carboxylates. In this case, the *effective* binding constant to nickel ions is indeed higher for an oxygen-donor ligand than for a nitrogen-donor ligand, which is the reverse order to that of the overall stability constants.

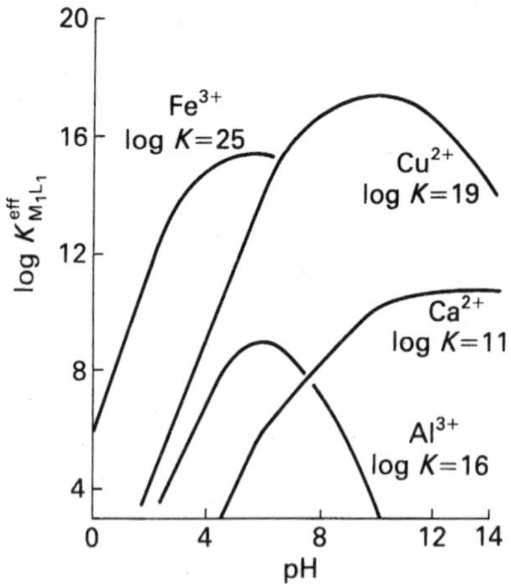

The effective stability constants at different pH values for EDTA complexes of four metal ions.

A similar type of limitation arises as the result of competitive binding by different elements M to a ligand L. The stability constants of their respective complexes may differ by a factor that largely overcomes differences in concentration and even small excesses of certain foreign elements over ligands may deprive some biological elements of their biological carriers or retaining ligands. In this respect, copper and mercury are particularly dangerous, since they usually form complexes of high stability with a wide range of ligands, and metal ions of elements such as Pb, Cd, Ni, Co, and Zn are also very competitive, especially for centres that bind through sulphur and/or nitrogen groups. Clearly, free metal ions of this kind must not be present except at very low levels, just as the proton concentration must be low for it also binds to these centres.

Finally, it must be noted that most of the stability constants available are for simple small-molecule ligands, which may or may not be good models for the sites of the metal ions in biological molecules and were not measured under identical conditions. As will be discussed later, there are several differences which will necessarily be reflected in the values of the stability constants for the real complexes with proteins, for example. To a first approximation we shall take the experimentally determined model-ligand stability constants of their respective complexes multiplied by a factor which converts them into the values valid for a protein which employs exactly the same set of donor atoms. This aspect must be borne in mind in later discussion and we shall include a special section on the influence of protein fold stability on the binding of metal ions (Chapter 6).

Given the variety of possible modes of uptake and incorporation of elements mentioned above we shall proceed by giving more details of simple equilibria (Section 2.5), and then by introducing complications in this analysis (Section 2.6 to 2.11) before turning to kinetic modes of retention or rejection (Section 2.12).

2.5 Element uptake at equilibrium: the selectivity of the uptake

We have seen that the uptake of metal ions M is almost invariably energized: consequently, it is not a true equilibrium process. However, if we assume a steady state in certain steps, we can treat the uptake problem as one in which the element M distributes between only two very different phases containing, say, L_3 and L_2 separately and between which only M itself can equilibrate. We call this 'equilibrium uptake' (Fig. 2.3) since the process of transport in Fig. 2.2 is presumed not to interfere. Selected uptake can then be dealt with through analysis of the binding constants of M in two different compartments: the outer aqueous phase and the inner aqueous phase. The uptake of M will depend only on the stability of ML_2 relatively to ML_3 and no kinetic considerations or irreversible energized steps will be invoked in this approximation. (Energy is in fact put into the system just because L_2 and L_3 are synthesized in a controlled way and kept separate, see Section 2.20).

On the other hand, since the values of K_{ML_3} in the outer phase in the absence of N or S donors do not vary very much from ion to ion (see Table 2.2), the discussion of selectivity can be centred just on the values of K_{ML_2} (or of K_{ML_1} for the uptake into a membrane assuming $D = 1$, see Fig. 2.2) and thus becomes a simple problem of co-ordination chemistry which in outline can be studied using model ligands.

Now, general coordination chemistry teaches us that the thermodynamic binding of cations to organic chelating agents in mononuclear complexes follows certain trends and that the different metal ions have preferences for particular donor atoms and for particular coordination geometries. Some general characteristics and the differential properties induced are shown in Table 2.5 to indicate directions in which we must be thinking. The ligand types and geometry preferences are summarized in Tables 2.6 and 2.7 and in Fig. 2.4.

Fortunately, the stability of the complexes formed by the different types of ions follow sequences which are, to a large degree, independent of the ligand when this ligand is built from an internally mobile organic molecule, i.e. where steric selectivity is minimal. For monovalent ions the binding constants K_{ML} are $Cu^+ > Na^+$, K^+ and, for divalent ions, one has the Irving–Williams stability order (Mg^{2+}, Ca^{2+}) $< Mn^{2+} < Fe^{2+} < Co^{2+} < Ni^{2+} < Cu^{2+} > Zn^{2+}$, again largely independent of the ligand. Few stability constants are available for

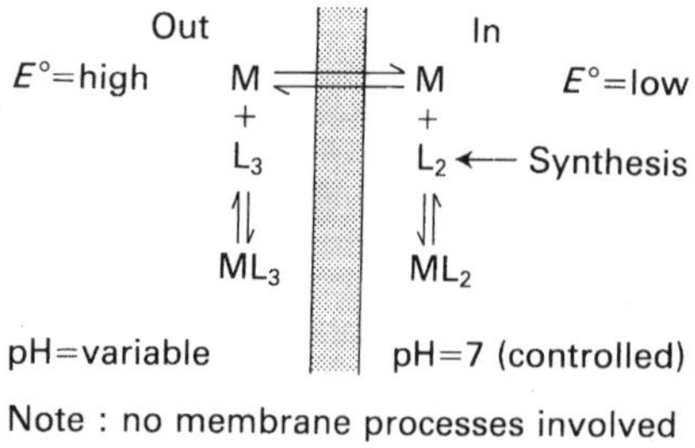

Fig. 2.3 A simplified scheme taken from Fig. 2.2 to illustrate uptake of M at equilibrium.

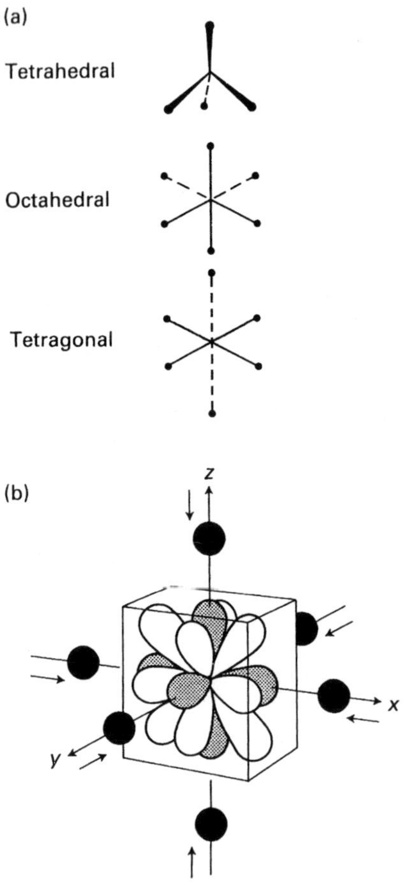

Fig. 2.4 (a) Common coordination geometries around a central metal ion. (b) d-orbitals in an octahedral field, where e_g (shaded) = σ-bonding and t_{2g} (unshaded) = π-bonding.

Table 2.5 Some general characteristics of metal ion complexes present in biological systems

Characteristics	Na^+, K^+	Mg^{2+}, Ca^{2+}	Zn^{2+}, Ni^{2+}	Fe, Cu, Co, Mo, Mn
Predominant oxidation state	+1	+2	+2	More than one state
Stability of the complexes	Very low	Low or Medium	High	High, but medium in the cases of Mn^{2+} and Fe^{2+}
Preferred donor atoms	Oxygen	Oxygen	Sulphur or nitrogen	Nitrogen or sulphur (and especially oxygen in the cases of some Fe and Mn complexes)
Mobility in biological media	Very mobile	Semi-mobile	Static, Ni; usually static, Zn	Static (semi-mobile in the cases of Mn^{2+} and Fe^{2+})

Table 2.6 Ligands preferred by different metal ions in simple coordination compounds

Metal ion	Ligands
Na^+, K^+	Oxygen-donor ligands, neutral or of low charge (–1)
Mg^{2+}	Carboxylate, phosphate, and polyphosphate (total charge >–2). N-donation (special)
Ca^{2+}	Carboxylate, phosphate (less than Mg^{2+}), some neutral O-donors
Mn^{2+}, Fe^{2+}	Carboxylate, phosphate, and nitrogen donors combined, (thiolate)
Fe^{2+} (special)	Unsaturated amines (particularly porphyrins)
Cr^{3+}, Mn^{3+}	Phenolate (e.g. tyrosine), hydroxamate, hydroxide
Fe^{3+}, Co^{3+}	Carboxylate, N-donors, polypyrroles, (thiolate)
Ni^{2+}	Thiolate (e.g. cysteine), unsaturated amines, polypyrroles
Cu^{2+}, Cu^+	Amines, ionized peptide $>N^-$, thiolate
Zn^{2+}	Amines, thiolate, carboxylate

The case of Fe^{2+} is especially complicated by spin-state changes and polymerization of mixed oxidation states.

Table 2.7 Preferred geometries in simple coordination compounds

Metal ion	Preferred geometries
Cu^{2+}, Mn^{3+}	Tetragonal >5-coordination>tetrahedral
Cu^+	Linear, trigonal, or tetrahedral
Ni^{2+}	Octahedral>others
Co^{2+}	Octahedral>tetrahedral>others
Zn^{2+}	Tetrahedral>octahedral>5-coordination
Mn^{2+}	Octahedral>others
Fe^{3+}, Co^{3+}, Cr^{3+}	Octahedral>others

*NB. In this table all metal ions are in high-spin states and liganding atoms are small O, N donors. S-donors favour lower co-ordination numbers. Ligand-field theory, that is polarization of and binding by the core electrons and orbitals of the metal ion (Fig. 2.4(b)), can explain the above observations; see inorganic chemistry textbooks in the 'Further reading'.

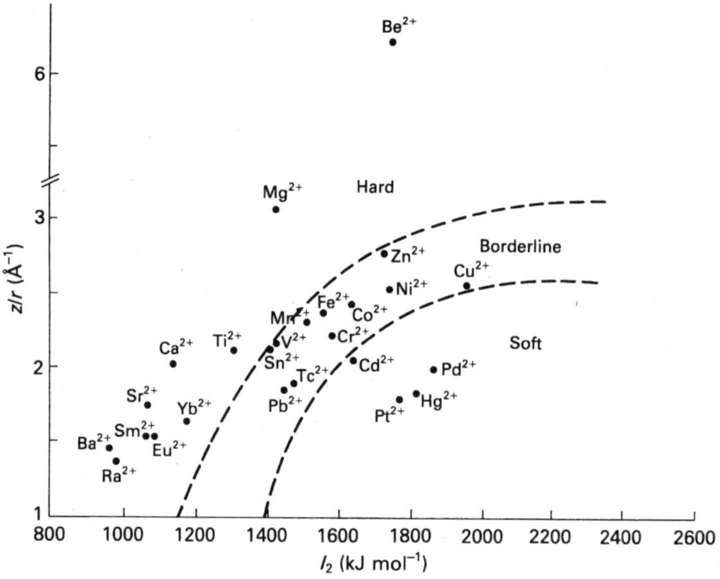

Fig. 2.5 A plot of z/r against I_2, ionization energy from M to M^{2+}, for some elements (see text). Note how the distinction between soft and hard ions depends on both parameters.

trivalent ions, but, in general, we expect and find the order Co^{3+}(low-spin)>Mn^{3+} >Fe^{3+} >Cr^{3+} >Al^{3+}.

These trends can be explained to a first approximation in terms of the ionic potential z/r (electrostatic binding) where z is the ionic charge and r the ionic radius of an ion, relative to I_n the ionization potential of M to M^{n}+, i.e. the electron acceptor power of M^{n}+. The plot of z/r against I_2 for divalent ions (Fig. 2.5) shows a clear separation of the so-called 'hard' (a-class) and 'soft' (b-class) ions (see also Table 2.8) and also generates the Irving–Williams' sta-

Table 2.8 Hard and soft acids and bases

Hard (class (a))	Borderline	Soft (class (b))
Acids		
H^+, Li^+, Na^+, K^+, Be^{2+}, Mg^{2+}, Ca^{2+}, BF_3, BCl_3, $B(OR)_3$, Al^{3+}, $AlCl_3$, $Al(CH_3)_3$, Sc^{3+}, Ti^{4+}, VO^{2+}, Cr^{3+}, Fe^{3+}, Co^{3+}	Fe^{2+}, Co^{2+}, Ni^{2+}, Cu^{2+}, Zn^{2+}, Rh^{3+}, $B(CH_3)_3$, R_3C^+, Pb^{2+}, Sn^{2+}	Cu^+, Ag^+, Au^+, Cd^{2+}, Hg^{2+}, Pt^{2+}, Pt^{4+}, $MoO_2^{2+}$$Pd^{2+}$
Bases		
NH_3, RNH_2, N_2H_4, H_2O, OH^-, O^{2-}, ROH, RO^-, CO_3^{2-}, SO_4^{2-}, ClO_4^-, F^-	$C_6H_5NH_2$, N_3^-, N_2, Br^-, Cl^-	H^-, R^-, C_2H_4, C_6H_6, CN^- CO, SCN^-, R_3P, R_2S, RSH, RS^-, I^-

NB Hardness is defined, see textbooks in the references, as $\eta_M = \frac{1}{2}(I - A_e)$ where I is the ionization potential of the state of M concerned and A_e is its electron affinity. I dominates for positive ions, whence the usefulness of figures such as Fig. 2.5, but in aqueous solution must be considered relative to z/r.

Ahrland, S., Chatt, J., and Davies, N.R. (1958). The relative affinities of ligand atoms for acceptor molecules and ions, *Quarterly Review*, **12**, 265–76.

Pearson, R.G. (1963). Hard and soft acids and bases, *Journal of the American Chemical Society*, **85**, 3533–9.

bility order (see Fig. 2.8). (Similar plots are readily obtained for other oxidation states.) For larger ions small differences exist between ionization potentials so that whereas ions of small size, when lowly charged, tend to be more 'b-class or soft' this difference disappears for larger ions, for example high-valent molybdenum binds well to RS^-.

While this plot helps to get a quick overall view of stability constant trends such as those in Table 2.5, a more detailed description is required if a deep understanding of the selectivity factors which operate in both chemistry and biology are to be appreciated. We must introduce for example stereochemical properties of ligands and preferred symmetry of sites, ligand–field effects, etc. for different metal ions.

The different ways of achieving selective thermodynamic binding, based on size, donor type, and shape of the ligand that we shall discuss, are, therefore:

1. selection by charge;
2. selection by ion size;
3. selection by liganding donor atom;
4. selection by preferential co-ordination geometry;
5. selection by spin-pairing stabilization;
6. selection by binding in clusters.

After discussing these thermodynamic factors we look at biological factors:

7. selection by control of concentration of M and L;
8. selection by transfer coefficients from water to proteins (or membranes).

Naturally, several of these mechanisms operate simultaneously and their separation in the discussion only aims at better identification of each possible basis of selection. While most of the discussion is based on model complex ion chemistry we shall refer time and again to biological molecules which illustrate the principles. Except where stated we refer to aqueous solutions.

2.5.1 Selection by charge type

In aerated aqueous media the elements of the first transition series are normally found in oxidation states +3 from scandium to iron and + 2 from cobalt to zinc (Fig. 1.9); hence these two groups can be differentiated by those ligands preferring ions of higher charge. It is found that hard singly charged ligands of small size prefer the highly charged, small, class 'a' (hard) cations and this means that appropriate biological ligands for the tri-positive ions are small first-row anions such as R_2N^-, RO^-, and F^-. Since F^- is uncommon in biology this ion cannot easily be used for capture devices. On the other hand, the concentration of R_2N^- will normally be small at pH ~ 7 since the pK_a of R_2NH is high. This leaves RO^-, which occurs in hydroxamates, enolates, phenolates, etc., as the adequate ligand to take up trivalent ions, e.g. Fe^{3+}, V^{3+}, and Mn^{3+}, preferentially. Chromium should also be captured by such anions, but there is yet no firm evidence for the occurrence of the corresponding species in biological systems. Aluminium is bound but more weakly and we find that, interestingly, aluminium is not taken up to any great degree. One of the interesting features here is the competition between organic ligands, e.g. phenolates, and inorganic hydroxide. Where the difference is small, hydrolysis to hydroxide precipitates prevents uptake, e.g. of aluminium, but this competition is affected by acidity, e.g. in acid rain conditions. At the beginning of the first row of the transition metals, scandium and titanium also are not taken up (Chapter 1) probably due to hydrolysis. Elements which would form even more highly charged cations are hydrolysed to anions. Vanadium, and molybdenum in the second row, are probably handled initially as oxoanions, VO_4^{3-} and MoO_4^{2-}. We return to a more detailed analysis of this problem in Chapter 17.

In conclusion highly negatively charged ligands having RO^- donors are useful in the selective uptake of a very few *trivalent* transition metal ions, e.g. Fe^{3+} and Mn^{3+}, while reduction to low oxidation states, e.g. Co^{2+}, Ni^{2+}, or greater relative hydrolysis of ions such as Al^{3+} removes these elements from competition for this type of ligand. Inside cells RS^- ligands assume greater importance and, surprisingly, are even found with trivalent metals, e.g. Fe^{3+}, see Section 2.10. Finally, we have to take into account competition from the doubly charged anions O^{2-} and S^{2-} which can lead on to polynuclear complexes or precipitation (see Section 2.19).

Structural considerations are to be considered with the nature of ligand donor centres and here the ligand size is not the only parameter since the way the ligand atoms are bonded together in multidentate chelates gives rise to problems of the stability of different rings, 5-,6-,7- and so on membered, and the bite angles subtended at the metal by the donor centres. The bite angle varies with ring size and its rigidity. We shall comment on the value of bite angles in the generation of reactivity in Section 4.3. but here we note that matching bite angles with the size or the stereo-specificity of metal ions (Section 2.5.5), gives rise to selective binding.

Note that this description applies largely to aerated solutions while in anaerobic conditions selection can be based on ions in reduced states, e.g. Cu^+. In fact, the reduction to Cu^+ gives a selectivity for copper of considerable magnitude since Cu^+ can be held strongly in complexes using only two linear donor centres.

2.5.2 Selection by ion size

The order of stability of different metal complexes would be that of the ionic potential of the metal ions z/r if covalence, steric hindrance, or other specific factors did not operate. Hence, for the ionic complexes of the alkali and alkaline-earth metals separately one would expect the following orders of stability to be obeyed from their ionic radii

$Li^+ > Na^+ > K^+ > Rb^+ > Cs^+$,
$Be^{2+} > Mg^{2+} > Ca^{2+} > Sr^{2+} > Ba^{2+}$.

Li^+	Be^{2+}
0.6	0.4
Na^+	Mg^{2+}
0.95	0.65
K^+	Ca^{2+}
1.33	1.00
Rb^+	Sr^{2+}
1.48	1.13
Cs^+	Ba^{2+}
1.69	1.35

Ion sizes (Å)

However, these orders are seldom found except with the smallest anions, such as F^- and OH^-, and many inversions of order occur, depending on the type of ligand used. The inversion is the more noticeable the larger the anion, e.g. SO_4^{2-} against OH^-. (NB $BaSO_4$ is insoluble but $MgSO_4$ is not.)

The simplest reason for the inversion of an order is of a stereochemical nature, which is readily understood with polydentate ligands which cannot fold around the smallest metal ions, e.g. Li^+, Be^{2+}, and Mg^{2+}, due to the crowding strain caused in the ligand on coordinating these ions. The ligands therefore do not displace water of hydration easily (water is a small ligand). We can analyse the problem on a semi-quantitative basis starting from the solubility of salts MX. The equilibrium governing selectivity is

$$M^{n+}\text{ (hydrate)} + X^{n-}\text{ (hydrate)} \rightleftharpoons MX\text{ (hydrate)}\downarrow + m\,H_2O.$$

Apart from the free energy gain of the water on the right, it is clear that the difference in electrostatic energy between the product and the reactant (ions of charge z^{n+} or z^{n-} and radius r) is related to a difference between individual ion hydration energies and lattice energies of MX salts,

$$\Delta G \propto C\left\{\frac{z^{n+}}{r_{\mathrm{M}}}+\frac{z^{n-}}{r_{\mathrm{X}}}\right\}-B\left\{\frac{z^{n+}z^{n-}}{d_{\mathrm{MX}}}\right\} \tag{2.1}$$

where d_{MX} of the lattice $= r_{\mathrm{M}} + r_{\mathrm{X}}$ to a first approximation and C and B are constants. By using C and B as fitting constants we are taking repulsions into account approximately. Simple examination of this function for a fixed r_{X} shows that at either very large or very small values of r_{M}, the hydration energy dominates. In a somewhat obvious way the lattice is most stable somewhere near the equality $r_{\mathrm{M}} = r_{\mathrm{X}}$. Detailed analysis allowing for $d_{\mathrm{MX}} \neq r_{\mathrm{M}} + r_{\mathrm{X}}$ only reduces the importance of the lattice term. This means that ions of approximately equal size pack together best. The solubility of salts can be discussed in detail using eqn (2.1) and it can be demonstrated that even the residual structural hydration of ions in lattices in the most stable form of MX(hydrate)$_n$ is decided by the radius ratio of anion and cation. As the anion gets larger relative to the cation there is an increasing probability that a hydrated cation, not the bare cation, will come out of solution with a given anion and that the hydrated salts are then the most insoluble, e.g. $BaSO_4$ and $SrSO_4$, but $CaSO_4 \cdot H_2O$ and $MgSO_4 \cdot 7\,H_2O$, which is to be contrasted with the series with a small anion, $Ba(OH)_2 \cdot 2H_2O$, but $Sr(OH)_2$, $Ca(OH)_2$, $Mg(OH)_2$, and $Be_nO(OH)_m$. In biology we find $BaSO_4$, $SrSO_4$, and $CaCO_3$. The data obviously reflect the drive toward matching of sizes of ions.

Next we shall consider the effect of radius ratio on the formation of ion pairs in solution. There is a close parallel between the above analyses of solubility and that of the reaction for complex ion or ion pair formation.

$$M^{n+}\ (\text{hydrate}) + X^{n-}\ (\text{hydrate}) \rightleftarrows$$
$$[M^{n+} \cdot X^{n-}]\,(H_2O)_y \text{ or } [M^{n+}\,(H_2O)_y \cdot X^{n-}] + mH_2O.$$

The equation for the formation energy of the complex is the same as that of eqn (2.1) but the constant B is different. In solution we expect, therefore, that size factors, of ions and water, will be important so that small anions and cations on one hand and large cations and anions on the other tend to associate. Additionally, in some complexes it will be the hydrated ion which dominates to give the species $M(H_2O)_yX$, whereas in others the bare ion will dominate to give $MX\,(H_2O)_y$. The difficulty in proving the point in solution lies in recognizing the species. We must go to spectroscopic studies in order to recognize them, but the central point is that ion-pairs always have structure-dependent energies as well as metal ion radii dependent energies. We have analysed the situation for small inorganic anions and cations, but the analysis is equally applicable to all sizes of charged molecules—substrates, proteins, and DNA, for example.

Selection by complexation based on size is greatly used by biological systems. The ligands synthesized are often cyclic and create a central cavity which fits most closely a particular cation size, e.g. either binding is $K^+ > Na^+$ or $Na^+ > K^+$. It has been possible to synthesize many similar compounds which are so tailored that they strongly favour the complexation of one particular ion, namely the so-called 'crown-ethers' and other macrocyclic ligands [Fig. 2.6(a), (b)]. There are well-documented parallel examples in bacterial antibiotics, e.g. nonactin and valinomycin [Fig. 2.6(c)] where the bactericidal action is due to the leakiness of membranes to K^+ or Na^+ through the presence of these ligands. [Note the ability to enter membranes (Fig. 2.2) as well as good binding]. Magnesium is taken up very effectively in a porphyrin-type cyclic ligand of fixed geometry and cavity size, chlorophyll, in clear contrast with calcium, which is excluded due to its large size, but it must be pointed out that the insertion of magnesium into this nitrogen heterocycle requires special conditions (see Section 2.12.3).

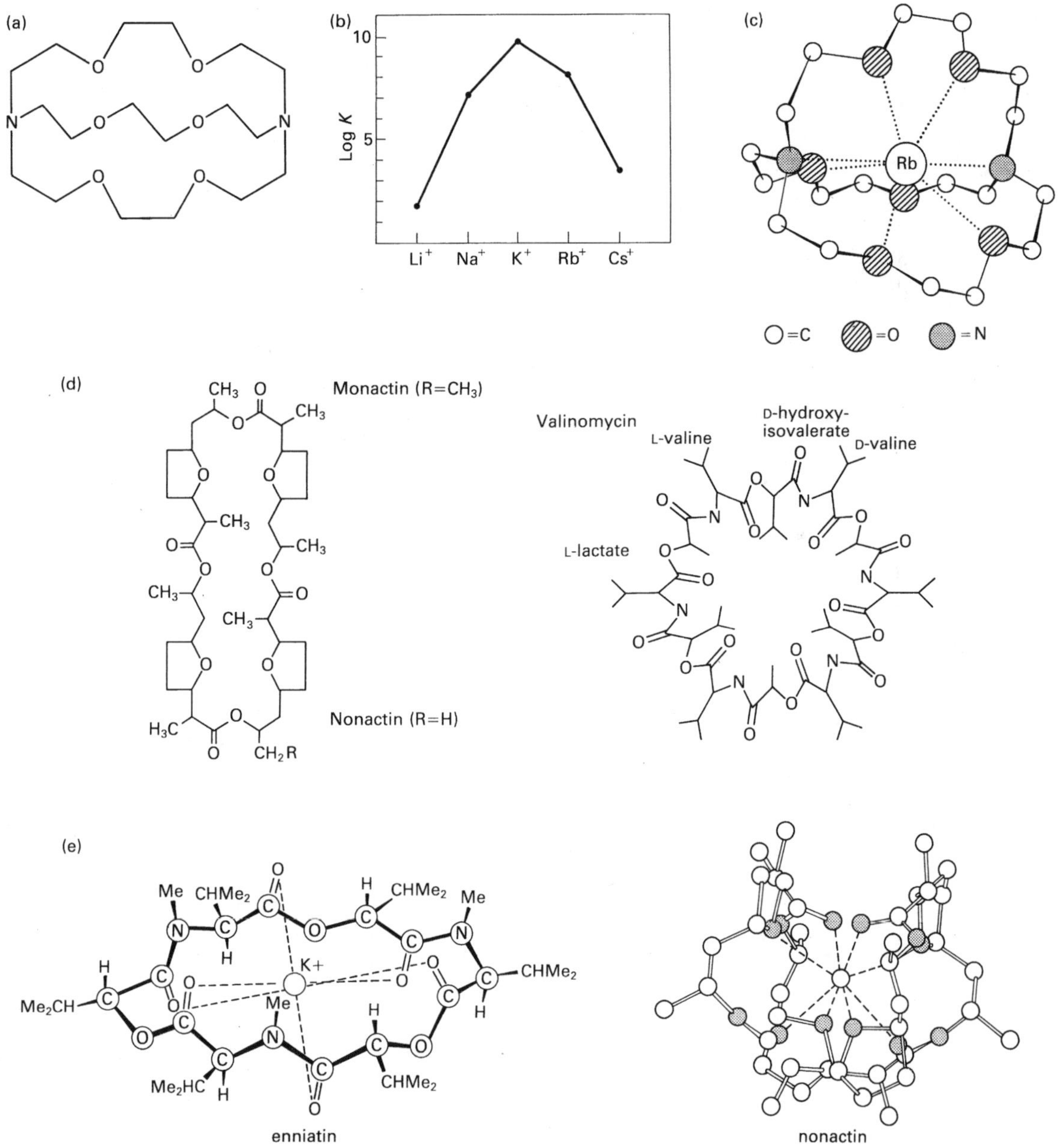

Fig. 2.6 Illustrations of biological ring chelates made from neutral O-donors for selective binding to ions such as K^+. (a) The cryptate (2.2.2)-diamine; (b) stability constants for complexes of the above cryptate with the group I cations; (c) structure of the rubidium cryptate-222 cation; (d) structures of two naturally occurring macrocyclic antibiotics, nonactin and valinomycin; (e) structures of the potassium complexes of naturally occurring macrocyclics enniatin B (D-hydroxy-isovaleric acid-*N*-methyl-L-valine) and nonactin.

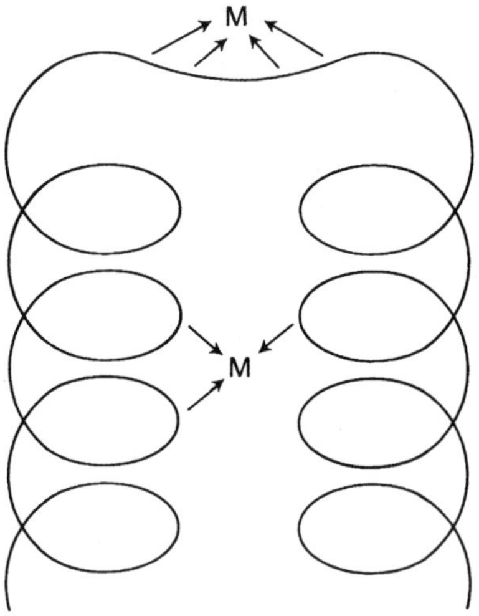

Fig. 2.7 A schematic representation of two ways in which a protein fold can generate sites of restricted size and geometry for any element M.

A second general method to generate size selectivity based on covalent ligand design is by protein folding. Proteins are large molecules and they can generate by folding different types of sites particularly suited for binding certain metals (and indeed many non-metals, substrates, cofactors, and co-enzymes) (Fig. 2.7). Here the fold strength is such that it provides a site (hole) which either: (1) can collapse around all ions and therefore is a better site for small ions; or (2) can only collapse to a particular large hole favouring large ions or even clusters. Typical examples are magnesium in phosphomutases (Chapter 9) or the calcium sites in many calcium-binding proteins, e.g. calmodulin, parvalbumin, thermolysin, etc. (Chapter 10). In most cases the proteins are folded in such a way as to bind preferentially one of these two metal ions virtually exclusively given the biochemical concentration conditions.

In many cases, however, size is not the only determinant factor and is complemented, even in the case of simple alkali metal ions, by the preference for a well-defined stereochemistry of the ligands (octahedral in the case of magnesium), the nature of the liganding atoms, and other effects. For instance, in phosphomutases coordination of magnesium appears to be by two carboxylates plus one imidazole group in an octahedral geometry where the metal ion retains three water molecules. Calcium binds relatively poorly to such a site. It prefers binding to carboxylate groups, together with carbonyl groups, in irregular geometry. This difference is seen in inorganic chemistry too. Discrimination between the two metals is not achieved only by the difference in ionic size, but clearly depends on stereochemistry and ligand–donor nature. We return to these problems in Chapters 8 to 10.

Finally, size selectivity is sensitive and can affect ions of quite similar size, e.g. compare the stability data with EDTA-like ligands for Mn^{2+} (0.75 Å radius) and Zn^{2+} (0.60 Å radius) (see Chapter 14, Fig. 14.2). This is also important in considerations of redox potentials where ion sizes change with charge.

In conclusion, ligands can be synthesized such that the constraints on the hole size which they generate or can generate are a powerful selectivity factor.

2.5.3 Combination of radius and charge effect

Clearly, a description of selectivity based on electrostatics for ions of both different charge and size needs a quantitative analysis of both the attractive (z/r) and repulsive energies. Such a treatment cannot be given here without interrupting the general consideration of factors involved in selective ion uptake. We shall therefore take up the whole matter of the electrostatic selective interactions in Chapters 8–10 where we analyse the biological chemistry of group 1 and 2 cations more thoroughly. In particular, it will be important to notice that the problem of understanding in biology is made much more complex by the occurrence of organic cations, e.g. RNH_3^+, which compete for the anionic centres (Chapters 8 and 9).

Fig. 2.8 A plot of binding constants, log K, for a series of single ligands with the divalent ions from Mn^{2+} to Zn^{2+}. The ligands are cysteine (○); ethylenediamine (▲); glycine (□); and oxalate (•).

2.5.4 Selection by liganding atom

This mechanism is based on the affinity of the a-group metal (hard) ions for oxygen (O)-donor ligands, whereas the b-group metal (soft) ions prefer to co-ordinate to sulphur (S) or nitrogen (N) donors. This topic is well dealt with in standard textbooks and reviews (see 'Further reading' section) and we will not discuss it in detail. The differences of stability between the complexes of the transition metals with ligands containing different donor atoms become greater the more polarizable the donors, i.e. they increase in the order

O ligands < N ligands < S ligands (see Fig. 2.8). This figure, which again illustrates the Irving–Williams' order of stability, shows that the divalent transition metals form complexes that may differ in their thermodynamic stability by several orders of magnitude if the liganding atoms are appropriately chosen.

Now we have pointed out that outside the cell most donors are of the O-donor variety and RSH ligands are oxidized away. This means that uptake, inside/outside selectivity, can be based on soft/hard discrimination and this would apply to all the metals and ligands equally if there were no control over ligand synthesis and metal uptake. Copper ions bind most strongly and, if copper were in excess over the total free N/S ligand concentrations, all the other divalent transition metals would be excluded from N/S centres. However, through biological control mechanisms (Section 4.4), particular ligands inside or outside the cells are likely to be present in only slight excess relative to particular metals, which are limited in availability. Both ligands and metal ions may also be restricted by selective pumping across membranes. Competition can now be such that each metal occupies its proper site and does not interfere with the distribution of those that form less stable complexes. The *effective* stability constants K_M^{eff} are operative in determining an adequate distribution. For this reason copper is bound by the best donor sites, e.g. 4(N) (as in superoxide dismutase) inside cells and 4(N/S) (as in plastocyanin-like proteins outside cells or in vesicles), overcoming competition with zinc which is more available. Once almost all copper is bound and there is little excess of its ligands, then for any site of lower donor power K_{Zn}^{eff} [Zn] will exceed K_{Cu}^{eff} [Cu]; hence the remaining 4N or N/S sites (or centres of two or three N/S donors, perhaps with one carboxylate ligand) bind zinc, e.g. carboxypeptidase or carbonic anhydrase. Nickel and cobalt are relatively rare elements, especially in advanced biological cells, so K_M^{eff} [M] for these metals is generally low and very special capture devices (vitamin B_{12} and factor F-430) are needed for them. (There are exceptional anaerobic environments and organisms for which this is not true of nickel, see below.) *Divalent* Mn and Fe, although abundant, do not bind strongly to any of these donors in sites of low coordination number and do not interfere in the distribution—they must be dealt with by biological systems in a different manner. They often bind to a variety of centres from four histidines (N) to a two-carboxylate and one-nitrogen centre but always in higher, often *octahedral*, coordination and they are then in fast exchange. Their selective geometric demands are fully utilized (Table 2.7). (We return in Section 2.5.7 to the ability to form mixed oxidation state clusters.) It will now be seen that combination of metal and ligands is dependent on the donor atom but also on the input (or exit) rates of metal ion to the cell and on the rates of synthesis and degradation of ligands. This is most important; it is a reminder that biological systems are kinetic states of matter.

The selectivity of organic O-donors relative to N- and S-donors is part of the 'a'/'b' or hard/soft classification of metal ions generally and does not just apply to mono and divalent ions. Amongst isolated trivalent ions, M^{3+}, only occasional affinity for organic S-donors is apparent within the metals of biological interest, e.g. Fe^{3+}, but note that this is not true for Tl^{3+}, Bi^{3+}, and the M^{3+} ions of later elements in the second and third transition metal series. Such competing ions are likely to be very poisonous since they can displace required divalent ions. We consider the case of copper in the Cu^+ state in Section 2.7.

In conclusion, selectivity by donor atom type is very powerful if the concentrations of ligand and metal are controlled.

2.5.5 Selection by preferential coordination geometry

In some cases, use can be made of the preference of certain metal ions for coordinating sites of particular geometry. The cases of zinc(II) and copper(II) are given first as an example. Both cations are only available in trace amounts, though zinc may be 100 times more readily available in soils and water than is copper. To counterbalance this, copper binds to most ligand centres at least 100 times more strongly than does zinc (see Section 2.5.4). Thus, log K_{M}^{eff} [M] is similar for both metals with many conformationally mobile ligands and we must ask how can selective uptake be achieved with over 99 per cent efficiency as is required in both copper and zinc enzymes. We believe that selectivity here is based upon their different polarizing power, the use of different donors, and their different stereochemical demands. [Here we ignore the influence of selection for copper(I), which is critical in some cases, and the enforced movement of copper and zinc into selected biological spaces (Chapter 3).]

(A)

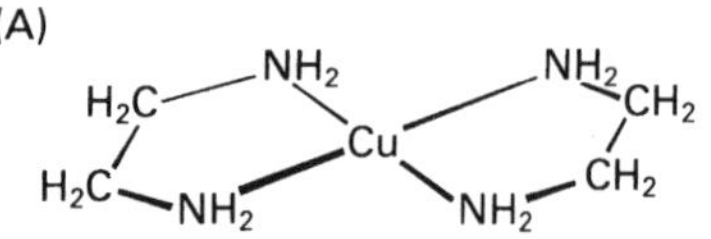

Tetragonal bis (ethylenediamine) Cu^{2+}

(B)

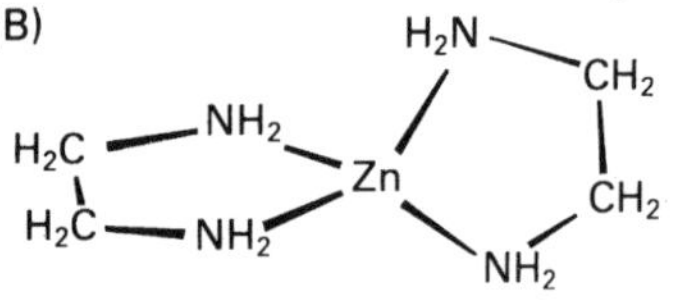

Tetrahedral bis (ethylenediamine) Zn^{2+}

(C)

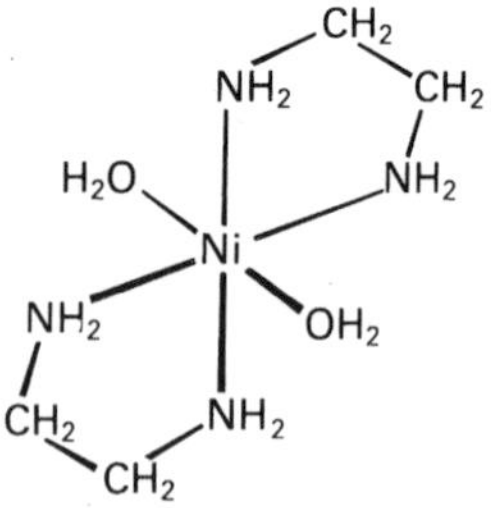

Octahedral di-aqua bis(ethylenediamine) Ni^{2+}

In the first place, copper(II) shows a remarkable ability not shared with zinc to form a bond at pH = 7 to peptide nitrogen in its ionized state. In this case the selectivity factor copper/zinc is perhaps as great as 10^4. Positive selection for copper is further enhanced by utilizing the preference of this metal for a tetragonal hole, (A), which arises from the ligand-field Jahn–Teller effect. An example of one of these effects is found in the copper-carrying plasma protein, an albumin. One or two of the metal–nitrogen bonds it forms are ionized amides while the geometry of the metal site must be that of a roughly square planar, four-nitrogen ligand, chelate. The site geometry is a property of the ligand conformation. (Here is an example of selectivity *outside* a cell which has few RS^- ligands.) Selectivity inside cells is largely based on Cu(I) bonding.

Positive selectivity for zinc can be based on its different demand now for tetrahedral geometry, (B), with nitrogen or sulphur ligands, and many examples of this are known in important biological zinc sites, e.g. in carbonic anhydrase and alcohol dehydrogenase.

A fine example of simultaneous zinc and copper selectivity appears in the blue-green superoxide dismutase: the zinc site is close to tetrahedral and the copper site is a distorted tetragonal hole (Fig. 15.8). Both sites have nitrogen bases as ligands and one of these (for copper) is an ionized nitrogen of imidazole.

There is a second pair of ions in the first-row transition series where strong advantage can be taken of different preferences for their coordination geometry: Mn^{3+} and Fe^{3+}. The first ion prefers tetragonal sites, like Cu^{2+}, whereas Fe^{3+} prefers tetrahedral or octahedral, (C), coordinating geometries. Due to the higher charge of the ions the preferred ligands in these cases are the phenolates of tyrosines and OH^-. Examples of the two situations are found in a second type of superoxide dismutase (SOD), the wine-red Mn^{3+}-SOD and the yellow Fe^{3+}-SOD. The dioxygen-release manganese centres of chloroplasts may select Mn against Fe in this way too, but we must be careful—biology can select by preferential pumping. A third case of advantage being taken by preferences for selected stereochemistry is that of Cu^+ in reducing solutions; i.e. in cells. Here, advantage may lie in the ability of this ion to have two linear bonds.

There is a most peculiar feature of the association of particular stereochemistries, ligand field stabilization, and functional use in biology. The two elements from the first transition metal series that in hydrates and oxide rocks favour most strongly an octahedral geometry are Cr^{3+} and Ni^{2+}, (C). We know that both are relatively available in the sea (Fig. 17.1), but neither has an important role in advanced, aerobic, organisms in marked contrast with Mn, Fe, Co, Cu, and Zn. It is as if the strong preference for a particular octahedral geometry and the possession of virtually no redox chemistry led to the exclusion of these elements. However, in primitive anaerobic organisms nickel is used in special ways in part associated with quite different geometries and oxidation states. We shall return to these problems of choice of metal ion in biology later since they may well concern the kinetics of reactions rather than thermodynamic selectivity (Chapter 4).

While the metal ions have preferred sites, it is the organic ligands which generate holes of particular sizes and shapes in biology. An example from model chemistry is the formation from three molecules of 1,10-phenanthroline (D) of an octahedral arrangement around low-spin Fe^{2+}, see margin, but the sterically restricted formation from two molecules of 2,9-dimethyl-1, 10-phenanthroline (E) to give a tetrahedron around Cu^{+} (Section 2.7). These two reagents are greatly used to analyse for Fe^{2+} and Cu^{+} selectively and they illustrate the parallel selectivity in biology by haem proteins for Fe^{2+}, octahedral, and plastocyanin for Cu^{+}, close to tetrahedral.

(D)

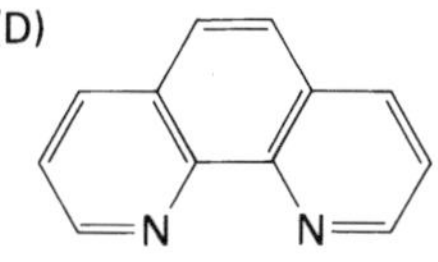

(E)

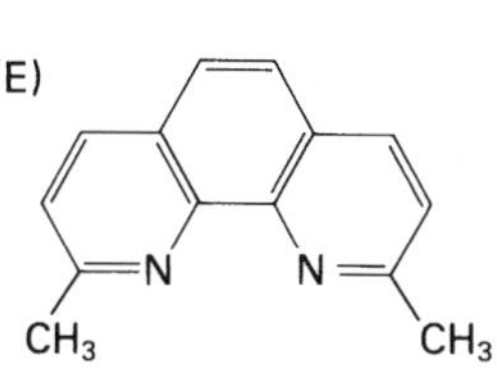

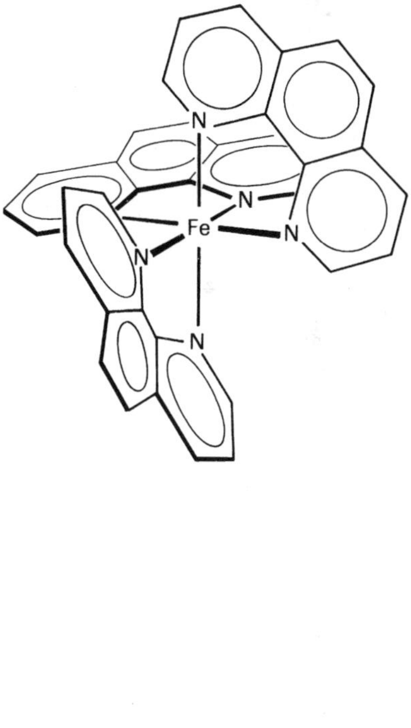

We have preferred to illustrate the ligand-field polarization of the metal ions rather than give a more detailed theoretical outline since this is available in many books (see the 'Further reading' section). The theory, well supported by experiment, shows that factors of up to 10^3 in stability can arise from a matching of ligand structures with energetically preferred metal ion polarizations (see Table 2.7).

Combining stereochemical preferences and types of ligand we can build a table of selectivity in which we list the preference of four of the ions considered for different sites (Table 2.9). Note that Mn^{2+} prefers octahedral co-ordination of N, O mixed donors (Table 2.9) to a greater degree than Zn^{2+} in a competition in which the supply of ligands is limited.

In conclusion, metal ion polarizability (ligand-field stereochemical effects) generates considerable selectivity on top of charge, size, and ligand polarizability differences.

2.5.6 Selection by spin-pairing stabilization

For transition metal ions there is a further possibility to increase selectivity, based on a spin-state change of a metal on binding to some ligands. The spin states, i.e. the way electrons are packed in 3d orbitals in the presence of different ligands, will generally be found to be as in Tables 2.10 and 2.11. Notice in Fig. 2.9 that changes of spin state affect ion size, and this is put to considerable use in biology, e.g. in haemoglobin (Chapter 13).

Table 2.9 Table of selectivity for Cu^{2+}, Zn^{2+}, Mn^{3+}, and Fe^{3+} (selection by adequate combination of donor atom and geometry of site)

Geometry of site	**Donor atom**			
	Phenolate RO^-	**Ionized peptide >N^-**	**Thiolate RS^-**	**Nitrogen⩾N or ⩾N, O**
Tetrahedral	Fe^{3+}	—	Zn^{2+}, Fe^{3+}, Cu^{+}	Zn^{2+}
Tetragonal	Mn^{3+}	Cu^{2+}	Cu^{2+}	Cu^{2+}
Octahedral	Fe^{3+}	—	(Mo(V))	Mn^{2+}, Fe^{2+}

Table 2.10 Usual spin states for some metal ions

Metal	**M^{2+}**	**M^{3+}**
Mn	High-spin (d^5)	High-spin (d^4)
Fe	Low-spin, high-spin (d^6)	High-spin (d^5)
Co	High-spin (d^7)	Low-spin (d^6)
Ni	High-spin (d^8)	Low-spin (d^7)

Table 2.11 Spin-state switches for some metal complexes

Cation	d-electrons	High-spin	Low-spin
Mn^{2+}	5	The vast majority	CN^-
Fe^{2+}	6	NH_3, H_2O, Cl	CN^-, NO_2^-, phen, cytochromes
Co^{2+}	7	The vast majority	CN^-, vitamin B_{12}
Ni^{2+}	8	The vast majority	CN^-, F-430
Cr^{3+}	3	All	None
Mn^{3+}	4	The vast majority	CN^-, phen
Fe^{3+}	5	H_2O, F^-, Cl^-	CN^-, phen, cytochromes
Co^{3+}	6	F^-	All the others
Ni^{3+}	7	O-donors	CN^-, many

phen is 1, 10 orthophenanthroline. Note the change to low-spin, paired *d* electron, states gains σ -acceptor and π-donor strength offsetting spin-pairing energy.

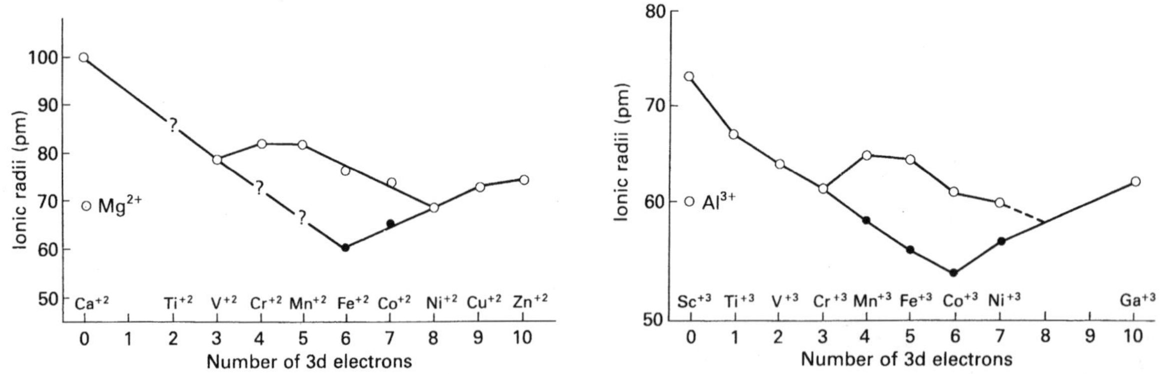

Fig. 2.9 Radii of the divalent ions Ca^{2+} to Zn^{2+} (left) and trivalent ions from Sc^{3+} to Ga^{3+} (right) as a function of the number of 3d electrons. Low-spin ions are shown as solid circles. [From Shannon, R. D. and Prewitt, C.T. (1970). *Acta Crystallographica*, **B26**, 1076]. 1 Å = 100 pm.

The selection of iron in preference to copper in porphyrins could perhaps be based on the low-spin stabilization of Fe^{2+} and the same may be said for the selection of cobalt in vitamin B_{12} in the presence of other divalent metals. However, later we shall show that kinetic effects are much more likely to dominate these incorporations. An important point again is that the selectivity factor will only operate if some metal ion concentrations, e.g. Cu^{2+}, are lowered by their removal in a complex with a controlled supply of a more powerful S, N ligand donor.

It is then expected that strong N/S-donor ligands will bind low-spin $Co^{3+} > Mn^{3+} > Fe^{3+}$ ($> Cr^{3+}$) and this may also apply with RO^- ligands, while the order for low-spin divalent ions will be $Fe^{2+} > Co^{2+} > Mn^{2+}$. Model ligands such as 1,10 orthophenanthroline can be used to illustrate spin-state changes (Table 2.11).

2.5.7 Selection by binding in clusters

The binding of metallic elements to proteins developed in a further way. In their higher oxidation states the hydrolysis of ions at pH ~ 7 is quite extensive and some low molecular mass species of hydrolysed units occur. These states will be present in all surface waters decreasing the availability of the simple ions. The most likely species to be present are Fe_2O^{4+}, $Cu_2(OH)_2^{2+}$, and Mn_2O^{4+}, highly charged cations with small, hard, anions. All three dimeric units can form complexes and, in fact, these dimer centres are present in some proteins where, in the course of evolution, they have undergone subtle changes so as to become of functional value both outside and inside cells. Interesting examples are those of haemerythrin, containing a Fe_2 dimer linked by one O^{2-} and two carboxylate bridges (Fig. 12.11), the copper dimer in copper oxidases (Fig. 15.1), and the manganese units in the dioxygen-producing enzyme of photosynthesis (Chapter 14). We shall find many other examples of the use of small groups of metal ions in hydrolytic enzymes.

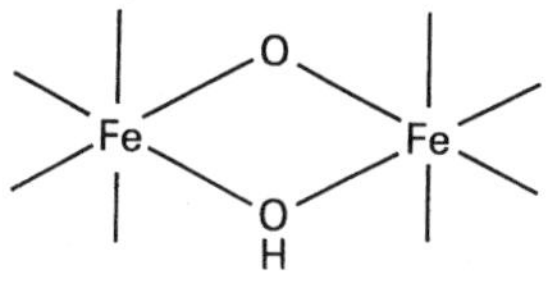

Oxo-hydroxo cluster

Inside cells (and also outside cells during the early history of life), sulphide ions are also present, and can react exactly as the hydroxide ions do, to give polysulphide units, e.g. $Fe_4S_4^{4+}$ based on linked tetrahedra of iron in the Fe^{3+} or both the Fe^{3+} and Fe^{2+} states, but not Fe^{2+} alone. These units can also bind to small molecules or proteins and widely known examples are those of the Fe_2, Fe_3, and Fe_4 units of ferredoxins, where S^{2-} acts as a bridge and cysteinate (RS^-) as the protein ligand. The selectivity between oxide and sulphide in dimers is in favour of sulphide in the order (Mo) > Fe^{3+} > Mn (see Chapter 1) as a consequence of the changing ratio of $I/(z/r)$ (see Fig. 2.5). Mn–S units have not been found in proteins. There is also the possibility of formation of mixed high-valence clusters and this may be the basis for the finding of Mo–Fe–S and V–Fe–S species in nitrogenase and for the formation of some nickel centres. Generally, clusters are stabilized by highly charged cations since they involve highly charged anions, O^{2-} and S^{2-}, and in biology are only observed in proteins (see Section 2.6).

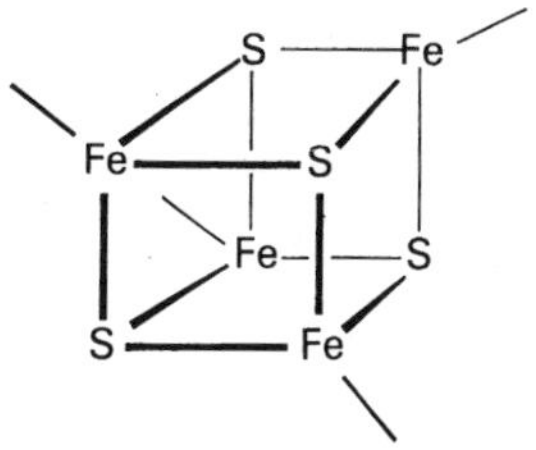

Fe_4S_4 cluster

Note that this binding of metal ions in pairs, or even more at a time, is naturally opposed by entropic considerations, but the unfavourable term in the association is greatly reduced if the binding organic anionic centres are already built into a preformer, e.g. a protein, such that the protein is largely folded into the correct conformation for the uptake of the clusters. This is a gross size selection much as we have seen to occur on a smaller scale for simple metal ions. This type of cooperativity effect will be discussed later in a wider context. The selection here also depends on the highly charged states of the cations (see Section 2.5.1) and clearly operates in favour of strong binding to organic anions; in this sense it assists $Fe^{3+} = Mn^{3+} > Cu^{2+}$. Clusters are usually found bound to anion protein sidechains. Note that the trace element zinc is excluded from this type of site since it only forms weak clusters.

2.6 Hydrolysis: hydroxides and oxenes

Cluster formation is a type of hydrolysis (sulpholysis). As stated in Section 2.5.1, in the case of trivalent and more highly charged cations there is competition between complex ion formation with organic ligands and hydrolysis. Furthermore, the hydrolysis is of two kinds leading to hydroxy-species and, ultimately, hydroxide polymers and precipitates, and oxo(oxene) species which can also take part in polymerization as oxides (see Table 2.12). These reactions allow the appearance in biology of a large variety of shapes and sizes

Table 2.12 Hydrolysis and sulpholysis of ions

General formulae	M^{2+} (M(II))	M^{3+} (M(III))	M^{4+} (M (IV))	M^{5+} (M (V))	M^{6+} (M(VI))
	Simple ions	MO^+	MO^{2+}	MO^{3+}	MO_2^{2+}
		M-O-M	M-O-M	MO_2^+	MO_4^{2-}
		M-S-M		MO_4	MS_4^{2-}
Specific examples	Mg^{2+}, Zn^{2+}	Fe^{3+}	MnO^{2+}	MoO^{3+}	MoO_2^{2+}
		Fe-O-Fe	FeO^{2+}	MoO_2^+	MoS_4^{2-}
		Mn-OH	VO^{2+}	$MoOS^+$	$MoOS^{2+}$

There are also many mixed states in polymers where there are internal O or S bridges and terminal groups. Note that S replaces O more readily down any group of the periodic table and increasingly toward the end of the transition metal series and in later groups

of oxygen-bound metal ions which must be held in controlled amounts. The reaction scheme is

$$\begin{array}{ccccc} M + H_2O & \rightleftharpoons & MOH(+H^+) & \rightleftharpoons & MO(+H^+) \\ & & \downarrow\uparrow & & \downarrow\uparrow \\ & & [M(OH)]_n & & [MO]_n \end{array}$$

As the pH of a solution of Fe^{3+} is raised, the amount of Fe_2O^{4+} rises rapidly as does the amount of a variety of oxy-/hydroxy-polymers such as $Fe(OH)_3$, ferrihydrite, FeO(OH), e.g. goethite, and haematite, Fe_2O_3. There is interconversion of the solids with ageing. We shall find these and other hydroxy/oxide precipitates in organisms in Chapters 12, 14 and 17.

A contrasting behaviour is that of V(IV) which has a variety of forms in biological media including VO^{2+} and V^{4+} in complexes. (For comparison we have the well-known inorganic species UO_2^{2+}, which is related to U^{6+}.) The formation of oxene species can also occur in well separated stages so that we may write for Mo(VI), for example,

$$MoO^{4+} \rightleftharpoons MoO_2^{2+} \rightleftharpoons MoO_3\downarrow \rightleftharpoons MoO_4^{2-}$$

with hydroxy-variations.

Whereas with the uranyl ion, and to some extent in vanadyl, VO^{2+}, chemistry, most reactions favour the maintenance of oxene units, in the case of molybdenum the different oxo groups can be replaced by organic chelating agents so that Mo(VI) chemistry in biology extends from units in which there are no oxene units to all the other possible chemistries of MoO^{4+}, MoO_2^{2+} and MoO_4^{2-}. The units can all be selectively recognized by proteins as we shall see in Chapter 17. We shall also find that the oxo group of VO^{2+} may be replaced by ligands in some biological species.

As already stated, oxygen is only one element which can engage in this type of chemistry; outstandingly in the inner compartments of biological systems today (and in early history in the environment) another available element is sulphur which can give the states M(SH) or MS. There is then competition between the deprotonation of H_2O and that of H_2S, which is also biased by redox reactions in the cellular compartments. Several factors are important in this case: (1) the dissociation constants of H_2S are much lower, making SH^- and S^{2-} much more available; (2) however, the concentration of H_2O exceeds that of the maximum H_2S concentration by some 10^5 and in conventional aerobic biological media this figure approaches perhaps 10^{10}, when there is very little H_2S; (3) the relative affinities of metal ions for sulphur- and oxygen-donor ligands (H_2S and H_2O) are metal-selective. In the presence of reasonable amounts of H_2S we know these affinities from the old student group analysis tables (sulphide affinities in precipitates for divalent ions)

Hg, Cu, Cd, Pb > Ni, Co, Zn > Mn, Fe > Mg, Ca.

(This is, in part, the Irving–Williams stability series.) For higher valent ions the sulphide preference is

Mo, As, Sb, Bi, Sn > V, Fe > Mn > Al.

Although this background to the oxide and sulphide chemistry is useful, it will not give more than general guidance for an understanding of the complexity of biological reactions, but it allows us to note with no great sense of surprise that the oxo- and sulpho-complexes we meet in biology are only of the kinds

FeO^{2+}, MoO^{4+}, MnO^{2+}, VO_2^+, VO^{2+}, and possibly CuO^+ (mononuclear), $[Fe_2O]^{4+}$, $[Mn_2O]^{4+}$, and possibly $[Cu_2O]^{2+}$, MoS^{4+}, Fe_nS_n, $(Fe, Mo)S_n$, $(Fe, V)S_n$

where only Mo or V carry terminal S and n is variable perhaps from 2 up to 6–8.

The difficulty of understanding the selective appearance of the different types of hydrolysis product and non-hydrolysed ions does not lie only in the inorganic chemistry but in the *controlled evolved* selection of proteins to incorporate the different shapes and sizes of these single sites and clusters (see Chapter 12). One general rule appears to be that oxide species appear with O- and N-donors from proteins and have high co-ordination numbers while sulphide species appear with S-donors and have low co-ordination numbers.

Finally, we note that it is relatively rare to observe sulphide precipitates in biology, but there are anaerobic species which contain them.

At this stage we hope to have introduced all the possible ways in which bonding aids selectivity of combination, and we draw attention to the fact that we have not had to discuss the major donor atoms, C and P, of organometallic chemistry. Very much has been learned from the studies on small model complexes.

2.7 Selective control of oxidation states of metals

One of the major tools of the analytical chemist is to use the redox potential of the medium to control differentially the redox states of metal ions under analysis. Since this control involves compartments in biology (see Chapter 3) only a few illustrative examples are given here.

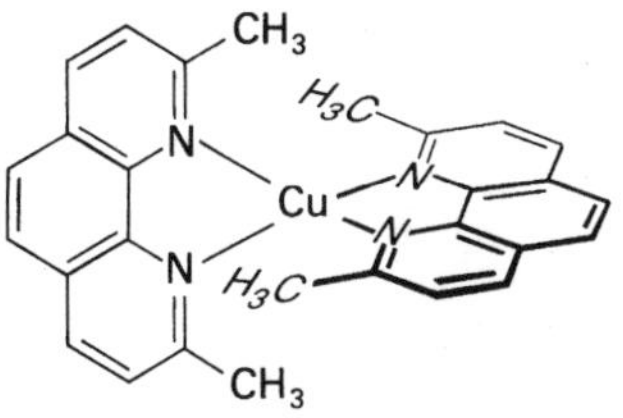

The easiest way in which to incorporate copper selectively into organic ligands is to reduce this metal ion to Cu(I). In analytical chemistry the use of reagents such as 2,9-dimethyl orthophenanthroline in reducing conditions allows selective estimation of copper(I). It is the combined use of reducing conditions, the strong binding of Cu(I) to nitrogen donors, plus the demand for a special geometry by the Cu(I) ion, see margin, which makes for *this selection*. Alternatively, Cu^+ can be bound by two strong donors linearly coordinated, which is an option not open to any other available metal ion in organisms. We shall see that biological systems have taken full advantage of this combination of factors in copper proteins.

A second example which we gave earlier was the use of anionic chelating agents to scavenge for iron(III) outside cells. The redox conditions and the selective reaction based on high charge effectively excludes divalent ions of metals such as Cu, Co, Ni, and Zn from phenolate complexes.

These manipulations of elements that depend on recognition of particular redox states require us to note carefully the redox conditions in compartments and to have knowledge of the redox potentials in different complexes. The potentials are not analysed separately in this book since they follow from the redox potentials in the oxidation state diagrams of Chapter 1 and the stability constants of the complexes of the individual states using the equation

$$E = E^0 + \frac{RT}{n\mathfrak{F}} \ln K_{red} / K_{ox} \quad \text{when } [M_{ox}L] = [M_{red}L]$$

where K_{red} and K_{ox} are the stability constants of the complexes of the two states concerned, and E^o is the standard redox potential of the couple.

2.8 Selection by control of concentration of M and L

In the following sections situations in which model chemistry is less helpful are analysed. We have seen that selectivity can be greatly enhanced by control over [M] in the expression K^{eff}_{ML}[M]. There may be conditions in which it will be advantageous if sites that are not truly specific will have to capture certain elements from among others which are more available so as to generate some special biological (chemical) function. A possible mechanism, certainly used by biological systems, is that of 'sequestering' the more common elements preferentially by using an even more powerful second ligand. The concentration of the 'interfering' elements will be lowered and their product K^{eff}[M] for the given site will be made smaller than that for the element to be captured. Let us look a little more closely at the nature of the discussion in Section 2.4.

Disregarding acidity effects for the sake of simplicity, i.e. assuming that the proton does not compete for the site, the desired situation can be represented as

$M_1 + M_2 + L_1 + L_2 \rightarrow M_1L_1 + M_2L_2$,

M_1L_2—excluded by lack of M_1, removed by L_1,

M_2L_1—excluded by lack of L_1, removed by M_1.

The generation of this circumstance can be approached in terms of effective stability constants

$$K^{eff}_{M_1L_1} = \frac{[M_1L_1]}{[M_1]^*[L_1]^*}$$

where $[M_1]^*$ is the concentration of M_1 *not bound* to L_1 and $[L_1]^*$ is the concentration of L_1 *not bound* to M_1, i.e. in each case it is a sum which one can easily derive as

$$[M_1]^* = [M_1]\,\{1 + K_{M_1L_2}[L_2]\},$$
$$[L_1]^* = [L_1]\,\{1 + K_{M_2L_1}[M_2]\}.$$

If binding of M_2 to L_1 is to be avoided, we have to decrease $[M_2]$ to such an extent that $K_{M_2L_1} \cdot [M_2]$ becomes smaller than say 0.01 (to be neglected in the expression for $[L_1]^*$). This can be done by preferential binding of M_2 to L_2. In this case, since

$$[M_2] = \frac{[M_2L_2]}{[L_2]\cdot K_{M_2L_2}},$$

one obtains

$$\frac{K_{M_2L_1}\cdot[M_2L_2]}{K_{M_2L_2}\cdot[L_2]} < 10^{-2}.$$

Now, if L_2 is in excess relatively to M_2, e.g. twice as abundant as M_2, it is enough that $K_{M_2L_2} > 10^2 K_{M_2L_1}$ to decrease $[M_2]$ to a level such that its competition with M_1 for L_1 will be less than 1 per cent. Larger ratios in the constants will allow the same effect with just a small excess of L_2 over M_2.

To avoid binding of M_1 by L_2 the same reasoning can be applied and one obtains

$$\frac{K_{M_1L_2}\cdot[M_1L_1]}{K_{M_1L_1}\cdot[L_1]} < 10^{-2},$$

and the problem is now related to the degree of saturation of L_1 by M_1 or to the differences in the values of the constants.

In general, for approximately equal concentrations of M_1 and M_2 and controlled small excesses of L_1 and L_2, selection giving only M_1L_1 and M_2L_2 will be achieved up to 99 per cent accuracy by sites differing by 10^2 in the ratio of their stability constants, i.e.

$$\frac{K_{M_2L_2}}{M_{M_2L_1}} > 10^2 \quad \text{and} \quad \frac{K_{M_1L_2}}{K_{M_1L_1}} < 10^2.$$

This requires that the order of ligand preference by different metals is different. In fact, this is the case for ions such as Mg^{2+} and Ca^{2+} compared with Cu^{2+}, as can be seen from Table 2.4 where Mg^{2+} and Ca^{2+} prefer O-donors but Cu^{2+} prefers N or S donors. Now, however, Cu^{2+} and Ni^{2+} both prefer N-donors over O-donors and here it is necessary that the stability constants are such that there is a higher selectivity in favour of Cu^{2+} with any one ligand in order to bind copper almost exclusively, see margin. Figure 2.8 shows that this condition will easily be achieved in biological systems using N/S ligands. Therefore, selection is dependent on the restrictions on concentrations of metal ions relative to ligands as well as on stability constants of the respective metal complexes.

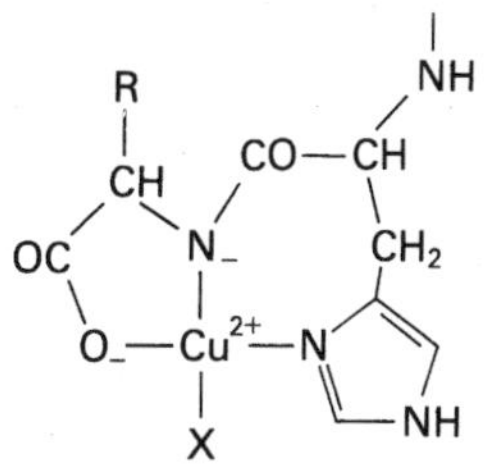

An example of a selective anionic nitrogen ligand for Cu^{2+} in a tetragonal field

Finally, M_2 can be H^+ when the approach given earlier in the chapter on effective stability constants reveals whether M_1 or H^+ will bind to any given L. The acidity of a compartment is critical.

This amounts to the statement that, *provided a cell limits free ligand (protein) and free metal ion concentrations (by controlled feedback mechanisms acting on synthesis and hydrolysis of the ligand and on the uptake and rejection of the metal ion, respectively), the selectivity of metal ion/ligand combination can be well-nigh complete given known stability constants*. This proviso will prove of the greatest possible interest to us and concerns genetic control of protein expression itself dependent on metal ion access to cells (see Section 2.20 and Chapter 6).

An example of control of concentration to achieve selectivity is that of the blood plasma, where the metal ions are removed one by one by powerful lig-

Table 2.13 Carriers in blood plasma

Metal ion	Free M	Carrier	Basis of selectivity
Iron (and manganese)	10^{-16}	Transferrin	Phenolates bind M^{3+} octahedral complex
Copper	10^{-14}	An albumin	Ionized peptide >N⁻ binds Cu^{2+} in tetragonal complex
Zinc	10^{-8}	An albumin	N(S) donors in tetrahedral geometry bind Zn^{2+}
Cobalt (as vitamin B_{12})	$<10^{-10}$	Carrier protein	The B_{12} sidechains are recognized
Calcium	10^{-3}	Phosphoproteins (when needed)	Weak binding for most metals
Magnesium, sodium potassium, (calcium)	10^{-3}–10^{-1}	None	Freely circulating as ions

ands so as to avoid mutual interferences (see Table 2.13). Thus, copper is removed first by an albumin containing an ionized peptide bond (after complexation). Iron is oxidized and the blood-protein transferrin picks-up the Fe^{3+} ion using phenolate binding groups. Zinc is also removed by an albumin which may have three or four neutral N-donors in a tetrahedral geometry. Manganese, Mn^{3+}, may be also transported by transferrin (same site as Fe^{3+}). Cobalt as Co^{3+} is carried as vitamin B_{12} (insertion control; Section 2.12.3); sodium, potassium, magnesium, and, to some extent, calcium are left free in the blood solution. If extra Ca^{2+} is required, it cannot travel as free Ca^{2+} in the blood since an increase in concentration would exceed the solubility products of its phosphate or carbonate. There is then a special calcium carrier protein with phosphorylated sidechains, which is generated under hormonal control. If Cu, Fe, Mn, and Zn had not been removed these ions would bind strongly to the calcium sites and the process would be impossible. We see here that it must be the combination of kinetic limitations on ligand synthesis together with inorganic chemical selectivity which dominate observed combinations. In all cases it is K^{eff} which we must consider. The consequences for biology are very considerable. Many states of disease relate to failures in these control processes.

Beside these thermodynamic binding considerations together with the limitations on synthesis, we shall need to note below the ability of cells to selectively pump elements in and out of cell compartments. [In turn, this means that we must analyse compartmentalization (Chapter 3), kinetics and energetics of transfer (Chapter 4), and biological synthesis of ligands (Chapter 6). The fact that pumps can be used to transfer any of the components becomes a dominant factor over selectivity since they control, within synthesis, the concentration terms.]

2.8.1 Availability and predicted cytoplasmic free ion concentrations

Before we enter into discussion of more complicated modes of transfer from the environment to the cytoplasm it is very important to recognize a peculiarity that would be generated by the establishment of thermodynamic equilibra-

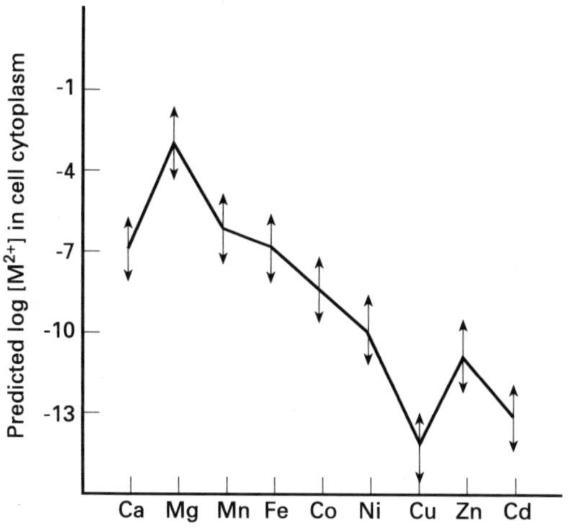

Fig. 2.10 Estimates of the free metal ion concentration in the cell cytoplasm assuming that the proteins that are produced in restricted amounts bind in simple thermodynamic equilibrium and using model complexes to assess binding constants.

tion between *selective element binding by the variety of organic ligands as seen in model studies* (Fig. 2.8) *if they were in the cytoplasm*, as opposed to the complexing of the *inorganic ligands* in the abiotic environment (Fig. 1.13). This use of such stability constants of model complexes in fact allows us to postulate that Fig. 2.10 gives us the likely variation of free metal ion concentrations in the cytoplasm of all cells assuming that each free metal ion concentration is buffered to an equilibrium concentration opposite the presence of restricted amounts of free ligand, and therefore of total metal, designed for selective uptake of the metal and controlled by genes (see Section 2.20). The free metal ion concentrations in the sea in aerobic and anaerobic (H_2S) conditions for a variety of ions are shown in Fig. 1.13. There is a large drop from Mg^{2+} to Mn^{2+} in *aerobic* conditions and then only a slight variation largely due to hydroxide precipitation or binding in silicates. In *anaerobic* conditions the variation along the series is hugely increased due to sulphide selectivity, see Fig. 1.13. This change, which occurred about 1.5×10^9 years ago, we believe to be one of the major causes of evolution, Chapter 20. In a cell cytoplasm we imagine that ligands using any combination of the donor types of ligand, O→, N→, and S→, are available in both environmental circumstances. The variation in free metal ion Fig. 2.10, deduced using model studies, is therefore intermediate between those of the aerobic and the anaerobic curve of Fig. 1.13. In all cases it is closely the inverse of the Irving–Williams stability constant series, see Fig. 2.8, with but minor exceptions. In fact, this steep pattern of variation of free element levels along the series *in the cytoplasm* is close to one characteristic of all living cells, anaerobic or aerobic (see Chapter 20). (Note, however, that a cell can also contain strongly bound metal ions which may or may not be in equilibrium with free ions, but this only accentuates the differences in Fig. 2.10. We deal with kinetically controlled incorporation in Section 2.12.) The total (not now the free) element concentrations in cells and organisms can vary independently from the free cytoplasmic concentrations by the production of binding agents or pumping into vesicles which has different possibilities in aerobic as opposed to anaerobic conditions. We shall follow these possibilities in much of Part II and Part III. We wish to have a name for the profile of the analytical element content of cell compartments, free or combined, and we shall use the words free and total 'metallome' to match the concepts of genome and proteome. We return to this description in Section 2.10 and especially in Chapter 20.

2.9 Selection by transfer coefficients from water to proteins (or membranes)

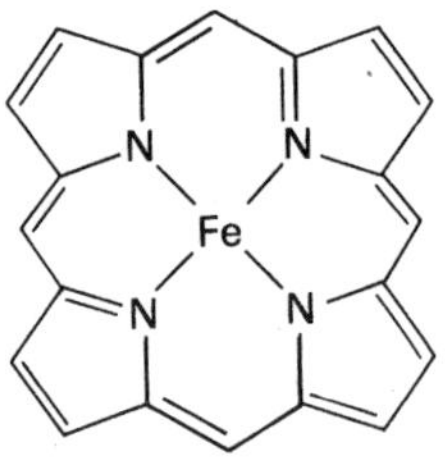

Iron porphyrin. Particular porphyrins have multiple ring substitutions. The protein-binding groups are attached above and below the Fe.

So far, the discussions on selectivity rest, for the most part, on the assumption that a protein behaves like a model complexing agent made from small monodentate ligands even in a membrane. In fact, a protein adjusts its fold around a metal ion and the changes in protein fold energy are likely to be commensurate with the metal-ligand bond energy. A compromise in coordination numbers, bond lengths, and bond angles is forced upon the metal and the fold of the protein imposing a constraint on both. The result can be a distorted energized situation to generate an activity which has been called an 'entatic', or constrained state of the metal ion. This free energy constraint is not just a matter of energetic effects in the first coordination sphere of the metal, since the fold generates an environment for the whole complex ML, where L is the equivalent set of metal ligands in a model complex. The energies of transfer of a simple complex ML from an aqueous medium (aq) to the inside of a protein (in a membrane or not) of lower dielectric constant can be very large and also selective, e.g. for M^{2+} against M^{3+}. The overall stability of a metal-protein complex is then related to a binding constant log $K_{ML(aq)}$ for the free unlinked amino acids that compose the binding groups by a transfer coefficient p of ML from water into a real protein. This involves local constraint in ML and a considerable interaction of ML with second- and third-sphere neighbours, which can be favourable or unfavourable, and we now show that p can generate selectivity just as much as log $K_{ML(aq)}$ through long-range energetics.

This can be illustrated by reference to the uptake of haem (iron), i.e. a special ML, into haem proteins, see aside (A) and (B). It is known that the haem can be bound in different ways (see Table 2.14) and these generate changes of the normal redox potential over a range of more than 600 mV. Hence, different proteins can selectively distinguish complexes of Fe^{3+} from Fe^{2+}, and therefore, in principle, of any M^{3+} from any M^{2+} by a factor of more than 10^{10} simply by changes of *one* or at most *two* ligands, keeping the other four fixed. Even without change of ligands, however, Table 2.14 shows that protein folds can distinguish Fe^{3+} from Fe^{2+} by a factor of 10^5, e.g. in peroxidase and myoglobin; this results just from a change of binding or 'solvation' within the protein, i.e. the redox potential of the iron couple is in part controlled by the degree of exposure of haem to solvent water and to protein hydrophobic sidechains. In effect this result implies that a selectivity of any M^{3+} from any

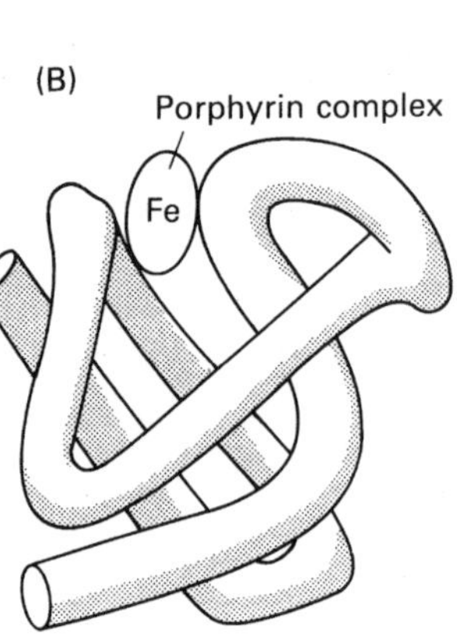

(A) The constrained (entatic) coordination geometry of iron (II) in a protein such as that shown in (B).

Table 2.14 Complexes of iron porphyrin IX with proteins

Protein	Binding groups	E° (mV)
Myoglobin	His or His, H_2O	> + 100
Haemoglobin	His or His, H_2O	> + 100
Cytochrome b_5	His, His	0
Cytochrome b_{562}	His, Met	< + 100
Peroxidase	His or His, H_2O	–200
Catalase	Tyr or Tyr, H_2O	–500

M^{2+} can be obtained by a change of internal dielectric constant, i.e. that of the protein. Similar selectivity applies to the partition of ML complexes into organic phases, e.g. membranes.

We must not be surprised, therefore, if sometimes M^{3+} and $(M')^{2+}$, e.g. Fe^{3+} and Zn^{2+}, are selected by the same coordination partners in the same geometric environment but by different protein sequences. This is observed in rubredoxin (Fe) and alcohol dehydrogenase (Zn) where the metal ions have four thiolate ligands. Similar factors can select Cu^{+} against Zn^{2+} and again we know that iron-sulphur proteins select between $|Fe_4S_4|^{+}$, $|Fe_4S_4|^{2+}$, or $|Fe_4S_4|^{3+}$, and also select between $|Fe_4S_4|$, $|Fe_3S_4|$, $|Fe_2S_2|$, and Fe all bound to the same tetrahedral group of four thiolate ligands, see Chapter 12. The protein fold decides the complex that is captured using many factors. Now, if it were so that different metals favoured different types of sulphide complexes M_nS_m, then proteins could also select between them, e.g. in the uptake of Mo–Fe–S as opposed to Fe–S species. Even when $n = m$ the size and charge of M_nS_m will be sufficiently different that several metals can appear in the same general coordination sphere selected by different proteins on the basis of folding energy, i.e. 'p' can be very discriminating.

It also follows that the fold itself will be dependent on the exact metal or metal complex, so that the surface of the protein can vary depending on the metal or metal complex and, of course, this dependence is variable being related to the strength of the protein fold. Here there is total co-operativity between metal ion geometry and protein fold. This may affect considerably the distribution coefficient D of M (in a protein) between an aqueous and an organic phase which is again a means of achieving selectivity. Examples are the folding of metallothionein (different for Cu, Cd, and Zn) and their passage through membranes. Of course, the metalloprotein may be recognized by other proteins as well as by membranes based on this protein fold difference. Here we begin to see how a particular ML may be positioned in a cell after uptake (Chapter 3).

Finally, we note that where the protein folding has open to it several conformations of somewhat equal free energy then different binding sites on a protein become interactive, cooperative, or anticooperative. The possibility of the adjustment of one site by another is called an *allosteric* or *rack interaction*, e.g. as is found in dioxygen uptake by haemoglobin. The different sites can be for a metal ion and, say, two proteins, when metal/protein A and protein A/protein B build a communication network dependent on the binding and on the redox state of the metal ion.

Before discussing this aspect in greater detail, especially its kinetic implications, we must stress again that in the previous considerations we have often made the important assumption that a given ligand (protein) is in only slight excess over the essential trace metal ion that it binds. This will only arise in biology if the synthesis of the protein is stopped once this small excess is achieved and there must therefore be a requirement for feedback homeostasis which effectively buffers free [M] and free [L] which may only be overridden locally by hormonal effects. The feedback becomes specific to M if there are different folds for different metal ions. We will come later to this aspect since we believe it contains the kernel of why element chemistry has so dominant a role to play in biological activity.

The above discussion of selectivity principles can obviously be applied to effects upon redox potentials between any two oxidation states and it applies equally to anions (see Section 2.15).

2.9.1 Soft and hard acids, solvents, and extraction

The classification of 'soft' and 'hard' acids and bases given in Table 2.8 and Fig. 2.5 is for aqueous solutions as far as ions are concerned. It is therefore solvent-dependent, and depends not just on electron affinity but also ionic size. Free ions behave quite generally more as 'a'-class (apparently *hard*) the lower the dielectric constant is, i.e. in organic solvents. However, when we consider systems of two immiscible solvents and the partition of complexes between them

$$\underset{\text{aqueous}}{M^+ + X^-} \rightleftarrows \underset{\text{aqueous}}{MX} \rightleftarrows \underset{\text{organic solvent}}{MX}$$

it is usually the case that MX in the organic phase is relatively more strongly favoured the more 'b' class the anion and the cation. The electrostatic controls over these equilibria are obviously considerable. Now, the interior of a protein or of a membrane is more like an organic solvent than an aqueous solution and the anion, X, may be trapped there so that 'b' class behaviour of all metal elements may be more prominent in proteins than expected from aqueous solution equilibria. There is a selectivity term here too in that complexes of one metal ion can have a different partition coefficient, D, from others. Thus metal ions can be phase-separated.

Ligands too can be partitioned into different media, so that metal/ligand combinations can be found in membranes as well as in water. Again ligands and/or metals can be moved across membranes so as to be present only in certain compartments. Proteins have 'leader' sequences which direct them to selected places in cells.

2.10 Selection at surfaces and in precipitates

Cells possess many surface sites for binding metal ions and molecules generally, where the binding depends on the concentration of the metal ion but the site has no concentration dependence though the number of sites employed is important. The major examples are membranes, large immobile particles of proteins, DNA/RNA, and polysaccharides. Here we include proteins in membranes such as pumps and channels. In these cases, which are undoubtedly of equal importance to those of free, soluble ligands, the binding equation can be readily deduced for sites of equal affinity since the problem is the same as for adsorption following the Langmuir isotherm where binding depends directly upon concentration of the material adsorbed, e.g. a metal ion, M. The binding constant is then given by the free metal ion concentration at 50 per cent saturation of the sites.

$$K_B = 1/[M]_{50\%}$$

The variation in K_B has the same dependence upon the specific nature of the site and selected metal ion as for free ligand binding. Hence all the above sections on selectivity are still applicable.

A further consideration follows when we treat conditions for simultaneous occupation of a surface site and binding by a free ligand. Consider, with the equation for surface binding, that for K_c the binding constant for a freely soluble ligand, X, i.e.

$$K_c = \frac{|MX|}{|M||X|}.$$

If we assume that both the surface and X need to be bound equally, then the requirement is [MX] ≈ [X] when $K = 1/[M]_{50\%}$.

[This condition can be deduced for solution equilibria alone if the system containing X is to respond optimally to [MX] and to [X] while reflecting changes in [M]. In fact this result is that commonly used in devising buffering solutions. For example, in the case of M = H^+ the best buffering by HX and X is where the pH = pK_a of X, i.e. at $[H^+]_{50\%}$. Biological cells are held at around pH = 7 close to the pK_a of phosphates, HCO_3^- and histidine].

If the substance, metal ion plus anion or molecule, form a cooperative solid phase (see Section 2.18) such as a new liquid or a precipitate, then there is a single critical concentration, C, at which the new phase forms. This critical concentration is related to a binding constant in the new phase, $K_S = 1/C$. For example—2.303 $RT \log C$ is the free energy difference between the material in the different phases which is given by 2.303 $RT \log K_s$. Clearly, the formation of a new phase restricts the system to an absolute critical and constant concentration.

We believe these observations are crutial to the integrated functioning of cells (see Section 7.6 and in particular Chapter 20) since they have simultaneously surface free solution and solid state binding agents. We turn to some examples of binding to surfaces such as pumps and channels.

2.11 The selectivity of channels

The existence of selective ion channels in the membranes is the basis of a special case of transport that is under thermodynamic control. To be selective a channel does not have to bind an element M; it is enough that it has the same radius as that of the (hydrated) ion, since being somewhat mobile and flexible molecules, proteins can simulate 'holes' of variable size and perform a 'sieving' action, allowing the free flow of certain ions while preventing that of others (Fig. 2.11). The sieving does not require that the interior of the channel is not hydrated and evidence suggests that the channel diameter for some ions is large, say 5–10 Å diameter, yet it still sieves (see Chapters 5 and 8). Selectivity, in this case, is shown by the fact that there are highly discriminatory potassium channels, sodium channels, calcium channels, and even proton channels (as well as various anion channels). Typical highly selective channels are the potassium channels of nerves, which show a selectivity Li < Na ≪ K > Rb > Cs, and the calcium channels which have selectivity Mg ≪ Ca > Sr > Ba (see Chapters 8 and 9). Since for these ions the energy at a site is a function of the inverse of the radius, it is clear that a channel hole can only match one ion closely for which there can be no real binding since it can pass unimpeded. This, in effect, is the same radius ratio effect that controls binding to chelating agents, but here there is a further problem, that of mobility, which requires a different process ensuring free movement *in* and *out* of the cell. Selection is based in part on repulsion, not on attraction: smaller ions do not cross the channels since they are too *big* to enter—they carry more water molecules and cannot be dehydrated except at the considerable expense of energy; larger ions cannot enter due to their size. Selection is then very effective and for these reasons Na^+ and Ca^{2+} channels exclude K^+, Li^+, Mg^{2+}, i.e. larger and smaller ions independently from charge considerations, by factors of, say, 10- to 100-fold. The lanthanides block the calcium channels, and barium blocks the potassium channels, showing that there is some interaction with channel surfaces since $La^{3+} = Ca^{2+}$ and $Ba^{2+} = K^+$ in size. In practice many channels are

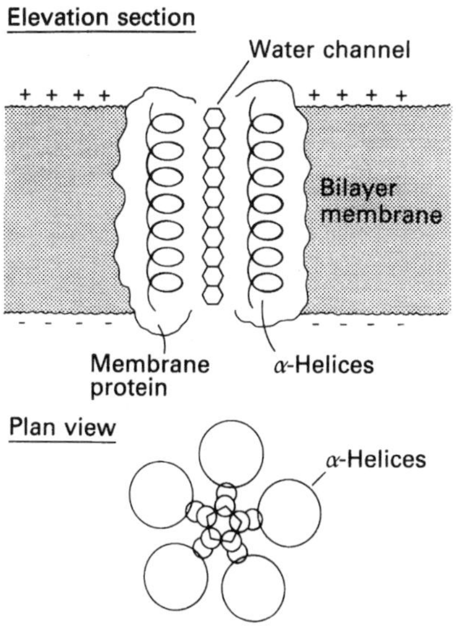

Fig. 2.11 A simplified view of a channel in a membrane. In real systems the channels can be in open or closed states and they are gated in various ways, e.g. electrically or conformationally, see Morowitz and Williams in further reading.

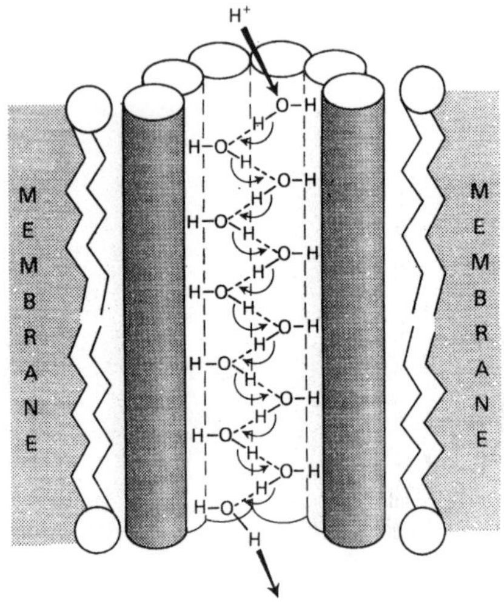

Fig. 2.12 The possible H-bond network for a proton channel or proton wire, see Morowitz and Williams in further reading.

controlled by *gates*, kinetic devices allowing or preventing free flow. The 'gates' may be charge movements in the channels due to external electrostatic potentials.

The case of the proton is clearly different from that of other cations since it binds easily to several side-groups of proteins. A unique possibility is that of successive protonation-deprotonation of water molecules (or other OH groups) inside a protein channel (Fig. 2.12). The exact network of hydrogen is not confined to linked H_2O; it could be any H-bond system which is continuous and can perform the necessary motions. Such channels are extremely important in energy capture devices (Chapter 5). Usually we do not think in terms of selectivity for protons, but biological systems have an outstanding need to make such a selection. The acidity of different biological compartments is an example of the result of such membrane selection.

Some channels act by exchanging ions both of the same and different charge types. Typical exchangers are H^+/Na^+, Na^+/Ca^{2+}, and Cl^-/OH^-.

2.12 Kinetic effects and control

The next step in our analysis of the uptake of metal ions will focus on the rate of movement of M from the external to the internal aqueous phase (see Figs 2.1 and 2.2). There are several possibilities which we consider in turn. These considerations could well oppose the thermodynamic assumptions made earlier (see Fig. 2.3), but we consider that it is more probable that they will act so as to enhance the thermodynamic selection within a steady state.

2.12.1 The distribution coefficient *D* for ML_1

The first limitation on this movement is the restriction upon the concentration of M in the membrane due to a generally low distribution coefficient *D* for the ML_1 complex. This reduces the rate at which ML_2 can form (Fig. 2.2). However, the effect can be made more selective if there is a special membrane protein, a receptor, that recognizes L_1 virtually only when it is bound to a particular metal ion M through stereochemical factors. The rate at which the metal crosses the membrane depends on the existence of such transport systems that may be refined to the level of protein–protein interactions and which may, through highly specific receptor recognition processes, act so as to restrict the distribution of different metals in different cells and cell compartments. An example of this is the special uptake of iron from transferrin into bone marrow cells upon which we will comment later. There is also a most intriguing system of transport proteins, called Ton B, for the movement of metal chelates, such as vitamin B_{12} molecules and siderophores, into *E. coli* across its membranes (see Chapter 12).

A particular case arises when an adventitious ligand L_1 is sent out of the cell to aid in the capture or sequestration and removal of some elements. The availability of M to L_1 is then dependent on the controlled synthesis of L_1, which is also a kinetic device. In higher organisms, L_1 can be a trap for an element which removes it into one organ, e.g. the kidney, so that it is not available to another. We describe in Chapter 11 the particular case of iron. Once again feedback control can determine that, once a cell has a sufficiency of M, it no longer synthesizes the ligand for capturing this metal ion. The distribution rate now depends on the rate of synthesis and cycling of the ligand, L_1.

2.12.2 Kinetics of transport

Although the thermodynamic binding features of ML_1 and ML_2 (Fig. 2.2) are very similar, a characteristic of transport systems, as opposed to those of retaining centres, is that the former must bind and release M reasonably quickly since transport must be continuous. The release of M may be achieved by lowering the oxidation state of the metal (to decrease the binding constant), by changes of pH (to decrease the effective binding constant), by changes in the conformation of the protein (to change the co-ordination site of the metal), or by hydrolysis of the ligand. We now know that certain metal ions are transported in aqueous phases as well as in membranes, e.g. see copper chaperones in Chapter 15. As stated in Section 2.9.1, a special feature of proteins is that they may have 'leader' sequences which ensure that they can cross membranes.

2.12.3 Insertion in pre-formed holes or chelating rings

A particular method of selection by kinetic control may also be considered: that of insertion into a pre-formed hole or chelating ring, from which exchange of the metal ion concerned is virtually impossible. This reaction is not rapid, due to steric constraints, as mentioned before. Consider the case where, while the binding of M to L_1 is governed by thermodynamic factors, it could well be that the rates of transfer of M from ML_1 to L_2 are M-dependent and, in fact, L_1 could merely act to control those rates

$$\mathrm{M}+\mathrm{L}_1 \underset{k'_{\mathrm{ML}}}{\overset{k_{\mathrm{ML}}}{\rightleftarrows}} \underset{\text{initial uptake}}{\mathrm{ML}_1} \xrightarrow[+\mathrm{L}_2]{k} \underset{\text{final uptake}}{\mathrm{ML}_2} + \mathrm{L}_1.$$

An example makes clear the selectivity that can be introduced by this mechanism. The insertion of magnesium in the chlorin ring of chlorophyll cannot be made directly *in vitro* in the presence of other free metal ions, like copper, zinc, or even nickel and cobalt. It is hard to see how any reasonably devised intermediate step could prevent the latter ions from entering preferentially a strong porphyrin-type ligand like chlorin at *equilibrium* (see Fig. 2.8 for the relative binding strengths).

Now, let us suppose that the step $ML_1 \rightarrow ML_2$ is irreversible and that the kinetic constants k are very different for different M. This could arise in part because the transition metals are complexed by other ligands (L_4) and the concentration of their free ions are low compared to that of magnesium, which is then complexed to a larger degree to a relatively poor ligand L_1, e.g. an O-donor protein site such as ATP, or because the MgL_1 complex is specifically recognized by L_2. It could even be that any ML_1 complex of the transition metals is not recognized due to conformational factors within a carefully modulated, even energized, protein transfer system. As a consequence of one or another of these factors, k would be much higher for the reaction $MgL_1 \rightarrow MgL_2$ than for any transition metal ion and, since the step is irreversible, this reaction would hinder the possibility of occurrence of all the other metal ions in ML_2. In different words, those weak-binding metal ions, which are in excess over ligands but react rapidly, can block the entry of ions that bind better to L_2 in an absolute sense, provided the kinetics of transfer are suitably arranged. It is a striking observation that, while *in vivo* the insertion is performed correctly giving magnesium chlorophyll, *in vitro* it is very difficult to avoid the production of zinc chlorin when this metal is present even in trace amounts.

Table 2.15 Polypyrrole chelating agents and their selection of metal ions

Chelating agent	Metal	Binding site
Chlorin	Mg	Chlorophyll proteins
Protoporphyrin	Fe	Haemoglobin, cytochromes
Corrin	Co	Cobalamin (vitamin B_{12}) proteins
(Nirrin) F-430	Ni	Factor F-430 proteins
Uroporphyrin	Cu	Turacin (pigment), free(?)

In fact it seems likely that the correct insertion of each metal ion with less than 0.1 per cent error is achieved for magnesium in chlorin, iron in protoporphyrin, cobalt in corrin, nickel in F-430 (nirrin), and copper in uroporphyrin (see Table 2.15). This result could only be achieved by kinetic control based on specific recognition of ML_1 complexes (Fig. 2.13). Thus the design of these ligands is to prevent easy access and so control associations.

Once in its new site, the complex ML_2 is recognized specifically and bound by a protein in such a way that it is still harder to exchange the metal M (see Fig. 2.7). In effect, ML_2 is almost a new 'element' in that it allows Fe, Co, Ni, Cu, and Mg to be used specifically in new ways. As proteins developed to bind these special complexes so the use of the elements expanded: it allowed mononuclear ring-chelate species of Fe^{2+} (and Fe^{3+}) in different spin states, of cobalt and nickel in at least two oxidation states and in low spin states, and of magnesium chlorophyll in membrane proteins, just as it was possible to retain copper, zinc, and molybdenum directly bound to the side-groups of the proteins. Previous to this development intracellular iron was overwhelmingly bound as Fe_nS_n complexes, i.e. mixed valence clusters, cobalt might not have been captured, and magnesium was associated only with rapidly equilibrating reactions of phosphate in water. Very large functional advantages accrue from the design of these small, relatively rigid molecule chelates, as will be seen later.

The selectivity conditions described in this section require that the different chelating rings produced for the different metal ions are synthesized in controlled amounts. Once more, there must be feedback from synthesized product to the genes.

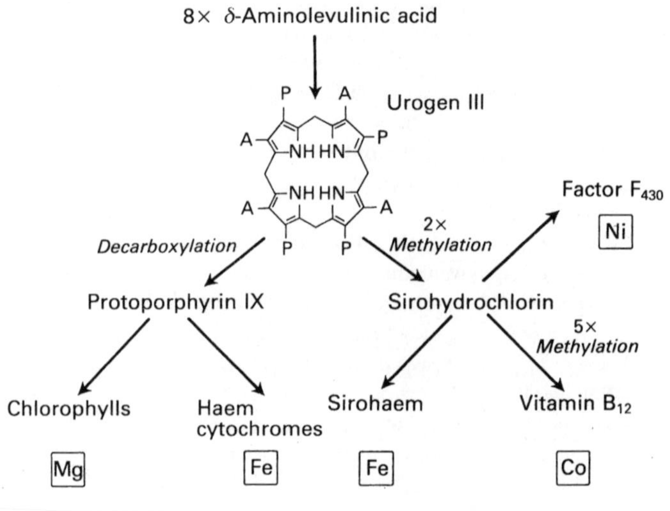

Fig. 2.13 The diverse ring chelates produced for different metal ions from one precursor.

2.12.4 **Energy coupling to the selective movement of M (gates and pumps)**

Energy can be applied to a membrane in many ways; in the first, energy can exert non-specific effects due to the fields, electrical potential differences, across membranes. These fields can be produced by light or metabolism and they can be general over the whole of the membrane or localized. The membrane here is nothing else but a restriction on diffusion, while the energy input separates negative and positive charges across it. Either general neutralization of these charges or general exchange of like charges across the membrane can produce energized uptake and will give a concentration gradient of perhaps several orders of magnitude. A particular pathway in the membrane, a channel, makes the device specific to, say, the proton, the sodium ion, or the calcium ion, but the membrane here is passive even though it has structure and a transmembrane charge gradient.

An alternative is to involve the membrane components in the transport activity, i.e. to put energy into the membrane structure. For example, a membrane protein can change its state in a cyclic fashion under the impact of energy. The general case is as follows: a structure E in a membrane accepts *n* M (or *n* ML_1) from an outer aqueous space with a binding constant $D \cdot K_{aq} \cdot p$. Energy is now applied to change E such that E′ is no longer open to the outer phase but is open to the inner phase, and such that E′ now binds *n* M (or *n* ML_1) with a different constant $D' \cdot K'_{aq} \cdot p'$ which is not large enough to retain these species. They are, therefore, released to the inner phase and E reverts to its initial state. Energy losses are involved in the cycle and the transport is irreversible in a thermodynamic sense (heat lost), but the pathway can be reversed by suitable energy input. Now, the direction of this pumping is extremely important since it will decide the concentration in a given cell compartment of an ion engaging in competitive binding. Generally the cytoplasm of a cell rejects Na^+, Ca^{2+}, Ni^{2+}, Cu, Zn^{2+}, and many poisonous elements. Examples are given throughout this book. In general, proteins depend on the supply of energy from pyrophosphate of ATP to force their changes in conformation in membranes. Figure 2.14 gives an example.

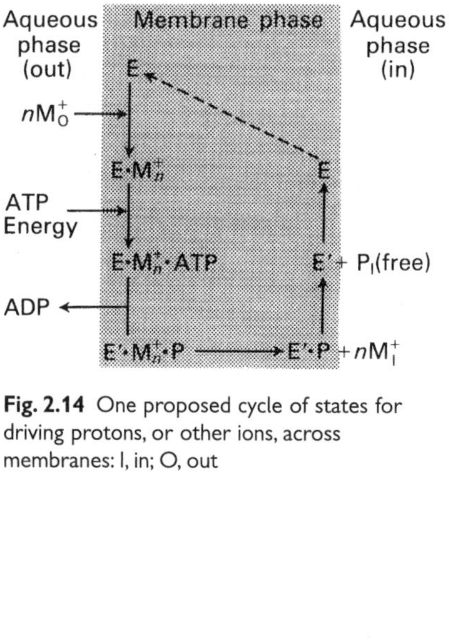

Fig. 2.14 One proposed cycle of states for driving protons, or other ions, across membranes: I, in; O, out

2.12.5 **A summary of the kinetics of uptake of metal ions**

From what has been said we now have the following rates to consider:

1. in-flow of M to the environment of the cell, which can be assisted by a certain ligand L_1;
2. out-flow of M from the environment of the cell, which can be assisted by another ligand L′;
3. transfer in and out of cells involving properties of ML_1, *K,p,D*, and pores and gates, all of which can be energized;
4. (irreversible) binding of M by L_2 inside the cell;

Later we shall have to add the possibility that M reacts with a group L to give a precipitate which may or may not be in equilibrium with free M and free L.

The actual capture of M by a cell is then no longer a problem of equilibrium thermodynamics, although binding constants are relevant and partially determinant. Rather we see that element uptake and rejection is a matter of a steady state in which paths and rates of various processes are as important as uptake equilibria. Remember too that a higher cell has many internal membranes so that different metals are concentrated by transfers in different vesicles,

Table 2.16 Pumped gradients of metals and their complexes

Metal	Inward to cytoplasm	Outward to	Organelle or vesicle
Fe^{3+}	Ferroxamines, etc. Transferrin	(Ferritins?) Citrate	 Mitochondria
Co^{2+}	Vitamin B_{12}	?	
Ca^{2+}	(Free ion)	Free ion	Reticula, outside
Na^{+}	(Free ion)	Free ion	Outside
Mg^{2+}	Free ion	(Free ion)	(Outside)
H^{+}	Free ion and many ligands	Free ion or with many ligands	Vesicles, outside
K^{+}	Free ion	(Free ion)	
Cd^{2+}	?	Metallothionein	Outside
HPO_4^{2-}	Free ion	Several organic phosphates	Mitochondria
Cl^{-}	Free ion	Free ion	Outside
Cu^{2+}	Albumin	Metallothionein	Outside
Zn^{2+}	Albumin	?	(Vesicles)

organelles and so on. Table 2.16 gives some examples. This brings us again to the topic of homeostasis, but we will deal with it in a separate chapter (Chapter 20 and see Section 2.19).

2.13 Rejection of metal ions

Before describing how anions are selected we must return to Fig. 2.1 to examine rejection. Rejection of elements M can occur in three ways:

1. pumping of M through membranes to external fluids or to the air either in the form of ions or special molecules (reversal of Fig. 2.14);
2. pumping into storage vesicles, vacuoles, again as ions or compounds;
3. precipitation as solids in vesicles which may be rejected as pellets or retained.

Clearly, the principles behind rejection are very similar to those for uptake, but we do need to repeat that the operation is very selective. Note that Na^{+}, Cu^{+}, Ca^{2+} and many other M^{2+} and maybe all M^{3+} (Al, Sc, lanthanides) ions are rejected from the cytoplasm.

We have now considered all the many ways in which metal ions can be handled and we shall see how these principles apply to particular metal ions in Chapters 8–19. Similar principles apply to non-metals and we give next an outline of their uptake.

2.14 General aspects of the uptake of essential non-metals

The trapping of non-metals by biological systems must also depend on the use of membrane restrictions, thermodynamic binding traps, and kinetically energized binding traps and steps. However, unlike ions, neutral molecules such as N_2, O_2, H_2O, CO_2, NH_3, $Si(OH)_4$, and $B(OH)_3$ carry N, O, H, C, Si, and B to all parts of a cell and cannot be prevented from free diffusion across membranes. Of these, compounds such as $B(OH)_3$ are readily trapped by binding to cis-diols (see margin), probably to specific polysaccharides, giving moderately labile condensation products.

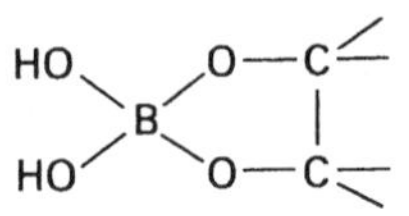

Apparently similar reactions can be written for $Si(OH)_4$ or any weak dibasic acid and we find these structures for phosphorus in cyclic adenosine phosphate. We also note the occurrence of boron in the natural compound boromycin and it could well be that silicon in mucopolysaccharides is another example of this type of binding. In this context an element such as molybdenum, visualized as $Mo(OH)_6$, is similar to $Si(OH)_4$ (Chapter 17). The ring structures decrease in thermodynamic stability, in the order B > Si > P, so that, for the most part, the later non-metal molecular hydroxides are retained by hydrogen bonding rather than covalent complexing, at least initially.

The other molecules, H_2, N_2, O_2, H_2O, CO_2 and NH_3, are less reactive and some need to be activated, reduced, or oxidized, before they can be bound. Their uptake is controlled by use of energy and kinetics of reactions and not by thermodynamics (see Fig. 1.8). After the initial capture, they usually pass into small molecule intermediates, such as charged amino acids and sugars, which remain trapped in cells by the membranes. These intermediates are *covalent* (as opposed to ionic) kinetic traps and, as small charged molecules, can be retained behind membranes before further synthesis leads to polymer formation.

Among these small non-metal molecules, dioxygen and dinitrogen are first absorbed at metal centres. Organic molecules can hardly activate them. For O_2 several types of centre are known, such as haemoglobin (Fe(II)), haemerythrin (Fe(II) . . . Fe(II)), and haemocyanin (Cu(I) . . . Cu(I)). In the last two cases oxygen is held as O_2^{2-} while in haemoglobin it binds reversibly as O_2, although there is partial charge transfer from iron to give a formal $O_2^{\bullet-}$ state. If oxygen is to be incorporated, the final capture of the element involves the attack on a substrate by O_2 activated at a metallo-enzyme centre similar to the dioxygen carrier centres to give oxygen containing organic molecules. In some cases O_2 is reduced to superoxide and then to peroxide, which can also be utilized in various reactions or taken down to H_2O, the last stage of reduction. The uptake of N_2 is known to be by the iron-molybdenum enzyme nitrogenase which carries out the reaction

$$N_2 + 6H_2O + 6e \xrightarrow[\text{molecules}]{\text{many ATP}} 2NH_3 + 6OH^-$$

The reduction is very energy-expensive in that it uses many ATP molecules. It is effectively irreversible through this use of energy, but also because the resulting ammonia goes on to be incorporated in organic molecules by condensation reactions. The question why the more active O_2 is not taken up by the N_2-absorbing sites is puzzling and still unsolved. Perhaps O_2 is removed first and its access to the site is prevented, or N_2 is indeed picked up by a special centre with little or no affinity at all for dioxygen, but this is not very common in chemistry, except for compounds involving metals that are not used in biology, e.g. ruthenium.

As to CO_2, it is rapidly converted to bicarbonate by the zinc-enzyme carbonic anhydrase or reacts directly with an activated organic molecule. The incorporation is again energy-driven by ATP. In this case there is no interfer-

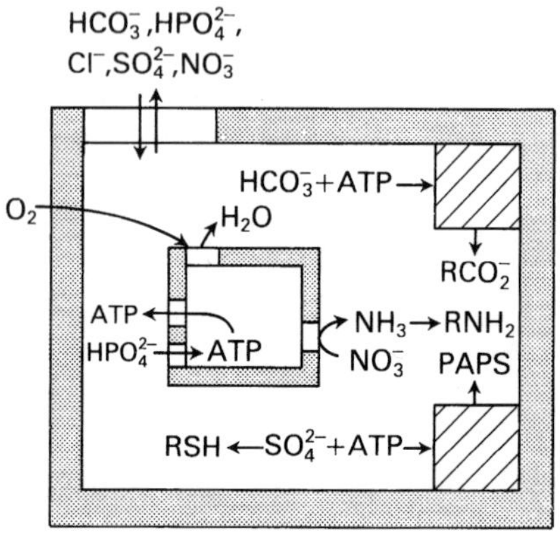

Fig. 2.15 Some of the incorporation movements of anions as they pass membranes or are activated into kinetic traps by enzymes. PAPS: phospho adenosine phospho sulphate.

ence by O_2 or N_2, since CO_2 is a quite different molecule with a quadrupole structure that allows a different type of binding. After incorporation, the CO_2 is reduced to states equivalent to HCHO, $-CH_2-$, etc. in organic compounds. Incorporation of $-CH_3$ from CH_4 involves typical organometallic chemistry (see Chapter 16).

Some of the above and all the other non-metals are also available to biological systems as simple anions, e.g. F^-, Cl^-, Br^-, I^-, or oxyanions such as CO_3^{2-}, HCO_3^-, $H_3SiO_4^-$, NO_3^-, $H_2PO_4^-$, HPO_4^{2-}, $B(OH)_4^-$, SO_4^{2-}, and SeO_4^{2-}. Molybdenum and vanadium can also be included in this group as MoO_4^{2-} and VO_4^{3-} and the principles of uptake for all these elements are similar. Naturally, the ions cannot cross membranes without carriers, and, once captured, they can be retained by such physical separation devices. The problem of capturing these non-metal ions by biological systems is therefore not very different from that of the capture of metals. We shall need to look for anion pumps, exchangers, and channels and their selectivity principles, as well as at simple thermodynamic control over incorporation (see Fig. 2.15). Once inside a cell several of them can be reduced and differential non-metal incorporation is then greatly assisted. (Note that apart from essential uptake cells must reject poisonous anions, e.g. $HAsO_4^{2-}$ and here the selective direction, in or out, of pumps is a critical factor).

2.15 Mechanisms of selection of anions based on thermodynamic properties

We will consider two major selection mechanisms based on 'equilibrium' binding properties in order to shorten discussion which should follow the full pattern of possibilities outlined for cations:

1. Selection by size and charge, taking into account the overwhelming influence of the hydration of anions on their binding to proteins.
2. Selection by differences in binding affinity for different types of cationic centres, e.g. H^+, $-NH_3^+$, M^{n+} (compare –OH, RNH_2, RS^-, etc, as binding centres for cations).

Table 2.17 Pauling's crystal radii of common anions

Ion	Radius (Ä)
F^-	1.36
Cl^-	1.81
Br^-	1.95
I^-	2.16
O^{2-}	1.40
S^{2-}	1.84
Se^{2-}	1.98
Te^{2-}	2.21
OH^-	1.40
SH^-	1.95

Source: Pauling, L. (1960). *The nature of the chemical bond* (3rd edn). Cornell University Press, Ithaca, New York.

2.15.1 Selection by charge and size; the hydration of anions and their binding to proteins

The various anions of group 17 and oxyanions of groups 14 to 16 show considerable variation in size (see Tables 2.17 and 2.18). F^-, Cl^-, Br^-, and I^-, for example, should be readily separated by the correct design of absorbing holes or channels. There is good reason to suppose that the associated thermodynamic selectivity can outweigh the vast differences in abundance and availability of these elements. On the basis of similar sizes only the anions F^- and OH^- or Br^- and SH^- show competition.

There are, however, some differences as to the nature of the binding sites for anions relative to those used in separating cations. First, anions are negatively charged and, secondly, anions have a relatively lower charge density as they are large ($F^- \cong K^+$). This means that, since it is difficult to build a molecule with a *positively* charged dipole surface (compare cyclic polyethers), anions must bind to clustered, positively charged groups, such as the ammonium groups $-NH_3^+$, the guanidinium group, and metal ions, and/or through very extensive hydrogen bonding. Rare cases are known in which there occurs binding at hydrophobic centres or regions containing $-OCH_3$, $-SCH_3$, $-CH_2-$, phenyl, etc., despite the fact they are of some polarity. Generally, these centres will only take up anions if they are surrounded by a 'buried' positive charge generated by the fold of the protein. This type of binding to buried charges is possible just because the charge density of anions is low. A deeper understanding of these observations comes from the balance of energy effects involved in the hydration of the anion and of the binding site in the interaction of these two entities, i.e. the energetics of the reaction (as for cations; Section 2.5.2).

$$X^{n-} \cdot (H_2O)_x + \text{site}^+ \cdot (H_2O)_y \rightarrow \text{site}\ {}^+X^{n-} \cdot (H_2O)_z + (x + y - z)H_2O.$$

This balance can indeed generate orders which differ from those simply based on ionic sizes or potentials. The basic electrostatic factors that regulate it are those that regulate processes such as the preferential precipitation of large cations by large anions or small cations by small anions, e.g. Cs^+ by ClO_4^- but Li^+ by F^-.

The net free energy change in binding to a site from aqueous solution including terms other than electrostatic can be written as

$$\Delta G_{\text{binding}} = \Delta G_{\text{elect}} + (\text{any } \Delta G_{\text{specific}}) - \Delta G_{\text{hyd}} \quad (2.2)$$

Here $\Delta G_{\text{binding}}$ is the net free energy balance between the bound (hydrated) state and the free hydrated forms; ΔG_{elect}, is the free energy change corresponding to the electrostatic interactions; $\Delta G_{\text{specific}}$ is the contribution to the free energy change derived from other interactions, e.g. covalent binding, hydrogen-bonding, stereochemical or other effects; and ΔG_{hyd} is the algebraic sum of the different hydration energies of the free and bound species, symbolized by $(x + y - z)$ in the above reaction.

Although the different terms are difficult to evaluate, it can be seen from eqn (2.2) that different trends of $\Delta G_{\text{binding}}$ can be generated depending on the relative values of the separate terms. Generally speaking, hydration opposes binding if the binding sites are hydrophobic and the less hydrated anions will then be bound preferentially. The so-called 'lyotropic' or Hofmeister series is obtained, which is just the reverse of the hydration free energy, i.e. in binding strength

$$I^- > Br^- > Cl^- > F^-,$$
$$ClO_3^- > BrO_3^- > IO_3^- > CO_3^{2-} > HPO_4^{2-}.$$

The non-polar sites in proteins which give such series often contain (buried) charged histidine residues or, not so frequently, arginine and lysine residues. (Notice how these inverted orders compare with the inverted orders of radii for cation binding, Section 2.5.)

However, in other cases, the opposite order is found, i.e. the more hydrated anions, like SO_4^{2-} and $P_2O_7^{4-}$, are exactly those which are bound preferentially and they are often the substrates of corresponding enzymes. This occurs when the binding sites are polar and heavily hydrated. X-ray studies have shown that, quite often, these regions contain arginine and lysine amino groups.

Table 2.18 Thermochemical radii of common anions

Ion	Radius (Å)
CO_3^{2-}	1.85
HCO_3^-	1.63
SiO_4^{2-}	2.40
BO_3^{3-}	1.91
MoO_4^{2-}	2.54
NO_3^-	1.89
PO_4^{3-}	2.38
AsO_4^{3-}	2.48
SbO_4^{3-}	2.60
OH^-	1.40
SO_4^{2-}	2.30
SeO_4^{2-}	2.3
TeO_4^{2}	2.54
ClO_4^-	2.36
IO_4^-	2.49

Source: Waddington, T.C. (1959). *Advances in inorganic chemistry and radiochemistry*, Vol. 1, p. 180. Academic Press, London.

2.15.2 Selection of anions by differences in binding affinity for different types of cationic centres

The affinity of anions for different types of cationic centres can be illustrated by data from the binding of free aqueous anions to free aqueous cations. Two different types of behaviour can be recognized: for class ‘a’ (hard) cations the order of binding by anions is $F^- > Cl^- > Br^- > I^-$ and $O^{2-} > S^{2-}$; for class ‘b’ (soft) cations the order is the reverse, i.e. they bind to the heavier donors rather than to the lighter ones (Table 2.8). As stated in Section 2.5, the order of binding of anions to class ‘a’ metal ions is overwhelmingly due to electrostatic terms, but binding to class ‘b’ metal ions is driven by stronger covalence, while taking ion size into account, (see Fig. 2.5).

Now, the main cations concerned in biological systems are Mg^{2+}, Ca^{2+}, Cu^{2+}, Cu^+, Zn^{2+}, Co^{2+}, Co^{3+}, Fe^{2+}, Fe^{3+}, Mn^{2+}, Mn^{3+}, but only Cu^+ is a class ‘b’ cation. Na^+ and K^+ cannot be considered as probable anion binding sites since they form only very weak complexes and are highly mobile, but they may have a role in some specific cases (see the end of this section).

Thus, for free aqueous ions, Cu^+ will amplify the lyotropic effect, while all other hydrated cations should oppose it to different degrees

$$Mg^{2+} > Ca^{2+} \text{ and } Mg^{2+} > Mn^{2+} > Zn^{2+} > Cu^{2+}.$$

When we consider the metal ions in proteins the situation is found to be very different. Consider:

1. highly charged positive sites in water-free regions or regions of restricted access of water in a protein, e.g. carbonic anhydrase where Zn^{2+} is bound to neutral ligands only but the metal is deep in a narrow pocket;
2. more lowly charged positive sites in highly hydrated sites, e.g. carboxypeptidase where Zn^{2+} is in an open pocket;
3. lowly-charged sites, e.g. in alcohol dehydrogenase where the metal (zinc) is bound in a pocket to two R-S$^-$ groups or in the haem-proteins where Fe^{2+} is similarly neutralized by the two negative charges of the ligand;
4. copper sites which utilize O_2 in oxidation reactions and go through an OH^- bound state, e.g. caeruloplasmin.

The third type of site will bind anions very poorly, as there is little electrostatic effect, and it is to be expected that binding agents for such sites will be neutral molecules which bind to metal ions, e.g. imidazole, or those anions that form covalent bounds, e.g. RS^-, but not any of the usual ions of the Hofmeister series. The substrate of alcohol dehydrogenase is C_2H_5OH.

The site (1), zinc in carbonic anhydrase, was referred to in Section 2.14. Being in a deep pocket, which can only be of low hydration, it gives the lyotropic series, e.g.

$$I^- > Br^- > Cl^- > F^-: \quad CNO^- > SCN^- > N_3^- > ClO_4^- > NO_3^- > CH_3COO^-.$$

This is what we would expect for a class ‘b’ cation which zinc is *not*; hence the protein pocket has changed the balance of energy changes involved and it is site hydrophobicity that determines the order of binding rather than the nature of the aqueous metal ion. The separation of metal ions into ‘a’ and ‘b’ classes is solvent-dependent (protein is here the effective solvent). Carbonic anhydrase can therefore act as a good Cl^- binding protein even in the presence of either SO_4^{2-} or HPO_4^{2-}. Sites (2) and (4) show the opposite properties. We now expect binding to follow the order of the ionic potential, dominated by overall favourable entropy effects. Actually, we observe frequently intermediate behaviour, depending on the degree of hydration of the metal in its site. Examples are the already mentioned zinc centre in carboxypeptidase and the single- and multi-copper centres of caeruloplasmin and laccase. The observed binding orders to these latter centres are

Carboxypeptidase: $F^- \cong Cl^- \cong Br^-$,

Laccase: $F^- > N_3^- > SCN^- > CNO^- \gg Cl^-$.

Note that in the case of laccase binding is to Cu(II), not to Cu(I) that does not bind the anion.

Typical of the zinc centres is that they bind simple carboxylate anions weakly, but this binding can be increased greatly by increasing the hydrophobicity of the sidechain, e.g. β-phenylpropionate $\gg$ acetate.

There is still one case of the binding of anions to proteins which deserves special attention: that of simultaneous uptake of an anion and a cation. The cations concerned must be those in fast exchange between their protein binding sites and the free aqueous state, i.e. Mg^{2+}, Ca^{2+}, K^+, Na^+, and certain small organic cations. The most frequently met example is that of the binding of polyphosphates such as ATP and ADP together with Mg^{2+}. The binding appears to be to very hydrophylic protein regions containing additional positive charges due to basic sidechains of the proteins and can be very selective since the combined requirements of the anion and the cation have to be satisfied simultaneously.

2.15.3 Kinetic binding traps for anions, with or without accompanying redox reactions

We have already seen that some neutral molecules can be trapped by condensation reactions with various organic compounds forming kinetically stable covalent bonds, e.g. $B(OH)_3$ which condenses with cis-diols (see naturally-occurring boromycin in margin). Other neutral molecules must be activated, reduced, or oxidized before they can bind to other compounds through kinetically stable covalent bonds. These can be considered together with the non-metal elements that are captured as anions rather than neutral molecules.

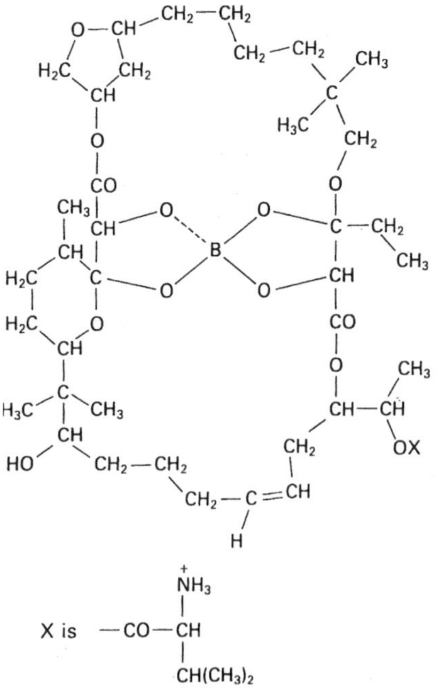

Boromycin

The phosphate anions can only be retained weakly by ionic interaction with positively charged groups, e.g. $-NH_3^+$, but they can be retained in a *not too stable* (energized) kinetic trap by condensing with –OH groups

$$HPO_4^{2-} + HOR \rightarrow R\text{-}O\text{-}PO_3^{2-} + H_2O.$$

In these forms phosphate is transported, e.g. as sugar phosphate, stored, e.g. as polyphosphate and ATP, and transferred to other molecules, simple or polymerized, to give a variety of substances, e.g. RNA and DNA. Among the other anions, only SO_4^{2-} can, to a lesser extent, give similar condensation products, but only as the simple ion, not polymerized.

In the above sections describing the behaviour of anions we have set aside the binding of sulphide. This anion was generally available in the early conditions of evolution and is still plentiful in anaerobic zones. It binds overwhelmingly and selectively in biology to metal centres, especially Fe and Mo, often in metal clusters. The nature of the competition between this anion and oxide has been described in Chapter 1 and Section 2.5.7. The selectivity of its reactions is re-introduced here since we wish to point out that in aerobic systems the successive stages of uptake of sulphur must start from the chemistry of sulphate and go through many steps to sulphide incorporation. In the next chapter we shall see that the part of space in which the element is found decides its oxidation state and its incorporation mode.

2.16 Redox incorporation of anions

Inside cells, SO_4^{2-} is reduced and sulphur is retained in sulphide, S^{2-}, the thiol R-SH, or the disulphide -S-S- form. The same reactions occur with SeO_4^{2-}, which is bound as selenethiol, and with NO_3^-, which is incorporated as an amino group or other nitrogen-containing molecule. In the halide group, F^- and Cl^- are usually handled as such, but I^- and Br^-, once taken into a cell, can be selectively absorbed by enzymes and the fact that they can be much more easily oxidized than Cl^- permits the reactions $Br^- \rightarrow Br$ and $I^- \rightarrow I$ in the presence of Cl^-, e.g. by the action of many peroxidases. The reactive radicals so produced attack organic compounds readily, e.g. phenols, hence Br and I are inserted covalently linked into organic molecules. A good example is the thyroid hormone. In some cases the free halogen can be formed and liberated, e.g. I_2 in some seaweeds, and it is said that this (or perhaps the volatile halogenated hydrocarbons formed by insertion of bromine and iodine in organic molecules together with some $(CH_3)_2S$ accounts for the smell of the sea-shore (not ozone).

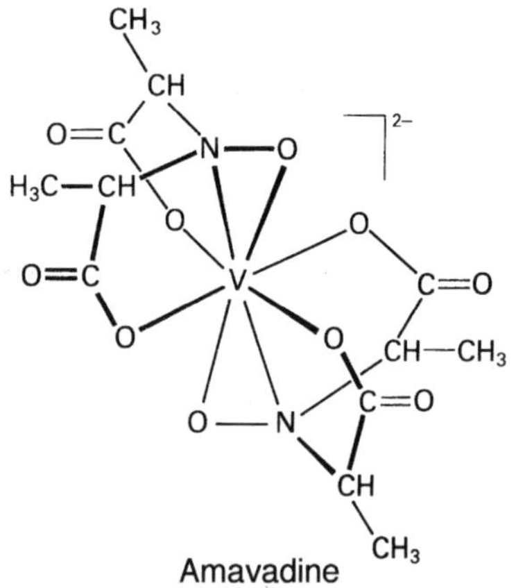

Amavadine

Finally, there are the cases in which reduction can change an anion into a cation. The uptake of molybdenum and of vanadium initially as MoO_4^{2-} and HVO_4^{2-}, respectively, i.e. in a non-metal form, is based on this principle. They are then reduced to oxocations or simple cations and bound into proteins or cofactors. Oxovanadium(IV), VO^{2+}, compares with nickel in its complexes with polyaminocarboxylic acids and binds strongly enough to N/O donors at pH ~ 7, but with certain strong complexing agents it can lose the oxo-group to give V(IV) complexes. The example known, the vanadium-containing toadstool *Amanita muscaria*, contains this species, V(IV), bound to an N-hydroxy derivative of iminodiacetic acid forming a very stable octacoordinated complex 'amavadine,' in which the N-hydroxy group is ionized and binds the metal, see margin. The unusual accumulation of vanadium in tunicates as V(III) or VO^{2+}, from the low concentrations of HVO_4^{2-} found in the environment, is probably due to this change in oxidation state, so that there is no 'saturation' inside cells relatively to the species available outside even when no internal ligand appears to be present and the retention of the ion (V^{3+} or VO^{2+}) is ensured by storage in vesicles (see Chapter 17). The case of molybdenum is very different.

The reactions of molybdenum are quite idiosyncratic since as a metal element it is the closest to the non-metals. Thus, its chemistry can be considered starting from MoO_4^{2-} (SO_4^{2-}). On entering cellular systems the hydroxide form of this ion $MoO_2(OH)_4^{2-}$ (formally written) can react not just with OH groups to form MoO_2^{2+} bound species (compare UO_2^{2+}) but, since it is a soft 'b'-class metal ion, it can give

$$MoO_2(OH)_4^{2-} + 2RSH \longrightarrow [MoO_2(OH)_2(RS)_2]^{2-} + 2H_2O.$$

Now, the more stable forms of this anion are probably $[MoO_3(RS)_2]^{2-}$. Here there is a resemblance to non-metals in that these compounds can be quite stable kinetically if $(RS)_2$ is a chelate. In fact molybdenum appears to be trans-

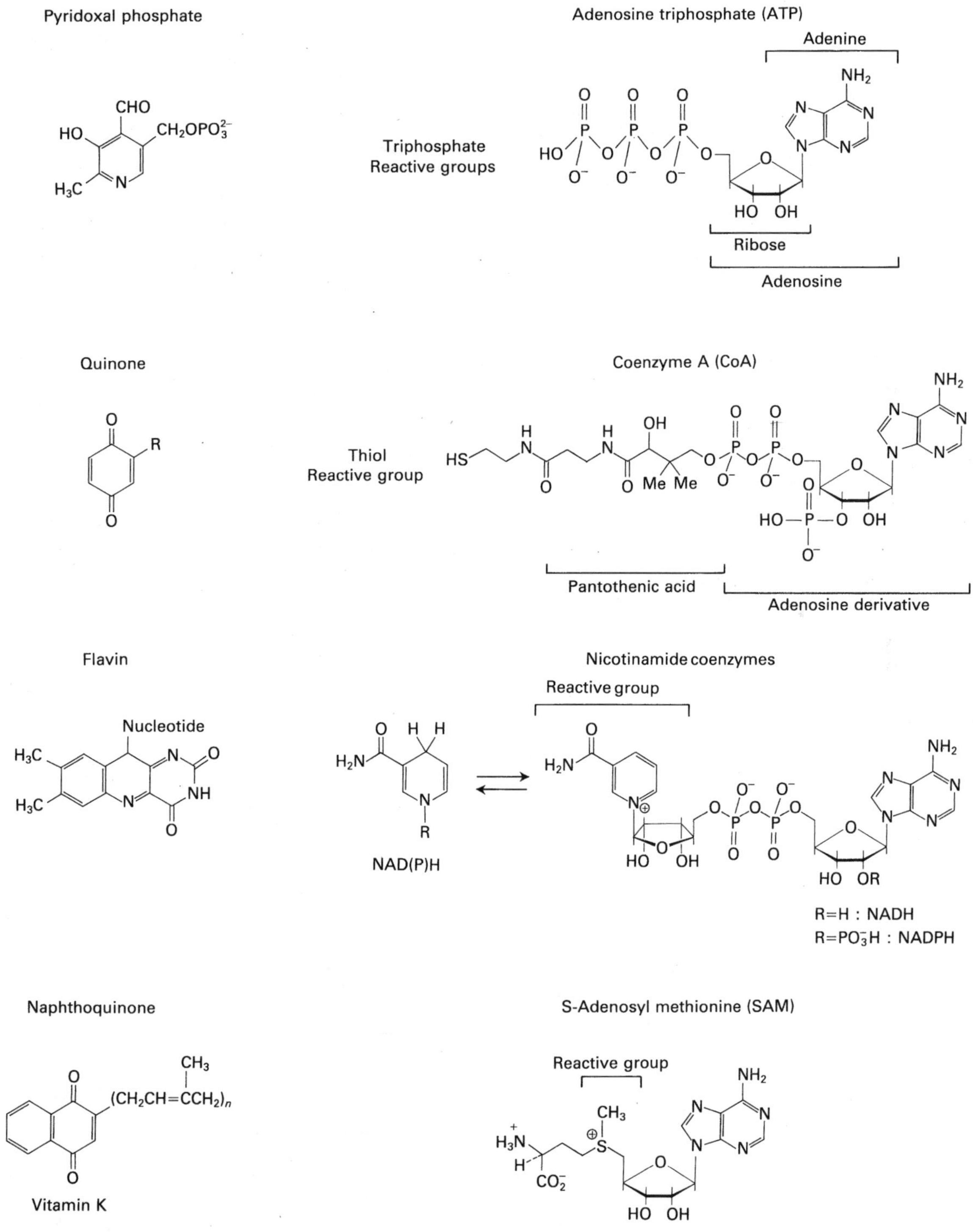

Fig. 2.16 The formula of the organic moieties which act as the coenzymes for carrying amines, phosphate, acetyl, hydride, and methyl groups.

ported in biology as the cofactor complexed metal with only *two* (covalent) bonds (Fig. 17.4). This is in contrast with common metal ions, which need three or more bonds, and with sulphate, which can be transported with but one covalent link, ADP-[OSO_3^-]. The intermediate nature of molybdenum chemistry is fully exploited in biology (see Chapter 17).

2.17 Coenzymes

Table 2.19 Some coenzymes which carry non-metals

Non-metal	Coenzyme or cofactor
H	NADH, NADPH, Flavin Quinones
P	ATP (NTP), IP_3
C	CoA, vitamin B_{12}
S	PAPS

While we were describing metal ions we stated that in some cases there were kinetic traps which removed the element making it into an effective 'extra' element. Examples were haem, chlorin, and vitamin B_{12}. Some literature refers to these bound entities as *cofactors*. When we look at the corresponding systems for non-metals, often called *coenzymes* (Table 2.19 and Fig. 2.16), we observe two broad differences from metal containing cofactors: (1) the kinetically stable unit holds the element by but one bond leaving it very accessible for transfer by catalysts, enzymes, which recognize the whole coenzyme and not the element: for example, phosphate is bound in the five nucleotide triphosphates, NTP (adenine, uridine, cytidine, guanidine, and thymidine) and hence phosphate can enter five selective transport (of phosphate) systems; (2) the coenzymes may be freely mobile or at least of restricted mobility, e.g. the NTP molecules. In addition, notice that each element, H, C, S, etc. is carried by coenzymes which belong to it alone, so that pathways of each are not open to blocking by the coenzymes of another element. Recognition is by the coenzyme tail and not by the element—contrast calcium and compare cobalt. The need for these properties in several different coenzymes is that the elements H, C, N, P, and S are parts of many synthetic substrate pathways as well as having group activities and these pathways must not meet haphazardly. The chemical form of the transported element is interesting: phosphorus always as phosphate, hydrogen as H^-, but carbon as $-COCH_3$, $-CH_3$, -sugar, etc. and nitrogen in amino-acids (glutamate). At the same time there are coenzymes which are immobile holders of non-metals, such as flavins, biotin, and pyridoxal phosphate. We shall return to the use of different coenzymes in different pathways in Chapters 3 and 18 since they, like metal ions, are differentially concentrated in space.

2.18 Precipitates in compartments: cation and anion cooperativity

We have left the description of precipitation as an uptake mode until last since precipitation requires uptake of X^- (the anion) and M^+ (the cation) of the MX precipitate into the same (vesicle) compartment. The selection here is by solubility product (S.P.), which can be appreciated by reference to the data for sparingly soluble salts found in Table 2.20. The selectivity of anion for cation in biology is precisely as expected from considerations of inorganic reactions, e.g. $BaSO_4$, $CaCO_3$, $Fe(OH)_3$, $SiO(OH)_2$; however, biological selection of a solid state goes much deeper in that there is choice amongst the allotropic forms of one salt, e.g. calcite or aragonite, both $CaCO_3$, and there is also a choice of shape and size of crystallites.

The solubility products in Table 2.20 hold a message and a warning for us when we inspect biological systems. It is undoubtedly true that free Fe^{3+} and Ca^{2+} are very close to their solubility limits as hydroxides and carbonates/phosphates, respectively. These ions, whose roles are dominant in biology, may well control in part homeostasis of cell metabolism since their precipitation reactions link them to the pH of the solution as well as to the levels of critical anions such as phosphate and carbonate. Note that such a precipitate buffers free cation and anion concentrations and these must match all other exchangeable bindings (see Section 2.10, and Chapter 20).

$$Fe^{3+} + 3H_2O \rightleftharpoons Fe(OH)_3 + 3H^+,$$
$$2Ca^{2+} + H_2O + HPO_4^{2-} \rightleftharpoons Ca_2(OH)PO_4 + 2H^+,$$
$$Ca^{2+} + HCO_3^- \rightleftharpoons CaCO_3 + H^+.$$

The message is clear that we must not look upon biological minerals as just inert structural deposits: they must be made and sustained in communication with the levels of free ion concentrations.

The fact that biological systems run with concentration levels of several pairs of ions close to their solubility products has obvious dangers too. The

Table 2.20 Solubility-product constants (S.P.)

Substance	Formula	Solubility product
Aluminium hydroxide	$Al(OH)_3$	2×10^{-32}
Barium carbonate	$BaCO_3$	5.1×10^{-9}
Barium oxalate	BaC_2O_4	2.3×10^{-8}
Barium sulphate	$BaSO_4$	1.3×10^{-10}
Cadmium hydroxide	$Cd(OH)_2$	5.9×10^{-15}
Cadmium oxalate	CdC_2O_4	9×10^{-8}
Cadmium sulphide	CdS	2×10^{-28}
Calcium carbonate	$CaCO_3$	4.8×10^{-9}
Calcium fluoride	CaF_2	4.9×10^{-11}
Calcium oxalate	CaC_2O_4	2.3×10^{-9}
Calcium sulphate	$CaSO_4$	2.6×10^{-5}
Copper(II) hydroxide	$Cu(OH)_2$	1.6×10^{-19}
Copper(II) sulphide	CuS	6×10^{-36}
Iron(II) hydroxide	$Fe(OH)_2$	8×10^{-16}
Iron(II) sulphide	FeS	6×10^{-18}
Iron(III) hydroxide	$Fe(OH)_3$	4×10^{-38}
Magnesium ammonium phosphate	$MgNH_4PO_4$	3×10^{-15}
Magnesium carbonate	$MgCO_3$	1×10^{-5}
Magnesium hydroxide	$Mg(OH)_2$	1.8×10^{-11}
Magnesium oxalate	MgC_2O_4	8.6×10^{-5}
Manganese(II) hydroxide	$Mn(OH)_2$	1.9×10^{-15}
Manganese(II) sulphide	MnS	3×10^{-16}
Strontium oxalate	SrC_2O_4	5.6×10^{-8}
Strontium sulphate	$SrSO_4$	3.2×10^{-7}
Zinc hydroxide	$Zn(OH)_2$	1.2×10^{-17}
Zinc oxalate	ZnC_2O_4	7.5×10^{-9}
Zinc sulphide	ZnS	4.5×10^{-24}

Data from *Stability Constants* (1964). Spec. Pub No. 7. The Chemical Society. London Note the preferred use of Roman labels of oxidation states.

precipitation of calcium salts is a major hazard in old age and deposits of iron oxides in haemosiderin are frequent in treatments of people with blood diseases. There has been also some suggestion of a correlation between aluminium silicate deposits and Alzheimer's dementia but this is probably a secondary effect. It will be necessary for us to see how in normal cellular systems the precipitation reactions are used and controlled so that diseased states are avoided for long periods of life.

Finally, in an atmosphere of H_2S the removal of elements such as copper and the much reduced concentration of zinc gave easier access of iron(II) and III to thiolate ligands, Fig. 1.12.

There is an intermediate condition between complex ion formation and precipitation termed condensation. Here a large number of cations bind simultaneously to an anionic polymer causing it to fold lightly, condense, but remain soluble. An example is the binding of various inorganic and organic cations to DNA, see Sections 6.20 and 8.8.

2.19 The application of equilibria and equilibrium exchange to carriers, buffers, pumps, enzymes, and gene regulation

Because of the nature of biological cells, the separation of elements in spatial regions (see next chapter) requires that suitable proteins must be produced *in situ* which are capable of binding as buffers, or as enzymes, or as transporters (carriers) moving the elements either through reversible thermodynamic binding and/or by kinetic transfer across membranes, pumps, or through the cell cytoplasm. These proteins, which must act cooperatively, can only be produced in quantitatively correct amounts by feed-back genetic instructions (see Sections 2.21 and 6.13). Here we describe some general principles of the binding properties of the proteins. These brief observations make the value of the study of model systems more obvious before we tackle particular biological cases in more depth.

2.19.1 Carriers

Carriers for metal ions are present in cells and external to cells of both higher and lower organisms. It is important to understand the selectivity of their binding and exchange properties and the corresponding rates of equilibration. Outside the cytoplasm as well as exchanging they must also have binding constants commensurate with the availability of the particular ion concerned since energy cannot be applied to binding or release. Equally inside the cell they must have binding constants closely similar to the inverse of the free ion concentration if they are to bind at equilibrium in the cell (see Sections 2.10, 6.13.4, and 20.3.3). The exchange of the ions from carriers is obviously a requirement, but all of them can either exchange spontaneously at equilibrium, or in the cell after an energy dependent process, or directly and only with a receptor. For exchange equilibrium binding strengths must be similar. There are very different types of carriers for different elements and in different cells or even in the same cell. Of course, a small bacterial cell has less use for carriers for distribution of ions which are at relatively high concentration, say $>10^{-6}$ M. We shall find that some ions equilibrate readily, but others are carried in kinetic traps (see Table 2.21). The kinetics of element transfer from carrier to carrier will be described in Chapter 4. Overall carriers give a path-

Table 2.21 Equilibrium carriers of ions (approximate stability constant, K).

Ion	Carrier outside	Carrier inside
Na^+, K^+, Cl^-, Mg^{2+}	None	None
MoO_4^{2-}	Mod A (10^6)	Mod C (10^6)
Ca^{2+}	Phosphoprotein (10^3)	Parvalbumin (10^8)* Calbindins (10^7)*
Fe^{3+}	Siderophores (>10^{17}) Citrate, transferrin, etc.	None (NB Ferritin) $[Fe^{3+}] = 10^{-17}M$, $[Fe^{2+}] = 10^{-7}M$
Cu^{2+}	Caeruloplasmin (>10^{14})†	Various (>10^{14} (Cu^+))†
Ni^{2+}	Albumin (>10^{12})† Histidine (10^8)	Histidine containing proteins (>10^{10})†
Zn^{2+}	Albumin (10^8)	Metallothioneins (>10^{10})

*Restricted exchange with free ions (see Chapters 10 to 12, 15).

†Virtually no exchange with free ions (see Chapters 13, 14, 16 to 19).

way for an element similar to the metabolic pathways for organic compounds. (Carriers are sometimes called *chaperones*, but this is unfortunate since a chaperone is a protein folding agent by definition.) A carrier can insert an element into a site irreversibly.

2.19.2 **Ion buffers and stores**

The buffering of ions is important when the supply to the cell or the extracellular fluid concentration is in danger of falling low, so that there is a back-up concentration of an element available to meet contingencies. There is, therefore, no need to buffer common ions such as Na^+, K^+, Cl^-, and Mg^{2+}, the concentration of which is high in the environment and which can be readily managed using inwardly and outwardly directed pumps suitably located and energized, see Section 2.19.3. However, a cell, especially a larger cell, may need a back-up store of other ions, certainly of iron and even of calcium, at equilibrium with the free ion. A simple buffer is an ML complex, but a more effective buffer is one in which there is cooperative binding, e.g. M_nL, where n may be as high as say 10. Such a buffer has a steep equilibrium relationship between the per cent bound and the per cent free, so that it buffers in a more sensitive manner. A further increase in cooperativity is found in buffers which are precipitates, e.g. $Fe(OH)_3$, when the MX combination equilibrates through

Table 2.22 Ion buffers (approximate stability constants, *K*)

Element	Extracellular	Intracellular
Ca^{2+}	Bone (10^3) Protein (Calsequestrin) in Vesicle (10^{3-4})	(Parvalbumin?) (10^7)
Fe^{3+}	?	Ferritin (10^{37})*.ppt
Cu^+	?	Metallothionein (>10^{12})
Zn^{2+}	Vesicle (10^5)	Metallothionein (>10^{10})

*Solubility product

a solubility product. Here M (and X) are buffered (trapped) independently from concentration above the solubility product. Some examples of buffers in biological systems are given in Table 2.22. As explained in Section 2.10, the best buffering is at 50 per cent occupancy of a binding site, (see also Chapter 20).

A buffer can control pH, log M, or redox potential. Two of these may be found in combination, e.g a feed-back pH/redox potential for hydrogen or the feed-back Fe^{2+}+/Fe^{3+} redox potential with $Fe(OH)_3$ precipitation.

2.19.3 Pumps

Pumps transfer metal ions selectively across compartmental membranes using energy. The pumps for different elements can all be similarly energized by ATP, but can also have very different affinities (Table 2.23). Many pumps are simultaneously (1) equilibrium binding devices, but with different binding constants *on each side of the membrane*, and (2) energized kinetic uptake devices *across the membrane*. The binding constants on a given side must be roughly equal to the *reversible* binding constants of carriers, enzymes, surfaces, precipitates and transcription factors on the same side of the membrane if all are to exchange, but they may also equilibrate with free [M].

$$\text{(Outside) } M^+ \underset{}{\overset{K_1}{\rightleftarrows}} \text{ Pump } \underset{}{\overset{K_2}{\rightleftarrows}} M^+ \text{ (Inside).}$$

To limit the input or output of an element M^+ the pump must be switched off when the element concentration has reached the desired level by feed-back from the metal ion concentration. The best control is at 50 per cent occupancy.

2.19.4 Enzymes, structural proteins and transcription factors

Some enzymes and structural proteins also bind ions reversibly or exchange at equilibrium so that their binding constants must also be similar to those of the corresponding carriers, buffers, and pumps. On the other hand, many enzymes bind ions irreversibly and often the ion is inserted with the help of a 'chaperone' protein (see Section 2.12). Finally, there are other roles for ion/protein binding in relaying feed-back information to the genes, so-called transcription factors (see Section 2.20 and 6.13) and structural proteins. Of course, it is important to know if these selective ion protein complexes exist at equilibrium with free ions or not, (see Section 6.13). Table 2.24, an abbreviat-

Table 2.23 ATP-pump binding constants (stability constant, *K*)

	Cytoplasmic	Vesicular or external	Direction from cytoplasm
Na^+	10^{+2}	10^{+1}	OUT
K^+	10^{+1}	10^{+2}	IN
Mg^{2+}	10^{+3}	$>10^{+3}$	IN
Ca^{2+}	$> 10^{+7}$	10^{+3}	OUT
Zn^{2+}	$> 10^{+10}$*	10^{+6}†	OUT
Cu^+	$> 10^{+14}$*	10^{+10}(?)*	OUT

* Probably not in equilibrium with free M^{n+}.

† Special vesicles have high free zinc.

NB Iron does not appear to use an ATP pump.

Table 2.24 Elements taken up and incorporated at equilibrium and/or by kinetic insertion

Equilibrium interaction	Kinetic insertion
H^+, Cl^-, SO_4^{2-}, HPO_4^{2-}	H, C, N, O, S
Na^+, K^+, Mg^{2+}, Ca^{2+}	Se, P, I
Mn^{2+}, Fe^{2+}, (Zn^{2+})	Co^{2+}, Ni^{2+}, (Zn^{2+}), Cu^+
MoO_4^{2-}, WO_4^{2-}	Mo, W
	Fe (haem), Co(B_{12})

NB All elements may be kinetically controlled in their uptake by pumps. Elements in brackets are only present in minor amounts.

ed version of Table 2.21, summarizes the two main situations of the chemical elements in biological systems.

Note that a carrier, a buffer, a pump, an enzyme, structural proteins and a transcription factor, see below, can only equilibrate and be reasonably occupied, if the *binding* constants to a given M are all similar and selective, but they may be loaded in a given sequence according to the rate of *on* and *off* reactions and the concentration of M in a given locality. Exchange between proteins depends upon similar considerations.

It follows from the study of the binding of model complexes in this chapter that to achieve selectivity these binding constants, approximately at a given value for all the proteins of one metal ion, must be in the order

$$Mg^{2+} < Mn^{2+} < Fe^{2+} < Co^{2+} < Ni^{2+} < Cu^{2+}\ (Cu^+) > Zn^{2+}$$

if all the metal ions are to be managed. [This does not include the non-exchanging sites such as chlorin (Mg) and porphyrin (Fe), see Fig. 2.13]. The implication is that binding ligands have to be generated selectively so as to give a profile close to Fig. 2.10 (see also Chapter 20).

2.20 Genetic regulation—an introduction to control of ligand (protein) concentration

The premiss in the above analysis is that the ligand (proteins in cells) concentration must be managed. Cell syntheses are under the influence of messages to genes, which we define as particular sequences of nucleotides in their DNA. The genes, with the assistance of enzymes and the intermediate help of RNA, manage the production of proteins, and in turn the proteins control uptake and insertion of chemical elements into coenzymes and enzymes in particular places in the cell. Not all the genes give readable messages which can be transcribed as mRNA (where m stands for messenger) and then translated to proteins in ribosomes since many of them only act to adjust transcription. The production of proteins can also be managed at the level of translation (see Section 6.10). It is therefore central to an understanding of biological inorganic chemistry *in vivo* to appreciate first the way in which the regulation of protein synthesis for element binding works. For this purpose, in addition to a reading frame there have to be attachment points on the DNA for the 'reading machine', the enzyme RNA polymerase, and the so-called transcription factors. It is through the transcription factors which bind the element and sense its concentration, directly or indirectly, and which command the polymerases to start reading the frame, that a feed-back, Ⓕ, relationship

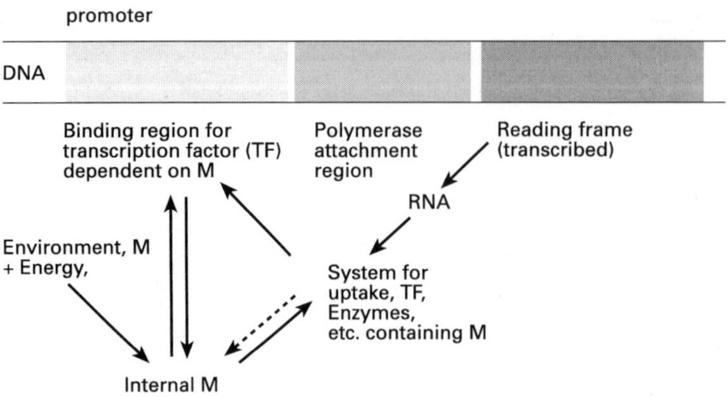

Fig. 2.17 The genetic structure for management of protein production according to need. The transcription factor TF binds to the promoter region of the DNA. TF prevents the polymerase from making RNAs and then proteins when there is an excessive supply of M but stimulates synthesis when [M] is adequate.

between element content of the cell and DNA expression is built. At its very simplest, the DNA then has three regions (Fig. 2.17).

Hence, which part of the DNA is expressed is not obvious from the DNA sequence, but equally important, *the amount of expression of a given metalloprotein is dependent upon the concentration of M either in free or bound form.* Thus, the quantitative cell content is a function both of DNA and the element content of the cell since one cannot operate without the other. In addition, a cell cannot operate without an energy input for pumping and synthesis, and this input must be connected to and controlled by a feed-back relationship to the element content of the cell and DNA (Fig. 2.17).

Overall, the management of cell constituents is very complicated, and in an account compatible with this book we must simplify. We have done this by giving first (Chapter 3) a detailed structure of cells and their compartments. We then analyse element distribution in the compartments before considering, in Chapter 4, rates of movement and reactions of cellular chemicals. In Chapter 5 we discuss the capture, transduction, and use of energy in biological systems, and in Chapter 6 we consider metal binding to biological macromolecules and come again to genes transcription in somewhat greater detail. Chapter 7 gives a first look at how a cell operates as a system. Specific elements are considered in Part II, while we return to integrated cell activity in Part III.

2.21 Concluding remarks: element handling in biological systems

The above discussion on selectivity provides most of the chemical information needed to understand how the different biological elements are captured and which kind of device is used to retain them inside cells or to keep them outside cells. We have considered metals and non-metals, cationic, anionic, and neutral species, thermodynamic and kinetic effects, and we have also briefly discussed some problems related to binding by proteins and in membranes and genetic regulation of concentration. It is, therefore, appropriate to check now some actual situations in biological systems against the predictions and we will do this first by examining the traffic of the biological elements in a generalized cell and its environment. A simplified model is that of Fig. 2.18 where only general trends are considered for the sake of simplicity.

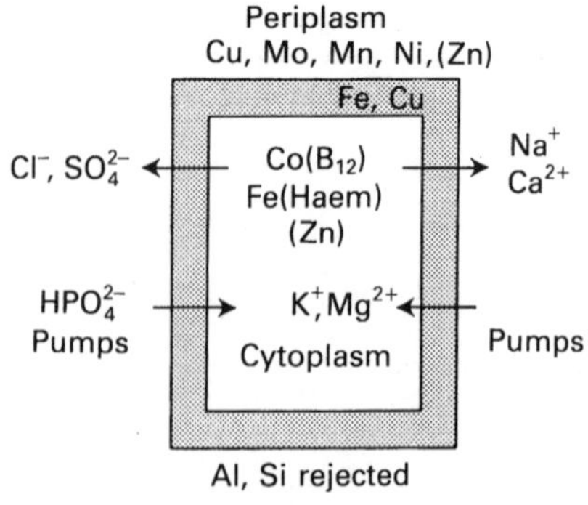

Fig. 2.18 The distribution of some 16 elements in a present-day prokaryote cell with no inner compartments.

Table 2.25 Traps for different biological elements

Element	Trap
Na^+, K^+, Mg^{2+}, Ca^{2+}, Cl^- ($H_2PO_4^-$, HPO_4^{2-}, SO_4^{2-})	Physical retention of ions by membranes (fast chemical exchange)
Cu^{2+}, Zn^{2+}, Ni^{2+}, Mo, Si, B, Fe^{3+}, Mn^{3+}	Thermodynamically and/or kinetically stable binding to polymers, proteins, and polysaccharides (slow exchange)
Fe^{2+}, Co^{2+}, Ni^{2+}, Mg^{2+}	Additional gated, thermodynamic and kinetic traps in small chelates, e.g. polypyrroles bound to proteins
H, C, N, O, P, S, Se, I, Br	Kinetic retention in thermodynamically unstable small or large molecules by covalent bonds
Si, Ca, P, C, S, F	Precipitates (biominerals), slowly exchanging

Figure 2.18 shows that some chemical elements are captured by cells in which they are retained by thermodynamically stable binding or by kinetic devices (covalent bonding, membranes, precipitates, etc.). Some other elements are rejected by the cells and remain in the external medium as free ions, combined with transporting ligands or as solid phases.

In their various situations, these elements follow the principles extensively discussed, which concern the type of traps adopted to retain them, the preferential binding to some atoms, and the geometry of the binding sites.

Table 2.26 Biological elements: some binding situations in biological systems

Binding atom (water excluded)	Element	Examples (with nearest simple geometries of site, which is usually distorted)
S only	Fe	Ferredoxin (tetrahedral)
	Zn	Alcohol dehydrogenase (tetrahedral)
	Cu	Atx carrier (linear)
S, O	Mo	Molybdenum cofactor
N, S	Cu	Plastocyanin (tetrahedral)
	Fe	Cytochrome *c* (octahedral)
	Zn	Alcohol dehydrogenase (tetrahedral)
N only	Fe	Cytochromes (octahedral)
	Cu	Superoxide dismutase (tetragonal)
	Zn	Carbonic anhydrase (tetrahedral)
	Mg	Chlorophyll (square-planar)
N, O	Zn	Carboxypeptidase (tetrahedral)
	Mn	Concanavalin (octahedral)
	Fe	Haemerythrin (octahedral)
O only	Ca, Mg	Calmodulin (Ca, 7/8 coordinate ionic binding)
	Mg	ATP 6-coordinate (octahedral)
	P, Si, B, S	XO_4 groups condensed with ROH, e.g. in polysaccharides (tetrahedral)
N, C	Co	Coenzyme B_{12} (octahedral)
C only	S, Se, I	Cysteine(S), selenocysteine(Se), thyroxine(I): single covalent bonds

Table 2.25 and 2.26 exemplify these situations. The combination of the different characteristics, affinities for certain atoms, preferences for determined geometries and all possible devices for retention and rejection available in biological systems, allows each element to be organized in the biological space in very specific patterns and thus acquire also very specific functional significance; role overlap occurs, but to a very limited extent, and closer examination usually reveals that either prevalent local conditions dictate duplications (e.g. the Fe^{3+}, Mn^{3+}, and Cu^{2+}/Zn^{2+} superoxide dismutases) or that they may be derived from the coexistence of different evolutionary stages (probably the case of V(V) and Fe-peroxidases and that of V- and Mo-nitrogenases). In evolution the availability of many elements, especially those which changed oxidation state, such as Mo and Cu, became useful in ways previously only open to other elements such as W and Fe. Examples of both situations will be discussed later in this book, in the chapters on specific elements.

Further reading

Cotton, F.A. Wilkinson, G. Murillo, C.A., and Bochmann, M. (1999). *Advanced inorganic chemistry* (6th edn.) John Wiley, New York.

De Bolster, M.W.G. (1997) *Pure and Applied Chemistry* (Glossary of terms used in bioinorganic chemistry) **69**, 1251–303.

Jensen, W.B. (1980) *The Lewis acid–base concepts.* John Wiley, New York.

Martell, A.E. and Hancock, R.D. (1996). *Metal complexes in aqueous solution.* Plenum Press, New York.

Mingos, D.M.P. (1998). *Essential trends in inorganic chemistry.* Oxford University Press, Oxford.

Morowitz H.J. (1978). Proton semiconductors and energy transduction. *American Journal of Physiology*, **235**, R99–R114.

Nelson, N. (1999). Metal ion transporters and homeostasis. *The EMBO Journal*, **18**, 4361–71

Phillips, C.S.G. and Williams, R.J.P. (1966). *Inorganic chemistry.* Oxford University Press, Oxford.

Radisky, B. and Kaplan, J. (1999). Regulations of metal ion transport across the yeast plasma membrane. *Journal of Biological Chemistry*, **274**, 4481–4.

Ringbom, A *in* Kolthoff, L.M., and Elving, P.J. (eds) (1978). *Treatise on analytical chemistry*, Vol. I. Interscience, New York.

Sarkar, B. (ed.) (1998). *Metals and genetics.* Plenum Press, New York

Schwarzenbach, G. and Flashka, H. (1969). *Complexometric titrations* (translation by H.M.N.H. Irving). Methuen, London.

Shriver, D.F. and Atkins, P.W. (1999). *Inorganic chemistry* (3rd edn). W.H. Freeman, New York.

Williams, R.J.P. (1993). The history of proton driven ATP formation. *Bioscience Reports*, **13**, 191–206.

Williams, R.J.P. and Fraústo da Silva J.J.R. (2000). The distribution of elements in cells. *Coordination Chemistry Reviews*, **200–2**, 247–348

CHAPTER 3

Physical separations of elements: compartments and zones in biology

3.1 General aspects

In order to understand the observed binding of the chemical elements, metals and non-metals, to particular chemical sites in biological systems, we have discussed three factors: (1) the limitations on the concentration of chemical elements imposed by their abundance and availability in the environment; (2) the possibility of adjusting this availability in a cell by pumping an element or compound across a single membrane into or out of its inner aqueous phase; (3) the intrinsic stability, thermodynamic and kinetic, of the association of the chemical elements with particular 'ligand' frameworks which have 'designed' structures, either in solution or in a condensed phase.

Whereas in primitive cells there was apparently only one internal aqueous compartment, i.e. one surrounding cytoplasmic membrane, and, in some cases, a periplasmic (outer) controlled spatial compartment, evolutionary pressures led to progressive modifications which included: (1) the differentiation of internal compartments by extra membranes in cells [Fig. 3.1(a) gives a real example and Fig. 3.1(b) is a simplified diagram for discussion purposes]; (2) the functional specialization of cells together with the development of

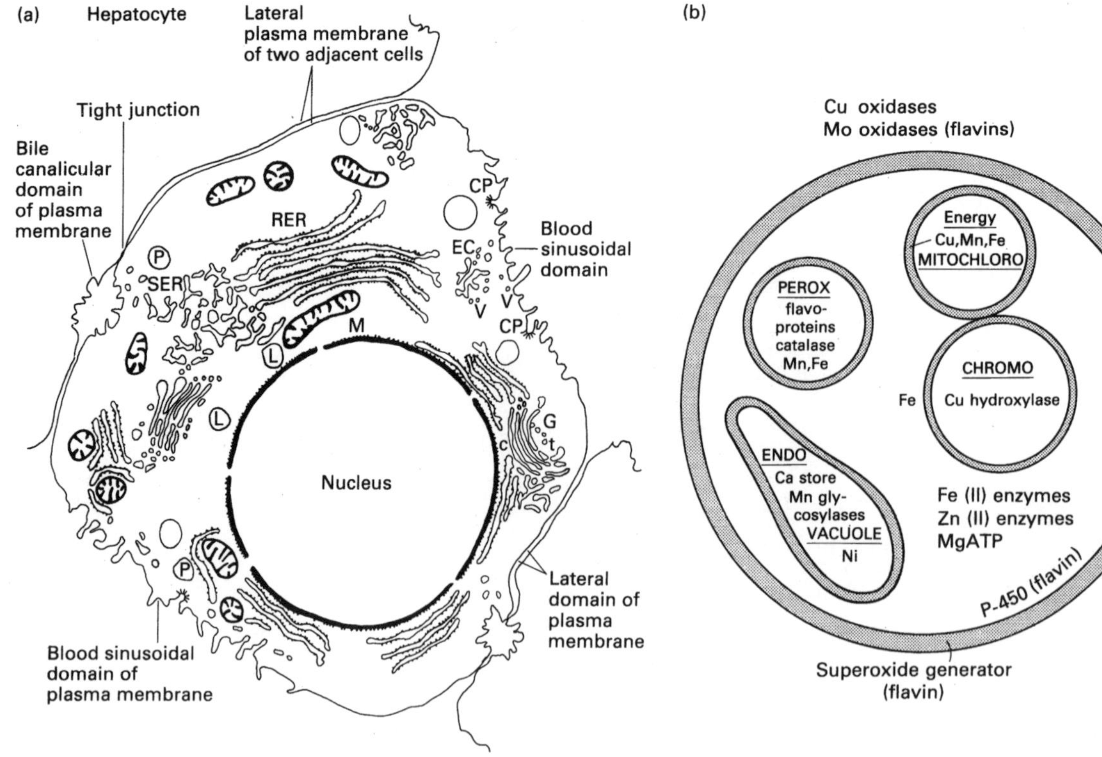

Fig. 3.1 (a) Section through an hepatocyte illustrating the various organelles and membrane systems. EC, endocytic compartment; G, Golgi apparatus (c = cis face; t = trans face); L, lysosome; M, mitochondrion; P, peroxisome; RER, rough endoplasmic reticulum; SER, smooth endoplasmic reticulum; V, vesicle; CP, coated pit. (b) A schematic indication of some of the different membrane separated compartments in an advanced cell. PEROX is a peroxisome; MITOCHLORO is either a mitochondrion or a chloroplast; CHROMO is a vesicle of, say, the chromaffin granule; ENDO is a reticulum, e.g. the endoplasmic reticulum. Other compartments are lysosomes, vacuoles, calcisomes and so on. Localized metal concentrations are shown. These figures are of transverse sections. To appreciate a cell fully it is necessary to have serial plane sections parallel to those along the z-direction.

multicellular organisms. In the present chapter we will be concerned mainly with the first of these aspects, but we must notice immediately that even in cells without internal, membrane-limited, compartments there must be some kind of internal organization; otherwise it would be practically impossible to ensure that the thousands of reactions which take place there could proceed without confusion. 'Modern' cells are complex cooperative organizations, with various internal divisions, some of which give simple physical meaning to our idea of 'compartment' since they have essentially different aqueous phases separated by membranes, but there are others where the rates of diffusion and reaction define *zones* less precisely. The inside of a cell is a network of filaments and vesicles, not space open to free diffusion. Special proteins and elements are bound in particular locations and we must not forget that each membrane is a defined zone or series of zones (see Section 3.4) with its own complement of proteins. Cytoplasmic solutions are quite strong solutions of proteins (see Aside).

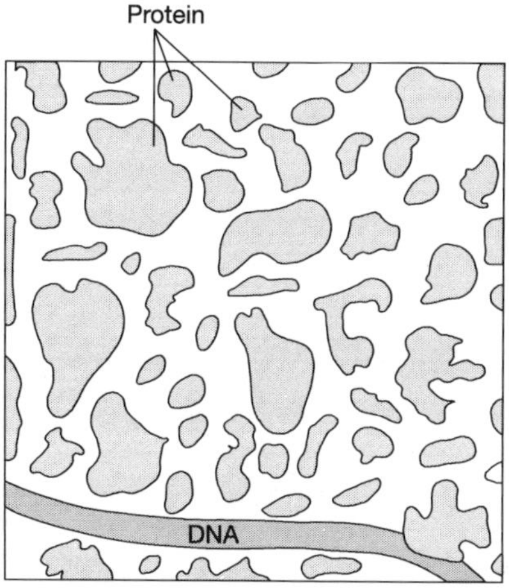

A picture of proteins in the cytoplasm with DNA running through it. (From Goodsell, see references.)

3.2 The nature of compartments

We start this description with the two major organelles. These are long, weaving compartments which may be remnants of prokaryote cells trapped inside a eukaryote cell. (Note that a eukaryote cell has a nuclear membrane enclosing its DNA.) The two are the mitochondrion (Fig. 3.2) and the chloroplast (Fig. 3.3). Each has its own DNA, not enclosed, and its own system of membranes. The mitochondrion has two membranes separated by a periplasmic space as in a prokaryote: the chloroplast has two outer membranes and an inner vesicle, the thylakoid. The inner membrane of the mitochondrion carries out oxidative phosphorylation; the thylakoid membrane is responsible for light capture and energy transduction. In mitochondria the metabolism is largely that of fats and the citric acid cycle and is oxidative; in chloroplasts the reactions generate free dioxygen and reductive incorporation of CO_2. Both organelles generate ATP; indeed they are the major energy-generating compartments.

Fig. 3.2(a) Early diagram of a mitochrondrion; see also Figs 3.1(a) and 3.2(b) and 3.11.

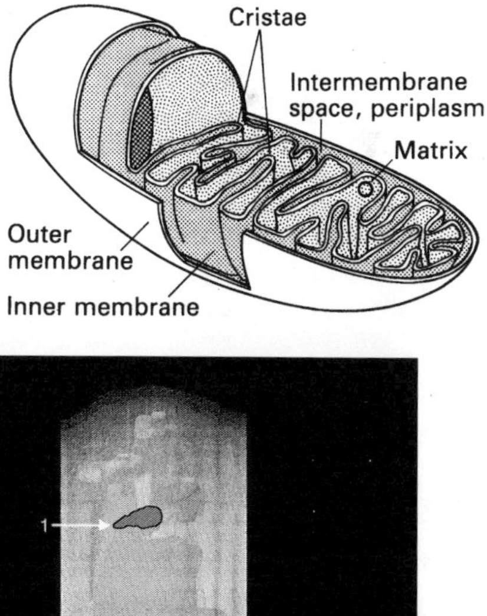

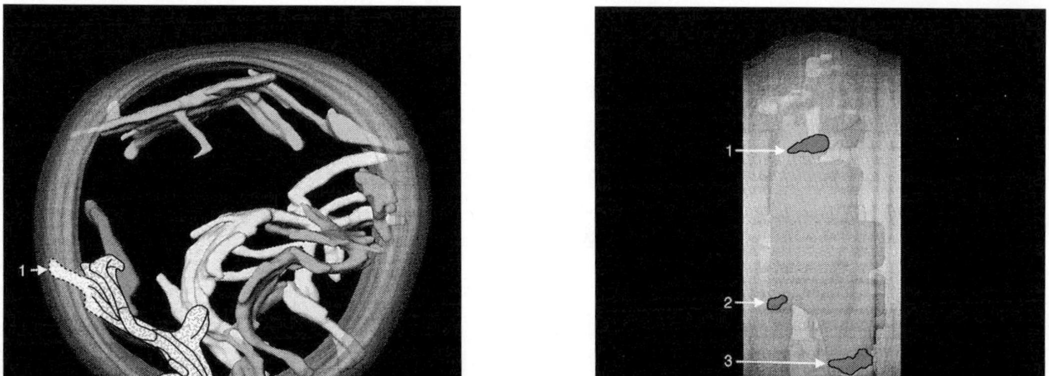

Fig. 3.2(b) Two views of the structures of a mitochondrion. A long roughly cylindrical sector of a mitochondrion is seen end-on in (i), revealing the cristae (several, differently shaded) running across and down it and joining to the intramembrane space. In (ii) the cylindrical sector of the mitochondrion is viewed parallel to its long axis. The arrows 1, 2, and 3 cross-located the two views and show one large cristae. 1 cm is equal to 0.5 μm. The extreme lengths and convolutions of the cristae do not permit a delocalised activity of the organelle, especially since it makes local connections to other vesicles. These pictures were provided by Dr Carmen Mannella to whom we are indebted. They are published with his permission.

At one time mitochondria and chloroplasts were thought to be small organelles, roughly elliptical in all three dimensions. Today we know that these organelles are elliptical in the *xy* plane (100 nM) but in the *z*-direction they can be extremely elongated bodies which weave through the cell. The inner membranes form tubes of very great length which are often cross-connected. It is inevitable that the organelles do not behave as a unit but in localized zones to produce ATP where needed. Here and there they make local close contact with other membrane enclosed systems. This is extremely important for *localized* cell activities. When we discuss bioenergetics (Chapter 5) we shall return to the problems of local domains and kinetic controls.

Other membrane-isolated compartments within cells contain no DNA; they are often used for storage or delivery of chemicals. There are a multitude of such vesicles containing transmitters, hormones, digestive proteins, and so on, according to cell type. Vesicles such as lysosomes contain enzymes that break down polymers and supply the cell with basic chemicals, while peroxisomes generate H_2O_2 and generally drive certain oxidative reactions, but they may also release enzymes externally. Pervading many cells is a reticulum which acts as a communicating network and on stimulation releases, for example, calcium from its internal store; it is usually a very extensive tube-like zone. Finally, there is the Golgi apparatus in which many proteins are modified before export (note especially glycosylation and sulphation). The Golgi apparatus also directs vesicles to different parts of a cell (see Section 3.5.4).

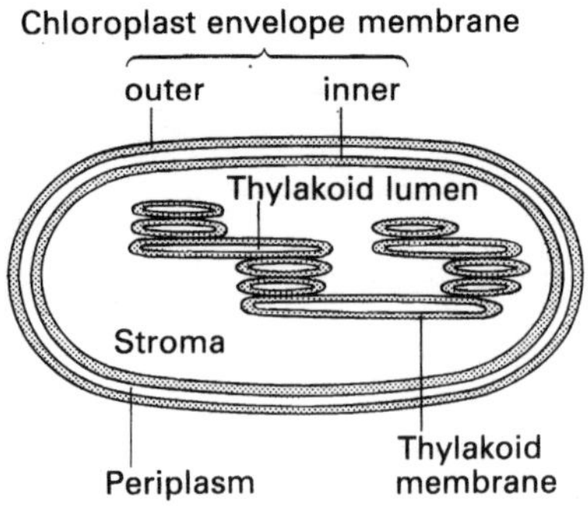

Fig. 3.3 Diagram of chloroplast; see also Fig. 3.10.

3.3 The chemical solutions and physical states of compartments

Across the membranes, as across cells, there is traffic of ions, ligands (including proteins), and complexes. Energy can be applied to these processes or to a whole phase by adjusting, for example, the pH or the redox potential in a compartment. If we take a reference compartment to be at a standard state pH = 7 and at a redox potential equivalent to that of O_2/H_2O, i.e. +0.8 V relative to the standard hydrogen electrode, then we would expect that the oxidation states and, perhaps through evolutionary origins, even the concentrations of the different aquo ions in this compartment would be similar to those in sea water. The extracellular fluids of blood are not too far from this condition. The situation in any cell compartment will be changed by ion pumping, as described in Chapter 2, by pH control (that is, by metabolism or by proton pumping) and redox control, by oxidative or reductive metabolism, or in other ways, which may even include mechanical pressure, as described in Section 3.4, so each compartment will be 'energized' relative to the reference state in many ways.

Furthermore, since the ions considered will combine with different ligands L and their functional value depends on this combination in most cases, we must look into the factors that regulate the availability and concentration of L in the various compartments. Here we find that some of these factors are common to those for the 'energization' of the metals, namely pH and redox potential. For example, the state of a ligand carrying two thiol groups can be $(\text{-SH})_2$, $(\text{-S}^-)_2$, or -S-S-, depending on the values of those two variables in the compartment into which the compound is pumped. In addition, every ligand, including any protein, is synthesized in variable amounts by cell and organelle metabolism and may be pumped into a given compartment. Finally, it is observed that some of the ligands L are in fixed positions, say, in a particle or on a filamentous construction or a membrane, so that ML is part of a surface and not free in solution. The position of ML in a cell is connected to function in addition to the controlled chemical combination itself.

The amount of ML formed in each compartment and its standing concentration are, therefore, decided by five new factors to be added to those stressed in Chapter 2 and at the beginning of Section 3.1:

1. the values of pH and redox potential of the phase in a particular compartment, affecting the states of both M and L;
2. the limitation to the delivery of M and L into that compartment, including the possibility of its local synthesis (note that L can be a protein);
3. the further possible movement of the complex ML from that compartment to other parts of the cellular organization (Fig. 3.4);
4. binding to surfaces, when we may well be interested in bound ML more than free ML;
5. feedback mechanisms from the concentration of M and L to their production especially to the genetic apparatus (see Chapter 6).

Given the multiplicity of factors controlling the siting of a given metal in any biological compartment it is not possible, with the present state of knowledge, to deduce where it will be found or where its different complexes are to be placed. We may, however, be able to work backwards from the knowledge of where the free ions or their complexes are found to the reasons for their respective distribution and then try to relate this distribution to the functions performed.

A first striking fact, which is becoming more and more apparent from recent data on the localization of the various elements in biological systems, is the

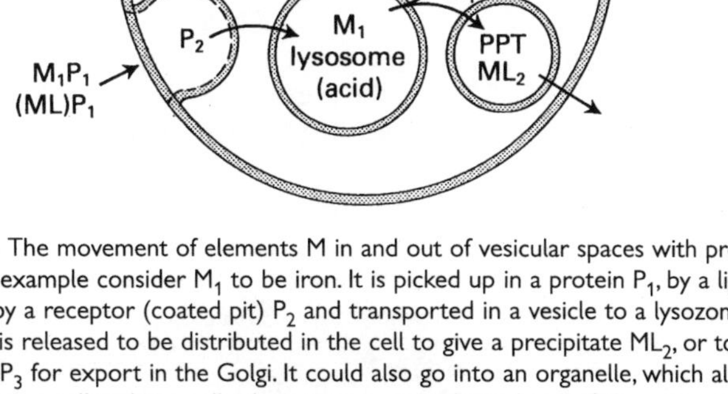

Fig. 3.4 The movement of elements M in and out of vesicular spaces with proteins P. As an example consider M_1 to be iron. It is picked up in a protein P_1, by a ligand L, and/or by a receptor (coated pit) P_2 and transported in a vesicle to a lysozome. Here it is released to be distributed in the cell to give a precipitate ML_2, or to a protein P_3 for export in the Golgi. It could also go into an organelle, which almost behaves as a cell within a cell, where it is processed into haem. Other elements M_2 enter the cell through pumping of ions. The whole arrangement may well have three-dimensional organization, directed flow, and time dependence. Within the cell cytoplasm copper and zinc circulate using metallothionein and iron circulates in ferritin. Both carriers, metallothionein and ferritin, are then homeostatic proteins. Of course, the synthesis of all the proteins must be commensurate and so they are under DNA regulation (Section 6.10).

asymmetry of their distribution. Figure 3.1(b) is a simplified description of some of the compartments. We have already observed that sodium and calcium are rejected from the cell cytoplasm whereas potassium and magnesium are concentrated in it, but similar observations can be made for the trace elements and more detailed information about compartments is now available for the oligoelements referred to above. For example, manganese is generally located in organelles and reticula, including the Golgi vesicles, but not in cytoplasm; in plant cells it is either in the vacuole or in the chloroplast organelle, but perhaps not so markedly in the mitochondria as in animal cells. Zinc, like calcium and manganese, is also sometimes strongly concentrated in special vesicular or reticular spaces, but not necessarily in the same compartments as manganese. Examples of these special spaces are the vesicles of pollen cells and the acrosome of male gametes; in addition, zinc is also found in the cytosol. Again the extensive involvement of this metal in hydrolytic (digestive) enzymes clearly requires that these are not active in the cells to avoid their self-digestion; hence this zinc and its proteins are in vesicles and are rejected at some stage to extracellular space. Bound iron is usually concentrated in special membranes, such as those of mitochondrial and thylakoid organelles, and, when inside the cytoplasm of cells, much is in the form of ferritin, which can be considered to be a special compartment of its own, where a protein wraps around $Fe(OH)_3$ solids like a membrane. Bound copper and perhaps molybdenum are largely extracellular (see Chapters 15 and 17), and some of their metallo-enzymes are found between double membranes, as in the periplasmic space of bacteria, e.g. in *E. coli* [see Fig. 3.1(b)], or organelles.

Amongst non-metals, the essential elements chlorine and silicon are largely rejected from cells, while elements such as phosphorus, sulphur (as sulphide but not sulphate), and selenium are accumulated. There is also a group of elements which have little or no biological function and which biology seems to exclude. They are group 3, 4, 13 and 14 elements, such as Al, Sc, La, Ge, Ti, (Sn), and Pb, all of which could have been accumulated if a useful function had been found for them. Surprisingly, some organisms take up large amounts of barium or strontium and deposit them in quite distinct cell vesicles as sulphates.

The different elements are moved in very different ways to different places by the use of metabolic energy and all these movements are kinetically controlled, so that their spatial distribution in a cell has to be described by a map of local, energized, steady-state concentrations, selective for each element, which relates to the 'wanted' function. This distribution is part of evolutionary 'design' and is contrary to expectation from thermodynamic pressures. The fact that the distributions are energized means that they are time-dependent, being susceptible to fluctuations of environmental conditions, such as light intensity cycles, oxygen tension variations, availability of food, reproductive or other binding cycles, and triggering of activity. The fluctuating movement of elements and its adjustment is much more obvious for some elements than for others. Thus, H^+, Na^+, K^+, Ca^{2+}, Mg^{2+}, Cl^-, or HPO_4^{2-} are mobile, or at most semistatic in the cases of Ca^{2+}, Mg^{2+}, HPO_4^{2-}, or even H^+, while other elements, namely some transition metals such as iron, are predominately static as ML complexes, and their function does indeed rest, very largely, in a fixed compartment or even in a localized surface zone of a membrane or of a cell interior.

From what has been said it becomes more and more evident that, if the full use of the elements in biology is to be appreciated, we must grapple with the description of the nature of biological space and with the disposition of elements in that space, for it is this organization which causes the flow of material as well as the uptake, storage, and transduction of energy. Since these are time-dependent phenomena, space and time relationships are essential knowledge in biology. Our immediate need is to describe biological space more thoroughly. We shall do so by reference to single cells, but in complex organisms, such as man, the element distribution is organ- as well as cell-specific. The intractable problem which this imposes will be approached only here and there in the present chapter.

One peculiarity of membranes is that there are considerable differences in composition. The membranes of archaebacteria appear to be largely based on long-chain esters. The coming of dioxygen introduced cholesterol into the membranes of eukaryotes giving the membrane extra strength.

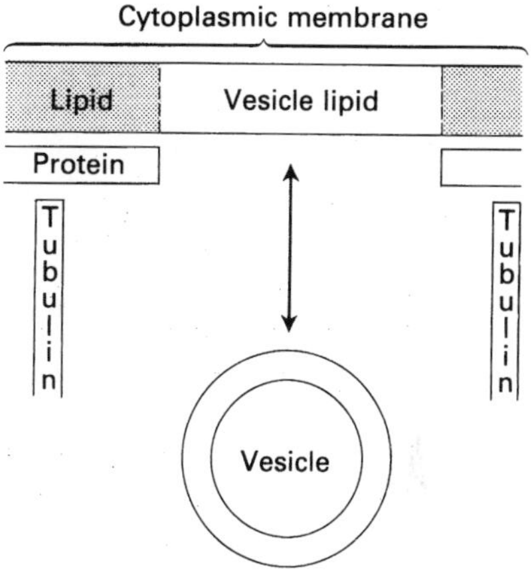

Fig. 3.5 A schematic representation of the relationship between the cytoplasmic membrane, an internal membrane (vesicle), and the protein network. One series of proteins, e.g. spectrins, lies under the outer membrane, bound to it, while another series, here called tubulins but including a variety of filamentous structures, crisscrosses the cell and holds vesicles in place. The vesicle membrane may form part of the outer membrane reversibly.

3.4 The role and nature of membranes

Biological systems have some obvious ways of dividing elements into spatial patterns using membranes:

1. capture in aqueous phases by membrane enclosures;
2. capture by proteins at localities either on or in membranes or filaments;
3. precipitation of salts in vesicles.

We now look more closely at the nature of the membranes related to all three. Biological membranes not only divide space but they determine mobility (in the membrane) and incorporate a number of energy-using devices that can move elements from one compartment to another either as simple ions, small chelates, protein complexes, or even crystallites. The energy-using devices are proteins which act as pumps fixed in the membrane or in a system of membranes, whereby the membrane can even bud into the cell to form a new vesicle or coalesce with an existing membrane to create a new compartment (see P_2 in Fig. 3.4).

However, a cytoplasmic membrane is not just a lipid barrier, but is part of a total cellular structure through a network of proteins in the cell. These proteins lie both under the surface of the lipid membrane as a network and perpendicular to it in the form of filaments of which tubulin is one example. In Fig. 3.5 we give a scheme which may be useful in thinking about this description. Notice that the proteins are tension fibres and have directional quality; the shape of the membrane is to some degree decided by the tension in the protein filaments. However, the system is also interactive, through these filaments, with internal vesicles (see Fig. 3.5 and Chapter 6).

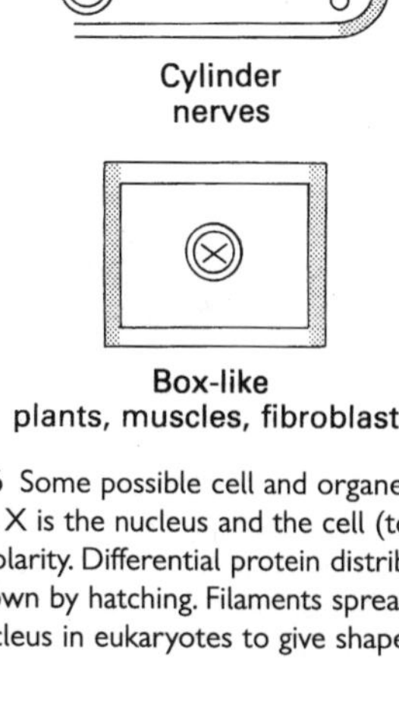

Fig. 3.6 Some possible cell and organelle shapes. X is the nucleus and the cell (top) may have polarity. Differential protein distributions are shown by hatching. Filaments spread from the nucleus in eukaryotes to give shape.

Obviously, due to the asymmetry of tension from the internal filaments, the outer membrane has a varying radius of curvature (Fig. 3.6), and from this fact alone there will be a lateral distribution of its chemical contents. This follows from the fact that all components in the membrane migrate to lower surface tension overall. Different parts of the membrane accumulate macromolecules plus their combined elements differentially; hence, they are ordered phases but with no repeats in the manner of a crystal: outside surfaces differ from inside surfaces and left is distinguished from right. Cell or organelle membranes are therefore asymmetric three-dimensional phases which undergo shape changes from one moment to another according to the local energization of tension.

Now, some cells confine the cytoplasmic membrane within an outer wall-like structure, for example many bacterial and plant cells. The shapes of their outer membrane may well be controlled by this outer wall, although complex invaginations are frequently observed. Other cells have a very flexible outer coat of polysaccharide and their shapes are open to external as well as internal forces. Their membranes are affected by the forces on the outer polymers, i.e. *connective tissue*, and the cell shape is now dependent on a steady-state relationship between the local pressures produced by both the internal and the external matrices. Such cells often show long extensions as villi or dendrites, as in nerve cells. The cell and its outer membrane are then interactive with the internal milieu and the external environment. This is frequently true for animal cells. It is against this background that we study the nature of membrane phases themselves and element distributions in them; detailed examples are given in Section 3.5.

3.5 Different types of membrane

The membrane described so far is for the most part the cytoplasmic membrane of higher cells or that of the organelles, mitochondria and chloroplasts. These membranes are connected to internal filaments (Fig. 3.5). Inside higher cells there are quite different vesicle membranes [Figs 3.1(a) and 3.4] which have no internal protein fibres. It is often even difficult to be sure whether the internal space behind these membranes connects with the extracellular space or not. In some cases this does not happen, e.g. the Golgi apparatus, but the various so-called reticula may or may not be so connected. Again there are vesicles which are formed from the outer membrane and which introduce external molecules or particles into the cell by phagocytosis, and other internal vesicles which reject chemicals from cells. Both of their membranes can be formed from local zones of the outer membrane (Fig. 3.5) and again the common feature of the *internal* spaces inside these membranes is that they do not have a fibrillar proteinaceous structure. They are limited by the forces imposed on them by the *internal* filaments of the cytoplasm, e.g. tubulin (Fig. 3.5), by the pumping of molecules into them, and only rarely by internal constructions not attached to the membrane. These may be based on polysaccharides or glycoproteins. The latter biopolymers are often related to export processes, to biomineralization, or to the sending of messages.

To deal with such complexity one must again try to divide the subject matter, i.e. the description of limitations on distribution of elements by different membranes, and we have chosen the following items:

1. internal and external biopolymer assemblies related to membranes;
2. lateral membrane organization and in-membrane flow;
3. transmembrane organization and flow;
4. the position of vesicles in cells.

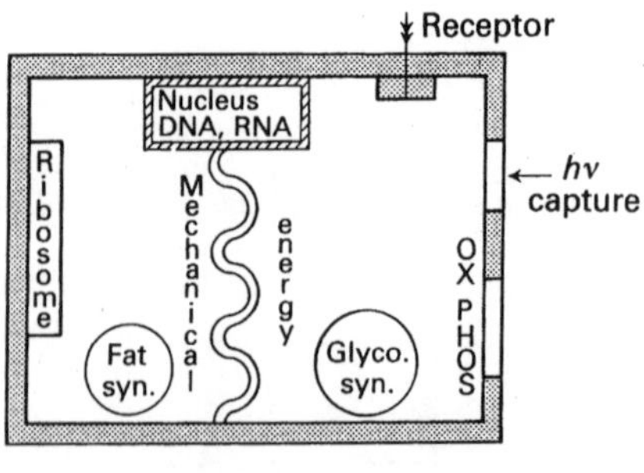

Fig. 3.7 A representation of some large particles in a cell. Some are free floating in the cytoplasm and some are in or on membranes. They are distributed in different membrane-limited compartments. The nucleus has a membrane in eukaryotes only and in these cells the energy capture devices are in organelles.

3.5.1 Internal and external particles and their relationship to membranes

Large particles, many of which are unrestricted by membranes, assemble both inside and outside cells. Here we include not just the filamentous structures described in Section 3.5, but also 'free-floating' units, such as enzymes associated with particular sequences of reactions, e.g. fatty acid synthesis (Fig. 3.7) and proteasomes for protein degradation. Inside cells, the most familiar large particles are those formed by RNA and DNA, i.e. ribosomes and nuclei (Fig. 3.7). These phosphate-rich polymers are held together by special basic proteins, e.g. histones, by magnesium ions, and by some monovalent ions. The DNA is connected to some of the filaments and is confined in eukaryotes by a special 'nuclear' membrane. One kind of RNA (ribosomal) is often connected to certain reticula membranes. In the course of a cell cycle these particles may form and divide, hence we know that once again we are examining a dynamic set of particles with an interaction with the membrane via filaments. The positioning of particles is very different in prokaryotes and eukaryotes.

As already stated, apart from the large particles there are, both inside and outside cells, large filamentous structures. Inside the cell the structures are made only from proteins and typical of these are the tubulins, vincins, and actin/myosin filaments (see Table 3.1). They have associated with them localized enzymes which assist in energy-driven changes of the filaments (Fig. 3.8). Outside the cells, associated glycoproteins of great complexity and length are to be found; they form both a surround for single-cell organisms and also the soft tissue matrix which holds together, often with the support of biominerals, the multicellular organisms. Typical proteins are collagens, but there are many others, including the proteoglycans. However, many of the glycoproteins are sulphated, as in myelinated tissues of nerves. Sulphated polymers are not found inside cells. Very many of the intra- and extracellular filaments are firmly attached to membranes (Fig. 3.7).

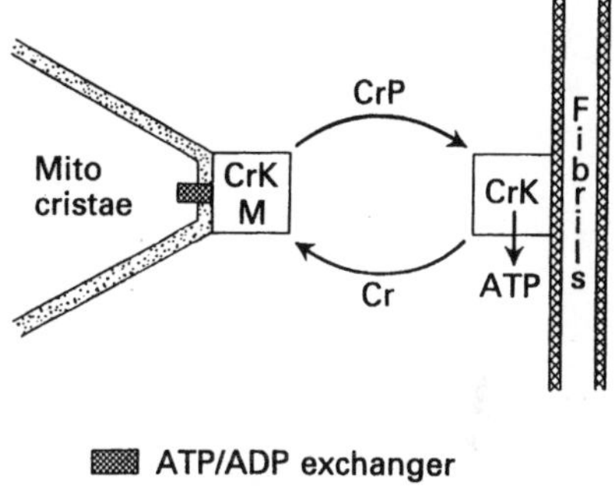

Fig. 3.8 The localized transfer of an energy-carrying chemical, creatine phosphate (CrP), from a mitochondrial cristae (Mito cristae) membrane (Fig. 3.2) organization to a fibril organization for contraction. Two different creatine kinase (CrK) enzymes act in different locations.

The chemical elements associated with these structures are highly specific; inside cells magnesium (and probably zinc) is deeply involved with the fibrous structures, their organization, energization, and synthesis; outside the cells, the major element involved is probably calcium. Most importantly, the actual synthesis and degradation of the extracellular matrices of many if not all cells is associated with the enzymatic activity of *extracellular* copper oxidases and zinc proteases. The constant balance of filament production and cross-linking, i.e. energized synthesis and degradation both inside and outside cells, is all part of the dynamic particulate organization of multicellular organisms much as it is a part of the division cycles of cells. Growth and repair are constant activities.

Filaments are organized in space, so that they give a cell vectorial quality within the aqueous matrix. The cell's metabolism governs their formation, but also exposes them to mechanical stress and strain, which is the basis of contractile action as in muscle. This action, which is directional, is again related to patterns of element flow.

If a cell is to maintain shape for a period and then adjust to another shape, it follows that the filamentous network inside as well as outside the cell must

Table 3.1 Some filamentous structures of cells*

Filament	Function
Actomyosin	Contractile protein which can be used to shape a cell or to exert force between cell units
Tubulin	Reversibly formed filament for structure and vesicle transfer and cell shape
Intermediate filaments	So-called because of their intermediate thickness: function uncertain
Collagen	Filamentous structure outside cells—connective tissue

*See Fig. 3.5.

have only temporary stability. The way in which this could be managed in cells is shown in Chapter 6, Fig. 6.14.

3.5.2 Lateral membrane organization and in-membrane flow

A biological membrane has to be described as a three-dimensional object with proteins embedded in it or protruding from it (see Fig. 3.9). Consideration of the inside/outside organization and ordering across the membrane is very important for the discussion of electron diffusion and ion or molecule movement *across* membranes. But the lateral organization of proteins must also be taken into account and this leads to the consideration of the *curvature* of the membrane (Figs 3.6 and 3.9). (This in turn is associated with the capture of certain proteins and to the flow of vesicles as already described, see Fig. 3.4). It is perhaps unnecessary to stress that all these aspects relate directly to the localization of the metals in the cell space since many of the proteins in the membranes are indeed metalloproteins. We will examine more closely this question by considering briefly three specific examples of membranes: that of the thylakoid of the chloroplast; that of the mitochondria; and that of certain epithelial cells.

The thylakoid membranes have a rather large set of incorporated proteins, largely metalloproteins, which occupy more than 50 per cent of the membrane volume; they include the whole apparatus of two photosystems, PS I and PS II, and the energy-capturing device, the ATP-synthetase. The relation of the different units has often been represented represented as a diagram—the Z diagram of photosynthesis which is supposed to indicate a path of electron flow (see Chapter 5, Fig. 5.3). Today much more is known of the topology of the membrane than this diagram describes (see Figs 3.3 and 3.10). For example, it has been found that, after exposure to light, the photosystem PS I and

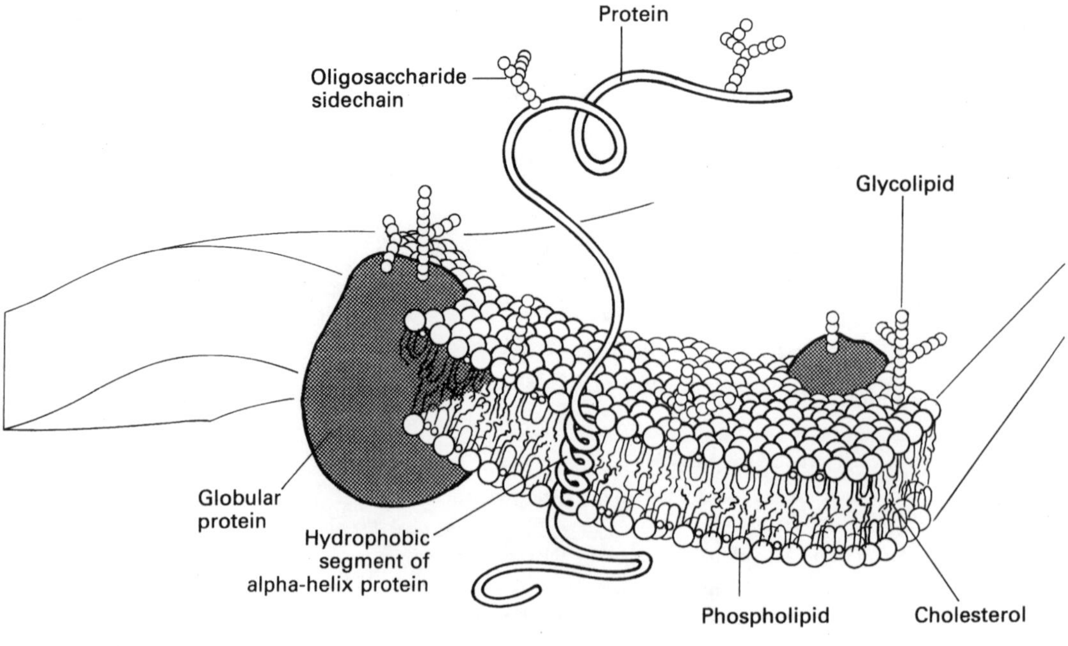

Fig. 3.9 The organization of proteins in membranes.

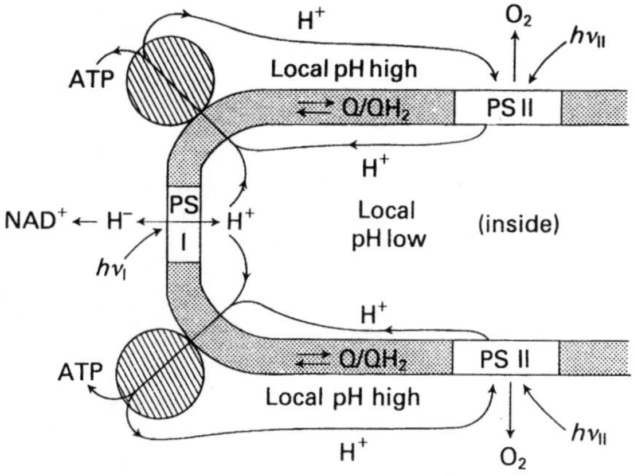

Fig. 3.10 A description of the distribution of particles, PS I and PS II (photosystems), and proteins in and on thylakoid membranes of chloroplasts. Q represents quinone and QH_2 hydroquinone.

the ATP-synthetase are in non-oppressed, curved, and exposed regions of the membrane, while PS II is in the oppressed region. This means that there are different microcircuits of electron and proton flow and of ATP generation laterally disposed along the membrane as well as across it (Fig. 3.10). The exact location of each metalloprotein involved (and consequently each metal) is not known, but it is suspected that many of metalloproteins face the aqueous phase of the chloroplast and not that of the thylakoid. The centre of the membrane itself is full of magnesium chlorophyll and the lateral organization places manganese (apparently bound to a 30 000 Da protein) in the oppressed regions (PS II), but iron is more evenly distributed. The case of plastocyanin, the only copper protein involved, is curious since it is found inside the thylakoid, perhaps assisting the electronic communication between PS I and PS II. This may be due to the fact that besides having the right redox potential this protein can withstand the low pH values (of the order of 4) generated in this compartment, whereas other metalloproteins may dissociate there.

The case of the mitochondria is similar, though the internal pH is not low, but here the lateral organization is not yet known with the same clarity although the proteins themselves have been more thoroughly studied. Figures 3.2(b) and 3.11 summarize the present knowledge of the essential structural features of the mitochondrial membrane system and again it is found that a large part of it is made up of proteins which are overall in excess of lipids. Besides the proteins of the interiors of the membranes in Figs 3.10 and 3.11, many other intraorganelle proteins exist, not bound inside the membrane but bound to it and to other proteins, extending more than 100 Å both ways from the 40 Å-thick lipid bilayer. One feature of this organization is that the inward movement of proteins, and perhaps of small molecules ML, is across the zone where the outer and inner membranes come together (Fig. 3.11). There is a special transfer unit here for iron complexes and vitamin B_{12} (Co), among other elements. We return again to mitochondrial organization when the tunnelling of energy is described in Chapter 5, and see also Fig. 3.8.

The time-dependent organization of these organelles is related to the local light intensity (thylakoids) and oxygen concentration (mitochondria); both factors cause the organelles to be energized and it is this energization that generates the lateral positioning of the different proteins, as well as that of the ATP-synthetase. Note that here it is not radial filaments which cause organization but interactions between proteins in membranes. Effectively these

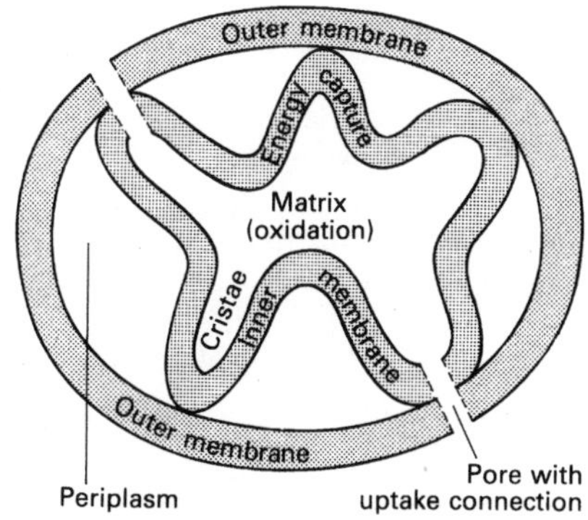

Fig. 3.11 The mitochondria have two membranes. Many of their proteins are made outside the cell cytoplasm and then are placed in different compartments of the organelle. The membranes of the organelle have shape and different proteins are also positioned in lateral organization. [See Frey and Mannella in the references for a more detailed description, and Fig. 3.2(b).] The distribution of charge is also differential both across and along membranes.

changes are massive invaginations of the inner membrane and similar invaginations are often seen in prokaryote cytoplasmic membranes. This lateral organization occurs together with vertical (transmembrane) movements of ions locally, so that ion gradients occur *across* and *along* membranes. As referred to before, the cations include Na^+, H^+, Mg^{2+} and the anions include OH^-, Cl^-, and HPO_4^{2-}; their lateral gradients cause ionic currents along membrane surfaces. Similar local eddy currents are well-known on nerve surfaces, but are probably common to all cells and organelle surfaces once energized. We discuss these features in more detail in Chapter 5.

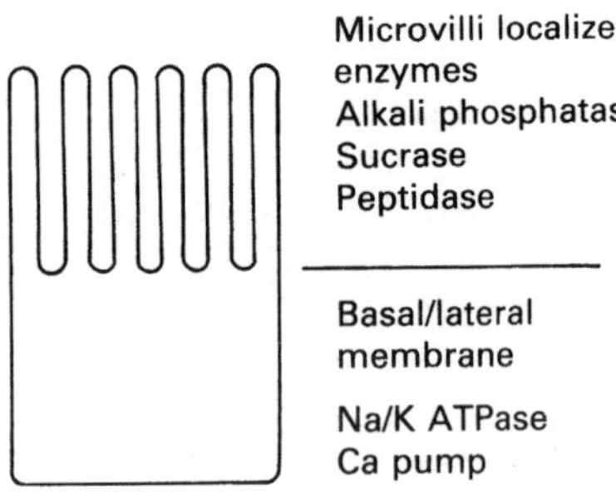

Fig. 3.12 The distributions of some proteins in highly-shaped epithelial cells.

Another and very different example of the differential placing of elements in proteins in a cell with a particular shape is illustrated in Fig. 3.12. There we show an epithelial cell. The bottom of the cell, the basolateral membrane, has one set of pumps for moving ions into and across the cell while the finger-like protrusions at the top have quite a different complement of pumps moving material in or out of this region. Flow from top to bottom of the cell is inevitable and no doubt there are special internal carrier systems to assist the transport. It is particularly important to observe the distinction between invaginations (Fig. 3.11) and extensive villi (Fig. 3.12). The shape of the epithelial cell is decided by long filaments inside the villi.

We next need to examine the lateral membrane organization in relation to the extremely anisotropic distribution of the large particles and the matrix of filaments inside and outside the cell. The interaction between the two is inevitable since energy production (ATP) is laterally different and it is ATP which energizes fibres. The tension in the filamentous structure is then connected to cell and vesicle morphology or shape and the membrane is a pattern of differentiated activity. External to the cell there will therefore also be energized organization. Quite generally, the outside of a cell becomes a pattern of energy flow of ions and tensions in filaments and this pattern could well be part of the morphogenic field, i.e. the force which decides the interrelationship of cells to one another. Of course, transmembrane currents are also related to those fields.

We have stressed in this section the organization of proteins, and elements in proteins, in membranes, but we shall see in Chapter 6 that the transverse and lateral distributions of lipids and polysaccharides are equally important.

3.5.3 Transmembrane organization and flow

Transmembrane organization is immediately apparent in cells from the fact that Na^+ and Ca^{2+} are largely pumped out of cells or into vesicles, whereas K^+ and Mg^{2+} are retained inside, so that, in a steady state, the Ca^{2+} gradient across the cytoplasmic and many vesicle membranes is about 10 000. Examples of vesicles that can hold high concentrations of Ca^{2+} are the sarcoplasmic reticulum of muscles (in fact reticula, generally), the calcisomes, and the various synaptic vesicles of nerves; it is frequently the case that these types of vesicles also contain high concentrations of calcium-binding proteins, e.g. calsequestrin and chromagranin A. When a local electrical pulse (Na^+/K^+) reaches a 'gate' for calcium, which corresponds to the opening of a hole in a membrane, this ion flows suddenly into the cell and triggers a predetermined reaction. Subsequently, it must be removed by pumping through the same or, more frequently, another hole; hence a current of Ca^{2+} always flows along a path in a cell, not just in and out of it, and the size of this current is time-dependent. This activity is localized by the position of the channels and pumps in the membranes. While all the simple cations have gradients across

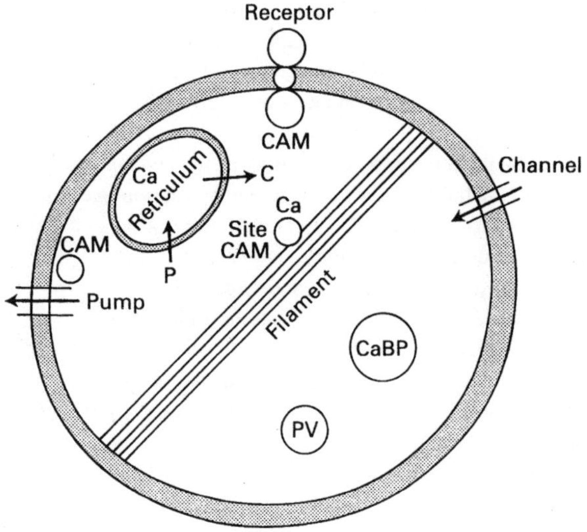

Fig. 3.13 The ways in which calcium distribution is affected and the ways in which its distribution affects cell organization. Calcium is pulsed into the cytoplasm by chemical or electrical stimulation through channels (C). It is distributed or retained by proteins, parvalbumin (PV) and calbindin (CaBP). It affects the cell through action on enzymes and filaments via calmodulins (CAM) and is removed by pumps (P). The disposition of many if not all these units is organized.

cells so do many anions but while chloride and sulphate are concentrated outside cells phosphate is usually concentrated inside.

These many ion gradients, localized or generally across cell membranes, imply that there are transmembrane asymmetries of large molecules, some of which will be partly dependent upon the gradients themselves. In Fig. 3.13 we show a receptor on the outside connected to an internal calcium-binding protein (CAM), a directional pump, attachment points for filaments, and a locally placed vesicle. All these features are related to transmembrane differences in protein constructs, in lipid distribution, and in polysaccharide positioning. In particular virtually all polysaccharides are on the outside of membranes. In Chapter 6 we shall discuss these distributions in more detail.

In parallel with the protein systems in membranes for the movement of calcium are particles for the transport of many other species and elements. The case of hydrogen (and other non-metals) is different from those of the alkali or the alkaline-earth ions; in its flow across membranes hydrogen can move as an ion or in the covalently bound form and, as such, it is pumped in many cells. Examples of hydrogen-containing covalent compounds pumped into cells are various simple sugars, carboxylic acids, or the coenzyme NADH. Inside some compartments it can be oxidized or ionized to the proton and so it may exist at a given free ionic level (as H_3O^+), different in each compartment, as is easily recognized by pH measurements. It also exists in many bound states which *do not* equilibrate with the ionic state (e.g. NADH) and this can be recognized by measuring the effective redox potential by a redox electrode or a reducible dye. Notice that the fact that hydrogen is common in many organic molecules where it is covalently bound must not be allowed to hide from us that these are HL compounds comparable with ML compounds.

To finalize these comments it is worthwhile to point out that, while many elements are moved about in proteins and others are transferred by the pumping of simple cations or anions, there is also the possibility of concentrating given elements in particular parts of the space by the movement of small ion chelates. Major examples are those of Fe(citrate), Fe(haem), Fe(hydroxamic acids), Co(vitamin B_{12}). These metal ion 'carriers' are similar to coenzymes in many respects and there must be receptors that can recognize such molecules specifically and then transport them across membranes in a time-dependent process. Two or more forms of one element then act as different kinetic entities even in the same pool, e.g. haem iron acts separately from free Fe^{2+} (see Table 7.4). Thus, the movement of some elements can take place in different forms; iron, for instance, enters cells as an iron complex, is processed by insertion catalysts in mitochondria, and much of it returns to the cytoplasm as iron-porphyrin. Similarly, phosphorus enters as free phosphate, is processed by a capture catalyst, ATP-synthetase, and remains as ATP. Both of these activities still leave the mitochondria with a quite high concentration of the free elemental forms—iron as a cation and phosphate as an anion. Other elements, such as calcium and manganese, are pumped into the mitochondria out of the cytoplasm so they too are considerably higher in this compartment, rather than in the cytoplasm. Although it is not possible, yet, to determine accurately the concentrations of *free* Cu, Fe, and Mn, the redox elements, in the cytoplasm of eukaryotes, it is likely that they will be present in low concentration to protect the exposed DNA from oxidative mutation. This problem does not arise with zinc and this is one of the reasons why this element is used almost exclusively in cytoplasmic acid-base catalysts.

A special case which is worth mentioning here is that of vanadium which is accumulated in the vacuoles of specialized cells in the blood of certain tunicates; recent results show that vanadium enters the cell as vanadate through the phosphate channel and is reduced to VO^{2+} and V^{3+} apparently by a set of pigments related to catechol called tunichromes. In these cationic states vanadium cannot leave the cells and accumulates there, apparently against the concentration gradient, see Chapter 17.

Finally, note again that the production of all proteins to be sent to any compartment is managed by feed-back controls acting on the genetic apparatus (see Chapter 6).

3.5.4 Flow of vesicles in cells

Whereas the compartmental handling of the elements Na, K, Mg, Ca, Cl, S, P, C, N, H, and, probably, Mo and Si, in advanced cells, and of all these elements plus Fe and Mn in prokaryotes and organelles, such as mitochondria, is often through the use of directed pumps (either for the element in some simple ionic state or as a small molecule, such as Fe-citrate or ATP), another device for the handling of the transition elements in higher eukaryotic cells is through the use of specialized complexing proteins transported in vesicles. Three possibilities can be considered for the flow of vesicles: one from the outer membrane of a cell to the large endosomal vesicular compartment (Fig. 3.4); a second from the Golgi membrane to the outer membrane (Fig. 3.14); and a third of flow along the tubulin structures running inside a cell (Fig. 3.15). In the first case, the outer membrane, which has receptors incorporated in it—the so-called 'coated pits'—can pick up certain metalloproteins and then fold around the receptor/metalloprotein pair. On closing off the membrane it phagocytoses the coated pit (see Fig. 3.4). A new compartment is thereby formed, which flows to a large endosomal vesicular compartment where the membrane becomes divested from the metalloprotein and reforms the vesicle that flows back to the outer membrane. Since a portion of external fluid is also carried in the process, metal ions abundant in that fluid can also be transferred to the endosome, e.g. calcium, but this is not their normal method of transport.

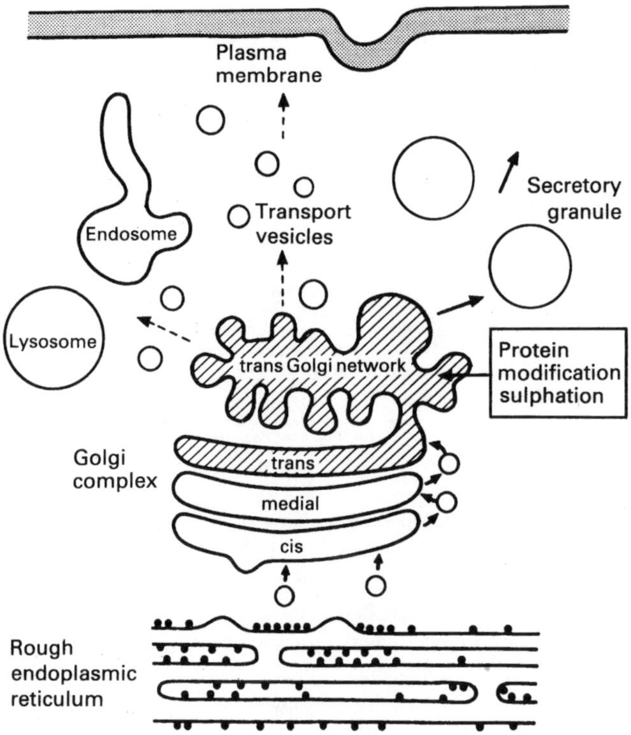

Fig. 3.14 The Golgi and the rough endoplasmic reticulum with extensive variation in protein and particle distributions and a variety of curvatures in the membranes. Some vesicles budding from the Golgi are shown, contrast Fig. 3.4. The positioning of the vesicles is controlled by filaments which are not shown.

The second circulation scheme involves the synthesis of proteins by ribosomes on membranes such that the proteins are excreted into an internal vesicle, e.g. into the Golgi apparatus (Fig. 3.14). The Golgi buds off vesicles continuously. These vesicles flow to the outer membrane, combine with it, and reject the proteins which are to be used in the external fluids of the organism, e.g. the lymph or bloodstream. They can also be discharged into *internal* compartments, as in the case of lysosomes, peroxisomes, etc. One should recall that much of what these internal compartments enclose is from 'outer' spaces, i.e. regions that are in many respects similar to the surroundings of the cells. The exported proteins may be designed to pick up a certain metal and then become the agonists of the receptors, to be again captured by a cell. Examples of this behaviour are those of transferrin (uptake of iron and manganese) and of the albumins (uptake of zinc and perhaps copper), both with known receptors, and that of the cobalamin carrier protein (uptake of cobalt), for which a receptor has been postulated.

Axonal transport along tubulin fibres is another device for the movement of vesicles. By careful time-lapse photography it has been shown that there is a continuous two-way traffic of different types of vesicles to the end and back of long cellular fingers (Fig. 3.15). It is considered that acetylcholine and adrenalin vesicles in nerve termini are moved in this way. (Interestingly, in the brain there are many nerve endings with vesicles which contain zinc at quite high levels. These zinc vesicle are very noticeable at the ends of the mossy fibres of the hippocampus.) This flow of vesicles along tubulins may well be the mode of transfer of silica, or other minerals, to the growth points of hard structures.

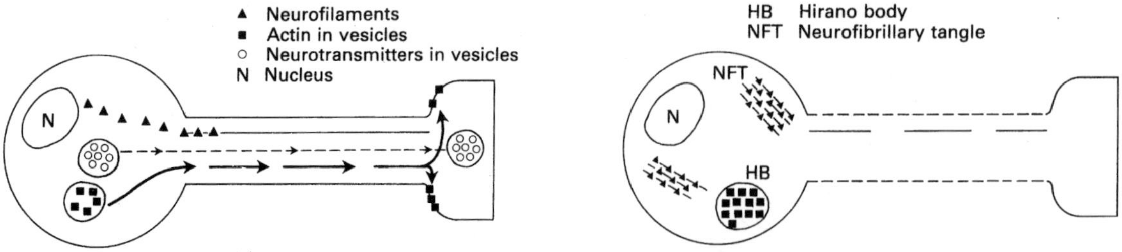

(a) Neurofilaments are produced in the nucleus and incorporated into the axon to form part of the vesicle transport system. Actin and the neurotransmitters are carried on this system in vesicles to the synapse and the growth cone where the actin is incorporated and the neurotransmitters are released.

(b) In the Alzheimer neuron, the neurofilaments have formed insoluble tangles around the nucleus. Axonal transport is stopped leading to neurotransmitted deficiencies at the synapse. Actin is no longer transported and collects in vesicles around the nucleus to form Hirano bodies.

Fig. 3.15 A comparison of the transport processes taking place in (a) a normal and (b) a diseased neuron illustrating vesicle incorporation and motion.

Failure of the transport of vesicles along tubulin filaments leads to an aberrant organization and the cell fails. This may be the case in Alzheimer's disease (Fig. 3.15) which has been associated with the incorrect distribution of elements such as Al, Si, and Ca, but is known to be complex.

Since the flow of protein-containing vesicles is a general process for the handling of the transition elements in their transport between compartments of higher eukaryotic cells, including the uptake and rejection of some of them, it is essential to recognize that the production of *receptor* proteins and of *export* proteins controls the spatial distribution of these elements. Since different cells produce different proteins, the pattern of cellular uptake of elements must be organ-specific; for example, liver absorbs much iron, but white muscle does not. It is equally important to realize that there is a time dependence of protein production; hence the uptake of transition metals can also be conditioned by need on certain occasions. Receptors for iron, for example, increase when cell division is stimulated and cancer cells are rich in such receptors for much longer periods of time than normal cells, a fact that suggests interesting lines of intervention to control their proliferation.

Only in passing do we have space to refer to the obvious flow of many cells from place to place in, for example, the bloodstream.

3.6 **Special solution conditions in vesicles**

As to the nature of the internal vesicle compartments three generalizations seem to apply: they are more acidic (pH <6); they are more oxidative; and they have a higher calcium content. The acidity arises from the fact that proton ATP-ases (pumps) are located in the membranes in such a way that they generate acid in the vesicle, and it may well be that these ATP-ases are associated with the 'coated pits'. [It is known (see Section 3.5.2) that proton ATP-ases accumulate in curved membrane regions of thylakoids and it is possible that they are responsible for the generation of the curved surfaces in these organelles and in the coated pits, favouring the formation of vesicles and their flow to and from the cytoplasm.]

The high redox potential arises from the absence of reducing materials and their enzymes; in lysosomes and peroxisomes, for instance, iron is held as Fe(III), not Fe(II), and manganese can be raised to the Mn(III) state. The Golgi apparatus retains sulphur as sulphate in contrast with the sulphide of the cytoplasm.

The high calcium content probably arises from the enclosing of the external fluids, but also from the activity of calcium ATP-ases which pump this element out of cells and into vesicles, since the outer membrane of the cells inverts on formation of a vesicle which then behaves as an inverted cell. This is the reason why the internal fluids of vesicles resemble the external fluids of the cell concerned.

The importance of the acidity and of the redox potential of the vesicle phases must not be missed; we have repeatedly mentioned that these factors can control the availability of certain ligands and their form; hence they introduce a limitation on the type of enzymes that can function in them. Zinc and indeed most *divalent* transition metal-protein complexes lose stability around pH 5 to 6, so that they are of no value in vesicles such as lysosomes, for example. The *trivalent* metal protein complexes can be made stable at a much lower pH and may be used in such conditions. Acid phosphatases in lysosomes, for instance, are Mn(III) and Fe(III) proteins, probably formed in these vesicles by apoproteins exported by the Golgi system.

It is also the acidity that releases the metals from their carrying proteins and releases the carrier from the receptor. The carrier returns to the outside of the cells, but the metal stays inside. An example is that of iron which is transported externally as transferrin, captured, and transported inside cells by a specific carrier and then used or stored, when in excess, on a special storage protein, ferritin, to which we have already referred. It is the escape of ferritin which leads to unwanted iron deposition external to cells (see Fig. 12.4).

The redox potential is also of the utmost consequence, as stressed before, and we will only note here that iron proteins are not very useful oxidases in external solutions or inside vesicles since the dissociation of iron caused by low pH is made irreversible by the oxidation of Fe(II) to free Fe(III). It is, therefore, the copper proteins which are of greatest value as redox enzymes *outside* cells and inside vesicles, since neither Cu(I) or Cu(II) will exchange from metalloproteins due to the high intrinsic stability of their complexes.

The further aspect of vesicles is that they allow the formation of inorganic deposits in cell compartments. Indeed, gross deposition of certain elements in certain specialized vesicles can lead to the formation of inorganic mineral phases, especially of Fe, Si, Ca, and P. These precipitates can be amorphous or crystalline, but may be given precise form ('pellets') by the proteins in the vesicles and, when exported, may be packed in regular solid structures, a process again favoured by other proteins that assist deposition of those solids. Bone, teeth, and shells are formed this way, as well as other solid structures, such as the gravity and magnetic sensors used by some species of animals. Curiously, this process is prone to loss of control in higher organisms, particularly in old age, so that the laying down of both iron and calcium minerals in an incorrect manner, especially in humans, is a major cause of ill health. Atherosclerosis, cataracts, haemosiderosis, etc. are all diseases related to incorrect movements of minerals, e.g. deposition in veins and arteries. Even Alzheimer's disease is associated with mineral deposits, although this appears to be a secondary event. Biological systems operate very close to saturation limits of many salts! The actual limits depend on such factors as the pH. The extracellular fluids of plants are at a low pH so that there is no precipitation of carbonates or phosphates but silica and oxalates can be precipitated.

3.7 Cooperativity of separations and localizations

We have seen that cell shape is connected to tubulin filament growth and we know that the shape will lead to segregation of units in membranes. If these two processes, tubulin growth and segregation, share a common energization device, then they become cooperative (or anticooperative). The positioning of vesicles is constrained by the filamentous, tubulin, structure of a cell; they move along the filaments, sometimes budding from outer membranes, sometimes coalescing with them. Clearly, the fate of vesicles is associated with the state of fibrils and outer membranes, and some coordination of activity is obviously required. It is highly probable that this coordination is dependent on the states of calcium and of phosphate (see Chapter 20).

3.8 Summary of metal ion positioning

We see now that the *position* of a metal ion, including its concentration gradient, in bound and unbound forms is controlled and must be absolutely critical to an understanding of its function in biology, and we return to this theme in Chapter 7. The picture is likely to be confusing until the distribution is established more firmly. To give those readers who have little acquaintance with the level of organizational complexity of biology a feeling for the situation we provide in Fig. 3.16 an indication of some of the zones into which different elements are placed. Table 3.2 is an illustration of how a variety of elements can occur in external or vesicle compartments of multicellular organisms.

We turn, finally, to a brief description of multicellular organization.

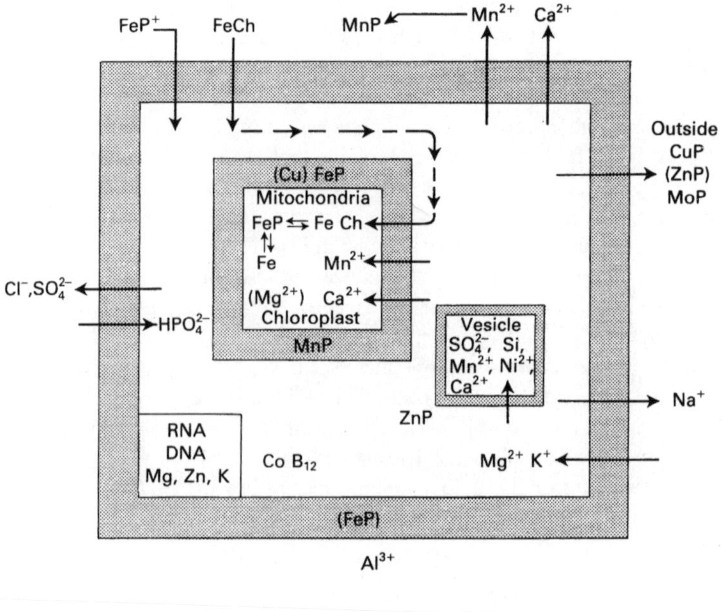

Fig. 3.16 A picture of some of the distributions of elements in eukaryotic cells. Iron is often in membranes while Cl^-, Ca^{2+}, and Na^+ are concentrated away from the cytoplasm; K^+ and Mg^{2+} concentrate inside cells, Mn^{2+} is to be found in vesicles, but copper is more usual outside cells. Zinc is everywhere and cobalt is rare everywhere. Aluminium may be rejected. P = a protein; Ch = chelating agent. Compare Fig. 2.16.

Table 3.2 Proteins (apoproteins) for external retention of metals

Caeruloplasmin	Cu	Oxidase in blood
Haemocyanin	Cu	Oxygen carrier in blood
Alkaline phosphatase	Zn	External surface of membranes
Xanthine oxidase	Mo	Oxidase in milk
Acid phosphatase	Fe, Mn	In lysosomes
Catalase	Fe	In peroxisomes
Hydroxylase (adrenal)	Cu	In chromaffin granule vesicles
Plastocyanin	Cu	In thylakoids
Urease	Ni	In plant vacuoles

3.9 Symbiosis and multicellular systems

The stress on the divisions within single cells must not be allowed to hide the differences in cells which are part of the same organism or are locked in symbiotic union. The most obvious peculiarity is the positioning of the nitrogen fixation system and, therefore, of much molybdenum, in root nodule bacteria while there is little or none of this activity or of molybdenum in roots. Symbiosis is very common (it may even be one of the main sources of evolution) and there are a multitude of ways of dividing activities and elements which parallel the division between mitochondria, chloroplasts, and cell cytoplasm. In fact, mitochondria and chloroplasts probably resulted from the capture by eukaryotes of prokarytes.

The organelles get most of their proteins from the cytoplasm but they make, using their own DNA, several essential proteins for themselves. They also take part in pathways such as those of the urea cycle, haem, and Fe_n S_n syntheses and the synthesis of one or two amino-acids (see Margulis in the references).

The creation of cells with different purposes in a single organism implies that element distribution is unique to a cell type and this is known to be the case. The most obvious is the distinction between red blood cells (iron mostly) and white cells (more zinc and little iron). We shall not discuss the problem of the distribution of trace elements here but will return to it in Chapters 7 to 19.

3.10 Spatial distribution of bulk non-metals

The spatial distribution of non-metals in some simple forms such as H_2O, CO_2, N_2, and O_2 is not limited by membranes, by binding, or by charge. In essence, these inert molecules form the liquid and gaseous atmospheres in which biology has developed. Here we are hardly concerned with these forms of the non-metals since they are not fixed in space; neither is it interesting to look at H, C, N, O in biopolymers since they are combined everywhere. Spatial distribution of H, C, N, and O becomes interesting when and where their small inert molecules are converted to reactive forms, i.e. where they are activated. It is now well known that N_2 goes immediately to NH_3, H_2O goes to bound hydride (NADPH), bound O atoms, and O_2 which is lost to the atmosphere, CO_2 goes to HCHO on its way down to $-CH_3$, etc. Once in reactive form, it becomes essential to move the non-metals about in such a way that they only react when and where they are required. In part this is managed by handling these elements in kinetically trapped covalent molecules, e.g. C/H in the carboxylic acids of the citric acid cycle (mitochondria) or as sugars in the Golgi, or the sugar phosphates of glycolysis (cytoplasm), N in amino acids or nucleotide bases (cytoplasm), etc. which are limited in their movements by membranes. In addition their distribution is managed, while still holding the elements in very water soluble forms so that they diffuse in any one aqueous compartment, e.g. C/H in sugars, by energizing their transport between compartments. This transport is observed to be by pumps or exchangers which move them through membranes, and it is very specific. For these reasons we become more concerned with the localization and transformation of specific compounds where the elements are incorporated into kinetically inert or somewhat labile forms as, for example, in C-H, rather than with the localization of elements H, C, N, or O as non-metals. This means that, for example, carboxylates of the Krebs cycle, amino acids, and phosphates of glycolysis, although they all contain C and H, often belong to quite separate C and H pools either in space and/or in kinetic connection. This is well known in metabolic and synthetic pathways of the biological systems, of course. Yet, despite this separateness of the metabolic pathways, they must all relate to one another for several reasons: first, all pathways have come from CO_2, H_2O, N_2—a few small primary molecules; second, they will all be degraded to these same molecules in an orderly manner so that energy and material are not wasted; third, the metabolic pools cannot be independent since amounts of proteins, fats, polynucleic acids, etc. are required in rather strictly controlled ratios in order to build an organism. The demand is then for an intercommunicating network based on additional kinetically controlled units which interact with the metabolic pathways distributors and controls, i.e. small basic carriers of the pools of –H, $–CH_3$, $–NH_2$, $–COCH_3$, $–OPO_3^{2-}$, etc. Naturally, these carriers can only distribute the basic units properly if they also act as (allosteric) controls over the networks of metabolic pathways. The essential carriers of the non-metals are the mobile coenzymes (see Section 2.17), and it is distribution of these carriers, frequently related to vitamins, to which we

Table 3.3 Transport and reaction systems for non-metals

Transported units	Mobile coenzyme (immobile)
H_2/reducing equivalents	NADH, NADPH, (flavin), OH_2, glutathione
CH_3–	Coenzyme M, S-adenosyl methionine (B_{12} coenzyme)
CO_2	(Biotin)
CH_3CO–	Coenzyme-A (lipoic acid) (F-430, Ni)
CH_3CHO	(Thiamin)
SO_4^{2-}	PAPS
PO_4^{3-}	NTP, ATP, etc. P_2, acyl phosphate
Sugars	UDP
NH_3	Glutamine (pyridoxal phosphate)

now turn. While doing this we must note which coenzymes are in a particular compartment and which go between compartments. Some coenzymes are not mobile (Table 3.3).

3.11 The spatial transfer of H, C, N, S: mobile and fixed coenzymes

3.11.1 H (and O) transport

There is no intracellular carrier for elemental oxygen, O, and it may well be that, since water is everywhere and dioxygen is a highly energized, mobile, species, extracellular distribution to cells only is necessary in multicellular organisms, being managed by a variety of dioxygen carriers, and the movement in cells is not controlled. The distribution of oxygen atoms after the element has been incorporated in specific localities by oxidases is then in combined form with carbon, for example in acetate, and oxygen atoms themselves are not transferred from one carbon to another. Even -OH as such does not have a carrier though it can be rearranged on carbon skeletons by isomerases. The distribution of H itself is managed by a series of carriers at different redox potentials which do not equilibrate directly with one another. At the low potential extreme (–0.5 V) there are the carriers NADH and NADPH which pick up H from organic molecules, such as malate, in aqueous phases. They transfer H, site selectively, to a variety of acceptors for synthetic purposes (NADPH) or to the membrane of the degradative respiratory cytochrome chain for energy production (NADH). Once the reducing equivalents enter the cytochrome chain, they remain membrane-bound or -associated until they react with O_2 to give H_2O, or perhaps in some systems with sulphur to give H_2S. While in the membrane two varieties of coenzyme handle H: fixed site reactivity is generated by flavins (see Fig. 3.1) and mobile transport of H is managed in the membrane phase by quinones, QH_2. Before passing the reducing H to dioxygen or sulphur, energy is retained in proton gradients (see Chapter 5). Remember that here we wish to observe the location and timed movement of H as an element, i.e. H^+, H^-, and $H^\bullet$ (including covalently held H).

Now, the H–transfer systems at –0.5 V (NADH, NADPH) in water are quite separate, in a kinetic sense, not only from the quinone QH_2 pools in membranes (0.0 V) but also from the external and internal redox couples which govern the equilibration of hydrogen reactions at somewhat higher potentials. Outside cells ascorbate, for example, is involved in maintaining a redox system with a potential above 0.0 V, while in the cytoplasm of cells the glutathione couple transports H at close to the same potential, 0.0 V. Because the couples NADPH/NADP, NADH/NAD, QH_2/Q, and 2RSH/R-S-S-R do not intercommunicate directly there is not a redox state of the cytoplasm in a cell, but at least four pairs of redox potential buffers acting on distinctly opposite series of different metabolic pools. The separation is maintained despite the fact that three of these pools lie in the same aqueous phase. The fact that there is no direct intercommunication between pools allows the energy of different C-H bonds to be used without equilibration and loss of energy as heat, and it allows specific networks of synthesis and degradation. A general transport of H between all systems would have been disastrous. The movement of H is then not just a matter of catalysis of substrates but of selective transport in selected pathways. The positioning of systems for hydrogen reactions in vesicles, on membranes and on filaments is clearly critical. (*All this activity is critically dependent upon the correct location of inorganic catalytic ions too.*)

Now, we have described the simplest form of transfer of H and before going forward to look at carbon, phosphorus, and nitrogen transfer we refer to H transfer in groups very briefly. Just as acetyl, CH_3CO, is an effective way of transferring O, it is also an effective way of transferring H (and C). Looking at the coenzymes in Table 3.3 we note that there are other coenzymes which transfer H with carbon, as $-CH_3$ for example, and this building of organic molecules from fragments in different pathways is central to metabolism.

3.11.2 Phosphate transport

Unlike O, H, and C, phosphate itself cannot gain direct access to cells and in this respect it is to be likened to inorganic cations. However, once it has been transferred across membranes by pumps it becomes locked into many kinetically stable, although not very stable, inorganic and organic compounds but it does not undergo redox state change. Note also that phosphate is not only transferred in small fragments but in larger molecular units, e.g. nucleotides (see Chapter 18).

The situation may next be compared with that of the hydrogen pools in the cytoplasm. The first step of synthesis involving phosphate is of ATP and then of GTP, UTP, CTP and other 'high' energy forms of phosphate from ATP. The exchange of energized phosphate is controlled such that each pool communicates to some extent with a separate reaction path, but also more directly to one another than the different forms of bound H. The networks of phosphorylation can then be separately blocked, although the pool of phosphate 'energy' is interlocked. This will be seen to become of consequence in later chapters.

The transfer of phosphate in different metabolic pools is also managed by a whole range of specific enzymes which recognize specific molecular species as is true for C, H, O, and N. Phosphorylated compounds apart from nucleotides are seldom moved across membranes.

3.11.3 Carbon transport

Carbon fragments, C_1 and C_2, are transported by very different sets of coenzymes which diffuse in the cytoplasm or operate at fixed sites. Thus, coenzyme A, methylmethionine CoA, vitamin B_{12}, and lipoic acid, all transport carbon. Notice that the carrier for acetyl-groups, CoA, is not a carrier for methyl groups, which methionyl CoA is. Methylation and acetylation must not be confused; they are different ways of moving C/H/O in combination. The interested chemist should take a biochemist's wall chart of metabolism and follow the way in which, say, C_1 fragments are introduced from CO_2, follow the path to $-CH_3$ groups, and then follow their subsequent distribution. The carbon has then to go into C_2, C_3, etc. units. C_1 incorporation into CH_3CO- fragments is an essential step. (Note in this connection the special pathways using permanently located Co and Ni coenzymes in archaebacteria (Chapter 16.)

The further incorporation of larger carbon units is managed differently for the individual chemicals. Fatty acid fragments are built up from acetyl CoA by successive C_2 (acetyl) additions which are retained by CoA. The control of the movement of units is very sophisticated. In particular one part of CoA, pantothenic acid, is suitable for swinging between a CH_3CO- uptake point and a delivery point during the addition of acetyl to a growing $(CH_2)_n$ chain. The end product is the synthesis of lipids. Other forms of movement of carbon include the varieties of carriers for fragments such as amino acids, saccharides, isopentyl groups and so on, and some of these systems are in special compartments, e.g. the Golgi apparatus for polysaccharide synthesis.

3.11.4 Nitrogen transport

Nitrogen is transported mainly as glutamine, which is the precursor of the amine group in most amino acids, and as pyridoxal. The amino acids are transported by tRNAs to the ribosomes for protein synthesis. There is a quite separate transfer of $-NH_2$ in the formation of lysine; in this case the coenzyme is pyridoxal. All this transport is confined to the cytoplasm.

Nitrogen is not incorporated in biological polymers outside the cytoplasm, unlike carbon or sulphur, both of which are also moved around in the Golgi apparatus.

3.11.5 Sulphur transport

Sulphur is transferred as sulphate using the coenzyme PAPS, see aside. The transfer is confined to the Golgi apparatus where sulphation occurs. Now, while sulphate is a dominant form of anionic sulphur outside cells, inside cells it is present largely as sulphide. Thus, the catalysts for sulphate reduction must be retained outside the cell cytoplasm. The transfer of sulphide is not understood yet but it may involve small Fe/S clusters that are mobile (see Chapters 12 and 19).

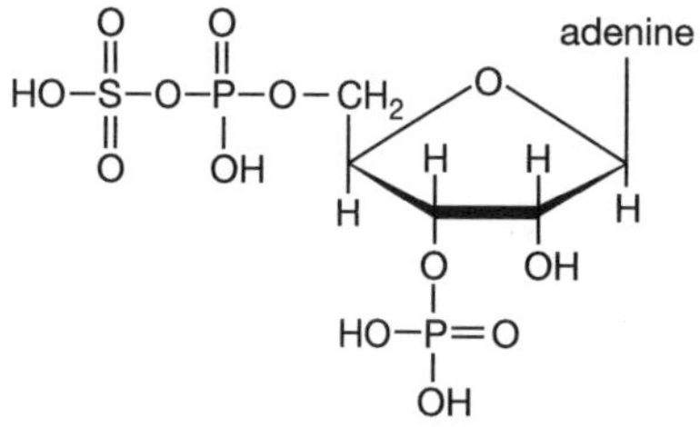

PAPS: phosphoro-adenosine-phosphosulphate

3.12 Summary of non-metal transport

As we have seen, the essential difference between metals and non-metals is increased kinetic inertness so that for non-metals chemical pathways are readily separated by placing catalysts in addition to using specific pumps. Many coenzymes are specially selected organic molecules which carry the units of H^-, HPO_4^{2-}, $-CH_3$, CH_3CO-, etc. in an activated condition so that later catalytic transfer is easy. Examples are ATP(HPO_4^{2-}), PAPS (SO_4^{2-}), and $-S-(COCH_3)$ in acetyl coenzyme A. However, there are also coenzymes that are located at fixed points in proteins, in contrast with those which are free and can traverse water or membranes on carriers; this is the case for ATP, which is required to enter or leave a large number of vesicle and organelle systems. There are also carriers for several metal ions and cofactors.

We conclude that biological non-metals, like the biological metals, may well be compartmentalized in very highly organized distribution networks. The fact that locally molecules may diffuse randomly in one phase does not mean that they are not following spatial paths out of the phase and into a designated second one. One general restriction is the membrane and another is the existence of points of reaction or transfer. The facts that, in membranes, pumps can be positioned specifically and, in aqueous phases, catalysts can be located on networks of filaments, means that distribution can be controlled. While it is easy to follow elements such as Ca^{2+}, which does not alter its chemical form as an ion, it is more difficult to follow carbon, or nitrogen, or even iron, because of their many possible oxidation states in kinetically controlled conditions. Even so, there are many kinds of devices controlling transfer to local organelles, reticula, and vesicles. We have seen that the vesicles themselves are transported in given ways too. The systematic movement of transmitter molecules to the ends of nerve fibres is not different from the movement and exocytosis of silica rods to form the *lorica* outside the cell of some protozoa. We shall return again and again to the dynamic organization of elements in biology.

While we have stressed the network involving the small organic molecules we must not fail to see that of necessity it involves the correct balance of inorganic elements and of proteins. The network of inorganic elements, which we have called the metallome, lies in the overall fixed ionic strength, free metal ions, metal ions bound to organic molecules, metallo-enzyme catalysts and metal ion transcription factors. Very clearly the whole of the cell works as a unity including the network of compartments. It is essential that this insight is not lost while we deal with individual reactions. Because of the intensive feedback there is organization which has been seen in DNA sequences, in proteins produced, in element uptake and in spatial distribution of compartments. These are all in communication. Therefore, while DNA and RNA analyses give us one way to look at organisms, their classification and their evolution, and protein or element composition give us others, compartmentation has given us the earliest and simplest method in the classifications of shape and form. It also allows a simple approach to evolution. All must be seen as aspects of chemical networking in which none is dominant. The network is not static, of course; it involves flow of two kinds. The one described in this chapter is transport between the environment and cellular compartments and back again. The second to be described in the next chapter, is a flow within chemical structures, i.e. patterns of transformation within chemical series, e.g. from $CO_2 + H_2O$ to a special variety of sugars and then back to $CO_2 + H_2O$.

3.13 Evolution of compartments and organization: summary

An understanding of biological inorganic chemistry depends not only upon the knowledge of how chemicals were introduced into organisms but also upon how compartments in and outside cells evolved.

The chemical elements in compounds are carefully inserted into different compartments to avoid mutual competition and to allow the development of novel properties. In part the insertions are localized along the membranes

Table 3.4 Membrane lateral local organization

Membrane	Organization
Thylakoid	Separation of photosystems I and II on energization with formation of convoluted membrane structure.
Mitochondria	Development of cristae and making of local inner/outer membrane connections for protein transport on energization. ATP-synthesis.
Osteoclasts	Formation of membrane fingers on contact with bone producing a localized acidified zone.
Epithelial cells	The separation of different enzymes in the differently shaped basolateral and microvilli zones of the plasma membrane.
Surface root cells	The indented membrane of the cells of the surface layer generates an external acid environment.
Embryo	The anisotropic positioning of Na^+/K^+ ATP-ase at early embryo stages allows shape to develop and nervous tissue to form.

(Table 3.4). At first cells of archaea and eubacteria had at most two aqueous compartments: an inner cytoplasm and a periplasm separated by two all encompassing membranes from the environment. These are the prokaryotes. The next development we know of is that of single cell eukaryotes which had a nuclear compartment and several vesicular compartments including organelles supported by filaments (see Fig. 3.5). Now it is not clear in what order these compartments appeared, so that single-cell eukaryotes cover a great variety of organisms. The next step in evolution was the combination of these eukaryote cells into multicellular organisms which includes primitive fungi, plants and animals. Only chemical messages could go long distances between cells. However, a new further development in animals was that of the nervous system which allowed an electrolytic connection between cells. The relationship of the environment as an external unconnected part of space to the internal spaces of the organism began to change until today when animals with brains, especially man, have learnt to use the external environment to make compartments and communication networks of value to themselves. We stress in Chapters 7 and 20 how this evolution of the use of space has been coincidental with the development of the use of the chemical elements, (see Williams and Fraústo da Silva in references).

Further reading

This chapter contains material which is found in many biochemistry textbooks, but without the emphasis on distribution of the elements as in this book. The books which follow are suggested for general reading. While reading these books we ask the reader to note which elements are involved in any process described and where the process occurs in cells.

Armstrong, F.B. (1989). *Biochemistry* (3rd edn). Oxford University Press, Oxford.

Bergethon, P.R. (1998). *The physical basis of biochemistry: the foundations of molecular biophysics*. Springer-Verlag, New York.

Frey, T. G. and Mannella, C.A. (2000). The structure of mitochondria *Trends in biochemical sciences*, **25**, 319–24.

Goodsell, D.S. (1998). *The machinery of life*. Springer-Verlag, New York.

Lodish, H., Baltimore, D., Berk, A., Zipurski, S. L., Matsudaira, P., and Darnell, J. (1995). *Molecular cell biology* (3rd edn). Scientific American Books, New York.

Margulis, L. (1999). *Symbiotic planet*. Basic Books, New York.

Metzler, D.E. (2001). *Biochemistry: the chemical reactions of living cells* (2nd edn). Academic Press, New York.

Stryer, L. (1995). *Biochemistry* (4th edn). W. H. Freeman, New York.

Williams, R.J.P. and Fraústo da Silva, J.J.R. (1999). *Bringing chemistry to life*. Oxford University Press, Oxford.

CHAPTER 4

Kinetic considerations of chemical reactions, catalysis, and control

4.1 Introduction

In Chapters 1 to 3 we have described the chemical elements of life, the principles behind their chemical speciation, and the observed compartmentalization of biological systems. Now, in their special places, the elements are used in a functional sense, e.g. to provide structure and to assist chemical transformation. Before we can look at their functional value in *biology* we need to be acquainted with their *chemical kinetic* potential and we shall do this in two steps. In this chapter we consider the basic rules of chemical transformations, i.e. the kinetic controls and barriers for ion and compound reactions which can be utilized by organisms and see how they are dealt with locally, but we do not wish to involve biological considerations in any depth. Then, together with the basic rules for the thermodynamic and kinetically controlled combination of the elements and for their distribution in compartments, outlined in the previous chapters, we can turn, in Chapter 5, to the possible ways in which energy can be put into biological systems in order to see how, in principle, the full functional potential of the chemical elements in biology, including catalysis, control, signaling, and structure, can be realized in the appropriate place and at particular times. This is not the same problem as that of the ways of developing the functional potential of elements in chemistry, since in chemistry we have a much wider range of variation of solvent, of pressure, of temperature, of phase removal, and so on. In biology there is instead an energized organization in the control over the distribution of elements. Fundamentally, of course, there is no difference from the deployment of elements in compounds by man and his spatial and energetic manipulation of that chemistry by, for example, phase transfer and temperature controls, e.g. in a chemical factory. Biology is just so much more particularized and complex on a microscale since all activities are in the same cell or organism, but, in essence, properly developed biology (and chemistry) seek to ensure that every atom has the correct partners and controls, in the appropriate zone, for function at the right time.

4.2 Chemical transformations

In this book there are only a few basic kinetic aspects of chemistry which we have a need to analyse in relation to biological activity, namely, those referring to:

1. acid-base reactions: hydrolysis;
2. the nature of ion flow: signalling and control, electrolytics;

3. electron-transfer reactions, electronics;
4. atom-transfer reactions (redox reactions);
5. group-transfer reactions;
6. free radical reactions;
7. conformational changes.

These aspects will be considered in the next sections, but some preliminary considerations on catalysis in general are necessary.

First of all, catalysis by itself can only assist a down-hill reaction, so that we shall need to consider how to incorporate or retain energy in a catalysed system. In part this means that the catalyst will not just activate a molecule, but will protect it from adventitious attack. The need for such organized and energized catalysts is seen if we consider the synthesis of any condensation polymer (high-energy reaction in water) as opposed to its degradation, e.g. in digestion. Since the two opposed reactions cannot occur effectively in the same site there must be catalysts for the forward and back reactions which are quite distinct, differently controlled, energized if necessary, and occur in different places (see Aside). The polymerization reactions require energy (e.g. from pyrophosphate) and a series of catalysts, Cat (2), which are placed inside cells. DNA and RNA are synthesized in the nucleus, proteins in ribosomes, and polysaccharides in the Golgi. The opposed degradation reactions need to be very largely prevented in cells. Down-hill protein degradation can be activated in cells in proteasomes by using unfolding agents which need ATP. Degradation is more general outside cells and is accelerated in digestion for example. In each case, catalysts (1) and (2) are enzymes differently placed and stored in space but with one fundamental feature in common: they are acid–base catalysts and not redox catalysts. In acid–base catalysis there is a rearrangement of atoms without changes of oxidation state. A very similar problem concerns the direction of redox reactions involved in both synthesis and degradation (see Section 4.8).

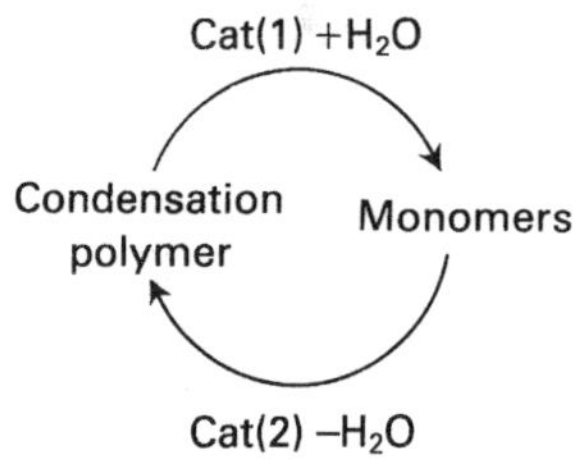

There is one group of biological polymeric molecules which we have not mentioned above—the lipids. The fully reduced carbon chains of biology are generated by condensation followed by reduction

$$CH_3COR + CH_3COOH \rightarrow CH_3COCH_2COR + H_2O,$$
$$CH_3COCH_2COR + 2H_2 \rightarrow CH_3CH_2CH_2COR + H_2O,$$

where R is a retaining centre on an enzyme. Naturally, two different catalysts are required, acid–base in the first reaction and redox in the second. Notice that these and many other biological polymers are formed by removing water, an unfavourable reaction in aqueous solution, as stated.

This very simple introduction to some living processes has shown that in cells two basic reactions must proceed at *controlled* rates and in chosen directions: (1) acid–base; and (2) redox reactions. We need to note the character of these reactions, to know how to control them, i.e. to catalyse and to stop them, and then to see that they go in a specific direction in space, all of which may require energy.

4.3 The nature of acid–base reactions: hydrolysis and condensation

Consider the reversible reaction of ester formation

$$RCOOH + R'OH \rightleftharpoons RCOOR' + H_2O \qquad (4.1)$$

The carbon atoms remain associated with the same numbers of oxygen and hydrogen atoms—there is no change of oxidation state—but the atoms are rearranged. The need for catalysis arises from the stability of the different carbon–hydrogen, oxygen–hydrogen, and carbon-oxygen bonds. To get reaction, the R′OH group must attack the >CO group of the acid, and naïvely we can see that this will be assisted by higher energy states, such as

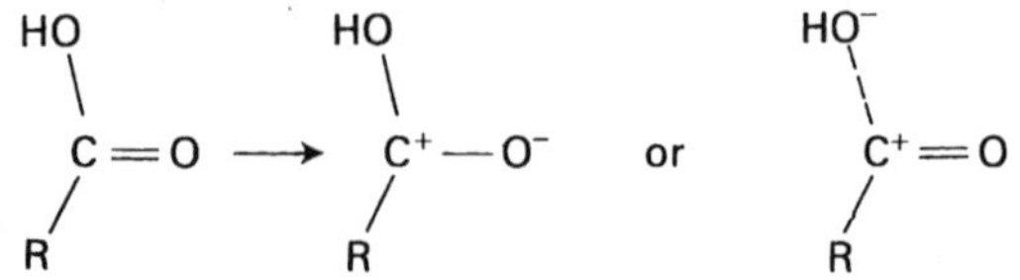
(4.2)

when we can expect that the C^+ is attacked by the negative polarity of the oxygen atom of R'OH to give

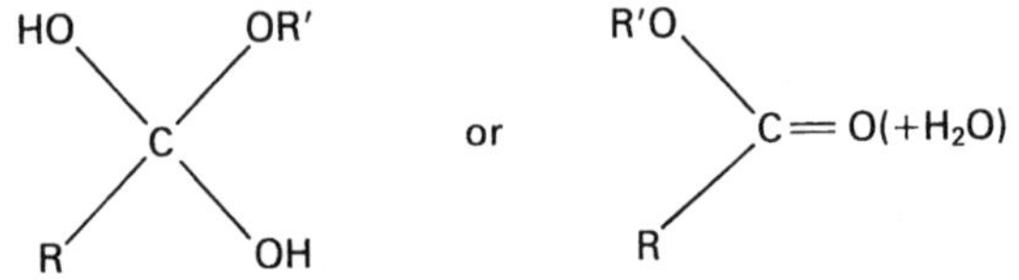
(4.3)

Overall loss of water (condensation) gives the product in both cases. The pattern is the same for all condensations mentioned above. Immediately we observe that catalysts must act by binding a molecule so as to favour vulnerable states (Fig. 4.1), the intermediates, and transition states. One way to achieve this is to induce strain in the substrate simply by deforming bonds, while using binding energy to induce that strain. This induced strain is, in principle, independent of the chemical nature of the binding in that it is just a mechanical effect. To strain a molecule, a large number of matched contacts is useful and surfaces of proteins (large molecules) are ideal for straining *large* substrates. If there is an electrostatic component of the binding, it can be used to help to induce charge separation [see eqn (4.2)] as well as to induce mechanical strain. Such a catalyst is an acid–base catalyst since acid and base centres are distinguished by their affinity for negatively charged atoms (acids are positive centres) or positively charged atoms (bases are negative centres). Clearly, the proton or a metal ion with high negative charge affinity, e.g: Zn^{2+}, will be a good acid, and an anion, such as methoxide or thiolate, will be a good base. In order to control condensation and hydrolysis, biology has developed a series of acid–base catalysts within proteins, i.e. enzymes. We shall see that the useful character of inorganic elements in these catalysts arises especially where the substrate is a *small* molecule or when the attack needs to be indiscriminate and fast. The metal ion is a site of great attacking power which is useful in both synthesis and degradation (Table 7.3).

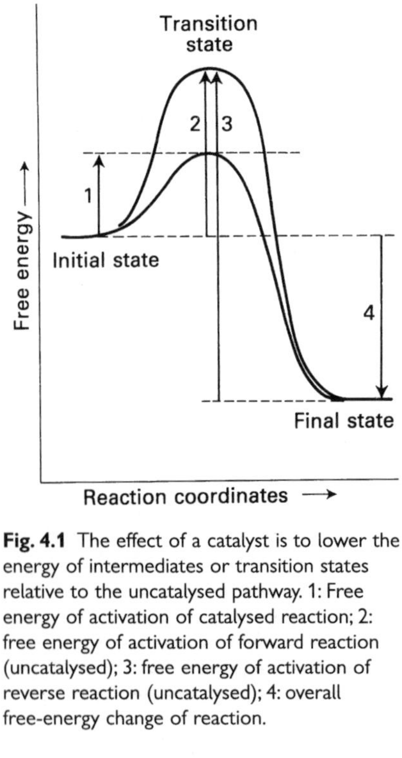

Fig. 4.1 The effect of a catalyst is to lower the energy of intermediates or transition states relative to the uncatalysed pathway. 1: Free energy of activation of catalysed reaction; 2: free energy of activation of forward reaction (uncatalysed); 3: free energy of activation of reverse reaction (uncatalysed); 4: overall free-energy change of reaction.

Now, acid attack depends on the electronic deforming (polarizing) power of an ion as well as on its electrostatic binding (it is permissible to refer to this as partial covalent bond formation). In Chapter 2 we observed that the Irving–Williams series, i.e.

$$(Mg^{2+}, Ca^{2+}) < Mn^{2+} < Fe^{2+} < Co^{2+} < Ni^{2+} < Cu^{2+} > Zn^{2+},$$

was a series of decreasing radius, increasing electrostatic effect, and increasing electron affinity, polarizing effect, or covalence.

This order is the observed order of acid catalysis by these ions in model acid–base reactions (see Fig. 4.2); however, in enzymes, the order, even when metals are available or bound in the same enzyme site, is never this one (Fig. 4.2) and is different in different compartments, due to control over concentrations of these cations, and in different enzymes, due to selective incorporation and control of conformation, as we shall see in the following sections.

Amongst bases, the common attack through –OH, –SH, or $-NH_2$ may give rise to relatively longer-lasting 'covalent' intermediates; the problem then is not just attack but the leaving ability to give products. In both respects proteins are 'designed' to assist the organic catalytic acts. The bases chosen by biology are not always conventional, but are often the conjugate base of a weak proton acid which is especially reactive within the protein (enzyme) matrix; a good example is the –OH group of an alcohol, e.g. serine.

This analysis of acid–base catalysed reactions has avoided the detailed description of the precise way in which a base like RS^- or an acid like Zn^{2+} can polarize an ester bond. The obvious mode is the formation of an intermediate so that in the reaction the base RS^- actually becomes acylated and the acid Zn^{2+} becomes complexed to the attacking anion (OH^- or RO^- above). The required property of the catalyst is that it must be a good attacking group, able to form bonds quickly, and must also be a good leaving group, i.e. it must form unstable bonds. Note that there is an obvious barrier to these reactions (Fig. 4.1), which needs to be overcome; within proteins there are special ways of increasing the attacking power of atoms or groups of atoms, for example zinc and alcohol –OH groups (of serine), which we shall describe in Chapter 6 under the name of the 'entatic state', or the constrained state. These features are now well known in model compounds designed to mimic enzyme sites. We must be careful with such descriptions of the catalysis of single steps, such as 'attack', since many reactions require several steps and it is the overall rate which is of greatest concern while only certain steps need heightened catalytic strength.

Let us seek very simple features of acid–base catalysis. The involvement of the cycling catalyst centre is in forming and breaking substrate bonds to itself so, as mentioned already, it must be a good attacking and a good leaving group. Consider the simplest case of an acid group which is to be hydrolysed by M^+, an attacking and product-leaving catalytic centre:

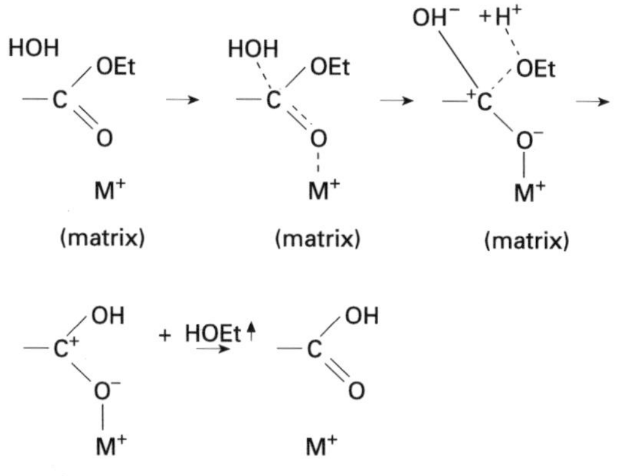

The properties of M^+ are clear: (a) It must be open-sided, accessible to substrate; (b) In the matrix it must be of high electron affinity while the matrix must retain it; (c) It must be made into a good leaving group, enabling it to

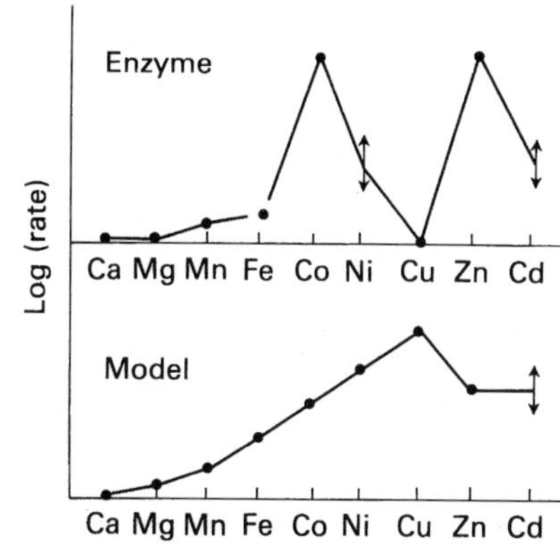

Fig. 4.2 (Lower) the order of hydrolysis of a simple ester by a series of Lewis acids and (upper) the order of hydrolysis by the same Lewis acids substituted in a zinc enzyme, carbonic anhydrase. Obviously, Lewis acid strength is not dominant in enzymes.

switch coordination to substrate or product, and this may involve change of coordination number, while it remains of fixed coordination to the fixed matrix; (d) the exchange of substrates and products must be fast, but loss of M^+ must be avoided; (e) extensive adjustments of ligand matrix links to the metal are also to be avoided since they would require energy.

For a multi-step reaction it is unlikely that any given constrained state of the catalytic atom is the best for a particular step since the best state for each step has to have somewhat easy flexibility to allow readjustment of ligand bonds during the reaction pathway to the next step. Many models have shown this to be true. A good example are the multistep reactions using terpyridyl complexes of palladium (see Groen et al. in the references).

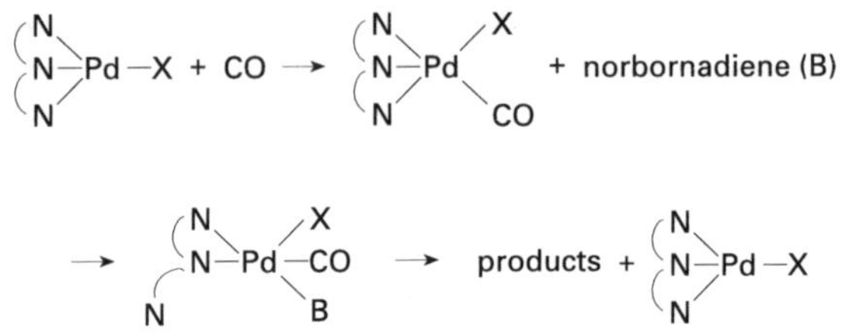

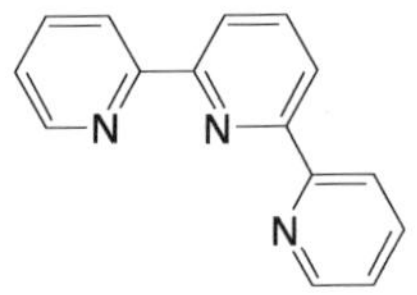

Terpyridyl

The use of terpyridyl, see aside, introduces the general point that such a tripodal ligand of constrained geometry can have only one chelate ring, not two, with a good 'bite' angle to the metal (see Section 2.5.5). This aids the ease of ligand–metal rearrangement. Advantage of this has been taken again in model redox chemistry involving the release of O_2 form manganese centres (Section 14.5) and it may illustrate a general way in which metal sites in proteins can be adjusted to give overall advantage to the free-energy of a multi-step process. The flexibility of coordination around central atoms is illustrated in the table (see aside); note that carbon has little flexibility due to activation energy barriers as in CH_4 or CO_2. It has been suggested that such ligands as terpyridyl should be called 'hemilabile', that is, half-labile.

While this description is clearly satisfactory for a downhill acid–base hydrolysis, it does not permit the reaction to go uphill in energy. All esters and amides are unstable in water and a way of protecting the active site for their syntheses and many of their other reactions has to be found, which avoids attack by the solvent. For uphill syntheses energy input is necessary, e.g. in the synthesis of pyrophosphate from two phosphates. We return to group transfer below and to energy input in Chapter 5.

Flexibility of reaction sites

Central atom	Coordination
PO_4^{3-} (P)	(3) 4,5
R_2S (S)	2(3)
MoO_4^{2-} (Mo)	4,5,6
Zn^{2+}	4,5(6)
Ca^{2+}	(6)7,8
Cu^+	2,3,4
Fe(haem)	5,6

NB. Carbon changes to 3 or 5 from 4-bonded with difficulty

Combinations of acids and bases in correct spatial organization opposite substrates obviously improves catalytic power since the activation event involves polarization of at least one bond, i.e. there are two centres to attack (eqn (4.2)). Again the use of two acid centres, e.g. metal ions, can be valuable since one can bind the substrate in a useful orientation while the other generates attack either directly or through a bound base, e.g. hydroxide. This point is made since it has been observed that many hydrolytic and synthetizing enzymes have two metal ions at the active site, e.g. zinc and magnesium in alkaline phosphatase and two zinc ions in phospholipase C. Model catalysts have been used to show the great enhancements of reactivity of two metal ions at active sites over one. We shall not describe the mechanisms of acid–base catalysis by enzymes in more detail until we have examined limitations of biology in Chapter 7.

4.4 The hydrolysis of proteins, RNA, DNA, and other polymers

While a major concern of acid–base catalysis is with small molecules, (glycolysis, and amide and ester hydrolyses), a section is added here to make sure that we remember a peculiarity of biological systems. The 'unstable' polymers of biology, which we have described in Chapters 2 and 3 as a framework for the reactions and movement of the elements, are constantly turned over. The turnover, comprising synthesis and degradation, is an effective way of generating a steady state concentration of a polymer. Acid–base catalysts control both these rates, as well as the supply of monomers and energy for synthesis. It will then be important to notice the parts of space in which the polymers and their acid–base catalysts are placed. Obviously, some zones are mainly digestive, e.g. the lysosomes, some are synthesizing, e.g. the ribosomes and the nuclei, and some are apparently neutral, e.g. the cytoplasm.

There are two factors which modify any general picture of hydrolysis rates of polymers from the above account of model reactions. One is the strength of the fold of the polymer, which varies from almost random coil to tightly folded and even cross-linked, and the second is the location of the hydrolytic catalysts. The random coil polymers are relatively easily digested, even in the cytoplasm, but the cross-linked proteins are not. However, the stabilizing cross-linking of proteins is reversible due to release of metals, redox reactions that break –S–S–bridges and so on. There is then a steady state of protein concentration dependent on X, where X can be a metal ion or a redox process giving an –S–S–bridge. The scheme shows that there may be little free apoprotein in a solution of a protein, a cross-linking agent, and a protease. This situation is made the more secure by the feedback control to DNA or RNA over protein synthesis by X, e.g. a metal ion, and its uptake. The limitation on free protein is a very important control over selective binding (Section 2.8). Similar considerations apply to RNA: messenger RNA is continuously made and digested; ribosomal RNA and DNA are in protected states.

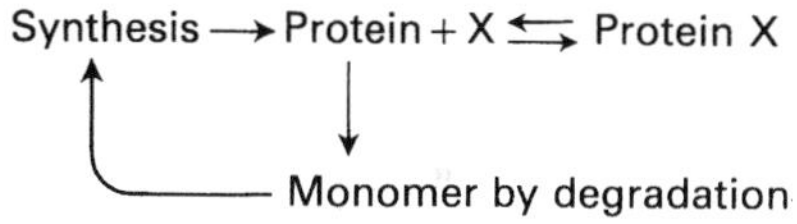

Now, as stated, the degradation process and the synthetic process need catalysts which must be kept apart, i.e. zoned, for obvious reasons (see Chapter 3). Proteins which are not strongly stabilized do not last long in the bloodstream of animals, for example, since there are proteases present everywhere outside cells. A major activator of catalysts of hydrolysis is external calcium. When excessive calcium enters a cell it even switches on degradation in the cytoplasm, which corresponds to cell suicide, *apoptosis*, aided by a specially activated protease in a particle, the proteasome, and by the use of unfolding proteins, ubiquitins. Magnesium rather than calcium is more concerned with the hydrolysis of phosphate-containing polymers, and again this is rapid outside cells, e.g. in digestion of DNA and RNA.

The function of the proton in acid-catalysed enzymic reactions is more to the fore in zones which are acidic, pH < 7. The obvious ones are parts of the extracellular digestive tract of multicellular organisms and lysosomal vesicles. Such zones often have *active* metal-containing enzymes for hydrolysis too. Clearly, we cannot speak of general principles of acid–base or redox catalysis in organisms without reference to a compartment and its content at a given time. Just as we shall find that certain elements, Zn, Ca, and Mg, are overwhelmingly used in acid–base catalysis, so we shall find that certain compartments are for acid–base reactions and not redox reactions.

However, some zones are designed to be oxidative as well as hydrolytic. Here–S–S–bridge formation, as well as calcium cross-links, protect certain proteins from hydrolysis, e.g. outside cells. Internally the cytoplasm of the cell is reductive and has less protection.

4.5 The nature of ion flow

4.5.1 Signals and controls

Acidic and basic ions are obviously important in the catalysis of chemical change, but they are equally important in signalling and control. The transfer of information from site to site between cells necessitates a signalling system. This need be no more than the passage of current, the flow of charge, and, in particular, the flow of ions. (The comparable particle in man's systems is the electron.) However, the kinetics of electrostatic flow, referring now to movement along or across cell membranes and cellular systems, must be selective, as in any transmission system, so that there must be allowed and forbidden flow paths for different biological ions. A selective carrier can be very useful, but speed of flow is also essential. Moreover, there has to be a receptor for the message, which eventually will be transcribed into a chemical change. All this means that in addition to the rate of movement from site to site, the rates of binding and release of the messenger ions or molecules have to be analysed. We then must consider a series of steps where an ion A flows to, binds, and then leaves a receptor B:

(1) A is released	(generation step);
(2) A flows to B	(rate of diffusion);
(3) A binds to B $\rightarrow$ AB	(rate of association);
(4) AB undergoes conformation change	(rearrangement rate);
(5) A leaves B $\rightarrow$ A + B	(dissociation rate);
(6) A is hydrolysed	(destruction step).

All these rate processes are not just conventional catalyses of acid–base reactions; they involve transport in space and time, i.e. they can be vectorial. However, quite frequently, we can relate rates of diffusion and/or rates of association to steps in acid–base catalysis.

For example, the movement of the ion A may well require the release of water from a solvation sphere of a cation M^+ or an anion X^-. The leaving of water is an acid-base dissociation

$$H_2O \rightarrow M^+ \rightarrow H_2O \text{ (base)} + M^+ \text{ (acid)}.$$
$$OH_2 \rightarrow X^- \rightarrow H_2O \text{ (acid)} + X^- \text{ (base)}.$$

These reactions and their reversals may well be required for action at a receptor or as an ion passes through a membrane and could well be vitally dependent upon ion selection. We shall then need to know the speed of water release since it will be important in message transmission, in control, in catalysis and in other processes.

Finally, binding involves mechanical rearrangement of the receptor ligand B (e.g. a protein) when bound to A as AB. We must observe the kinetics of rearrangement of molecules here. The distribution rates of strain energy are involved (step (3)), since the metal binding pulls on a protein. The control of surfaces and the control of mobility gives proteins extreme value as channels, gates, triggers, and so on.

There is little doubt that H^+ and OH^- are the most useful acid and base, respectively, not only in catalysis but in ion migration since they have the fastest diffusion rates. They are very valuable in organism, but we must remember that the biological pH is 7.0, which is extremely restrictive for the concentration of these species. Function in biology does not follow easily from chemical principles without very careful examination of the restrictions which do not allow wide variation of parameters. Since the physiological pH is close to neutrality (~7), H^+ and OH^- are 10^6 to 10^5 times less available than Na^+ and K^+ and, therefore, despite their higher mobility, they are at least 10^3 times less useful as charge carriers. For this reason, K^+, Na^+, Cl^-, and even Ca^{2+} are preferred in biology's electrolytic circuits.

We turn now to some details of acid–base reactions in ion flow.

4.5.2 **Ion migration rates: electrolytics**

The migration of an ion in a membrane and, more especially, its movement to and from water will be partly dependent on the rate at which water leaves and enters the coordination sphere of the ion. For anions this rate is almost invariably very fast, of the order of 10^9–10^{10} per second. For cations the rate decreases from this limit for H^+, Na^-, and K^+, down via the large divalent and trivalent ions, Ba^{2+}, Sr^{2+}, Ca^{2-}, and La^{3+} (10^7–10^9 per second), to the smaller divalent ions, Mg^{2+} (10^5 per second), to the smallest highly charged ions, Al^{3+} (1 per second) (see Fig. 4.3). The rates of this exchange are one major factor, apart from concentration, in the selection of Cl^-, Na^+, K^+, and Ca^{2+} as biological ion current carriers. The second factor is binding, which is minimal for Cl^-, Na^+, and K^+, so that on off rates to organic matrices remain very fast (see Section 4.6).

Two factors clearly need to be carefully distinguished now in the discussion of the selective passage of ions across membranes. There may be thermodynamic barriers to entry into the channels, as discussed in Chapter 2, but there can also be differences in rates of diffusion in a channel. When we measure

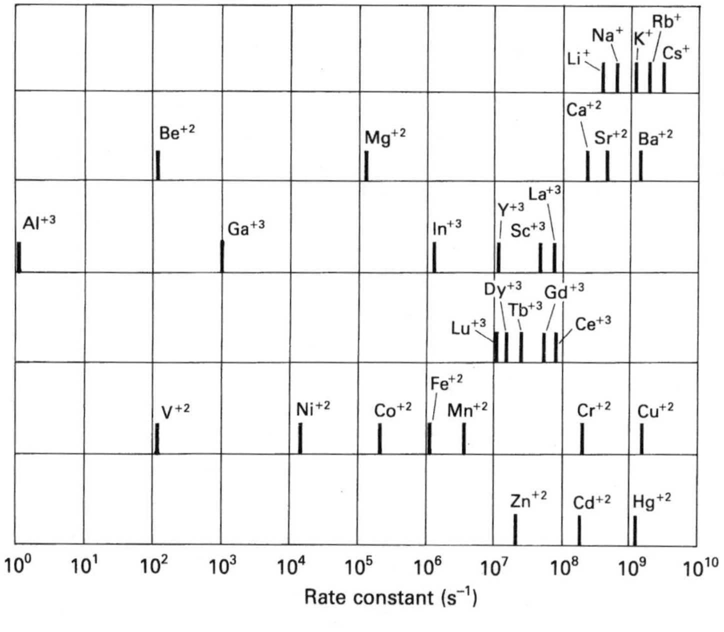

Fig. 4.3 Characteristic rate constants (s^{-1}) for substitution of inner-sphere H_2O of various aqua ions. *Note.* The substitution rates of water in complexes $ML(H_2O)_n$ will depend on the symmetry of the complex. [(Reproduced with permission from Frey, C. M. and Stuehr, J. (1974). Kinetics of metal ion interactions with nucleotides and base free phosphates. In *Metal ions in biological systems*. Vol. 1 (ed. H. Sigel), p. 69. Marcel Dekker, New York.]

Table 4.1 Permeability ratios for two types of channel

	Li^+	Na^+	K^+
Sodium channel	0.95	1.00	0.05
Potassium channel	0.02	0.02	1.00

currents, i.e. permeabilities, through membranes in the presence of mixtures of ions, we study a combination of these factors which are locked in the permeability ratios (Table 4.1), which differ for channels designed to allow passage of say, Na^+ predominantly or K^+ predominantly. Many of the channels appear to be of considerably larger diameter than the bare ions and it could well be that some of the channel walls are lined with water molecules (Fig. 2.11); these water molecules may reorganize differently as different ions pass through them (see Chapter 8 for more details).

In order to establish a net flow it is necessary to generate an initial out-of-balance configuration of ion concentration or of field gradient, for example, across membranes. To achieve the concentration or field gradient, some charged species, ions, must be pumped across the membrane. The rate of pumping depends on the rate of energy input and the capacity of the pump. The design of pumps for H^+, Na^+, K^+, Ca^{2+} and Mg^{2+} amongst many other moieties is a critical acid-base kinetic problem in biology (see Chapters 8 and 9). The selectivity of the rate of pumping is quite different from the thermodynamic selectivity of ions entering channels, although it too depends in part on interactions between ions and water molecules. The actual transfer step in a pump requires also a conformation change of a protein, which is of necessity slow, about 10^2 to 10^3 s^{-1}, and this motion of the protein may or may not be allowed, depending on the size of the ion in the pump channel.

A very special case of ion migration is that of the proton since as stated it can diffuse in an H-bond network by rotational/translational diffusion of water molecules. This mode of very rapid diffusion is characterized by a hop distance of some 0.5 to 1.0 Å (Table 4.2); the equivalent motion of OH^- is easily written. Proton channels readily exclude all other ions since no other ion can follow such an H-bond path, Fig. 2.11. An important consideration will be the coupling of proton channel movement to electron movement in proton pumps which, if they are to conserve energy, must be 'gated' (see Section 4.7 and Chapter 5). An impression of a proton network within a protein is given in Fig. 4.4 which shows the interior of cytochrome *c*. an electron transfer protein.

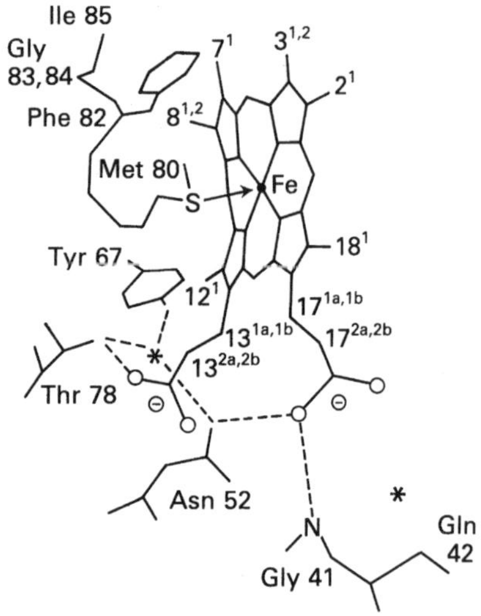

Fig. 4.4 A diagram of the interior of part of the redox protein cytochrome *c*, to illustrate the coupling of proton networks in proteins with electron transfer to their metal ions. On redox reaction, Fe(II) to Fe(III), there is a change in the hydrogen-bond network associated with Gly-41, Asn-52, and Thr-78. Such changes could alter rates of proton flow through the network; *represents a water molecule, see references in Williams in further reading.

It is now known that during energy transduction (see Section 4.7), the flows of current in the membranes of both photosystems and the chain of oxidative phosphorylation are not electronic alone but are due to proton mobility in the membrane phase. There is a similar flow in bacteriorhodopsin [see Luecke *et al.* (1999) in 'Further reading'] which uses light and no electron flow to create a proton gradient. In other words, this is a proton photocell, not an electron photocell. The proton flow uses some protein sidechains but frequently chains of water molecules are employed (see Fig. 2.12). The movement of the protons or of other ions can be gated by conformational changes involving cooperative helix–helix rearrangements (see Fig 2.11 and Section 4.14), which can break or make H-bond connectivity, including that to water within the protein (see

Table 4.2 Proton and electron transport

	Protons	**Electrons**
Hop distance (Å)	0.5 to 1.0 Å	<15 Å
Site	Bases	Metal ions, aromatics
Activation energy	pK_a change Relaxation, *R* Rotational diffusion	Redox energy, ΔE Relaxation, *R*

Fig. 4.4). Analysis by site specific mutation of proteins could be very misleading, since it could alter long-range H-bond networks.

There is one other type of current carrier: small organic cations, such as acetylcholine (AcH) and glutamate, which can be very selectively recognized by receptors to build a specific current chain, e.g.

$$\underset{\text{(Cell A)}}{\text{Message } Na^+/K^+} \rightarrow \text{Release of X (AcH)} \rightarrow \underset{\text{(Cell B)}}{\text{Message } Na^+/K^+}$$

Here X is an organic carrier of current which selects for a special B-type cell from amongst many others. It is the specificity of these current carriers which makes slow complicated (parallel) biological circuits so different from the fast electronic circuits of man, e.g. in computers. However, the use of the slower ion movements allows diversity in the current carriers themselves and each current carrier can function specifically at any point in the circuit since it can be differentially recognized. This allows very sophisticated information distribution.

4.5.3 Exchange

Exchange includes not just exchange between free ions and bound ions, inorganic or organic, or exchange across membrane barriers, but also direct exchange between two sites of binding. For example, a carrier may donate ions directly to a pump or to a final site. This means that the free ion may not be involved in the pathway through a cell. Obviously, if there is exchange with the free ion from both sites or even only from one then the system of exchange will equilibrate generally. Now, very generally we know the rates of exchange of metal ions from sites and this allows us to tabulate those free metal ions and metal complexes which will exchange easily and those which will not (see Aside). Fast exchange across membranes is an extremely valuable way of using an initially energised gradient of, say, H^+ (or Na^+) in the uptake (or rejection) of other ions (inorganic or organic) to or from cells. Ion uptake systems will be examined in this light.

Table of exchange

Fast exchange	**Slow exchange**
Na^+, K^+, Cl^-	(Co^{2+}) Ni^{2+}, (Zn^{2+})
Ca^{2+}, Mg^{2+}	Fe^3, haem, vitamin B_{12}
Substrates	Mo-cofactors

4.6 On/off reactions: control systems

Control demands that, as well as diffusing rapidly, the messenger system should bind, change the conformation at the site of binding, and subsequently should leave as rapidly as possible so as to allow the system to relax after information has been received. Thus we require the rate constant, k_{off}, to be as large as is compatible with the ability to switch the device. Now, the energy capacity to switch a protein conformation will depend on binding strength, where the binding constant is

$$K = \frac{k_{on}}{k_{off}}.$$

Biological switches need to be fast and their speed appears to be controlled by the time scale limit at which proteins can rearrange since proteins must have reasonably stable structures. This limit seems to be close to 10^{-3} s. Thus k_{off} should ideally be 10^3 s^{-1}, when K can be no larger than about 10^6 to 10^7 since k_{on} is diffusion-limited at 10^9–10^{10} s^{-1} for all ions or molecules. Ions able to bind with a k_{on} as high as possible and K around 10^6 make therefore the fastest and consequently most powerful controls. The Ca^{2+} ion meets these criteria wonderfully well, while K^+ and Na^+ fail on grounds of poor binding, except in ring chelates, where on and off rates are slow, Mg^{2+} fails on the grounds of its

Table 4.3 Idealized binding and rate constants

Ion	k_{on} (s^{-1})	k_{off}(s^{-1})	K	Function
Na^+, K^+, Cl^-	$>10^9$	$>10^6$	$<10^3$	Electrolytic message
Ca^{2+}	10^9	10^3	10^6	Mechanical trigger
Mg^{2+}	10^5	10^2	10^3	Phosphate transfer
Zn^{2+}, Fe^{2+}	10^8	10^{-2}	10^{10}	'Hormone' communication
Cu^{2+}	$>10^8$	$<10^{-7}$	$>10^{15}$	No exchange
C, H, N, O	Covalent, enzymic control			Structure building
HPO_4^{2-} (RPO_4^{2-})	10^9	10^6(?)	10^3	Trigger
Protein PO_4 (phosphorylation)	Slow	Slow	Weak	Kinetic control faster than C, N, O bond breaking
H^+	$>10^{10}$	Slow to 10^{10}	Huge range	Catalysis, energy store
OH^-	$>10^{10}$	Slow to 10^{10}	Huge range	Catalysis, energy store

slow on and off reactions, and Ni^{2+} has even slower on rates. Not surprisingly therefore, it is found that calcium is a universal biological trigger. We shall see how these on/off reactions communicate through protein matrices in Chapters 6 and 10.

For slower controls, resembling hormones rather than transmitters, there are several possible candidates: Mg^{2+} [on rate slow (10^5 s^{-1}), weak binding but relatively high concentration (mM range)], Zn^{2+} (on rate faster but stronger binding, i.e. much slower off rate, and present in lower concentrations as a free ion), Fe^{2+} and Mn^{2+} (on rates fast, stronger binding, but present in lower concentrations and redox active). Clearly, such elements are not useful in triggering (10^{-3} s time scale) but they can be used in regulation (10^3 s time scale). The very strong binding of ions such as Cu^{2+} and Fe^{3+} makes them almost permanent parts of a system once bound, so they are of lower value in either process. Table 4.3 summarizes some of the important on/off rate constants.

Amongst anions, phosphate and phosphate derivatives match calcium in control function since they diffuse rapidly and can bind quite well. (In a simple-minded sense, HPO_4^{2-} is negative calcium.) The organic ions are like acetylcholine in current selectivity (see Section 4.5.2) but, like organic molecules generally, they can be metabolized and it is hard to maintain their concentrations steady, especially outside cells, without wasting metabolic energy. Binding of organic molecules can be quite strong and selective due to their large surfaces. In this case, it is the hydrolysis of a weak ester or anhydride which is used for control. In Chapter 20 we look at the general way in which calcium plus phosphate compounds inside cells act cooperatively in control devices.

While discussing molecular ion movement we must not forget the diffusion and binding of larger organic ions, even up to the size of proteins. It appears to be a general observation that barriers to on/off reactions, apart from binding constants, are related to the polarities of the interacting surfaces. Polar surfaces come together and part rapidly; non-polar surfaces stick together in water.

4.7 Electron-transfer reactions: electronics

Electron-transfer reactions are of considerable importance in cells. Perhaps the most important of these reactions occur in the electron transfer chains of energy transduction systems in mitochondria and chloroplasts, where electron transfer is coupled to proton transfer. Their electron transfer centres are now known and have been well described (see Chapter 5).

Electron-transfer depends on the approach of an electron donor to an acceptor. There are two cases:

1. Long-range or outer sphere electron transfer, when the coordination sphere of both donor and acceptor is altered to only a minor degree during reaction: the intervening medium is not of great consequence.
2. Inner sphere electron-transfer, where the donor and acceptor form a direct chemical bond by changing their coordination spheres.

In biology, the first case is used more generally and especially for making electronic conductors, i.e. biological wires. Direct electron-transfer in an inner sphere complex is often reserved for difficult chemical transformations, i.e. for catalytic steps in enzymes.

Long-range electron migration in biology is via hop conduction, i.e. the electron goes from one metal ion to another over a limited distance. A satisfactory tunelling theory of the rates of electron-transfer has been developed, particularly by Marcus. The Marcus equation (see 'Further reading') is

$$\ln k = \ln \nu_{\mathrm{ET}} \frac{(\lambda - \Delta E_{\mathrm{m}})^2}{4k_{\mathrm{B}}\lambda T}$$

where ΔE_{m} is the redox potential difference between source and sink; λ is the relaxation energy due to the small motions required before electron-transfer; k_{B} is the Boltzmann constant; and ν_{ET} is the limiting rate when $\lambda = \Delta E_m$.

We must then consider ΔE_{m}, the redox potential difference between the source and the sink of the electrons, which is the driving force, and the relaxation energy, λ. Obviously, the only chemical elements suitable for electron transfer devices must have at least two oxidation states (Fig. 1.9). Hidden in the term ν_{ET} is the very important dependence on distance of the overall rate.

The effect of distance on unimolecular electron-transfer rates ν_{ET} has been clearly demonstrated in model complexes of many metals held apart on a rigid organic framework (see the 'Further reading' section). The rate constants fall exponentially (by a factor of approximately 10^2 for each 3 Å increase in distance); hence it is expected (and found) that hops of 15 Å are at the upper limit of usefulness in biology. These inferences from models have been thoroughly demonstrated in the electron-transfer of the reaction centre of photosynthetic bacteria and in model systems, see reference to Dutton and especially Gray (see Aside). Figure 4.5 illustrates this in photosystem I in which electrons excited at the chlorophyll dimer $(BChl)_2$ hop successively to other chlorins (BCI) or phaeophytins (BPh) and then to quinones, Q, while new electrons are fed in from cytochromes, Cyt. The medium, here a variety of protein groups, appears to have no particular influence and all redox reaction potentials are below + 0.6 V. Photosystem II is very similar except in the section at very high oxidizing potentials between $(Chl)_2$ and the Mn (for O_2 production) when a tyrosine radical intervenes. Tyrosine, tryptophan, or histidine radicals can only be really useful in redox reactions in the range above + 0.75 V. The study of almost every electron hop in this wiring diagram follows expectation based on model reactions (see Fig. 4.5).

We shall consider later in more detail the value of inserting unsaturated groups in the chain between, but not bound to, the metal ions. Bound unsatu-

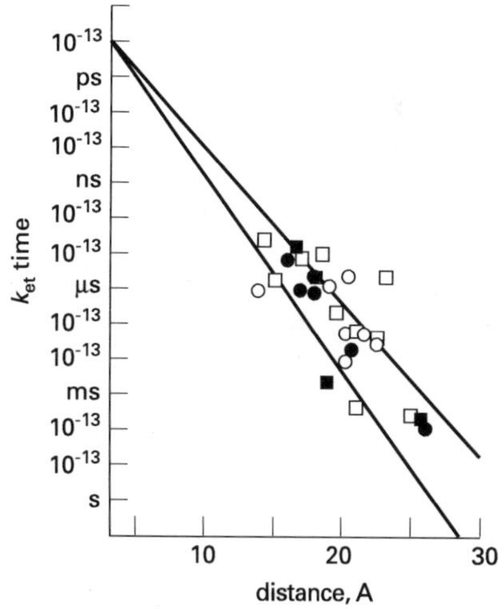

Tunnelling timetable for four proteins: Ru-modified azurin data (filled circles), β-sheet; Ru-modified cyt *c* data (open circles) ill-defined, α-helical; Ru-modified myoglobin data (filled squares), α-helical; Ru-modified cytochrome b_{562} data (open squares), α-helical. Electron transfer is from a differently placed bound Ru-dipyridyl complex to the biological metal centre. Data from J.R. Winkler, *et al.* see references. Note however that a single straight line through all the points rather than different lines for different sets of proteins seems to be appropriate.

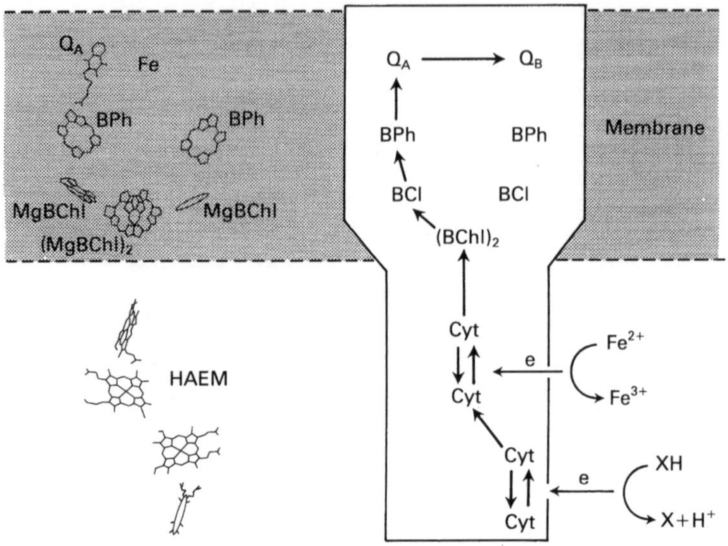

Fig. 4.5 The electron transfer scheme of the bacterial reaction centre equivalent to photosystem I. Cyt, cytochrome; BChl, chlorophyll; Q, menaquinone; BPh, phaeophytin.

rated ligands, e.g. porphyrins, spread the orbitals of the metal, but keep it protected from substitution reactions. The remaining restrictions on the rate of electron transfer are hidden in relaxation energies λ, given a fixed redox potential difference which helps to decide the direction of flow.

Of course, the nature of the medium between centres may have some influence and rather than use a general bulk dielectric to describe this medium some authors prefer to use a full statistical quantum mechanical treatment of the separating organic protein medium. The first procedure gives a simple sense of the effect of distance and at present there is no clear-cut evidence in favour of a more detailed model except in one or two high redox potential systems (see aside on page 119 and Williams in references).

All electron-transfer processes involve a change in charge locally. There is then a change in the electrostatic interaction with the environment and in certain circumstances, e.g. a bare ion going from $M^{2+} \rightarrow M^{3+}$ in water, there will be a considerable change in bond energy and in the entropy of the water. This would require a high reorganization energy hindering transfer, since the states, source and sink, must be thermally adjusted before the electron can move. The term λ would then be caused to be disadvantageously large; apparently this is observed in some model reactions. Biological electron transfer need not suffer from this problem because the electron-transfer centres can be held relatively rigidly in the correct structures to make optimal electron transfer in proteins and out of water. We shall look at their structures later, since it is the nature of the protein matrix which is central to biological electron transfer. Notice that here the electron transfer is a case of a single step reaction when the catalyst centre for transfer can be optimally designed (see Section 4.3).

In a protein, the change from one oxidation state to another, involving relaxation of the surrounds of the electron transfer site, could therefore be simply vibrational adjustments. Alternatively, it could involve quite major conformational movements, but in this case energy is used and properties other than simple electron conduction can be achieved and the coupled conformational change has to occur separately in time from electron transfer. The former situation at a series of hop centres would generate a 'wire' where electron transfer is not coupled to any other process; the latter case would correspond to a coupled process in which electron transfer was not permitted unless a certain conformational change had occurred or was followed immediately by a major relaxation of nuclei. The major relaxation could then cause other events, such as movement of ions (see Fig. 4.4). This reaction is called 'gating' and the transfer of energy is called 'energy transduction'. It is possible that electron transfer is more strongly coupled to the lattice through H-bond changes which can also be linked to tunnelling. This may well be part of the mechanism of coupling between in-put energy and proton transfer, (see Chapter 5).

The above applies to all those reactions in which the co-ordination sphere of the metal ion is not changed. Obviously, then, it applies particularly to the metal ions which have all their coordination sites filled by ligands which are sidechains of proteins. In the case of open-sided metal ions there is the possibility that the non-protein ligand, water or any other ligand, could be lost, and this change in structure prior to and/or following electron transfer means that the relaxation term may again be large so that coupling to these steps can occur. We find that wiring 'circuits' for rapid electron transfer in cells are built from the closed coordination-sphere metal ions and that the rate is limited at the end of a chain by open-sided catalysts (Fig. 4.6). These are the ones which bind dioxygen, hydrogen peroxide, or even dihydrogen so as to couple electron transfer to reactions which effectively are the energy generators of an electronic (chemical) battery (see Chapter 5). This applies to a relatively highly organized system—almost a solid state device—but there is also the possibility of molecular diffusion-based devices, e.g. liquid three-, two-, or one-dimensional molecular flows, for carrying electrons between sites. All

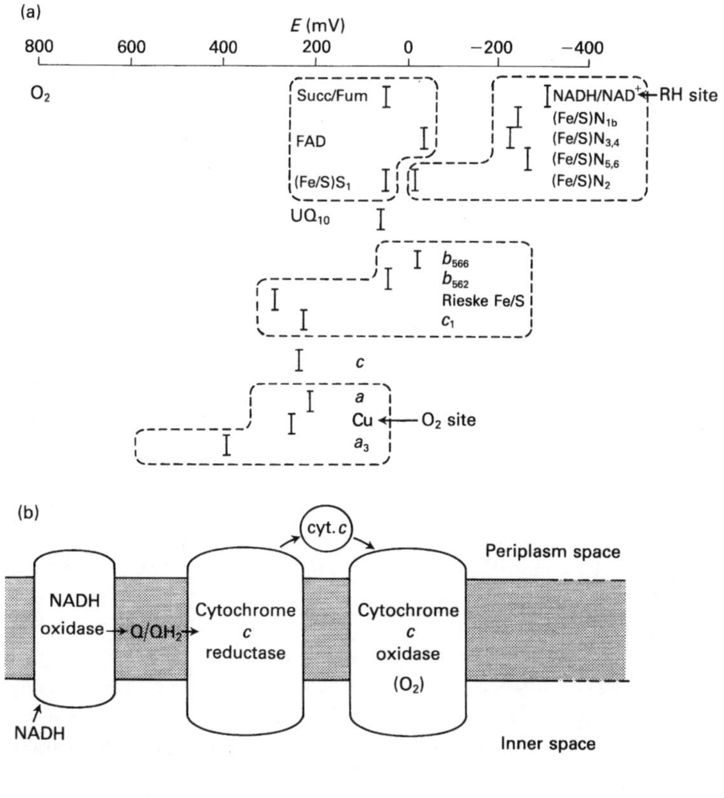

Fig. 4.6 The mitochondrial respiratory chain. (a) The sequence of redox centres given in the diagram is based on the *E* values obtained with de-energized heart mitochondria. Cytochrome *c* links the cytochrome *c* reductase and cytochrome *c* oxidase complexes. (b) The arrangement of cytochrome *c* with respect to the positions of cytochrome reductase and cytochrome *c* oxidase is not well defined. The reductase and oxidase complexes span the inner mitochondrial membrane with cytochrome *c* loosely attached to the outside of the membrane. The open-sided reaction centres are at the beginning (RH) and end (O_2), *a* and *b* refer to different cytochromes.

these possibilities are put to use in biology.

As to the material used, it is curious to note that man makes his wires from metals such as copper; biology makes hop conductors from metal ions embedded in insulating material, the proteins. The choice of metal is not limited except by the availability of two oxidation states at the appropriate potential. Nearly all transition metals in a number of oxidation states could be so manipulated, but it is observed that iron is virtually the only metal used. It could well be that this was an accidental result of the availability of iron in the early evolution of life. Especially interesting is the use of Fe_nS_n *clusters*, which contain high-spin iron, but relaxation is minimized by the delocalization of the electrons giving more effective centres than one metal atom units and this is also true of haem iron. Besides iron, only copper is used, and much less commonly, in aerobic systems which were a later development.

Before concluding we must note that a control protein or a coupling device must have a physical character different from a catalyst in that the former must change shape easily. In Chapter 6 we return to these problems.

4.8 Redox potentials of complexes

It is useful to discuss the thermodynamic redox potential change associated with electron transfer here since it is the driving force referred to in the preceding section. The redox potentials are given by the gradients of lines in Figs 1.7 and 1.8 and are those for free ions in aqueous solutions at pH = 7 and in the absence of other ligands. If we give this redox potential the symbol $E°$, then in the presence of ligands the value changes. Consider two free aquo ions M^{II} and M^{III}; if they both form complexes ML with the same ligand, the redox potential changes from that of the hydrated ions by the strength of binding in ML when for $[M^{II}L] = [M^{III}L]$

$$E = E° + \frac{RT}{\mathfrak{F}} \ln K^{II} / K^{III}$$

where K^{II} and K^{III} are the stability constants of $M^{II}L$ and $M^{III}L$, respectively (see Section 2.7). The use of a great variety of ligands to manipulate redox potentials is of the greatest importance since, for example, the assembly of metal ion complexes into a series within a membrane (Fig. 4.6) is the way in which biology makes a wire. The wire has the best current flow with least loss of energy if the redox potentials of the couples in the wire are ordered in sequence and the rate of electron transfer is also controlled by the difference of redox potential (see Section 4.7). By adequate selection of ligands L, usually proteins, the redox potential of iron in different couples may extend from −0.5 to +0.4 V.

Using model systems it has been easy to show how increase in σ-donor power of ligands tends to decrease the redox potential of a transition-metal couple. On the other hand, increase in π-acceptor power of ligands tends to stabilize lower oxidation states and increase the redox potential. The use of saturated and unsaturated ligands based on nitrogen donors readily illustrates these points, but it is not so easy to explain the properties of, e.g. S-, P-, and As-based donors in inorganic chemistry, since in these cases the π/σ effects are very mixed. Especially noticeable are the effects on redox potentials of larger (heavier) transition metals which are little affected by such complexing agents, e.g. molybdenum.

Finally, we must consider a series of ligands which induce spin-state changes in, say, the Fe^{2+} and Fe^{3+} ions. Models show that, since the energy for spin-pairing is less for Fe^{2+} than for Fe^{3+}, increase in π-bonding together with σ-bonding will drive Fe^{2+} low-spin first and, therefore, raise the redox potential as the total ligand field strength increases, but at still stronger ligand fields the Fe^{3+} state becomes low-spin and the redox potential falls (see Section 2.5.6 and 2.7). The example of the iron porphyrin proteins illustrates the point (Fig. 13.2).

The redox potentials of organic molecules in the electron transfer chains are matched with those of the metal ions (see Fig. 4.6). At low potential, pyridine nucleotide and flavin coenzymes have potentials around –0.5 to –0.3 V, i.e. near those of Fe_n/S_n couples; at intermediate potentials, around 0.0 V, there are quinone coenzyme couples. Of course, dioxygen has a higher potential, + 0.8 V at pH = 7, and is associated with haem and copper couples. A feature of these three non-metal systems is that redox change is necessarily associated with acid-base protonation reactions:

$$Q + 2e + 2H^+ \longrightarrow QH_2,$$
$$\tfrac{1}{2}O_2 + 2e + 2H^+ \longrightarrow H_2O,$$
$$\text{Flavin} + e + H^+ \longrightarrow \text{Flavin H}^\bullet,$$
$$NAD^+ + 2e + H^+ \longrightarrow NADH.$$

This covalent hydrogen binding is fundamental to the energy capture devices.

In enzymic reactions, the oxidation of organic molecules can be achieved so as to give organic radicals of high redox potential using couples such as FeO(IV)/Fe(III) (about + 1.0 V). Groups which can then be oxidized are phenols and indoles, and these oxidations are greatly used to detoxify, to protect, and to strengthen connective tissue. It is also possible to use the oxidation of these groups to radicals reversibly when they are part of a side-chain of an amino acid in a protein (Table 4.4). They may be quite generally valuable in electron transfer through proteins at high redox potential.

Table 4.4 Organic radicals in enzymes

Enzyme	Radical
Ribonucleotide reductase	Tyrosine, glycine, B_{12}–CH_2 of enzyme
Photo-reaction centre	Tyrosine, phaeophytin, chlorophyll
Cytochrome c peroxidase	Tryptophan
Galactose oxidase	Tyrosine
Prostaglandin H synthetase	Tyrosine
Glucose oxidase	Tyrosine
Cytochrome oxidase	Tyrosine

4.9 Atom transfer reactions

While acid-base and electron transfer catalyses are major activities within cells, a third equally important process is oxidation and reduction by atom transfer; an obvious need in biology is to transform CO_2 to $HCHO + H_2O$, N_2 to NH_3, and so on. The catalytic act may well be the uptake on one site of an atom or ion, H^-, O, etc., which is obtained from, say, H_2 (NADPH) or O_2 (FeO in an oxidase), and its transfer to a carbon centre, e.g.

$$CH_4 + O \rightarrow CH_3OH.$$

When an oxygen atom or H^- are added to a compound we term this a 'two-electron oxidation or reduction' (by atom transfer). Some other examples are

$$ClO^- + SO_3^{2-} \rightarrow Cl^- + SO_4^{2-}$$

and

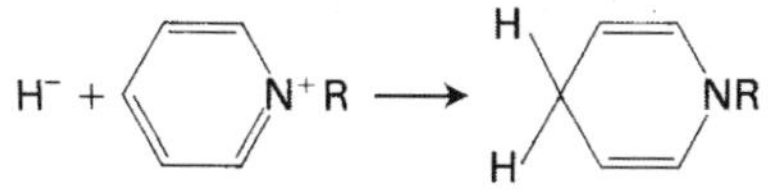

These reactions require the activation of oxygen or hydrogen and the usual mechanisms in chemistry require the creation of an intermediate such as FeO or CoO or Fe-H, Ir-H, etc. on the surfaces of metals or in organometallic compounds. These metal intermediates are often seen in biological systems too, but we must observe the very extensive use of non-metal coenzymes described in Chapter 3. Very frequently, the use of a metal is associated with very small redox substrates as in acid-base catalysts (see Table 7.3).

Here we must observe that while oxidation may be downhill thermodynamically, the reduction of say CO_2 to CH_3OH requires energy. Thus oxidation and reduction generally in cells are not reversible and use different catalysts (see Section 4.3).

In order to make atom transfer (as opposed to electron transfer) easier, two conditions must be met: (1) the central atom must be able to make and break its immediate atomic links, i.e. it is an inner sphere reaction; and (2) the change of the central atom formal oxidation state has to be two. A typical reaction would be

$$ClO_4^- + R_2Se \rightarrow R_2SeO + ClO_3^- .$$

Referring to oxygen atom transfer we find that this is difficult in the first row of the periodic table but easier down each group given that redox potentials permit the reaction (see Fig. 1.8). Thus SeO_2 gives O more readily than does SO_2. Elements confined to one oxidation state do not take part in these reactions, so that they are possible properties of C, N, (O), S, Se, and a variety of transition metals, but not of Si, P, and Zn, for example. Notice that there is a potential danger with the transition-metal ions since they are likely to confuse one- and two-electron reaction paths; Fe and Se peroxidases (Chapters 12 and 18) are good examples to contrast for their different value.

A typical two-electron transfer in hydrogen chemistry is H^- transfer, and once again a shift of two in effective oxidation state is required. For these reactions biological systems resort to complex coenzymes of the pyridine nucleotide and flavin classes or to the use of transition metals. Here cells have not generated many of the equivalent electron-rich series of metal compounds for catalysing these reactions which are well-known in organo-metallic chemistry; the best they do is to use nickel or iron sulphide complexes as we shall see.

Many very fine catalysts made by man use quite heavy metals, e.g. Rh, Ir, Pt, which biology has not managed to obtain. Instead, biological systems have manipulated the first-row transition metal ions and especially their spin states to induce reactivity (Fig. 4.7 and see Section 4.11). This heightened reactivity in the catalyst atoms is required for the activation of O_2, NO_3^-, NO_2^-, SO_4^{2-}, and H_2 species among others.

While the atom addition reactions of such reagents as dioxygen and dihydrogen are clearly described, by definition, as oxidation and reduction, the further classification of reactions is not so simple. 'Oxidation' is still adequate for the description of reactions of good electrophilic reagents, such as the halogens and sulphur, but is not so easily applied to species based on carbon or nitrogen. For example, if two ethylene molecules combine to form cyclobutane we do not call this oxidation and we do not say that the oxidation state of carbon is changed, but we do describe the reaction of ethylene with dioxygen in these terms since a cyclic ethylene peroxide is formed. The problem is well-known (and irresolvable) and arises from the definition of oxidation states. We shall not attempt any such classification of reactions here, but observe that all reactions are catalysed by re-adjustment of electron density, although it is not always obvious which electron adjustments have been made, due to a variety of factors. Similarly, the transfer of groups, like the transfer of O-atoms, is often loosely described by referring to types of 'oxidative' reactions of catalysts.

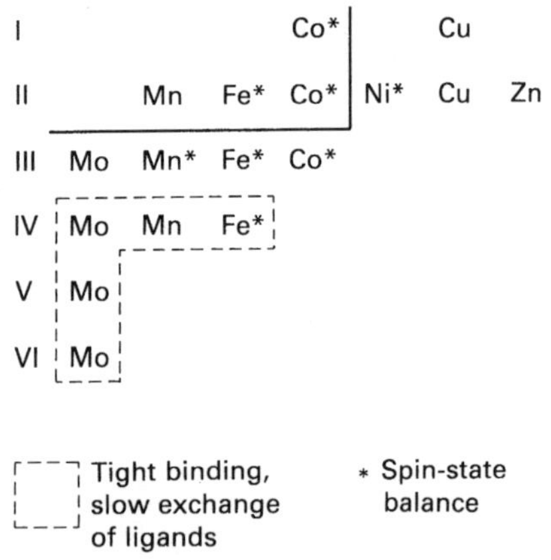

Fig. 4.7 The ranges of available oxidation and spin states of different elements which, together with their energies (redox potentials), generate different efficacies in catalytic acts. Biology has selected elements on this basis, but by incorporating the elements in special ligands and proteins has enhanced their functions both in chemistry and spatial effect.

4.10 Group transfer

Atoms other than H or O which can be transferred usefully are the halogens, as X^+, but a much more profitable series of reactions for biological systems is the transfer of groups such as CH_3^+, (CH_3^-), CH_3CO^+, and, in fact, any such similar ion (or unit). The preparation of group transfer agents is also a major preoccupation of organic and organometallic chemists; most of this chemistry is today outside the reach of organisms since it requires non-aqueous solvents and inert atmospheres, but it may well have been that in very primitive organisms this path was used (see Section 4.11). Instead, biology has now a series of coenzymes and metal centres as well as side chains of amino acids for such reactions; as expected, it is often heavy elements, especially sulphur, which are used because they make good attacking and leaving groups, e.g. in coenzyme A (Fig. 2.16), but lighter atoms such as N can also be activated, especially in unsaturated molecules, to perform these functions, e.g. in biotin or imidazole. For those who wish to think more deeply about such processes and relate acid-base and redox functions of atomic centres it is of interest to compare the release and uptake of H^+ with that of CH_3^+ or CH_3CO^+ in detail. Here we give a brief summary only, referring readers to books on bio-organic reactions.

It is conventional to separate reactions such as

$$H_2PO_4^- \rightarrow H^+ + HPO_4^{2-} \tag{4.4}$$

from

$$CH_3COPO_4^{2-} \rightarrow CH_3CO^+ + PO_4^{3-} \tag{4.5}$$

and from

$$CH_3COSR \rightarrow CH_3CO^+ + RS^-, \tag{4.6}$$

although they are all electron-pair reactions. In fact, it must be generally true that a leaving base in an acid base reaction such as (4.4), which has a pK_a in the range 4–12, will function as a positive ion group transfer reagent. Obvious examples are imidazole, phosphate, thiolate, and carboxylate. The big difference in considerations of catalysed group transfer is that while proton and simple ion reactions of type (4.4) come very rapidly to thermodynamic equilibrium, reactions (4.5) and (4.6) do not, so that kinetic pathways, dependent on catalysts, decide which reaction will occur. The mobile coenzymes of biology for group transfer are usually based on reaction (4.5) or (4.6) (cases of CoA, lipoic acid, NTP) or upon a well designed aromatic moiety which does not equilibrate rapidly, e.g. NADH in hydride transfer.

In summary, non-metal atoms suitable for group transfer are particularly S and Se, followed by N, while, amongst metals, Mo, Mn, V, and Fe are more suitable for transferring O, and Co and Ni are more useful for the transfer of H^- or CH_3^-.

4.11 The nature of transition-metal centres in catalysis

In describing metal ion functions we have not looked at all their kinetic capabilities; we have only examined the Lewis acid properties of the metal ions, which are a function of the energy of unoccupied exposed orbitals of any

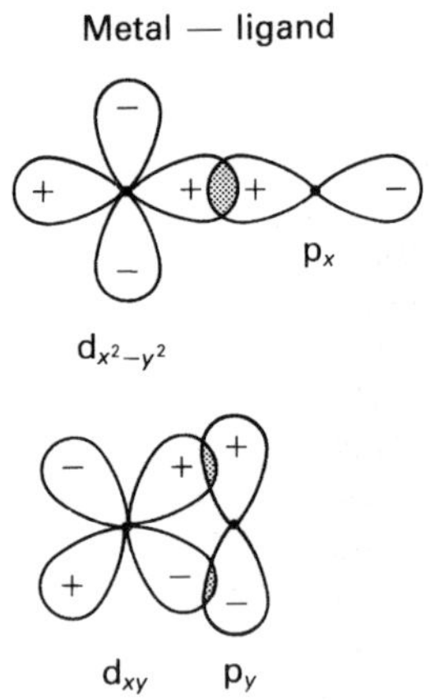

Fig. 4.8 Simple picture of the overlap between metal d orbitals and ligand p orbitals to give σ- and π-interactions. d_{z^2} orbitals also give σ-bonds; d_{yz} and d_{xz} give π-bonds.

element or ion. This term is additional to electrostatic polarization due simply to the charge on an ion. High-affinity hidden orbitals are not effective, but the degree of their exposure depends critically on the metal in question, on the other ligands of the metal, on the spin states generated in a complex (Fig. 4.7), and on the spatial characteristics of the complex. When considering ions such as Mg^{2+}, Ca^{2+}, and Zn^{2+} some of these considerations are of little importance since these ions have a hidden spherical core, but in the transition metals the more exposed 3d core orbitals are polarized, as in Fig. 4.8, for octahedral symmetry. The two less stable orbitals are now in the direction of the donor lone pairs of a ligand, while the three more stable orbitals can form π-bonds to it. Naturally, it is the lower three which are occupied preferentially by the ions' own electrons. Usually, in biological systems, we only need to discuss ions with five or more 3d electrons, i.e. metal ions from Mn(II), d^5, onwards, which tend to be π-donors from the highest occupied molecular orbitals (HOMO), e.g. d_{xy}, but σ-acceptors to the lowest unoccupied molecular orbitals (LUMO), e.g. $d_{x^2-y^2}$. Examples are low-spin $3d^6$ (Fe(II), Co(III)). As the number of 3d electrons increases further, the σ-orbitals become occupied and the metal ion can become a σ-donor of two electrons (Ni(II) d^8, Co(I) d^8). Thus, a metal ion is not just a σ-acid, but can be a π-donor, or a σ-base. Clearly, while acidic functions are closely paralleled by the activity of the proton, which shares with transition metal ions the property of acidity not just due to electrostatics, other properties are quite different. Hence, the metal ion, while binding via σ-acceptor properties, can totally adjust the reactivity of the bound group by acting as a π-donor. These 'basic' properties of transition metals allow them to bind and to activate unsaturated small substrates such as CO, O_2, NO, etc. The 'basic' properties can be called into use after the acidic properties of the metal ion in that, after forming the acid base complex, the metal centre (with a lower formal charge) may alter its role and act as a π-base to unsaturated ligands. In this way, a reaction can proceed from a planar d^8 system to a pentagonal prism and then to an octahedral system, e.g.

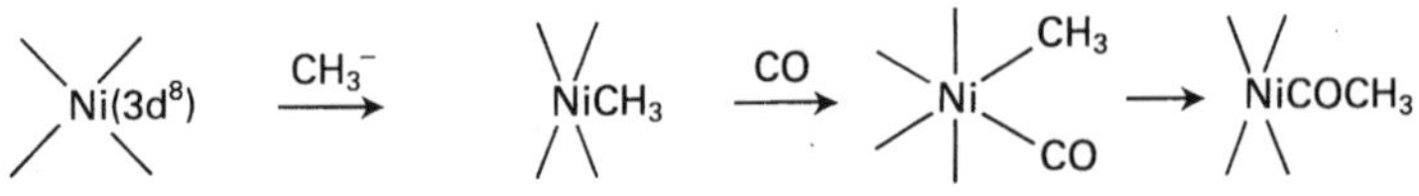

with subsequent atom rearrangements. Such reactions are well known on metal centres in organometallic chemistry.

In the description of these reactions and their catalysis a rather loose language has again developed. For example, if we consider the reaction of Ni(II) in a square planar geometry as a d^8 ion and add to it e.g. Cl_2 such that the nickel becomes octahedral, we may describe this as 'oxidative addition' since Ni can now be seen as a Ni(IV) d^6 ion with two chloride ligands. By the same reasoning, the reverse reaction is 'reductive elimination'. It is clear that many syntheses and degradations can proceed by these pathways. Notice that in reactions such as these oxidative additions nickel may well have been the element of choice, especially early in evolution. Reference to metallo-organic chemistry texts may lead one to think that some form of organometallic catalysis pre-dated life.

The above description has concentrated upon octahedral complexes, but a similar description applies to other structures. Additionally, the 3d shell acts as a partial protection for free radical intermediates; indeed, while all half-filled orbitals, e.g. in high-spin Mn(II), $3d^5$, are free radicals, they only give rise to free-radical reactions readily if the d orbitals are exposed. Electrons are clearly more reactive in σ than in π orbitals so that $3d^7$ (low-spin) cobalt in vitamin B_{12} and $3d^9$ copper cations in enzymes readily initiate radical pathways.

In this same context we note again that too rigid a limitation on conformational change at a centre could prove to be disadvantageous to catalysis, see Section 4.3. Most reactions other than electron transfer require that the catalyst flows through a series of states as the substrate is transformed into the products, see Aside. On the other hand, fast simple electron transfer reactions require this structural switching to be minimal. If follows that cations which are extremely stabilized by polarization in a particular geometry are not likely to be of great functional value. Here we can compare ions such as Fe^{3+} (d^5) and Zn^{2+} (d^{10}) with Cr^{3+} (d^3) and Ni^{2+} (d^8) noting that the first two, which do not have strong preferences for coordination symmetry, differ strikingly from the other two, which are strongly constrained in octahedral sites. There is here a correlation with the general absence of Ni^{2+} (except for urease, see Section 16.6.) and Cr^{3+} (possibly involved in one or two proteins related to sugar uptake control) from aerobic biology and the suspicion arises that they are deliberately avoided, even rejected. Given that this is only true of aerobic conditions, there is the further probability that, in the presence of sulphide, nickel chemistry is much more flexible, for example in archaebacteria, but this sulphide chemistry is not open to Cr(III). Chromium is not found in archaebacteria.

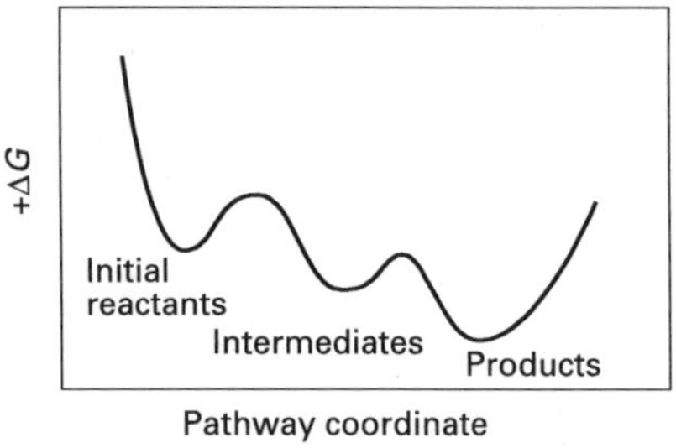

An example of a free energy profile of a reaction which proceeds in steps.

4.12 **Free radical reactions**

The major part of the above mechanistic account of the central reactions of chemistry, except the discussion of electron transfer, concentrates on two-electron reactions, e.g. transfer of ions such as H^-, CH_3^+, CH_3^-, OH^-, and H^+ or of units such as CO and O_2 inserted into molecules. There remains the possibility that, for example, $CH_3^{\bullet}$ will be transferred. Figure 4.9 illustrates the likelihood of radical transfer from potential catalyst elements as against CH_3^+ or CH_3^- transfer. We see that as polarity is removed from the substrate so $CH_3^{\bullet}$ transfer (and similarly $H^{\bullet}$, etc. transfer) becomes more probable. This is not to say that radical pathways become generally easier; rather ionic pathways become very difficult. Radical pathways will be assisted by an intermediate catalyst if the central catalytic atom can undergo one-electron reactions easily. In this case, there are three possibilities:

1. the use of transition-metal ions;
2. the use of aromatic molecules such as quinones and flavins and, with more difficulty, tyrosine and tryptophan;
3. the use of heavy elements, where ionization energy differences are small, e.g. lead.

Rather than elaborate here, we simply state that certain reactions such as those of H_2, $R-(CH_2)_n-CH_3$, terpenes, and even sugars, i.e. molecules obviously not open to ionic reactions, do take place more readily through radical pathways. The usual metal ions involved in the catalysed steps are Mn, Fe, Co, Ni, and Cu; Mo and Se are more usually employed in two-electron reactions. The situation will be discussed in several later chapters (see Table 4.5). Finally note that $H^{\bullet}$ transfer can be by tunnelling.

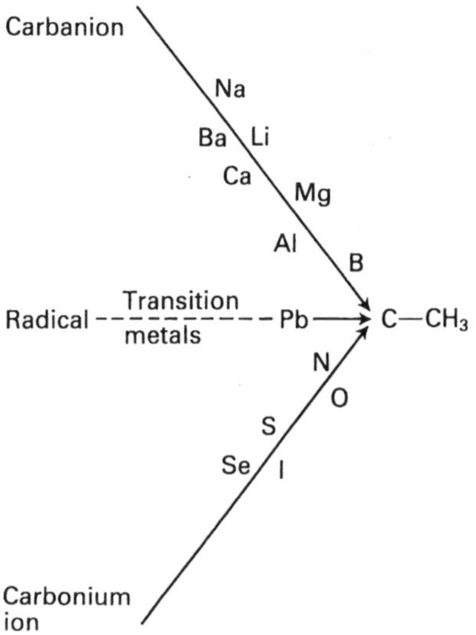

Fig. 4.9 The relative control over the type of leaving group derived from CH_3 or H, i.e. X^+, X^-, and $X^{\bullet}$. Carbanions obviously come from group 1, 2 and 13 metal methyls and carbonium ions from S, Se, I. Radicals can be generated by elements more able to undergo one-electron reactions and they may be the lowest pathway of dissociation for atoms of closely similar electronegativity, e.g. H_3C-H.

Table 4.5 Free radical systems in biology

Value	System	Elements
Free radicals for polymerization	Construction of cuticles, lignins, etc. Protective pigmentation	Cu
Detoxification (removal)	(a) In cytoplasm: superoxide dismutase (b) Outside cells: ascorbate oxidase (c) In membranes: radical reactions	Cu, Mn, Fe Cu (Vitamin E)
Protection	Reactions of leucocytes in vesicles	(Flavins)
Special reactions in cells (local traps)	Sugar reductions (for DNA) Sugar rearrangements P-450 oxidations	Co, Fe, Mn Co, Fe, Mn Fe

We observe again the need to consider other factors than just catalysis. A biological system, especially DNA, is very vulnerable to free radicals; hence cytoplasmic catalyses must not release products which are free radicals. Reactions are then controlled in space (see Section 4.14) and partly by localization of the catalyst.

Having dealt with the main kinetic mechanism we need now to make some additional points which apply to most, if not all, the different reaction paths mentioned.

4.13 Sizes of atoms, stereochemistry, and reaction paths

The intermediates of reactions may be formed by dissociative or associative paths. It is generally thought that, for a given geometry of the initial compound, the dissociative path is favoured by the smaller centres, while larger centres favour associative intermediates. Given that bond strengths are usually greater for smaller atoms the implications are that:

1. reactions of saturated small atoms, e.g. carbon, are slow;
2. reactions of saturated larger atoms can go by associative mechanisms, e.g. Si, P, S.

Examination of standard texts provides many examples.

The dissociative and associative paths involve different stereochemical demands and here the polarization of the inner core of electrons (3d) of the transition-metal ions plays a role. Figure 4.10 shows the crystal-field energy changes for trivalent ions from Sc^{3+} to Co^{3+} for the two steps (the crystal-field energy change parallels the change of polarization of the 3d core). The bottom part of the figure shows the observed ranges of rate constants for these ions in reactions by either pathway. It is obvious that some metal ions, e.g. Co^{3+}, cannot be used as catalyst centres since they do not bind or release groups easily, while others are excellent, e.g. Fe^{3+}.

The adjustment of rate processes by bound ligands has both electronic and steric components and is too complicated for further analysis here.

4.13.1 Selective binding to protein (DNA) sites in pathways

We have not stressed the problem of the selectivity of biological catalysts although this is of extreme importance. In fact, selectivity is a common

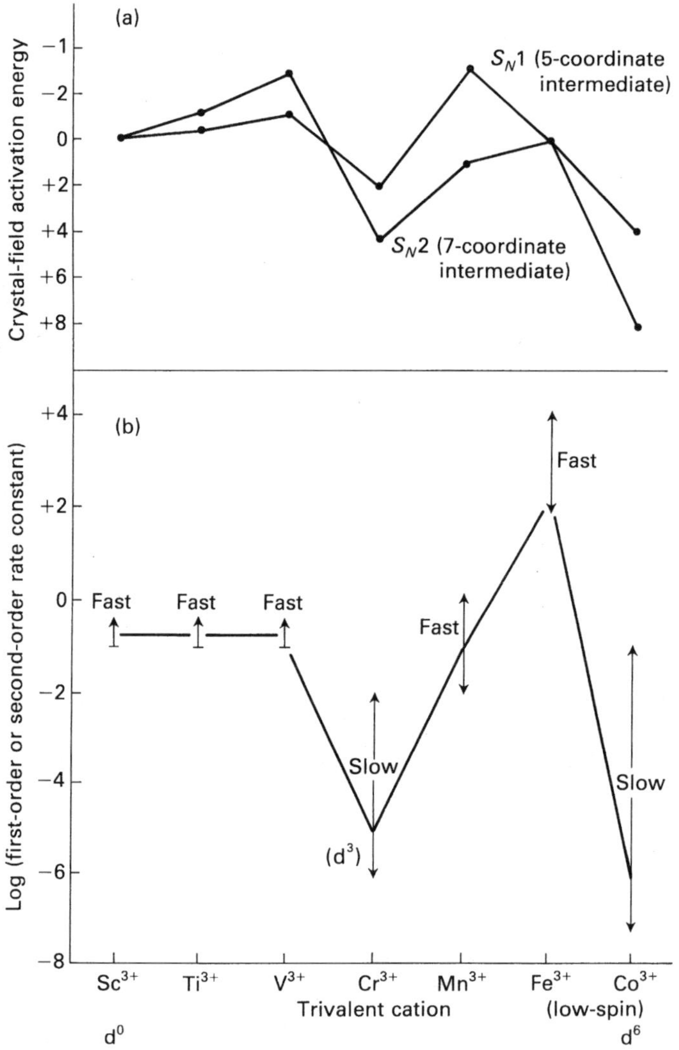

Fig. 4.10 The crystal field stabilization energy on going from a 6-coordinate complex to either a 5-coordinate (dissociative reaction) or a 7-coordinate (associative reaction) intermediate. S_N1 and S_N2 denote nucleophilic substitutions via unimolecular or bimolecular reaction paths, respectively. (b) An impression of observed rates of substitution showing how the energy in (a) affects reactivity depending on the number of d-electrons in the ion.

requirement in all the reactions of biology since individual catalysts are designed to be part of pathways. There may be as many as ten catalysts in the pathway so that each one must give only the product for the next step. In fact, this principle is more important than the establishment of the best or fastest catalyst. Hence enzymes are not likely to be the best catalyst for a given step. Again, the catalytic activity in the chain needs to be controlled so that products in the pathway or at its end do not accumulate. Hence series of controlled (by feed-back) enzymes generate the flow of chemicals along precise pathways which have a rate commensurate with flow in other pathways. There is therefore a communication network based often on units common to more than one pathway, e.g. units such as ATP, NADH, Ca^{2+}, and Mg^{2+}. Any reference to a biochemical textbook will reveal how selective enzymes are due to shaped surfaces of particular sidechains and the fact that the surfaces are optically active. DNA must also bind transcription proteins selectively (see Chapter 6).

4.14 The creation of local small spaces: mechanical devices for transfer reactions

Discussions of mechanism in inorganic and organic chemistry usually leave aside mechanical devices; the reactants are seen as just diffusing together randomly. Biological catalysts can behave just in this way, acting as a relatively rigid recognition surface for partners. This is the standard approach to digestive enzymes, for example proteases. A very different reaction route is also found in which the process of binding alters the catalyst in a mechanical sense; a simple example is a hinge-bending enzyme (Chapter 6). The reaction sequence, where E represents the enzyme, A and B the reactants, and C and D the products, is

1. $E + A \rightarrow E'A$;
2. $E + B \rightarrow E''B$;
3. $E'A + B \rightarrow E'''AB$;
4. $E''B + A \rightarrow E'''AB$;
5. $E'''AB \rightarrow E''''CD$;
6. $E''''CD \rightarrow C + D + E$.

The different enzyme conformational states are represented by dashes.

The last step (6) could be re-written as a series of steps to follow steps (1) to (4) replacing C and D for A and B, respectively. The series of reactions shows that the enzyme is a mobile polymer. If steps (1) and (3) only and not (2) and (4) are permitted, there is a required order of addition of reactants to the catalyst. If the way back from E'''' can be arranged to be via conformations very different from E′, E″, and E‴, then the enzyme cycles rather than retraces its steps and energy is required in such a one-way flow (see Chapter 5). Pumps are of this kind; there is a big difference between a two state (open $\rightleftarrows$ shut) reversible equilibrating system and a multistate cycle, which can be energy driven to avoid reversibility that allows synthesis of unstable compounds, as discussed in Section 4.2.

Now, let us examine what can be achieved by avoiding a random set of conformations. If we take the case of the desired reaction

$$ATP + X \rightleftarrows XP + ADP$$

where ATP is just an organic triphosphate, A-P-P-P, this is in competition with hydrolysis, which is energetically much more favourable

$$ATP + H_2O \rightarrow P_i + ADP$$

where P_i is free inorganic phosphate. Hence, the binding of ATP must not be such that it is activated without the presence of X. It may also happen that the reaction

$$XP + H_2O \rightarrow X + P_i$$

is favourable, so neither the reactant ATP nor the product XP must be bound initially in an activated state. The required transfer of P from ATP to X or of P from XP to ADP must follow only if X and ATP or XP and ADP are together and activated after a conformation change forms a local pocket which eliminates water. Since ATP is an energy store, the reaction retains energy in the products instead of losing it to hydrolysis. The above case is one of random substrate addition; an example is that of phosphoglycerate kinase. Consider next the case of attack by a reagent which, no matter what the conformation of the protein, is very aggressive, e.g. oxygen bound to iron not in a carrier protein, such as haemoglobin, but in a detoxifying aggressive oxidase, such as cytochrome P-450. Oxidative attack on the protein itself is now the risk, but this possibility can be removed if step (1), the binding of the substrate A to be attacked, is required before B. In this case previous binding of A allows O_2 to see A only in its site of binding and protects the enzyme from O_2, i.e. steps (1) and (3) are allowed, but steps (2) and (4) are forbidden. This is a required order mechanism which opens a space for a reagent.

These mechanical movements in proteins require 'hinges'. Later we shall attempt to show that the enzymes which have hinges can use β-sheets for their recognition sites, but that the hinges themselves are based on α-helices (Chapter 6). There are two classes of such hinges, simple and energized (compare the rising hinge of a door). The latter stores restoring energy after movement.

We turn next to the final mechanical trick using helical motions—the irreversible ratchet. If during a series of reactions there is an unstable intermediate then the destruction of the intermediate loses energy from the reaction system. It is possible then to have two kinds of pump; one which transfers energy from one system to another without deliberate loss (a balanced pump) and another which during its cycle loses energy. The second will not run reversibly. Examples are given in Chapter 8. The movement of ions in channels may well depend on such mechanical properties of proteins and so may the synthesis of ATP from proton gradients. Finally, this is the mode of transfer of the energy of ATP to the filamentous systems which maintain tension.

So far in this chapter we have considered kinetics as a problem in a free aqueous solution, so that reaction rates are expressed in terms of rate constants and concentrations per unit volume. In fact, as seen before, a biological cell is criss-crossed by fibrous proteins and each cell has many membranes; hence molecules attached to the fibres or membranes can diffuse along a one-dimensional path, while those in a membrane can diffuse on a two-dimensional matrix. Because of the attachment of the fibres to the membrane there are many ways in which movement can be controlled. At one time such ideas were considered fanciful, but it is clear today that, as well as channelled diffusion, there are such phenomena as the diffusion of proteins along DNA and in membranes, and the diffusion of vesicles along fibres in nerve cells. It is also likely that ion diffusion on charged surfaces is extremely important.

4.14.1 Rates of conformational changes

A further possibility is that a reaction in any matrix involves a cooperative change of a large number of atoms. Such a reaction may be a local cooperative change or a global concerted change in a protein much as in a phase change. In this chapter we have mentioned such changes many times with reference to conformations. If a protein is treated as a mini-phase then the description of kinetic alterations of structure has to be in terms of allosteric switches of state since a protein is not static internally. As a protein is a very small phase and such switches may involve small movements of, for example, several helices or small numbers of water molecules, there is no reason to suppose that co-operative change cannot be fast. Mechanistic considerations here then have the same complexity as those of other phases for which descriptions use thermodynamic functions of state. A typical problem of this kind is protein folding when only two states may be seen. The part of this problem which concerns the chemical elements is their incorporation during folding, which can be all important for their eventual localization. Another area for study of possible phase changes is in redox processes. An example is haemoglobin on dioxygen uptake and another possible case is gating of proton transfer in cytochrome oxidase. There then is a connection between shape of a system and its reactivity. However, it would not be wise to limit our analysis of these changes to just two rigid structures. There is some evidence from NMR that all proteins undergo small but extensive changes even on binding an electron, a proton, or a small substrate.

4.15 Networks: kinetic circuits

A man-made electronic circuit is based on the flow of electrons and various flow-control (kinetic) devices can be put into the circuit such as resistors, transistors, condensers and switches. Moreover, it can be self-controlling through its own fields, magnetic or electric, devices, e.g. the ringing of a door bell. An electrolytic circuit is similar, but the current carriers are ions, not electrons, the medium is a solvent (channel), not a wire, and so on. We shall see that biological circuits are never based on electrons alone, but they can be based on electrons plus ions especially the proton. More usually biological circuits are based upon Na^+, K^+, Cl^-, H^+, and Ca^{2+} ions and of course upon a range of flowing small or even large (carrier) molecules. An example is phosphate in that it is used in energy transfer and signalling (see Sections 18.2 and 18.3), but here transfer is through carriers such as ATP, not free phosphate. However, circuits can be of any material or energy and need not use charge. In this book we are concerned to know what other circuits are possible. Consider an element such as nickel. It needs to be guided to the correct active site. This is managed by letting it diffuse in compounds, which sounds haphazard until it is remembered that each nickel carrier may pass the nickel to the next member of the chain *selectively* without equilibrating with the free nickel ion. Such specific transport by high affinity carriers is possible for ions such as Ni^{2+}, Cr^{3+}, Cu^{2+}, and Cu^+ but is increasingly difficult for ions such as Zn^{2+} and Co^{2+}. Finally Fe^{2+} and Mn^{2+} exchange with their free ions from most sites so that we return toward circuits based on free ions (as for potassium) rather than on exchange of bound ions. There will then be cases where one element is part of more than one circuit. For example free iron circuits are quite distinct from those of haem iron. We shall return to these topics in Chapter 20.

4.15.1 Control over networks of reactions

When we turn to the kinetics of reactions in cells we observe that catalysis of chemical transformation, much like the physical movement of uptake and transfer across membranes (Chapter 3), is under control so that the homeostasis of cells is maintained. The implication is that enzymes as well as pumps are under feed-back restriction

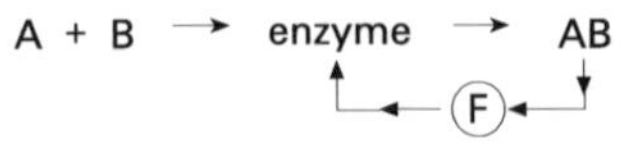

where Ⓕ indicates that as AB increases in concentration it inhibits the enzyme. Now we shall see that rather than dealing with single reactions and their control a cell controls reactions in sequences, metabolic paths (see Fig. 5.10 for an example). The path is linked to uptake and to other paths of

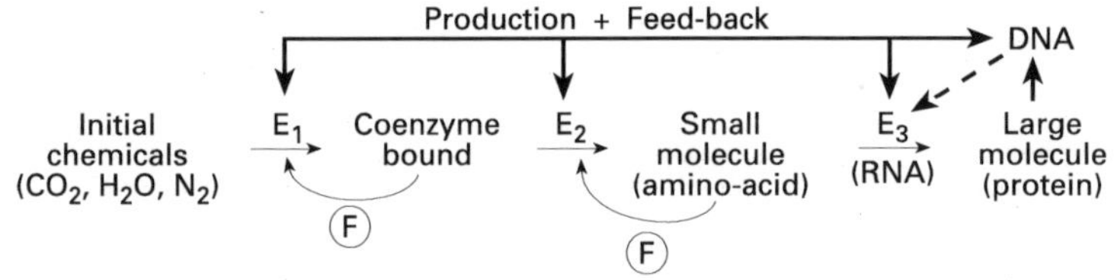

Fig. 4.11 The rates of production of any compound in a cell are under feed-back control Ⓕ. This feed-back can be upon the synthesis of RNA or a protein, but it can also be upon an enzyme reaction by large or small units of flow, see Section 7.5 and Fig. 7.3.

necessity, so that the whole has a functional purpose. It is then found that the feed-back control is exerted to some degree on all the steps in the whole of the cell. This level of control applies to the synthesis and degradation of the enzymes themselves through messenger systems that connect to RNA and DNA (Chapter 6, and see Fig. 4.11). Yet again, the synthesis of these very nucleotides is linked to the availability of the elements. The whole is therefore a unity of reactions so that an organism has its particular timed sequence of chemical actions allowing it to be characterized by its system *behaviour* as well as its DNA, its protein content, its element content and its shape. We will come again to this subject in Chapter 20.

4.16 **Summary**

This chapter has described in brief outline the kinetic constraints and controls in chemical systems with only brief reference to living organisms to keep the topic in the context of the book.

The kinetically valuable features of the use of elements other than C, H, N, O, S, and P which we have discussed are:

1. strong acid-base catalysis, e.g. by Zn^{2+};
2. rapid mobility with weak binding in electrostatic currents (Na^+, K^+, Cl^-);
3. rapid exchange with moderate binding (Ca^{2+}, Mg^{2+});
4. electron transfer using Mn, Fe, Co, Ni, Cu, Mo;
5. ease of atom transfer from Se, Mo and to some degree from Mn, Fe, Co, Ni, Cu;
6. potential for free radical reactions (many transition metals);
7. the control of networks of reactions (many elements).

We must next look at an area where biological chemistry is particularly unlike normal chemistry. This lies in the use of energy for synthesis, for organisation and to maintain flow, which is the subject of Chapter 5. We shall then be able to return to the way in which essential elements are exploited functionally in biology and, in Chapter 7, we shall show how the chemical functions of the elements, now not just the catalytic activity but also the ability to transfer electrolytic charge and electrons, and to participate in structures, are utilized in biology. This returns us to the way in which elements are partitioned in space. It might be difficult for a chemist to say why he should choose iron rather than copper as an oxidation catalyst in a given process since both could, in principle, do the same job. Biological systems use both iron and copper in oxidative catalysis, but their roles are very well separated in space so that their total function is very different. This is equally true of acid-base catalysts, where Zn^{2+} is used as a general acid with the proton, but we find that for particular reactions Mg^{2+} is preferred, e.g. phosphate transfer, while other metal ions are most important in special zones, for example Fe^{3+} and Mn^{3+} in lysosomes, and Ca^{2+} outside cells. Again Na^+ and K^+ are biologically totally distinct and are not similar ions as in chemistry.

Further reading

There are many standard textbooks which describe the kinetics of organic compounds and catalysis by enzymes.

Atkins, P.W. (1998). *Physical chemistry* (6th edn). Oxford University Press, Oxford.

Basolo, F. and Pearson, R.G. (1958). *Mechanisms of inorganic reactions*. John Wiley, New York.

Connors, K.A. (1992). *Chemical kinetics in solution*. VCH, New York.

Cramer, W.A. and Knaff, D.B. (1990). *Energy transduction in biological membranes*. Springer-Verlag New York.

Eigen, M. (1960). Exchange of water from around cations. *Zeitschrift für Electrochemie*, **64**, 115–30.

Groen, J.H., De Jong, B.J., Ernsting, J.M., Van Leuwen, P.W.N.M., Vriese, K., Smets, W.J. and Spek, A.L. (1999). Synthesis and reactivity of some palladium complexes. *Journal of Organo-metallic Chemistry*, **573**, 3–13.

Harrison, L.G. (1993). *Kinetic theory of living patterns*. Cambridge University Press, Cambridge.

Johnson, M.K., King, R.B., Kurtz, D.M., Kutal, C., Norton, M.L. and Scott, R.A. (eds) (1990). *Electron transfer in biology and the solid state*, Advances in Chemistry Series, No. 226. American Chemical Society, Washington, DC.

Kuznetsov, A.M. and Ulstrup, J. (1999). *Electron transfer in chemistry and biology*. John Wiley, Chichester, UK.

Laidler, K.J. (1987). *Chemical kinetics*. Harper & Row, New York.

Luecke, H., Shobert, B., Richter, H.T., Cartailler, J.P., and Lanly, J.K. (1999). Structure of bacteriorhodopsin at 2.0 Å resolution. *Science*, **286**, 255–60.

Marcus, R.A. and Sutin, N. (1985). Electron transfers in chemistry and biology. *Biochemica Biophysica Acta*, **811**, 265–322.

Miller, C. (2000). Ion channels. *Current Opinion in Chemistry and Biology*, **4**, 148–51.

Moser, C.C., Keske, J.M., Warucke, K., Farid, R.S., and Dutton, P.L. (1992). The nature of biological electron transfer. *Nature*, **355**, 786–802.

Ponce, A., Gray, H.B., and Winkler, J.R. (2000). Electron tunneling through water. *Journal of the American Chemical Society*, **122**, 8187–191.

Norman, R.O.C. and Hill, H.A.O. (eds) (1986). *Inorganic and organic radicals*. Proceedings of a discussion. The Royal Society, London.

Packer, L. (ed.) (1986). *Protons and water*. Methods in Enzymology, Vol. 22. Academic Press, New York.

Purves, W.K., Orians, G.H., Heller, H.C., and Sadava, D. (1998). *Life—the science of biology* (5th edn), Chapters 6–8. Associates, Inc. and W.H. Freeman, Sunderland, Mass.

Taube, H. (1970). *Electron transfer reactions in solution*. Academic Press, London.

Winkler, J.R., Di Bilio, J.A., Farrow, N.A., Richards, J.H., and Gray H.B. (1999). Electron tunneling in biological molecules. *Pure and Applied Chemistry*, **71**, 1753–64.

Williams, R.J.P. (1999). Electron transfer and proton coupling in proteins. *Journal of Solid State Chemistry*, **145**, 488–95.

Williams, R.J.P. and Fraústo da Silva, J.R.R. (1999). *Bringing chemistry to life*. Oxford University Press, Oxford.

Note: Structure and Bonding (1991), **75**, contains a series of articles concerning electron transfer in proteins, pp. 1–224. *Journal of Biological Inorganic Chemistry*, (1997) **2**, 373–404 contains a series of papers summarizing recent research on electron-transfer reactions.

CHAPTER 5

Energy in biological systems and hydrogen biochemistry

5.1 Introduction

As we have seen in the previous chapters, the second major requirement for living organisms, besides the raw materials, ultimately chemical elements, is *energy* made available in appropriate, usable forms. In this chapter we examine which forms of energy are available for life and which forms are easily harnessed or stored for later use and transducible into other convenient forms, e.g. chemical bond energy. The parallels with man-made systems for generating useful energy are not sufficiently useful for us to employ such models but the reader should be aware of the difficulties of utilizing available energy sources efficiently both by man and by organisms. Despite the difference which will become apparent, captured energy generally can only be contained by cellular systems in the same way as man contains it in chemical and physical storage devices Some of the obvious static forms follow.

1. **Electrical storage.** Storage due to charge separation, as in a condenser (electron store), or at a boundary between aqueous phases where diffusion, which would neutralize charge, is prevented, i.e. individual ion stores.
2. **Concentration storage.** Here a compound be it a salt, NaCl, or an organic molecule is held in solution and prevented from diffusion by a membrane. Such junctions occur in electrolytic concentration cells.
3. **Light energy storage (capture).** Light is absorbed locally in an excited pigmented layer before going to heat or another form of energy, e.g. electrical, chemical, conformational, or mechanical energy.
4. **Chemical bond energy storage.** Some combinations of chemicals are less favourable than others as is illustrated by the use of fuels by man. The combination of C with H (CH_4) separately from O (as O_2) is less favourable than bound C with O and bound H with O. The basis of the petrol engine is

$$CH_4 + 2O_2 \rightarrow CO_2 + 2H_2O + \text{energy}.$$

The redox states of carbon and other elements in Figs 1.7 and 1.8 are plotted on an energy scale; this scale of chemical energy ignores combinations of elements other than with H and with O and refers to fixed unit concentration conditions for each, i.e. unit gas pressure or unit concentration in water. We obtain the energy equivalent of a chemical change per mole from differences along the ordinates of these figures; the slope is the chemical potential of unit change of redox state, i.e. the redox potential. The difference in redox potential between two couples is an energy source that can be used by organisms, but they can also obtain energy from chemicals in many different ways.

5. **Energy storage in a field.** Life on earth is in a gravitational field and movement away from the earth, i.e. in this field, costs energy. Therefore, objects held high above the earth store energy even if this energy is usually trivial, and ignored in chemistry. It is not trivial in physical and biological systems. Of much greater importance are the electric fields in cells.
6. **Energy due to tension storage (mechanical energy).** Storage of energy in a stretched rubber band is an obvious example. Biological systems are full of deformable string-like objects, e.g. muscle fibres and ligaments, and very similar fibres (filaments) are present in most cells. We can extend this description to two-dimensional surfaces, e.g. of balloons. All cell membranes or containing walls of biology are in balloon-like tension, e.g. the skin. Deformation energy is often contained in the confined conformational or configurational energy of polymers.

It would certainly be useful if we could give quantitative expression to some of these energy stores. Using ΔG to sum the free energy change per mole, that is the available energy for work on going from one state to another, we define $\Delta G°$ of a chemical as its energy in a defined state relative to its component elements in defined elemental states in which they sense no field, are not in tension except being under unit pressure, are at unit concentration (if in solution), and are at 25°C. Change of concentration to a value [M], where M is any element, different from unity causes a change in ΔG given by the expression

$$\Delta G = \Delta G° + RT \text{ In } [\mathrm{M}].$$

Changes of other factors extend the equation to fields and tension

$$\Delta G = \Delta G° + RT \text{ In } [\mathrm{M}] + \text{energy due to physical effects}.$$

(A high negative ΔG represents a highly stable state by convention.) Change of chemical combination is then considered in the term $\Delta G°$, which contains the sum of the internal energies of all the chemical bonds in the molecule relative to those in the separated elements, as well as entropy differences.

Biological systems accumulate energy in order to create or develop ordered chemical constructs, so they have to adjust $\Delta G°$, [M], and the physical condition of a large number of chemical elements within compounds. The starting condition of the elements before they enter biological systems differs from element to element (see Chapter 1). Most required non-metals outside biology are present as stable and available simple gases, N_2, O_2, and CO_2, or as molecules or ions in solution at reasonable concentration, H_2O, HCO_3^-, NO_3^-, SO_4^{2-}, and Cl^-. These are basic food stuffs for many unicellular systems and plants. This high availability applies to some metals too, e.g. Na^+, K^+, Mg^{2+}, and Ca^{2+}, but for the rest of non-metals and many metals their concentration, [M], is very low. The biological problem is to get these elements out of these relatively stable conditions (large negative ΔG values) into the constructs we observe in biology. The major consideration is then transfer, from one kind of energy to another. This is achieved in various ways as described in the following sections.

5.2 Biological systems and light

Organisms depend very largely on sunlight energy, $h\nu$ (Fig. 5.1). Light is absorbed by membrane pigments, e.g. green (Mg) chlorophyll, and, as in a

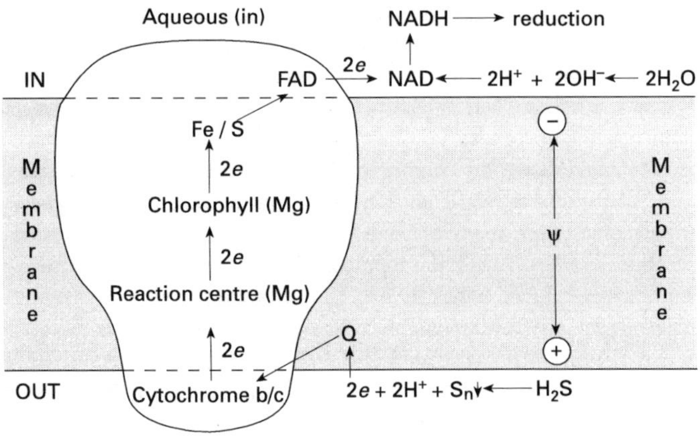

Fig. 5.1(a) The most primitive type of photosystem in green sulphur bacteria which deposits elementary sulphur.

man-made photocell, this energy is converted to a local gradient of electronic charge, ψ; that is, an electronic potential energy. Chemically, this is seen as a change of oxidation state between, say, two compounds of hydrogen, (or of two metal ions) on opposite sides of a membrane [Fig. 5.1(a).] One centre becomes more positive, the other more negative. The electron/positive charge separation can then be converted to freely diffusing entities by moving the charges to unbound molecules or ions in the two solutions on opposite sides of the membrane [Fig. 5.1(a)]. The most primitive example is that of green bacteria which have a system shown in Fig. 5.1(a) which resembles photosystem (PS) I of the chloroplasts of plants today. Here the transfer of charge generates reducing equivalents via NAD on the inside of the cell and oxidizing equivalents as sulphur from sulphide on the outside of the cell. Simultaneously a proton gradient is created across the membrane (acid outside) and this gradient is an energy store to be used subsequently to make ATP (see below).

Another scheme, seen in purple bacteria, uses cyclic photosynthesis and more closely parallels photosystem (PS) II of modern plants. It includes the Q-cycle which is simplified in Fig. 5.1(b). In this case we may write

$$\text{Light } (h\nu) \rightarrow \begin{array}{c}\text{membrane}\\ \text{bound } \psi\end{array} \xrightarrow{(\mathrm{QH_2\ cycle})} \begin{array}{c}\text{free}\\ \mathrm{H^+_{out}} + \mathrm{OH^-_{in}}\end{array}$$

where the subscripts ‘in’ and ‘out’ represent relative separation in space.

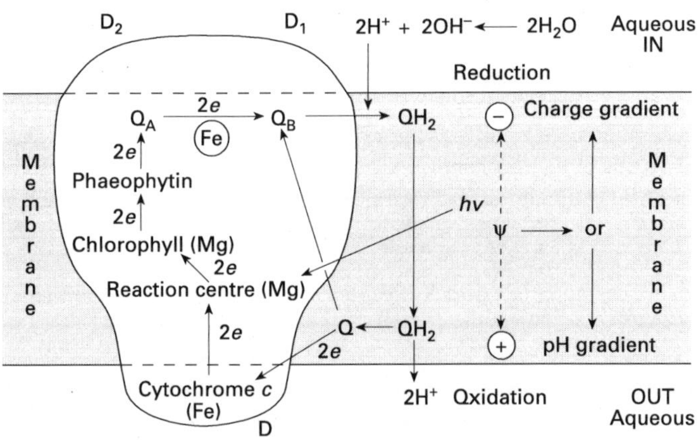

Fig. 5.1(b) The reactions of the photosystem of purple bacteria. Light drives electrons and hence the separation of charge across the membrane to give an electrical potential, ψ. The separated charges can rest on metal ions or organic molecules. At the edges of the membrane the charges can migrate to the aqueous phase to give a proton gradient and the chemical species generated can be used to drive organic reactions. Note the metals used. Q, quinone, which cycles in the membrane. This system is more like photosystem II of chloroplasts today (see below).

The succession of steps can be followed in Fig. 5.1(b). 'Membrane bound ψ is initially formed as in photosystem I by the action of light on a *reaction centre* of two chlorophyll (Mg) molecules, Chl_2, to give an excited diradical state; $Chl^{+}Chl^{-}$. The electron passes from Chl^{-} to another single chlorophyll molecule and then in succession to single phaeophytin and quinone molecules before H^+ moves into the membrane of the bacteria from water leaving negative charge as the OH^- in the *inner* aqueous phase of the bacteria. Meanwhile, the initial positive charge of Chl^+, has been moved from this site in the opposite direction by the input of electrons from a cytochrome (or a multicytochrome unit), which in turn gets electrons from QH_2 leaving the positive charge as H^+ in the *outer* aqueous phase. Thus ψ covers all intermediate dispositions of in-membrane potentials before the H^+/OH^- gradient appears in the aqueous phases (inside and outside). There are obvious similarities between parts of these two schemes of Fig. 5.1. The cyclic nature of the system II follows from the flow of Q. The electron transfers from site to site have been thoroughly analysed (see Section 4.7). The proton transfer path is more difficult to study, but in part it involves series of water molecules (see Williams and Morowitz in the references). Notice that the switch is from physical energy as light, to a physical field (electronic potential) localized on atoms, to a chemical transformation ($\Delta G°$ change), and finally to a concentration change of H^- (QH_2) and H^+ ($RT\ln[H^+]$ change), Fig. 5.1).

In the chloroplasts today the electrons of photosystem II (see purple bacteria) go to Q_A/Q_B but not just in cyclic fashion since these quinones are coupled to non-cyclic electrons from H_2O giving O_2 as in Fig. 5.2. Reducing equivalents from PSI go to NADP giving NADPH and then sugars, the ammonia for amino-acids, and so on. That is a further, now favourable, ΔG change. Such intermediate small molecules are required by all metabolic systems and these intermediates already hold elements at a considerable higher (less negative) chemical free energy; making polymers still further increases the positive ΔG change and an extra source of energy from ATP is required (see Sections 4.2 and 5.5). Alternatively, CO_2 could be reduced to CH_4 or heavier hydrocarbons, e.g. oils and stored gas. In chloroplasts the two photosystems are combined (Fig. 5.3).

We see that today in addition to using light, $h\nu$, for direct reductive metabolism, chloroplasts use metal ions in a second photosystem as oxidizing reagents, e.g. for the production of dioxygen of the air which is a 'waste' energized product or rather a remote biological energy store (Section 5.3). Manganese is the metal used to make dioxygen from water in biology (see Section 14.5); hence oxidation states of manganese are the intermediates (Figs 5.2 and 5.3), i.e. the light is used to drive the change $2Mn^{2+} \rightarrow 2Mn^{4+}$ followed by

$$2Mn^{4+} + 2H_2O \rightarrow 2Mn^{2+} + 4H^+ + O_2$$

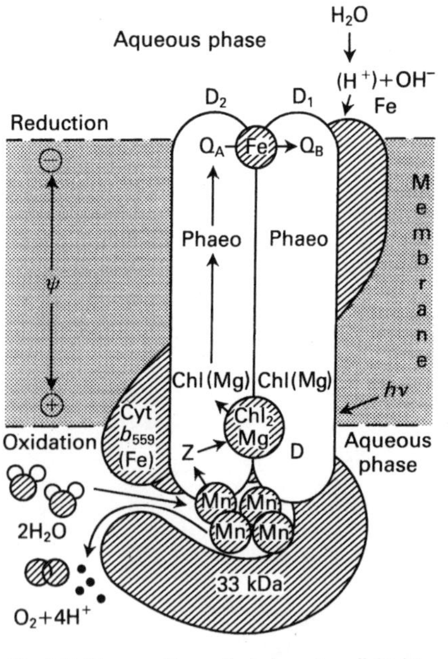

Fig. 5.2 The reactions of photosystem II. Light drives electrons and hence the separation of charge across a membrane to give an electrical potential, ψ. Because the system works at higher redox potentials than photosystem I the charge separation is used to generate dioxygen as a product as well as a proton gradient and reducing equivalents. Note the metals used. Chl, chlorophyll; Phaeo, phaeophytin; Q, quinone; Z, a tyrosine radical. The figure is schematic.

The combination of the two photosystems, Figs 5.1 and 5.2, is shown schematically in Fig. 5.3 but their real spatial distribution is complicated (see Chapter 3). The storage of O_2 in the air plus hydrocarbons such as oil and gas, CH_4, in the ground (an alternative reduced form of CO_2 is $(CHOH)_n$, cellulose, e.g. in plants) by the combination of the two photosystems (see Fig. 5.3) can be looked upon just as ΔG storage of separated oxidizing and reducing compounds in chemicals, as far as living systems are concerned, since they can combust (eat) oils or cellulose with O_2. The reaction of O_2 with these reduced chemicals then gives a second supply of oxidative energy in mitochondria:

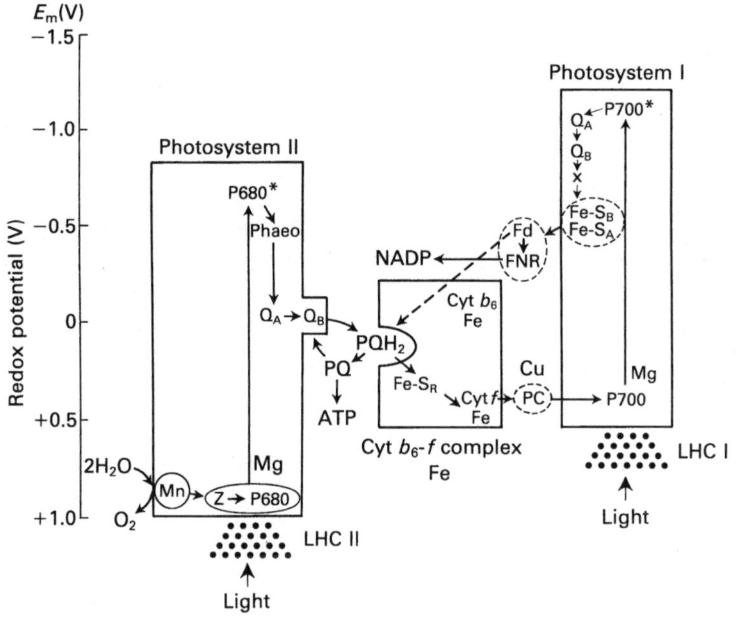

Fig. 5.3 The combination of photosystem I and photosystem II shown on a redox energy scale and indicating their interaction (but see Chapter 3). Z, a tyrosine radical; P and P*, ground and excited states of reaction centre chlorophyll; Cyt, cytochrome; PC, plastocyanin; Fd, ferredoxin; Q, quinone. Note the different uses of four metal ions. Fe-S_A, Fe-S_B, and Fe-S_R are different iron-sulphur proteins. LHC is a light harvesting system, FNR is ferredoxin reductase.

$$[CHOH] + O_2 \rightarrow CO_2 + H_2O + \text{energy}$$
$$CH_4 + 2O_2 \rightarrow CO_2 + 2H_2O + \text{energy}$$

The energy is transduced into ATP, (see Section 5.5), and then energy driven reactions produce intermediates for chemical synthesis either in plants or animals.

5.3 **Oxidative energy: mitochondria**

The reducing compounds, e.g. bound H_2 in organic compounds, and the oxidizing compounds made from light, O_2, can be applied to opposite sides of a separate bacterial or mitochondrial membrane, which is different from the membrane used in light energy capture, and their reactions then can lead back to electrical ψ (electron gradients) (Figs 5.4 and 5.6).

In this case the dioxygen is first bound to an Fe, Cu centre which it oxidises so generating charge distribution $[FeCu]^+O_2^-$ inside the inner membrane, in an enzyme called cytochrome oxidase, also known as particle IV (Fig. 5.4 and see Section 12.6). Here we describe the case of the mitochondria (see Fig. 3.2). The dioxygen itself is finally reduced to water by the flow of electrons starting from the oxidation of a C/H compound (which was produced from metabolism of CO_2 to sugars by the above photosystem) generating initially the intermediate NADH. The flow of electrons from NADH to O_2 generates energy which is used to create the flow of protons from inside through particles I, III, and IV, so as to give a proton gradient across the membrane (compare the systems in Fig. 5.1). The electron flow is complicated in that the initial oxidation of NADH in complex I of the mitochondrial chain, (Fig. 5.4) is followed by flow of reducing equivalents to a quinone ($Q \rightarrow QH_2$) and then from QH_2 to cytochrome c (via particle III) and finally to O_2 in particle IV. The free energy difference between O_2 and the NADH in this case is converted within three particles to a total protonic potential ψ (Fig. 5.4). For

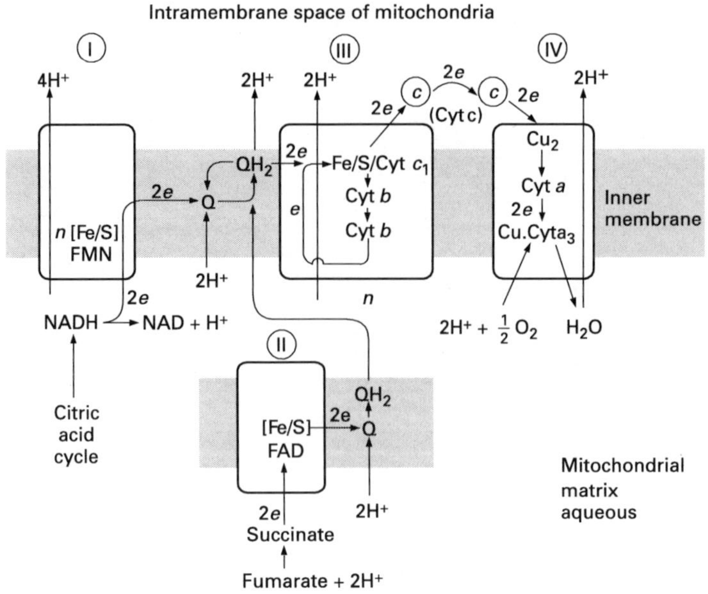

Fig. 5.4 The mitochondrial redox chain for making proton gradients from the oxidation of NADH with O_2. There are four particles: I, NADH reductase (see Section 12.2); II succinate dehydrogenase/Q-reductase; III, Quinone cytochrome *c* reductase (see Section 13.6.4); IV, cytochrome oxidase, (see Section 13.9.1).

one reduced C/H compound, succinate, the reduction potential is closer to that of O_2, and electrons from its oxidation enter the chain of e/H^+ reactions via particle II and then, leaving particle I to one side, go directly to particle III giving a lower energy yield (Fig. 5.4). Apart from the generation of ψ directly from (dislocated) chemical reactions across membranes,

$$4H^+ + O_2 \xrightarrow{4e} [4H^+ \text{ out } + 4OH^- \text{ in}] + 2H_2O$$

and reduced C/H compounds → oxidized C/H compounds + H^+ (out) gradient, there is pumping of H^+ to the outside of the inner membrane of the mitochondrion by the mechanical action of the enzymes. Thus, the total gradient of charge and pH is made up of many separate reactions to secure a high efficiency of energy conversion from the chemical free energy difference of organic compounds relative to dioxygen. Note that both e and H^+ travel in the membrane before any charge separation appears in the aqueous phases.

Today there is much detailed knowledge of the structures of all the four particles, but in particular there are complete X-ray structures for particles III and IV in different states (see Chapters 12 and 13). The understanding of these states remains incomplete, for although the location of electrons is readily uncovered with the help of spectroscopy the paths of protons are obscure. Model studies of cytochrome *c* (Sections 4.7 and 13.6) and what is known of structure itself indicate how water molecules as well as proteins in the membrane can be important in the pathways, [see Figs 4.4 and 5.6(a)]. Now, a further difficulty is encountered in that the enzymes act as mechanical pumping devices. The relationship of parts of the fold to other parts, or of subunits to other subunits, is adjustable to greater or lesser degrees. These switches of conformation are a necessary part of the pumped movement of protons, which requires gating. At present the structural resolution is not sufficient to give a clear picture of the way in which the molecular machinery works. Some idea of the workings can be obtained from the study of bacteriorhodopsin which generates a proton gradient by proton (not electron) diffusion in certain bacterial membranes. Here it is known that the small helix/helix conformational changes gate proton movement, which follows a path along certain ionizable groups and water molecules (see Lueke *et al.*, in the references). In this book we shall not give the detailed structures of any of these complicated proteins (see references) since we are more concerned with principles of energy capture, but we note the extreme conformational changes involved in particle III, the bc_1 complex, reactions.

One very interesting feature of oxidative phosphorylation (Fig. 5.4) is that starting from the organic C/H compounds (NADH) the electrons and protons move through a very long chain. There are some 20–30 Fe atoms in Fe/S clusters, 5 haem units, and 3 Cu atoms apart from coenzymes, NADH, flavin and quinones. The most primitive part, which excludes particle IV, has no Cu and, possibly, used sulphur originally as an oxidant in particle I. This Fe/S system contrasts with the complexity of the earliest photosystem, see the Mg chlorophyll units together with several haem units in photosynthesis system I and Cu or haem and 4Mn associated with photosynthesis system II, and with the haem and copper units in particles III and IV of oxidative phosphorylation. The different roles of trace inorganic elements in energy capture is quite remarkable and suggestive of an evolutionary sequence from the simplest involving Fe/S then Mn to the most complex involving Cu. The two modes of energy generation, light and oxidation of hydrocarbons, are closely similar in that both produce gradients of potential and then of H^+, and both use some similar components in membranes (Fig. 5.5).

The above account of energy capture cannot be that of very primitive life since it is (1) extremely organized even in the simplest bacteria of Fig. 5.1; (2) in Fig. 5.1 we also see that very complex synthetic molecules are used, for example chlorophyll and haem. In Section 5.11 we shall look again at possible sources of energy and their capture at the earliest times of life's existence. Here we note the probable evolutionary sequence green bacteria, purple bacteria, combined photosystems in bacteria, chloroplasts, mitochondria. It may well be that chloroplasts derive from cyanobacteria and mitochondria are derived from *Paracoccus denitrificans*.

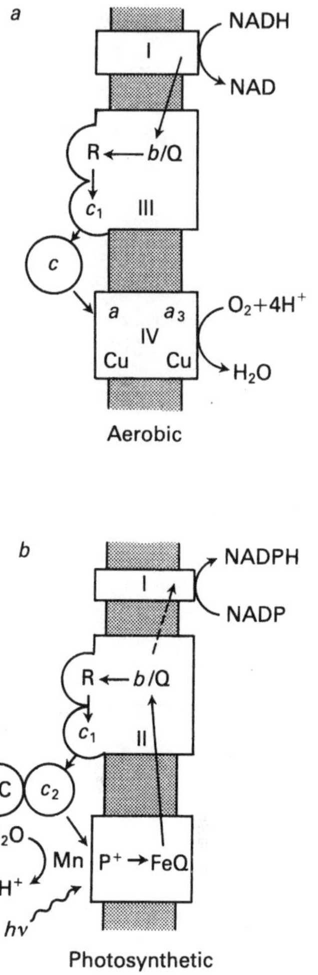

Fig. 5.5 The parallels between the electron transfer, middle sections, of photosynthesis, below, and of oxidative phosphorylation, above, are illustrated. *a*, *b*, and *c* represent cytochromes *a*, *b* and *c* respectively, R is a high potential iron protein. P is a photocentre, PC is plastocyanin, and Q are quinones. Details are not necessary here.

5.4 **Proton migration coupled to redox reactions**

The question of proton as opposed to electron flow is, as stated, very difficult to analyse. There are two separate problems—proton flow within proteins and proton flow along membranes. In Section 3.2 we have described the most recent knowledge of organelle membrane structure and in Section 4.5.2 we have considered proton flow using both free water and surfaces. Putting the two together and given a driving force proton flow may well be on or almost in membranes rather than in the bulk fluid. Considering the distribution of the sources and sinks of protons it may also be the case that *local* surface diffusion, see Fig. 5.6(a), dominates any equilibration of the vast networks of membrane tubes in mitochondria or chloroplasts. We must be aware too that protons can tunnel over short distances and, therefore, the electron tunnelling processes in proteins may be directly coupled to proton movements.

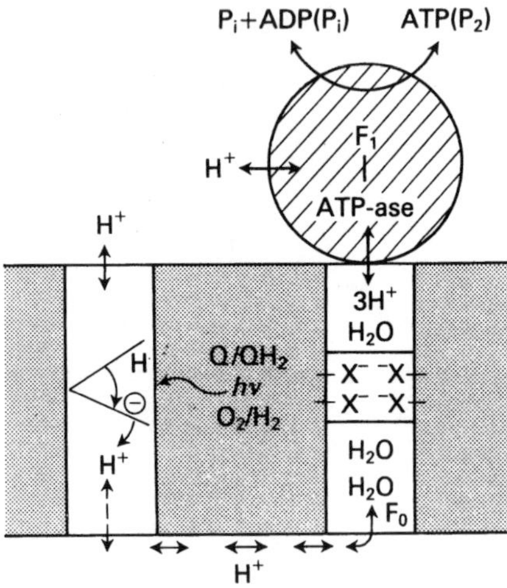

Fig. 5.6(a) Outside the pathway of energized protons, generated by the photocell or the fuel cell of Figs 5.1–5.4, leads to the ATP synthetase unit F_1 where ATP is made from ADP and P_i. The nature of the F_0 unit is to provide a proton channel and a transducer of the energy from proton gradients to stored mechanical energy which in turn drives the rotation of F_1 so that it absorbs ADP + P_i but releases ATP in a cyclic series of events. Follow the arrows from left to right but note that the system is reversible. X^- is a carboxylate anion of a protein sidechain. The left-hand channel is a generalized picture of the production of proton gradients. [See *Science* (1999), vol. 286, p. 1687–8, 1200–4, 1722–4 and *Trends in Biochemical Science* (1997), **22**, 420–3 and Williams (1961) in references].

5.5 The coupling of gradients to ATP formation

Apart from the creation of gradients we have also described a mechanism for accumulating energy in compounds, i.e. as free energy of oxidized and reduced elements which are ordered in space across membranes. The basis is oxidation-reduction reactions and it is achieved by the organized assembly of essential elements (Fe and Mn, particularly) in membranes, as in Figs 5.1, 5.2, and 5.4; this mechanism is common to most if not all living systems. We still need another procedure since biological polymers are not made by redox reactions but by condensation reactions which remove water

$$n[\mathrm{HOROH}] \rightarrow [\mathrm{OR}]_n + n\mathrm{H_2O}.$$

We know that this is achieved in biology by using a very small condensation polymer, pyrophosphate $P_2O_7^{4-}$ (in ATP) (see Fig. 5.6(a)), which can be represented as P ~ P where ~ stresses the unstable bond.

Pyrophosphate can drive condensation of all biopolymerizations since it is a drying agent, e.g.

$$P_2O_7^{4-} + 2(\mathrm{HOROH}) \rightarrow \mathrm{HOROROH} + 2\mathrm{HPO_4^{2-}}$$

written here for sugar reactions, but a very similar reaction can be written for amino-acid or nucleotide polymerizations. Thus, the free energy change $\Delta G°$ for pyrophosphate going to two phosphates is great enough to drive the polymerization, but where does pyrophosphate (ATP) come from? It has been clearly shown that the reaction

$$2\mathrm{HPO_4^{2-}} \rightarrow P_2O_7^{4-} + H_2O$$

is driven by the electrolytic (pH) gradient of H^+/OH^- made across biological membranes described in Section 5.3. We shall illustrate the principles behind this process later but the basic transformation takes place in a membrane. After the charge separation has generated ψ, a membrane potential, or a pH gradient confined by a membrane, charge in the form of H^+ is allowed to flow back down a special channel for H^+ ions [Fig. 5.6(a)]. As it flows through the channel, the H^+ ions use their energy, i.e. ψ or ΔpH to drive conformational changes in the proteins of the channel [Fig. 5.6(b)] (i.e. gradient energy goes to mechanical energy) and, in turn, this mechanical energy is transduced into chemical bond energy forcing the condensation of two inorganic phosphates to give pyrophosphate in free solution

$$P_i + P_i \rightarrow P_2 \text{ [see Fig. 5.7]}$$

Actually in cells this condensation is of a special diphosphate, adenosine diphosphate (ADP) with P_i to form adenosine triphosphate (ATP)

$$\mathrm{ADP} + P_i \rightarrow \mathrm{ATP}.$$

Furthermore Mg^{2+} is required for the reaction.

The H^+ flows in a circuit driven by light or redox reaction (Fig. 5.6) and connects systems dissipating this energy in the production of the condensation reagent ATP. The equivalence of the energy storage modes on reversal of flow is shown in Fig. 5.7. A more detailed description of the functioning of the ATP-synthetase based on detailed structures [see Fig. 5.6(b)] is given in the references, but note that it must operate so as not to hydrolyse pyrophosphate, i.e. it is not strictly reversible (Fig. 5.7).

The exact mechanism of ATP formation in ATP-synthetase is still somewhat uncertain but there is a strong suggestion that the $\alpha\beta$ units which make the ATP and release it are open to conformational change due to the proton motion in the c-units driving a rotation of the γ-unit within the $[\alpha\beta]_3$ set, [Fig. 5.6(b)]. As γ-rotates it opens one $\alpha\beta$ unit for ATP release and simultaneously allows ADP + P_i to bind to the second $\alpha\beta$ pair and generates ATP in the third pair. The rotational motion then begins again to make further ATP molecules. It is the active sites in the $\alpha\beta$ unit which have hinge-bending open/closed conformational switches driven by energy to release ATP while ATP itself forms from ADP + P_i with little energy input due to binding energy differences. The motion has to be locally cyclic to prevent simple reversal, i.e.:

$$\text{Protein} + \text{ADP} + \text{Pi (bound)} \longrightarrow \begin{matrix}\text{Protein} +\\ \text{ATP bound}\end{matrix} \longrightarrow \text{ATP released} + \text{protein}$$

(the released protein returns to the start of the cycle)

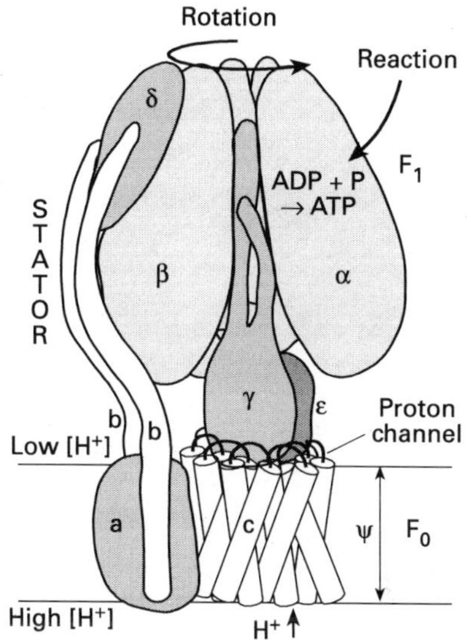

Fig. 5.6(b) The ATP synthetase machine. Energised protons entering at the bottom, F_0, drive the rotation of the central region so as to force ADP + P to form ATP in the F_1 region. ([From Junge *et al.* (1997). *Trends in Biochemical Science*, **22**, 420–3.)

The action resembles that of the flagella-motor of bacteria, (Fig. 5.14). There must be a similar but simpler device in the primitive bacteria which make ATP, or rather pyrophosphate, from proton gradients, using a single protein, pyrophosphate synthetase. Here a rotational trick is probably replaced by an oscillator of the kind shown in Fig. 5.8. Note that, as we have stressed repeatedly, the major motions are of helical segments, see Williams (1993) in further reading.

Two points remain. The pyrophosphate ΔG° energy can also be used in the reverse sense to drive gradients, since this is just the reverse of its synthesis (Fig. 5.8). Indeed pyrophosphate (ATP) is often used to accumulate and distribute elements and compounds all over the cell (see Fig. 5.6 and Chapter 6), but the chemical energy can also be transferred to other nucleotides, GTP and UTP (generally written NTP), which are used selectively in different metabolic synthesis pathways for the same general purposes as ATP. This selective use of phosphate energy parallels the production of NADPH for synthesis but of NADH for degradation in (high-energy) hydride transfer (Chapter 3).

The second point to observe is that in the above account the inorganic ions and elements dominate the several processes subsequent to the absorption of light. This chemistry is of redox changes of metal ion (electronic potentials and currents), changes in ion concentration gradients (electrolytic potentials and currents), and the polymerization of phosphate. It is for this reason that Fig. 1.3 was provided so early in this book. Naturally, without proteins the organization could not be built, but without the inorganic elements the energy capture could not operate. Dominant also in all this biological chemistry is the function of hydrogen as the go-between from light and electrical energy to pyrophosphate; hydrogen is indeed the only element that can carry out this role (see Chapter 1, Section 1.2).

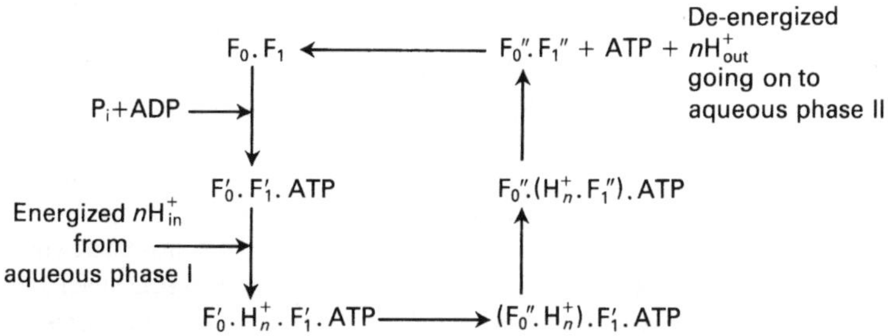

Fig. 5.7 A set of intermediates in the transfer of energy from an ion gradient to the synthesis of ATP or the reverse. Curiously, in some sense, the steps of transfer of energy in the drive of muscle or bacterial flagellae may be similar to that in pumps. (see Fig. 5.13).

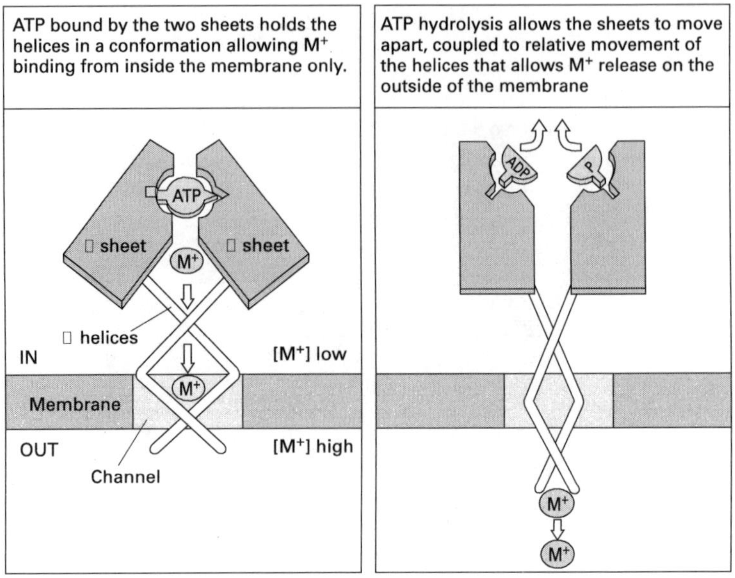

Fig. 5.8 A stripped down version of the way in which the dynamics of helices connect to reactions on β sheets in pumps and ATP synthetase (from pyrophosphate). The process is again cyclic and irreversible.

5.6 **Anaerobes in the dark**

Some biological systems do not use light and do not use dioxygen in energy generation. If we are to understand these anaerobic energy sources in the dark then we must start from a compound with one given chemical energy, $\Delta G°$, and transform it without changing the average redox states of carbon but producing energy. The usual source is the reaction of sugars, e.g.

$$[HCHO]_n \rightarrow n/3\ CH_3CHOHCOOH \text{ (lactic acid)}$$

which does not involve a general oxidation or reduction of carbon, i.e. although there is an internal oxidation-reduction of this element its average oxidation number remains the same (here it is zero).

The process can be looked upon, schematically, as

$$-\underset{\large OH}{\underset{|}{CH}}-\underset{\large OH}{\underset{|}{CH}}-CHO \longrightarrow -CH_2-\underset{\large OH}{\underset{|}{CH}}-CO_2H$$

Here, every third carbon is oxidized and one carbon is reduced. If we take the oxidation states of all carbons in glucose to be zero, the two carbons of lactic acid involved in the reaction are in oxidation states –3 and +3, respectively. This is then an internal disproportionation, oxidation reduction reaction, driven by a favourable $\Delta G°$ change.

Indeed, if we look at the differences in $\Delta G°$ between the reactants and the products in this type of reaction (Fig. 1.7), one finds that lactic acid must lie lower than $[HCHO]_n$ and, therefore, this reaction of sugars can be used to make pyrophosphate (ATP) in a coupled series of reactions involving phosphorylated intermediates, as shown in Fig. 5.9. This series of organic chemical transformations is called 'glycolysis'.

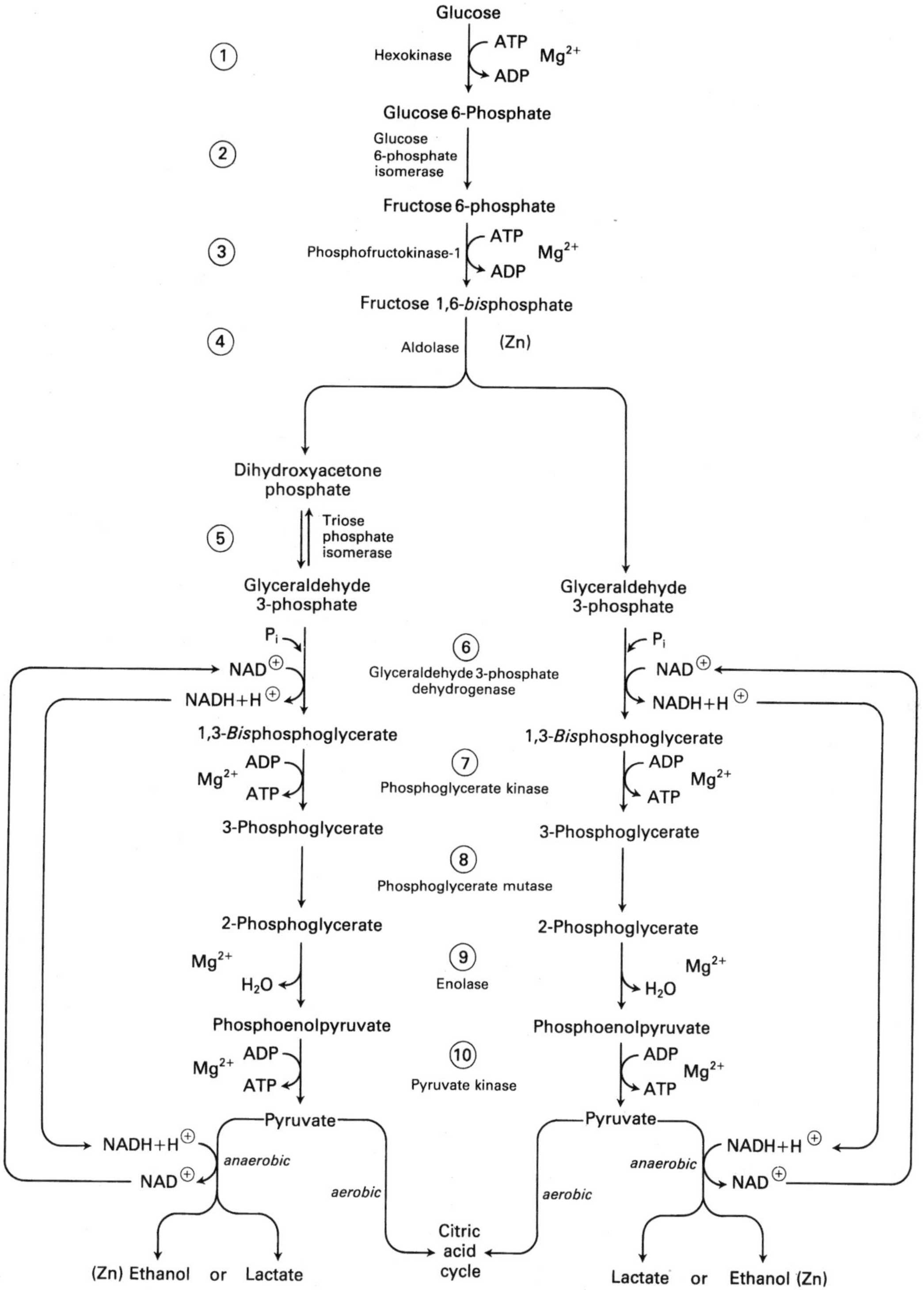

Fig. 5.9 The anaerobic breakdown of carbohydrates in the cytoplasm of cells. Notice the uses of the inorganic elements Mg and *Zn* while control is exercised by Ca and phosphates as shown in Chapter 10. Details of the pathway are readily available in biochemistry textbooks.

Given that the phosphorylation of intermediates and the formation of ATP require energy, i.e.

$$P{-}OH + HO{-}P \xrightarrow{+\Delta G^{\circ}} P{-}O{-}P + H_2O$$

$$(ADP + P_i \xrightarrow{+\Delta G^{\circ}} ATP + H_2O),$$

and, since the free energy change on going from $[HCHO]_n$ to $CH_3CHOHCOOH$ is small, the final yield of ATP is rather limited—just two ATP molecules per molecule of glucose going to pyruvate (see Fig. 5.9).

Notice how poor an energy source this is compared to the aerobic process (in the respiratory chain, see Figs 4.6 and 5.10)

$$[HCHO]_n + nO_2 \rightarrow nCO_2 + nH_2O$$

which requires the different set of transformations of Fig. 5.10, called the 'citric acid cycle', leading to a proton gradient that is able to generate a much larger amount of ATP, now a total maximum of 38 ATP molecules for each molecule of glucose transformed. In Table 5.1 we compare the yield of the two processes which is considerably different, but notice also the entirely different chemistry and *the different essential inorganic elements used* (Figs 5.9 and 5.10).

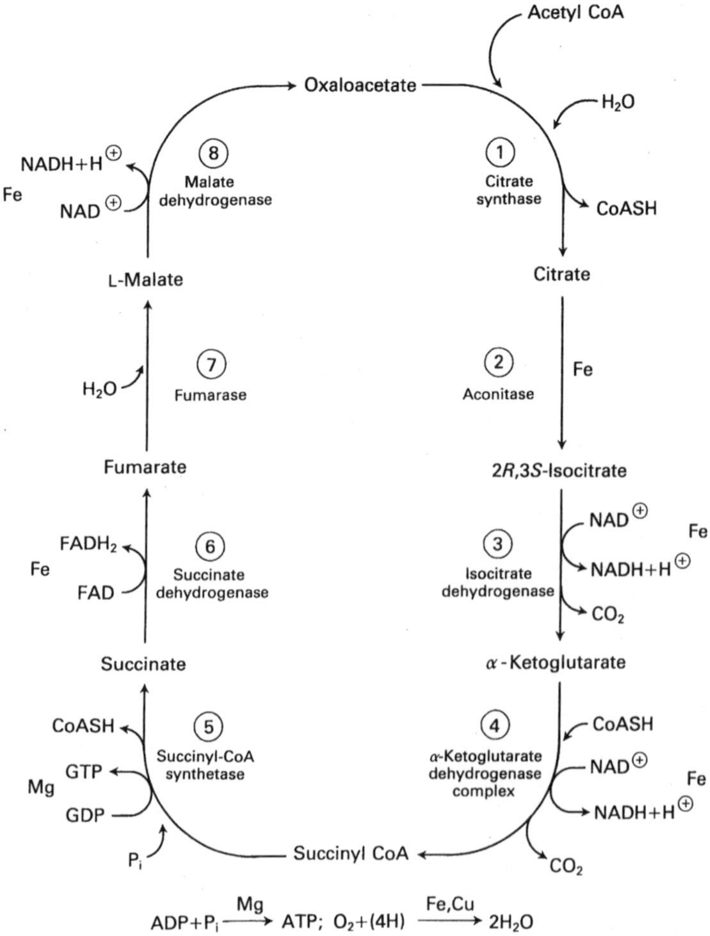

Fig. 5.10 The aerobic breakdown of carbohydrates in the mitochondria of eukaryotes (the citric acid cycle). Notice the uses of the inorganic elements Fe, Cu, and Mg, while control is exercised by Ca (see Chapter 10). Details of the pathway are readily available in biochemistry textbooks, see references in 'Further reading'. The NADH feeds oxidative phosphorylation, (see Fig. 5.4).

Table 5.1 The production of metabolic energy—a comparison of the anaerobic and the aerobic processes

Processes	**Anaerobic glycolysis**	**Glycolysis + respiration**
Free energy change in the reaction (kcal)	47	686
Number of ATP molecules produced	2	38
Free energy change stored in ATP (kcal)	15.6	296
Yield of conversion (%)	33	43.2

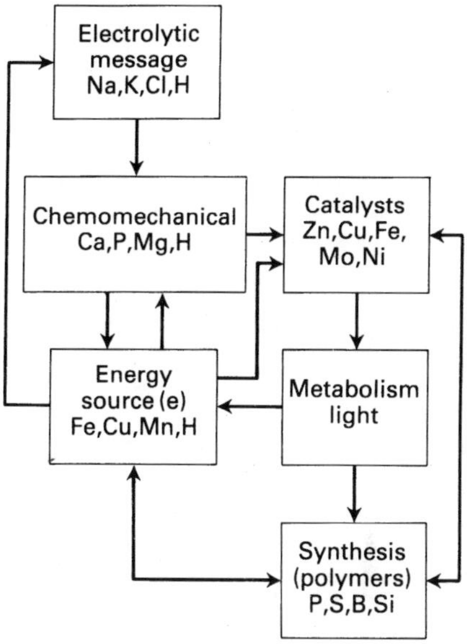

Fig. 5.11 The interrelationship of the functions of the elements is given here to stress the intimate connection between them and energy. This diagram is only an aid to thinking about the ways in which energy and metabolism have to be linked frequently using inorganic chemical elements. C, N, O are excluded.

There are reasons to believe that glycolysis was a very early metabolic pathway in the history of life. Whether it preceded energy from transformation of minerals is a moot point and there are, of course, other ways of obtaining energy from changes in chemical states of several elements. Bacteria are known which use nitrogen or sulphur cycles, redox reactions of iron and manganese, and a multiplicity of carbon sources (see Section 5.11).

Now we can examine other metabolic cycles or reaction pathways in the same way, looking for their connection to energy generation or use, noting where they occur in the cell, and observing the functional roles of inorganic elements in catalysis and control apart from their direct participation in reactants. The reader is asked to do this for himself so that he gains the feeling for the unavoidable interaction between pathways, controls, and energy. It is important that the essential involvement of many elements is observed and for this reason we give a reminder scheme of those elements which are most prominent in different functions, Fig. 5.11. In this book we wish to keep the inorganic elements in the forefront of the mind.

5.7 Compartments, energy, and metabolism

One of the features of complex cells, eukaryotes, is the separation of metabolic pathways in compartments. The most obvious are the three we have been examining, photosynthesis in chloroplasts, oxidative phosphorylation in mitochondria, and glycolysis in the cytoplasm. They all synthesize ATP for the further metabolic steps which generally consume energy either just in ion or molecule transport through membranes or in molecular syntheses. So far in this account ATP was considered to act only on the cytoplasmic side of the membrane. Virtually all compartments have pumps associated with them, but compartments such as the Golgi apparatus (Fig. 3.14) engage in glycosylation (polysaccharide synthesis) and sulphation, for example, which are processes that require energy; hence the nucleotide triphosphates themselves must be pumped into the Golgi. There are other cell compartments which are degradative, e.g. lysosomes and peroxisomes, whose names explain their functions. They have no need of internal energy from pyrophosphate. Yet others, such as the outside of cells and vesicles are just deposits, even of solids such as $CaCO_3$, SiO_2, FeO(OH), and so on. These different compartments may be energized only by their own reactions, not by ATP, but they are set at a pH or a redox potential poise which is different from that of the cytoplasm. Lysosomes are usually at pH <6, as are plant vacuoles, and ATP is used externally to pump protons and other ions into them. When we look at the outside

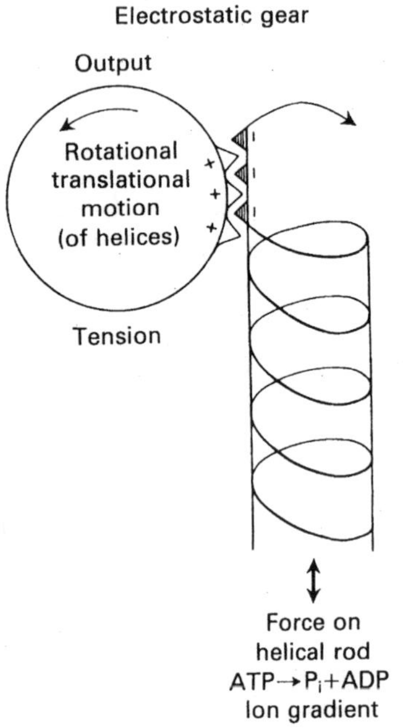

Fig. 5.12 A diagrammatic representation of the general way in which drive is transferred from one helical system to another. The pumps in membranes as well as mechanical drive in muscles depend on the adjustment of protein/protein surfaces due to the flow of chemical bond or ion gradient energy. This type of gear is similar to that in the ATP-synthetase [Fig. 5.6(b)].

of cells we find no pyrophosphate; hence energy must be obtained in a quite different way from transferred chemicals, i.e. from chemical ΔG° directly (see Chapter 20). All the compartments contain special proteins, so energy is always used in the transport of proteins through membranes since all proteins originate from ribosomes.

5.8 Organization and tension

Now we have seen how biology can obtain energy from different sources and in many different forms. A final form is tension in filaments that controls the shape of organisms and allows movement, [see (6) in Section 5.1]. A shaped object in biology is usually an energized object. Once again it is the hydrolysis of pyrophosphate (ATP) which is used to drive conformational and configuration changes in polymers to cause constantly held tension and to allow movement. Note that Mg^{2+} ions are always used in these ATP reactions.

Tension in biological polymers can be achieved in different ways. The first is by the use of mechanical displacement of units accompanied by conformational changes in polymers; this is called the sliding filament model of muscle action shown in Fig. 10.12 and generalized in Fig. 5.12. Another elementary device of this kind is seen in rubber where stretching restricts configurational entropy. A second mode for the creation of tension is to drive (reversible) polymerization in selected directions; the diagram of polymerization of tubulin monomers shows the basic idea (Fig. 6.14). The polymer is built from shaped units, therefore it can only change length in a given direction and as it does so it stretches or relaxes the membrane to which it is connected: the tension develops in the containing membrane and any fibrous molecules attached to this membrane. An elastic rubber is very different in that it can be deformed in all directions. Polymerization with a direction may well be very significant in defining biological shape and it is involved in many processes such as cell division and nerve cell extension. This, in turn, may be the basis of memory—a record of external observations reconstructed in a spatial internal pattern, the brain.

5.9 Motion of organisms

The motion of organisms requires what is effectively an engine. The development from whip-lash flagellae of unicellular organisms to muscle movement cannot be covered here, nor can we describe the apparatus in any detail. We need to note certain features of it, however. Energy is required and dissipated as heat, and in large part this energy comes from ATP or from ion gradients. The machinery to which it is applied must be a mechanical device which moves in a controlled manner. The devices are not all the same, but they depend upon cyclical operations as in simple man-made pumps and engines, (Fig. 5.13) Finally, they must have controls such that they can drive vectorially and with adjustable speeds. The involvement of various inorganic ions, such as Mg^{2+}, H^{+}, Ca^{2+}, and phosphates, like the machinery itself, is still the subject of much research and it is therefore only possible here to lead the reader who is interested to references.

5.10 Local storage of elements and energy

Once energy has been devoted to constructing a biological cell in the form of compartments, the next step in our discussion is to see how it is used to trap

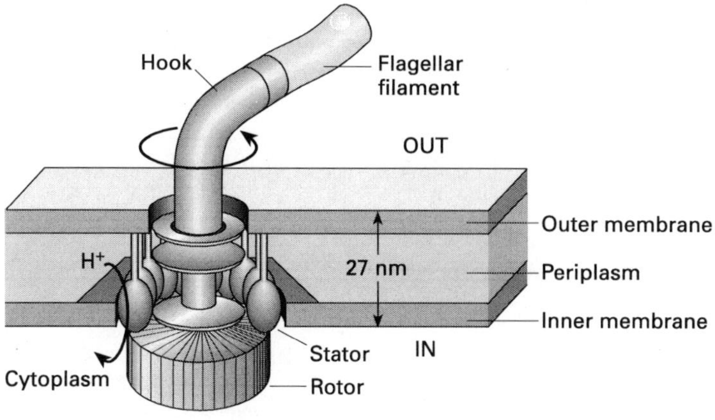

Fig. 5.13 Flagellar motor rotor-stator architecture (schematic) depicting location of the transmembrane ring particle and cytoplasmic motor modules, thought to be rotor and stator components, respectively, as denoted, within the flagellar base. [After Khan, S. (1997). *Biochimica et Biophysica Acta*, **1332**, 91.]

elements and molecules, including proteins, in these compartments. Only if the elements are properly placed in a physical (Chapter 3) as well as a chemical sense (Chapter 2) will the biological system work. The discussion is, of course, strangely circular in that without organization no new organization can be generated, but we do not ask here how the first ordering (physical or chemical) was achieved and became a biological system needing all the components of Fig. 1.3.

The different energy storage systems of biology are steady state stores maintained by the constant activity of cells. For example, the concentration of ATP is kept very closely constant at a fixed value, slightly different in different cells. All the storage devices in this kinetic scheme can only remain in a fixed relationship, and thereby cellular activity and shape remain constant, if they all communicate amongst themselves. The cell must then have an information network (always energized) which maintains this homeostasis (Fig. 5.11). We shall see that calcium, the proton, and phosphate play major roles, but that they are all maintained in steady states far from equilibrium with their surroundings. While calcium is held in concentration gradients, i.e. a $RT \ln[M]$ energy store, and phosphate is largely held in kinetically stable but thermodynamically unstable chemical bond systems, i.e. a ΔG° store, the proton (or hydrogen) is held in a variety of $RT \ln[M]$, ψ, and ΔG° stores. The marvel of hydrogen chemistry is this easy switching from $RT \ln[M]$ to ΔG° energy or to ψ field energy and the reverse.

There is, therefore, the following connection of energy types in a homeostatic transient state between light and heat:

Light $\rightarrow$ [electronic $\leftrightarrows$ electrolytic $\leftrightarrows$ mechanical $\leftrightarrows$ chemical ($P_2O_7^{4-}$)] $\rightarrow$ heat.

This is the essence of biological energy capture and use which remains largely related to inorganic element properties since

Light capture	$\rightarrow$ (Magnesium) chlorophyll proteins,
Electron transfer	$\rightarrow$ Iron, copper proteins
Electrolytic current	$\rightarrow$ Proton flow (Na^+, K^+, Cl^-),
Condensation	$\rightarrow$ Phosphate to pyrophosphate (magnesium required).

While stressing these aspects of energy capture and use, we must not overlook the fact that the electron transfer is also used directly in biological chemistry so that the separation of oxidizing from reducing equivalents, starting from water (or H_2S), causes the reduction or oxidation of metal ions or that of compounds of C, N, etc. It is only after these stages that organic syntheses can start. It is essential for an understanding of living systems that the reader realizes that much though the system uses organic molecules and much though inorganic elements are present only in trace amounts, there is no way energy capture or catalysis can proceed without both. Despite the contradiction in words, life requires inorganic elements.

5.11 **Primitive sources of energy**

Sections 5.2 and 5.3 described the energy of cells obtained from either of the two photosystems or dioxygen much as is found today. In fact energy capture could not have started in these ways since any of the three main processes described requires very much relatively recent organization (Fig. 5.1 to 5.4). We can imagine that the origin of the earth or the action of light could have developed more primitive energy sources in compounds of raised ΔG on the surface of the earth, before life started see Chapter 20, and we need then to consider how a primitive organization evolved to manage the use of this extra free-energy. Such organisms are called chemoautotrophs (or heterotrophs). Many accounts exist of such possible reactions, which vary from the direct production of energised vesicles to the use of energized clay or other mineral surfaces, e.g. iron sulphide which can undergo the reaction (see Wachterhäuser in the references).

$$\underset{}{FeS} + H_2S \rightarrow \underset{\text{pyrite}}{FeS_2} + H_2 + \text{energy}$$

This reaction, when suitably linked to some organic chemical transformations produces reducing H_2 in bound form for the initiation of carbon chemistry. In short we can write for example:

$$H_2S \rightarrow S + \text{bound hydrogen } (H_2)$$
$$\text{Bound hydrogen } (H_2) + CO \rightarrow HCHO \text{ (formaldehyde)}$$

The bound hydrogen is found first in nicotinamide adenine di-nucleotide (NADH), a reduced coenzyme that connects to several different redox pathways and feed-back loops. It is in all cells to this day. From formaldehyde and ammonia almost every required organic chemical can be synthesized.

Alternatively the reducing H_2 could have been reacted with any oxidizing agent present, such as free sulphur, to give energy as electrochemical gradients (of H^+) as is known in some archaea

$$H_2 + S_n \rightarrow H_2S + \text{energy}$$

Or course, all these possible reactions needed a membrane (a compartment), catalysts for synthesis and an agent for chemical energy transduction. The only effective catalysts are transition metals such as nickel, iron, and cobalt, when the reactions can be written as

$$\begin{array}{ccc} (H_2) & & CO \\ | & & | \\ Ni & \ldots & Ni \end{array} \rightarrow Ni \ldots Ni + HCHO$$

OR

$$\begin{array}{ccc} (H_2) & & CO \\ | & & | \\ Ni & \ldots & Fe \end{array} \rightarrow Ni \ldots Fe + HCHO$$

The subsequent energy capture as ATP involves more complex steps as in Fig. 5.5.

Another possible early source of chemical energy, which could also lead to the development of proton or charge gradients was due to the presence of nitrogen oxides. Lightning could have generated quite considerable amounts of nitric oxide from the photoreaction of dinitrogen with water:

$$N_2 + 2H_2O \rightarrow 2NO + 2H_2$$

Although NO is a strong oxidizing agent it may have avoided being reduced immediately since it can bind to Fe^{2+} ions and to sulphides and therefore be temporarily stabilized. The bound NO would then have been an electron acceptor and so possibly linked to reducing agents across membranes with the production of proton or charge gradients. The initial enzymes using NO could have been based either on Fe^{2+} or RS^-, or both, in Fe/S clusters. These centres were present in association with cells and their membranes in very early life-forms. (Note that some present-day nitrogen oxide metabolism involves copper enzymes but this was necessarily a later development, see Chapter 1.)

Table 5.2 lists these and other primitive energy capture systems, giving pyrophosphate, ATP (or other NTPs) or other compounds with 'high energy' bounds, e.g. thioesters (see the next section).

Table 5.2 Evolutionary sequence of energy sources

1. (FeS + H_2S), minerals, H_2/S, H_2S/S_n, (NO).
2. Light
3. SO_4^{2-}
4. Oxides of nitrogen
5. H_2O/O_2

5.11.1 **Sulphur compounds as energy transfer agents**

Although not so obviously involved in the energetics of cells today, where nucleotide triphosphates such as ATP are dominant, there is one other molecular construct of high positive ΔG relative to stabler conditions, namely the –S–CO– bond of thioacids and thioesters. Note that many syntheses based on the transfer of acyl fragments go via thioacyl units, for example using coenzyme A (see Chapters 3, 7, and 19).

There has been the suggestion that the earliest bioenergetic controls worked through this 'high energy' bond, which may have preceded the 'high energy' pyrophosphate bond, although the reverse is also a likely possibility. Thioacid and thioester bonds could have been made on metal sulphide surfaces by the reaction of CO, CH_4, and H_2S, schematically

$$2MS + CO + CH_4 \rightarrow \overset{H_2\uparrow\ +}{2MS - CO - CH_3} \xrightarrow{+\,2H_2S} 2MS + 2CH_3CO{\cdot}SH + H_2.$$

They may also have been formed in cells from carboxylic acids and thiols at high temperature and low pH:

$$R\,COOH + R'SH \rightarrow R'S - CO - R + H_2O.$$

Once available thioacids and thioesters could be used as energized carriers to drive reactions (see de Duve in the references).

5.12 **Energy networks and summary**

At the end of each of the first four chapters we have stressed the way in which the survival of a cell is a complicated problem concerning homeostasis. Here homeostasis refers not just to a cell at a given condition in its life's history but also to a given pattern of reactions including growth at fixed shape. A cell's condition is not a static concept. Since there is constant transfer and reaction of chemicals, many synthesised in an unstable state, energy is constantly required in a multiplicity of activities. Hence energy use has to be controlled so that pathways are kept in harmony. Once again this is managed by having feed-back and feed-forward controls over the application of energy in many steps. The gradients of ions across membranes and organic syntheses then become closely related in particular ways to the level of nucleotide triphosphates. Hence, there is a network of energy production and distribution just as there is a network of material distribution and a network control of material movement and chemical transformation, much of it largely based upon nucleotide triphosphates. In this book we have so far largely concentrated upon the chemical elements in this network showing the absolute demand for trace elements, especially metals. This has been strikingly obvious in this chapter on energy capture. To see the full extent of the interaction with synthesis we must now turn to the nature of the polymers in cells which have a very important role in establishing and maintaining all the networks, including those of the trace elements. We are not able to state how these feed-back networks began nor how they became coded in DNA, but today we begin to see how the code is connected to the material environment generating with energy input coded activity rather than the rather poorly organised chemistry which must have started life. Chapter 6 therefore, will be largely a simple outline of bio-organic polymer chemistry which we then must put together with biological inorganic chemistry to begin to see the cellular essential requirements (see Chapter 7).

The energy network must be seen in the context of the genome, the proteome and the metallome. In fact, each of these reflects the energy status of the energy in the cell as is the case for any system but cannot be understood from the sequence of DNA. As far as this book is concerned we trust that the reader will recognise the deep involvement of the metallome in all aspects of bio-energetics.

Further descriptions of the proteins in the energy capture machinery described in this chapter are given in Chapters 12 to 15.

Further reading

The general ideas on the nature of energy capture in biological systems are to be found in standard books such as the following.

Cramer, W.A. and Knaff, D. (1990). *Energy transduction in biological membranes*. Springer-Verlag, Berlin.

de Duve, C. (1991). *Blue print for a cell.* Neil Patterson Publishers. Burlington (North Carolina), Chapter 7 (thioester world).

Harold, P.M. (1986). *The vital force—a study of bioenergetics*. W.H. Freeman, New York.

Lodish, H., Baltimore, D., Berk, A., Zipurski, S.L., Matsudaira, P. and Darnell, J. (1995). *Molecular cell biology*. Scientific American Books, Chapters 17, 18. W. H. Freeman, New York.

Morowitz, H.J. (1978). Proton semiconductors and energy tranduction. *American Journal of Physiology*, **235**, R99–R114.

Nicholls, D.G. (1982). *Bioenergetics*. Academic Press, London.

Stryer, L. (1995). *Biochemistry* (4th edn), W.H. Freeman, New York.

Vallee, C.A., Solomon, E.P., Martin, C.E., Martin, D.W., Berg, R.B., and Davis, P.W. (1989). *Biology* (2nd edn) Chapters 6–8. Saunders College Publishing, Philadelphia.

Details on the structure of ATP synthase, bacteriorhodopsin and the photosynthetic reaction centre can be found in the following papers and references in them. Structures of the proteins are described in Chapters 12 and 13.

Deisenhofer, J., Epp, O., Miki, K., Huber, R., and Michel, H. (1985). Structure of the photosynthetic reaction centre. *Nature*, **318**, 618–27.

Fillingane, R.H. (1999). Molecular rotary motors. *Science*, **286**, 1687–8.

Luecke, H. Shobert, B., Richter, H.T., Cartailler, J.-P., and Lanly, J.K. (1999). Structure of bacteriorhodopsin at 2.0 Å resolution. *Science*, **286**, 255–60.

Sambongi, Y., Iko, Y., Tanabe, M., Omote, H. Iwamoto-Kihara, A., Ueda, I., Yanagida, T. Wada, Y. and Futai, M. (1999). Mechanical rotation of the c subunit oligomer in ATP synthase. *Science*, **286**, 1722–4.

Stock, D., Leslie, A.G.W., and Walker, J.E. (1999). Molecular architecture of the rotary motor in ATP synthase. *Science*, **286**, 1700–4.

The development of the theories of the involvement of proton gradients follows from three articles:

Mitchell, P. (1961). Coupling of phosphorylation to electron and hydrogen transfer by a chemi-osmotic mechanism. *Nature*, **191**, 144–8.

Williams, R.J.P. (1961). The functions of chains of catalysts. *Journal of Theoretical Biology*, **1**, 1–13.

Williams, R.J.P. (1993). The history of proton driven ATP formation. *Bioscience Reports*, **13**, 191–206.

CHAPTER 6

The role of biological macromolecules and polymers

PART A *Proteins and nucleic acids*

6.1 Introduction

In the previous chapters we described the elements, the physical-chemical basis of their uptake, their movements and distribution inside and outside cells, the chemical properties on which their functions are based, and the energy required to generate biological systems containing these elements. In the discussion it became clear that most biological elements, particularly the transition metals and many non-metals, have their differential effects through binding to proteins, either permanently, in catalysis, electron transfer, etc., or transitorily, in activation, control, or triggering. Even sodium and potassium pass through protein channels in membranes. For this reason, proteins may be considered, from the point of view of inorganic biological chemistry, to be the most important biological macromolecules. Although we have already stressed the co-operative action of metals and proteins and referred to the need for mutual adjustments on binding, we did not consider in any detail the structural character of protein structures nor the stability of their folds. We must now look at these properties and then go on to their dynamics which have developed to fulfil the various functions they perform. The importance of structural adjustment and conformational mobility is becoming more and more evident and justifies the inclusion in this text of a large part of a chapter dedicated to the analysis of the structure and dynamics of *proteins*, an analysis in sufficient depth to enable the understanding of their functional specialization even when metals are not involved, at least directly. While stressing the role of these polymers, other biological macromolecules, nucleic acids, polysaccharides, and membrane lipids, must not be overlooked. It is true that they hold a less obvious position in the context of the biochemistry of the elements due to the fact that their element related functions are of lower consequence (saccharides and lipids) or are less known (nucleotides) Nevertheless, these biopolymers are overwhelmingly important biological molecules and an effort has been made later in this chapter to consider their role in an analogous perspective. We have previously mentioned the role of *polysaccharides*, especially when combined with proteins, in assisting the recognition of the cell environment and we have also discussed in some detail how the bi-layer *lipid* biological membranes divide space, restrict the flow of materials into and out of cells and cell vesicles, and assist the organization of chains of metalloproteins and enzymes embedded in them (Chapter 3). The other class of polymers, *nucleic acids* within DNA and RNA, are, of course, determinants of the synthesis of proteins, partly through their interaction with metals, particularly Mg^{2+}. There are also control proteins which allow the influence of metal ions to reach the DNA and RNA through their structures (regulatory systems; see Sections 6.11 and 6.14). These aspects are well enough known to show that (inorganic) chemists would be well-advised to pay attention to all these biopolymers and to take into account also their particular structural and dynamical features.

There is a further reason for an examination of all these biological polymers. It is more and more necessary to see cellular or organism element activity in some integrated way with that of biological polymers rather than as isolated units. In this chapter we shall not be concerned just with isolated proteins, polynucleotides, lipids, and polysaccharides, but we shall try to describe their total functional potential when bound to elements such as metal ions to enable the discussion, in the next chapter, of integration of all components in a cell. We shall start each section on proteins, polynucleotides, polysaccharides, and lipids, respectively, with a simple description of isolated polymers and then we shall look at their roles in organized systems. This will require some reference to their distribution in compartments (see Chapter 3).

6.2 Protein composition and basic structure

As is well known, proteins are linear chains of α-amino acids connected by peptide bonds that usually exceed 50 units in length. In fact, *enzymes* are usually above 100 amino acids in length. Shorter chains are referred to as polypeptides and several of these have important control (hormonal) functions. There are some 20 natural amino acids (see Table 6.1, which also shows their potential for binding metal ions), and some modified versions of them also appear through, for example, methylation, carboxylation, or hydroxylation (Table 6.2). The amino acids can be divided roughly into four classes as shown in Table 6.3, and their relative presence in a protein largely conditions the structure and dynamics of that protein.

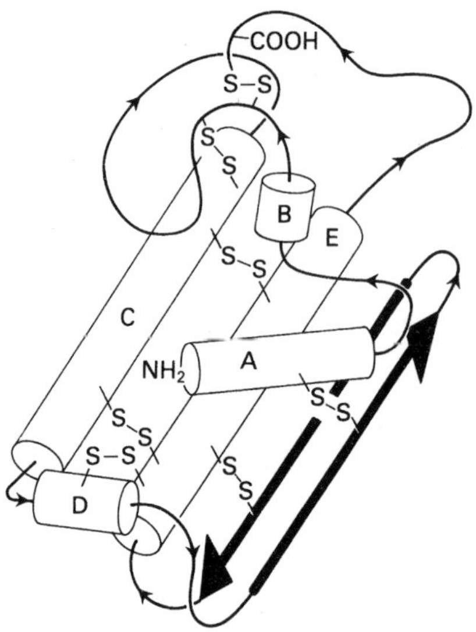

Fig. 6.1 An outline structure of a protein, here phospholipase A_2, showing α-helical runs of amino acids as cylinders, A to E, and anti-parallel β-sheet runs as heavy black arrows. Disulphide cross-links are shown, and runs of no secondary structure appear as thin lines. The structure is relatively immobile, and binds calcium in a constrained loop (see Chapter 10). (Reproduced with permission from Professor J. Drenth.)

The amino acids are incorporated into linear sequences in various proteins and the vast array of proteins arises from the huge number of possible combinations. The exact sequence of the amino acids corresponds to the 'primary structure' of the corresponding protein and this sequence largely determines all the other forms of internal association which lead to the so-called 'secondary', 'tertiary', and 'quaternary' structures of proteins. The 'secondary' structures (helices (A), parallel (B), and anti-parallel (C) β-sheets, and β-bends) arise from hydrogen bonding between the peptide links, i.e. the –CO–NH– units. The particular structure which arises is related to the nature of the amino-acid sidechains. The 'tertiary' structures arise from further folding due to interactions between sidechains, e.g. hydrophobic or electrostatic interactions or formation of covalent disulphide –S–S– bridges, and are, of course, specifically dependent upon the nature of the sidechains. In later sections we will see that other types of interactions, e.g. bonding to cofactors, metal ions, etc., can also lead to specific tertiary structures. The 'quaternary' structure refers to the form of association between several complete protein subunits, usually by specific surface interactions. Figures 6.1–6.3 exemplify most of the types of structures, but see Fig. 6.13 for a β-sheet barrel.

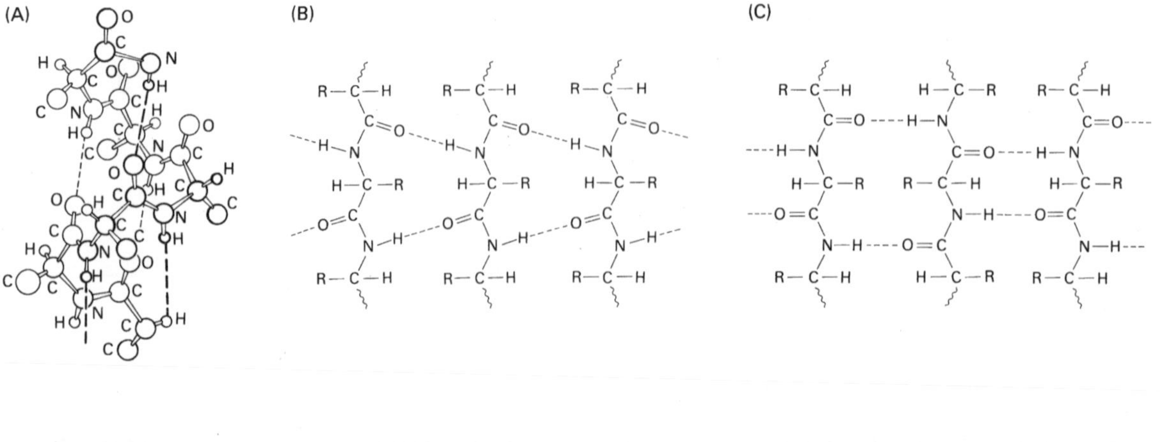

Table 6.1 The natural protein amino acids*

Name	3-letter symbol	1-letter symbol	Sidechain	Character	Metal-binding
Aspartic acid	Asp	D	$HOOC-CH_2-CH(COO^-)(NH_3^+)$	Acid hydrophilic	M^{2+}, M^{3+}
Glutamic acid	Glu	E	$HOOC-CH_2-CH_2-CH(COO^-)(NH_3^+)$	Acid hydrophilic	M^{2+}
Tyrosine	Tyr	Y	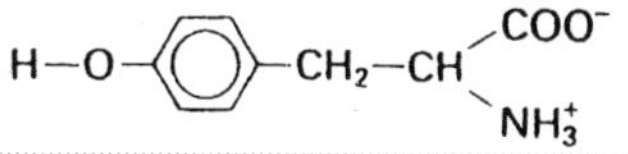	Neutral	Mn^{3+}, Fe^{3+}, (Cu^{2+})
Alanine	Ala	A	$CH_3-CH(COO^-)(NH_3^+)$	Neutral	—
Asparagine	Asn	N	$H_2N-CO-CH_2-CH(COO^-)(NH_3^+)$	Neutral	—
Cysteine	Cys	C‡	$HS-CH_2-CH(COO^-)(NH_3^+)$	Neutral	Zn, Cu Fe, Ni Mo
Glutamine	Gln	Q	$H_2N-CO-CH_2-CH_2-CH(COO^-)(NH_3^+)$	Neutral	—
Serine	Ser	S	$HO-CH_2-CH(COO^-)(NH_3^+)$	Neutral	Mg
Threonine	Thr	T	$CH_3-CH(OH)-CH(COO^-)(NH_3^+)$	Neutral	Mg
Histidine	His	H	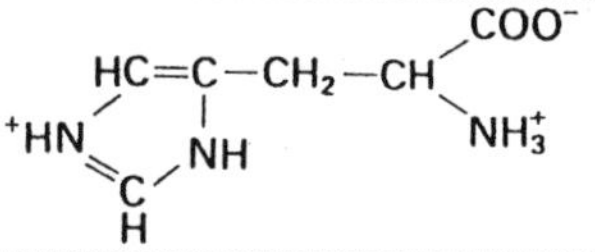	Basic hydrophilic	Zn, Cu Mn, Fe, Ni †
Arginine	Arg	R	$H_2N-C(=\overset{+}{N}H_2)-NH-CH_2-CH_2-CH_2-CH(COO^-)(NH_3^+)$	Basic hydrophilic	†
Lysine	Lys	K	$H_3\overset{+}{N}-CH_2-CH_2-CH_2-CH_2-CH(COO^-)(NH_3^+)$	Basic hydrophilic	†
Glycine	Gly	G	$H-CH(COO^-)(NH_3^+)$	Non-polar hydrophobic	—
Isoleucine	Ile	I	$CH_3-CH_2-CH(CH_3)-CH(COO^-)(NH_3^+)$	Non-polar hydrophobic	—
Leucine	Leu	L	$(CH_3)_2CH-CH_2-CH(COO^-)(NH_3^+)$	Non-polar hydrophobic	—

Table 6.1 The natural protein amino acids* (contd)

Name	3-letter symbol	1-letter symbol	Sidechain	Character	Metal-binding
Methionine	Met	M‡	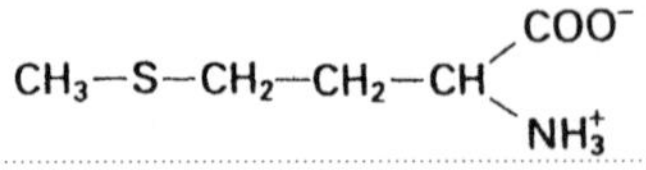	Non-polar hydrophobic	Cu, Fe
Phenylalanine	Phe	F	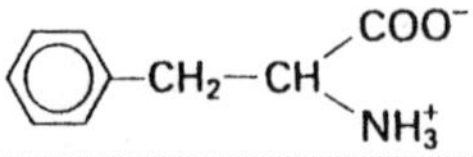	Non-polar hydrophobic	—
Proline	Pro	P	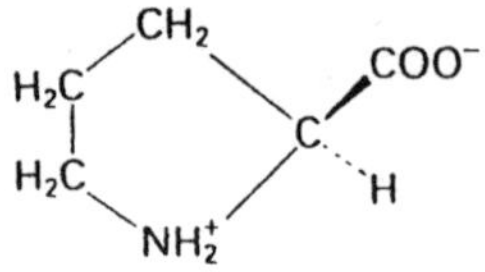	Non-polar hydrophobic	—
Tryptophan	Trp	W	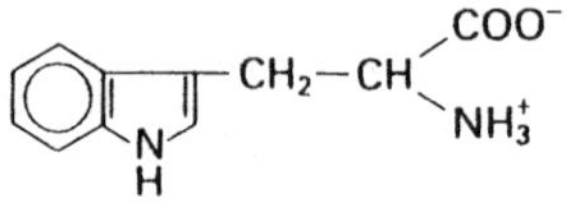	Non-polar hydrophobic	—
Valine	Val	V	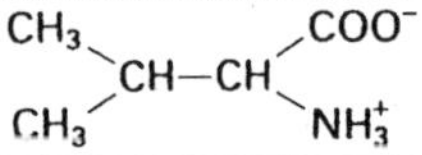	Non-polar hydrophobic	— —

NB Carbonyl groups of main- or sidechain may bind M.

*Ionic forms, predominant under physiological conditions at pH 6–7.

†Anion-binding.

‡Se can replace S in these coded amino-acids.

Table 6.2 Some modified and unusual amino-acids

Amino acid	Metal-binding	Metal involvement in synthesis
γ-Carboxy-glutamate (Gla)	Ca^{2+}	None
Hydroxy-aspartate	Ca^{2+}	Fe?
Hydroxyproline	(Bone)	Fe
Hydroxylysine	(Bone)	Fe, Cu
(Methyl)$_n$lysine	None	None
Selenocysteine	W, Ni	None
Selenomethionine	?	?

NB Several amino-acids can be phosphorylated and some can be glycosylated.

Table 6.3 Classes of amino acids

Charged hydrophilic	Intermediate (hydrophilic)	Hydrophobic	'Structural'
Lysine	Serine	Leucine, isoleucine	Glycine
Arginine	Histidine	Valine, alanine	Proline
Glutamate	Glutamine	Phenylalanine	
Aspartate	Asparagine	Threonine, tyrosine	
(Histidine$^+$)		Tryptophan, (proline),	
		Cysteine methionine	
		(and selenium analogues)	

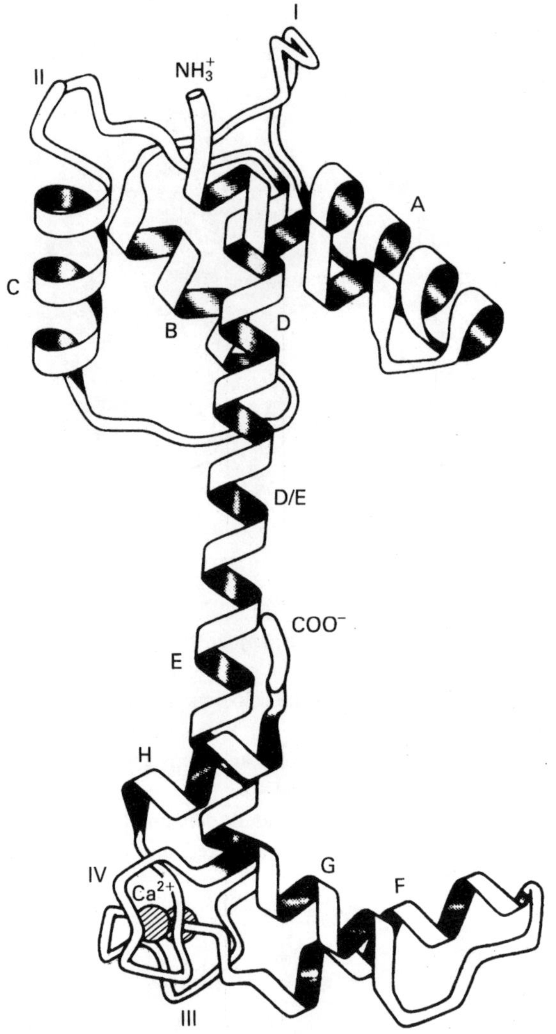

Fig. 6.2 The structure of a simple helical protein, troponin C, and see calmodulin (Chapter 10. Fig. 10.7). Note the absence of any extensive β-structure and that the helices, A to G, are relatively mobile. The calcium sites I to IV have mobility, but in crystals only the bottom two are occupied. (After M. W. G. James and C.E. Bugg.) The helix D/E is not fully formed in solution and the bottom and top lobes are not fixed as here.

These structures correspond to a static picture of the proteins, i.e. to a certain particular conformation for each one, but we are further interested in conformational changes and structural dynamics since it is through these processes and properties that each protein acts during its function, whether this is catalysis, control, triggering, or some other. A simple list of some important classes of mobility is given in Table 6.4.

To understand the corresponding relationship to functional mechanisms it is necessary first to clarify the meaning of 'conformational change' in the context of protein dynamics and for this we must realize that proteins are not rigid bodies. Indeed, even the most structured proteins show some mobility and there are others that are indeed very mobile. As a matter of fact, the range of 'mobility' (a concept in which we include all the rate processes within proteins) is such that at one extreme a change of conformation may mean something like a phase change, the so-called R (Relaxed) $\rightleftarrows$ T (Tense) switch, or a change in molecular structure like a cis $\rightleftarrows$ trans isomerization (relatively slow processes), while at the other extreme it can only be described as a fast

Table 6.4 Mobility in proteins

Unit of mobility	Example
Sidechain	
Aromatic ring	Flipping of ring Flapping of ring
Aliphatic chain	Rotation about several bonds possible
Aliphatic ring	Cis/trans isomerism (proline)
Segment	Motion within β-bends
Domain	Motion of termini Concerted motion of helices Partial unfolding
Whole protein	Unfolding/folding Rotational tumbling Diffusion free or on a membrane surface or on DNA, etc.

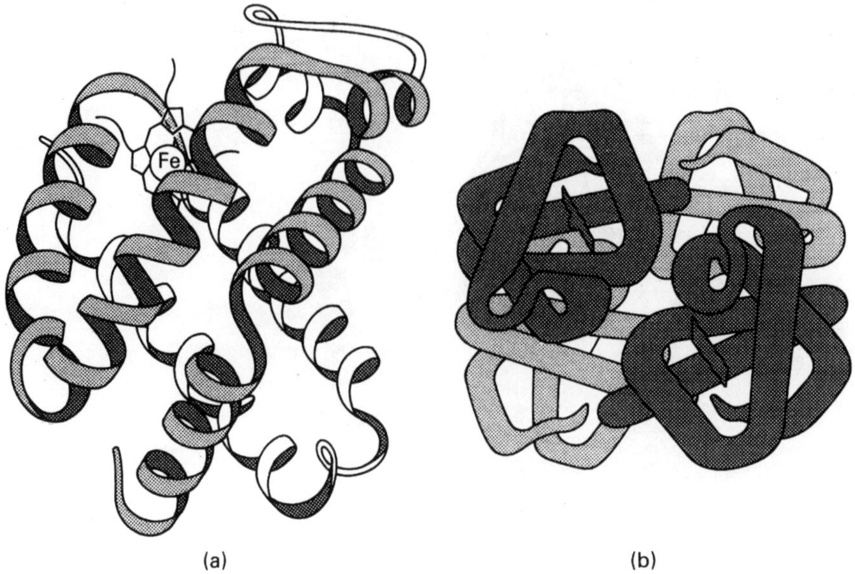

Fig. 6.3 (a) The subunit of haemoglobin.
(b) Quaternary structure of haemoglobin from four multihelical peptide chains ($\alpha_2\beta_2$). (After M. F. Perutz.) The structure can undergo considerable conformational changes. Note that it contains no β-sheet.

equilibration over many conformers of similar energy or, in terms of statistical thermodynamics, a variation in the population of states within a large ensemble, whose appraisal requires the calculation of the corresponding partition functions.

Figure 6.4 exemplifies the two extreme situations and an intermediate one. In Fig. 6.4(a) there is only one ground state (free energy minimum) for the ordered (native) and one for the disordered (denatured) form; in Fig. 6.4(c) there is no sharp distinction between ordered and disordered forms; in

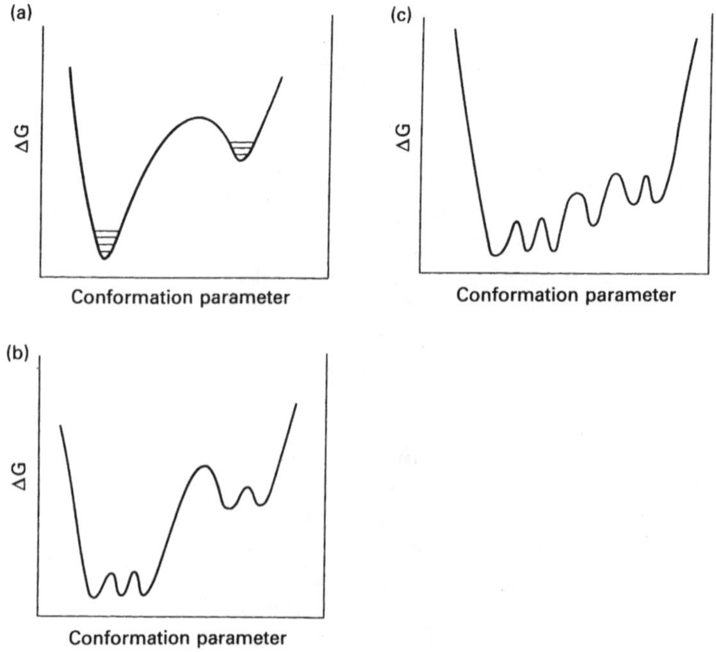

Fig. 6.4 The relative values of the free energy of different folds, ΔG, and its dependence on conformation for different theoretical classes of protein. In (a) there is only one deep free energy minimum for the ordered (native) and one for the disordered (denatured) form. In (b) local motion within these two states gives rise to several conformations associated with the ordered and disordered forms. In (c) there is no sharp distinction between ordered and disordered forms. The abscissae are labelled 'conformation parameter' to indicate that a change from one state to another is made by altering the fold. The horizontal lines in (a) illustrate states which are associated with motions of sidechains. Examples are (a) phospholipase A_2, (b) calmodulin, (c) metallothionein, metal-free.

Table 6.5 Protein conformational motions related to known or possible functions

Functional property	Protein motion
Ion movement in channel	Very fast local bond movement around an ion 'site'
Electron transfer	Fast relaxation of conformation around site: small changes
Enzyme catalysis	Movement of sidechains with substrate transformation: fast binding adjustments (induced fit)
Allosteric cooperativity	Segmental shift, especially of helices: relatively slow
Triggering by ions	Segmental shifts, especially of helices: intermediate rate
Binding of surfaces	Adjustment of surface groups: fast
Two-dimensional diffusion on membranes or DNA	Adjustment of surface groups, e.g. lysine: fast, helix adjustments: intermediate or slow
Folding	The whole structure is dependent on metal ion binding in some cases: relatively slow
Covalent attachment	Phosphorylation is particularly important: intermediate rate of local change

Fast here implies $>10^5\ s^{-1}$, intermediate 10 to $10^5 s^{-1}$, and slow $<10 s^{-1}$

Fig. 6.4(b) there are several conformations associated both with the partly ordered and disordered forms which are clearly separated by energy barriers but differ little in energy. Binding of molecules or metal ions can switch the fold which is most stable.

It should be noted that even when the change of conformation is (apparently) just from one conformation to another, as in the case of Fig. 6.4(a), corresponding to a very structured and rigid protein, it is highly improbable that the change is a single-step process. The probable nature of such a readjustment is that an initial change at one point in the structure spreads through small steps into the protein to other distant regions. This has been confirmed by many studies of mobility either of the protein backbone or its sidechains, using various techniques, with particular emphasis on nuclear magnetic resonance (NMR). One of the most interesting results of these studies was the establishment of definite correlations between the conformation motions of several proteins and their known or possible functions (see Table 6.5 for various examples). The number and nature of the different cases studied is sufficient to confirm that not only structure (a static concept) but also conformational changes and their rates, i.e. structural dynamics, are essential data to understand fully protein function. The discussion can be extended from the highly structured proteins all the way to random coil proteins to which we return. Of great interest is the conformational change of the proteins as they pass through a series of membranes (see Fig. 3.4), which requires special *leader sequences* of amino-acids and a prevention of full folding until a predetermined moment. A problem of great concern is the finding, that some proteins can exist in two different conformations which have different properties and may be differentially inherited. These proteins, called *prions*, can affect the normal metabolism of an organism and cause diseases such as the 'mad cow' disease or BSE (bovine spongiform encephalopathy).

6.3 The protein fold and the internal motions in solution

Our knowledge of protein folds comes mainly from X-ray diffraction studies of protein crystals and from studies in solution using NMR methods. The essential structures of proteins up to 20 000 Da (mol. wt) have been completely determined in solution.

Some uncertainty exists with regard to the dispositions of sidechains, even in those proteins which form crystals (and many will not), and fluctuations are found to occur about one determined average structure, forming a family with related conformations of sidechains. In crystals these can be measured by crystallographic B-factors, which in effect measure vibrational/rotational motion, but it is clear that there is much more mobility in solution.

The mobility of the protein chain in solution increases continuously with the temperature for all proteins, but the rate of this increase varies from case to case. It can also vary within a protein from segment to segment and is influenced by the pH, ionic strength, and, of course, by binding to other molecules or ions. Naturally, the active sites of enzymes must have relatively low mobility, since they must recognize a specific substrate, and the surfaces in contact with water must be much more mobile than the interior due to the interactions with mobile solvent and both inorganic and organic ions. It is usually the case that enzyme sites are pockets, concave surfaces, where, naturally enough, conformational changes are much more limited than on the open convex surfaces. Limited mobility is an aid to specific recognition, visualized in the lock and key concept, but mobility is required to aid speed of recognition, induced fit or even hand in glove fitting. (see Fig. 6.15).

The influence of the temperature is easy to understand, but it may take different forms as illustrated in Fig. 6.5. In this figure two classes of proteins are considered: for one of them, A, there is a small increase in mobility (corresponding to a decrease in order) until a large sudden cooperative change is observed and the protein is denatured. This behaviour is typical of globular proteins, e.g. phospholipase A_2 [see Figs 6.1 and 6.4(a)] For the second protein, B, there is a gradual increase in mobility, but not sharp change is observed at any temperature. Troponin C (Fig. 6.2) free from calcium is an example of this type of behaviour. This class of protein is obviously represented at the opposite extreme, (c), of Fig. 6.4. Finally, there is a group of proteins which does not fold at room temperature (Table 6.6). Notice that troponin C is an all α-helical protein while phospholipase A_2 has a β-sheet

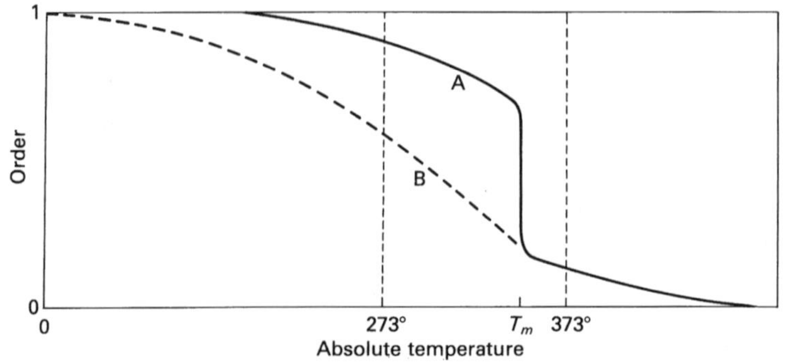

Fig. 6.5 The variation of order with temperature in two classes of protein, where order is defined as unity for a protein which occupies only one, the most stable, conformation and is zero for a random coil. In class A there is increasing but very local disorder with temperature until the onset of melting. In class B, disorder increases continuously and is more widespread at all low temperatures. A third class of protein (not shown) will have very little order at all temperature of biological interest.

Table 6.6 Some proteins of little tertiary structure studied by NMR

Protein	Characteristics in solution
Chromogranin A	Loose cross-linking net
Calcium-binding proteins of Troponin C type	Loosely structured proteins especially at T >50°C with marked changes on binding calcium
Glycophorin	Membrane protein with little or no tertiary structure in solution
Phosvitin and tooth phosphoproteins	Structural protein with no tertiary fold
Histones	Bind to DNA; in solution some segments are not folded
Lipoxygenase-I	Region remote from active site behaves as random coil
Myelin sheath protein	Poorly structured in aqueous solution
Glucagon	Very nearly a random coil in solution
Metallothionein	No fold as apoprotein: folded with metals
Osteocalcin	Little or no fold in the absence of calcium
Cytochrome *c*	Little fold in the absence of haem

core and is heavily –S–S– cross-linked. Thus troponin C is a trigger protein while phospholipase A_2 is an enzyme. We also observe that Fig. 6.5 typifies *phase changes* of bulk materials known as first (A) and second (B) order phase changes. Troponin C undergoes a second-order phase change.

As to the sidechain groups of proteins we expect that there will be rapid rotation of all methyl groups, flipping rotation of aromatic and of some aliphatic groups, flapping motions of groups such as proline and coenzymes, and swinging motions of some larger groups such as tryptophan (see Fig. 6.6).

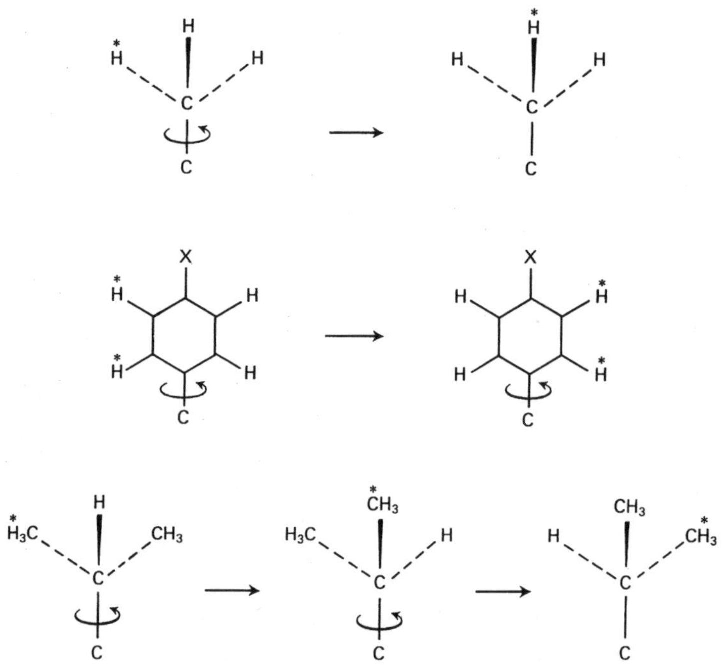

Fig. 6.6 Some internal motions of protein sidechains.

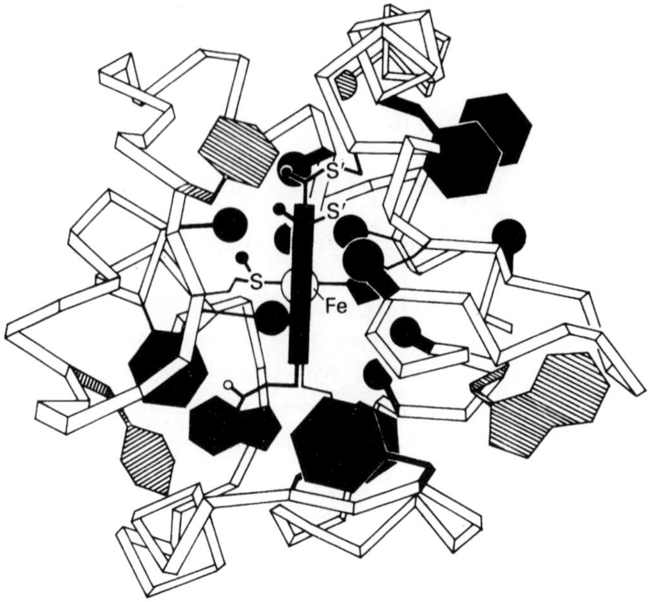

Fig. 6.7 The fold of cytochrome *c* showing the haem (black rectangle) linked by two thioether bridges (S′) into the chain. The shaded residues have very restricted motion while the hatched residues are more mobile (see also Fig. 4.4).

Naturally, the motions of all these groups will be more or less limited by the folding of the protein and their location in the chain; accordingly, the motion activation energies vary from 5 to over 100 kJ mol^{-1}. Cytochrome *c*, Fig. 6.7, is an example of large differences in mobility of differently lcoated residues.

6.3.1 The folding process

We cannot describe folding of proteins here except in outline. The folding process in a cell is often controlled by the location in which it is made. Synthesis occurs on ribosomal RNA and as the protein appears it may be automatically passed through a membrane, if the ribosome rests on that membrane, or released into the cytoplasm. The release process starts, of course, from the N-terminus. As the protein appears it may be modified, glycosylated and sulphated (in the Golgi), or it may pick up metal ions. The final folding may be prevented (or assisted) by binding to other special proteins, so-called 'chaperones', while steps are taken to make sure it does not fold incorrectly and is correctly modified. The precise way in which a protein folds probably differs from one protein to another. Since the overall process can be likened to a phase change, all the kinetic difficulties in understanding associated with phase changes are inherent in the protein folding process. One description depends on collapse through a great variety of states down a funnel-shaped reduction in the number of states to the folded form which occupies relatively few states [see Fig. 6.4(a)].

6.4 The amino-acid composition of specific proteins

The amino acids can be divided in four classes as shown in Table 6.3 and it is to be expected that the relative percentages of each group in each protein, particularly of the charged (hydrophilic) and of the hydrophobic amino acids, will have a profound influence on its properties. Indeed, a high percentage of charged amino acids is likely to lower the stability of folded structures, increase hydration, and increase the mobility of the proteins (which can be further increased if extra hydrophilic chains are incorporated into the proteins through glycosylation). Thus, we may suppose generally that there will be a relationship between the protein-chain mobility and the percentage of hydrophilic groups, but we must also ask about those proteins which contain large numbers of oppositely charged amino acids and are virtually neutral–will they fold tightly and be rigid or will they be more mobile? A high relative percentage of hydrophobic amino acids will, of course, have an influence opposite to that of charged amino acids and one expects to find this in the most structured proteins. However, should such proteins become embedded in membranes, the variety of states open to them may well change dramatically. Again binding to metal ions may constrain mobility and protect proteins from hydrolysis. It is clear that much of evolution to a function is dependent on the development of protein composition and sequences.

The ratio of hydrophobic to charged amino acids may therefore be examined as a first approximation to judge the character of the proteins in water. Table 6.7 gives the amino acid composition of a series of proteins from the well structured globular enzymes at the top of the table to the non-structured, almost random coil, control or structural proteins at the bottom, while Fig. 6.8 shows the distribution of those proteins with different ratios of hydrophobic to charged amino-acids. The separation is clear at the extremes: to the right one finds the less mobile; to the left the most mobile proteins. Enzymes, which are relatively rigid, are, accordingly, found to the right, whereas most control proteins, which are relatively mobile, are to the left side. Inhibitor proteins fall together with enzymes; they are rigid blockers. Some proteins, however, appear in unexpected positions, e.g. the cytochromes, which have a content of charged amino acids apparently higher than expected for the function they perform and are located at the left side of Fig. 6.8. Note that cytochrome *c* does not fold fully until hydrophobic haem is bound.

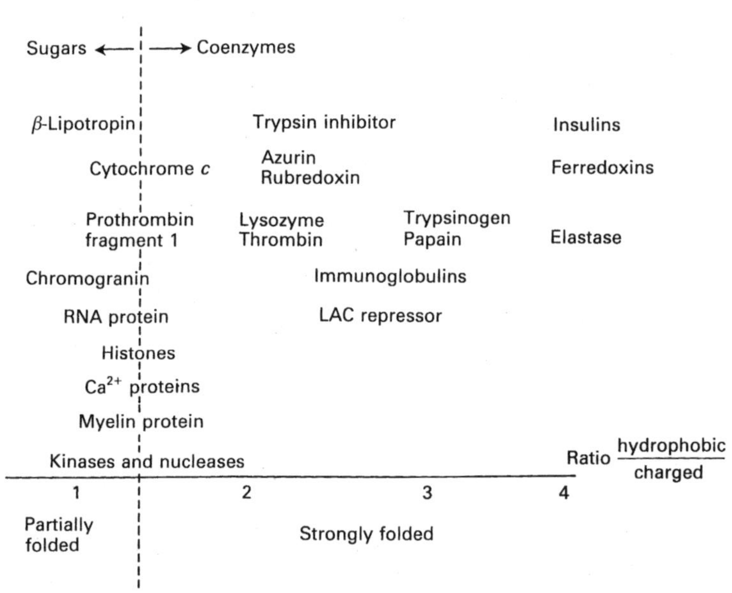

Fig. 6.8 The distribution of proteins with different ratios of hydrophobic to charged amino acids. Note the division between most enzymes, to the right, and most structural or control proteins, to the left. Proteins to the right of the dotted line have much lower mobility than those to the left.

The case of cytochrome *c* (Fig. 6.7), is common to other proteins that strongly bind some hydrophobic cofactor such as haem, flavin, or even some sterols. Without the cofactor they may be only loosely folded, but when cofactors bind they become tightly folded, i.e. the cofactor acts as a strong folding agent and may be considered as equivalent to several hydrophobic groups contributing to the shaping of hydrophobic platforms within the protein structure. Folding will be even tighter when stabilized by chemical cross-links, as happens with some haem proteins such as cytochrome *c*. It may be that fitting large units inside proteins requires an initially mobile sequence.

Now it must be noted that proteins are not uniform in charge or hydrophobic group distribution and may have some segments that are predominantly hydrophobic and others that are hydrophilic, which merely states that local sequences as well as overall composition are important especially for larger proteins which can have different domains.

It is the sequence which determines whether the fold is dominantly β-sheet or α-helical, thus before we enter a more detailed description of the properties of the different kinds of segments or domains we need to know which amino-acids or amino-acid sequences favour β-sheets as opposed to α-helical structures. Unfortunately, although there is some bias toward β-sheets due to the presence of aspartate and asparagine, for example, and some bias toward α-helices due to the presence of lysine and leucine, theory is not yet able to give an accurate prediction of secondary structure. What is clear is that when the tertiary fold is known certain properties are associated with the different segments. There is low relative movement of strands in a β-sheet while individual helices can move on one another (Fig. 6.9). Helices can move on the surface of sheets and sheets can move on one another. An example makes clear the value of the different domain or segment constructions and their mobilities.

Kinases, for example, are bi-lobed and have highly charged, more mobile, central linker regions, and these features may explain the apparent exceptional behaviour to the general distribution of enzymes related to their amino-acid content in Fig. 6.8. Since their primary effect is to transfer phosphate from ATP, kinases must bind not only this species but also the substrate and cations such as Mg^{2+} (see Fig. 9.3) but must not activate the substrates separately, e.g. ATP must not be hydrolysed. Conformational control of reaction could well be as represented in Fig. 6.9(a) in which a V-shaped bi-lobed structure has a hinge-bending action of the central linker region upon substrate binding [see Fig. 6.9(b) for details]. This movement brings together the two substrates on sites X and Y and the catalytic group Z, thus promoting the reaction between X and Y only in the closed state; the original conformation is re-established after reaction and the products are expelled. (Hinge bending of this kind is a functional necessity if reagents have to be protected during reaction.) Another example of this hinge-bending movement in catalysis is that of cytochrome P-450, when the substrates, some organic molecule, RH, e.g. a sterol, and O_2 (bound to the catalytic group, the haem) are brought together through closing of the V-shaped enzyme molecule by sequential reactions. The sterol must be bound *before* O_2 is activated since otherwise the haem.O_2 complex could attack the protein itself. Thus sterol binding engenders a change in the haem iron region, allowing O_2 to bind. The products, H_2O and ROH, are expelled after the reaction, and the V-shaped structure opens again in a redox cycle for binding a new set of centres of the substrates [see Fig. 6.9(c)]. We find that the binding regions of these proteins are usually β-sheet domains while the hinges are mobile often α-helix domains, see for example vitamin B_{12} enzymes. Another example now of

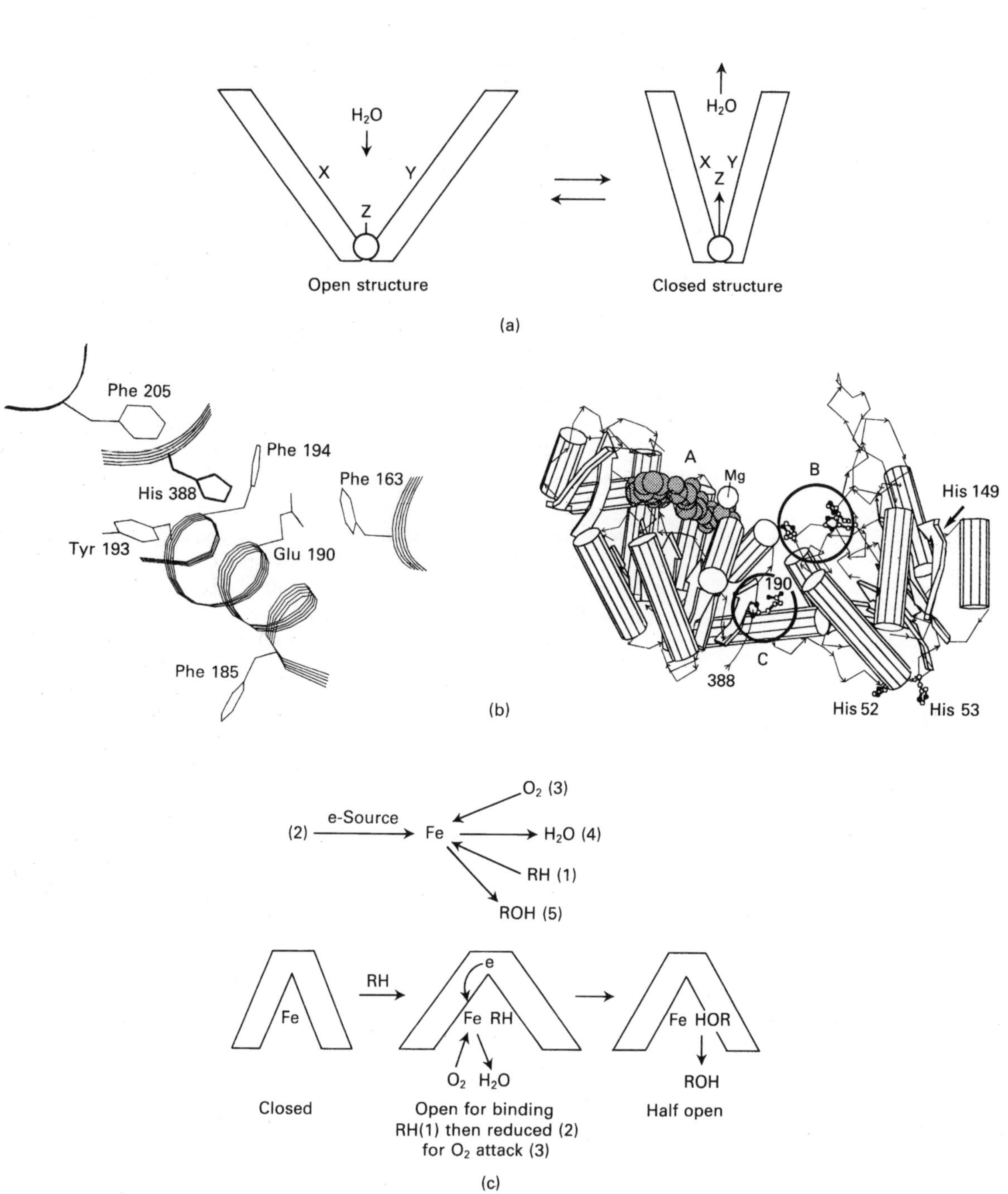

Fig. 6.9 (a) A schematic representation of kinases showing the V-shaped structures before substrate binding and the hinge action on substrate binding. The latter brings together the two substrates on site X and Y and the catalytic group Z. Water is expelled in the closed form. Much conformational mobility is required in such an enzyme near the hinge.
(b) The open structure of phosphoglycerate kinase (right) showing ATP.Mg (shaded) binding on one β-sheet domain at A; the binding site for 3-phosphoglycerate at circle B with three histidines on the other β-domain; and the looser hinge region at circle C. C is shown in more detail to the left of (b). (After C.C.F. Blake and H.C. Watson.) Helices carrying Tyr 190 and His 52 are part of the more mobile hinge region, C. (c) The proposed opening and shutting of P-450. The enzyme is not cross-linked and is helical. Reaction occurs in the half-shut state and via a further open state when product is lost. The reaction step order, numbered, is shown above.

extreme hinge movement is in particle III, the bc_1-complex, of mitochondria, (see Section 5.3).

Of course, these considerations apply in aqueous solution and other factors must be considered in lipid bilayers. Sequences of some 20 *hydrophobic* amino acids are usually associated with mobile helices *in membranes* and may alternate with hydrophilic (charged) sequences in aqueous phases making for channels of helices in the bilayers (Chapter 8). Segmental mobility in membranes will require movement between hydrophobic contacts of the helices and the functions of the few charged amino-acids in the membrane helices will now be associated with active sites for fast exchanging ion-binding, e.g. H^+, Na^+, Ca^{2+}. The regions outside the membrane often contain β-sheet domains for the recognition of specific channel activators for binding energizing chemicals such as ATP. The combination of helical membrane segments and β-sheet containing aqueous domains is found in pumps, in ATP-synthase, in receptors and in certain enzymes.

We must enquire next about the role of the 'structural' amino acids, glycine and proline, since those of intermediate character need no particular comment. Most proteins in Table 6.7 contain 9–17 per cent glycine plus proline and these amino acids frequently occur at *turns* which link β-strands or α-helices. A much higher glycine plus proline content is found in collagen, and proteins with such high ratios of the two amino acids are often termed 'collagen-like'; they have little tertiary structure and form special quaternary structures, e.g. long multiple interwoven helices (Fig. 6.10). In collagen there is incorporated the hydroxyproline and hydroxylysine of Table 6.2, and the latter is made into cross-links by the action of metallo-enzymes. Such

Table 6.7 Amino-acid composition (per cent) of some proteins and corresponding hydrophobic/charged amino-acid ratios

Protein	Charged (C)	Hydrophobic (H)	Ratio H/C	Pro + gly
Elastase (pig)	10	53	5.3	13
Trypsinogen (cattle)	13	47	4.3	15
Papain (papaya)	16	52	3.2	18
Ribonuclease (pig)	20	48	2.4	9
Alcohol dehydrogenase (horse)	22	54	2.5	17
Lysozyme (man)	23	51	2.2	10
Immunoglobulin, kappa chain	19	51	2.6	11
Collagen (cattle) α-l-chain	13	46	3.5	54
Lac repressor (*E. coli*)	18	54	3.0	10
Azurin (*P. fluorescens*)	21	50	2.4	12
Flavodoxin (*P. elsdenii*)	22	52	2.5	15
Myoglobin (horse)	27	46	1.7	12
Cytochrome *c* (man)*	30	46	1.5	16
Cytochrome c_3 (*D. gigas*)*	27	43	1.6	14
Myelin membrane protein*	25	38	1.5	22
Histone H.2 (trout)*	27	48	1.7	14
Parvalbumin (hake)*	31	50	1.6	11
Ribosomal protein (*E. coli*)*	33	44	1.3	10
Troponin-C (rabbit)*	40	42	1.1	9
Chromogranin A*	~44	41	~0.9	10

*Proteins which are partially or not folded unless bound by cofactor.

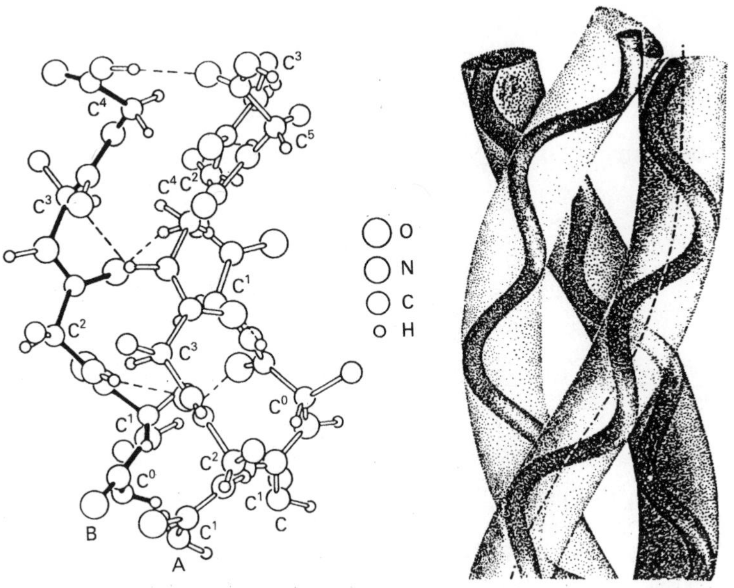

Fig. 6.10 Model of the collagen triple helix with the structure of the intermolecular H-bonds. Three left-handed single strands wind themselves around one another in a right-handed spiral. For steric reasons every third residue must be glycine.

filaments are not enzymes, not control devices, but structural elements inside and outside cells. Smaller groups of prolines, now next to residues larger than glycine, even such as alanine or even an acidic amino-acid such as aspartic acid, do not allow triple helix formation. In this case the effect of a series of prolines with other amino-acids can be to create a somewhat mobile multijointed rod. In the rod a very limited number of conformers are attainable, so that it can act as a swinging arm for the transfer of chemicals. Indeed, this is the way in which fatty acid synthetases are presumed to work as one $-CH_2CO-$ unit after another is added to a growing chain. An over-simplified survey of many of the above points is given in Table 6.8.

The importance of changes of tertiary structure and the possibility of using domains in different confirmational relationship under the control of substrates or metal ions is common not only to intracellular proteins but also to the activation of many extracellular proteins. We shall have occasion to refer to blood-clotting proteins (Section 10.18), to integrins, and to many cell–cell network proteins (Section 9.12). The discussion can be extended to the interaction between different subunits when the changes of conformation of one subunit are relayed to another in quaternary structures (e.g. in haemoglobin Fig. 6.3). or to the relay of protein conformation change to other polymers, e.g. to DNA, by transcription factors (Section 6.14). Finally, the formation of quaternary association itself can be under conformation control. Thus, two 'subunits' may be dissociated in the absence of a triggering molecule which forces association. The signalling of several membrane proteins is thought to depend on this dissociation/association equilibrium.

Table 6.8 Major secondary structure properties

β-sheets	**α-helical bundles**
Relatively rigid	Relatively mobile
Excellent recognition	Good recognition
Excellent catalysts	Poor catalysts
Most simple enzymes	Signalling
Platform constructions	Mechanical machinery

↘ ↙

Complex enzymes
(pumps, kinases, receptors)

6.5 **Structural proteins and mechanical devices**

Because of their lack of a precisely defined function those proteins that are not enzymes are least discussed. In fact, through their mechanical, not chemical, properties they control cell shape, contractile systems, cell division, and so on, Table 6.9. Frequently, these proteins are made from single- or multiple-strand helices. They extend as intricate networks both in the body of the cytoplasm,

Table 6.9 Some protein mechanical devices in cells

Mechanical device	Example
Gate	Ion or molecule pumps Electron transfer in energy transduction
Swinging arm	Fatty acid synthesis—movement of acyl-groups to growing chain
Swinging domain (Hinge)	Kinases: the transfer of phosphate from ATP to substrate
Rotator	ATP-synthetase [see Fig. 5.6(b)] Bacterial flagella
Helical rod	Helicase enzyme for undoing DNA
Ratchet	Movement of vesicles in tubulin

e.g. actomyosin and other filaments, and under the cell membrane (Fig. 3.5). They may pass through the membrane in places. We leave aside the contact with the outside matrices until we have described polysaccharides. Inside cells they contact and hold in place organelles, vesicles, and the nucleus. As a consequence of the composition and nature of these proteins, different cell types have different shapes and different roles. Some cells have multiple long protruding villi which move easily, e.g. epithelial cells, some are elongated and relatively rigid, e.g. muscle cells, and some have discoid shape, e.g. red blood cells. These proteins isolated on their own do not show their full functional value since: (1) they connect to one another; (2) in the cell they are in energized 'stretched' states; and (3) they are controlled by the exact ionic medium. It is through these proteins and their ability to adjust their structures that there is mechanical communication between in-cell, in-membrane, and external events (see Figs 6.25 and 6.30). They usually have both anion, e.g. ATP, and cation, Mg^{2+}, binding sites.

6.6 Matching of proteins and organic and inorganic ions

So far the examination of proteins in terms of their amino-acid composition indicates that there are several major factors affecting the ability to maintain a fold: one is the ratio of hydrophobic to charged amino-acids; another is the percentage of glycine plus proline; a third is the local sequence; a fourth is the ability to form helices or sheets and the fifth is the location in a membrane or in an aqueous solution. Naturally, the possibility of cross-linking, e.g. through –S–S– bridges, is also a major factor to consider that depends upon the amino-acid composition (Table 6.8). Not directly dependent upon the amino-acid composition, although related to it, one must also consider the possible effects of the interaction of opposite charges in a buried hydrophobic region and, particularly important, binding to ions, prosthetic groups, and other cofactors, which also give cross-linking and may lead to the folding of even heavily charged proteins (Table 6.9). We shall see that calcium and, to a lesser degree, zinc are extremely important in controlling and stabilizing folds.

At the same time, as the charges in the sequence of the proteins interact with one another internally and with salts, on the surface of the protein they interact with other proteins, RNA, DNA, membranes, and polysaccharides. Where these polymer-polymer interactions are electrostatic they are very sensitive to

Table 6.10 Some -S-S- bridged proteins and their mobilities

Protein	Number of bridges	Mobility
Lysozyme	4	Relatively stable fold, $\alpha\beta$ type; high internal mobility; hinge motion
Phospholipase A_2	7	Stable fold, $\alpha\beta$ type; high internal mobility
Prothrombin kringle	2	More mobile fold; higher internal mobility
Toxins EGF-peptide proteinase inhibitors	4	Restricted internal mobility, β-type

NB In all cases loops are the most mobile regions.

salts, either through general salt effects or through specific ion-binding. This means that many functions of proteins can be regulated by altering general salt levels or by changing the concentration of particular ions. Biological cells, e.g. of nerves or muscle, are then capable of regulation by dissipating controlled ion gradients, directly or via membrane potentials, especially by control of the flow of K^+, Na^+, H^+, Ca^{2+}, and Mg^{2+} amongst cations, and of Cl^-, HCO_3^-, $HOPO_3^{2-}$, and OH^- amongst anions. It is becoming increasingly clear that these ions act by changing the relative conformation of the mobile proteins. Interestingly, the cations are found to be in competition with polyamines, e.g. spermine, or polyamine-like surfaces built from small basic proteins, e.g. histones. These interactions are very important in nuclei (DNA) and in ribosomes (RNA). We note, in particular, the competition between many different charged species which must exist at the surface of membranes, at several points along highly charged polymers such as DNA and RNA, and on the highly charged structural proteins. It is inherent in this interaction that rolling contacts between the polymers rather than exact fitting of rigid shapes are made to allow rapidity of structural change.

Although the roles of calcium are frequently stressed, the importance of a variety of other ions, organic and inorganic, must not be overlooked. Many effectors, transmitters or hormones, are small, highly mobile, *organic* molecules with charged end-groups, e.g. acetylcholine, γ-aminobutyric acid, adrenalin, cAMP, etc., and the binding of these small molecules is often extremely strong. The same can be said for various drugs and toxins, e.g. chloropromazine and the neurotoxin from Cape cobra, that also have charged centres. It would seem that, especially through the use of organic cations and anions with particular shapes, biological systems have at their disposal subtle ways of adjusting charge-charge interactions and, through these, various metabolic processes, but very frequently these binding reactions affect or are affected by inorganic ions. *Since all the inorganic ion concentrations are energized (Chapter 5) so are the states of all proteins in cells.*

Table 6.11 Metal cross-linking agents

Metal	Protein	Cross-link
Zn	Alcohol dehydrogenase	$(RS^-)_4$
Zn	Zine fingers	$(RS^-)_4$
Zn	Carbonic anhydrase	$(imid)_3$
Fe	Rubredoxin	$(RS^-)_4$
FeS_2Fe	Ferredoxin	$(RS^-)_4$
Fe_4S_4	Ferredoxin	$(RS^-)_4$
Fe	Cytochrome c	(imid) (thioether)
Cu	Azurin	(imid) (RS^-) (thioether)
Ca	Parvalbumin	$(—COO^-)_4$

6.7 Enzymes

Enzymes generally have the following features: (1) they select for binding and action from all the molecules in their own compartment down to the level of one type of molecule in a thousand; (2) they act as catalysts often giving catalytic factors of as much as 10^{10}; (3) their surfaces are prochiral. While these are general rules, several enzymes, especially digestive enzymes, are much

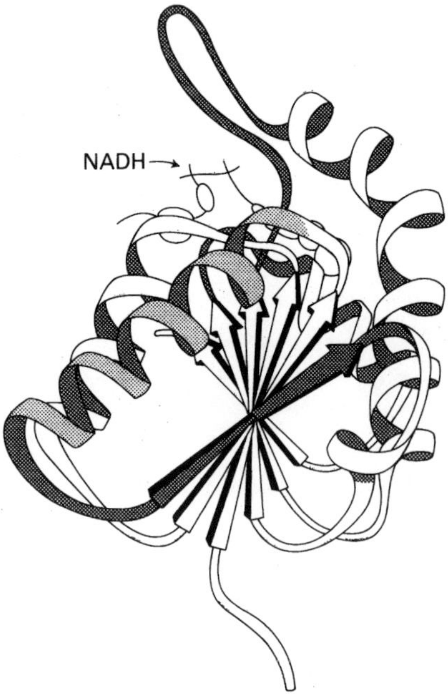

Fig. 6.11 Lactate dehydrogenase has an NADH-binding domain which is shown as a β-structure with connecting helices. (After M.G. Rossman).

less selective. Given the above discussion of mobility and selectivity opposite a variety of structures we cannot be surprised that most enzyme sites are in concave zones on the surface (grooves) of folds based on β-sheets since this is an ideal way in which to develop selective and strong binding, i.e. ensure recognition (Fig. 6.11), while avoiding adjustment of the chains.

In this chapter we do not attempt to explain the action of any particular enzyme but wish to draw attention to the general problem of catalysis by folded proteins, especially those containing metal ions. We refer to metal ions not just because they fit the general content of the book but because by their nature they allow a more detailed examination of the activity of a reaction centre, using a huge variety of spectroscopic tools, than is possible for enzymes lacking metal ions. We can see the nature of the complexity through the following simple description.

The substrate or substrates are bound and in this process a large number of interactions come into play. The semi-mobile enzyme surface is adjusted to make particular long-and short-range bindings, but there will also be repulsive energy changes and changes in hydration which will result in considerable entropy alterations involving the enzyme itself and the substrates. Going forward from the binding step, which may have distorted the substrate(s), toward the transition state, i.e. towards product(s) (Haldane-Pauling postulates), we need to consider the attacking groups which will form the intermediates (Fig. 6.12, and see Fig. 4.1). In hydrolytic reactions (acid/base) these are usually polar centres such as basic RS^- or RO^- (either protonated or not) or acid proton-carrying centres. A proton transfer from enzyme to substrate or the reverse is a very usual step in such a reaction path and it is often possible to analyse but difficult to describe the intermediates. It has been observed that the active site-attacking groups are usually more aggressive than the common run of RS^- or RO^- sidechains or protons in proteins which are not at active sites. To explain this a series of reasons has been invoked involving solvation, energy distribution of bonds, and so on (see Section 6.8.2). Even using crystallographic structures it has proved to be very difficult to give precise details

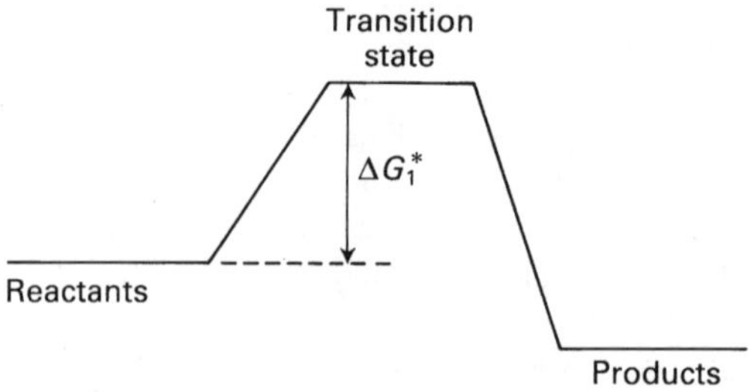

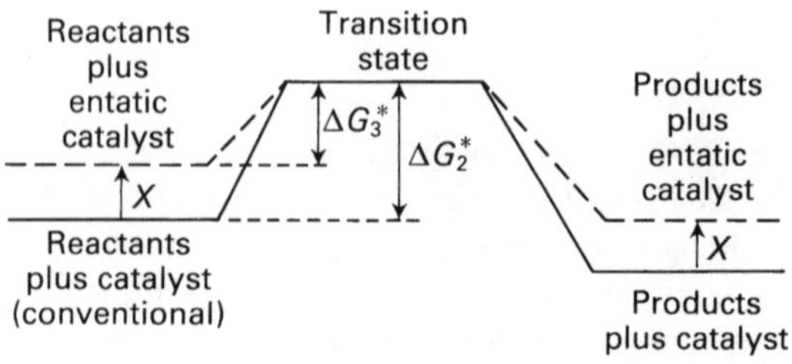

Fig. 6.12 Above, the normal activation free energy, ΔG_1^*, in a diagram for a one-step reaction (see Chapter 4). Underneath is the catalysed path for a normal catalyst with reduced activation free energy, ΔG_2^* ($<\Delta G_1^*$), and the path for an entatic catalyst, ΔG_3^*, which shows that the catalyst site is 'energized' at all stages of the reaction relative to a normal catalyst by the strain energy (X) imposed by the protein fold on the catalytic groups.

of these activation steps. Much as in solution reaction kinetics, the solvent, here the surface of the enzyme in the concave enzyme pockets of active sites, can behave peculiarly in mobility as well as directly as a catalyst. Again, there are many points of contact between protein and substrate and no doubt strain is induced in large substrates. Throughout this description we have had great difficulty in following the processes which we describe because our spectroscopic tools are limited. (Notice that no reference has yet been made to leaving groups simply because they must be the same as those which we call attacking groups since we can write the reactions of a catalyst in terms of either the forward or the back reactions; it is not useful to make a catalyst from a group which becomes unable to leave an intermediate.)

Perhaps the most obvious step for the purposes of this book is to look next at those enzyme reactions which require metal ions as opposed to those that do not. Two simplifying features emerge. The substrates are often very small molecules and the states of metals can be followed by spectral means. Table 7.3 supplies a list of very small substrates and their enzymes. One feature of biological (and man-made) catalysts is that all small molecules, such as N_2, CH_4, NO, CO, H_2O_2, and, generally speaking, O_2 and H_2, are activated by metal ions. Such small molecules can hardly be activated except by electronic perturbation through metal ions since they cannot bind to any, never mind many, organic sidechains on the surfaces of proteins. Metal ions are the most aggressive reagents, i.e. attacking groups, in the chemical (biological) armoury. Many of them are excellent Lewis acids since they can carry two or more positive charges on one small atom and remain of high electron affinity; this is a direct consequence especially of the underlying 'd'-shells of electrons. They can also create radicals easily due to their one-electron chemistry—again a property of the 'd'-shells. One essential function of metal ion chemistry is then that metals give very localized attacking power to a protein (see Chapter 4). In passing, notice that there is a risk too if this attacking power is not extremely well controlled since the metal/substrate complex might attack the protein; here lies the importance of localization of metal ions. Larger substrates do not, in general, need to have enzymes with very powerful attacking groups since bond strain can be induced by multiple points of attachment using the binding energy.

A factor which is particularly useful in the study of metallo-enzymes is that the metal can be replaced often without loss of activity (see Fig. 4.2). This site-specific substitution can introduce a spectroscopic probe, e.g. Co(II) for Zn(II), or extend the range of techniques used, e.g. lanthanide ions for calcium or Mn(II) ions for magnesium. Such isomorphous replacements, compare D for H, are extremely powerful.

6.8 States of metal ions in proteins

6.8.1 General aspects

Early in this book we stressed that metal ions in a protein, especially an enzyme, had properties modulated by the protein or coenzyme organic moiety to which they are bound. Where the modulation of electronic or geometric state was a distortion of these states from those normally observed in inorganic chemistry, the modulation being obviously of functional value, Vallee and Williams (see 'Further reading') described the metal ion as being in an 'entatic' (under tension) state, but an easier description is to call it a *constrained* state of free energy. One example is sufficient here, since the point will arise in several later chapters.

To clarity this concept of a constrained state one may imagine that the metallo-enzyme in its ground state has a kind of permanently 'energized' catalytic atom, so that, as well as lowering the energy of the transition state for the catalysed reactants as with ordinary catalysts, the 'starting' energy of the metal protein ground-state complex is at a level closer to that of the catalyst's own transition state through adjustment of the bound metal. Figure 6.12 compares the two situations for normal and constrained catalysts. An example is given in the next section. Notice that energization here is not just a matter of metal-protein bond energy strain; it is an overall free energy term and so includes, for example, exclusion of solvation molecules. It is then due to a general misfitting or *constraining* of the metal ion away from the best energy for it and its possible ligands. The protein plus the metal ion are then in a compromised 'best' condition for catalysis.

6.8.2 The copper blue centres

The nature of the deeply blue coloured centres of such proteins as azurin and plastocyanin (Fig. 6.13) has been the subject of intensive examination by very many methods that we do not intend to summarize here, but it very quickly became apparent (1960–65) that the coordination geometries of the copper(II) blue sites were unusual; their outstanding properties were the high redox potentials, the curious absorption spectra, and the intriguing electron paramagnetic resonance (EPR) signals. The prediction was made that these sites had a constrained, in free energy, geometry for Cu(II) so as to reduce the rearrangement energy on going from Cu(II) to Cu(I); it then becomes apparent that the protein was designed for binding Cu(I) rather than Cu(II). These views were beautifully proven by the X-ray crystal structure determinations of Freeman and Jensen, who established that the protein fold was a β-barrel which formed the binding site and which was virtually unaffected by removal or addition of the metal; in other words, the metal is captured in a designated hole in the protein (Fig. 6.13) and the hole generates a constrained state for Cu(II). All subsequent work in solution really establishes this position in detail. The design ensures that the coordination sphere and the metal form an excellent electron transfer site of controlled potential and minimal relaxation energy. The complex formed by the constrained site itself would surely have had a very high electron transfer rate, say 10^9 s^{-1}, to a second small donor/acceptor, but for the fact that it is somewhat hidden in the protein by an enclosing strand of peptide (see Chapter 4). In fact, this very fast rate of electron transfer is sacrificed to gain: (1) a large electron transfer distance, about 10–15 Å, the beginnings of the making of a spatial construction like a wire but based on a sequence of hops; (2) an external surface near the electron transfer site to provide a recognition zone for a very limited number of protein partners, to or from which electrons are required to go (Table 13.1); (3) a protected site away from such molecules as H_2O, O_2 and CN^-. Thus the electron transfer rate becomes optimized together with other functions at a rate constant (protein-protein) of around 10^4 s^{-1}. The proteins control one other factor: the location of these copper blue proteins in an appropriate cell compartment and assembly. The most important point to note here is that the protein is a β-barrel and it is this that confers the rigidity and supports the constrained state of the metal site. Some further examples are given in Table 6.12. While we concentrate on metal ions in Table 6.12, sidechains of proteins of organic cofactors can be forced into similarly activated environments by the effect of the fold and its energy demands, e.g. tryosine free radicals. We stress that it is the virtue of the metal ions that they allow detailed knowledge of the energetics of enzyme sites.

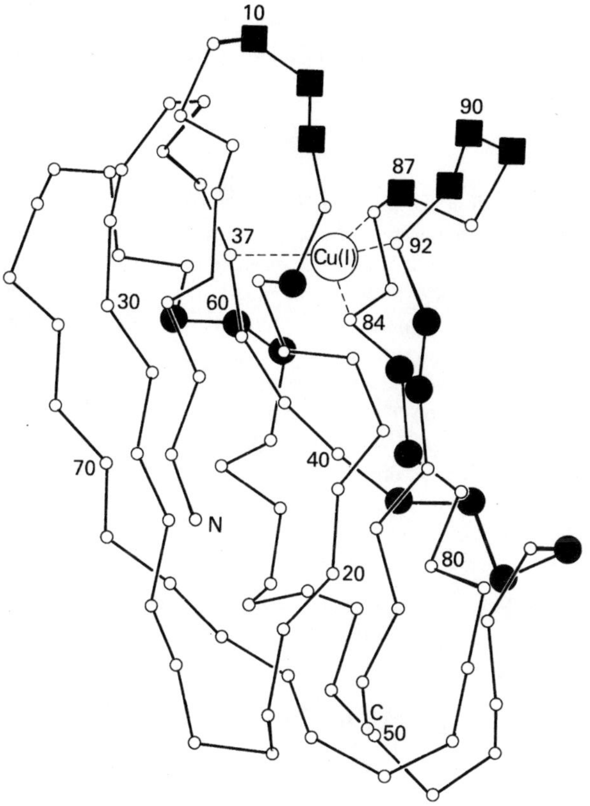

Fig. 6.13 The β-barrel fold of the copper protein plastocyanin, which is a rigid structure designed to fit Cu(I) but not Cu(II) that becomes constrained in the rigid site (see Fig. 6.12). Note the difference from the protein mechanical rack of haemoglobin which is made from helices which can move and share strain with the metal ion. (After H.C. Freeman and L. Jensen.) The filled squares and circles show binding zones for donor or acceptor proteins. [See also references in Grey, Malmström and Williams, see further reading].

Table 6.12 Examples of constrained states in metallo-enzymes and metalloproteins

Metallo-enzyme	Metal	Donors	Structure of active site
Carboxypeptidase	Zn(II)	2 imidazoles 1 carboxylate 1 water	Pentacoordinated, with bidentate carboxylate
Alcohol dehydrogenase	Zn(II)	1 imidazole 2 thiolates 1 water	Distorted tetrahedral
'Blue copper' site, e.g. in plastocyanin	Cu(II) Cu(I)	2 histidines 1 cysteine 1 methionine	Distorted tetrahedral, with a very long methionine–copper(II) bond
Cytochrome *c*	Fe(II) Fe(III)	Porphyrin 1 imidazole 1 methionine	Distorted octahedron, strained axial bonds in Fe(III); see also cytochrome *b*
Some ferredoxins	Fe(II, III)	4 cysteines 4 inorganic sulphur atoms (S^{2-})	Distorted cubane type cluster (Fe_4S_4); a great variety of structures
Vitamin B_{12}	Co(III)	Corrin 1 benzimidazole	Open-sided distorted octahedron (the cobalt–carbon bond is broken when cobalt is Co(II))

6.8.3 Metalloproteins in multiple-step reactions

The above description of the value of a constrained state applies to single step reactions and their reversal such as

$$M + e \rightleftarrows M^-$$
$$M + O_2 \rightleftarrows MO_2$$

Most enzyme reactions require several steps at the same site, for example in hydrolysis the substrate needs to be bound, then go through an intermediate and products must leave. It is then not possible to have a single constrained state as the best catalytic centre since it could lead to the trapping of a particular intermediate. To this degree the active site must allow mobility of the groups to be attacked and it may well be advantageous in a metallo-enzyme for the metal ion itself to undergo small changes in its coordination with the protein during the reaction. These changes allow the series of steps to have an overall *optimal free energy of activation for the whole reaction profile*. We must consider the types of protein and metal ion most suitable for such catalytic action.

We can take urease as an example, since this allows us to discuss a two-metal ion cluster reaction also. The overall reaction is

$$H_2O + CO(NH_2)_2 \rightarrow CO_2 + 2NH_3.$$

The requirement for activation is polarization of two substrates, H_2O approaching as $OH^- \ldots H^+$, to bound polarized urea,

$$^-O - C^+\begin{matrix}\diagup NH_2 \\ \diagdown NH_2\end{matrix}$$

, to give $(HO)_2\underset{\underset{NH}{\|}}{C}(NH_3)$ so that while the atoms of urea undergo rapid switches in conformation the two nickel ions at the active site (see Section 16.6.1) have to bind a series of molecules, based on H_2O or $CO(NH_2)_2$ initially, of different conformational demands and hence space-filling at the active site. At first sight the nickel(II) ion would appear to be an unlikely catalyst since its coordination in model systems is dominated by its preferred octahedral stereochemistry (Section 2.5), and hence slow exchange of ligands. Here the constrained (yet here somewhat mobile) chelation of a metal at an active site of an enzyme is shown by the peculiarities of the nickel site of urease (see Fig. 16.6). Of the two nickel ions one is in a four coordinate pseudo-tetrahedral and the other ion a five-coordinate trigonal geometry. Both are quite unusual geometries and undoubtedly assist activity. It is noticeable that they are linked by a carbamate ion which is a fine charge relay. While we do not know the intermediates which the enzyme uses, we can picture them as adjustments of charge on the nickel ion (A) as it polarizes the water molecule and simultaneously affects the exact position of the nickel ion (B) which carries the urea molecule through the charge relay. The two groups $(OH^- \ldots H^+)$ and $CO(NH_2)_2$ must be allowed to come together after binding. There are undoubtedly small changes in the positions of the coordinating atoms. Note especially that the carbamate attached to the ε-N-atom is joined to a saturated chain of lysine (see also Rubisco, Section 9.7). It will take very exact structural studies of the dynamics of the reaction to unravel the path, but we can be certain that the peculiar constrained states of the two nickel ions are part of an adjustable set of overall variations to generate an 'optimal' overall free energy path. Particular interest centres here on the use of more than one metal ion at a reaction site as this seems to be common to a large percentage of metallo-enzymes for both hydrolytic and redox reactions. Obviously, in general, metal ions which strongly favour particular geometries, e.g. Ni^{2+}, may fall into energy traps, while Zn^{2+} will not. Zn^{2+} is then a better general catalyst for multiple step processes such as hydrolyses. Again cobalt(III), low-spin (d^6), is not very useful in a sequence of redox reactions involving both electron and atom transfer, while iron(III) (d^5) is more versatile as indeed is Cu(I). We shall see how well these ideas apply in Part II, where we shall also see how combinations of metal ions are more effective than single ion sites.

In this account the only mobility required is very local and may be very small so that an enzyme based on a β-sheet fold with an active site on the loops can be extremely valuable, contrast the discussion in Section 6.4. The fact that metal ions are forced into unusual constrained sites has now been shown in examples of not just Cu and Ni enzymes but also of Fe, Fe/S, haem, Zn, and Mo enzymes (see the chapters on the individual elements).

From what has been said earlier it must be clear that immobile β-structures will support the constrained states most readily. When we come to very helical proteins (not usually enzymes) we find a somewhat different situation in that the metal-binding and the apoprotein fold are not compatible, so that strain in free energy, relieved somewhat by helix–helix motions, is induced *in both* on binding, e.g. in calmodulin. This is a *rack* action (see below, Fig. 6.15). A combination of a constrained state and a rack is possible, as in haemoglobin where the porphyrin takes the place of the rigid β-sheet. The porphyrin holds high-spin iron, Fe(II), in constrained (out-of-plane) state, but constraint at the metal is relieved on binding O_2; the iron goes low-spin and moves into the plane of the haem, while by rack action the strain is induced in the protein helices, see vitamin B_{12} enzymes, chapter 16. There must be no confusion between final constrained states and rack action which generates such states in some cases [see Fig. 6.15 and Gray, Malmström and Williams in 'Further reading'].

6.8.4 **Protein–protein assemblies and metal ions**

When proteins come together they may form a well defined structure, as in the reaction centre for photosynthesis in Fig. 4.5, and in haemoglobin, Fig. 6.3, or a loose association. The value of the well defined organization is that it gives very specific direction to the movements of reactants and products, for example, to electron flow. In particular, buried metal ions for electron transfer can be in organized arrays. The value of looser association is that assembly and disassembly are rapid. The need for fast exchange between proteins is obvious in such functions as:

1. scavenging, e.g. cytochrome c scavenges the periplasmic space for electrons from several electron donor enzymes;
2. flexible development and contraction of villi, e.g. tubulin forms rapidly reversible filamentous structures (Fig. 6.14), which stretch out or contract the membranes (Fig. 3.5);
3. control, such as that exerted by calmodulin;
4. dissociation/association of membrane surface receptors, e.g. the integrins (Section 9.12) or EGF segments of sensors (Section 10.18.2).

The different types of organization and its rigidity depend upon the type of amino-acid and metal ion involved at the junction. Oppositely charged amino acids on two protein surfaces are likely to generate highly mobile electrostatic joints, while hydrophobic matching is more likely to give rise to more rigid structures. Amongst metal ions we must expect that inside cells the role of

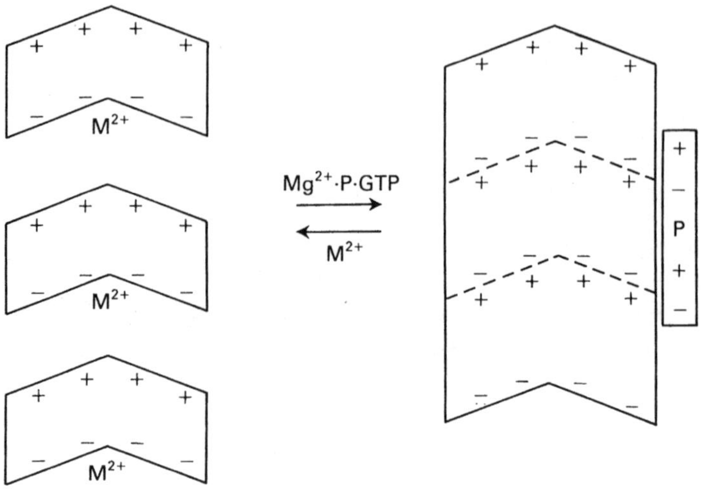

Fig. 6.14 The assembly of tubulin monomers, probably quite rigid units, into filaments due to the presence of specific concentrations of metal ions, e.g. Mg^{2+}, and GTP, then followed by hydrolysis of GTP (energy dissipation). Dissociation may be aided by Ca^{2+}. The whole assembly may require additional proteins, P. Compare integrins and membrane sensors.

magnesium will be important at protein contacts, whereas outside cells calcium has the dominant functional significance and provides greater mobility.

Figure 6.15 gives summary diagrams of the way in which groups X (and here we are most frequently involved with metal ions) can be affected or can affect proteins so that the combination generates a particular function. However, X can be any added molecule, such as a cofactor, including not only those which contain metals such as haem, Chapter 13, vitamin B_{12}, Chapter 16, and chlorophyll, Chapter 9, but also coenzymes such as flavin and NADH. Again X could be a substrate, but then there is the further problem of considering its transformation, not just its binding. Some examples amongst many of different kinds apart from those described in detail above include, the hinge-binding of iron and sulphate uptake proteins (Section 12.3); the transfer of electrons in nitrogenase involving changes in domain interaction due to ATP binding and hydrolysis (Section 12.6), conformational switches in ion-pumps and energy capture devices (Sections 4.5 and 5.4); and changes in Ca^{2+} trigger proteins on binding to say kinases (Sections 10.8 to 10.11).

6.8.5 Protein degradation

While protein synthesis is under the control of DNA and RNA and properly belongs in Sections 6.13 and 6.14, and protein folding has been described in Section 6.3, we have not examined protein degradation which is under cytoplasmic cellular control. The central steps of protein degradation appear to be an initial unfolding by binding to other proteins called ubiquitins, which are therefore related to chaperones in function, followed by a degradation in a large particle, a *proteasome*, which recognises the ubiquitin-bound protein and uses enzymes, *proteases*, to degrade it. The process requires ATP, which may well be concerned with the energy input to unfolding, first through covalent bond formation, a thiol-ester, between ubiquitin and the protein, and then in the interaction with the proteasome particle.

The proteasome is a 700 kilodalton particle which requires an activating 600 kilodalton regulatory protein complex. The protease activity based on threonine residues is quite unusual. No metal ions are involved, except for Mg^{2+} with ATP in several steps.

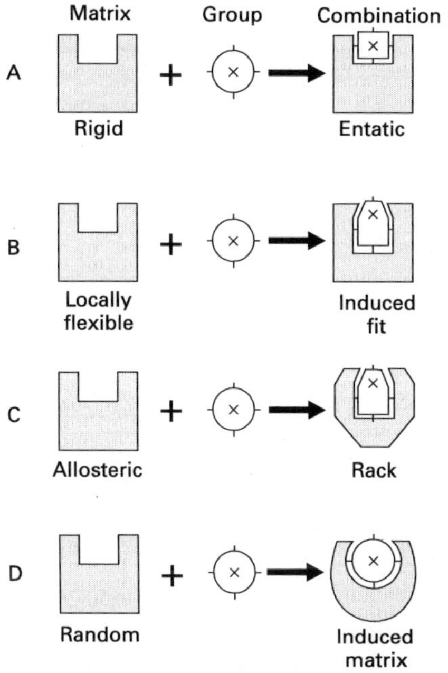

Fig. 6.15 *Schematic illustration of the differences between the ways in which the properties of a group X, which can be an intrinsic amino-acid of a protein or an extrinsic factor such as a metal ion or a coenzymes, can be structurally energized by and/or can energize a protein structure.* In (A) the protein structure is a rigid frame and gives rise only to an entatic constrained (in free energy) state of the group X through misfitting of protein structure to the desired geometry of the group. The group geometry and/or its electronic state may both be energized and it may be left with an open site. This permits optimal features for catalysis. In (D) the group X dictates the fold of the protein while it remains unchanged except for its solvation. If the group X, when so bound, has residual open sites it is effectively energized (and differently *constrained* relatively to A), but its surrounds are ill-defined and mobile so that, while it may bind substrates effectively, it is not likely to be especially catalytic, although it could be an effective carrier. In (B) small changes of *local* conformation in the protein and more considerable changes of the group X on combination, leading to so-called *induced fitting*, are shown and can also be associated with various kinds of energization at X, i.e. we may again call the group constrained and it can generate catalysis. In (C) longer range more extensive conformational protein changes are associated with allosteric effects and maybe with stretching of X. The case of stretching has been called a *rack* and here X is more likely to be effective in cooperative binding than in catalysis, and constrained energization of X is likely to be lower. Combinations of different groups X in complicated protein reaction sites generate the different features to different degrees, (see Gray, Malmström, and Williams in 'Further reading').

6.9 Summary of proteins: the proteome

In the first part of this chapter we have described different kinds of proteins and the way in which proteins and inorganic elements combine functionally; we emphasized that it is the protein which provides *structure* within the organization of biology. However, it should always be kept in mind that: (1) the whole organization of a protein is energized by the physical conditions, temperature and pressure, by the input of energy due to ion and other gradients, i.e. electrostatic fields, and by chemical binding of cations, especially Mg^{2+} and Ca^{2+}, as well as of phosphate; and (2) the protein's position in space is also energized. Chapters 8–20 will amplify this combination and organization of elements and polymers. Within structure there is a variety of dynamics associated with different activities. As stated helical segments are more mobile than β-sheets; the former give rise to many mechanical devices, e.g. pumps and signal systems, while the latter are more used structurally and at active sites of enzymes. These generalizations apply to metal binding sites.

Finally, all the proteins in a cell taken together form its *proteome*. The proteome is just becoming open to study. Though it is expressed via the *genome* (see next section) it is a different concept in that the genome has but one variable, its sequence, while the proteome relates to concentration and compartmental quantification as well as to sequence. The proteome is a name of a *system* not of a set of *molecules*. Moreover, both the genome and the proteome are not independent from the cells' third system, that of the metal ion (and small non-metal molecule content) which we refer to as the *metallome*.

6.10 **Nucleic-acid composition and outline structure**

The linear polymeric structures of deoxyribonucleic acid (DNA) and ribonucleic acid (RNA) are closely related. They are both condensation polymers of ribose phosphates which carry four purine and pyrimidine bases. The difference in the backbones lies only in the reduction of one, 2′CHOH group of the ribose of RNA to >CH_2 in DNA, but this does appear to make double-helical strands more possible in DNA. This extremely important reduction is controlled by one of the three metal ions, Fe(III), Mn(III), or Co(III), in a quite extraordinary set of enzymes which generate a free radical pathway in the cytoplasm of cells (see Chapter 12). The backbones of the polymers carry the bases shown in Fig. 6.16 and these sequences in DNA are the basis of the genetic code. Of course, there is a close relationship between the bases of any RNA derived from a DNA by transcription and the bases of the DNA itself. DNA has both the function of replication and of RNA production and the

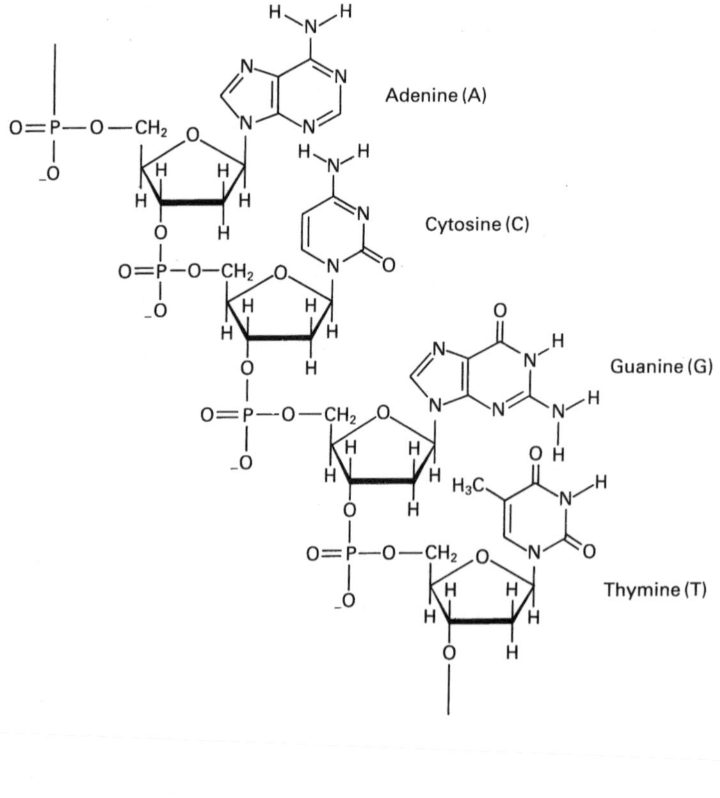

Fig. 6.16 A short stretch of single-stranded DNA giving the formulae of four bases.

Table 6.13 Metal enzymes and proteins related to DNA and RNA production

Enzymes	Metal ion	Function
Ribonucleotide reductase	Fe,* Mn, Co(B_{12})	Ribose→deoxyribose
Transcriptional factors	Zn, Fe, Cu	Control DNA→RNA
RNA polymerases (also reverse transcriptase)	Zn, Mg	Condensation polymerization
Various NTP-utilizing systems	Mg	Bioenergetic driving forces

*Two kinds: Fe_n/S_n (primitive), Fe_2O (recent).

enzymes which carry out the polymerization reactions are often, but not always, zinc enzymes. Moreover, the production of the bases is under the control of a zinc-requiring enzyme, aspartate transcarbamylase. The RNA is translated to proteins via the ribosome machinery and magnesium enzymes play a considerable role in this process. The DNA in eukaryotes is partly bound to structural proteins such as *histones* in nucleosomes, see aside, while the RNA forms large particles, e.g. the ribosomes, which contain a specific set of some twenty proteins. Note also that the DNA in eukaryotes is seen at the time of cell division in the form of chromosomes, separate units of DNA which appear to have arisen through duplication. Finally, the control of transcription and translation rests with regulatory transcription and translational factor proteins, many of which in eukaryotes are zinc metalloproteins (*not* enzymes) (Table 6.13). In this part of the chapter we shall not be concerned with any of these metal-containing proteins but with the polymeric nucleic acids themselves and their metal ion and possibly small cation binding sites. We shall look at the nature of the nucleotide monomers more closely while describing saccharides since the basic monomer of RNA and DNA is a phosphated saccharide. The origin of the bases is described by Eschenmoser, see Further reading.

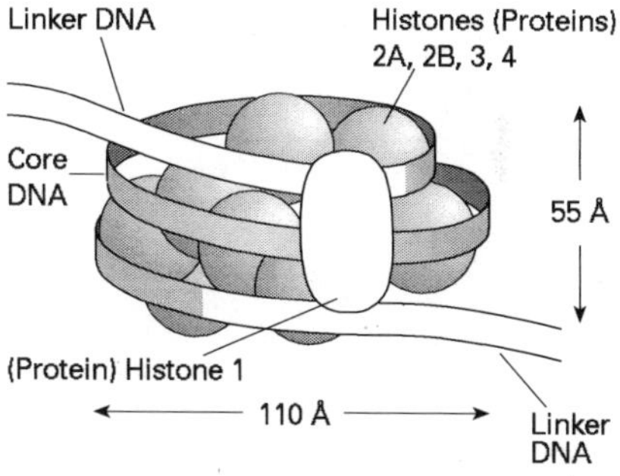

The nucleosome structure

Some of the known structures of the DNA and RNA molecules are illustrated in Fig. 6.17 The structures have a very different character from that of proteins in that they do not fold easily since they are polyelectrolytes. However, DNA must be able to assemble and disassemble quaternary double-helical structures rapidly in order to replicate. Single-stranded DNA (RNA) is a very mobile polymer both at the phosphate and at the ribose. The grooves of DNA provide selective specificity since the groove width and chemical structure are dependent on the local base sequences. The grooves are recognized by helical sections of proteins and by a great variety of chemicals (drugs) some of which are metal chelates. Specific base reactions are known, for example, with anticancer drugs such as alkylating agents, and reagents such as copper, iron or platinum complexes. They can be chosen so that they bind in the groove and generate destructive reactions of great value in the foot-printing of DNA.

In the double-helical structures there is little mobility of the sidechains since they are all cross-linked by H-bonds but, in principle, the five-membered ribose rings can flip between conformers very easily and the obvious way to pull apart the two chains is to use a zipper-like action employing the mobility of both the rings and phosphate ester joints. The single-stranded polymers are much more readily hydrolysed by enzymes.

(*Note*. In what follows we shall often refer to *poly*nucleotides, synonymously with nucleic acids, to avoid confusion with *mono*nucleotides.)

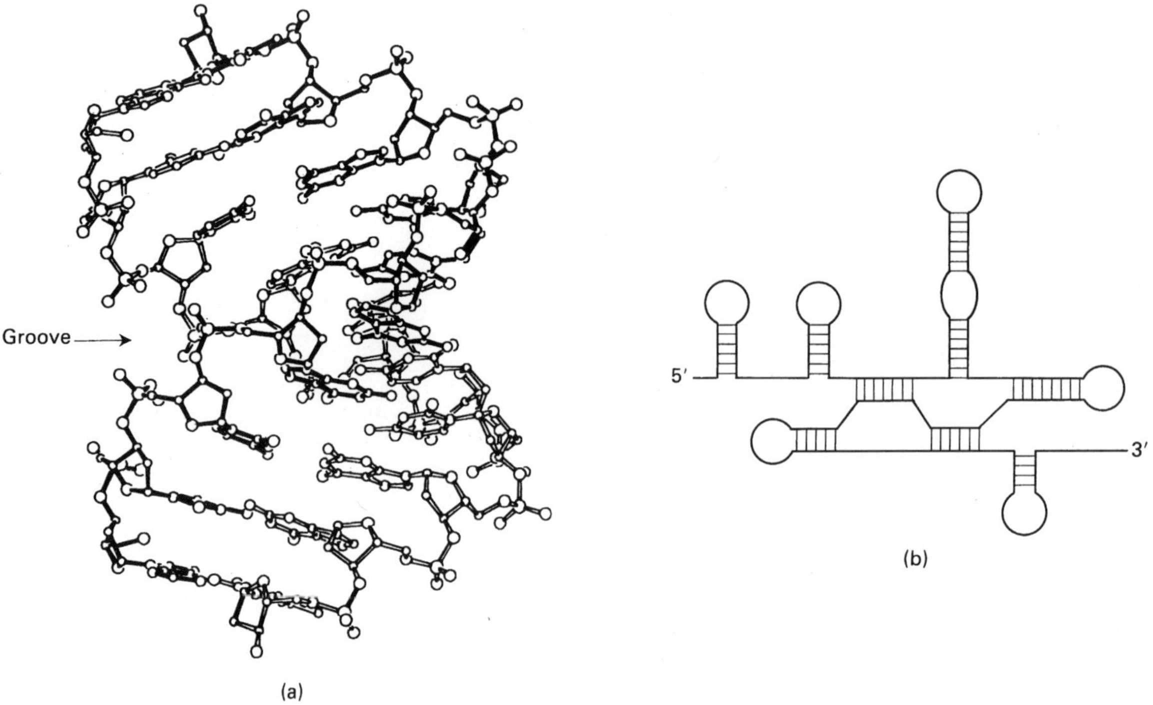

Fig. 6.17 (a) The fold of DNA. (After A. Rich.) (b) The outline of alternating double helices and loops in a stretch of ribosomal RNA (see Fig. 6.18 for detail).

6.11 Metal-ion binding to polynucleotides

The outstanding features of a single-stranded RNA or DNA (note RNA in particular), assuming a random fold, as a potential centre for ion binding can be simply listed.

1. *The diester phosphate with the sugar moiety*. This is a very weak centre and it is doubtful if there will be binding of effective affinity by one such site to any cation. The charges are then simply screened by long-range effects familiar in all polyelectrolyte solutions, i.e. by Na^+, K^+, Mg^{2+} in organisms. However, organic cations, such as spermine, behave differently.
2. *The nitrogen centres of the bases*. Again, in a random polymer, these are not likely to be binding sites for cations except for reactions with very heavy metals such as Hg^{2+}. Aromatic bases of low pK_a are poor donors and most transition metals can only form stable complexes with them through chelation.

Thus, free single-stranded polynucleotides are of little interest to us in the context of the present book. However, such single-stranded polymers can bind to proteins and become locally structured, so metal ion chemistry again becomes involved and undoubtedly this is the case in the complex units of ribosomes (see Section 6.12) and during transcription.

The centres on *double-stranded helical* DNA and single-stranded RNA folded into *double helical segments* are similar, but now the organized surfaces become of interest since the double strand (Fig. 6.17) has potential sites for metal-ion interaction. Thus the major and minor grooves, which have base-specific widths, expose both the ribose phosphate moieties and parts of the bases. These organized centres of charge and N-donors (bases) can bind metal ions, but the phosphate anions can also bind polyamines and small sequences of single-strand polynucleotides (i.e. anions), so that the exact form of a double helix structure and its curvature may well be controlled by molecules or ions in the grooves. We shall turn to the competition and cooperative effects between them later (see Chapter 9).

There are potential co-operative structures which are not double-stranded but can be created irregularly along the chains if the chain alternates between double and single (or triple) strands. Such a structure is seen in transfer RNA (tRNA) (Fig. 6.18), in larger RNA (Fig. 6.17), and is the basis of the cruciform of DNA (Fig. 6.19). The stability of these folded forms is again cation-dependent. The cooperativity here involves directly Mg^{2+} or polyamine binding. These structures may well be critical in the expression of function.

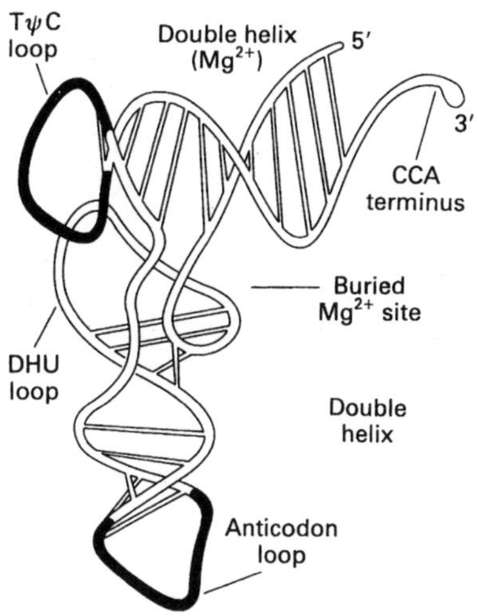

Fig. 6.18 Schematic diagram of the three-dimensional structure of a tRNA. (After A. Klug and A. Rich.)

6.11.1 Cavities and metal-ion binding in DNA

The reader is reminded of the discussion of cavity size and cation binding which was given in Chapter 2. There we saw that there were many factors controlling binding to a cavity which we could simplify under the headings:

1. the cavity size, treating it as an empty circle or hollow sphere of given radius;
2. the lining atoms of the cavity which act as donors;
3. the support structure behind the cavity.

There are also known exceptional specific features, e.g. at the telomeres* (K^+)

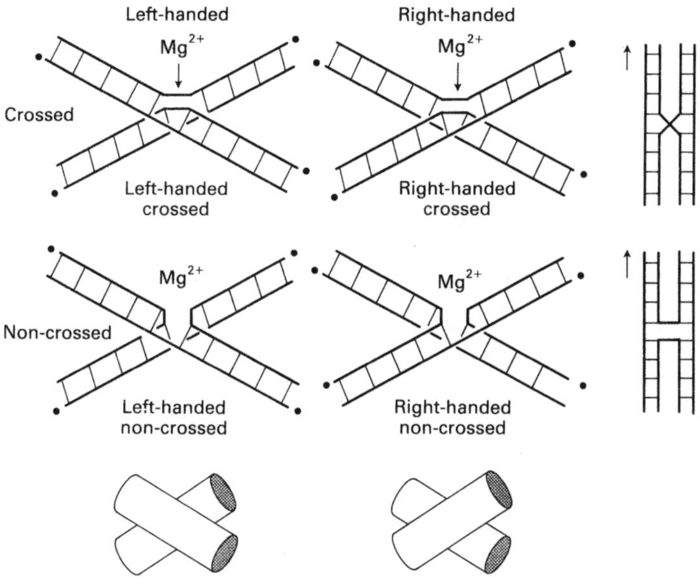

Fig. 6.19 Possible topology of strands and helices in the stacked X structure of the four-way DNA junction. The strands can be drawn as non-crossed or crossed, generating an antiparallel or parallel alignment of the sequences respectively (indicated by the arrows). Because of the non-planarity of the X, the helices can cross in either a right-handed or a left-handed form (the convention for the handedness indicated is based on the definition of dihedral angles). The helices at the bottom are stippled. The combination of these variables generates four possible structures. The 5′ termini of strands are indicated by filled circles. (After a diagram by Dr D. Lilley.)

*Telomeres are specialized structures at the 5′ end of chromosomes which contain a series of repeats of non-coding hexanucleotide sequences, e.g. TTAGGG . . . in humans. This is a device to overcome the problem of replication of DNA at the extremes due to the lack of space (this problem does not occur in the circular DNA of bacteria).

and at functions or cross-overs (Mg^{2+}) (see Fig. 6.19 and under the chapters on specific elements).

Now it is known that DNA (RNA) must be relatively mobile in all its interesting biological forms since, for the most part, these polymers must pass through a reading (translational, transcriptional) machinery; the construction as phosphate ribose esters aids this mobility. The sizes of the cavities these polymers can form, as seen in grooves (Fig. 6.17) and crucifixes (Fig. 6.19), are unlikely to have strict limitations; hence many cations and polyamines can bind but not very specifically or strongly. In cells, however, only Mg^{2+} and polyamines are present in sufficient quantity (Chapter 9). We need therefore to consider fast equilibria of the kind

$$\left.\begin{array}{l}\text{Open structure with}\\ \text{large cavity for}\\ \text{very large cation}\end{array}\right\} \underset{\text{polyamine}}{\overset{Mg^{2+}}{\rightleftarrows}} \left\{\begin{array}{l}\text{Closed structure with}\\ \text{small cavity for}\\ \text{small cation}\end{array}\right.$$

To a first approximation Mg^{2+} appears to vary but little in concentration in cells; if this variation can be neglected, then the degree of folding DNA/RNA is controlled by association with polyamines, such as spermines and histones, and selective transcription factor proteins.

While Mg^{2+} tends to force nucleic acids towards crucifix forms, both spermine and histones appear to stabilize condensed states of regular DNA and, since Na^+ and K^+ do not bind, except at telomeres, they stabilize very open ionic surfaces. In all of these interactions the relative selectivity of interaction with small molecules and metal ions is very slight. There is then little question of enzyme activity except for RNA (with RNA as substrate) where the fold has peculiarities (Fig. 6.18). Specificity develops only over very considerable surface area, e.g. in binding of promoter proteins. We return to these problems in Chapter 9 which treats magnesium biochemistry.

The bases in the folded forms offer the possibility of reaction (binding) to many heavy metal ions. This field of research was strikingly awakened by Rosenberg's discovery of the cisplatinum drugs. Here binding is to N(7) of guanidine bases. Undoubtedly many drugs in this and related families remain to be found in man's attempt to fight cancer and viral diseases. At the same time we must notice that other heavy metal ions will also bind to the bases, but this represents damage to DNA and could either be useful or cause mutational activity which must include cancer itself. Here the relationship between the value of drugs and the dangers of chemicals is close.

6.11.2 DNA sequences and mutation

The DNA sequences contain the coded information for the cell. It is the development of these sequences that records evolution as we understand it. The rearrangement of sequences and the changes of the bases by mutation are well recognized mechanisms for the trial and error search by biological systems in their efforts to evolve greater 'fitness'. It is found that whole genes can be inverted in their arrangement, and genes can 'dance' along a chromosome in evolution. Note immediately that there must be a 'driving force' for the search for fitness, in which environmental changes play a major role. After all, we must ask: 'fitness' for what? The details of the involvement of simple elements in these changes of sequence are not clear but it is known that elements such as Mn, Fe, and Cu through their oxidizing actions can cause DNA breakdown and then mutations. It appears that all biological systems reduce exposure of their DNA to damaging reagents such as $O_2^{\bullet-}$, H_2O_2, and $OH^{\bullet}$, produced by these oxidation catalysts, to a large degree. In this respect eukaryotes have the most powerful controls in that they use a superoxide dismutase in the cytoplasm which equilibrates only with an extremely low level of metal ions (Cu, Zn) while the prokaryotes use other enzymes (Fe, Mn) which involve a higher risk. Moreover, eukaryotes keep the reactions of H_2O_2 in special vesicles. In multicellular organisms the protection against mutation, and therefore against cancer, has a high priority. Development of complex organisms is often extremely slow allowing control.

6.11.3 Completed genomes

The genomes of more than twenty species have now been completed. This allows a comparison of genome development from archaebacteria, eubacteria e.g. *E. coli*, Yeast, *C. elegans* (a worm), and fruit fly to man. Such comparisons show the ways in which metabolism has or has not developed, how signalling and control have evolved, and how spatial division is managed, amongst many other features. However, we must remember that by itself a genome is only a one-dimensional linear sequence with *qualitative* connection to a cell's organization. A cell is a chemical system needing *quantitative* understanding which is a multidimensional, not a linear, problem. The dimensions include the concentrations of components, spatial division and energy input. Even so we shall use knowledge of the genome to show evolutionary trends. Here following 'fingerprints' of inorganic element binding over time is simpler than following organic chemical development. Hence, for example, we shall look for calcium EF hands, zinc fingers, haem-binding motifs and Fe_nS_n thiolate sequences as they arise and increase in kind and number in the genome throughout the major divisions of evolution (see the separate sections in Chapters 8–19 and Chapter 20).

6.12 Nucleic acids and proteins

Major particles in cells are nuclei (DNA) and ribosomes (RNA). They contain folded forms of both proteins and polynucleotides but it is the proteins that control the shapes of nucleosomes, DNA units of about 200 base pairs, and ribosomes, RNA combinations of high molecular weight. Each of these units has associated with it an unknown complement of cations, especially Mg^{2+}. There are present also two special classes of proteins which can activate the transcription or translation machinery; these proteins bind in the grooves of DNA but hold RNA in a complex form (Fig. 6.20). For these activities there needs to be a series of factors and enzymes and we shall note the particular role of iron (and haem) and of zinc in the DNA transcription systems. It is obviously necessary to have a communication network amongst all the systems. This is especially obvious in cell division before which proteins and DNA have to be replicated. The division process itself requires filamentous structures of yet other proteins to be made mechanically active at a precise time. The filaments bind both to the DNA and to the membranes. There is a deep involvement of the simultaneous directed use of energy and the completion of a synthetic cycle. How is this integration managed? We do not know yet, but Mg^{2+} and NTP must be involved.

Again we repeat that it is necessary to realize that the organization of DNA and RNA is not rigid but energized and dynamic. There are general structural principles, but these are temperature-, pressure-, electric-field-, and

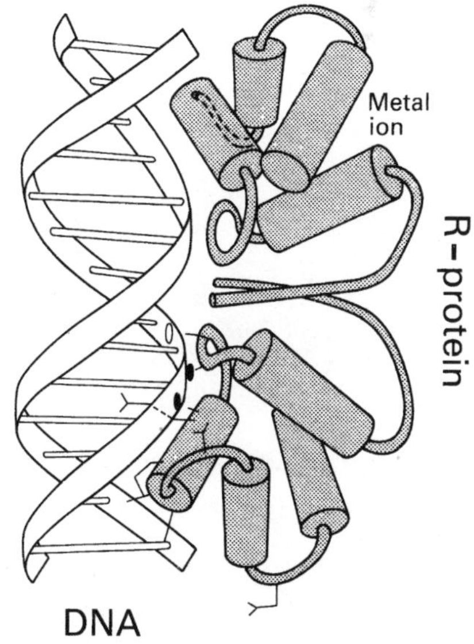

Fig. 6.20 The way in which a helical set of two protein segments could fit into the grooves of DNA. Note the value of mobile helices. Several metal-binding regulatory proteins may be of this kind (see Section 10.7). The structure is from the works of B.W. Matthews and T.A. Steitz

ion-concentration-dependent and are linked into the organization of proteins and, as we shall see, of membranes and even of polysaccharides. DNA in particular has to be a dynamic structure, more like a river than a lake, since it has to flow through changing structures all the time.

6.13 Controlled synthesis and degradation of biopolymers

An important part of our understanding of biological systems concerns the rates of synthesis or repair of proteins and nucleic acids and the rates of their degradation. The polymers are produced in response to messages; in fact, even the monomers are not wastefully generated. A full appreciation of the interactive system which is required to generate such control is possible only when we see that: (1) because DNA is a double-stranded helix, adenosine and thymine as well as cytidine and guanine must be produced in appropriate amounts; (2) similar statements can be made about proteins and RNA: all polymers demand a somewhat equal production of a great variety of monomers; (3) no nucleic acid or protein can be made without generation of appropriate amounts of ATP, GTP, etc. for synthetic energy. Again associated elements must be collected in correct proportions, so there has to be over-riding chemical integration within the system as well as an appropriate means of generating particular polymers and the physicochemical conditions of their existence. In a steady growing system this implies a homeostatic set of rates of production and degradation which can only be achieved if feed-back systems regulate the synthesis by start and stop signals and/or if the excesses to requirement are destroyed by a feed back self-hydrolysis. All this activity demands controlled flow of information to and from the central organizing unit—the DNA.

6.14 Genetic control

It is accepted today that the production of proteins follows a given synthesis sequence: (1) *transcription* of the DNA to give, after processing, various kinds of RNA: ribosomal RNA, transfer RNA, and messenger RNA; (2) the messenger RNA is then *translated* into proteins on the ribosome (RNA + proteins). It is the DNA which contains the inherited genes which relate to protein or to RNA production. In this book we are not greatly concerned with DNA or RNA themselves since, as described above, they do not interact with many inorganic elements or simple compounds directly. Exceptions are K^+, Mg^{2+}, and small polyamines, but of course DNA binds many proteins, as does RNA, and proteins bind metals. The major interest is, therefore, the way in which DNA, through RNA, produces proteins. To appreciate this regulation we need to give a brief description of DNA functions in terms of genes, where the major concern is with those genes which generate the proteins. It must be understood that there are large regions of DNA, especially in higher organisms, that are not transcribed at all, and their functions, which may be structural, are not known today, and there are also considerable regions of DNA just for binding a variety of proteins.

Turning to genes for protein production we have already mentioned (Section 2.21) that each gene is composite in that in its simplest form it has three regions: (i) a sequence of bases which in a triplet code reads out the sequence of amino-acids of the protein, although reading may be interrupted

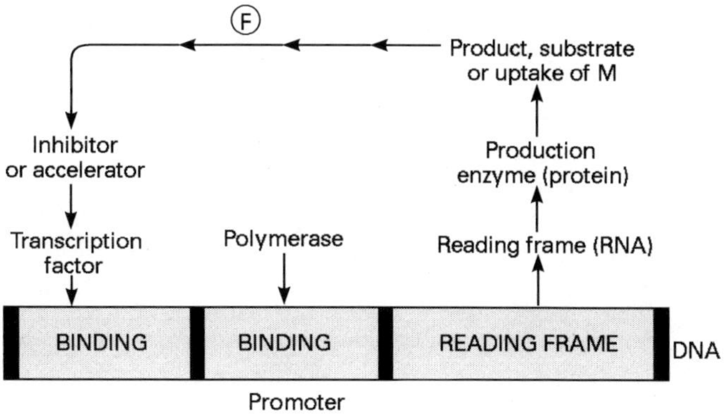

Fig. 6.21 A simplified version of a genetic construct which requires not only a reading frame to generate the m-RNA and then the protein but requires a binding region for the polymerase machinery and an instruction region for the binding of a transcription factor, both of which must be bound in appropriate form before reading can occur. The form of the transcription factor which may interact by feedback Ⓕ with an element, M, can be adjusted (by [M]) to stop reading. The whole is necessary for cell operations not just for that of DNA.

by so-called intron regions which are not translated; (ii) a sequence* for binding the protein reading machinery, RNA polymerases, which transcribe DNA virtually into m-RNA before proteins are produced (this is called the promoter region); (iii) a sequence to which a control factor, usually a protein, binds which only allows the machinery to work when this 'transcription factor' is bound and in the appropriate conformation (Fig. 6.21). The reader must realize that in practice the apparatus may be much more complicated with multiple transcription factors for even one translated gene. Our concern is with the relationship between final gene expression in the form of a protein and the involvement of elements. Apart from the facts already mentioned that K^+ and Mg^{2+} are involved with all RNA and DNA structures and Zn^{2+} and Mg^{2+} are required for many polymerases, our major concern is therefore with transcription factors. It is the transcription factors which bind many elements or small compounds, and then in the bound mode these factors allow or prevent the production of proteins (see Fig. 6.21) many of which function as their own pumps, enzymes, carriers, buffers and even transcription factors themselves. In essence, transcription factors are the feed-back controls related to those elements which are required together with proteins and small compounds to generate the cell machinery, see Aside. Thus DNA, RNA, and the cell contents, including metal ions and organic molecules all belong to one system which has to be inherited, as a whole, and this system is not just bare DNA which is inert. By itself DNA does not give the essential proteins, the proteome, or metals, the metallome, content of a cell.

Activators and Repressors

(a) Activator, MerR- like for Hg, Zn, Pb, Co, Cu

(b) Repressors, Ars R- like for As, Zn, Cd, Hg

NB. Sensors are known for Cu, Ag, Cd, Ni

Now, genes may not be controlled by transcription factors and promoters in isolation. In prokaryotes genes are organized in *operons* which contain several genes which are read cooperatively (see Fig. 6.22). In eukaryotes, however,

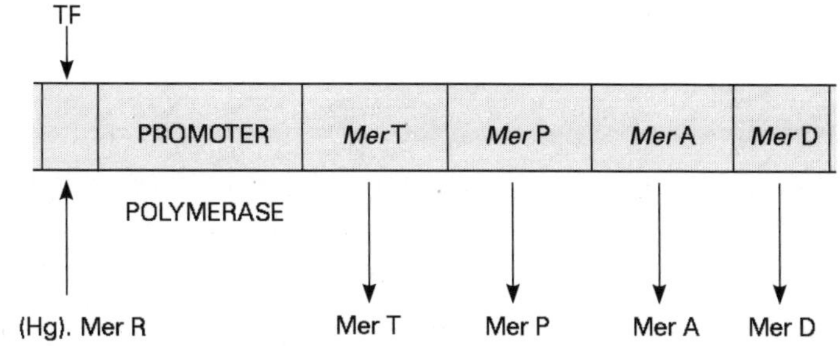

Fig. 6.22 The mercury resistance operon. The presence of mercury (Hg) and its binding to the transcription factor Mer.R activates the polymerase and causes reading of *Mer* T, P, A, and D generating the proteins for rejection of mercury.

* Usually the sequence of nucleotides is TATAA, commonly called a TATA box.

Table 6.14 Sizes of genomes

Organism	No. of bases (Mb)	No. of genes (estimate)
Mycoplasma genitalia	0.58	470*
E. coli	4.3	4000*
Yeast	13.5	5–10 000†
C. elegans (worm)	100.0	13 000
Fruit fly (drosophila)	165.0	15 000
Arabidopsis thaliana	70.0	~10 000
Human	3000.0	35–45 000‡

*No introns
†Few introns
‡97 per cent non-coding DNA

each gene has its own promoter site and there are many more genes and introns (see Table 6.14). Whereas 80 per cent of the genome of *E. coli* and of organelles generates RNA products, and then proteins, this may drop to as low as 10 per cent in some higher organisms (see Table 6.14). Prokaryote cells may contain small circular DNA in *plasmids* as well as the major DNA. Plasmids (see Aside) often carry drug and poisonous metal resistant protein factors. The DNA of higher organisms is arranged in chromosomes but we do not understand the reasons for the nature of the division of genes between chromosomes and the relationship to protein production as yet. It is expected that this will be elucidated in the next few years.

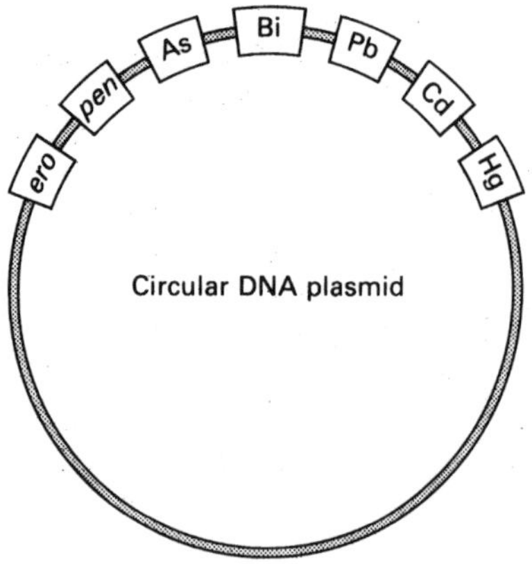

Parts of the DNA may be replicated and these new pieces may be inserted elsewhere. These pieces are called *transposons*. One way of responding to stress is for the stress to increase mutation rates, another is to initiate transposon formation and insertion. We cannot go into detail here, but one obvious stress during evolution has been the continuous change of the environment including element availability. It is uncertain how these stresses have forced adjustments on cells including their DNA, but they undoubtedly were and are one major determinant of evolution.

There are, therefore, many complications which should be kept in mind while examining the nature of DNA in relationship to the observed characteristics of an organism. For example:

1. The genetic code is somewhat different in different organisms; at the present time there is no logical explanation for the particular code(s) we observe.
2. The code is not read in a strictly linear way and there are different ways of reading in different organisms.
3. Some bases are chemically modified and this affects their reading, hence once again DNA is in a feed-back relationship with its own proteins, for example the methylation enzymes.
4. It is possible to reverse reading from RNA to DNA.
5. Genes can be transferred across species boundaries.
6. Translation as well as transcription has controls which differ in prokaryotes from those in eukaryotes.

Given these problems we cannot go directly from DNA information to the biological chemistry of the elements although they are certainly in feed-back and feed-forward relationship. We shall be able to draw attention to interactions between the two under each element in Part II and to indicate ways of thinking about whole systems in cells in Chapter 20. Note that the power of DNA sequence analysis is that it is directly informative about the type of protein in a cell and is extremely valuable for discerning evolutionary pathways. However, it does not indicate the need for quantitative production of proteins, nor the uptake of any element, nor of energy requirements. It is knowledge of the whole machinery of a cell that will explain why life exists (see Fig. 1.3), and it is here that biological inorganic chemistry is so central since it links the problem of survival to the environment.

6.14.1 A note on gene expression in multicellular organisms

While genes are expressed in bacteria according to external circumstances, a good example being a facultative aerobe which can live using one set of genes under anaerobic and another under aerobic conditions, multicellular organisms have differentiated cells in which some genes are permanently switched off and specific differentiated cells have little flexibility. The major purpose of a cell in an adult organism is not to multiply but to serve the whole organism while surviving. A muscle cell serves a mechanical function, a brain cell an electrical function and a gland cell serves to supply hormones. All the cells of the same organism have the same DNA, but the DNA is read differently in each case. (This does not mean to say that in artificial circumstances any or all of the genes cannot be re-activated.) It is during development that the cells differentiate. Hence, starting from one genetic inheritance in one cell, a fertilised egg, of a multicellular organism, the *expression* of genes changes in particular cell locations as the number of cells increases. The cause of change is the position of the cell on repeated division since all cells cannot be in exactly the same relationship when packed together after the eight cell stage. Gene expression is then influenced by the environment, initially physically, but increasingly chemically, using signals from neighbour cells. For the purposes of this book we note that to some degree the environmental factors are reflected in the way in which different metal ions are concentrated in the different cells (see Sections 20.5 to 20.8). The final steady state of cells in the adult may take years to develop, as in man, and may fail in a poor environment. For example, humankind do not go through puberty if the diet is too low in zinc, embryonic development depends on the supply of iodine for the thyroid hormone, and so on.

Now, the data in Section 20.14 show that during development the pumping of ions into and out of cells changes, so that cell gradients of different cells are also different. For example, neural plate development, at one stage in embryogenesis, requires activation of the sodium pump so as to give a threefold lower sodium in cells. Incorrect sodium levels yield altered brain structure. We see that ion levels back interact with the DNA, so that gene expression is continuously changing with the cells internal and external environment. Other factors affecting expression are physical fields (gravity) and light intensity, especially in plants. Thus the genes of a multicellular organism interact with inherited and non-inherited external factors. The full degree to which this happens in human development during early life is unknown.

6.14.2 Regulation: constitutive and induced protein production using genes

For operational purposes it is useful for us to separate genes into classes, *constitutive* and *induced*, no matter if in practice the distinction we can make is not so clear-cut.

A constitutive gene is responsible for the regulated production of a protein which is essential for the well-being of the cell in all normal circumstances. For example, some proteins are produced in a controlled fashion in concert with the cell cycle, but for most of the other constitutive proteins the reading of the constitutive gene must be switched off so that excess protein is not generated. The usual regulation process is one in which a given substance, a protein or metal ion or a small molecule generated directly or indirectly by the gene and which must first be produced to a certain reasonable concentration, builds up in concentration so that eventually it becomes a feed-back inhibition of production by binding to a transcription factor (see Figs 6.21 and 6.22). This feed-back inhibition, called *repression*, is necessary to ensure that energy is not wasted and a given protein is not produced excessively. Several transcription factors may be present and control expression of one protein. Since quite a high concentration of the protein is normally required (and acts at a given time) a weak restrictive effect (binding) on DNA by the active participant in repression, that is, by the transcription factor or by substrate or metal ion binding to a transcription factor is desirable (see Fig. 6.21). Notice that a transcription factor itself must be constitutively produced often by cell cycle limitation.

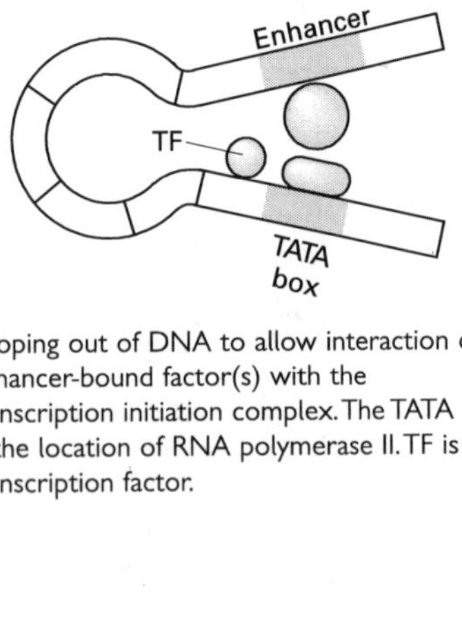

Looping out of DNA to allow interaction of enhancer-bound factor(s) with the transcription initiation complex. The TATA box is the location of RNA polymerase II. TF is a transcription factor.

The situation is made somewhat more complex in that when supply of an essential substance runs low it pays the cell to increase the production of the proteins for uptake of that substance so that the transcription of relevant genes may be *enhanced* by binding of a different factor not bound to the substance to be supplied but to some later product of metabolism. This is an example of *positive feed-forward* (see Aside).

An inducible gene is responsible for looking after a cell in unusual circumstances, so that expression of its protein is not required normally. The gene is repressed by some particular transcription factor protein which binds to it strongly. However, under stress of some kind, the effect of the stress on the cell leads to a signal which removes repression and leads the gene to express the inducible system so as to relieve the stress. The stress itself or the stressed condition of the cell, its adjusted metabolism, informs the genome when the induced system should be switched on. In the absence of the stress the system is not noticeably expressed. Clearly, the switching on of an inducible system may well have to be very sensitive to the inducer (or poison) so that a high-binding constant of the inducer to the protein transcription factor, which by itself represses the expression of the DNA, is required in the feed-forward activation of transcription. Note that since the repression is due to binding of a repressor there will always be some production of a protein since binding constants are not infinitely large. While many genes are repressed under particular circumstances there are also positive regulatory genes which enhance gene expression.

Examples help to see the way the genes and their regulation work (see Table 6.15). In *E. coli* the glycolytic pathway enzymes are produced constitutively in tune with the cell cycle. Amino-acid synthesis is also essential for the survival of bacterial cells, hence enzymes for amino-acid synthesis are also constitutive, however they suffer end-product repression. Taking the examples

Table 6.15 Definitions of genes with examples in *E. coli*

Definition example	Pathway example	Metal distribution
Constitutive or constitutive with repression	Glycolytic path and early H^+/ATP synthase Anaerobic metabolism	Na^+/H^+ pump K^+ channel Ca^{2+}/H^+ exchange Cl^-/H^+ exchange
Constitutive but with (end-product) repression	Amino acid synthesis Nucleotide synthesis	Ca^{2+} channel(?) Ca^{2+} ATP-ase(?)
Induced	Organic drugs Oxidative metabolism	Cd^{2+}, Hg^{2+}, Pb^{2+} pumps (outward)

of histidine and tryptophan: the synthesis by the enzymes of these two amino-acids has to fit requirements for the production of say 1000 proteins. However, there must not be excess of either amino-acid, or of any other, over requirement, hence both histidine and tryptophan themselves as they build up in concentration begin to bind to transcription factors which then stop further synthesis of the enzymes for the production of these very amino-acids. This is *negative feed-back*, or *end-product repression*, and requires a relatively weak-binding constant, $<10^6$, so that free amino-acid levels never fall below this optimal value. Notice that a transcription factor itself may have to be constitutively produced.

Consider next a man-made drug which is not readily metabolized and the metabolic products of which are valueless. After suitable mutations of genes pre-existing the ability to remove the drug, there is generated a gene for destruction of the drug. Now the gene product, an enzyme, has to be produced in effective amounts only in the presence of the drug although the gene is now a part of the genome. (The gene is only inherited in gametes of sexual organisms). Provided that the drug binds to a transcription factor with a high affinity and that only this combination binds to DNA there is only effective production of the gene product when the drug is present. The drug induces the enzyme by *positive feed-forward* (see Aside). In bacteria such inducible genes are often on a separate plasmid (see above, Section 6.14). (Note that it may not be concluded that absolutely no copies of an 'inducible' protein are produced in the absence of inducer.) All of this account of gene regulation is to be found in more detail in specialized references (see 'Further reading').

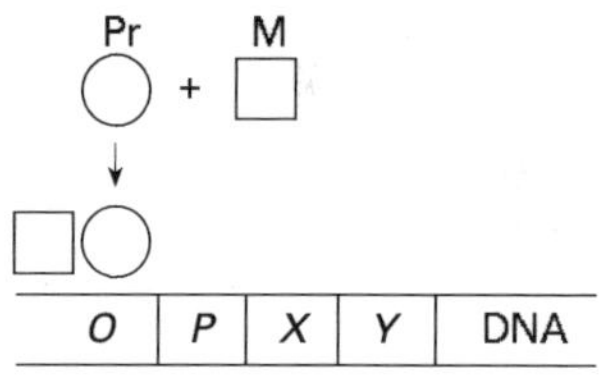

Positive (induced) control is managed by binding of an agent, M, to a protein, Pr, which activates, at *O*, the polymerase, *P*, to read genes *X* and *Y*.

Now we are interested in this book in the handling of the chemical elements by cells and hence we need to know which elements are looked after constitutively and which by induced systems. In many cases it is found that a cell has to be flexible in that it has to adjust gene activity so as to control levels of proteins and associated metal ions and non-metal elements in cells in close harmony opposite a variable environment. This means that both repression and acceleration, feed-back and feed-forward systems, are required over the element uptake/rejection as well as over the protein concentrations. To meet this requirement, uptake and rejection systems for elements coexist in the endeavour to maintain homeostasis. Individual elements must therefore be associated with selective binding to transcription factor proteins. It would be interesting to put this discussion on as quantitative basis as possible, and we have to remember that some required elements are poisonous in the cytoplasm at very low levels (Table 6.16). We expect these elements,

Table 6.16 Levels of required metal ions which can be poisonous in the cytoplasm

Poisonous concentration (M)	10^{-1}	10^{-3}	10^{-5}	10^{-6}	10^{-10}
Metal ion	Na^+, Mg^{2+}		Ca^{2+} Mn^{2+},	Fe^{2+}	Zn^{2+}, Cu^{2+}
					Ni^{2+}, Co^{2+}

NB Many metal ions are poisonous at almost any concentration e.g. Hg^{2+}, Pb^{2+}, Ag^+.

e.g. Hg^{2+}, to be selectively handled by strong induced binding while required elements such as K^+ or Mg^{2+} will be handled by weak constitutive binding. The difficulty of a general treatment is obvious, but we hope that a brief outline will be useful.

6.14.3 Constitutive and induced genes for elements

We consider in the first instance bacterial cells, in particular *E. coli.* An element, for example a metal ion, M_i, is required at selected level in cells but excess is damaging. The complexity of the uptake is shown in Fig. 6.23. There has to be adequate production of constitutive uptake systems, inward channels, pumps or exchangers, accepting sites and so on, especially if the element is in short supply, and means of switching off the production when an adequate concentration has been reached. It is also necessary to pump the metal ion outwards if the element can gain access to the cell to a degree larger than required, when outward pumps must be produced also in some regulated mode. The rate of production of the pump proteins, MP_1(out) and MP_3(in), can be inhibited or enhanced as discussed above.

It can be the case that on entering the cell the element M_i is bound first to a carrier MP_4 and eventually placed in a functional unit, ME_1 etc., for example an enzyme, while also interacting with a transcription factor, TF (Fig. 6.23). In all cases the expression rate of the protein production during the growth cycle takes the same form, so that free metal ion (and its associated proteins) is allowed to reach a certain level but is then restricted by metal ion binding to transcription factors depending on the values of the stability constants of the corresponding complexes, small for Mg^{2+}, large for Zn^{2+}. *Thus through genes the proteins are produced commensurately with the element concentration so that there will be related free M_i* and free apoproteins, compare the incorporation of histidine described above. The production of any of the proteins is eventually stopped by feed-back inhibition (see Fig. 6.21) but in some cases it can be enhanced over basic constitutive rates by other transcription factors which bind the element.

At this point it is worth repeating (see Section 2.19) that if all the above proteins, transcription factors, carriers, pumps, and enzymes, are to bind to an element in exchange equilibrium either with one another or with the free element then for a given element there has to be a similar binding constant to each of the proteins of Fig. 6.23. This is central to the concept of a metallome.

Now even when the free metal ion does not equilibrate with the proteins in the cytoplasm it is still necessary for the carriers, the transcription factors, the connected part of the pumps, and the final receptor protein or enzyme to have similar binding energy unless external energy (ATP) is applied to exchange. Thus we can still relate the system for an element to a notional equilibrium constant even though exchange with free ions may not occur (see

Table 6.17 Cytoplasmic and extracellular proteins in equilibria

Metal ion	Protein	Estimated **equilibrium K**	Ligands
(A) Equilibria between complexed and free ions			
Ca^{2+}	Calmodulin	10^7 feed-back to pump	O
	Calbindin	10^7 carrier	O
	ATP-ase pump	10^7	O
Mg^{2+}	MgATP	~10^3 feed-back to pump	O
Mn^{2+}	Fur	10^6 transcription factor	O(N)
Fe^{2+}	Fur	10^7 transcription factor	O/N
	Aconitase	10^7 transcription factor	Fe_3S_4
	Fnr	10^7 transcription factor	Fe_3S_4
	DCT1/Nramp 2	(?) carrier	O/N
(B) Equilibria between complexes			
Co^{2+}	(?)	10^8 carrier/pump	N(S) ?
Ni^{2+}	(?)	10^9 carrier/pump	N ?
Cu^{+}	Ace-link	>>10^{12} transcription factor	RS^- (cluster)
	ATP-ase pump	>>10^{12}	RS^-
	Metallothionein	>>10^{12} carrier(?)	RS^-
Zn^{2+}	Zinc-fingers	>10^{11} transcription factor	RS^- (imid)
	ATP-ase pump	>10^{11}	(N) RS^-(N)
	Metallothionein	10^{11} carrier(?)	RS^-
	Zur	10^{11} (?) transcription factor	N(RS^-)
Fe^{3+}	Transferrin	10^{21} carrier	ϕO^-/N
Zn^{2+}	Albumin	10^{12} carrier	N
Cu^{2+}	Albumin	>10^{15} carrier	N/N^-
Ca^{2+}	Phosphoprotein	10^3 carrier	O

The examples below the dashed line are of extracellular carriers

Table 6.17). The advantage of protein–protein exchange as the mode of transfer in a cell is that protein/protein recognition rather than direct selective element recognition (binding) can decide which element is involved. Of course, at some early stage in the extracellular fluids or environment direct selection of the free element must occur. *For strong binding the implication is that in a steady state all the proteins are overwhelmingly bound to M_i* and there can then be very little free M_i (notional or real) and very little free protein, TF P, E, *or* C, *when the cell is running optimally*. The sum of the concentrations of all of these proteins is effectively stoichiometrically related to the bound M_i. Clearly, this will be the case for the more poisonous metal ions of Table 6.16. We can see how this can be achieved in the case of mercury ions by having all its genes under the control of one mercury binding transcription factor (see Fig. 6.22).

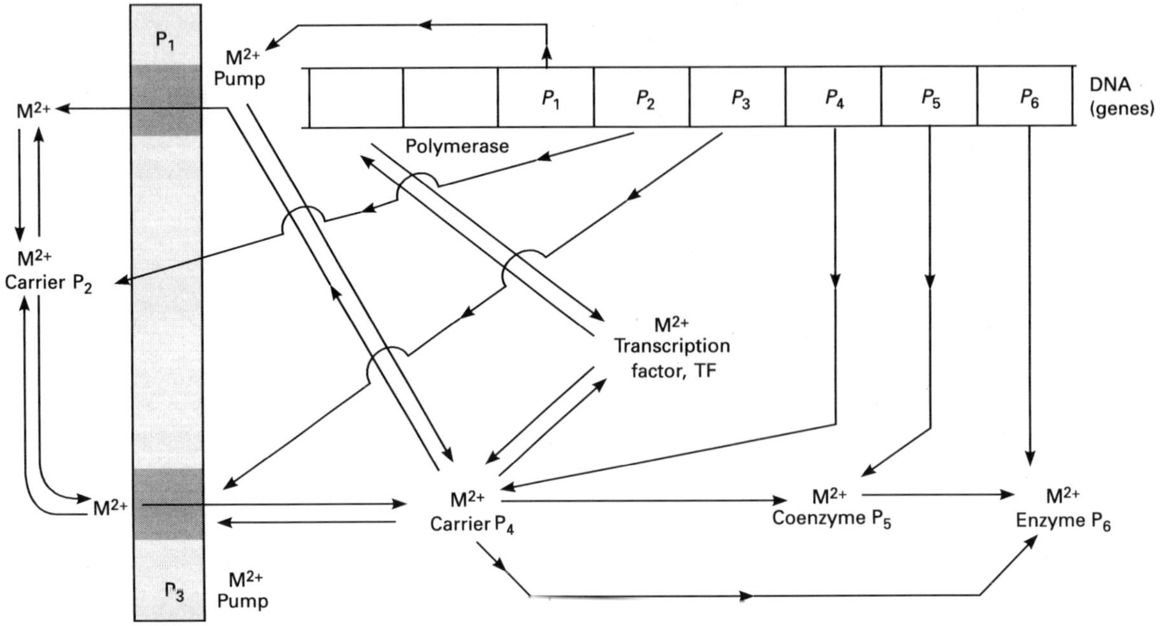

Fig. 6.23 The uptake proteins which may well be required in the management of any metal (or other element), M^{2+}. All the proteins, P_1 to P_6, and any enzymes, described in earlier chapters and in the chapters to follow are expressed by *controlled* production via DNA and RNA (see Fig. 6.21). The result is a cooperative interaction of all the units in the figure.

6.15 **Summary of genetic control and regulation of elements at 'equilibrium'**

All the above discussion uses *equilibrium* binding constants, *K*, which may control activity or regulation of synthesis of a protein. As stated, all *K* values must be closely similar for the handling of one element M_i in a bound state opposite free M_i to ensure equal activity. It is then very important to know which processes can be safely described as being at equilibrium with free ions. This returns us to the problem of describing the modes of incorporation or rejection of given elements, since some incorporations are not under equilibrium thermodynamic control but are under kinetic exchange control, as we have seen. Before doing so we give a table of elements the incorporation of which is constitutive, rather than induced, in *E. coli*, and may be generally so (Table 6.18).

While we see that for any one metal ion the binding constants to its selected proteins must all be rather similar for all reactions which equilibrate with a given free M_i, it is also imperative that under the conditions that exist in the cell no other metal ion, M_2, binds to these same sites. One of the severe difficulties of control and regulation that is of necessity at equilibrium or close to it, is that such selectivity is required. We have indicated how this is possible largely through studies of model complexes in Chapter 2, that is through the application of 'conditional' (or effective) equilibrium constants with feed-

Table 6.18 Constitutive and induced systems for element incorporation or rejection in many cells

Constitutive Incorporation or partial rejection (essential elements)	H, C, N, O, for essential organic metabolism Na, Mg, (Si), P, S, Cl, K, Ca, (V) (Cr), Mn, Fe, Co, (Ni), (Cu) Zn Mo, Se, I
Induced rejection (poisons)	(Al), As, Cd, Ba, most heavy elements after molybdenum in the periodic table NB Many foreign organic molecules, drugs.

Rejection can be by metabolism when an enzyme, not a pump, is induced.

back control of ligand production, even when a cell has to handle at least ten similar elements simultaneously, all of those listed under constitutive in Table 6.17, in fact (see Sections 2.4 and 2.8). A final implication of this tight feed-back control over equilibrium concentrations of free ions is that, to a first approximation, given the values of total metal–protein concentrations and binding constants, *the free ion concentration of each metal ion is approximately the inverse of its binding constant to the proteins present*. We know from model studies the probable values of these binding constants (Fig. 2.10). The question arises as to how closely cell systems follow this prediction (see Chapter 20). In any event, taking the nature of proteins and the selectivity of metal ion binding and activities required together with the use of energy in pumping, their distribution, or one closely like it, probably represents a fundamental feature of the cytoplasm of all life.

Clearly we must determine these constants in cell compartments. However, one conclusion is that the genetic code, the genome, and the cells protein content, the proteome, was and is limited by the element content of cells which has environmental, chemical implications.

When we turn in the next section to the final two types of biological polymers, fatty acids $(CH_2)_n$ and polysaccharides, then these principles of interaction with elements remain valid but the synthesis becomes less exacting, their direct role in control appears to be less, and especially in the case of their phospholipids the problem is not now referable to the properties of single molecules but only to *lipid phases*. The lipids form the solvent of a phase, much as water molecules do, and the polysaccharides are even more difficult to describe in structural molecular terms. Since they are all polymers we describe them first in terms of single molecules, but we shall see that phase concepts are of great value in the analysis of their functions.

PART B *Polysaccharides and lipids*

6.16 Introduction

The importance of saccharides and lipids to our understanding of the biological chemistry of the elements is somewhat less than that of the importance of proteins and nucleotides. Nevertheless, we must appreciate some of their properties since these molecules are critical for the building of cell structures and compartments, membranes and connective tissue. On the whole the value of their interaction with various elements lies in the general electrostatic field effects and concentration gradients rather than with specific interactions. These ionic gradients and electrostatic fields are largely confined to H^+, Na^+, K^+, Mg^{2+}, and Ca^{2+} chemistry amongst cations and to that of Cl^- amongst anions, and they are absolutely critical to the well-being of an organism. To some degree, of course, these interactions act as cross-links stabilizing mainly the external parts of cells.

6.17 Introduction to polysaccharides

Polysaccharides in living systems (Table 6.19 and see Fig. 6.24) have five main roles: (1) in the formation of plant *skeletons* (cellulose) and cell walls; (2) in connective tissue of animals e.g. chitin; (3) in generating recognizable features on external proteins or on the outer surfaces of cells; (4) in mucus, adhesive and cohesive fluids, etc., many of which either aid of inhibit flow; and (5) in C/H chemical and energy storage (glycogen). In the present chapter we will be concerned mostly with the first four roles. Note that much of this material now appears under the heading 'Glycobiology'.

In most organisms the outside (exo- and ectocellular) regions of the cell have the carbohydrates associated with proteins (i.e. glycoproteins and proteoglycans) and lipids (i.e. glycolipids) or in the form of hydrated polysaccharide gels. In unicellular organisms the combination of these macromolecules, polypeptides and polysaccharides, gives rise to cell walls, protective polymers, receptors, and enzymes, both in membranes and in the periplasmic and extracellular spaces. In multicellular organisms, while the same diversity exists, the mixed (hetero) and homo polymers also form the structural network which holds the cell–cell organization together (Table 6.19). (For a review on saccharide structures and functions, see 'Further reading'.) The primary structure of these polysaccharides is often very complex since their constituent monosaccharides can be linked in many different ways. Consequently, the potential information encoded into a polysaccharide via its primary sequence and three-dimensional structure is considerable allowing very selective interaction with proteins. We shall find that the interactions of the different polysaccharides with inorganic ions are quite different from those with proteins and that these interactions have a less selective nature of their own. In part this is due to the very different locality of their synthesis and use (outside the cytoplasm) from proteins, in part from their lower level of folded structures given greater structural covalent diversity, but mainly due to the absence of strong chelating centres, especially those based on S- or N-donors. It is the latter which removes the possibility of binding to trace elements.

Table 6.19 Important polysaccharides of some living organisms

Organisms	Polysaccharides	Composition	Biological function	Functional groups other than –OH
Algae	Fucoidan	D-glucose, D-mannose, D, L-galactose, L-rhamnose, L-fucose, L-xylose, L-arabinose, D-glucuronic acids, D-galacturonic acid, L-3,6-anhydro-galactose	Extracellular mucilage	$-COO^-$ –O– $-O-SO_3^-$
	Cellulose	β-(1→4)-D-glucan	Cell wall	–
	Xylan		Cell wall	–
	Mannan		Cell wall	–
	Alginic acid	D-mannuronic acid	Intercellular space in marine algae	$-COO^-$
Higher plants	Cellulose	β(1→4)D-glucan	Cell wall	–
	Hemicellulose	β(1→4)D-xylan, xyloglucan, arabinogalactan, arabinoxylan, etc.	Cell wall	–
	Pectin	α(1→4) D-galacturonan, arabinogalactane rhamnogalacturonan, etc.	Cell wall of young tissues	$-COO^-$
	Gums	Uronic acid, hexoses, pentoses	Secretion products	$-COO^-$
	Mucilages	Galacturonic acid	Cell wall, nutrient reserve	$-COO^-$
Animals	Chitin	β(1→4)*N*-acetylglucosamine	Invertebrate skeletons	(a) $-NH-COCH_3$
	Glycogen	α(1→4) D-glucose with ramifications in α(1→6)	Nutrient reserve	–
	Hyaluronic acid	D-glucuronic acid, *N*-acetyl-D-glucosamine	Cell cohesion	(a) $-NH-COCH_3$ (b) $-COO^-$
	Chondroitin	D-glucuronic acid, *N*-acetyl-D-galactosamine	Cartilage tissue	(a) $-NH-COCH_3$ (b) $-COO^-$
	Chondrotin sulphate	Idem with galactosamine sulphated at positions 4 or 6	Cartilage tissue	(a) $-NH-COCH_3$ (b) $-COO^-$ (c) $-O-SO_3^-$
	Glycoproteins	Mannose, galactose, glucuronic acid, glucosamine, galactosamine, sialic acid	Cartilage tissue (many others)	(b) $-COO^-$ (d) $-NH_2$

NB Some of the polysaccharides may be phosphorylated. They and carboxylated derivatives bind metal ions.

6.18 The backbones and sidechains of polysaccharides

A description of polysaccharides is given in general textbooks of biochemistry. We give some typical structures of linked hexoses in Fig. 6.24 including their backbones.

Polysaccharides are quite variable in sidechain structure but the sidechains are relatively small (see Fig. 6.24). Generally, the sidechains help to make the polymer more water-soluble. Moreover, they are not so mobile as in proteins in the sense that they are not connected via CH_2-units to the main chain, the

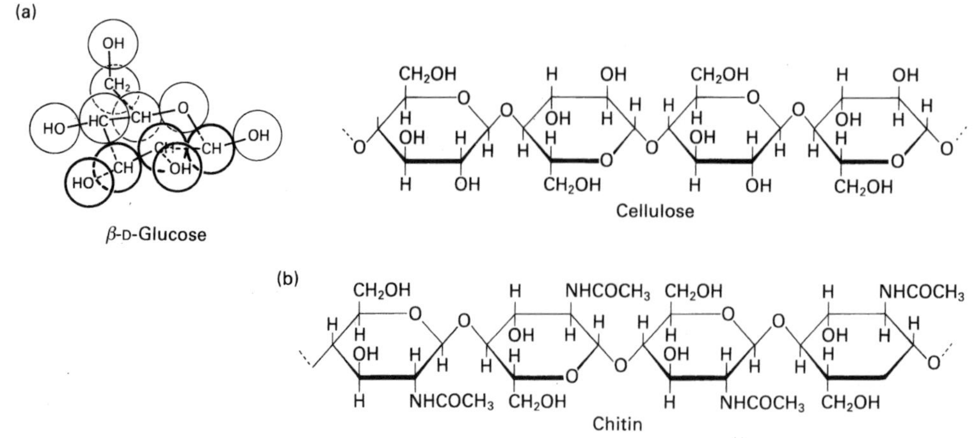

Fig. 6.24 (a) Cellulose; (b) chitin.

single exception being the β–CH_2OH unit attached to carbon 5 in pyranoses. Of the two faces to each saccharide ring, one is generally more hydrophilic than the other due to the greater number of hydroxyl sidechains on one side of the ring than on the other; this is especially marked in monosaccharides such as mannose. The hydrophobicity of a polysaccharide is then based on *one face* of the hexose ring, i.e. is a property of the backbone and not of the sidechains, in marked contrast with proteins and polynucleotides. Thus, on the whole, any fold of a polysaccharide is based on weak main-chain/main-chain interactions and tertiary structures are not strong. They are such then, that they can be very selectively recognized by large surfaces, but again they do not supply good selective *recognition* centres for small molecules or ions. In some cases the side chains such as–NH–CO–CH_3 do link together in antiparallel β-sheets to give strong structures, e.g. in the probiscus of some insects, see chitin [Fig. 6.24 (b)]. However, very many polysaccharides carry several negative charges, –CO_2^- and –OSO_3^-, and repel one another making open mesh or entangled structures in which diffusion is easy. This is a desirable property in that small molecules and ions must reach all cells easily through the extracellular matrices.

6.19 Glycoproteins and cell surface packing

Glycoproteins generally have short polysaccharide chains of some sugar moieties attached to the proteins, frequently at –OH or –NH_2 sidechains The attachment of sugars to proteins occurs in the endoplasmic and Golgi compartments, and labelling by short polysaccharides is part of the traffic-directing of proteins to their destinations. Some glycoproteins, e.g. calrectin, act as chaperones in these compartments assisting folding of metalloproteins, e.g. transferrin and caeruloplasmin.

One way in which expanded regions of polysaccharides attached to proteins or lipids (Fig. 6.25) in membranes may be responsible for, or related to, volume restrictions is through spatial packing which relates to shapes of cells. The function of different polysaccharides which are on the outer surface of membrane proteins could then be to assist localization of protein activities in zones along the inside as well as the outside of the membrane; all these associations will be affected by cation, mainly calcium, binding, if the surfaces are

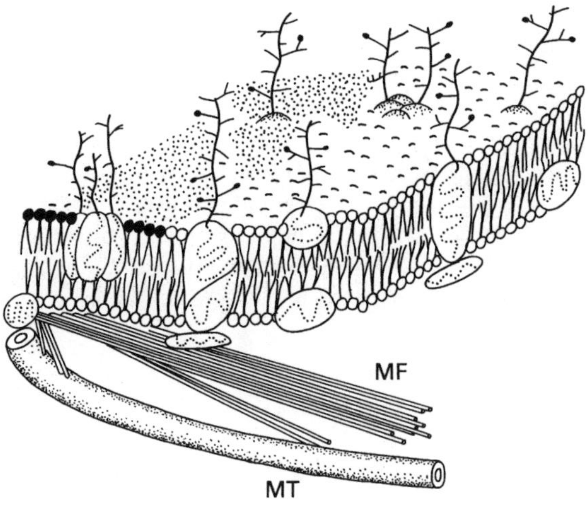

Fig. 6.25 Membrane protein networks and contact with membrane glycoproteins: MT, microtubules; MF microfilaments The precise organization is thought to be critically dependent on the level of Ca^{2+} and Mg^{2+} ions in the cells and in the surrounding medium.

negatively charged (Figs 6.25 and 6.26). They are also a major source of cell–cell recognition and ordering, and potentially of message transmission (see also Fig. 3.9).

6.20 **Interaction of polysaccharides with metal ions**

Now, just because some of these polysaccharides, especially those which are sulphated, are *anionic* gels, or sols, they are extremely dependent upon the inorganic salt medium in which they reside. Two types of interaction affect their volume:

1. the general ion effects (Debye–Hückel–Onsager);
2. the specific binding (cation) effects (Bjerrum).

It is here that the inorganic chemistry of biological systems again becomes important.

General theories of ion interaction with polyelectrolytes have been developed (see Kotyk *et al.* in 'Further reading'). These interactions can cause collapse (called condensation) of a widely dispersed solution of an anionic polymer, a sol, down to a very small volume, a gel, over a very small concentration range of cation. Furthermore, like all other attractive forces, they give rise to equations of state with terms parallel to those for condensable non-ideal gases. We note that such condensations can be selective again to the calcium ion in particular. While the calcium interactions may be between pairs of polysaccharide units as in Fig. 6.26, they frequently are involved in combinations between proteins and saccharides where one acts as a receptor for the saccharide extensions of a glycoprotein. These protein receptors are often labelled *lectins* and it is the calcium ion which controls this extracellular binding. The consequence of the binding can be that a message is sent into a cell informing it of the presence of the glycoprotein, or it can result in structural changes and stabilization of cell–cell organization.

At this stage of the discussion it is clear that polysaccharides were 'designed' to interact with cations to very different degrees so that the full potential value of their structural variety was developed. Their selectivity can be based especially upon the incorporation of different sidechains. For

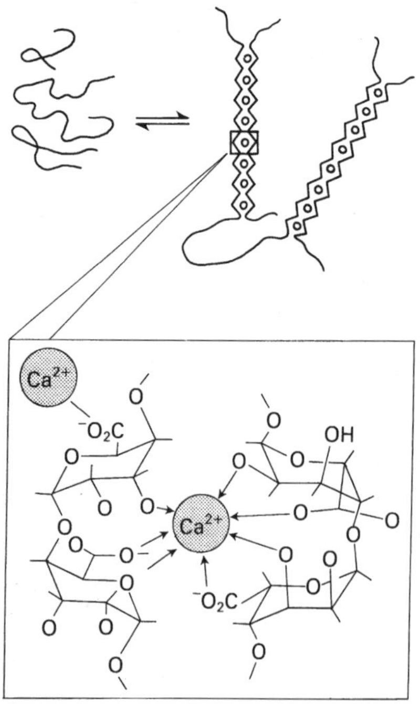

Fig. 6.26 Schematic representation of association of polyglucuronate sequences by chelation of Ca^{2+}, the 'egg-box model'. Oxygen atoms coordinated to calcium are shown by arrows.

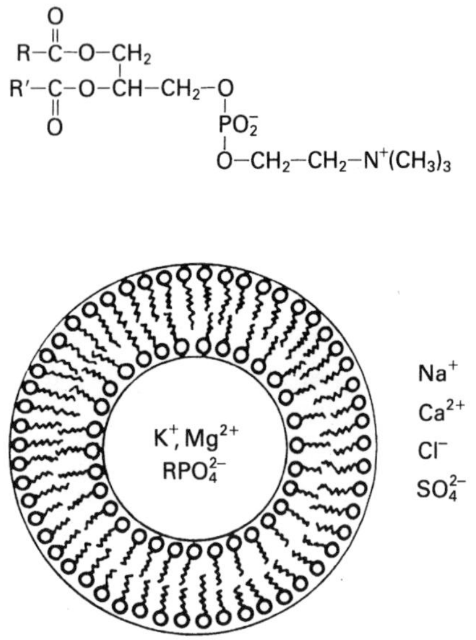

Fig. 6.27 (*Below*) A diagram of the bilipid layer membrane of a vesicle or a cell with (*above*) a typical lipid, phosphatidylcholine. The ions surrounding and inside a cell are shown, but the distribution is reversed in vesicles.

example there were various developments in evolution for interaction with inorganic materials: (1) very low density of charged sidechains, as in celluloses, with virtually no role for cations but generating a potential interaction with hydrated silica particles, or possibly semi-crystalline calcium salts, e.g. carbonates, which tend to carry low negative surface charge as seen in statoliths and corals; (2) increasing anionic density due to side chain carboxylation, where such polymer chains repel but may have considerable binding interactions with single cations; (3) further increasing anionic density, but now including sulphation (a late development in evolution), which has little interaction with cations and can help to give an open mesh allowing free diffusion.

6.21 Properties of lipids

6.21.1 Lipids in membranes as macromolecular assemblies

Lipid molecules in themselves are polymers based on such chains as $-(CH_2)_n-$ and obviously, such chains do not become involved with the binding of inorganic elements; they are very loosely held to one another and the membranes of biology are liquid rather than solid barriers between aqueous solutions. However, the headgroups of the lipids are polar and are exposed to ions. For the most part we shall not be concerned with single molecules of fatty acids or phospholipids but with the separated membrane phase which they form. This phase has a repeating pattern of headgroups in water, such as saccharide $-CO_2^-$ or $-(RO)_2PO_2^-$, and such headgroup repeats are reminiscent of the grooves of DNA and have the same interest for us. We shall therefore treat the membrane surface as if it were a two-dimensional, very large, mobile, fluctuating macromolecule carrying a considerable electrostatic potential, and proceed to the discussion of membrane organization and dynamics and the effect of cations and anions upon them. First, we must see the complications in such an approach by examining the molecules involved in some more detail.

6.21.2 Composition and general functions of lipids

As is now clear, the basic framework of the biological membranes (see Chapter 3) consists of double layers of lipid molecules held together mostly by hydrophobic van der Waal's forces (see Fig. 6.27). Embedded in these double layers one finds several 'integral' proteins (i.e. non-dissociable without disruption of the membrane) with carbohydrate chains attached to them and protruding from their external, extracellular surface as also happens with glycolipids discussed later in this section. The relative percentages of these three components of membranes (lipids, proteins, and carbohydrates) vary considerably from one membrane type to another; usually one finds 25–65 per cent lipids, 25–62 per cent proteins, and about 10 per cent carbohydrates, see Fig. 6.25, which is similar to the accepted model of the biological membrane, based on that proposed in 1972 by Singer and Nicholson, in which the various components are represented. The arrangement is not static but in some regions the proteins are highly organized in complexes. Besides the 'integral' proteins, membranes also contain 'peripheral' proteins and glycoproteins associated with their internal and external surfaces respectively.

All membrane lipids are amphipatic, i.e. they have a hydrophilic head group and a hydrophobic (double) tail in their molecules. Most of them are phosphoglycerides, i.e. derivatives of glycerol with two of its three hydroxyl groups esterified with fatty acids and the third esterified with a phosphorylated molecule which may be choline, ethanolamine, serine, or inositol. The fatty acid residues are long hydrocarbon chains, generally with 12 to 18 carbon atoms, which can be saturated or unsaturated, usually with one or two double bonds (see Fig. 6.27). These membrane lipids are, therefore, diglycerides, and differ from storage fats (non-membrane lipids) that are triglycerides, i.e. have the three hydroxyl groups of glycerol esterified with fatty acids. The other lipids that occur in biological membranes are cholesterol, (a disc-shaped molecule with four fused rings. and an hydroxyl group) and some derivatives of sphingosine, an amino alcohol with a long unsaturated C_{15} hydrocarbon chain in which the amino group is linked to a fatty acid residue. When the hydrophilic head group of these lipids is phosphorylcholine the resulting derivative is sphingomyelin; when it is a carbohydrate one obtains a glycolipid, either a cerebroside (with a single saccharide) or a ganglioside (with an oligosaccharide). It is clear that the different lipids confer different properties on a membrane. However, one common property of all these membranes is their fluidity or mobility.

The mobility of the membrane proteins (or lipids, in some cases) may be affected by binding to the membranes, e.g. of Ca^{2+}, protonated polyamines, or other peripheral proteins. In some cases this may lead to the formation of 'clusters', 'patches', or 'caps' (see Fig. 6.28).

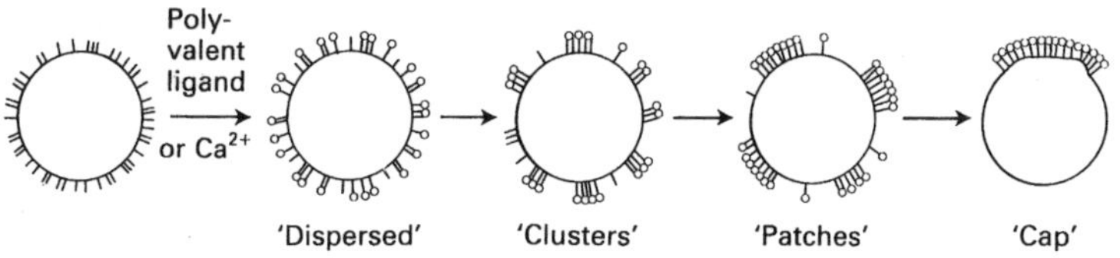

Fig. 6.28 Formation of clusters, patches, or caps driven by cations, e.g. Ca^{2+}, or cationic organic compounds.

6.22 Ion association with lipid surfaces

Since membrane surfaces are negative they will accumulate cations through both a general salt effect and specific binding. These are very hard to distinguish since the types of group exposed on the membrane surface, O-donors, have little selective binding for different cations. We expect binding to be important, however, since the charge density on the surface of a membrane is high. We also expect polyamines or protein surfaces which closely resemble polyamines, e.g. those of polylysine and cytochrome *c*, to bind well to these membranes. As in the case of DNA a competition exists between organic and inorganic cations for the surfaces.

$$M^{2+} \text{ surface} + \text{polyamine} \rightleftarrows \text{polyamine surface} + M^{2+}.$$

The modulation of the interaction, as in the case of DNA, can be managed: (1) by altering $[M^{2+}]$ (pumps); (2) by synthesis of the polyamine; and (3) by modifying the surface, for example, by phosphorylation.

6.22.1 The influences of selected inorganic elements

We have seen that ions are pumped across the membranes so that the lipid bilayer faces a gradient from the cytoplasm of high K^+ (Mg^{2+}) and phosphate anions, which is opposed by high Na^+, Ca^{2+}, (H^+), and Cl^- ions in both the outer bathing solution and in most vesicles. These energized gradients can interact with the lipids, so that the distribution of lipids across a membrane is energized by the chemical potentials (not just electric potentials) at the surfaces. The strength of the interaction is very like that discussed in Section 6.20. Undoubtedly, the most effective ion involved in this interaction is the calcium ion. Most of the charged saccharides attached to the lipids also carry negative charge and as such they assist calcium interaction. In so far as there is lateral discrimination of ionic concentrations due to steady-state flow of ions along membranes there will be lateral organization of lipid molecules also. Another form of energization is phosphorylation of the proteins on the inside of the bilayer; for example such phosphorylation is known to control the organization of protein complexes in the thylakoid membrane.

We assume that there is no direct binding of trace elements or of monovalent ions to the membranes.

6.22.2 Potentials on membrane surfaces

The fact that positive ions are unequally distributed in the aqueous phases on either side of membranes gives an inner to outer *solution* potential across membranes (see Chapter 8). Quite different, though interactive, is the potential which arises from the (anion) surface charge disposed differentially across the 50 Å of the membrane. The interaction of this *surface* potential with the *free* (not bound) ions in solution is usually handled by Gouy–Chapman theory (see Kotyk *et al.*, in 'Further reading'), so that there is a summed potential. If the membrane is not isotropic, local areas have local surface potentials and we must then always be aware of the possible development of local eddy currents through holes in membranes while a steady state is maintained by ion pumping.

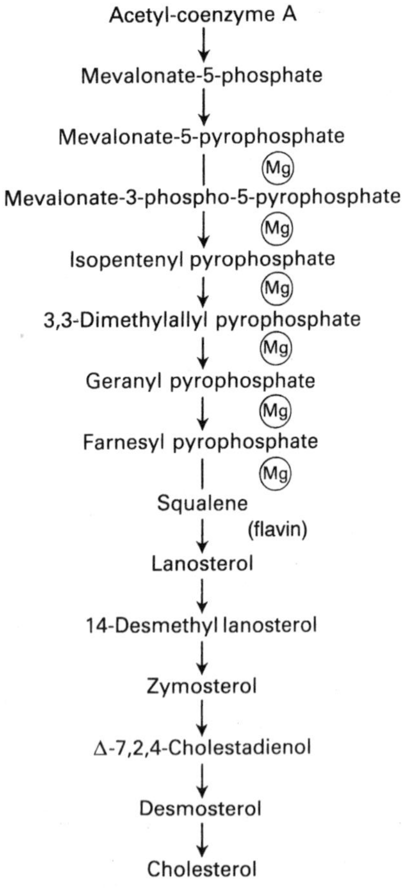

Cholesterol biosynthetic pathway

6.23 Essential fatty acids

Animals cannot synthesize all fatty acids. Of particular interest are the linoleic acids, which they require in their diet. Their release as free fatty acid glycerides from inositol phosphate derivatives is an extremely important mode of signalling associated with Ca^{2+} and kinase C. They are also precursors of arachidonic acid and then of the prostaglandins. The syntheses of the latter require iron enzymes and dioxygen.

There is a grossly intriguing feature of the secondary metabolism described in this section, namely, that it is so dependent on dioxygen and iron. Behind the hormonal activities of sterols and prostaglandins lies a deep level of homeostatic and morphological control, that of iron.

A further important point not to be missed in the context of this book is the way many essential molecules have interlinked syntheses, for example related to isopentenyl pyrophosphate, the reactions of which require magnesium, see aside, which leads to ubiquinone, cholesterol, and dolichol.

6.24 Summary of biopolymers

The nature of the three groups of large biological polymers, proteins, nucleic acids, and polysaccharides, based upon a description of the way in which their monomers are built into sequences and the structures which then developed, can be said to be similar. By contrast when we turned to situations in which the properties of the polymers, overwhelmingly the lipids but also to some extent the polysaccharides, they were described better by bulk (equations of state) not molecular properties. This approach was found to be particularly suitable for the description of fatty acids in lipid membranes. The fatty acid polymers often separate zones of protein action and create an organic phase as well as separated aqueous compartments for reactions, but they cannot be usefully described through structures of isolated molecules.

All the biological polymers are invariably energized in more than one way. They are energized in the obvious sense that their chemical bonds are unstable to hydrolysis. Additionally, a polymer is energized by the way it is placed and contained in space—polysaccharides outside cells, for example. In the local space in which it is found the polymer is obviously in a gravitational field but may also be in an electrical field. Such fields can separate lipids across membranes and affect the conformation of a flexible polymer, e.g. DNA. While any polymer distribution is energized, so is the distribution of all interacting polymers, small molecules, and ions by pumps and metabolism. The polymer, through its interactions is then additionally energized. Finally, since the energized polymer is attached to other energized polymers, the whole system cannot relax so that the polymer has to some degree a configurational energization. Some, if not all, of these factors are seen in the most obvious features of biological molecules—they are integrated in an organization.

Now these interactions are dependent on the presence of elements other than those of organic chemistry. First, we note that proteins are largely made from H, C, N, O, S, nucleic acids from H, C, N, O, P, and polysaccharides and lipids from H, C, O. The surfaces generated by these polymers give rise to very different ways of binding 'inorganic' ions as well as organic molecules. Each and every one of the binding ions has an energized distribution so that all the polymers are energized differentially. The polynucleotides and polysaccharides give a narrow range of electrostatic interactions, e.g. with Na^+, K^+, Ca^{2+}, and Mg^{2+}. Membrane lipid two-dimensional surfaces come into this category too. Proteins bind a vastly richer variety of metal ions and anions. Finally, there is covalent modification which is only really different from cation binding in its kinetics. Thus all the polymers can be phosphorylated, amidated, methylated, and so on. Apart from the thermodynamic differences kinetic distinctions arise as follows and each class of polymer can be modified:

Simple cation and anion binding	**Phosphorylation (esters)**	**Methylation**
Electrostatic	(Covalent)	Covalent
Fast	Moderately slow	Very slow

Organism do not understand chemical bonding, but they behave as if they do since evolution is a teacher *par excellence* and, following the functional value of chemical modification of polymers, we see how all biological polymers are wonderfully kinetically controlled as if they were guided by expert hands.

Before concluding, we must stress again that the chemical and physical combination of the four types of polymer has produced a network of interactive energized phases. It is essential that we do not allow analyses of individual extracted molecules or of artificial vesicles, i.e. highly localized analyses, to deceive us into thinking that molecular structure holds the key to biological activity. The structures are valid important starting points, no more. Mobile energized molecules and semiliquid phases are organized in many very different ways and need different thinking concerning systems, see Chapter 20.

Hence we must adjust our thinking so that we can incorporate knowledge concerning both single molecules and selected elements within the system of cooperative activity within cells and organisms. In this book we are concerned to show that this organized activity requires not only appreciation of the functioning of proteins, nucleotides, polysaccharides, and lipids as described in this chapter, but also the interactive role of some twenty essential chemical elements. We shall therefore go forward in Chapter 7 to an outline of this cellular chemistry but a fuller analysis has to be put off until after we have provided information concerning the specific roles of each of these twenty elements in Chapters 8 to 19.

A note on small organic molecules in cells

The description of the cell content of glycoproteins is incorporated into the proteome, but that of the polysaccharides and lipids is not. It is likely that their presence is not independent of the concentration of small organic molecules and enzymes for their synthesis. This leaves the description of cellular activity generally as largely covered by the genome, the proteome and the metallome, but with missing sets of chemicals, those of lipids, sacharides and the small organic molecules. The small molecules range from amino-acids, bases, sugar molecules and molecules such as cholesterol and glutathione as well as coenzymes. We also need to know their concentration and distribution since they link to the proteome, the genome, the metallome and to the environmental levels of non-metals such as C, H, N, O, S, Se, and P, but we cannot describe them further in this book.

Acknowledgement
We thank Dr. E. Owen of Oxford University Press for help with Sections 6.10 to 6.15.

Further reading

Cells and proteins

Ballou, D.P. ed., (1999) *Essays in biochemistry*, Vol. 34. *Metalloproteins*. Portland Press, London.

Blackstock, W. and Mann, M. eds. (2000). Proteomics. *Trends guides* (Elsevier), 1–51.

Cooper, G.M. (1997). *Cell—a molecular approach*. ASM Press, Washington DC.

Dinner, A.R., Sali, A., Smith, L. J., Dobson, C.M., and Karplus, M. (2000). Protein folding. *Trends in Biochemical Science*, **25**, 331–39.

Ferscht, A. (1985). *Enzyme structure and mechanism* (2nd edn) W. H. Freeman, New York.

Gray, H.B., Malmström, B.G., and Williams, R.J.P. (2000). Copper coordination in blue proteins. *Journal of Biological Inorganic Chemistry*, **5**, 551–9.

Kotyk, A., Jamaček, K., and Korita, J. (1998). *Biophysical chemistry of membrane functions*. J. Wiley, Chichester, UK.

Lodish, H., Baltimore, D., Bark, A., Zipiveski, S.L., Matsudaira, P., and Darnell, J. (1995). *Molecular cell biology* (3rd edn.). Scientific American Books, New York.

Pandey, A. and Mann, M. (2000). Proteomics to study genes and genomes. *Nature*, **405**, 837–46.

See also *Science* (1999) **286**, 1881–1905 on protein folding and assembly. Protein structures are available in data bases.

Nucleotides

Brown, T.A. (1998). *Genetics* (3rd edn). Stanley Thornes, Cheltenham.

Cech, T.R. (2000). The ribosome is a ribozyme. *Science*, **289**, 878–9.

Eschenmoser, A. (1994). In *Pioneering ideas in physical chemical science* W. Fleischhecker, and T. Schoenfeld. (eds). Plenum Press, New York.

Lewin, B. (1997). *Genes VI*. Oxford University Press, Oxford.

Scott, W.G. and Klug, A. (1996). Ribozymes and catalysis. *Trends in Biochemical Sciences*, **21**, 220–4.

Trends in Biochemical Sciences (1996) (ed. R.D. Kornberg). **21**, 325–55. (Volume devoted to RNA polymerase and transcription control.)

NB There is a huge amount of material in data bases covering the genomes of some thirty species.

Polysaccharides

Drickamer, K. (1999). C-type lectin-like domains. *Current Opinion in Structural Biology*, **9**, 585–90.

Rudd, P. and Dwek, R.A. (1997). Glycosylation: heterogeneity and the 3D structure of proteins. *Critical Reviews*, **32**, 1–100.

Wormald, M.R. and Dwek, R.A., (1999). Glycoproteins: glycan presentation and protein-fold stability. *Structure*, **7**, 155–60.

Lipids

Betteridge, D.J. (ed.) (1996). *Lipids: current perspectives*. Martin Dunitz, London.

Hemming, F.W. and Hawthorne, J.N. (1996). *Lipid analysis*. Bios Scientific Publishers, Oxford.

Vance, J.E. (ed.) (1987). *Biochemistry of lipids and membranes*. Benjamin/Cummings, Menlo Park, California.

The nature of states of polymers is difficult to describe in thermodynamic language since their configurational and thermal entropies and energies are extremely complex. This applies to proteins, even the surfaces of folded proteins, to polysaccharides, to fats and to extended molecules such as DNA and RNA. All of them come under the heading of 'Soft Materials' (see references below) and have to be analysed by systems, not molecular language — see Chapter 20. Important features of these polymers appear not only in general bulk concepts such as exclusion volume and entanglements but in properties of molecules in close range contact (as in the van der Waals equation). Their dynamics are also extremely important in controlling viscosity and rates of recognition of surfaces. We gave a beginner's approach to some of these topics in the first edition of this book but it has been excluded here as a subject which awaits further development in terms of ensembles of molecules and, in particular in the context of this book, of the interaction of the polymer molecules with elementary ions.

Doi, M. and Edwards S.F. (1986). *The theory of polymer dynamics*. Oxford University Press, Oxford.

Nagasawa, M. (1988). *Molecular conformation and dynamics of macromolecules in condensed systems*. Elsevier, Amsterdam.

CHAPTER 7

The functional value of the chemical elements in biological systems

7.1 Introduction

The major aim of this book is to describe the requirements by any living system for a series of chemical elements and the functional value they have within organisms. We can do this in two different but complementary ways, showing what each element, taken separately, does in living organisms on the one hand, and illustrating, on the other hand, how the whole of cellular activity is based upon the cooperative functioning of these many elements. In two recent books (see references) we have stressed the cooperative activity of the elements while in this book our main objective is to show what each element taken separately gives to the whole. Even in the latter there are two different considerations; we need both to describe the individual molecules which involve a particular element (this is the conventional molecular approach) and to examine the controlled way in which all such molecules are produced and used. However, this could still leave the unfortunate impression (in this book) that the living process could be understood by piecing together the separate activities of the elements whereas we believe that life is an emergent property of the *concerted, organized action* of all the elements involved. Hence, while in Part II, Chapter 8–19, we shall provide an appropriate background for each element, in this chapter, and in much more detail in Chapter 20, we shall indicate the overall involvement of the chemical elements in life.

Before going to their overall survey of the biological chemistry of the elements it is appropriate for us also to give in this chapter the conclusions to be drawn from the preceding six chapters. In Chapter 1 we introduced the twenty or so elements which are essential for life, discussed their environmental availabilities and the fact that only certain combinations of them in cells led to living systems of high survival value (see Aside, p. 26), that is, as seen in viable organisms. In Chapter 2 we went on to examine the potential modes of incorporation of the elements in compounds and complexes, separating binding at equilibrium from kinetic trapping particularly as seen in model systems. We indicated that in organisms proteins were overwhelmingly involved in both. Chapter 3 was concerned with the description of cells at all levels of evolution, noting that higher organisms had more and more compartmental divisions separated by membranes. Different compartments in the higher organisms could be in local contact with one another, so that 'channels' of communication both in cells and between cells were observed forming networks where the cells and their internal compartments were supported by filamentous structures. Now, these chapters give a view of the static distributions and combinations of elements in divided space. However, life is overwhelmingly concerned with the turn-over of chemicals, so that in Chapter 4 it was necessary to illustrate the principles of the rate controlling steps of transformation of chemicals, to show how selected chemical elements could aid the transformations, for example as catalysts. We also indicated how elements could assist communication and stabilise structures and how the whole set of biological reactions was in physical flow in and between cells. In a final section of that chapter the way in which *networks* of reactions involving feedback processes lead to control over reactions was indicated. A further clear requirement beyond that of particular chemical elements for many of the biological reactions in cells, particularly syntheses and physical flows, is for energy, ultimately obtained from the environment. Chapter 5 uncovered the way cells capture and use that energy. Since energy has to be managed and distributed in many pathways simultaneously it was again necessary for a cell to have a communicating *energy network* on top of the *material network* described in Chapter 4.

The complexity of the cellular systems is managed in part by the polymers and fatty substances described in Chapter 6. The fatty substances, lipids, form the membranes; proteins form other structures, the filaments, part of the membranes, and the active catalysts (enzymes) with or without metal ions: the polysaccharides are parts of receptors, sensors and connective tissue; the nucleic acids DNA carry the essential functional units of coded information, genes, and in transcribed form (RNA), which to some extent can be used to characterize an organism. This molecular coded information in the DNA is known as the *genome* while *the system of proteins* it codes and controls in part is the *proteome*. We noted in that chapter that knowledge of the DNA sequence by itself does not lead to an understanding of the living process nor in any simple way to a full appreciation of speciation or reproduction. Cellular life and reproduction are associated with the *network interaction* of the coded DNA (RNA) and the cell contents, machinery, and energy distribution which connect to the environment. While the cell contents of proteins are in part a reflection of the DNA code, the reading of the code is limited by the proteins themselves, feed-back, and their interaction with smaller units, including simple ions and molecules in the cell, and with the energy, so that DNA gives no quantitative indication of a cells' contents. The whole *system of a cell* also needs a description of a set of elementary chemicals taken from the environment. The system of element concentrations in cells we have called the *metallome* using language parallel to that used for the protein content, the proteome.

To summarize, the first six chapters have shown that the living process is a *network* of flow of material, energy and information in a structured cell or set of cells in contact with the environment. The total system is interactive and includes

1. Networks of chemical combinations some of which are in local equilibrium exchange, while others rely on kinetically controlled carriers;
2. compartmental material networks including filaments inside and outside cells;
3. energy networks;
4. information networks.

The cell is indeed a very complicated cooperative system. As already stated, as well as stressing this total cooperativity, we wish to emphasize particularly *the peculiarity of the role of the individual chemical elements and their compounds in this system*. Therefore even before going to the description of the biological role of each element, Chapter 8–19, we give in the following sections a reminder of the differences between the chemistry of the elements (Section 7.2) and their known individual functional value for living processes (Section 7.3). We shall then be in such a position that we can introduce the main features of the living process in terms of element uses (section 7.4) rather than in terms of DNA and proteins, as seen in Chapter 6. A particular feature is the flow of material and of energy, leading to organized patterns in space for each element separately.

We have mentioned that not only can the elements be energized through their concentrations, e.g. by uptake into compartments, but that many of the molecular forms in which they appear in living organisms are thermodynamically unstable, e.g. phosphate in ATP and magnesium in chlorophyll. Furthermore, there is always intensive interaction with the polymers in the cell. While in no way suggesting that biological systems can invent a chemistry totally foreign to *in vitro* laboratory chemistry, these considerations do suggest that biological chemistry might be unorthodox. This goes beyond the obvious feature that biological organic chemistry, unlike most of man's organic chemistry, is carried out in the presence of water, to the idea that the 'elemental' chemistry in biology is sometimes not easily appreciated from *in vitro* behaviour. Once we have provided a brief impression of integrated cellular systems in this chapter (Section 7.5) we shall there upon give in chapters 8–19 not just the nature of the biological molecules involving each element, but the particular manner in which each element is linked to the genome and to its set of proteins within a cell's activities. It is the purpose of Chapter 20 to integrate these separated element systems into the unity which is a cell. Much more than an understanding of molecular chemistry is required.

7.2 Major chemical properties of elements in aqueous solutions

A very brief general summary of the chemical properties of metals and non-metals, a reminder of what was discussed in Chapter 2 and 4, is conveniently given here.

Taking metals first, it may be said that the great advantages of their chemistry in aqueous solution are:

1. the transport of positive charge by stable ions;
2. ready, often strong, and selective interactions with organic ligands;
3. easy bond breaking of ionic bonds in low positive oxidation states;
4. considerable polarizing power of the ions of high electron affinity, i.e. Lewis acidity in later transition and group 12–15 metals;
5. readily adjustable bond directions and lengths of some but not all elements;
6. easy change of stable oxidation states of some transition elements, i.e. easy one-electron chemistry;

7. in high oxidation states, some properties closely related to those of non-metals, e.g. slower ligand exchange but still readily activated exchange for atom transfer;
8. good π-electron donor properties of heavier elements in low oxidation states.

The advantages of non-metal chemistry in water are in quite different directions:

1. the transport of negative charge by stable anions, e.g. Cl^- and HPO_4^{2-};
2. strong kinetic control of homolytic and heterolytic breaking and making of covalent bonds which are frequently thermodynamically unstable;
3. weak but selective interaction between polar groups, including organic sidechains of proteins, through H-bonds, anions and cations;
4. high electronegativity, especially of the light non-metals, providing good Lewis bases;
5. noticeably fixed bond lengths and bond angles, giving stereochemistries of high precision, especially for light elements;
6. difficult single unit changes of oxidation state and thus a restricted (but important) radical chemistry since one-electron non-metal reactions are rarely easy or easily controlled.

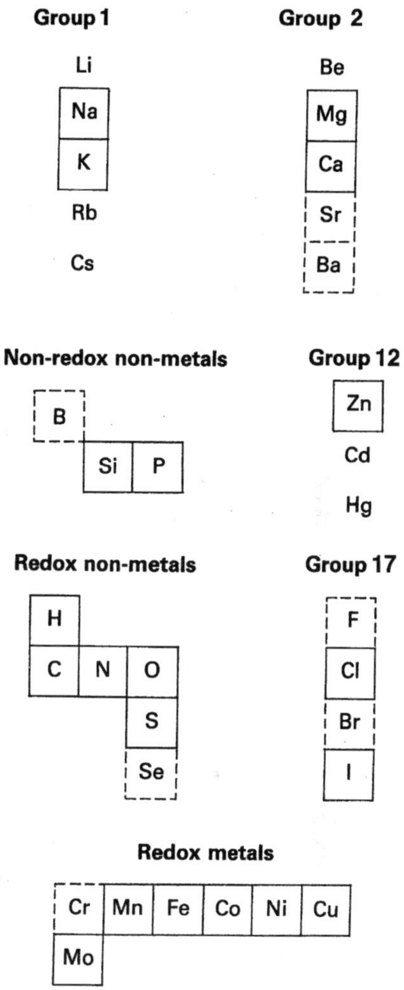

Undoubtedly the reader may wish to add one or two points to these two lists to make them more comprehensive but what is immediately obvious is that generally elements in different groups of the periodic table, see Aside, have different properties while elements in the same group have similar properties whereas at the same time combinations of elements from certain regions of the period have very different properties from other combinations.

For example, in the presence of water, non-metals are the basis of large, mostly thermodynamically unstable but kinetically stable compounds, i.e. all major polymers and substrates, while metals are the best electron-transfer agents, the best local catalyst atoms, and amongst the conducting devices via ionic diffusion they provide a multitude of agents. These statements are independent from biology but we see that any organism needs both non-metals and metals because it has to build structures, requires catalysts, and needs communication systems. While it is extremely dangerous to overgeneralize, we can provide a guide table of chemical reactivity from these observations (see Table 7.1) and we can ask what kind of elements do we need to take part in or

Table 7.1 Examples of chemical reactivities

Extremely difficult reactions	Slow reactions	Fast reactions
$[CH_2]_n$ reactions in all classes	Ester hydrolysis, amide hydrolysis	Dissociation of ion pairs, M^+X^- and $M^{2+}X^-$
Rearrangement of sugars Inversion of configuration at carbon atoms	Many cis-trans conformational changes	Proton reactions of strong and moderately weak acids
Atom (group) transfer from weakly polar compounds	Atom (group) transfer from more polar compounds	Electron transfer M_1 to M_2 across short distances

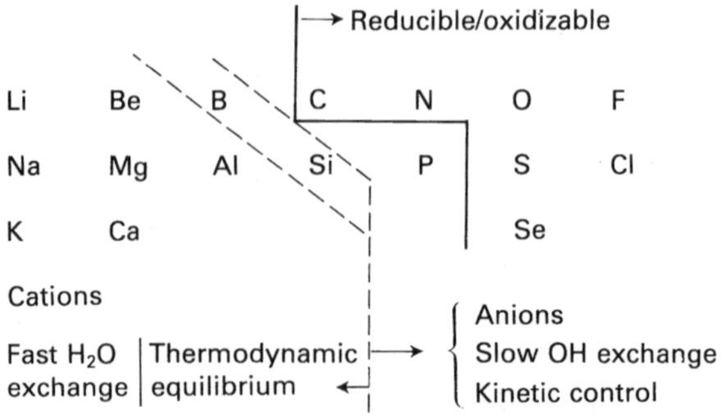

Fig. 7.1 A simple division of part of the periodic table stressing the separation in biology between elements: (1) those which are found as cations from those found as anions; (2) those which have a variety of redox states from those that do not; and (3) those which exchange ligands or groups rapidly from the first coordination sphere from those that do not. B and Si are intermediate in many ways.

to activate these reactions. In addition, Fig. 7.1 gives an easy-to-read overview of some of the different aqueous solution chemistries of the *early elements* in the periodic table and it is clear that they give rise to all sorts of covalent and ionic molecules but inadequate catalysts, especially for redox reactions. If good catalyst centres are required we need to turn to some of the heavier elements and observe also that in every group of the periodic table heavier elements react faster than lighter elements in the same type of compound.

All the above are points made from knowledge strictly from *chemistry*; the immediate next step in this discussion is to use the *biochemical* data obtained by analysis usually of *extracted* compounds to get a first impression of how the elements are used functionally in biology, but note that such biochemical analysis is to some degree destructive of biological systems. This discussion cannot take into account to any degree that a biological system is an organized, energized, system *in permanent flow* (Section 4.15).

7.3 Biochemical functions of the chemical elements

It is useful at this stage to summarize the biochemical roles that the different elements perform (Table 7.2). A systematic examination of these roles by biochemical methods has been made possible by remarkable increases in our knowledge over the last 50 years. More powerful methods of chemical analysis have been developed and practice and theory have progressed enough to allow detailed investigation of the structure, dynamics, and reactions of many isolated large molecules. Naturally, most of the biochemical molecules are organic compounds of the bulk elements–carbon, hydrogen, oxygen, and nitrogen–sometimes with sulphur and phosphate groups and in a few cases with some covalently combined minor non-metal elements, e.g. selenium. These elements are required for the major monomers and polymers of biology, that is, the central constructs of life. It is not very useful to classify any of these compounds under individual elements and it is usual to classify them under active groupings, e.g. acids, bases, aldehydes, ketones, amino-acids, and so on. However, in catalytically active enzymes we frequently find other (inorganic) elements; in fact, some one-third of all enzymes need metal ions, but now their activities are selectively associated with the element itself. There are also biochemical materials that are fully or predominantly inorganic, e.g. bones, teeth, shells, etc. Furthermore, there are quite high, and precisely main-

Table 7.2 Main biochemical functions of the chemical elements

General function	Elements	Chemical form	Examples of specific functions
Structural functions (biological compounds and support structures)	H, O, C, N, P, S, Si, B, F, Ca, (Mg), (Zn)	Combined in organic compounds or sparingly soluble inorganic compounds	Components of biological molecules; formation of tissues, membranes, skeletons, shells, teeth, internal structures, etc.
Electrochemical functions	H, Na, K, Cl, (Mg), (Ca), HPO_4^{2-}	As free ions	Transmission of messages (nerves); production of metabolic energy
Mechanical effects	Ca, (Mg), HPO_4^{2-}	As free ions exchanging with bound ions	Triggering of muscle contraction; lysis of vesicles
Acid-base catalysis	Zn, (Ni), (Fe), (Mn) –OH, SH,	Combined in enzymes	Food digestion (Zn); hydrolysis of urea (Ni); phosphate removal in acid media (Fe, Mn)
Redox catalysis	Fe, Cu, Mn, Mo, S, Se, (Co), (Ni), (V)	Combined in enzymes	Reactions with oxygen (Fe, Cu); oxygen production (Mn); nitrogen fixation (Mo); inhibition of lipid peroxidation (Se); reduction of nucleotides (Co); reactions with H_2 (Ni); bromoperoxidase activity (V)
Signalling to DNA (via transcription factors)	H, O, C, N, P, S, Mg Ca, Zn, Fe, Cu	As small molecules or ions	Binding agents to transcription factors
Various specific functions	Mg	In chlorophyll	Light harvesting in photosynthesis
	Fe, Cu	In proteins	Transport of oxygen
	I	Special covalent compounds	Hormonal action
	Fe, Ca, Si	Mineral compounds	Sensors (magnetic, gravitational)
	O_2, N_2, CO_2	Gases	Flotation (fish)

tained, concentrations of a variety of simple inorganic ions throughout biological systems. Some of these are essential as information carriers and in the coordination of activities. One thesis of this book is that biochemistry cannot be considered as just a part of organic chemistry, i.e. the chemistry of C, H, N, O, etc., since combinations of these elements alone do not represent the analytical make-up observed in a living system; on the contrary life seems to have evolved from a much wider use of elements in the periodic table. A few examples of the uses of inorganic elements, most of which have been referred to in the previous chapters, clearly show some, not all, of these functions, as revealed by biochemical analysis (Table 7.2).

1. The hard structures of animals are largely mineral, most frequently calcium salts (see Table 10.11 for examples of biological minerals). Of course, this is not the case for the plastic cuticles of insects or of the woody stems of many plants.
2. Electrolytic connection in and between cells, e.g. nerve messages, rests largely on the concentration gradients of sodium, potassium, and calcium amongst metal ions and chloride amongst non-metal ions.
3. Phosphate transfer, essential in bioenergetics, is almost invariably linked with magnesium reactions in enzymes.
4. Mechanical changes in cells are frequently initiated by calcium concentration changes.

5. Triggering in digestion, blood-clotting, fertilization, muscle contraction, cell-surface interactions, etc. also depend largely on calcium ions and their concentration changes.
6. The capture of light in photosynthesis depends on a magnesium-containing pigment, chlorophyll.
7. Electron transfer in energy transduction is very largely dependent on iron proteins.
8. Oxidative metabolism is almost always dependent on iron and copper catalysts.
9. The capture of combined nitrogen from nitrate or dinitrogen and of sulphur from sulphate relies amost totally on a molybdenum-containing catalyst centre, though there is some use of vanadium, or even iron alone.
10. The release of oxygen from water, that maintains our present atmosphere, is absolutely dependent on manganese (part of an O_2-evolving enzyme).
11. Transport of oxygen requires iron or copper proteins.
12. The synthesis of some DNA and RNA, the activation of water as a leaving or attacking unit in condensation and hydrolysis reactions, etc. depend on the intervention of more than 300 zinc (and magnesium) enzymes involved in biological reactions. (DNA synthesis also requires either cobalt, manganese, or iron in ribonucleotide reductase.)
13. Bound pyrophosphate (ATP with magnesium) is the major carrier of chemical free energy for driving reactions.
14. The role of the free proton is dominant as an intermediate in energy metabolism.
15. Element concentrations are specifically related to the regulation of their gene products, see especially the example of Fe^{2+} and Zn^{2+} proteins in Chapters 12–16, amongst metal ions, and phosphate amongst non-metal ions.
16. Selenium is deeply involved as an anti-oxidant in aerobes.

Again the reader will want to add other points, many of which we hope to cover from Chapter 8 onwards. We repeat that an immediately obvious point is that organisms use elements very much in accordance with expectation given their chemistry. However, use is equally clearly a biologically selected and controlled feature. Many different elements are required because there are many different *essential* tasks which are best performed by particular elements in constrained biological chemical molecules and compartments.

Glancing at the above list of activities of the various selected groups of elements one is struck first by the extent of the total involvement of inorganic elements in structural and mechanical functions, electrochemical control, acid–base and redox catalysis, generation, transmission, and storage of energy, transmission of messages, triggering effects, and so on, very much in line with the chemical possibilities for individual elements (compare Figs 7.1 and Table 7.2). A novel feature is that there is remarkably little in common between the functions of different elements (see Table 7.3). Such a complete separation of element function is not expected from chemistry judged directly from the periodic table, since neighbouring chemical elements especially in a group are often similar in properties, although some of these specialized

Table 7.3 Some specific metal ion catalyses

Small-molecule reactant	Metal ion	Examples
Glycols, ribose	Co in B_{12}, Fe	Rearrangements, reduction
CO_2, H_2O	Zn	Carbonic anhydrase
Phosphate esters	Zn Fe, Mn	Alkaline phosphatase Acid phosphatase
N_2	Mo(Fe)(V)	Nitrogenase
NO_3^-	Mo	Nitrate reductase
SO_4^{2-}	Mo	Sulphate reductase
CH_4, H_2	Ni(Fe)	Methanogenesis
$O_2 \rightarrow H_2O$	Fe Cu	Cytochrome oxidase Laccase, oxidases
O-insertion from O_2 (high redox potential)	Fe	Cytochrome P-450
SO_3^{2-}, NO_2^-	Fe	Reductases
$H_2O \rightarrow O_2$	Mn	Oxygen-generating system of photosynthesis
H_2O_2/Cl^-, Br^-, I^-	Fe(Se)(V)(Mn)	Catalase, peroxidases
H_2O/urea	Ni	Urease
Small peptides and esters	Zn	Hormone control (peptidases)
Ethanol	Zn	Dehydrogenase

biochemical roles are more or less obvious. For example copper, zinc, and iron are good Lewis acids, and each is used to some degree in cells; however, at first sight puzzlingly, zinc has been selected overwhelmingly above the others for this function. Again potassium is largely inside cells, and sodium outside cells, so there must be good *biological* (not chemical or biochemical) reasons for such an individualized choice which biochemical analysis by itself does not reveal. Some of these features will have arisen, of course, since biological evolution, but not chemistry, has been very restricted by the abundance and availability of the elements themselves in the universe and on earth (see Chapter 1). The changing availability on Earth has then allowed, perhaps driven (see Chapter 20), much of evolution.

This analysis of the selection of an element for a given function is even more complicated since we have seen in the previous chapters that an element may be present (selected) in living systems in different locations, different oxidation states, and different forms, with distinct properties and functions (Table 7.4). For instance, iron (II) is different from iron (III) and haem iron behaves as a separate entity from both in many different ways. This means that any attempt to relate biological functions in living systems to biochemical analysis of elements must take into account specific non-exchanging complex species in physiochemical environments. However, no matter how complicated we make this type of analytical description there are obviously some overriding principles of living systems which are peculiar to them. One such peculiarity is the fact that we usually find metalloproteins of living systems with a stoichiometric ratio, but apparently limited excess apoprotein and only a very limited fixed excess of free metal (Section 2.19). A different feature is

Table 7.4 Examples of separated functional pools of one element

Element	Pool 1	Pool 2
Mg	Free ions	Chlorophyll
Fe	Free ions	Haem
S	-S-S- and RS^-	Sulphate
Ni	Free ions	F-430
Zn	Free ions	Enzyme sites
H	NADH (degradation)	NADPH (synthesis)
P	ATP (glycolysis)	UTP (glycosylation)

the observation that the ratios of all the elements in a given cell are quite fixed. This applies to the concentrations of the free elements as well as those bound which allows us to describe a metallome in parallel with the proteome. A third is that a given living system is always in flow within a highly specific structure. We turn now to the opposite approach from that of individual element activities to the nature of organisms that is of biological systems.

7.4 The living process

We shall take it that all life known to us is based on similar physical and chemical features though of very different complexity. The major physical features largely described in previous chapters are:

1. Life is a controlled (chemical) *growth* process at pH = 7.0, temperature 300 ± 25 K with but few exceptions amongst archaebacteria.
2. Growth requires *energy*.
3. Life exists in a series of aqueous and membranous isolated *compartments*, which it makes (see Table 7.5).
4. Living systems are *organized*, develop, and then fail. The system *self-assembles* and *self-degrades*.
5. Such systems must *reproduce* to remain extant. It is cheaper and better to renew than to constantly repair.
6. Inevitably they grow in association with a wide variety of *elements in the environment*.
7. When they grow they grow in *well-defined* ways; there is *morphological* growth and development.

We can also make some further restrictions and largely chemical statements which must be kept in mind.

1. Living processes occur in fixed internal, aqueous, *salt* solutions which are retained by cells and their membranes.
2. Organisms (life) must capture energy in small molecule chemicals, i.e. within *unstable compounds*, or in *gradients* in order to generate their larger compounds.
3. The initial function of energy capture must be to take up or make the appropriate special *unstable monomers* for the synthesis of polymers, which also occurs in water. There then has to be a balanced set of monomers and polymers; hence, ions, molecules, and polymers are all in a *homeostatic* relationship.
4. Life depends on the continuous *synthesis and degradation* of unstable water-soluble as well as organic, lipid-soluble, *polymers*. (While we now know that reproduction today is dependent on DNA, we do not have to consider this finding as fundamental; in principle, there could be other polymers which would carry genetic information. Similar statements can be made about the functions of RNA, proteins, polysaccharides, and fats. The necessary feature is just a trapped, controlled, reproducible set of polymers which are of considerable but limited kinetic stability in water and which require energy to produce and a coded unit to reproduce them faithfully.)
5. The overall composition of a living system contains elements in closely fixed ratios.

6. The synthesized molecules must be placed in an *organized array* because *directed flow* of material is essential.
7. Control of growth requires a complex set of *messages* and *feedback controls*, which again demand flow which is constantly maintained.
8. Series of organized *catalysts* are necessary to interconvert chemicals at room temperature and this requires capture of various elements.
9. Life has evolved with time and the chemistry of the elements in life has been and still is an evolving development. This is necessary to allow advantage to be taken of environmental changes.

In summary organisms have a quantitative selection of chemicals in organised flow. Once again the reader may wish to add further points, but the essence is clear.

While the biochemical analysis in the previous section revealed some of the refinement of the chemistry, and this will need to be examined carefully when we describe the biological chemistry of specific elements in Chapters 8–19, there are in addition definite new features in living systems. Outstanding are the organization of flow at any one time (Table 7.5) and the time-dependence of the whole flow system, synthesis/degradation, and specific growth (Tables 7.6 and 7.7). There is clearly some integrating activity which has a large energy input. The function of an element or a compound in biology has to be seen in the context of where and when it combines and how it moves, is used, and is removed. In an effort to see how these features must force a very separate and almost individual chemistry upon each of the elements (already seen above to some extent by biochemical analysis) we shall examine some of the major requirements to build an organized system which can develop and reproduce.

Such a system must have a regulated stability and a 'channelled' flow at any time, and in this organization there must be integration of function (see Figs 7.2 and 7.3). In these figures all the activity pockets must feed back to

Table 7.5 Compartments and reactions in eukaryotes

Compartment	Reactions	Elements
Cytoplasm	Synthesis of DNA, RNA proteins, Oxidations, Glycolysis	Mg, Zn, K Fe, Co, Ni Ca, Mg
Mitochondria	Oxidative phosphorylation Citric acid cycle	Fe, (Cu), (Ca), (Mg)
Chloroplasts	Photophosphorylation Carbon assimilation	Mg (chlorin), Fe(Cu) Mg
Sarcoplasmic reticulum	None(?)	(Ca)
Golgi apparatus	Glycosylation Sulphation	Mn, (Ca)
Endoplasmic reticulum	$2RS^- \rightarrow R\text{-}S\text{-}S\text{-}R$ Glycosylation	(?) (Ca)
Vacuoles	Urea hydrolysis	Ni, (Mn)
Peroxisomes	Oxidative degradation	Fe, Mn
Circulating fluids	Hydrolysis oxidations	Zn, Ca, (Na) Cu

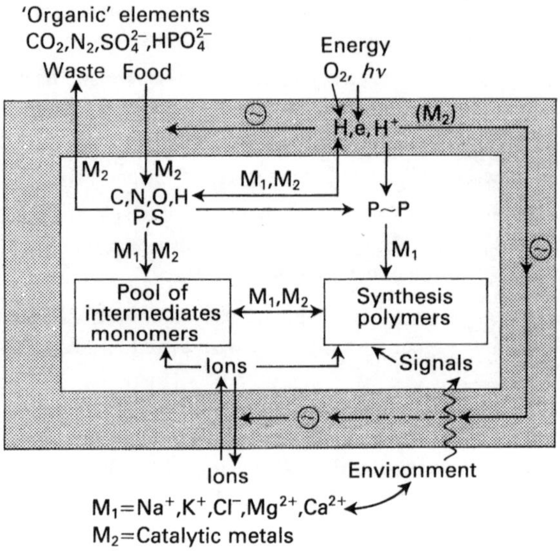

Fig. 7.2 A very simplified picture of the multitude of cellular activities in a prokaryote which have to be in mutual communication. The role of inorganic elements is stressed.

Table 7.6 Some basic element flow units of biology

Metabolite (element)	Reactant and control carriers (coenzymes)	Protein-catalysed reactions
Electron (e^-)	Metal ions, glutathione	Many, e.g. energy conversion, oxysugar reductase
Hydride (H)	Pyridine nucleotide (NAD)	Many, e.g. energy conversion
Methyl (C)	Methionine	Methylation
Acetate (C) (and other carbon fragments)	Coenzyme A	Fatty-acid synthesis, etc.
Phosphate (P)	Adenosine (guanosine) triphosphate (ATP)	Many, e.g. glycolysis and oxidative phosphorylation, DNA, and RNA synthesis
Nitrogen (N)	Pyridoxal phosphate	NH_3 exchange
Amino acid (N)	Adenosine (guanosine) monophosphate (AMP)	Ribosomal synthesis of proteins
Sugars (C, H, O)	Uridine diphosphate, dolicol phosphate	Many, e.g. glycoprotein formation
Nucleotides (C, N, O)	Nucleotide phosphates	RNA/DNA synthesis in nucleus
Hydride/electron (H^-/e^-)	Coenzyme Q (quinone)	Energy conversion (membrane)
Sulphate (S)	Adenosine phosphate-sulphate (APS)	Sulphonation
Energy (ΔG)	Triphosphates	Energy utilization
Light quanta ($h\upsilon$)	Chlorophyll	Energy capture (membrane)

Although there are carriers for H, C, P, N, S, and e^-, there are no simple carriers for O. Rather, this element is carried in carbon compounds or H_2O or O_2. Note different carriers for different oxidation states.

Table 7.7 Examples of diffusion-controlled zones

Process under control	**Control of diffusion**
Oxidative phosphorylation	Membrane curvature (mitochondrial)
Photo-capture	Membrane curvature PS I/PS II (separation of O_2 release)
Creatine kinase exchange	Muscle and mitochondrial relationship
Na^+/K^+ channelling	Nodes of Ranvier

one another so that directed network organization is of the very essence of biology. The regulation is based on DNA but note that the more complicated systems at least are segregated in species and do not mix (see Section 7.7), while DNA can be varied continuously in principle. We illustrate this point first with the nature of biological polymers and here it will be necessary to repeat in part some points made in previous chapters. [While we may believe that each different species has a different element content, its metallome, this is not easily related to functional specialization. However, the overriding major evolution of organization through prokaryotes, eukaryote single cells, multicellular organisms is readily related to the changes in the metallome in evolution and in the environment (see Table 7.11 and Chapter 20).]

7.5 The chemical flow in biology

7.5.1 The synthesis and degradation of the polymers of life

We ask first 'Is there a common feature amongst the biological polymers?' Chemically, the answer is that all the biological polymers are condensation polymers; indeed biological systems rarely use free radical polymerization since this is very difficult to control, while it is relatively easy to carry out condensation reactions in a very controlled step-by-step manner

$$AB + C \rightarrow ABC + H_2O.$$

Man achieves this in solid state synthesis of peptides and polynucleotides following Nature. The reactions are reversible also in a step-by-step manner and this is part of control (Chapter 4).

None of the condensation reactions is spontaneous, of course, so it is clear that they need some driving power. It is equally clear that biological systems have a need for a wide variety of catalysts for these syntheses. In order to appreciate these requirements we must see the close interrelationship of the whole activity. The overriding need is to build a set of polymers of four major types, see Chapter 6, from simple monomers:

1. polynucleotides from phosphate and sugars (and bases);
2. polypeptides from amino acids;
3. polysaccharides from sugars;
4. hydrocarbons from acetic acid, followed by reduction.

There are immediately three perplexing features: (1) the polymers are produced in balanced ratios, yet the exact mode of production is not the same for each, e.g. for proteins and DNA; (2) the products appear in position in space; (3) all the reactions use the same initial sources of energy. All the reactions are acid-base reactions, so that the catalytic elements involved are usually Mg, Ca, and Zn, and various triphosphates are used for energy. While some separation of the different syntheses is managed by the differential uses of ATP, UTP, CTP, and GTP in spatial organizations, there still remains the ultimate need for integrated feedback control. Furthermore, each set of the four polymers is processed in a specific machinery located in spatial compartments: polynucleotides in the nucleus, polyamino acids in ribosomes, polysaccharides in the Golgi, and fats in fatty acid synthesis particles. Communication between all these compartments is obviously required.

These features and the observed stoichiometric ratios of irreversibly bound coenzyme in enzymes and of metals in metalloproteins, as well as the general absence of large excess of apoprotein (or metal ion or free coenzyme, e.g. flavin) indicate that the synthesis must have a degradative counterpart, for example,

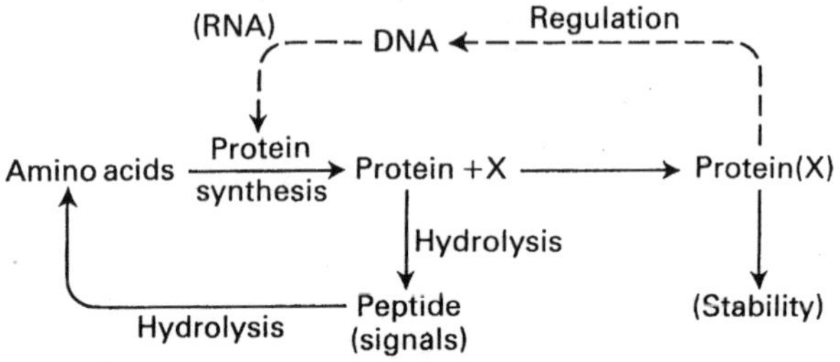

where X is either a metal ion, such as Zn, or a coenzyme such as flavin or ATP, and the broken arrows to and from DNA represent instructions or feedback of information. (see Section 4.15). (X may stabilize as well as aid signalling to DNA).

Synthesis and degradation lie at the heart of flow within the turnover of all macromolecules and feedback is clearly essential for balanced growth, Table 7.8. Moreover, the feedback is not just in one pool, e.g. proteins, but must be between pools, RNA and proteins, to maintain stability.

We now summarize our knowledge of the organization of the flow of polymers:

1. They are synthesized selectively and quantitatively in specific localities.
2. Starting from a locality they are transported in a particular manner across membranes, or in vesicles, or they are prevented from such movements.

Table 7.8 Feedback in polymer synthesis

Polymer	Some feedback systems
DNA (nucleus)	Transcarbamylase (Zn), ribonucleotide reductase (Fe) (Mn) (Co)
RNA (nucleus)	Controlled by similar chemistry to DNA and by DNA
Proteins (ribosomes)	Control at the level of ribosomes and DNA by regulating genes, hormones, Fe^{2+}, Zn^{2+}, etc.
Polysaccharides (Golgi)	Pumped supply of monomers from cytoplasm, Mn^{2+}
Glycolysis/glucogenesis	c-AMP, Ca^{2+}, ATP controlled proteins
Fatty acids	Redox controls sharing ATP, cAMP, Ca^{2+} with glycolysis, glucogenesis, and the citric acid cycle series of enzymes

All syntheses are controlled by material and energy supply and such ions as HPO_4^{2-} and Mg^{2+}.

3. They are deposited in a region assisted by local recognition.
4. They are turned over by destruction locally.
5. There is linked to them a communication network which relays back to the synthetic machinery information about excesses or deficiencies (see Table 7.8).
6. The information in the communication network is not made from polymers alone but much of the structural network itself, as well as the biological support structures, are part of it.
7. The polymers form the genetic code in DNA, the device for the transcription of the code, RNA, the protein framework and the filaments of the cell, and also the localized catalysts. The catalysts then locate monomer activity.
8. The lipid polymers form the membranes which make all functional compartments (Table 7.5).
9. Finally there is a condition of a cell which is its programmed death. The process, called *apoptosis*, is one of degradation of all polymers. A cell then *has an overall cycle* which reproduces the same *polymers quantitatively while keeping only DNA*, of necessity, to generate and repair the proteome and indirectly the total metallome.

The polymers are all produced from monomers and we therefore turn to the ways in which monomers are synthesized and degraded. We are looking at various parts of Figs 7.3 and 7.4.

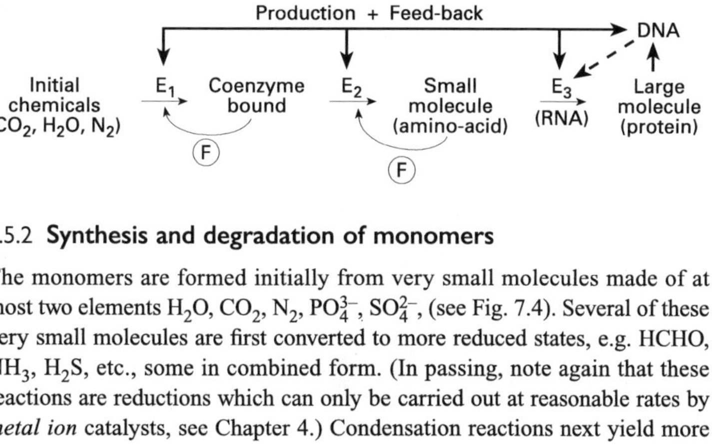

Fig. 7.3 A simple diagram of the feedback (F) relationships between the initial environmental chemicals the various products and DNA, the information centre. Here we also stress the connection to small molecules, see Section 4.15.

7.5.2 Synthesis and degradation of monomers

The monomers are formed initially from very small molecules made of at most two elements H_2O, CO_2, N_2, PO_4^{3-}, SO_4^{2-}, (see Fig. 7.4). Several of these very small molecules are first converted to more reduced states, e.g. HCHO, NH_3, H_2S, etc., some in combined form. (In passing, note again that these reactions are reductions which can only be carried out at reasonable rates by *metal ion* catalysts, see Chapter 4.) Condensation reactions next yield more complex molecular species. The production of these monomers is written here in formal, not real, reactions

$$\begin{aligned} 6\,HCH(OH)_2 &\rightarrow C_6H_{12}O_6 + 6H_2O \\ HCH(OH)_2 + NH_3 &\rightarrow HCH(NH_2)\,OH + H_2O, \\ HCH(OH)_2 + HPO_4^{2-} &\rightarrow HCHOH(OPO_3)^{2-} + H_2O. \end{aligned}$$

In this way amines, esters, and ethers are made. It is also possible to go on to make more complex small compounds and in series of reactions (shown in all biochemical textbooks and on wall charts in biochemical laboratories) to go further to amino acids, nucleotides, saccharides, and ketones, i.e. to the monomers needed for polymer syntheses (see Fig. 7.4). Some steps require further reactions and again and again catalysts are required. The catalysts differ, but the initial elementary components are the same for all the monomers

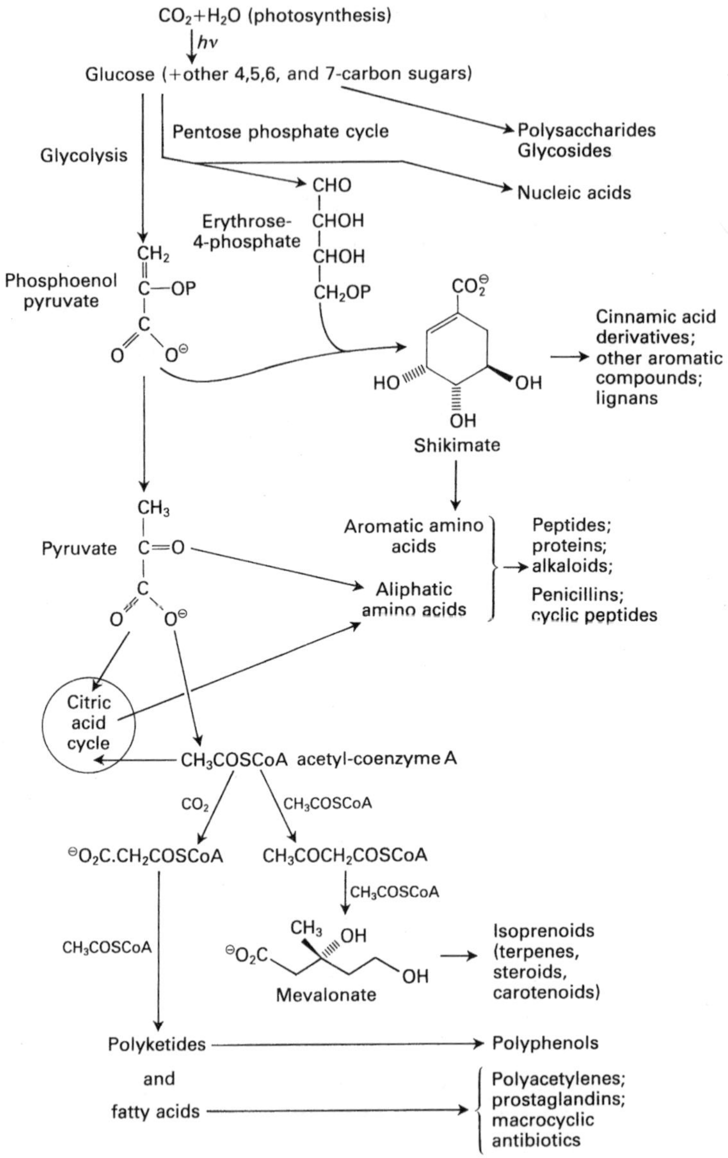

Fig. 7.4 A sketch of some of the required balanced connections in the substrate chemistry of cells. It is a revealing exercise to label the arrows with required metal ions for catalysis.

and the process, oxidation/reduction followed by condensation, is fixed. The network of all reactions then resembles a computerized circuit (see Fig. 20.4.).

However, if we take any set of monomers they have to be generated in particular ratios to meet synthetic demand based on the polymers. Hence all the monomers are in constitutive but feed-back regulation via the proteins which handle and transform them see (Section 6.14 and Fig. 7.3). The co-ordination of these syntheses must rest in DNA/RNA. At any time there exists a homeostasis but this changes with time during development and growth.

In order to carry out these reactions in an orderly controlled way the elements H, C, N, P, S must be packaged and sent to controlled destinations. One way cells achieve differential packaging, for example, of reducing equivalents, is through the action of the freely mobile coenzyme carriers

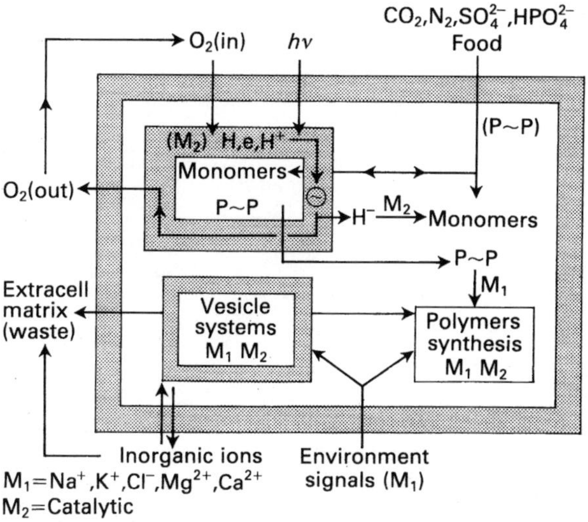

Fig. 7.5 A very simplified picture of a eukaryotic cell illustrating some of the complexity and sophistication needed to maintain a network of balanced interactions in a multicompartment cell.

NADH, NADPH, and glutathione. NADH is involved in the degradation of small substrates in both glycolysis and the citrate cycle, while NADPH carries hydrogen for incorporation in the synthetic paths of glucogenesis and photosynthesis. The path of incorporation is further split since glutathione, as well as NADPH, delivers hydrogen in specified paths of reductive synthesis. As mentioned before there is a parallel here with the differential transport and use of phosphate in different NTP molecules, of iron as free ion or as haem iron, and of magnesium in chlorophyll or free magnesium ions (Table 7.4.)

Now monomers, indeed substrates, occupy a curious position in that they are the source of synthesis but also very frequently the source of energy. We observed above that the digestion of polymers (outside cells or inside in proteasomes) was kept separate from their synthesis (inside cells) for very good reasons. The consumption of the monomers for other purposes than syntheses has also to occur inside cells and this consumption can give both energy and new intermediates. Clearly, the activities must also be separated; thus, we find oxidative degradation in mitochondria and sugar degradation (glycolysis) in the cytoplasm well separated by membranes just as was the handling of reducing equivalents well separated from synthesis by different coenzymes. The advantages of compartments also for transmitters and hormones and for stores of digestive enzymes becomes obvious. We repeat that all compartments must remain in communication with one another (Fig. 7.5). To generate many separate activities the monomers have to be pumped into compartments.

The implication is clearly that in order to achieve balanced synthesis the relative amounts of monomers flowing in different synthetic and degradative paths must be balanced and, as for the polymers, feedback control is essential from the different monomer pathways of Fig. 7.3. However, we also require a balanced supply of chemical elements of a greater variety than those used to make monomers for synthesis.

7.5.3 The uptake and loss of elements

As commented in previous sections, apart from incorporation into organic molecules, a large number of inorganic elements are used in communication, control, regulation, and catalysis and many of these elements have to be

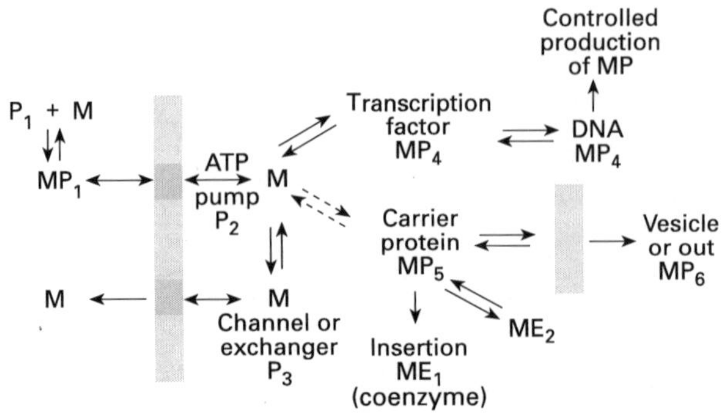

Fig. 7.6 A scheme for the uptake of a chemical element M into proteins, P, of various functional value and enzymes, E. Double arrows indicate exchange. Compare Fig. 6.23.

obtained in controlled amounts from outside the cell. There is then a balance utilizing many devices including synthesis of reagents and pumping into or out of compartments but all the time the elements are held in steady states of elevated or depressed concentration levels which are critical for all functions of cells (see Chapter 20). Here we include carefully controlled salt solutions containing Na^+, K^+, Mg^{2+}, Ca^{2+}, Cl^-, and so on both inside and outside the cell, e.g. in circulating fluids. All these ions are pumped and clearly the pumps must be under feedback control and linked to one another by exchangers.

The need to make special chemicals in special places means also that their catalysts have to be positioned, but we have seen in Chapter 4 that the catalysts for small molecules have to contain activating metal ions. Clearly, to make the catalysts, metal enzymes, the metal ion transport has to be as specialized as the polymer and monomer transport, and, as for the monomers, the metals must react in the required place. Not only is this the case but it is also necessary to ensure that redox active metals, e.g. Cu and Fe, do not get into certain other spaces for they, as ions, catalyse simultaneously both oxidation and hydrolysis indiscriminately. They must not interfere with the chemistry of zinc. There have to be 'metal-free' zones as well as prescribed metal-activated zones, so that detoxification must also be managed. The reason for the intensive element selection in Table 7.3 is becoming clear; each catalytic element is identified with idiosyncratic sets of reactions in a compartment so as to keep control partly through the secondary use of the element in communication systems but partly to avoid confusion in reaction pathways, and at the same time its free ion concentration is fixed relative to that of other ions (see Fig. 2.10).

In Section 6.14 we indicated how this control was managed by the feed-back relationship between the amount of an element in a cell and the production of its selective binding partners and its pumps, via the regulation of DNA and RNA activities (Fig. 7.6). Thus polymers, monomers and required elements are all coordinated qualitatively in uptake by the DNA code and quantitatively by feed-back to produce a material whole which self-assembles. Note particularly the predicted concentrations of the free elements, in Fig. 2.10 which together give what we have called the free metallome (see Chapter 20), Many elements are taken up constitutively while others required induced uptake (Section 6.14). There is one further requirement—that of energy.

7.5.4 The flow of energy in compounds

While building up a picture of the distribution of polymers, monomers, and of the chemical elements, we have insisted that all the polymers are unstable and have a limited kinetic stability and this is true of all the monomers too. The distribution of elements, ions, across membranes is also unstable. In fact the whole biological set of chemicals is unstable to hydrolysis, oxidation, and diffusion to remove concentration gradients, so that unless energy is constantly put into each part of the system it will degrade cell by cell, polymer by polymer, monomer by monomer, gradient by gradient. The whole biological set of chemicals is an evolved energized group with a timed life from synthetic to degradative steps connected to energy. Of necessity, the use of H, C, O, N, P, and S as elements and, indeed, of all inorganic elements is built into this covalent and kinetically regulated, energized, system. Another way of visualizing the situation is to note that the major cycles and pathways of primary metabolism are both synthetic and degradative, e.g. glycolysis, glucogenesis, and the citric acid cycle (Fig. 7.4). Degradation in one section of a pathway, giving energy, leads to components for synthesis in another to achieve balance in the production of all the many monomers and four types of polymers. It is required then, that always the system retains stored *energy* which is generated largely as NTP in particular parts of space. Thus, energy produced locally must flow and this flow has to be guided (Table 7.9). It must be continuously captured and lost: light always passes eventually to heat.

We have analysed these processes in our book *Bringing chemistry to life*, indicating that since all these energy using steps have to be operative simultaneously but to different degrees, feed-back control is an essential feature of energy uptake and use (see Chapter 20).

The complexity of the uses of energy, the fact that the whole system is constantly multiplying and growing or repairing, that it can maintain shape or alter it, and that the several operations are timed require a continuous monitoring, feed-back, communication and energy network. We have then, in every respect, a flowing energized chemistry within a dynamic organization. It is this operation with which our description of the biological chemistry of the elements (and of genes) must come to terms. When the energy flow stops, the order and the organisation of the system disintegrate and death occurs.

Energy transfer, as we showed in Chapter 5, is often in the form of NTP, phosphate transfer. Now, phosphate transfer is not involved just in syntheses of polymers, in energizing stress fibres, or in pumping ions and molecules, but through the process of phosphorylation of proteins it brings about major controls and regulations (see Chapter 6). Indeed, some one-third of in-cell proteins are phosphorylated and dephosphorylated regularly. The kinases which bring about phosphorylation interact mainly with ATP and GTP. However, the GTP and ATP connections are distinct (Table 7.9), just as NADPH and NADH transfer H in separate paths. Again, phosphorylation and

Table 7.9 The uses of pyrophosphate in four coenzymes

Coenzymes	Uses
ATP	Pumps of ions and most molecules Actomyosin tension generation Many kinases. Assimilation of CO_2, N_2 Fatty acid and glycolytic paths of metabolism
GTP	Protein synthesis, tubulin growth, many kinases (G-proteins)
UTP	Synthesis of polysaccharides
CTP	Choline synthesis

dephosphorylation (often using metal ions) of proteins are closely connected to transcription. Thus, phosphate *movement* is a fundamental part of energy transfer, communication and synthesis, and is linked to energy storage via NTP. All must be in a homeostatic balance at a given time. We have strong reason to believe that this coupling of transferable energy and controls via the reactions of phosphate is extremely primitive. (NB In Chapter 9 we show that in cells the controls exerted by these phosphate compounds are almost invariably magnesium complexes).

7.5.5 **Equilibria and kinetic balance**

In so far as a biological system is driven by energy to give flow patterns both in space and in chemical bond pathways, it is not even close to an equilibrium system (see Figs 7.4 and 7.6). However, in given aqueous compartments there are certain exchanges between units and binding sites which are fast compared with the overall flow (see Sections 2.10 and 2.19). Hence, there are localized equilibria. When we come in Part II of this book to discuss the roles in organisms of individual elements then we observe that some of these roles are due to equilibration while others are not. Thus, in the resting condition of cells most of the reactions of Na^+, K^+, Cl^-, Mg^{2+} and Ca^{2+} in a given compartment are in equilibrium while most of the bindings of Cu^+, Ni^{2+}, and Co^{2+} are not and distribution is carrier mediated. The difficulty then arises that we must describe each element differently, and unfortunately some of the rates of exchange are uncertain and of uncertain significance. In particular we shall find difficulties in the biological chemistry of Zn^{2+} and Cu^+. Again, amongst non-metals it is clear that the exchanges of C, N, O and S between compounds are slow, almost zero, and, while those of Cl^- are at equilibrium, the exchanges of H^+, due to the localization of flow in bio-energetics, may well not equilibrate within some compartments. However, while there is no overall equilibrium there is an overall steady-state or homeostasis in the resting or constant activity states of cells. As we now see clearly these states require a coupled input of energy. Just as particular equilibrium free energies characterize chemicals, particular homeostatic, kinetic states characterize organisms. Thus, particular selection of concentrations of elements arise in species but fortunately there is much in common between all species.

7.6 The integration of activity

7.6.1 **The biological selection of elements**

We have found a satisfactory way of appreciating the interrelations of the chemistry and biochemistry of elements and molecules in living systems from very detailed analytical observations, Fig. 7.7, but this leaves us still distant from an understanding of the specific functional roles of either the elements or the molecules within *biological* systems. A simple example illustrates the point. There is no obvious *chemical* reason why an oxidative catalyst for a given reaction should be made from copper rather than from iron. While in biology it is found, now and then, that either copper or iron can be used for a particular function, e.g. dioxygen transport or electron transport, it is also clear that, generally, copper and iron are used in quite different ways. If we are to understand the biological functions of the elements, we must look more closely at the selected roles the major 25 elements play against the background of the chemical and biochemical principles discussed in Chapters 1–5

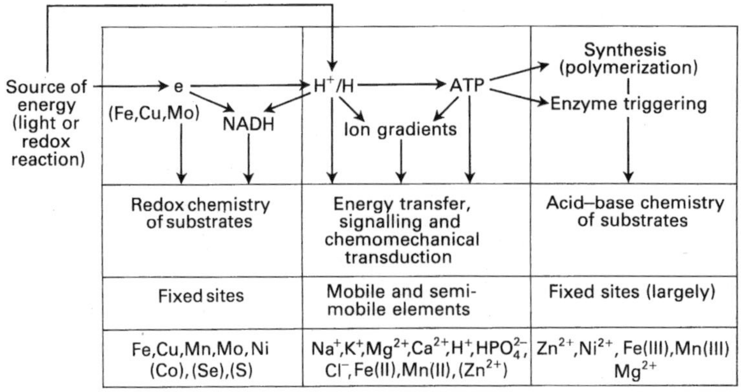

Fig. 7.7 Interrelation of some functions of the elements in biology. Controls are largely excluded.

and against the background of evolution, i.e. the way living systems have become organized. We must also look therefore at their selective association with and evolution of biological polymers, especially the proteome. It will be the purpose of Chapters 8–19 to show the roles of each element in detail both regarding the structure and properties of its free and combined forms and regarding the genetic, energetic and environmental influences acting on the elements in these forms.

7.6.2 Self-organization and flow patterns

The self-synthesis of an organism depends further on the mutual recognition of lipids and polymers so as to build membranes nuclei, ribosomes, and so on (see Chapter 20). While so doing they also pick up metal ions and organic cofactors so as to assemble the catalytic machinery of the cell. The polymers are forced into folded states by their sequences (Chapter 6) and we need to know how these sequences are related to activity and flow. We also need to observe that their surfaces come together to give the organization but that much organization has to be relatively flexible since the systems are internally mobile and grow. (We have seen the importance of electrostatic interactions in Chapter 6.)

The flow we have described so far has concerned the movement of polymers, monomers, elements, and energy through aqueous phases and across membranes into compartments. As we pointed out in Chapter 3, the compartments contain organized systems due to the ways in which filamentous structures self-assemble both along membranes and perpendicular to them, so that a cell has a shape. Now when such anisotropic organization evolved, flow of material and energy became more complex and this is especially true once organisms gained control of the external solutions of cells and built cell–cell organization. The flow which ceased to be random, developed into defined patterns and these patterns had a timed appearance. An obvious example is the flow of blood in the human body but, at a more microscopic level, flow will be caused by the anisotropic distribution of active systems along all membranes. Clearly, this flow becomes part of the information carrying machinery and part of the self-assembly so that the flow and the organization are an interactive whole. The growth tip of a fungus is a fine illustrative example of such a development (Fig. 7.8).

We go forward in Chapter 20 to look for the relationship between the fundamental *flow* of energy and the flow of material *within the self-organized*

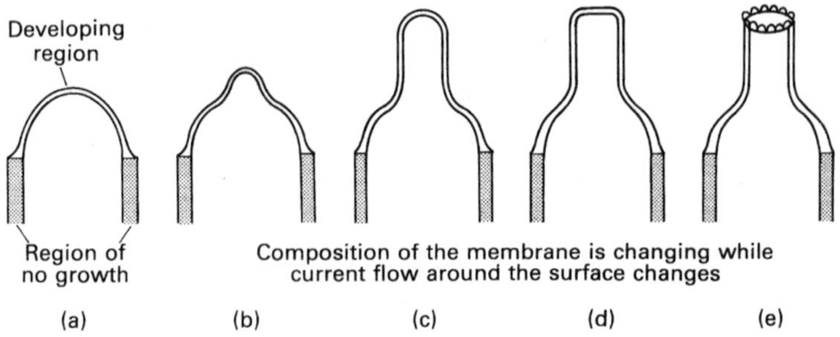

Fig. 7.8 Representation of the morphogenic changes taking place in the regenerating tip of *Acetabularia*. The sequence (a) through (e) shows the morphogenic changes from the initial state to the beginning of whorl formation. We wish to uncover how the development of the tip is related to the distribution of proteins in the membrane and of currents of ions which are known to flow at the tip. The current carriers are calcium ions here.

system. Energy capture, organized information and activity, like the chicken and the egg, had to develop simultaneously. We shall see that this means that 'organic' and 'inorganic' chemistry had to come together to make a living system since we have already demonstrated that organic materials on their own do not and could not have generated many of the flow systems we observe in catalysis and communication.

7.7 Conclusion

The above illustration of the growth of a given organism is an example of the way in which the integration of the whole system of chemical activities in a specific cell produces a resultant physical pattern in space. In the past the pattern, the shape of the organism, a systems property, was in fact used to define a species. Today we use sequences of *single* molecules, DNA, the genome, or part of the DNA, or the RNA, to define species. However, these two descriptive procedures do not lead to an understanding of the relationship between them nor to an understanding of why such species exist. The shape and *raison d'être* of an organism is an emergent property of the chemical system inside its cells. Such chemical systems (organisms) can be characterized only in name by either shape or the linear DNA but it requires quantitative analysis of the active chemistry in the cells to understand their functioning. One major step in this direction is the *quantitative* description of the proteins in all the organism's compartments, the proteome. *The proteome is a system of molecules* and is open to many variables, e.g. concentration, unlike the DNA. (Individual members of a species are not exactly alike.) The proteome itself does not define the cell system fully since for activity the cell needs a constant uptake *and concentration* of essential elementary material. Inside the cell compartments there is therefore *an essential quantitative level of free elements, the metallome, and of their bound forms, the total metallome* within which we include the small non-metal (organic) components required. Again these characteristics are not defined by DNA. Now we cannot hope to explore the metallomes in either of these ways for the twenty essential elements simultaneously, especially since we must do this compartment by compartment as for the proteome. What we can do is to describe the metallomes element by element (Chapters 8–19) and then to analyse the interactions between elements in the integrated cell system (Chapter 20). An intriguing feature is that all *the free metallomes of the cytoplasm* of all living organisms have much in common. We must discover why this is so. However, the organism as a whole with its many compartments is linked to an environment, which itself has changed and this fact must have consequences for any organism.

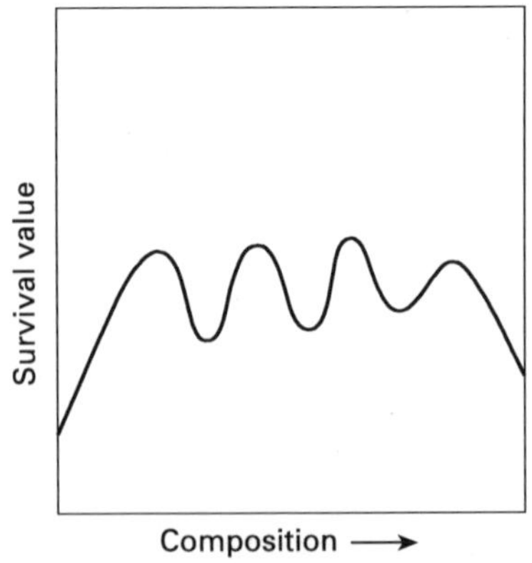

Fig. 7.9 A plot of survival value against chemical composition for organisms which compares with a plot of free energy against composition for a complicated mixture of chemical elements in compounds. The composition axis can either be of free elements or of total element content and each element requires a separate axis.

Genome $\leftrightarrow$ Proteome $\leftrightarrow$ Metallome $\leftrightarrow$ Environment
$\uparrow$ ________________________ $\uparrow$

The peculiarity of these systems (see Chapter 20) is that only certain free ion compositions in certain compartments give rise to persistent organization (species) which can survive long enough for it to reproduce successfully, (Fig. 7.9). This is another fundamental feature of life which we must attempt to appreciate since it can not be understood from DNA sequences.

Further reading

The main references for the novel features are given in our previous books and a recent paper: more detailed references are given in Chapter 20.

Williams, R.J.P. and Fraústo da Silva, J.J.R. (1997). *The natural selection of the chemical elements—the environment and life's chemistry*. Oxford University Press, Oxford.

Williams, R.J.P. and Fraústo da Silva, J.J.R. (1999). *Bringing chemistry to life—from matter to man*. Oxford University Press, Oxford.

Williams, R.J.P. and Fraústo da Silva, J.J.R. (2000). *Coord. Chem. Rev.*, **200–202**, 247–348.

PART II

The roles of individual elements in biology

CHAPTER 8

Sodium, potassium, and chlorine: osmotic control, electrolytic equilibria, and currents

8.1 Introduction

There are large amounts of sodium, potassium, and chloride ions in sea waters but not very much in fresh waters, yet life abounds in both media and to a very good approximation it holds the internal free ion concentrations of these elements constant, not only in cells but also in the circulating fluids of higher organisms (Table 8.1). Even small deviations from normal levels in man are recognized as symptoms of malfunction or disease. (Common salt is not to be despised within organisms.) The constancy of the levels can only be maintained in the biological organisms which dwell in the sea by rejecting much Na^+ and Cl^- while accumulating K^+. By way of contrast all organisms in fresh waters must accumulate Na^+ and Cl^- and they must also accumulate K^+. Again in multicellular organisms quite generally the cytoplasm of cells is relatively rich in K^+ and poor in Na^+ and Cl^- compared with the circulating fluids. Table 8.1 shows some data for the single-cell algae *Valonia* in sea water and for red blood cells surrounded by blood plasma. Obviously, all organisms expend vast energies in keeping these gradients and they all do so internally versus externally in a very similar way. The balances themselves are related to the general problem of homeostasis treated in Chapter 20. There can be little doubt that these three elements, Na, K, and Cl, are and always have been essential components of life (As a reminder, elements directly related to them in the periodic table are shown in the margin, on the next page.)

In this chapter we discuss two related functions: one connected to the concentration (activity) of the same ions, which is associated with the generation of gradients and electrical potential differences and is a determinant of

Table 8.1 Concentration of the main free cations and anions in living cells and their environment (mmol l^{-1})

System	Na^+	K^+	Ca^{2+}	Mg^{2+}	Cl^-	HPO_4^{2-}	HCO_3^-
Sea water	460	10	10	52	550	<1	30
Valonia	80	400	1.5	50	50	5	(10)?
Red-blood cells	11	92	10^{-4}	2.5	50	3	(10)?
Blood plasma	160	10	2	2	100	~3	30

the osmotic and charge stability of cells, and the other connected to the mobility of these ions (Section 4.5), and hence to the flow of electrolytic charge in living organisms. Only these three elements have simple electrolytic functions and occur in biological systems very largely in the form of monatomic free hydrated cations, Na^+, K^+ (and note H^+), or anions, Cl^-. All three ions bind weakly to all organic molecules (see Table 8.2 which gives an example of a very strong chelating agent but relatively weak binding at pH = 7) so that they can only be retained by membrane enclosures very much as in some electrochemical cells. Functional significance in the bound state is rare, but we must remember the binding by cyclic ligands (Section 2.5), which can also be used as drugs or poisons, and by DNA telomers, see below.

It is probable that the rejection of the most abundant anion, Cl^-, and cation, Na^+, was required early in the development of cells to maintain the osmotic balance necessary for cell stability. The inside of cells is of necessity rich in anionic organic molecules, for example DNA, RNA, lipids, and substrates, which implies that some inorganic materials had to be rejected to maintain osmotic balance with the sea. It was also necessary that some positive ions should neutralize the excess negatively charged groups of the biological organic molecules inside the cells since many are ionized at physiological pH, e.g. the carboxylate, phosphate, or sulphate groups of proteins, lipids, nucleic acids, or their constitutive units, as well as of any other compound present. While rejection of chloride partly compensates for this anion accumulation, residual sodium but largely potassium, and to a certain extent magnesium, in effect serve the function of neutralizing negative charge in cells.

Rejection and accumulation require movement across membranes but although the means by which the individual ions can traverse the membrane have already been referred to in the previous chapters (summarized in Fig. 8.1), some thermodynamic consequences of the processes were not discussed. We shall look first at the connection of the thermodynamic balances to the three main types of movement: (1) passive diffusion down the concentration gradient through (gated) channels; (2) active transport, 'pumping', against the concentration gradient coupled to an energy-yielding process, usually the hydrolysis of ATP; (3) ion exchange across membranes, e.g. Na^+ for H^+ or $2Na^+$ for Ca^{2+}. Passive diffusion is a spontaneous process driven by an increase of entropy; active transport results in a decrease of entropy and must be driven by an input of energy. Notice that the reverse of pumping is energy capture from pre-established gradients (H^+, Na^+) since it could lead to ATP formation from a cation gradient (see Chapter 5). A steady state of all three ions is maintained in all living systems and can be analysed using thermodynamic expressions but this implies that all movements are controlled. They are also related to the H^+ gradient.

8.2 Passive diffusion

The free-energy change, ΔG, that occurs during diffusion depends only on the difference of concentration and potentials between the compartments separated by the membrane, reaching zero at equilibrium. Here we leave aside potentials except those due to electrical fields.

For uncharged species ΔG, the free energy of transfer between two compartments of different concentrations, C, is given by

$$\Delta G = RT \ln(C_2 / C_1)$$

Group 1 elements

Li	(rare)
Na	common
K	common
Rb	(rare)
Cs	(rare)

Group 17 elements

F	(not uncommon)
Cl	common
Br	(rare)
I	(rare)

Elements related to Na, K, and Cl

Table 8.2 Binding constants K (l mol^{-1}) and substitution rate constants k at 25°C

$M^+ + EDTA^{4-} \underset{k_2}{\overset{k_1}{\rightleftarrows}} [M(EDTA)]^{3-}$

Cation	log k_1	log k_2	log K
Li^+	7.7	~4.0	~3.7
Na^+	7.7	~5.0	~2.7
K^+	7.8	~6.5	~1.3
Rb^+	8.2	~8.0	~0.2
Cs^+	8.3	~8.0	~0.3

$K = k_1/k_2$ = stability constants which are some 10^3 times lower at pH = 7.0.

$EDTA^{4-}$

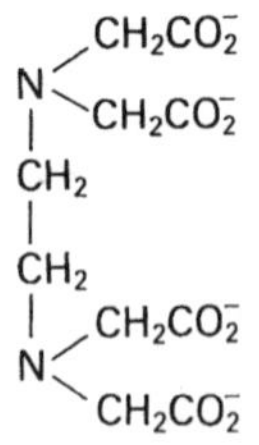

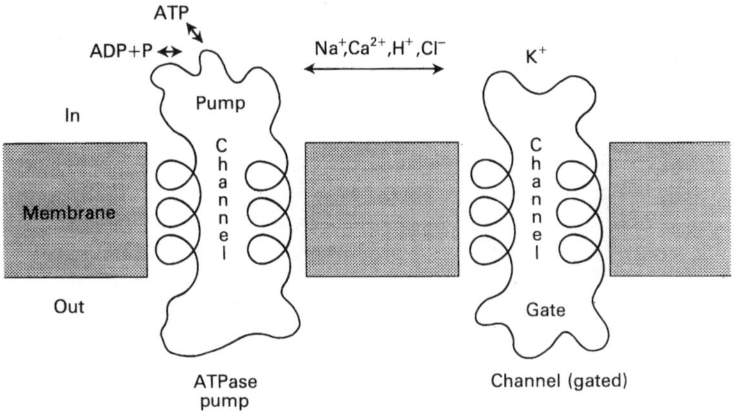

Fig. 8.1 Two of the ways in which the simplest cations (and anions) pass through membranes. On the right is a channel which may have a gate; it allows facilitated diffusion. On the left is an ATPase pump. The enzyme is remote from the channel here and it energizes the movement across the membrane usually pumping Na^+, Ca^{2+}, H^+, or Cl^- out of the cell. There are also specific combinations of gated channels which work as exchangers, e.g. the Na^+/Ca^{2+} exchanger. Note: the α-helical constructions allow segmental motion.

and it will be negative for processes down the concentration gradient and positive in the reverse direction, representing in this case an energy barrier to be overcome by coupling to an energy-yielding reaction.

For charged electrolytes, however, besides the concentration gradient one must also consider the overall charges in the two compartments, and the difference in free energy between them will then depend on the electric potential difference ψ generated across the membrane

$$\Delta G = RT\ln(C_2 / C_1) + z\psi\mathfrak{F} \tag{8.1}$$

where z is the charge of the ion considered and $\mathfrak{F}$ is the Faraday constant.

Now, biological membranes are not equally permeable to all species due to their particular composition and organization. Among the charged species, membranes are usually impermeable to the ionic charges of the amino-acid residues (in proteins or not) and to free or bound phosphate, relatively impermeable to sodium ions, to prevent lysis, and more permeable to the larger potassium or chloride ions (but permeability can vary greatly for different types of membranes or even for the same membrane under different conditions). The difference in permeability adjusts the distribution of diffusible ions, as described by Donnan in 1911, and causes what is called a 'Donnan equilibrium' and an electrolyte potential.

To understand this effect let us consider a 'cell' containing only potassium and chloride ions and a charged (anionic) protein, placed in an isotonic solution of potassium chloride

Inside (cell)		Outside
$[K^+] = 100$ mM		$[K^+] = 100$ mM
$[Cl^-] = 50$ mM		$[Cl^-] = 100$ mM
$[Prot^-] = 50$ mM		

(8.2)

The osmotic pressure of the two sides is equal (since it depends on the number of particles per unit volume), but the concentration gradients of the charged protein and of chloride are not. Since the membrane is permeable to chloride there will be a tendency for this ion to enter the cell. Its movement cannot be compensated by the diffusion of the charged protein in the opposite direction because the membrane is impermeable to this species, whence the diffusion of chloride must be accompanied by a simultaneous movement of potassium ions in the same direction to attempt to maintain the electroneutrality of each compartment.

The final situation can be derived from eqn (8.1), applied to both K^+ and Cl^-, i.e.

$$\Delta G_{K^+} = RT \ln \frac{[K^+]_{in}}{[K^+]_{out}} + z\Phi\psi = 0, \tag{8.3}$$

$$\Delta G_{Cl^-} = RT \ln \frac{[Cl^-]_{in}}{[Cl^-]_{out}} - z\Phi\psi = 0, \tag{8.4}$$

Since ψ is the same in eqns (8.2) and (8.3),

$$\frac{RT}{\mathfrak{F}} \ln \frac{[K^+]_{out}}{[K^+]_{in}} = \frac{RT}{\mathfrak{F}} \ln \frac{[Cl^-]_{in}}{[Cl^-]_{out}}. \tag{8.5}$$

It follows that $[K^+]_{in}[Cl^-]_{in} = [K^+]_{out}[Cl^-]_{out}$, where the brackets represent concentrations or, more accurately, 'activities' of the two ions. The equilibrium condition is, therefore, that of the equality inside and outside the cell of the product of the activities of potassium and chloride.

From this result and considering that $[K^+]_{in} = [Cl^-]_{in} + [Prot^-]_{in}$, one easily derives the final situation, given to the nearest whole number, as

Inside (cell)		Outside
$[K^+]$ = 114 mM		$[K^+]$ = 86 mM
$[Cl^-]$ = 64 mM		$[Cl^-]$ = 86 mM
$[Prot^-]$ = 50 mM		

($\leftarrow H_2O$) (8.6)

At this stage, further net movement but not exchange is prevented by the difference of potential generated across the membrane, which can be calculated from eqn (8.3) as

$$\psi = \frac{RT}{\mathfrak{F}} \ln \frac{[K^+]_{out}}{[K^+]_{in}} = 0.059 \log \frac{86}{114} \approx -7 \text{ mV}. \tag{8.7}$$

It must be said that the value obtained is only a rough indication since we have used several approximations, e.g. total concentrations instead of activities of *free* ions, and neglected volume variations and pH changes that may occur, etc. As to the volume variations the establishment of this 'Donnan' equilibrium has resulted in a net movement of particles into the 'cell' thereby raising its osmotic pressure; the inside becomes hypertonic to its environment and water will enter the cell increasing its volume [see scheme (8.6)]. (Remember, however, that plant cells and bacteria have a retaining outer wall to prevent such expansion.)

If in the above treatment there had been no diffusible ions outside and only protein$^-$ and K^+ ions (100 mM) initially in the cell, outward diffusion of K^+ would have left an unequilibrated K^+ potential at the membrane. However, we must include one more diffusible ion, the proton. The possible change in pH becomes clearer when considering the case of a negatively charged protein having K^+ as counter-ion on one side of a semipermeable membrane and fresh water on the other side. Naturally, the tendency will be for diffusion of K^+ out, but, since the protein cannot diffuse, the process would be prevented if it were not for the dissociation of water. This does indeed occur and H^+ tends to accumulate on the protein side of the membrane while the corresponding OH^- ions accumulate, along with K^+, on the side of the initially pure water until equilibrium is established. If there are other ions present, as in real biological systems, their effect has to be considered following similar lines of reasoning. Since Ca^{2+} and Mg^{2+} amongst the cations are much lower in concentration and have a high tendency to combine with biological ligands, the electric field potential will be mainly due to Na^+ and K^+ gradients (and to H^+ gradients in particular cases) but there are situations where the calcium potential is important.

This rather detailed description is given to show that from the very beginning of cellular enclosures there was of necessity a link between the simplest inorganic ions, Na^+, K^+, and Cl^-, the pH, and the electrolytic potentials of cells which are crucially important in living systems. In fact all cell membranes have a potential across the plasma membrane. Evolution has amplified these effects using energy to vastly increase the gradients by pumping (see *active* transport, Section 8.5).

There are some particular cases of passive diffusion in which the substance or electrolyte considered does not diffuse by itself but the movement is promoted or facilitated by carriers (see Section 2.5) or by proteins in the membrane (named channels, Fig. 8.2) in a way similar to the catalysis of a reaction, i.e. the protein has no effect on the thermodynamics of the system but acts on the kinetics of diffusion by lowering the energy of activation for transfer and accelerates the rate at which diffusion occurs (Fig. 8.1). As with enzymes, channels facilitate the process in both directions, depending on the concentration gradients, and can be inhibited both competitively and non-competitively. It is important to note that the walls of the channel may be coated with water molecules (Fig. 2.11). In fact, most cells maintain an open potassium channel much of the time but closed channels, if any, for all other ions. In certain cells the channels can be opened or closed by the application of a voltage, by the action of a messenger such as acetyl choline, by an energy carrier, for example ATP, or by another ion, e.g. Ca^{2+}.

Facilitated diffusion is particularly common for molecules formed in one compartment but needed in another, e.g. in the transfer of sugars and amino acids across cell membranes, but notice that facilitation is very selective. Facilitated diffusion can also be used to destroy cells. The introduction of a small ring chelate carrier (Fig. 2.6) or a protein, e.g. a colicin channel (Fig. 8.3) into a membrane can generate unwanted diffusion and kill a cell. These weapons are known both to bacteria and to drug chemists.

Let us suppose that the membrane is only permeable to one ion, e.g. K^+. If we place electrodes across the cell membrane we can follow the net current which flows as we change the applied potential. This potassium current is zero when no potential difference is applied despite any steady-state Donnan potential. The current at any applied potential will also be zero if we close the channels. In fact, we can only observe the potential of those ions which carry current. In the example given at the beginning of this section (expression (8.2)), there could be two extreme cases where the observed current was due to the K^+ current (potassium channels open) or where it was due to Cl^- ions (chloride channels open). (Physiologists do not use the simple thermodynamic equation (8.7) for membrane potentials but a related Nernst equation which includes the permeability of the ions; see 'Further reading'.) Now manipulation of channel opening and closing for different ions has led to profound developments in evolution, e.g. the central nervous system. Chloride channels are also used to modulate cellular behaviour. Defective chloride channels give rise to the disease of cystic fibrosis.

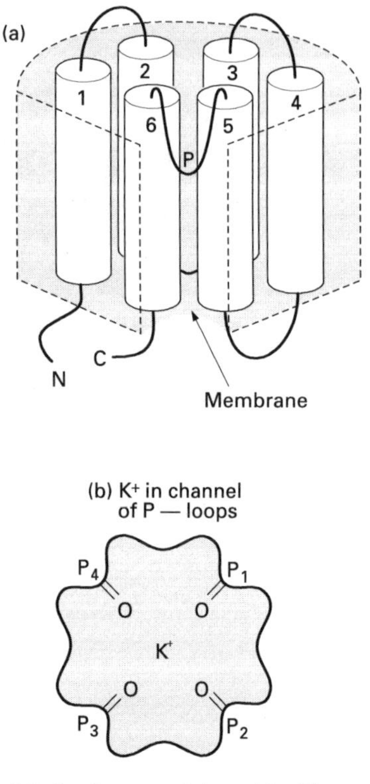

Fig. 8.2 A voltage-gated channel for K^+ consists of four subunits, each of which has six trans-membrane helices and an inward facing P-loop (a). Each P-loop lies in the central region of the four subunits and subtends into this 'pore' a series of carbonyl residues to give a lining of a succession of four roughly planar oxygen atoms which can bind potassium (b). Regions of the pore also have charged residues which can presumably alter conformation when a potential is applied so as to block access of K^+. A chemically modulated ion channel which binds a messenger or a drug could be blocked or opened by this binding, e.g. the acetyl choline receptor for the Na^+ channel. The channels we know are all based on multiple helices which can adjust mutually, compare calmodulins (see Fig. 10.7). The structure of the channel is known (see Doyle *et al.* in 'Further reading'). The figure is after Choe *et al.*, (see 'Further reading'). Na^+ and Ca^{2+} channel units have only four helices with P-loops but are similar.

8.3 Gated channels

The channels for simple diffusion may be held closed so that, in the absence of an effector molecule which opens the 'gate' of the channel, there is no ion

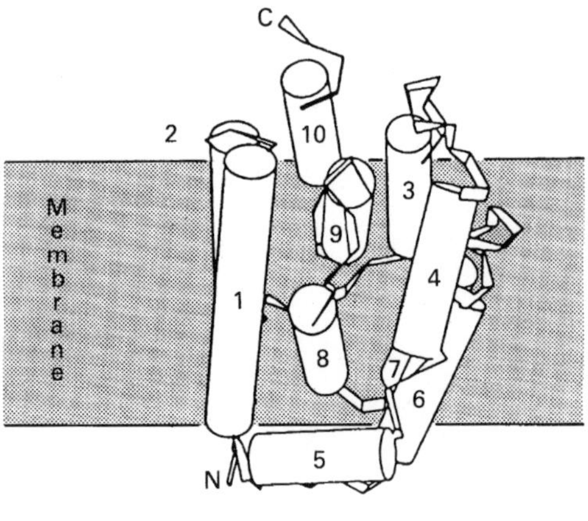

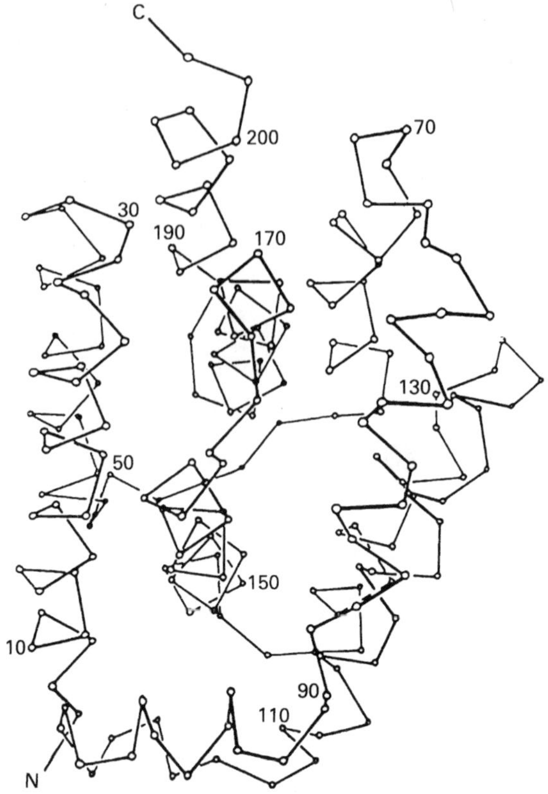

Fig. 8.3 A detailed description of a protein, colicin A, which can form a channel in a membrane. Notice the multiple helices (shown in detail on the right) which can be inserted into a membrane leaving a hole through it (Fig. 8.2). The structure is due to the work of G. Tsernoglou.

flow. The usual biological situation is that the gate has a receptor for an effector; on binding to the receptor the gate is opened by the binding energy which causes conformation changes. (Where these conformation changes involve movement of charges on the proteins in the membranes there is a charge drift or gating current *in* the channel without net flow *through* the channel.) Ideas of how the gated channels operate are being developed (see Section 8.5). The opening of a gate allows an influx of ions which can be used as a message either by changing the local electrical potential ψ or the local chemical concentration, *C*.

A gated channel can be held open or closed by a chemical, as just discussed, or by an applied voltage. The latter are called voltage-gated channels and are opened by the switching of the applied voltage.

Table 8.3 Relative permeabilities of the sodium and potassium channels†

Ion	Na^+ channel	K^+ channel
Na^+	1.00	<0.01
K^+	0.09	1.00
TI^+	0.33	2.30
NH_4^+	0.16	0.13
Guanidinium$^+$	0.13	<0.01
H_3O^+, $(H_9O_4)^+$	~250.0	Low?
$CH_3NH_3^+$	<0.01	<0.01

† Relative permeabilities are ratios to $Na^+ = 1.0$ for sodium channels and to $K^+ = 1.0$ for potassium channels.

8.4 **Channel selectivity and possible constructions**

The differences in selectivity between Na^+ and K^+ channels are remarkable (Table 8.3). Note that:

1. Only TI^+ approaches K^+ in permeability in the K^+ channel. No larger or smaller ion passes easily. (TI^+ is a useful probe of K^+ channels.)
2. The K^+ channel excludes $NH^+{}_4$, which has the same radius, better than does the Na^+ channel.
3. Large H-bonding spherical molecules, e.g. guanidinium, pass the Na^+ channel.
4. No methylated compounds pass through either channel.

We can propose an explanation for these facts. Though both channels are made from protein helices (Fig. 8.3), the Na^+ channel made by the proteins must be the larger. This is the inverse of expectation. It would appear from the data that its low selectivity, when open, could be due to the ease with which ions move in or out through a water layer which partly fills the channel. This would account for high permeability by H^+ and very large H-bonding ions which could displace water as they pass. (Note that in fact little H^+ does pass since it is at 10^{-7} M.)

The Na^+ ion may well move through its channel only partially hydrated. By way of contrast, the K^+ channel is less likely to be hydrated and matches the K^+ radius in size, Fig. 8.2. Only Tl^+ enters equally to potassium and selectivity is made to be remarkably high since, perhaps, it involves temporary ligation to somewhat polar protein sidechains, $Tl^+ > K^+$.

The overall impression is that the Na^+ ion has a very small binding constant $<1.0\ M^{-1}$ to the K^+-channel since the water molecules bind too strongly to Na^+ and cannot be replaced sufficiently by the designed set of carbonyls which fit K^+, see aside for relative sizes. The binding constant of potassium is close to $10^3\ M^{-1}$. This result is exactly that predicted from model studies; certain chelating agents displace water from K^+ more readily than from Na^+ (see Section 2.5.2, and see also Section 8.11 on DNA).

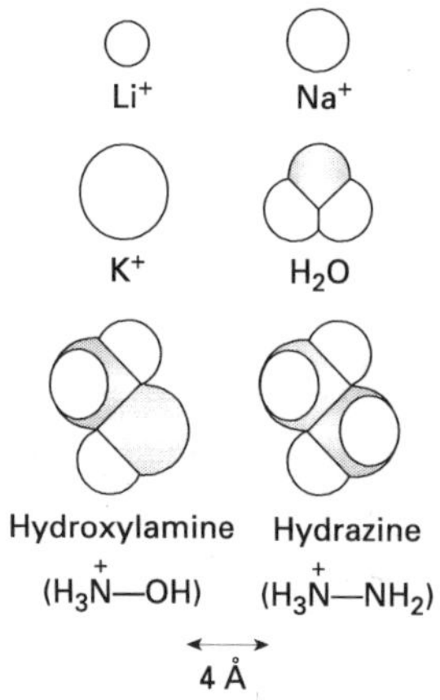

Some ion and molecule sizes

The difference between Na^+ and K^+ channels must be seen against the background of the very great difference in solubility of their salts with hydrophobic anions such as tetraphenyl borate, where $Na^+ > K^+$ by 10^6. Highly structured water could form a proton as opposed to an inorganic ion channel (see Section 2.11).

The description of a channel has been greatly aided by the study of peptides. It is clear that even a single short helical strand of polypeptide, such as gramicidin, can make a voltage-gated channel (Fig. 8.4). The details of such channels are not important here but the shape of the interior of these helices is, in that their exact shape decides the selectivity of ion movement. The interested reader should refer to the literature on the peptides gramicidin and alamethicin. The known features of non-synthetic channel proteins are consistent with a multiple set of helical strands forming a helical bundle across the membrane. Today synthetic peptides have been made which generate such helical channels. A typical composition is

$$^{+}H_3N\text{–}(Leu\text{–}Ser\text{–}Ser\text{–}Leu\text{–}Leu\text{–}Ser\text{–}Leu)_3\text{–}CONH_2.$$

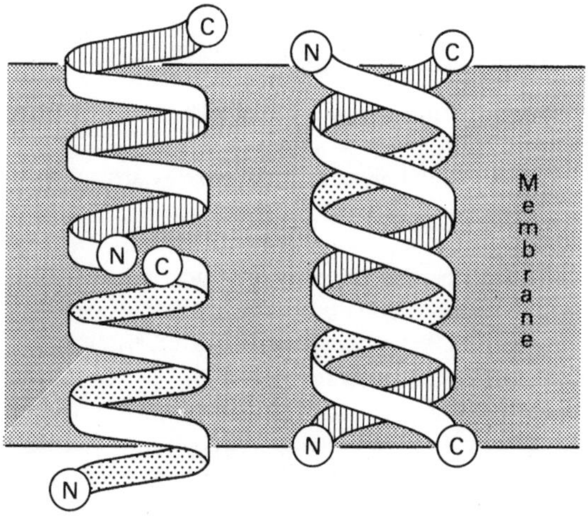

Fig. 8.4 The probable structures of peptides which form (gated) channels in membranes. Examples are given in the text. The channel helices are usually formed from 20 hydrophobic amino-acids.

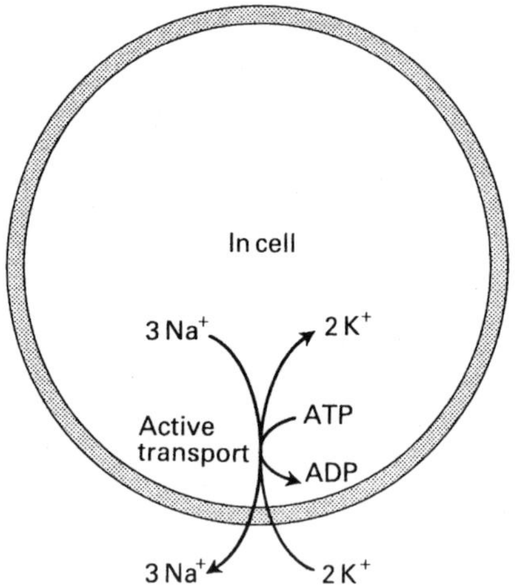

Fig. 8.5 The pumped movements of sodium ions out of and potassium ions into the cell. ATP hydrolysis is required for the movements of the ions against their gradients. The actual coupling is of two K^+ for every three Na^+ so that the pump is electrogenic, i.e. it develops charge and chemical gradients (see Fig. 8.6).

Notice that rates of movement of the ions, Na^+, K^+, and Cl^-, in these channels are close to the diffusion limit, i.e. 10^{10} s^{-1} for water exchange (Fig. 4.3), but that binding to such ligands as ethylenediaminetetracetate, EDTA, is not quite as fast (Table 8.2). Entry, exit, and passage through channels remains fast even if weak binding to the channel walls occurs.

Before proceeding, it may be helpful to give a brief reminder of certain structures of cells. The above-mentioned membrane is the cytoplasmic membrane, but the considerations of ion equilibrations across it through channels when some ions or molecules are trapped inside apply to all vesicle membranes inside cells as well as to the cytoplasmic membrane. Prevention of swelling is managed partly by ion balances, as described, and partly by tension either in the outer membrane or in filamentous structures of the cell. Cells have different structures to maintain stability. Plant cells and many unicellular organisms have a strong outer wall; animal cells, especially, have a network of inner fibres connected to a protein network under the cytoplasmic membrane (Section 3.5). Vesicles are controlled in shape by the filament networks in the cytoplasm. In this chapter we must note that the Na^+, K^+, and Cl^- balances are linked to the pressures in cells and thence to the mechanical tensions in various parts of an organism. These ions are also differently distributed in different compartments, e.g. the cerebrospinal fluid is very special. Why?

8.5 Active transport: pumps

Much larger Na^+/K^+ concentration gradients are found in biological systems than those one would expect from Donnan equilibrium considerations, and other mechanisms besides diffusion, facilitated or not, must be operative. For example, a typical K^+ concentration ratio for intra- and extracellular fluids may be of the order of 20:1 (see Table 8.1) and, since the membrane is permeable to potassium in the resting condition, potassium channel open, this gives an approximate potassium membrane potential of about –80 mV (inside negative relative to the environment). If the membrane channel is permeable to more than a single ion type, the effect of other ions must also be considered when examining observed potentials. Whereas the direction of the potassium gradient is that expected from the presence of ionized protein and other non-diffusible anions, e.g. organic phosphates inside a blood red cell, the sodium gradient is opposite from that expected and this can only happen by a process of active transport ('pumping') coupled to an energy-yielding reaction. If the energy input stops the transport ceases and sodium ions will enter the cell slowly through appropriate holes or channels. The measured distribution of sodium and potassium inside and outside the blood red cell corresponds to a steady-state condition in which sodium ions are continually moving in by diffusion, through leaks, and pumped out by active transport across specific routes, whereas potassium ions follow the reverse process (see Fig. 8.5). In the resting state of cells it is the sodium channel which is held closed, with the potassium channel open and the pump working.

The mechanisms of generation of this steady state include an enzyme with ATPase activity, i.e. promoting the hydrolysis of ATP, an energy-yielding reaction, which is active only in the presence of Na^+ and K^+ (as well as Mg^{2+}, which acts as a cofactor; see Fig. 8.1). The pathways for the ions in the pumps are formed by an association of proteins which cross the membrane, so that each kind of ion may be passed through a series of protein conformational changes to successively weaker and weaker binding sites as it traverses the

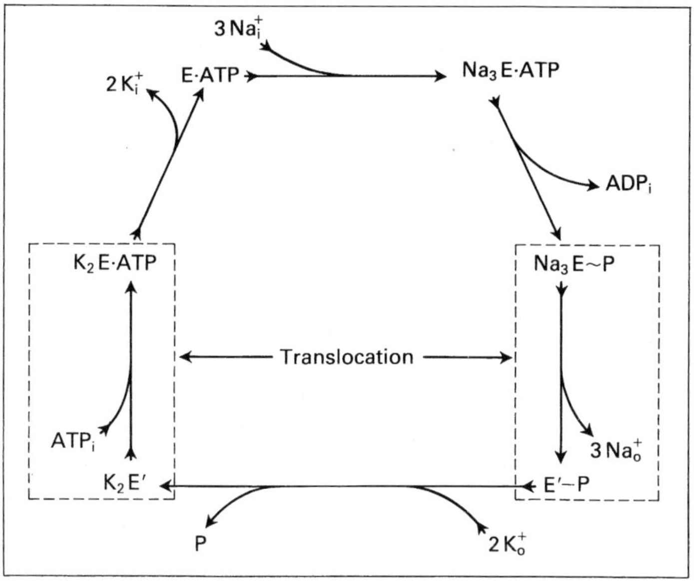

Fig. 8.6 Biochemical reaction scheme for active Na-K transport. Translocation by the enzyme, Na, K-ATPase, across the membrane corresponds also to the biochemical conformational changes of the enzyme from E (Na^+ inside the cell) to E′ (Na^+ outside the cell), or vice versa. The whole process could run backwards to make ATP from large ion gradients but the process would lose energy. ATP, adenosine triphosphate; ADP, adenosine diphosphate; P, inorganic phosphate. K^+_i and Na^+_i; and K^+_o and Na^+_o are internal and external potassium and sodium ions, respectively. Note that here and elsewhere ATP and ADP are really Mg^{2+}.ATP and Mg^{2+}.ADP.

membrane until being delivered (see Fig. 8.6). The probable motions within the pump are due to relative rotation/translation of helices as in Fig. 8.7 and it is these movements which provide adjustments to binding constants and to gates so as to generate flow. Halide pumps will be described in Chapter 18.

Pumping systems do not have to be electrically neutral. The Na^+/K^+ ATPase pump of animal cells moves $2K^+$ in for every $3Na^+$ moved out so that the membrane potential which is established is large and directly related to thc hydrolysis energy of ATP.

The establishment of concentration gradients, such as those of sodium and potassium, and, as we have seen, of H^+ (Chapters 1 and 5) provides a means by which free energy can be stored in a living cell very much as in an electrochemical cell. This energy can be made quickly available for various

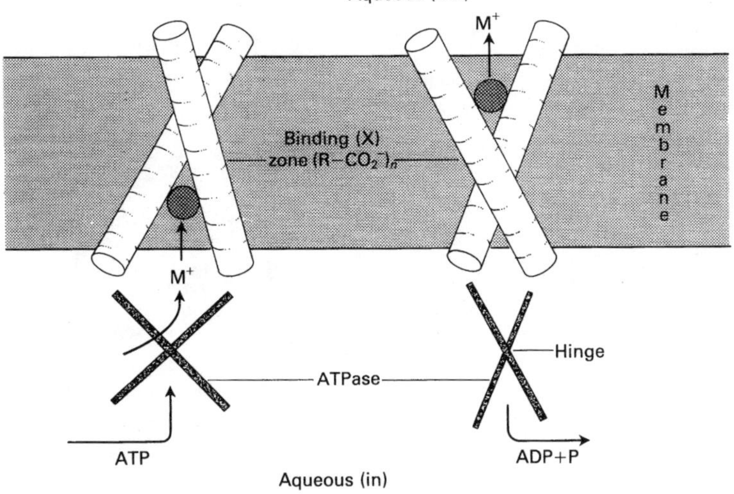

Fig. 8.7 A schematic model for an ATPase pump (or the ATP-synthase). The degradation of ATP to give an ion gradient takes place with movement of M^+ (H^+, Na^+, Ca^{2+}) as it flows from a low to a high concentration (see Chapter 5), left to right in the figure. There are a series of helices in the membrane forming a channel. Binding of nM^+ with high affinity at X, (RCO_2^-) (left) causes the helices to change conformation to a highly energized state (right) in which binding is much lower so that M^+ is released. The movement of the helices is transmitted via a groove-opening reaction which uses the energy of ATP and which releases ADP (right). Energy can be reversed in its flow to make ATP from a gradient (see Figs 10.13 and 5.6).

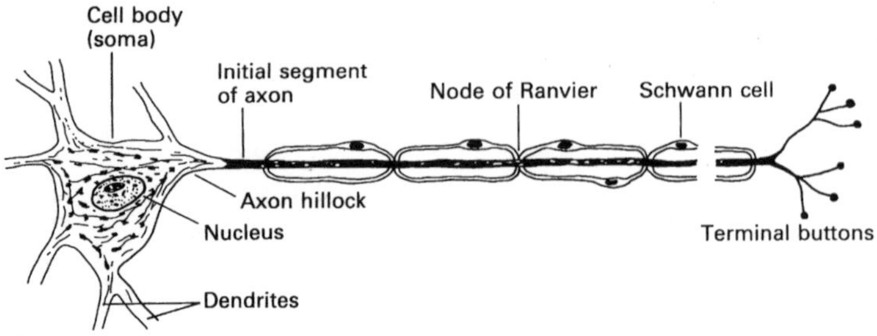

Fig. 8.8 Motor neuron with myelinated axon.

processes, e.g. for the transmission of impulses (messages) in nerve cells; for exchange reactions so as to accumulate or reject other chemicals, e.g. amino acids; for coupling with other chemical reactions, e.g. the 'high-energy' intermediates (ATP) for oxidative and photophosphorylation; for synthesis of new materials, e.g. in the process of replication of bacteria; and for motility motors. These links to simple ion gradients are of extreme importance in biology.

As stated, the potential recorded by placing electrodes across a membrane tells us which ions can flow across a membrane. Resting cells are kept highly impermeable to Na^+ and most other ions, e.g. Ca^{2+}, Cl^-, and HPO_4^{2-}, but they are permeable to K^+ in the sense that the specific K^+ channels are open. Remember there is no loss of K^+ despite the huge gradient since the potential ψ and the pumps oppose *net* flow, while allowing exchange. (There are cells of a specialized nature where the resting potential is not a potassium potential but the interested reader must consult a book on physiology to obtain insight.)

Through a sudden disturbance, e.g. a light touch on the skin, the K^+ channel in a nerve cell (Fig. 8.8) can be blocked and the Na^+ channel opened when the resulting flows of ions are as shown in Fig. 8.9. The measured Na^+ poten-

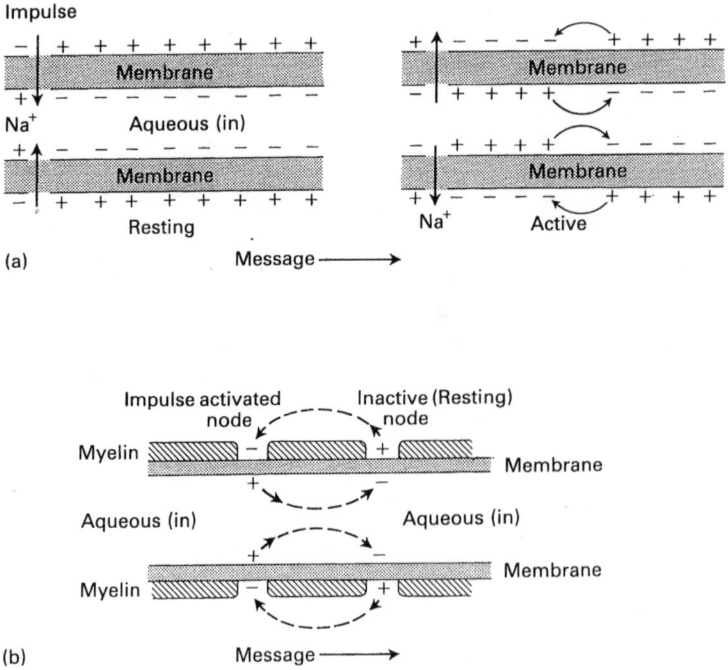

Fig. 8.9 Local current flow around an impulse in an axon, shown as a longitudinal slice. Note that current flow is represented as movement of positive charges. (a) The situation in non-myelinated nerves; (b) that in myelinated nerves.

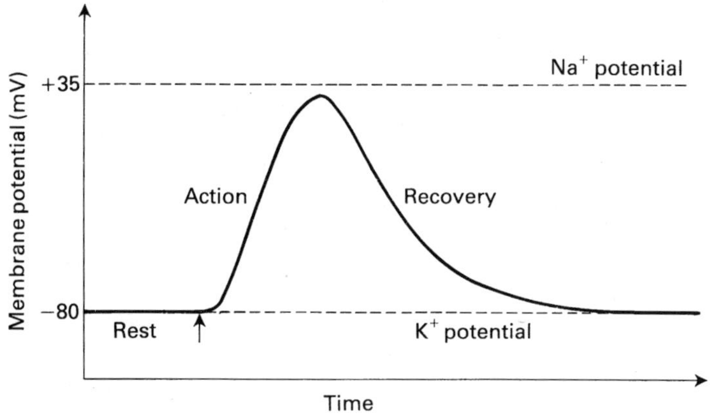

Fig. 8.10 The action potential at a nerve or muscle due to opening of Na^+ channels followed by their closing and the re-opening of potassium channels.

tial across the membrane is now similar but opposite in sign to the K^+ potential (Fig. 8.10). Moreover, the steady state is lost and a large net inward current of Na^+ will flow which disturbs the potential on the membrane locally. Immediately, the Na^+ channel is closed and the current reverts to a K^+ current. In nerve cells this alternation of potassium steady-state potentials with pulsed sodium potentials due to opening and closing of K^+ and Na^+ gates gives rise to a travelling wave of depolarization. This is the electrolytic message which is largely due to transverse movement of ions and local eddy currents generating a travelling 'Na^+ gate open/K^+ gate closed' message (Fig. 8.9). It is the travelling gating and ungating which is the current message. (At the end of the nerve cell the gates which are opened are those for Ca^{2+} ions or for organic chemical messengers since a new type of message is required; see Chapter 10.) This system is the basic process of nerve cells everywhere.

(While we have given a very simple account of relationships between membrane potentials and concentrations gradients in Sections 8.2 to 8.5, there are today extremely sophisticated descriptions of the relationships between membrane voltages, currents, concentration gradients, and, above all, permeabilities and their mechanisms. The interested reader is advised to read an appropriate physiology text; see 'Further reading'.)

In Fig. 8.11 we put side by side two types of circuitry. The first is man's electronic circuit. Man uses electrons in narrow zones of metal or semimetal devices surrounded by the broad insulator, air, or non-conductors such as plastics. Cells cannot make such materials and its only electronic circuits are very different (see Chapters 4 and 12). They carry very little current and are not used for message transmission at all. The major biological circuits are based on elec-

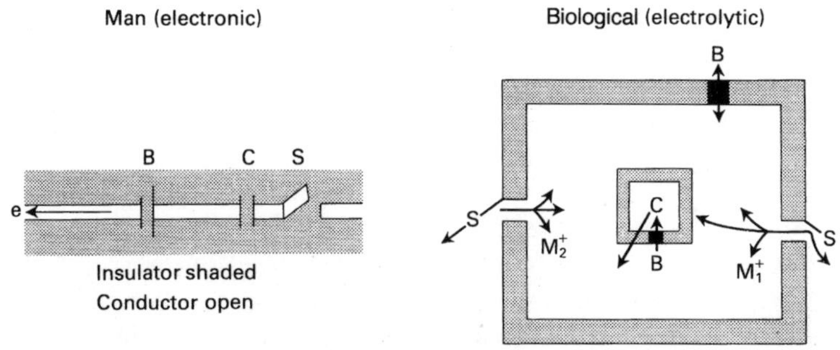

Fig. 8.11 An illustration of the way in which man's electronic circuits based on metals or semiconductors are related to biological electrolytic circuits. The insulator in the latter is the membrane: the battery in the membrane is described in Chapter 5, the switch is a gated channel, and the condenser is a vesicle store (see Chapter 9). Flow of charge is of the essence of biology. S, switch; B, power supply; C, condenser; e, electron; M, metal cation.

trolytics in elongated cell protrusions where the internal conducting aqueous phase is broad and the insulator layer narrow. However, the currents are not the flow of ions along tubes as might be thought but depolarization waves along the surfaces of the elongations. Such a circuit construction is possible when, as in cells, the current-carrying medium is water and when fatty acid (or glass) membranes are the insulators. Notice that in circuits the power is along the wire of man-made electronic devices but across all the membranes of the current-carrying devices of biology. Now these biological circuits can incorporate switches (gating systems) or condensers (large concentrations of ions locked in vesicles which may be suddenly released by triggering currents of ions) (Fig. 8.11). In fact, electrolytics has the same versatility as electronics. The evolution of such circuitry is seen at its best in the human brain. It must be remembered that the above account applies to animal nerve cells and that all cells have similar features but they are far from identical in circuitry.

8.6 The nature of selective pumps

Certain points from previous chapters can now be re-introduced. In the first instance, the manner of generating ion selectivity by rings of donor atoms can be analysed specifically for Na^+ against K^+ as was described in Section 2.5.2. The action of ion gating of pumps can also be referred to models of helix/helix motions in membranes. Driving the opening of pumps by binding of a ligand is not unlike the adjustment of the binding site of haemoglobin by allosteric effectors (Chapter 6) or movements in calmodulin (Chapter 10). A pump must introduce a one-way drive into the channel opening and closing. A simple picture is that of change from one closed configuration to another driven by ATP hydrolysis. In the cyclic movement the gating device is transferred from 'in front of' to 'behind' the ion to be pumped (Fig. 8.7). Simultaneously, the binding constant of the ion to the pump is reduced. The pump returns to its initial unloaded state via a low-energy state. This style of pumping is possible for any molecule or ion. Of course, exchangers must have two channels operating cooperatively and oppositely.

8.7 Building electrolytic circuits

8.7.1 The positions of pumps and channels

In a real cell the distribution of the pumps and channels for ions is not isotropic along the cell membrane so that there are zones where ions are taken in which are laterally separated from zones where ions are pumped out. This allows ion gradients to develop across rows of epithelial cells, for example (Fig. 3.12). Interestingly, it also allows current to flow via concentration gradients along a cell membrane which are laterally controlled. Another example of the positioning of the channels is that of the sodium channels of nerve cells which are found overwhelmingly at the nodes of Ranvier and not along the whole body of the cell (Fig. 8.8). This increases considerably the efficiency of message transmission.

8.7.2 The ATPase pumps: some general comments

We have described the Na^+/K^+ ATPase pump for making gradients of these ions and we have noted earlier that H^+ gradients and ATP formation are related reversibly at mitochondrial and chloroplast membranes. In the next

Table 8.4 Classes of ATPase membrane pumps

Class 1: no intermediate		**Class 2: phosphorylation**	
Mitochondrial	H^+	Cytoplasmic	H^+
Chloroplast	H^+	Vacuole	H^+
Prokaryote	H^+	Cytoplasmic	Na^+/K^+
Archaebacteria	Na^+	Cytoplasmic	Ca^{2+}
		Vesicular	Ca^{2+}
		Bacterial	Mg^{2+}?

Note. Class 2 proteins are single chains while class 1 proteins are made from multiple subunits. Class 1 pumps are often called P-type.

chapter we shall describe Ca^{2+} ATPases, but undoubtedly there are many other ATP-dependent movements of ions and compounds. In the analysis of these pumps it has been found that there are two kinds: (1) without formation of *covalent* intermediate protein phosphorylation; and (2) with such phosphorylation (Table 8.4). There may be quite considerable similarities in all the pumps in so far as the construction of the routes through them is concerned; all are based on cross-membrane helices that are mobile with respect to one another. The helical structures generate selectivity of the route. The coupling to ATP is through helix/helix motion (Fig. 8.7). The distinction between the two types of ATP-coupled transport lies in the ATPase and its connection to the helical channel. There is known to be close sequence homology between the pumps of class 2, i.e. Na^+ and Ca^{2+} pumps, for example. Here the last step is dephosphorylation of the protein (Fig. 8.6) and the loss of energy in this step leads to a high degree of irreversibility in the energetics. In class 1 pumps there is no phosphorylation step (see Fig. 5.12) and irreversible losses are just those associated with heat loss in any machine and may be small. The desirable differences between the two classes are obvious in that certain steps must not be readily reversed. These pumps are all part of biological electrolytic circuits.

The positioning of these pumps along membranes and the lateral currents induced cause concentration gradients which in one way or another interact with membranes and filaments. Ion gradients are not to be dissociated from considerations of cell shape. Micro-organization of electrical activity is then associated with internal and external macroscopic structure.

The connection between the shape of a cell and the positioning of pumps will be made by the interaction of the product of the pumps with the filamentous internal structure. At the same time, as the outer membrane is extended or contracted by the changed tension developed in the filaments, the curvature of the membrane changes. This change alters the positioning of the ATPase pumps in the new more curved or straightened regions of the membranes since the chemical occupancy is a component of the surface energy (see section 3.4). Pumps and channels therefore develop local current circuits in association with shapes and tensions in cells and the two are interactive. It is likely that the proton and calcium pumps have a larger part to play than the Na^+/K^+ systems since Ca^{2+} and H^+ interact more strongly with the filament structures (see Chapter 10).

Finally, in a steady state the particular pump exchanger or channel for an ion needs to be back connected to the concentration of that or another ion in the cell so that the activities are stopped selectively, or at least maintained at a fixed rate of circulation of ions. This gives an ion-selective homeostatic state maintained by *feed-back control*. Thus some potassium channels may be closed by Ca^{2+} and others by ATP concentrations. Feed-back control of all ion levels or flows is an essential feature of cells at rest. The resting condition is not one of inactivity but of fixed, usually low, circulation of ions. The maintenance of ionic strength largely due to K^+, Na^+ and Cl^- in cells is essential for the activities of enzymes and of DNA and RNA, and the conformation of membranes.

8.7.3 Exchangers

We have stressed the nature of pumps and channels and we saw that in some of the pumps more than one ion was involved. To keep the total system of ions under homeostatic control there are also regulating exchangers which are also under tight feedback control so that *relative* concentrations of ions are strictly managed. Examples are the Na^+/H^+, Na^+/Ca^+ and the $Na^+/K^+/Cl^-$ exchangers of various cells, which are specific for sets of ions and are controlled. It is observed that some of the exchangers are used to move organic molecules across cell membranes, e.g. the uptake of amino acids. This is another part of homeostasis since it ensures that there is a relationship between organic chemistry in the cell and inorganic balances. In this section of the book we wish to keep our attention on the connections between charged particles, potentials, and currents, turning only now and then to organic ions and metabolism. The exchangers are likely to be structured like pairs of cooperative gated channels.

8.7.4 Organic anions and cations as current carriers

Just as variations in organic chemistry could lead to the development of selective coenzymes such as NADH and NADPH or different nucleotide triphosphates, so variations on inorganic chemistry lead to different message carriers. Thus Mg^{2+} and Ca^{2+} are used sometimes in place of Na^+ and K^+ and HPO_4^{2-} can replace Cl^-. This diversity of current carriers makes it possible to develop fantastically complex parallel circuits since current carriers which are chemically different (note the singular nature of the electron in man's circuits) can be filtered by specific gated channels or pumps (Fig. 8.11). Now, there is an obvious limit to the variations on the theme of inorganic ions for making parallel processes. The variations are vastly increased by using organic ions and evolution could not fail to find these conductors. It is known that there is a huge variety of ionic organic messengers (and many non-ionic ones) which carry 'current'. The obvious ones use acetylcholine and adrenalin which carry single positive charges. Note that the ions are based on quarternary bases, which do not go through K^+ or Na^+ channels (Table 8.3) and cannot bind through H-bonds to proteins in a general way but flow as do the inorganic ions Na^+ and Cl^-. The number of possible parallel processes is virtually infinite. Organic reagents which act as blockers of organic current carriers are also an extremely important group of drugs.

Acetylcholine

$CH_3CO{\cdot}O—CH_2—CH_2—N^+(CH_3)_3$

Adrenalin

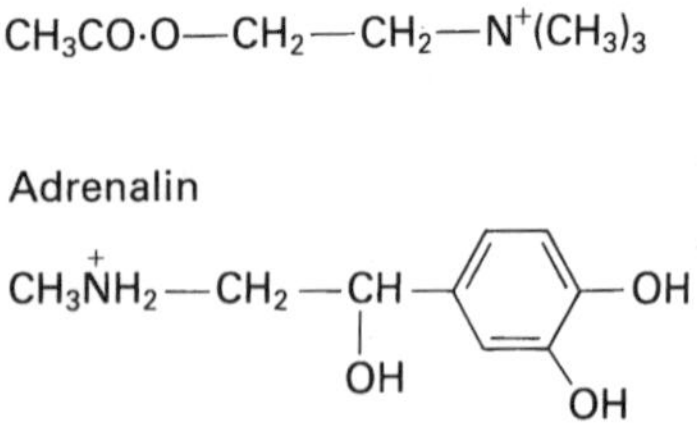

Glutamate

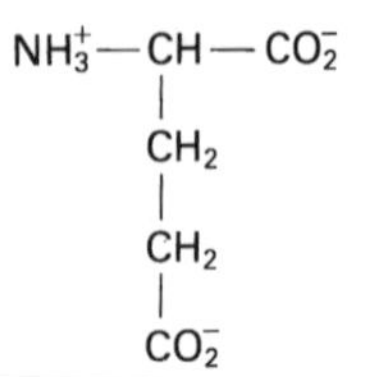

We shall see that, when the simple process of current-carrying has to be changed to current (message)-carrying plus mechanical effect at a target, then the organic ions through their size, shape, and chemical variation can and do bind at special sites and in this respect they are an advance on Na^+ and K^+ and are related more closely to Ca^{2+} electrolytic currents. In fact, we now stray from primary messages to secondary action and the inorganic chemistry switches from that of Na^+ and K^+ to that of Ca^{2+} and Mg^{2+} while the Cl^- chemistry switches to HPO_4^{2-} chemistry. We shall see that second organic messengers, for example after acetylcholine (a cation), utilize the organic chemistry of phosphate, e.g. cyclic AMP (cAMP), inside cells.

8.7.5 Currents of ions and morphogenic patterns: growth

The chemist understands very well the general idea of biological electrolytic currents and from the preceding sections he can see the way in which these currents are used in nerves to send messages. They are sodium/potassium currents. The physiologist is also aware of chloride and calcium currents in different tissues. There is, however, a wider issue than the general mechanism of ion-transfer in messages; there is the impact of the current carrier on the synthetic/degradative organic system. At some stage a current has to be transferred to a mechanical or chemical event if it is to serve a function. Before examining this point more closely we must look at where the circuits are. We know the topography of nerve circuits (see Section 8.7.1); we also know the microtopography of electronic currents in the membranes of chloroplasts and mitochondria (Chapter 3) but we do not know the external cell/cell or internal cell/vesicle currents at all well. In Section 3.5 local as well as general transmembrane ion flow was described. This means that there will be local as well as general lateral flow to rebuild the concentration. Now, if inlets for ions are placed along a surface at different distances from ion pumps then there will be a constant lateral current, dissipating energy, along a membrane surface both outside and inside the cell. Are such currents related to control over cell shape and do changes in the currents affect growth? There are now many observations which lead thought in this direction, see Fig. 7.8. Proton as well as metal ion currents have been observed. Is there a connection to morphogenic fields? The morphology of a living object is a dynamic state and it requires homeostatic control just as the chemical status of cellular metabolism must be controlled (see Chapter 20). Here it will not just be the inorganic ions which will be involved but a whole range of flowing organic messages.

8.8 Simple salts and the conditions of polyelectrolytes

Biological polymers are polyelectrolytes. Some, like DNA, have an extended almost linear series of charges while others, e.g. fats, give two-dimensional arrays, and yet others, proteins, give tightly curved surfaces often almost spherical. These polyelectrolyte surfaces are stabilized by the surrounding ionic atmosphere due mainly to Na^+, K^+, and Cl^-. The interactions are of two kinds, one quite general and the second quite selective. The subject is complex since the polyelectrolytes can interact with one another and with a huge array of soluble organic anions and cations. For this reason we give here only references to the problems in biology associated with surface potentials of polyelectrolytes and membranes. The nature of the polymer matrix of a cell is adapted to the controlled ionic conditions.

A particularly important consideration is the effect of cations on the anionic poly-electrolytes outside the cell. The anionic groups are of two kinds: the anions of strong acids such as $(RO)_2PO_2^-$ and $ROSO_3^-$ and the anions of weaker acids, RCO_2^-. All occur in the polysaccharides of connective tissue (see for example section 6.20). While none of these anions has a sufficient affinity for the major cation outside cells, Na^+, clusters of RCO_2^- could well bind it and become cross-linked. Clusters of the other anions remain ineffective and thus they are particularly important in keeping an open external matrix.

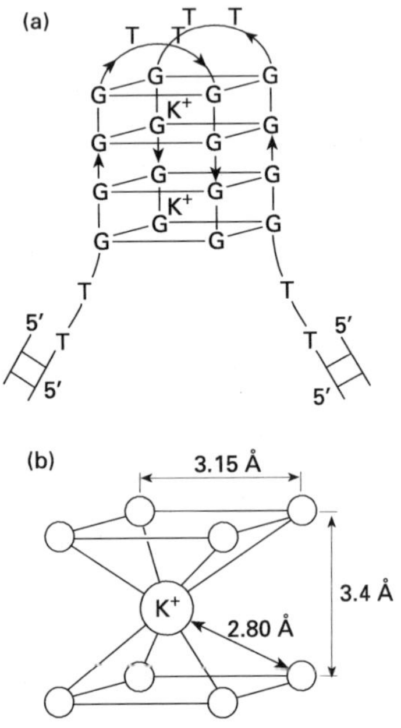

(a) Schematic DNA telomere structure.
(b) K^+ in 8-coordinate hole of telomere carbonyls.

8.8.1 Binding to DNA

While general electrostatic effects, ionic strength and condensation equilibria are important for DNA there are also some specific bindings—certainly of K^+. The telomere is a structure of DNA at its initial beginning and final end and is critical for complex replication. The composition, poly-G, of the tail leads in the presence of K^+ to a functional telomere; K^+-binding is shown in the aside. Now the binding of Na^+ is weaker and could have a different effect on some DNA structures, for example chromosomes are known to form 'puffs' in the presence of high Na^+, not K^+.

8.9 Enzymes requiring potassium

The genuine specific requirement of enzymes for Na^+ or K^+ is limited to a few special cases although the stability of all states (conformers) of monomers and polymers in cells is connected to the ionic strength of about 0.2 M. Here we do not need to refer further to the general balance of ionic conditions in a biological aqueous medium since it is clearly true that the reactions of most biological systems have some dependence on the ionic strength which is largely either NaCl (outside) or KCl (inside). A striking case of the latter is in the halobacteria where all the biological polymers have evolved to match high salt conditions. Putting such considerations aside there are a few enzymes which have a simple stoichiometric requirement for K^+, but the present authors know of none where Na^+ is required. The examples are listed in Table 8.5. The value of the dependence is not known.

Table 8.5 Some potassium-dependent enzymes

Enzyme	Ion dependence
Pyruvate kinase	$Tl^+ > K^+ > Rb^+ > Cs^+ > Na^+ > Li^+$
Dioldehydratase	$Tl^+ > NH_4^+ > K^+ > Rb^+ > Cs^+ > Na^+ > Li^+$
Na^+/K^+ ATPase ($K^\pm$ site)	$Tl^+ > K^+ > Rb^+ > Cs^+ > Na^+ > Li^+$

NB Tl^+ is an excellent probe (NMR and EM) for K^+ sites generally, K^+-sites are sometimes blocked by Ba^{2+}, an ion of the same size.

It is interesting to note that Tl^+ can substitute for K^+ and this substitution has been used to probe both ion channel and enzyme sites (see Table 8.3. and Table 11.13). The NH_4^+ ion does compete for some of these sites, but generally K^+ and NH_4^+ do not react at the same sites in cells. Much greater selectivity is found for the small organic cations, such as acetylcholine, which act as transmitters.

8.10 Ion genetics and networks

To maintain the concentration of K^+ in an *Escherichia coli* cell there is a constitutive genetic system. The level of K^+ in the cell controls the activity of a sensor kinase, KdpD, in the gene sequence. The kinase phosphorylates a response protein, KdpE. When K^+ is low KdpE acts as a transcription factor for the expression of proteins KdpA, B, and C which include a K^+ pump for moving K^+ into the cell. KdpE is switched off by phosphorylation due to K^+ bound KdpD.

A similar system of constitutive control is found in most if not all cells. The ABC transport system is known for a wide variety of other entities which must be transferred across membranes. The K^+ concentration relative to Na^+ affects certain enzymes and DNA/RNA as described above. Now the level of K^+ flow is linked to other flows not just through exchangers and pumps in single cells but through networks even acting between different cells in multicellular organisms (see Chapter 20).

8.10.1 K^+, Na^+, Cl^- circuits

In essence K^+, Na^+ and Cl^- circuits, Fig. 8.12, control in all cells of all organisms:

1. osmotic pressure;
2. electrolyte (charge) balance internally;
3. membrane potentials;
4. condensation of polyelectrolytes;
5. the required ionic strength for activity.

Hence these three ions are essential components of the free metallome of the cytoplasm and of those of vesicular compartments, but do not forget that the metallome, like the proteome, is different in these different compartments. In fact the free metallome of Na^+, K^+ and Cl^- is quite different in extracellular fluids from that in the cytoplasm of cells but it is also fixed.

The ion circuits run constantly since there is always a maintained membrane potential and some flow as the channels will open now and then, and the exchangers and pumps will also work depending on existing gradients and ATP availability. Total body levels of these ions in higher organisms are managed by the kidney and mineral-corticoids (sterols) (see Chapter 20).

In animals with a nerve and muscle system the circulation is highly active all the time, but note that nerves do not exist in early embryos, neither do muscles. Again, we draw attention to the special connection between the Na^+/K^+ ATP-ase and the development of the nerve system of animals. In fact during development of the fetus there is a special change in the amounts of this pump coincidental with the appearance of nerve tissue in the ectoderm.

The very complicated relationship between K^+ flow and that of other inorganic ions and organic molecules can be appreciated for example in the pancreas. Glucose levels affect ATP release externally from say cell A; ATP affects the K^+ channel of its receptor B cells in the pancreas and K^+ flow

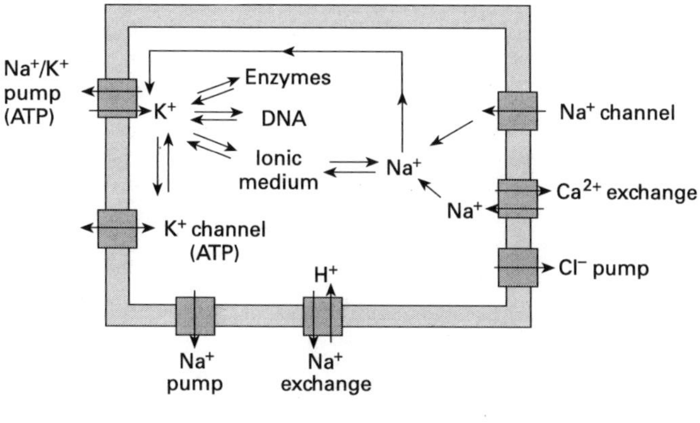

Fig. 8.12 Flow diagram for Na^+ and K^+ balances. Note that balances are maintained by connected requirements and by feed-back, which together fix the cell's ionic composition, part of the metallome.

adjusts the polarity of the cell B membrane; this change of polarization allows Ca^{2+} to enter cell B; Ca^{2+} then causes releases of insulin from cell B; insulin restores the level of glucose circulation through altered metabolism in many cells. This is a typical feed-back circuit.

8.10.2 The evolution of channels

The main known evolutionary change in ion proteins is the huge development and variety of channel proteins and pumps. Exchangers appear to change less and to be very primitive. Clearly, for nerve systems to work it is necessary to have gated K^+ and Na^+ channels, but many other cells have such channels and pumps. The first evidence of such K^+ channels is in *E. coli*, where there are three while in yeast there is a similar number. The worm *C. elegans* has some seventy different kinds of ion channel including some 20 K^+ voltage-gated channels and 6 chloride voltage gated channels. The number in the human genome exceeds one hundred. We shall refer to the evolution of pumps for different elements in Section 9.18 but the numbers are perplexingly large.

8.11 Summary

While nobody denies that life as we see it is linked to DNA, RNA, and proteins, it is rare indeed to see the statement that life is linked directly to sodium, potassium, and chloride chemistry. In some highly interactive manner and as the biopolymers evolved, not before or after them, the sodium/potassium balances had to be present since all the major polymers are anions and currents later became increasingly sophisticated. The dynamic homeostasis of these elements is now very precisely maintained and small deviations in levels of less than 1 per cent can be a symptom of a weakness in an organism. There is a deep-seated need to maintain these ion balances which interact with the states of the polymers via controls on osmotic pressure (affecting the mechanical stresses in the filaments of the cell), on electrical potentials (affecting the membranes), and on local states of polyelectrolyte polymers, both inside, including DNA, and outside cells. The temptation in the study of physiology is to look at fluctuations from the steady state (rest), thus addressing the problem of messages as the fundamental feature of Na^+/K^+ chemistry in life. It is not. The fundamental link is between two types of energized structure even at rest. One is represented by the polymerized states of organic molecules and a second by the resting concentration gradients of simple ions. The most direct way in which to see this link is through the chemical forms of energy—the bound pyrophosphate of NTP, e.g. ATP. Through NTP the whole of polymer biosynthesis is maintained and through NTP all gradients are generated. However, conversely, through ion gradients ATP can be made. This circle of interchange of energy from gradients (electrical or concentration), to weak covalent compounds (NTP), to biopolymers is the essence of life. We shall show that not only Na^+, K^+, and H^+ but also the other essential bulk and trace elements are involved. None of these conditions are static so that at all times currents flow. It is an essential feature of living systems as we know them that, in fact, current always flows. This is a part of homeostasis.

There is no sense in denying the fundamental need for metabolism of C, H, O and N compounds in organisms for synthesis but there is equally no sense in forgetting that sustaining any such activity depends on concentration and voltage controls so that an enclosed volume is stable physically. Evolution is built on organic synthesis but also on physical chemistry so that enforced introduction of control of these three inorganic elements, Na, K and Cl, in organisms led step by step to signals concerning the charges and concentrations of ions around spherical single cells, to signals along cells and from cell to cell in shaped organizations, to nerves and muscles, and eventually to electrolytic computers—the brain. For this reason, amongst many others, we wish to force to the notice of scientists that organisms are *systems* based on genomes, proteomes, and metallomes in a feed-back interactive organization (see Chapter 20). The *free* metallome of the cytoplasm in all organisms is 10^{-1}M K^+, 10^{-2}M Na^+ and 10^{-2}M Cl^-, but the total metallome of different organisms is very variable due to vesicle and extracellular fluid contents.

Further reading

General

Escherichia coli and Salmonella (2nd edn), ed. F.C. Neidhardt. American Society for Microbiology Press, Washington DC (1996). See sections on K^+, Na^+ and Cl^-.

Guyton, A.G. (1986). *Textbook of medical physiology* (7th edn). W.B. Saunders, Philadelphia.

Lamb, J.F., Ingram, C.G., Johnson, I.A., and Pitman, R.M. (1988). *Essentials of physiology* (2nd edn). Blackwell Scientific, Oxford.

Miller, C. (2000). Ion channels, *Current Opinion Chem. Biol.*, **4**, 148–51.

Miyazawa, A., Fujiyoshi, Y., Stowell, M., and Unwin N. (1999). Nicotinic acetylcholine receptor at 4.6Å resolution. *Journal of Molecular Biology*, **288**, 765–86.

Murchie, A.I.H., Aboul-Ela, F., Lauglan, G., Luisi, B., and Lilley, D.M.J. (1995). Structure of Guanine Tetraplexes Nucleic Acids. *Journal of Molecular Biology*, **9**, 143–64.

Russell, J.M. (2000). Sodium-potassium-chloride co-transport. *Physiology Reviews*, **80**, 211–76.

Sundquist, W.I. and Klug, A. (1989). Telomeric DNA structures. *Nature*, **342**, 825–9.

Williams, R.J.P. (1970). The biochemistry of sodium, potassium, magnesium and calcium. *Quarterly Reviews of the Chemical Society, London*, **24**, 331–60.

K+ channels

Choe, S., Kreusch, A., and Pfaffinger, P.J. (1999). Towards the three-dimensional structure of voltage-gated potassium channels. *Trends in Biochemical Science*, **24**, 345–9.

Doyle, D.A., Cabral, J.M., Pfuetzner, R.A., Kuo, A., Gulbis, J.M., Cohen, S.L., Chait, B.T., and MacKinnon, R. (1999). The structure of the potassium channel. Molecular basis of K^+ conduction and selectivity. *Science*, **280**, 69–77.

Na+ channels

Catterall, W.A. (2000). The structure and function of voltage-gated sodium channels. *Neuron*, **26**, 13–25.

Polyamines

Igarashi, K. and Kashivagi, K. (2000). Polyamines. *Biochemical and Biophysical Research Communications*, **271**, 559–64.

CHAPTER 9

The biological chemistry of magnesium: phosphate metabolism

9.1 Introduction

Magnesium in biological chemistry is a Cinderella element: we know its hidden power and personality only indirectly since we are unable to label and follow it in a sensitive manner. Analysis *in situ* by X-ray fluorescence using the electron microscope is difficult since the element is light and present in the millimolar range; a very useful radioactive isotope is not available; there are few good spectroscopic tools (contrast calcium); even atomic absorption spectral methods need care because of the interference by phosphate with which magnesium is usually associated in cells. In these circumstances a description of the biological role of magnesium is hazardous. Three features stand out, however. The first is the general involvement with phosphate compounds and phosphate metabolism, the second is the binding in chlorophyll pigments, the third is associated with enzymes concerned with some essential steps in carbon assimilation and metabolism. These three roles are quite different since in the first and third magnesium is a freely diffusible cation in aqueous solution involved not only in catalysis but in a multitude of controls, while in the second magnesium is an immobilized element in a light capture device in a membrane. We shall describe the first role before going on to chlorophyll.

Note first that the charge density of Mg^{2+}, radius 0.6 Å, is higher than that on any other freely available charged centre inside cells. As a consequence, it dominates binding to negative centres of high opposite (negative) charge density, e.g. pyrophosphate. This equilibrium chemistry has been explored in detail and we give in this chapter only an outline of observations, especially those concerning competition with positive polyamine centres. In order to bring this magnesium/phosphate chemistry into a biological context we must first observe the concentration of magnesium in various compartments.

9.2 The spatial distribution of magnesium

We begin with a description of magnesium ion levels outside cells. Magnesium is plentiful in the sea (5×10^{-2} M) but is at a relatively low level in fresh water ($<10^{-4}$ M). It is seen that magnesium, like sodium, must be rejected by organisms in the sea but pumped into organisms in fresh water to maintain the free magnesium concentration in the body fluids and in nearly all cells at around 10^{-3} M. However, the presence of bound magnesium outside many cells may well be essential also to maintain the integrity of their outer surfaces (see Section 9.6). Again magnesium in the outer fluids greatly affects the laying down of mineral phases. It is usual for calcite to be the major mineral of freshwater life but for aragonite to be the dominant mineral of life in sea water and we know, of course, that *in vitro* magnesium ions inhibit the growth of calcite much more strongly than the growth of aragonite. Magnesium is also a prominent minor element in bone and in calcitic shells.

Inside cells we have no clear knowledge of the distribution of magnesium other than to say that it is universally approximately 10^{-3} M. For a multicellular animal the bloodstream or extracellular fluids are also kept at about the same level (10^{-3}M) so that there is virtually no gradient of magnesium ions across the cytoplasmic membrane in a range of multicellular organisms. The function of magnesium at these membranes is then very different from that of calcium, sodium, or potassium ions. Magnesium does not act there as a current carrier or a trigger. An exception is that magnesium has an energy-dependent distribution across the thylakoid membrane where this metal ion may be sequestered in the thylakoid by phosphate and carboxylate groups at high pH. It is then quite likely that the acidification of the inside of the thylakoids on exposure to light leads to the simple displacement of bound magnesium by protons. As a consequence there is movement of the cation into the cytoplasm of the chloroplast and magnesium acts as a slow trigger or control over cytoplasmic photosynthetic reactions of plant cells. The cation may well stimulate carbon dioxide formation through activation of the enzyme ribulose bis-phosphate carboxylase, Rubisco. Note, however, that the free magnesium ion concentration always remains in the range 0.5–2.0 × 10^{-3} M.

The question arises next as to the magnesium levels in the Golgi and other cysternal internal vesicle compartments of higher cells. Although it is not certain, the authors believe that these compartments are relatively low in magnesium but high in calcium and manganese due to the positioning of cation pumps (Fig. 3.16). We need to look further at the comparative functions in compartments for the three divalent cations, magnesium, manganese and calcium, after we explore the chemistry of magnesium in more detail.

9.3 Magnesium chemistry

The magnesium ion is very different from the calcium ion. The differences are in structure, thermodynamic affinity for certain classes of ligand, and reaction rates. All the differences spring from the difference in radius (Mg^{2+}, 0.6 Å; Ca^{2+}, 0.95 Å) when measured in an octahedral environment of oxygen donor atoms. The first obvious feature is in the structure of compounds and complexes since it is the case that within the closed shell cations of the periodic table (groups 1,2,3), there is a sharp structural separation between the ions of radius less than 0.75 Å and those of larger radius (Fig. 9.1). This distinction puts a diagonal line in the periodic table, leaving Li^+, Mg^{2+}, Al^{3+}, and Sc^{3+} on

Table 9.1 Some typical magnesium salt structures

Salt	Mg(II) co-ordination no.	Mg-O distances (nm)	
		Minimum	Maximum
Mg hexa-antipyrine · $(ClO_4)_2$	6	0.206(Ca, 0.230)	0.212
$MgS_2O_3 \cdot 6\ H_2O$	6	0.205	0.212
$MgSO_4 \cdot H_2O$	6	0.204	0.209
$Mg(HPO_4) \cdot 6\ H_2O$	6	0.200	0.212
$Mg_2P_2O_7$	6	0.200	0.211
$Mg(CH_3CO_2)_2 \cdot 4\ H_2O$	6	0.200	0.210

See Tam and Williams (1985) in the 'Further reading' for sources of the data.

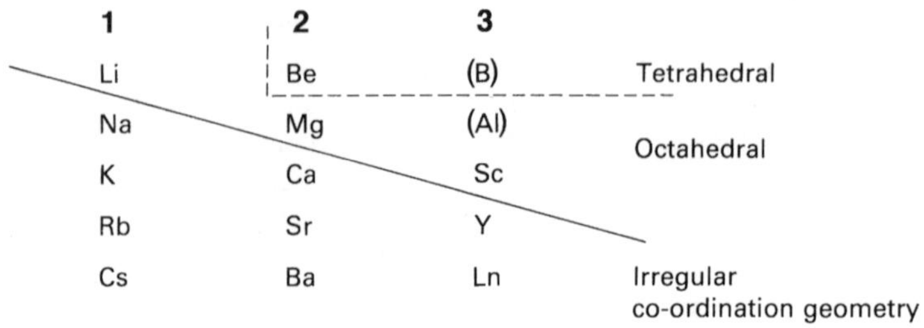

Fig. 9.1 A simple division of the periodic table to indicate some structural divisions associated with the diagonal relationship.

one side and Na^+, Ca^{2+}, and Y^{3+} (Ln^{3+}) on the other. The smaller ions are strictly octahedral in the symmetry of their complexes with, in the particular case of Mg^{2+}, very rare different geometries. The bond distance Mg-O is very close to 2.05 ± 0.05 Å in all cases (Table 9.1). The contrast is very strong with the irregular coordination geometry, bond angle, bond distance, and co-ordination number (7–10) for the larger cations including Ca^{2+} (see Table 10.3). The explanation is that in the case of the smaller cations the central field of the cation dominates the coordination sphere while in the cases of larger ions the second and perhaps third coordination spheres have a more important role to play so that a compromise, irregular structure results around, say, Ca^{2+}. (Note that Mn^{2+} is not too dissimilar in size from Ca^{2+}.)

The easy adjustment of the larger ions to steric crowding affects the relative thermodynamics of binding (see Chapter 2). In Table 9.2 we show that, as the number of coordinating groups increases and especially where these groups are weaker donors than water but are linked together through covalent bonds, i.e. chelating ligands, then the stability constants for calcium exceed those for magnesium. Similar features are seen in the chemistry of the lanthanide cations when compared with that of Al^{3+}. Even trends within the lanthanide series follow this expectation. The value of this difference of the biological chemistry of magnesium relative to that of calcium is apparent everywhere in biology. The difference appears in large opposed heats of formation of the complexes of ions such as magnesium which are only partially offset by entropy gains. However, Mg^{2+} is not a powerful enough acid to bind effectively to such hard centres as phenolates (pK_a = 11.0) at pH = 7. Thus, structural and pK_a factors force the strong association of Mg^{2+} with organic polyphosphates, highly charged anions, e.g. ATP^{4-}. At the same time the stability of Mg^{2+} complexes is insufficient for it to bind to the common small anions of metabolic pathways (Table 9.3). In proteins Mg^{2+} is often bound to one or more carboxylate sidechains. Only extremely rarely do we find imidazole (histidine) coordination, but a further feature of Mg^{2+} ion binding to proteins is that it sometimes binds to simple 'alcohol', ROH, and carbonyl, C=0, groups, but not to ether R_2O centres. We shall refer to this point later as the chemistry of enzymes indicates that Mg^{2+} is virtually the only cation to bind to these centres. This unusual binding site for a metal ion is seen for example in integrins (see Section 9.12). The binding is weak, log K = 3.5 to 4.0, and as in complexes with DNA and RNA much of the Mg^{2+} coordination sphere is made up from water molecules. We stress immediately that all known DNA polymerases have an active site requiring two Mg^{2+} ions in part bound to the protein and in part to the bases. Thus Mg^{2+} is an essential ion for all life-forms. In some enzyme reactions catalysed by magnesium the cation is

Table 9.2 Some stability constants, log *K*, of magnesium, calcium, and manganese complexes at zero ionic strength and T = 25°C

	Log *K*		
Ligand	Mg^{2+}	Ca^{2+}	Mn^{2+}
Acetate	0.8	0.8	~1.0
Glycine*	~3.0	1.5	3.5
Iminodiacetate⁺	3.7	3.4	
EDTA*	9.1	11.0	14.1
EGTA*	5.5	11.2	12.5
4,5 Dihydroxybenzene-1,3 disulphonate⁺	5.9	6.0	8.6

*Smaller donor centres, N or O, assist Mg^{2+} binding relative to Ca^{2+}, e.g. unsaturated amines and phenols. Data for pH = 7 are > 10^3 fold lower.

EGTA

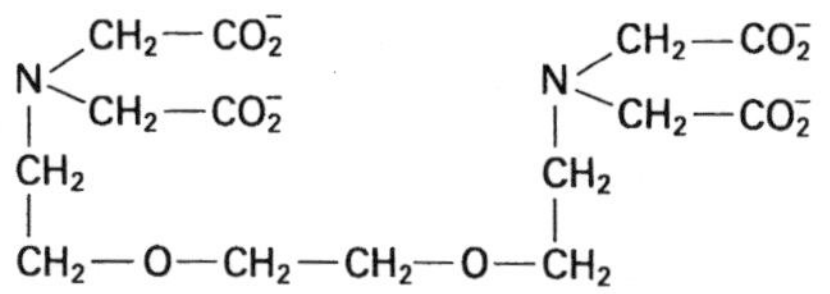

Table 9.3 Free substrate binding to Mg^{2+} and Ca^{2+} at pH = 7

No binding	**Binding**
(Simple carboxylates)$^-$	Oxalate^{2-} (?)
(Kreb's-cycle molecules)$^{2-}$	ADP^{3-} (other nucleotides)
(Amino acids)$^{+-}$	ATP^{4-} (other nucleotides)
(Mononucleotides)$^{2-}$	IP_3^{4-}, IP_4^{5-}, etc.
(Sugar phosphates)$^{2-}$	(see polynucleotides)
(Sugar sulphates)$^-$	Weak

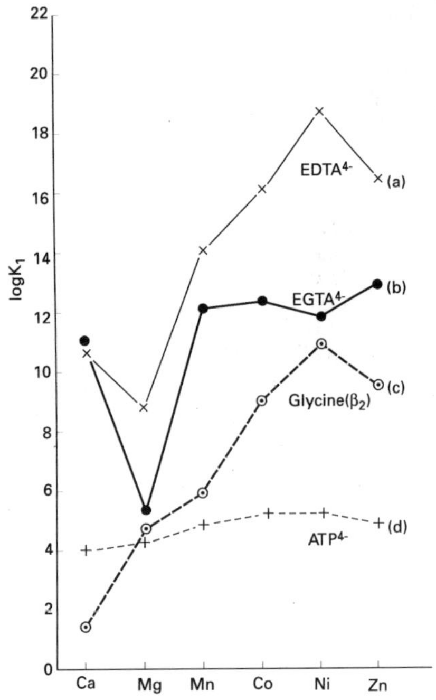

A plot of stability constants for the complexes of several elements indicating the low selectivity and stability of ATP^{4-} complexes which allows Mg^{2+} to bind in a competitive situation (see text).

largely bound to the substrate, e.g. to ATP in kinases, and only very weakly to the protein.

The reason for the combination of Mg^{2+} to selected weak donor sites despite the fact it binds more weakly than many other cations (see Aside) is as explained in Section 2.4, that binding depends on effective binding constants and not on the absolute constants as in the Aside. Thus the strength of binding and the number of binding molecules available in a cell at pH = 7 together with the controlled access of elements to cells removes all cations from cell cytoplasm to below 10^{-6}M except for Mg^{2+}. At these concentrations only Mg^{2+} can bind to ATP, for example (see Aside and Fig. 2.8).

There is a noticeable, though weak, tendency for the small cations, including Mg^{2+}, to be able to bind to nitrogen donors and here there is again a marked contrast with Ca^{2+} (this is a matter of the polarizing power of the cation) so that the possibility arises of designing a small-cavity ligand from nitrogen donors carrying negative charge and excluding calcium on two grounds: ligand donor type and cavity size. Such a ligand is the chlorin ring. We return to the difficulties of the incorporation of Mg^{2+} into this ring in Section 9.14.2, but the competition here is no longer from Ca^{2+} but from transition metal cations such as zinc. Protein surface cavities can be formed in principle to fit Mg^{2+} from O-donor ligands, but this is unusual.

The kinetic difference between calcium and magnesium, and, in fact, for all metals above as opposed to those below the line of Fig. 9.1, are due to packing more than to the total bond energy. As stated in Chapter 4, the exchange of water is approximately 10^3 times slower from around Mg^{2+} than from around Ca^{2+} (Table 9.4); the magnesium octahedron is not easily broken. The result, again well-known, is that calcium provides the major, fast, trigger ion of biology using quite strong binding in very rapid exchange, e.g. with calmodulin. It is easy to show that magnesium cannot act as such a fast trigger ion due to the slow on-rate of its reactions. However, the very use of calcium in this way means that it had first to be excluded from many parts of intracellular biological space in order to build a triggering gradient. Presumably, this exclusion allowed Mg^{2+} to take over the major role of anion charge neutralization in the cytoplasm, e.g. with ATP^{4-}. The binding constant here does not inhibit reaction rates. In general, magnesium is not held with a binding constant greater than 10^4 in biological systems, but there are some notable exceptions where function is hardly likely to be related to exchange. We shall see that where magnesium is used in control reactions it is a relatively slow control, e.g. in chloroplasts.

The manganese ion, Mn^{2+}, lies between magnesium and calcium in size, and can therefore be used as a spectroscopic probe for the properties of both of them, but it has a higher affinity for N-donors than either of them (Table 9.2). We return to the uses of Mn^{2+} chemistry as a probe in the study of

Table 9.4 Rates of water exchange

Fast (10^{-10} to 10^{-7} s)	**Medium (10^{-7} to 10^{-5} s)**	**Slow (>10^{-5} s)**
Na^+, K^+, Rb^+, Li^+	Mg^{2+}, Co^{2+}, Ni^{2+}	Al^{3+}, Fe^{3+}
Ca^{2+}, Sr^{2+}, Mn^{2+}		
Ln^{3+}, Cu^{2+}, Zn^{2+}, Fe^{2+}		

See Section 4.5.2.

Table 9.5 Some magnesium, calcium, manganese, and zinc enzymes* and their locations

Magnesium (see Chapter 9)	**Zinc**† (See Chapter 11)	**Calcium** (See Chapter 10)	**Manganese** (See Chapter 14)
Enzymes			
ATPases	Phospholipase C	Phospholipase A_2	Glycosylases
Kinases	Peptidases	Thermolysin	Photosystem II
Enolase	Alkaline phosphatase (plus Mg^{2+})	Saccharases	Arginase
Glutamine synthetase	Carbonic anhydrase	Thrombin	Mn-catalase
Compartments			
Cytoplasm	Extracellular	Extracellular	Golgi vesicle

*The metal ion is at the active site in all the examples cited. In some other enzymes the metal is probably at the active site but there is no definite evidence. Structural use of a metal away from the active site is not included in this table.
†Zinc is not so restricted to a particular compartment as other metal ions (see Chapter 11), but most of its hydrolytic enzymes operate outside cells.

Mg^{2+} systems later, but note here that Mn^{2+} appears to be largely excluded from the cytoplasm of higher cells and to be functional in organelles and in Golgi vesicles, so that competition between the ions within cells is slight.

Using the differences in size, biological systems remove both free manganese and calcium ions to levels below 10^{-7} M (see Chapters 10 and 14 respectively) while keeping the magnesium ion at about 10^{-3} M in the cytoplasm. The outward pumping of the first two ions concentrates them in reticula, e.g. sarcoplasmic and endoplasmic, and in Golgi bodies, and also outside cells. Therefore, of the divalent cations magnesium dominates the cytoplasm but is of less consequence inside reticula. We must look for the differential use of the three cations with different enzyme systems placed in different compartments (Table 9.5).

9.4 Magnesium pumping in cells

We need to consider how the concentration of Mg^{2+} is maintained in cells or indeed in body fluids. To do this we have used a general way of inspecting the functions of metal ions by writing a formal equation for all metal ion biological functions which come to equilibrium.

$$\text{Function} \propto K \cdot [\text{M}] \cdot \{\text{structure factors}\}.$$

We have described in Chapter 2 and in Section 9.3 the binding constants, K, and the structure factors, the latter containing kinetic effects, but we have not dealt in detail with the energized maintainance of the concentration, [M]. In the case of magnesium it is a general rule that, in a steady state of a metabolizing cell, the free [Mg^{2+}] is close to 10^{-3} M in the cytoplasm. This statement does not refer to the condition outside the cell. In fact, it appears that, no matter what the external concentration, from about 10^{-6} M to $10^{-1.3}$ M (the sea), the internal [Mg^{2+}] remains in the millimolar range in all cells. However, there are undoubtedly fluctuations when single (bacterial) cells are subjected to gross environmental changes.

Now, while higher organized systems maintain their cells in external circulating fluids of 10^{-3} M magnesium, unicellular systems cannot control the environment. As we have seen fresh water can be low in magnesium,

$<10^{-4}$ M, or even 10^{-6} M, yet unicellular organisms remain able to multiply. Below this level they enter stationary phase no matter what the level of other essential nutrients. (In fact, it appears that the cells become very run down in the virtual absence of Mg^{2+}.) Studies on *E. coli* established the nature of the required magnesium pumps. Their formation requires synthesis of special proteins, as shown by genetic studies, which are inserted in the cytoplasmic membrane. What we know about these proteins suggests that they resemble the calcium pumps but why they act selectively on magnesium is unknown. We are forced to imagine that they have special acceptor 'holes' which fit small divalent cations. Interestingly, nickel, which has a similar size and stereo-chemistry, may well enter cells via the magnesium pump. When considering pumping we have to keep in mind too that this is a kinetic energized process, so that information about binding constants alone could be misleading. There is the probability that weak binding, say $\log K = 4$, connected to an *energized* pump can lead to accumulation even from very dilute solution, 10^{-6} M, via two steps

$$\text{Mg} + \text{L}_1 \rightleftharpoons \text{MgL}_1 \xrightarrow{\text{E}} \text{Mg(inside)}$$

where L_1 is a ligand, E is a pump, probably an ATP-hydrolysing enzyme, and the last step is considered to be irreversible. Here there need not to be more than 1 per cent formation of MgL_1, with $K = 10^4\ M^{-1}$, so that pumping could be effective down to external $[Mg^{2+}]$ of 10^{-6} M. Again the selectivity of the pump may depend not on the selectivity of binding for magnesium but on the recognition of the size and shape of MgL_1 in the pump, E. This type of selection is related to the difference between an enzyme substrate and an enzyme inhibitor. Both may bind equally, but only the first flows along a reaction coordinate because of its shape and size. The selection of magnesium rather than calcium may depend on a similar principle.

Living systems in the sea must reject Mg^{2+} and their outward pumps must also use ATP, but at present they are not known; however, there is some evidence for outward pumping of magnesium. The chromaffin granule, a vesicle of the adrenal gland, is surrounded by an unusual membrane which maintains in the interior of the granule a pH of 5.8, a very high concentration of adrenalin >0.1 M, and ATP at about 0.025 M. (Note that this is not Mg · ATP which is everywhere present in the cytoplasm.) The level of cations present is extraordinary in that there is some calcium (10 mM) but virtually no magnesium (Table 9.6). To maintain such strange gradients the chromaffin granule must have a series of exclusion mechanisms since it is known that the contents of the granule are free in solution. By some device, magnesium is excluded from being a partner of ATP, as it is in the cytoplasm, even to a much greater degree than calcium is excluded. Once again, this is extraordinary considering that the ratio of magnesium to calcium in the cytoplasm is 10^4. It would appear that this membrane is not too unlike the plasma membrane (and the Golgi membrane?) in that there are outwardly directed pumps, i.e. from the cytoplasm, for H^+, Ca^{2+} and Mn^{2+}, while there are, if anything, inwardly directed pumps for Mg^{2+}, i.e. out of the vesicle but into the cytoplasm.

Table 9.6 The contents of chromaffin granules (mM)

Na	K	Mg	Ca	P	S	Cl
43	28	low	10	624	82	33

Other granules are very different. For example, in the platelet cells there are storage vesicles of pyrophosphate which has both calcium and magnesium with it as well as 5-hydroxytryptamine.

We return to the discussion of control over Mg^{2+} concentrations in cells in Section 9.18.

9.5 Very strong binding of magnesium

A very few cases of very strong binding of magnesium ions in biology are known besides the case of chlorophyll (Section 9.14). The major cases, which are well established, are in the ATP-synthetases of thylakoids and mitochondria and in the ATPases of muscle for which binding constants well in excess of 10^8 M^{-1} have been described. The function of magnesium in these enzymic systems cannot be related to turnover or exchange. Now the rigidity of the coordination sphere of this ion points to the possibility that it is being used to deliberately prevent motion in certain regions of organized systems. Thus, here Mg^{2+} has almost the opposite function to that of calcium and possibly zinc which, even where they form cross-links, will allow swivelling motions due to the easy distortion of their coordination spheres. However, although we do not know the nature of these sites, we see clearly that biological systems can build centres which have very high Mg^{2+} binding constants. Even within the design of organic ligands for Mg^{2+}, e.g. EDTA, very few show such binding strengths at pH = 7 (Table 9.2). It is probable that the very tight unusual binding of magnesium is a property of the protein fold and its energy.

There are two other compartments where tight binding of Mg^{2+} may be present—the ribosomes and the nucleus. Little is known about the magnitude of the binding constants but there are sites of multiple junctions in both RNA and DNA where Mg^{2+} appears to be strongly retained.

9.6 Magnesium in walls and membranes

While examining the divisions of biological space we need to observe that much of it is occupied by protein, polysaccharide, and lipid, especially in and surrounding membranes, and much magnesium is found in these membrane and wall zones. Magnesium is used here to cross-link the carboxylated and phosphorylated anionic polymers, and shares this role with calcium ions. The binding of the two is somewhat different, of course, and so they act cooperatively in stabilizing external structures of cells. This is a very general phenomenon and is such that strong chelating agents, such as $EDTA^{4-}$, are needed to disrupt cell wall and cell–cell structures. A major binding agent in the cell walls of *E. coli* is teichoic acid (Fig. 9.2).

9.7 Magnesium enzymes: magnesium and phosphates

While magnesium is not generally associated with carboxylate-containing substrates, binding to phosphate substrates is clearly an imperative function in all cells. However, the affinity of magnesium for phosphate or phosphate monoesters is not determinant at the concentration in cytoplasm, since it does not form complexes with simple phosphate anions, HPO_4^{2-} or RPO_4^{2-}, nor does it form insoluble salts with them (Section 9.3). In order to bind to Mg^{2+} further negative charges or dipoles are necessary in the molecule which carries the phosphate. The most striking examples are pyrophosphate and other condensed polyphosphates, e.g. ATP (Table 9.7), for which the binding constants are close to 10^4. In biology these phosphates are largely Mg^{2+} complexes, and most of the energy transfer due to NTPs and the phosphate

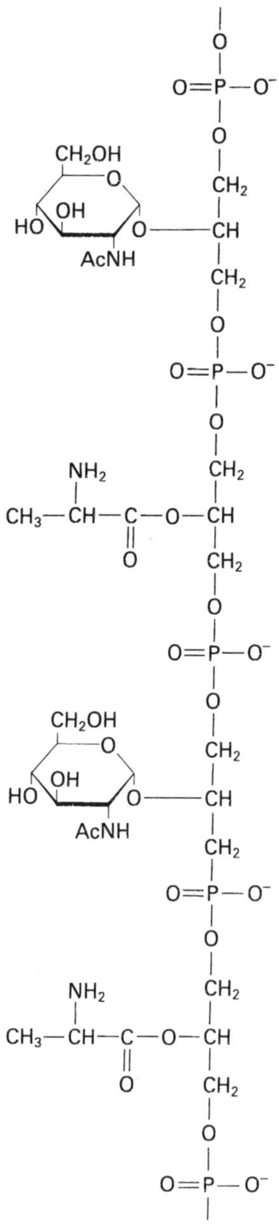

Fig. 9.2 Segment of a teichoic acid which has D-alanine and N-acetyl glucosamine residues. Note the glycerol phosphate backbone.

Table 9.7 Magnesium mononucleotide complexes

Mononucleotide complex	Example
Mg · adenosine triphosphate	Kinases, ATP-pumps, contraction
Mg · guanosine triphosphate	Protein synthesis, membrane responses, tubulin polymerization, elongation factors (ETu)
Mg · uridine diphosphate	Saccharide reactions and glycosylations
Mg · cytidine triphosphate	Choline synthesis (lecithins)

NB. NTP with Mg^{2+} ties together many degradative and synthetic routes with mechanical devices.

transfers of kinases depend on Mg^{2+}. Other important phosphates that could form magnesium complexes of high stability constant are the multisubstituted phosphates of sugars, for example, inositol (tri-), tetra-, penta-, and hexaphosphates. The binding to these phosphates rather than to carboxylate ligands, e.g. molecules of the citric acid cycle, arises through the high local charge density (2–) on more than one terminal group. As stated, the absence of competition for NTPs by other cations is due to the tight binding and low concentrations of them all inside cells (see Sections 2.17 and 9.3).

Now, molecules such as ATP are bound into proteins to a large degree through their organic adenine centres, often leaving the phosphate chain carrying its magnesium freely exposed or weakly bound. The activation of the triphosphates for transfer of the terminal phosphate is then brought about by using an anion centre on the protein, as shown in Figs 9.3 and 9.4. The figures indicate that magnesium bound between the β- and γ-phosphates protects the ATP, bound or free in solution, from hydrolysis, while on the enzyme surface the protection is removed from the γ-phosphate by movement of magnesium to the β- and α-phosphates when it is also bound by a protein carboxylate or other groups (Fig. 9.4). Here magnesium is a locally mobile cation between phosphates of ATP or on the protein surface even though absolute rates of movement can be quite slow. Catalysis also often involves minor migration of substrate groups relative to the attacking centres of the enzymes but also a major conformational groove-closing reaction (Chapter 6), which probably involves all the helices in Fig. 9.3. The full value of magnesium in nucleotide phosphate transfer reactions of the cytoplasm can only be appreciated by reference to detailed compilations of the mechanisms of the relevant enzymes (see 'Further reading'). These show that magnesium is a totally dominant cation in these biomechanical, NTP-dependent enzymic reactions and it is therefore an absolutely essential element. The enzymes involved extend from traditional kinases to the G-proteins involved in many receptors, the elongation factors, ATP-synthases (β-subunits), myosin heavy chain (muscle) ATPases, and the ras proteins (tyrosine kinases). The feeling that in this group of proteins there is a mobility device with which Mg^{2+} is involved and that this device is of extreme importance for the phosphate homeostasis is increased by the observation that single mutations in the ras proteins are associated with tumourogenesis. The mechanism of action of many of these proteins (enzymes, kinases, pumps, contractile devices, etc.) is likely to be fundamentally the same in that there is a site for NTP associated with a β-sheet domain and that this domain is linked via a helical hinge section to other domains (Figs 9.3 and 9.4, and see Fig. 6.9). The closing of the gap between domains, or its opening, can be coupled to many devices as in ATP pumps (Chapter 8). In this sense we have repeatedly used knowledge of the properties of

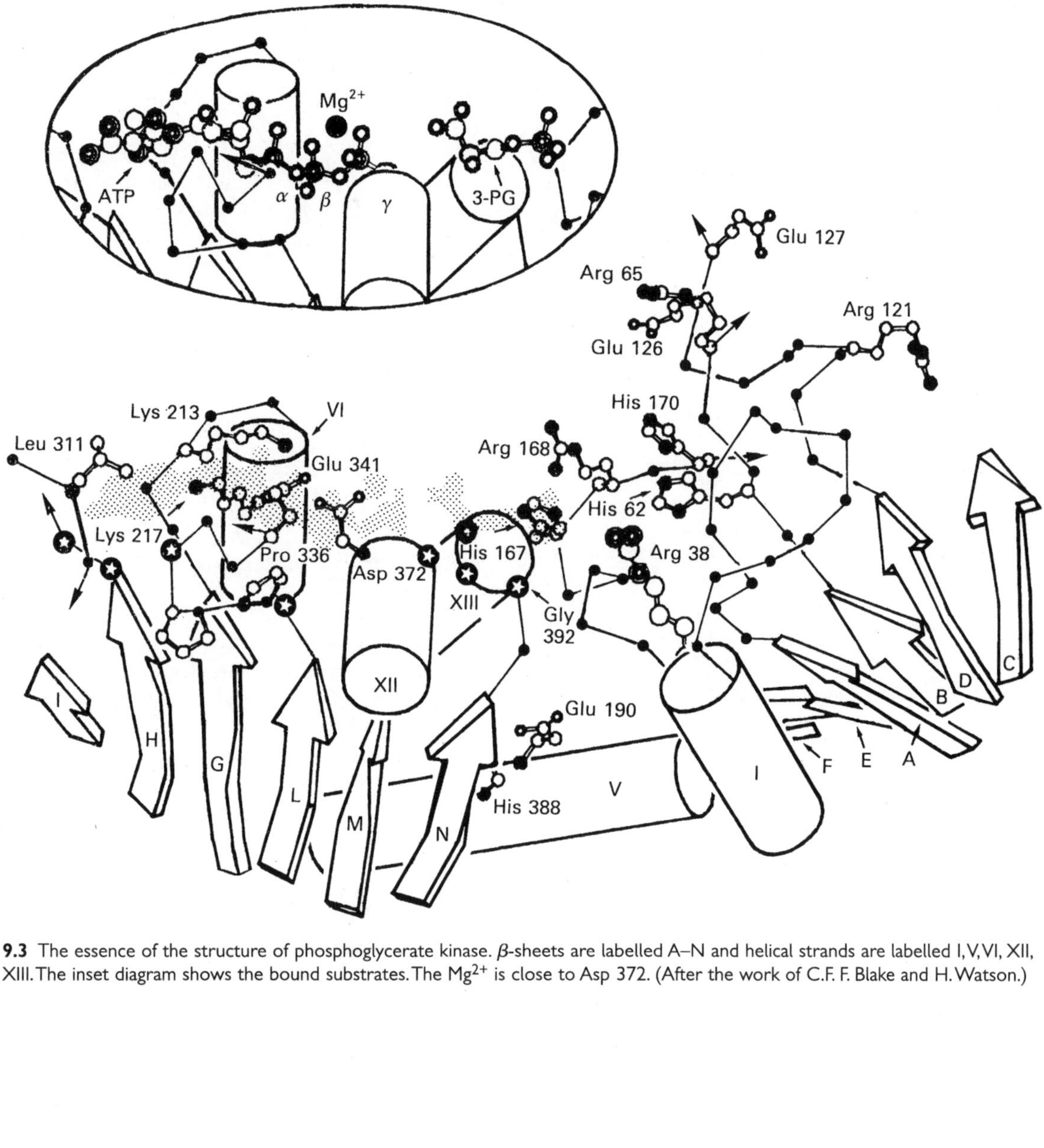

Fig. 9.3 The essence of the structure of phosphoglycerate kinase. β-sheets are labelled A–N and helical strands are labelled I, V, VI, XII, and XIII. The inset diagram shows the bound substrates. The Mg^{2+} is close to Asp 372. (After the work of C.F. F. Blake and H. Watson.)

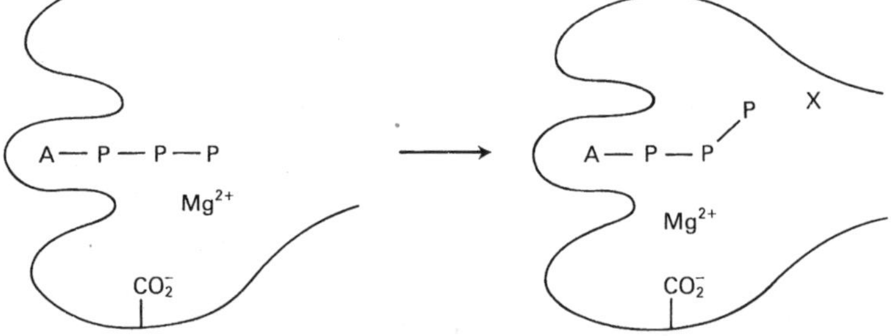

Fig. 9.4 A schematic representation of the movement of Mg^{2+} in a kinase and the groove-closing reaction that is required to bring substrate X near to the transferable phosphate at ATP.

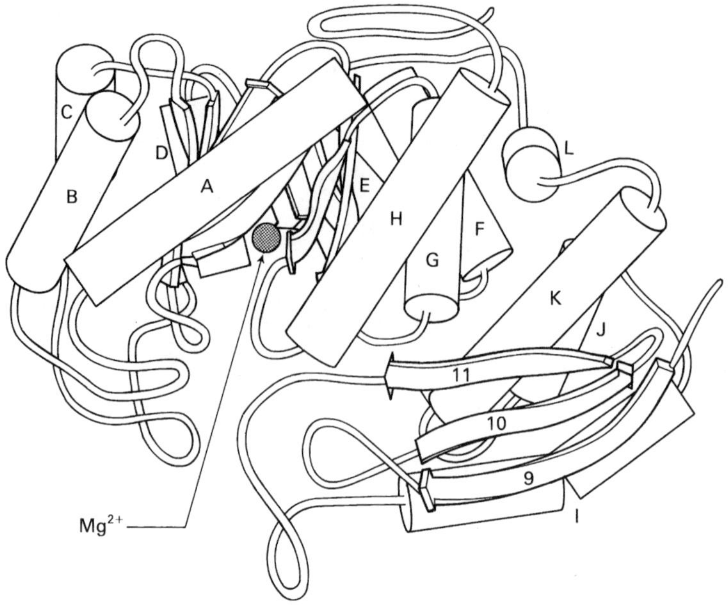

Fig. 9.5 The structure of yeast enolase after the work of J. M. Brewer. The Mg^{2+} ion is shown as a shaded cycle. The enzyme is mainly a β-barrel with linking helices.

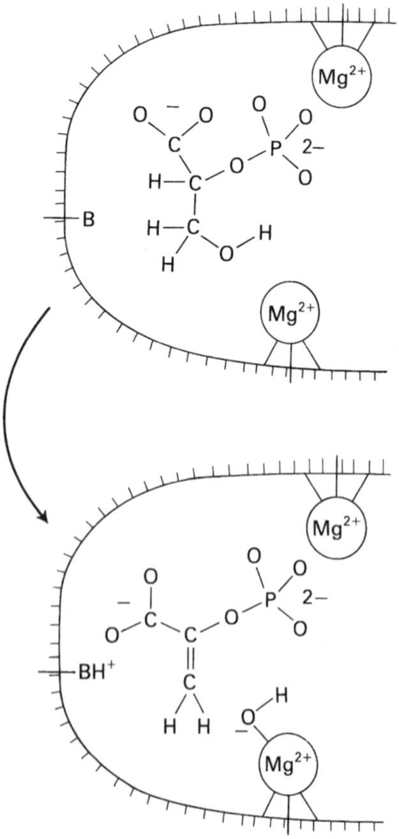

Fig. 9.6 The proposed mechanism of reaction of enolase. (After A. Mildvan.)

phosphoglycerate kinases, see Fig. 9.3, in order to appreciate the function of internal mobility in all of these magnesium-dependent biomechanical devices. The role of magnesium in this activity must never be overlooked.

The reactions of phosphate condensates are not confined to the organelles and the cytoplasm, that is, to the transfer of phosphate from the nucleotide triphosphates. There is, in the Golgi vesicles, the apparatus for sugar transfer using the reactions of UDP^{3-}. The binding constant of magnesium to the condensed pyrophosphate unit $[R\text{-}O\text{-}P\text{-}O\text{-}P]^{3-}$ is always about 10^4 so that, in the Golgi, where magnesium ion concentration may well be low, Mg^{2+} can still be bound. Manganese can substitute for magnesium and it is very difficult to know which enzymes in the Golgi compartments are magnesium-activated and which are manganese-activated.

There are many enzymes which do not involve the triphosphates and which are magnesium-requiring (Table 9.8). In some of these the substrate is a magnesium complex, e.g. isocitrate lyase, while in others the magnesium enzyme is the active species, e.g. enolase. The binding of magnesium in such enzymes is probably through two or three carboxylates and perhaps one imidazole (Table 9.9). Reference to the section on stability constants shows that such sites are selective against manganese in the cytoplasm where $[Mg^{2+}] = 1 \times 10^{-3}$ M and $[Mn^{2+}] \leqslant 10^{-8}$ M. We advise the reader to look at wall charts of metabolic pathways to see the importance of these magnesium enzymes. Notice the β-sheet constraints at the active site.

A case of the mechanism of action of magnesium in such an enzyme reaction is provided by enolase. A picture of the active site is shown in Fig. 9.5 and of the possible pathway of action in Fig. 9.6. Undoubtedly some enzymes, such as enolase, use two magnesium atoms, and in the particular case of pyruvate kinase one magnesium and one potassium, so that the way in which the two (three) metal ions work together could be very complex (see Chapters 4 and 11 on zinc-catalysed acid–base reactions). In the pyruvate reaction magnesium is thought to stabilize an intermediate enolate.

Table 9.8 Some examples of magnesium-requiring enzymes

Enzyme	Requirement
Kinases (generally)	Phosphate transfer from ATP · Mg
Adenylate cyclase	ATP · Mg is the substrate
ATPases	ATP · Mg is the substrate
Alkaline phosphatase	Activator of the enzyme but is not essential; Mg binds the enzyme
Enolase	Magnesium is bound to the enzyme and not directly to substrate
Isocitrate lyase	Mg · isocitrate is the substrate
Methyl aspartase	Magnesium is bound to the enzyme
Ribulose bisphosphate carboxylase	Magnesium binds to intermediates and enzyme
Myosin ATPase	Mg-ATP is the substrate
Nucleases	Activation of RNA or DNA hydrolysis
GTP-dependent enzymes	Hydrolysis of GTP

Table 9.9 Magnesium-binding sites in enzymes

Enzyme	Binding site
Several kinases	ATP, one protein carboxylate
Alkaline phosphatase	Possibly three metals at active site
Xylose isomerase	Sugar, two carboxylates
Enolase	Three carboxylates
Ribulose bisphosphate carboxylase	Two carboxylates, one histidine
Glutamine synthetase	Two carboxylates, one histidine

Several magnesium-dependent enzymes have more than one metal ion binding site (see Section 4.3). Some of the magnesium appears not as a catalyst but as a conformational control.

Magnesium is not only involved with phosphorylation of proteins by NTP, but is concerned with many protein phosphatases; see Kennelly and Potts (1999) in 'Further reading'. Both classes of enzyme are extremely primitive and have been involved in linking the control of proteome expression to the genome throughout evolution. The critical value of magnesium as part of the further link to the metallome must not be missed.

There are also several magnesium-dependent enzymes involved in the metabolism of other phosphates including inositol phosphates, but one of extreme importance is ribulose bisphosphate carboxylase (Rubisco) which is essential for the incorporation of CO_2 in plants. Possibly it is the most abundant of all enzymes. Mg^{2+} at the active site is bound by carboxylates and one lysine carbamate. In fact, CO_2 incorporation to give the carbamate depends on the initial reaction of CO_2 with this lysine, on the concentration of Mg^{2+} in the stroma of chloroplasts, as well as on the pH. It appears as if Mg^{2+} had an absolutely essential control role in many aspects of living cells from the very origin of life since CO_2 incorporation must have been a very early demand.

A few enzymes use Mg^{2+} to catalyse the isomerization of sugars. Here Mg^{2+} is bound to the protein by two carboxylates, but it binds to the sugars

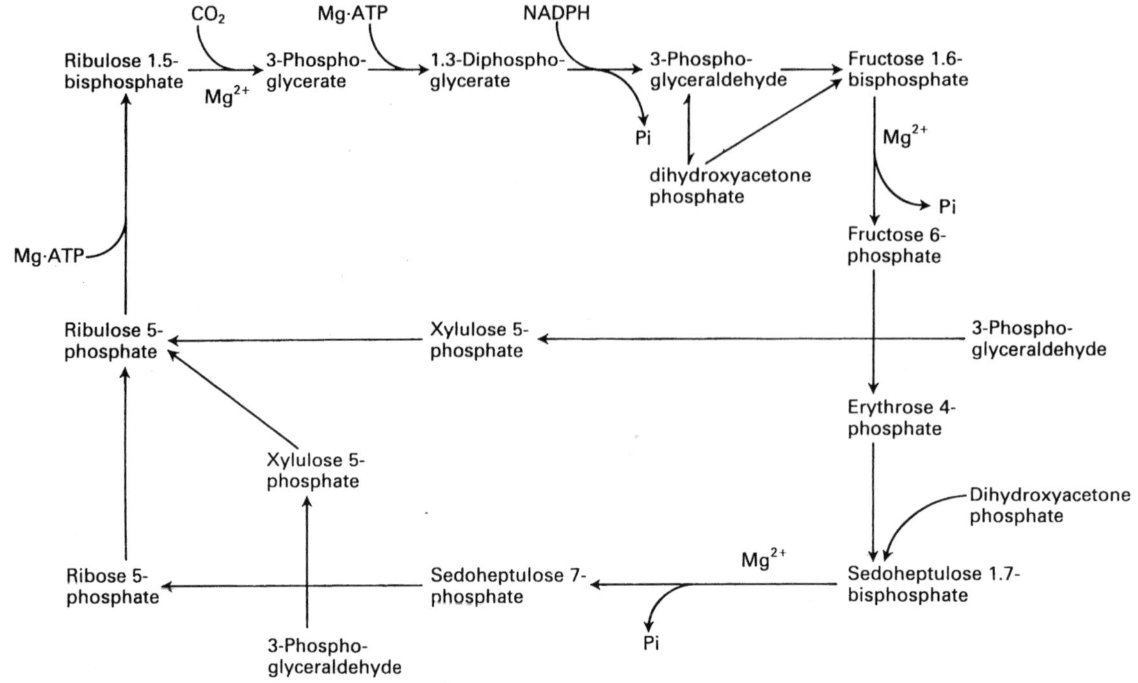

Fig. 9.7 Reactions of CO_2 assimilation and the role of magnesium.

through hydroxy residues. In all cases the cation retains some water molecules and remains octahedrally coordinated.

There are undoubtedly many more enzymes which use Mg^{2+} but which are not too well characterized and have not been included in our tables. An outstanding case is that of the vitamin B_1, thioamin pyrophosphate, dependent enzymes, while a second is in its use in all polymerase chain reactions which require a very well controlled Mg^{2+} concentration.

This section by no means gives a complete list of the deep involvement of magnesium in enzymatic reactions. One way of recognizing this is to examine the enzymes of glycolysis (Chapter 5), glucogenesis, the pentose shunt, and of carbon dioxide assimilation (Fig. 9.7). *In almost every major metabolic path in the cytoplasm magnesium is deeply involved but it is also associated with very much polynucleotide biochemistry* (see Section 9.9).

9.8 Magnesium and muscle cells

9.8.1 Activation of tension

The tension in fibres is of extreme importance for the shapes of single cells but it is best known in the study of muscle cells of multicellular organisms. Here a model called the sliding filament model is used to explain the development of tension. The fibres are shown in Fig. 9.8. On one fibre, labelled M for myosin, there is a headgroup which can bind in different ways in de-energized and energized states to the second fibre, actin (A). The myosin contains an ATPase. The hydrolysis of Mg · ATP drives the head of myosin through a series of conformational changes so that the energy released causes the filaments to slide and so to reduce the overall length of the muscle fibrillar

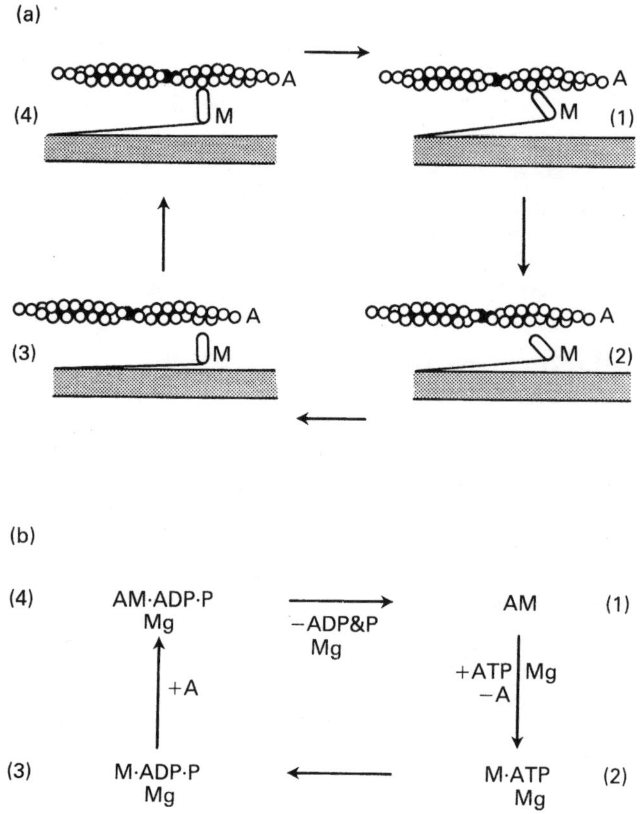

Fig. 9.8 The reactions of Mg ATP which drive contraction. The sliding filament series of steps is shown in (a) with the enzyme steps in (b) (see text). AM, actomyosin; A, actin; M, myosin. There is also a tightly bound Mg^{2+} within the actomyosin complex (see Section 9.5).

bundles (see Fig. 10.13). (The process in muscles cells is activated by calcium invasion of a cell.) For further details see books on physiology in the 'Further reading' to Chapter 8, but here notice the total dependence of the process on Mg^{2+} and that the motions in the headgroup of the myosin may be very similar to those of the ATPase pumps while the ATPase itself may have hinge movements like that of phosphoglycerate kinase.

9.8.2 Magnesium and muscle relaxation: calcium buffering

The fastest muscles require very rapid relaxation, i.e. very fast lowering of calcium ion concentrations after an activating pulse. The usual mode of relaxation of the pulse is the outward directed calcium pumps using the energy of Mg·ATP (see Chapter 10). A faster mode of relaxation involves the use of a soluble protein chelating agent for calcium, parvalbumin, which is designed in a special way. It has an absolute *higher* affinity for both calcium and magnesium than the trigger calcium proteins, troponin C and calmodulin. As a consequence, in the resting cell it is present as Mg · parvalbumin and is only able to accept calcium, from a calcium pulse, quite slowly, i.e. at the rate at which Mg^{2+} is released. The trigger proteins are metal-free before the pulse and accept calcium at the diffusion limit 10^{-9} s. Thus, despite a weaker affinity for calcium, troponin C and calmodulin take up calcium before parvalbumin since the latter only releases Mg^{2+} in some 10^{-3} s. The sequence of events for calcium binding following a pulse is then

$$Ca^{2+}(\text{in}) \rightarrow (\text{trigger proteins} \rightarrow \text{parvalbumins} \rightarrow Ca^{2+}\ \text{pump}) \rightarrow Ca^{2+}\ (\text{out})$$

Parvalbumin acts relatively quickly (10^{-3} s) reducing the calcium in the cell cytoplasm before calcium rejection by the pump, which has a turnover of only 100 s^{-1}. Very fast muscles are found to contain excess parvalbumin up to millimolar levels. If all this is a correct interpretation, the cycle of activation and relaxation of muscle is a wonderful example of the function of kinetic controls using calcium/magnesium ratios in receptor systems.

Another use of magnesium in a mechanical device is in the response regulatory protein of chemotaxis. The protein, CheY, self-phosphorylates but needs a Mg^{2+} ion. This ion is coordinated to two aspartate residues, one of which is phosphorylated with change of protein conformation.

9.9 Magnesium and polynucleotides

9.9.1 Structures of DNA and RNA particles

The determination of the structure of transfer RNA (tRNA) revealed that there are different classes of magnesium associated with this folded polynucleotide. It was already known, of course, that magnesium was often associated with polynucleotides but there was no knowledge of the coordination chemistry. Figure 6.18 shows that the magnesium is found in one conventional binding mode and also in one particularly interesting structural trap. The fold of tRNA is in considerable part just a conventional double-stranded helix and Mg^{2+} does not bind very well locally in the short grooves of this structure (see Section 6.11) at high salt concentration though it may be retained by the total surface potential. The fold also contains a half crucifix-like structure where three strands intersect. It is here that there is an interesting binding of Mg^{2+} in a partially dehydrated state. It would appear as if the stability of the tRNA fold, and hence its functioning, is Mg^{2+}-dependent. This is a very pretty example of cross-linking of oxyanion centres by a cation. Since the Mg^{2+} will not allow rapid rearrangement of its coordination sphere, this half crucifix is likely to be an essential maintained part of the fold during activity. The half crucifix form does connect the various essential exterior and functional parts of the tRNA in its transferring capacity. It is probable that certain regions of other RNAs and DNAs have similar folds and a parallel need for Mg^{2+} stabilization (Table 9.10; Fig. 6.19) but the Mg^{2+} binding here could be very strong due to the greater collection of negative charges. Of course, Mg^{2+} is required in all steps of transcription and translation, in large part through the involvement of nucleotide triphosphates. It is also known that Mg^{2+} stabilizes the bending of DNA into particular curved structures (Section 6.11).

Table 9.10 Magnesium requirements in polynucleotides

Species	Requirement
tRNA	At least one strong site and many weaker sites Fold stabilization Half-crucifix structures
rRNA	Stabilization of fold and association of 30S, 50S, and 70S subunits. Crucifix structures at attachment points?
mRNA	Attachment to ribosomes Crucifix structures at attachment points?
Ribozymes	Mg^{2+} acts to bind the RNA enzyme to the RNA substrate
DNA	Crucifix structures
DNA/RNA	polymerases
DNA/RNA	primases

9.9.2 Ribozymes

The discovery of catalytic activity associated with RNA and especially its connection with splicing reactions has led to speculation that these RNA enzymes, ribozymes, could have been some of the earliest catalysts in organisms. Be this as it may, it is noticeable that they, like many another enzyme-handling phosphate hydrolyses, are Mg^{2+}-dependent even *in vitro*. Other metal ions that are much stronger Lewis acids do not interfere. The reason could well be that these cations, such as copper, bind preferentially to the nitrogen bases of the RNA and the active metal ions are therefore $Mg^{2+} > Mn^{2+} > Ca^{2+}$. Nickel, Ni^{2+}, does work quite well so that a preferential size factor is also operative. In cells, Mg^{2+} is, of course, the only ion available for binding to the weak O-donor sites of the RNA since it is available at

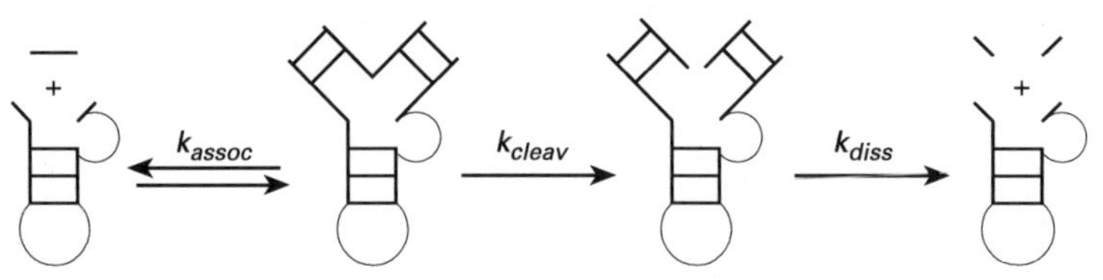

Minimal reaction scheme for hammerhead ribozymes

10^{-3} M. It is likely that the Mg^{2+} is regulatory too so that, if it falls below some 10^{-4} M, the ribozyme is inactive just as the ribosome is inactivated. There are several Mg^{2+} ions required for the cleavage of an RNA by a hammerhead ribozyme in the reaction shown above. This could be important in regulation of unicellular organisms.

9.10 Competition with polyamines

In the context of RNA and DNA it is interesting to compare Mg^{2+} chemistry with that of a wide range of other cations. The binding of a series of cations ranging from simple ions such as Mg^{2+} and $[Co(NH_3)_6]^{3+}$ to complex polyamines such as NH_3^+-$[(CH_2)_2NH_2]_m^{m+}$-$(CH_2)_n$-NH_3^+ (Fig. 9.9) with a series of anions ranging from HPO_4^{2-} and RCO_2^- to X-$(CH_2CHX)_m CH_2$-X (where X is carboxylate or phosphate) has been explored (see Fig. 9.10). The framework cations and anions are intended to be models for biological molecules such as spermines and histones on the one side (cationic) and DNA, carboxylated proteins, and sulphated polysaccharides on the other (anionic). Molecules studied in these model analyses have either mobile or rigid frameworks. A parallel study with encapsulating cyclic ligands and amines has been made, but here we are more concerned with open surfaces. The findings can be stated simply using the data and figures such as Fig. 9.10.

1. Where the anionic ligand has a high local charge density, e.g. $C_2O_4^{2-}$ and $P_2O_7^{4-}$, then its binding of small cations such as Mg^{2+} is much stronger than its binding of polyamines of any kind.
2. Where the charge distribution is spread out on the anion, but not too spread out, then binding of polyamines is the stronger.
3. All binding of polyamines to the anions is weak unless the charge product (anion charge times cation charge) exceeds 12.

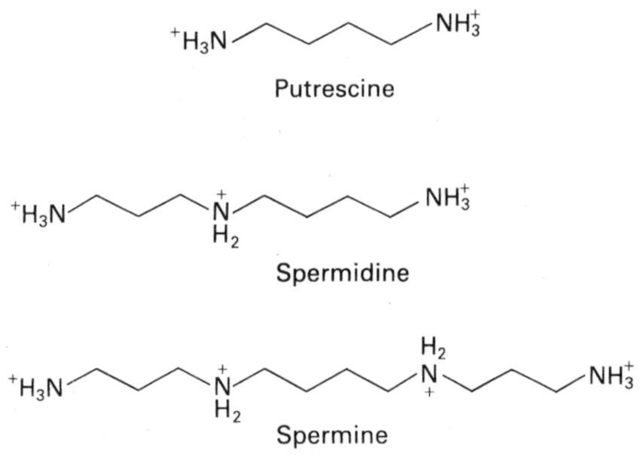

Fig. 9.9 The structures of some important polyamines.

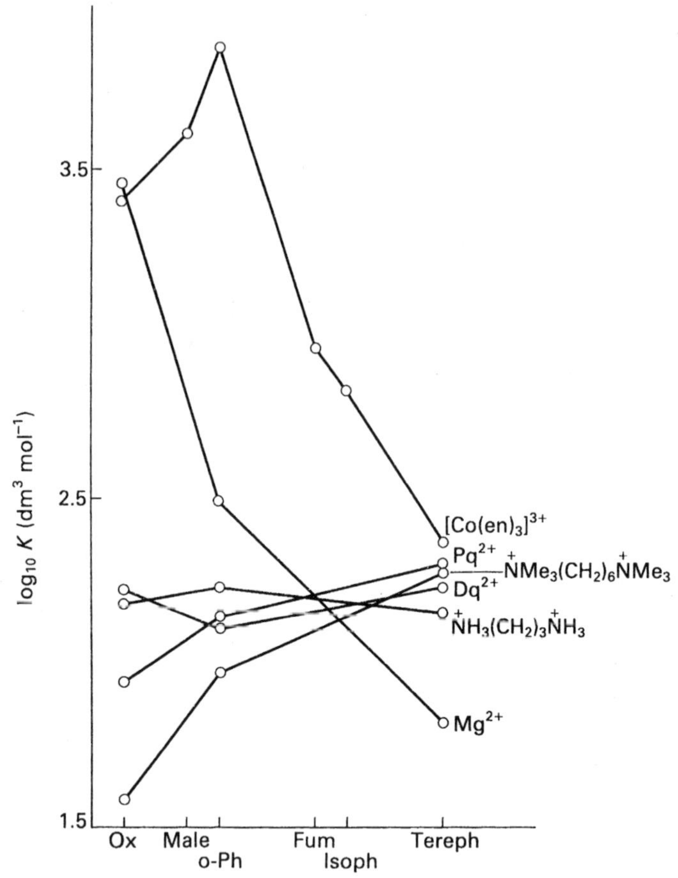

Fig. 9.10 The binding constants to dicarboxylate molecules by a variety of cations. As the negative charges are moved apart, Mg^{2+} binding falls rapidly but that of the organic amines remains steady or increases. Ox, oxalate; Male, maleate; *o*-Ph, orthophthalate; Fum, Fumarate; Isoph, metaphthalate; Tereph, paraphthalate; Pq^{2+}, paraquat; Dq^{2+}, diquat; en, ethylenediamine; Me, methyl.

4. Given a large number of spread-out charges, stereochemical factors become important in charge-matching and hydrophobic interactions between frameworks (Fig. 9.11).
5. Intermediate anion charge density leads to competition.
6. In general, $-OPO_3^{2-} > -CO_2^- > -OSO_3^-$ in binding to all cations.

(In passing we observe that, in a sense, calcium is intermediate in character between magnesium and the polyamines since it is large, but in cells it has little impact since it is only 10^{-7} M.)

This set of observations is important in that it leads us to an understanding of the roles of frameworks of positive charge as found on labile molecules such as spermine, or on more specific polyamine cation surfaces such as those on the surfaces of proteins like histones and cytochrome *c*. These polycations bind to matching spread-out polyanion surfaces, such as the diester phosphates in the grooves of DNA and RNA double helices, or they bind to the carboxylates on the surfaces of partner proteins such as cytochrome *c* oxidase. At equal concentrations magnesium ions cannot compete for such surfaces in the cytoplasm of cells, while the polyamine cations do not interfere with the binding of Mg^{2+} to ATP^{4-}. We see too that in the intermediate situation, given by only one carboxylate or perhaps two on an enzyme surface and a terminal phosphate $ROPO_3^{2-}$ on a substrate, there is the ready possibility for the use of a bridging Mg^{2+} ion at the active site of hydrolysis without competition from amines. Alternatively, if amino

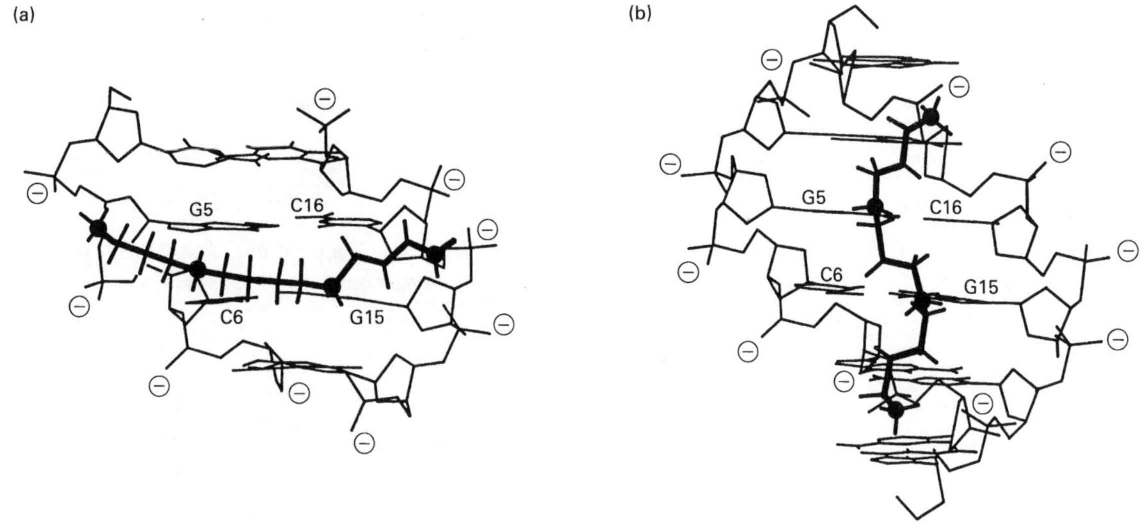

Fig. 9.11 The fitting of a polyamine into the groove of DNA seen so as to reveal the interaction with (a) the phosphate diester groups and (b) the stacked bases. NB Polyamine binding changes the structure of DNA.

groups of the protein side chains are suitably placed, the binding site for the anion can utilize the different selectivity of electrostatics while not involving Mg^{2+} (see 3-PG^{3-} in Fig. 9.3). It will then be the presence of special selective factors, say imidazole sidechains, and of relative charge density in proteins which will assist or prevent Mg^{2+} binding.

The preceding discussion has neglected the use of the hydrophobic framework of the polyamine cation (Fig. 9.9). Clearly, it is possible to develop part of the cationic molecule so that the interaction with a protein or DNA surface is enhanced. It is this component of organic molecules which adds a new dimension to the competition between organic cations and inorganic cations. The analysis of this competition shows that Mg^{2+} may well hold DNA, for example, in one conformation, while on synthesis of competing polyamines, say during a cell cycle, the Mg^{2+} is removed and DNA is switched both in structure and function. (NB There is the obviously related competition between Na^+, K^+ and organic cations such as acetylcholine for protein surfaces in pumps and receptors, see Chapter 8.)

Once again, notice that it is in the cytoplasm that the competition or cooperation between Mg^{2+} and organic polycations occurs and Ca^{2+} is not involved, but outside the cell it is often calcium and amino groups which act in competition or cooperatively in enzymes and other macromolecules.

Finally, sulphated polymers do not bind so well to cations so that by using the sulphate group organisms can produce free anionic surfaces as in sulphated polysaccharides (Chapter 6), which therefore make up a passive extracellular matrix.

The above discussion has concerned competition between organic molecules and inorganic cations in terms of stability constants. There is a way of modulating this competition in cells through metabolism. While Mg^{2+} concentration is relatively fixed in the cell cytoplasm, that of the polyamines can vary grossly during the cell cycle. There is then a possibility that the timed relative concentration of polyamines and magnesium is an important factor in cell division. The competition may also lead to the production of antibiotics since synthetic polyamines can be introduced into cells.

9.11 Polymeric equilibria: tubulins, DNA, and the cell cycle

So far the account of magnesium biochemistry in this chapter has dealt with simple equilibria, where L is any ligand,

$$Mg^{2+} + L \rightleftarrows MgL^{2+}$$

and the competition for L by Ca^{2+}, polyamines, or protons. There are further important reactions involving magnesium which are of the kind

$$nMg^{2+} + mL \rightleftarrows Mg_nL_m^{2n+}.$$

The analysis of such reactions shows that they can be extremely cooperative ($n > 5$) when we almost have a phase change as in a precipitation reaction

$$Mg^{2+} + L^{2-} \rightleftarrows MgL(\text{solid}).$$

Here an extremely small change in the concentration of magnesium switches the system from a one 'phase' to a two 'phase' system. An intermediate situation between the above is one in which there is a level of cooperativity where n is some number considerably greater than one. A plot of formation of $Mg_n\ L_m$ against magnesium concentration now shows a rapid change as [Mg] increases while the system remains soluble. In this case changes in magnesium concentration do not bring about an abrupt solubility change, but only a very rapid switch in the solution, sol $\leftrightarrows$ gel, conditions and this could apply to DNA and RNA structures. We now turn to a different case where we treat the development of a filament as if it were like a gel reaction.

The polymerization of one of the polymers, tubulin, associated with the filamentous network of a structured cell can be pictured as in Fig. 6.14. There is involved the monomer of tubulin, L, the cation, M^{2+}, and GTP, the nucleoside triphosphate. The formation of the polymer is linked to the energy state of the cell through GTP and the concentration levels of the cation M^{2+}. It is observed that, while magnesium aids polymerization, calcium opposes it.

The balance between these cation concentrations could decide the internal structure of cells but it is likely that $[Mg]^n$ is the effective agent since calcium in cells is so low and Mg·GTP is dominant. We now see that relatively small changes in [Mg] and [GTP] during the cell cycle could lead to formation or dissolution of the tubulin filaments. Again minor changes in the environment, such as stress on the filaments leading to their dissolution, will be poised opposite the level of $[Mg]^n$. We are suggesting here that, in part, Mg^{2+} controls the *mechanics* of the cell-division cycle. As stated in Section 9.10. this control could extend to features of DNA itself and the competition with polyamines might be involved.

9.12 Magnesium outside cells

The role of Mg^{2+} outside cells is not as well defined as that of calcium although both are held at very fixed free ion concentrations in higher animals and there equilibration with bone helps to buffer them. There are a few extracellular enzymes, notably alkaline phosphatase, which require Mg^{2+}, but perhaps the most interesting use of this cation is in integrins. Integrins are receptor proteins which hold cells together and can assist activation of responses inside the cell

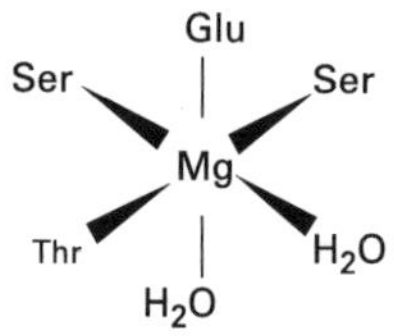

as they have connecting links through the membrane to domains inside the cell. The role of magnesium and/or calcium is to hold together the extracellular domains in a pattern which is then adjusted by the further binding of a foreign body or a new cell development. The binding groups on one domain appear to be two serines, a threonine and two water molecules, (see Aside) plus a further water which is replaced by a carboxylate of a neighbour domain. The response of integrins may arise during recognition of foreign cells and may then be the cause of activation of neutrophils to give rise to superoxide and peroxide attack on the foreign body. There are likely to be many other examples of Mg^{2+} binding to extracellular matrices as in bacterial cell walls.

9.12.1 Sol/gel equilibria and magnesium

The polymerization of polysaccharides and of glycoproteins, all of which are negatively charged and are frequently parts of the cell exterior, controls the stability of membranes and cell–cell interactions. These cation-dependent polymers are often parts of adhesive surfaces of various kinds in different organisms. The systems lose stability in the presence of metal-binding agents such as EDTA. The suggestion is frequently made that the exact structure of these open polymer meshes is dependent on condensation between 'sol' and 'gel' states. The switching between the two states is well known in model systems of intertwined polymers. Once again the switch sol/gel is dependent on very small changes in divalent ion concentrations, including $[Mg^{2+}]^n$ and $[Ca^{2+}]^n$. Particularly important here are the outer cross-linked membranes of Gram-negative bacteria which contain teichoic acids (Fig. 9.2).

The overall picture of these polymeric equilibria is that, differently from calcium-dependent second-messenger switches in the cytoplasm, which appear to be cooperative to only a small degree and which act very rapidly, magnesium-dependent switches are almost always phase-like changes dependent on $[Mg]^n$, where n is large, and are relatively slow.

9.13 Magnesium and lipids

The synthesis of many lipid-soluble molecules in organisms is based on the transfer of the isopentenyl unit. The unit, C_5, is carried by pyrophosphate, so that there is coupled to synthesis a loss of the pyrophosphate bond energy. The pyrophosphate moiety is synthesized from ATP. In this reaction, the part played by magnesium is at several levels from the initial formation of the pyrophosphate from Mg · ATP to the reactions of the pyrophosphate esters in condensation reactions. These reactions are on the route to the synthesis of terpenes and cholesterol.

Another use of magnesium is in the synthesis of choline-based lipids such as phosphatidylcholines (lecithins) where cytidine triphosphate is the energy source driving reaction.

9.14 Magnesium and chlorophyll

9.14.1 Chlorophyll binding in proteins

There are two questions of great interest in the association of magnesium with chlorophylls, as for example in Fig. 9.12, which gives the dominant molecule for light capture in biology. The first is what is magnesium for and the second is how is chlorophyll placed where it is. The requirement for magnesium

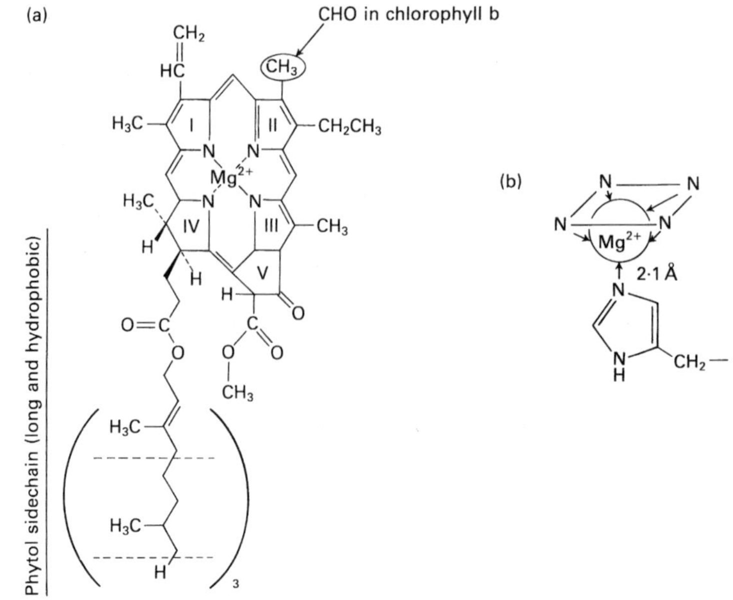

Fig. 9.12 (a) The formula of chlorophyls a and b and their binding in proteins; (b) the binding of a protein histidine to the magnesium of chlorophyll.

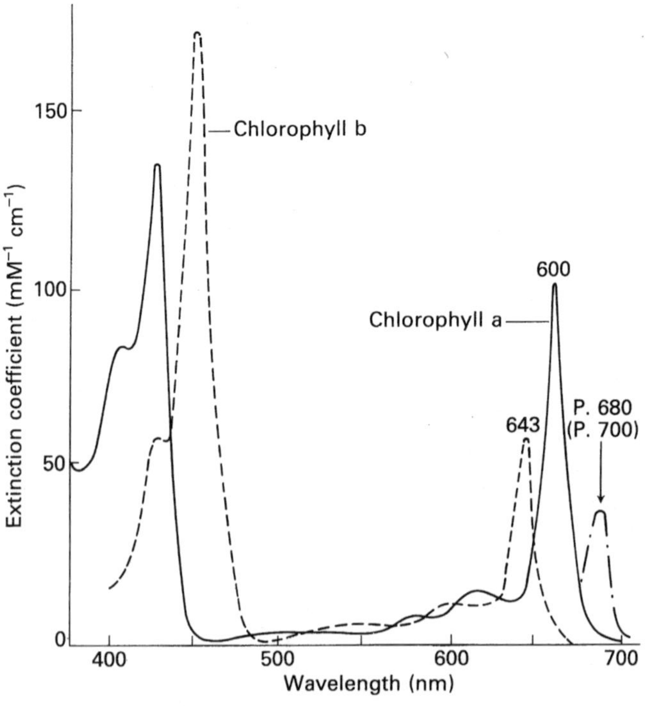

Fig. 9.13 The absorption spectra of chlorophylls and of the reaction centres, P-680 (P-700). Light is harvested by a series of chlorophyll a and b molecules (Fig. 9.14) and then radiation transfer to a special chlorophyll pair (P-680 or P-700) occurs which is the reaction centre for converting light to free-radical energy (Section 9.14.2).

rather than another metal ion in this light-capture molecule is perhaps due to the fact that an element of such low atomic number does not enhance fluorescence and does not itself have redox properties. Small size allows other ions to fit in the chlorin cavity, but fluorescence is much greater from zinc chlorins and unwanted redox properties are to be found for Mn, Co, Fe, Ni, and Cu heterocycles. While serving no obvious chemical function, magnesium does not interfere with the photo- and free-radical reactions of reaction-centre chlorins and clearly can assist them. Moreover, the presence of magnesium in the chlorin ring gives the ring an approximate tetragonal symmetry and hence enhances the molecular extinction coefficient around 650–700 nm relative to the lower-symmetry metal-free molecule while moving the absorption maximum to longer wavelengths. Figure 9.13 illustrates these properties.

The binding of magnesium in chlorin proteins is peculiar in two ways. The chlorin is an N-donor, and the further binding to the protein is through another N-donor from a histidine imidazole (Fig. 9.12). The coordination sphere is then five nitrogen atoms with the magnesium somewhat out of plane. It is difficult to believe that this arrangement could be a thermodynamically stable form of simple magnesium coordination chemistry in the presence of many free oxygen donors, e.g. from water. The chlorin/magnesium complex itself may well be stable to hydrolysis, as it is effectively an encapsulation by a ring dianion, but binding to such a site by ions such as Fe(II) or Zn(II) is so much more powerful that some selective insertion pathway has to be imagined (see Section 9.14.2). The binding to the imidazole of the protein in the authors' opinion is another encapsulation reaction but this time of the chlorophyll as a whole which is encapsulated by a helical *protein* in an '*organic* solvent' (Fig. 9.14; see Chapter 2). Effectively this is a solvent extraction reaction and no doubt it is aided by the large lipophilic tail of the chlorophyll molecule. The coordination chemistry is itself an interesting case of the domination of the fold energy of the protein over the ion coordination chemistry, but the positioning in space of the complex is a property of a special protein.

Now, as well as capturing light, the special chlorophyll molecules in a reaction-centre dimer, Fig. 9.13, generate the first step of energy transduction in that at this centre the photon generates a diradical and to and from it electrons flow to give the first membrane potential (Fig. 5.3) for the structure of photosystem I, while that of photosystem II is mentioned in note (2) on p. 417.

The chlorophyll itself has an interesting property in which it differs from porphyrins: there are no sidechain anionic carboxylate groups. It may well be that these groups that connect electron transfer reactions to proton transfer reactions in iron porphyrins are unwanted in a light-capture reaction centre (see Section 13.6.3). The evolution of all these pigmented molecules will be described in Chapter 20 (and see Chapter 5.).

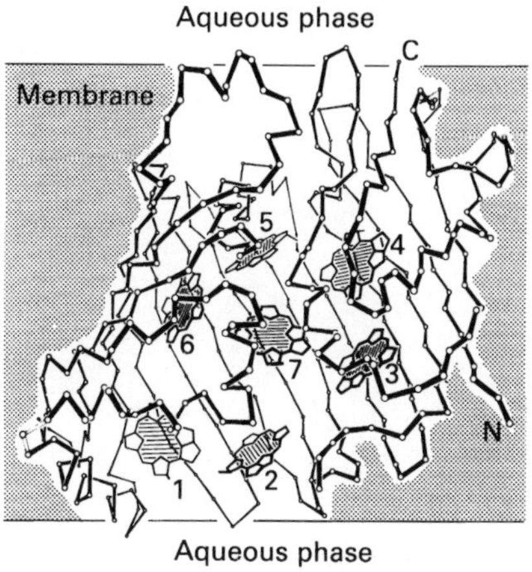

Fig. 9.14 The chlorophyll-containing protein for light harvesting which is in the chloroplast thylakoid membranes (After B. W. Matthews.)

9.14.2 Magnesium insertion in chlorophyll: chelatases

Magnesium ions are inserted into the chlorin ring through the action of a chelatase enzyme and without the enzyme this cation could not combine with chlorins nor could it overcome the much greater affinity of the chlorin site for the first-row transition metal ions, e.g. Zn^{2+}, Cu^{2+}, and Fe^{2+}. The requirement is for a series of reactions comparable with the reactions for the pumping of magnesium ions across membranes using ligands, L,

$$Mg^{2+} + L \rightleftharpoons MgL_1 \text{ (reversible)}$$

$$MgL_1 + L_2 \longrightarrow MgL_2 + L_1 \text{ (irreversible).}$$

The reaction scheme is such that magnesium is the preferred partner of L_1 and other cations M are excluded from binding to L_2, etc. by kinetically irreversible steps. We know, for example, that ATP binds Mg^{2+} but not other metal ions in cells since the product [M] [ATP]K, where K is a binding constant, is large for magnesium but small for most other ions. This is due to a similar stability constant K for most divalent ion binding to ATP (Table 2.2 and see Section 9.3) but vast differences in [M]. Most Zn^{2+}, Mn^{2+}, Cu^{2+}, and Fe^{2+} ions are removed by one of three mechanisms: (1) binding to N,S ligands of protein; (2) oxidation Fe^{2+} to Fe^{3+}; and (3) pumping into vesicles (see Chapters 2 and 3). The case is very different where L is a nitrogen donor such as chlorin whence it is essential to avoid direct *reversible* binding of Mg to chlorins.

Given that there is a very small amount of ML_1 other than MgL_1 this could still be dangerous in a transfer system but for the fact that MgL_1 must have a recognizably different structure. The structure, i.e. the size and stereo-chemistry of a complex, e.g. Mg · ATP, is a high degree selective to Mg^{2+}. The *recognition* of such a structure can be used in the kinetic control

$$MgL_1 + \text{chlorin} \longrightarrow \text{chlorin} \cdot Mg + L_1,$$

which is effectively irreversible due to the nature of encapsulation. We know that the enzymes for this last reaction exist, but *in vitro* even they usually incorporate Zn into chlorin in the presence of vast excesses of Mg^{2+}.

In these biologically interesting chlorophyll complexes in proteins magnesium is five-coordinate and the magnesium atom would appear to be just too large to fit the cavity of the chlorin. The misfit is very slight since model complexes are known where magnesium sits either in the plane of a porphyrin or about 0.3 Å out of the plane. The first group of complexes is six-coordinate, but they have unusually long 2.25 Å axial bonds while the second group is five-coordinate and the axial bond length is 2.1 Å which is normal. The Mg -N bond to the porphyrin nitrogen in the ring is about 2.05 Å and it could be that the magnesium has to squeeze into such a ring making for long axial bonds (Fig. 9.12).

All the magnesium chlorophyll pigments in membranes do not absorb at exactly the same wavelength (Fig. 9.13). Just as copper is tuned to its function of electron transfer (see Section 6.8.2) so the absorption of dye-stuffs, here chlorophyll, is tuned to the process of energy transfer. Here the tuning consists of modifying the electronic absorption spectrum so that it moves to longer wavelengths the closer in space the molecule is to the reaction centre, a chlorophyll dimer, which has its absorption bands at the longest wavelength. The system can be written as a photon cascade $h\nu \rightarrow$ chlorophyll a $\rightarrow$ chlorophyll b $\rightarrow$ reaction centre chlorophyll. The control of the absorption of the dye molecule—the same organic molecule in different environments—is an entatic (constrained) situation. In essence the protein has evolved to control the relative stability of the ground and excited spectroscopic states. It can do this by minor changes in the coordination bond-lengths of Mg^{2+}, by adjusting the planarity of the chlorin ring, or by altering the solvation by the sidechains of proteins. A comparison of this trick for a photon pipe with the adjustments around haem to generate an electron pipe (a wire) is intriguing (see Chapter 13).

As stated earlier, magnesium has another function in the chloroplast as a free ion. When light shines on the thylakoids the pH falls to around 4.0. The protons force Mg^{2+} out of the thylakoid while some chloride enters, and there is a rise of Mg^{2+} in the stroma of the chloroplast. Here magnesium activates the carboxylase rubisco, for incorporation of CO_2 into ribulose bisphosphate, a step in the Calvin cycle of photosynthesis (Fig. 9.15).

9.15 The use of manganese as a magnesium probe

The chemistries of magnesium and manganese as divalent ions are rather similar despite the difference in radii and in affinity for N-donor ligands. This

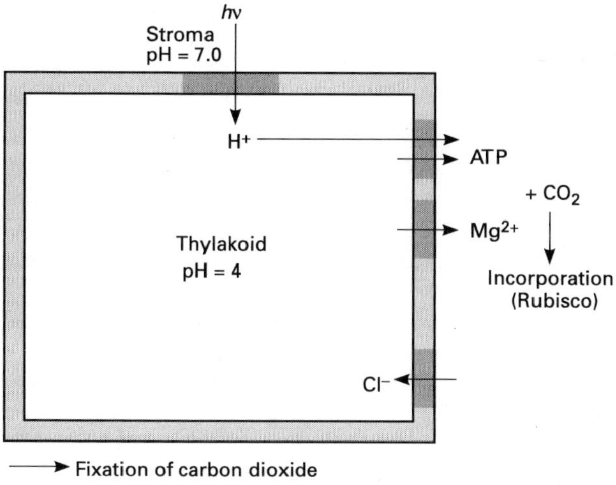

Fig. 9.15 Fixation of carbon dioxide.

similarity can be put to good use provided that the system under study has been shown to be *functionally* similar when one cation replaces the other. Manganese(II) is a good EPR reagent and a good NMR relaxation reagent. Other probes can be used; for example Cr(III) is used in Cleland's reagent, ATP(Cr), but has no activity; again the Ln(III) ions can be used but do not often show activity in Mg^{2+}-dependent systems. In those structural studies where $[Mg(H_2O)_6]^{2+}$ rather than the simple cation Mg^{2+} may be important, it is found that such cations as $[Co(NH_3)_6]^{3+}$, $[Cr(NH_3)_6]^{3+}$ and so on are useful substitutes. Particularly in the hands of Cohn and Mildvan Mn^{2+} has been the major valuable probe of the sites of enzyme activity of magnesium (see Fig. 9.16). Most important has been the use of Mn^{2+} in probing phosphate transfer reactions. It would appear that Mg^{2+} acts as a mild Lewis acid while performing the major role of aligning the anionic reactants in the enzyme site. If anything, Mg^{2+} is a structure-based catalyst not an attacking group. Where a stronger acid is required in order to break substrate bonds then it is usual to find a zinc metallo-enzyme rather than a magnesium metal-complex. Notice that the case of hydrolysis follows the order anhydrides > esters > amides. Magnesium is not used to assist amide hydrolysis which requires stronger acids (see peptidases and urease in Table 7.3).

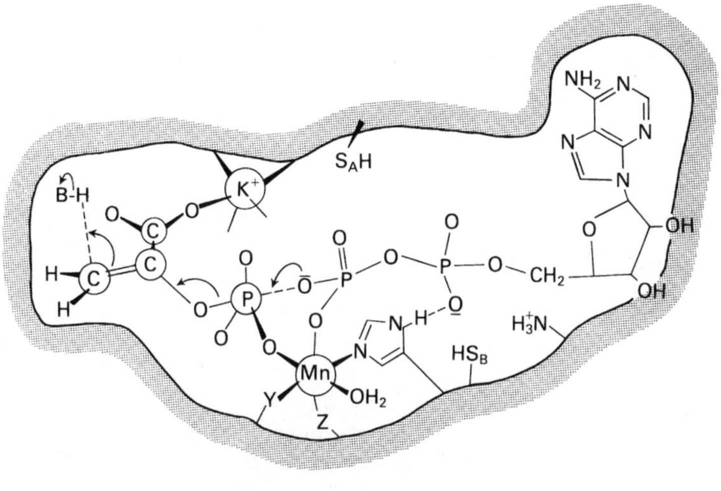

Fig. 9.16 An illustration of the use of Mn(II) as probe for Mg(II) activity in an enzyme (after the work of M. Cohn and A. Mildvan). The enzyme is pyruvate kinase. Other features of note are: (1) the use of two metal ions, here the second is potassium; (2) the necessary closed pocket; and (3) the coordination involving but one histidine and possibly two carboxylates Y and Z.

9.16 Interference of other cations with Mg^{2+} biochemistry

There is little doubt that in model systems Mg^{2+} can be replaced by a variety of cations such as Mn^{2+}, Co^{2+}, Ni^{2+} and Zn^{2+}. Sometimes Ca^{2+} will act as a replacement. The activity of the enzyme is often only moderately impaired since the main function of Mg^{2+} is to neutralize negative charge. In biological systems Mg^{2+} functions are not open to this type of replacement chemistry since the levels of all divalent cations are $>10^4$ times lower. Accidental contamination could still be a problem and here we must concern ourselves today with the polluting elements such as Al^{3+} and possibly Be^{2+}. There is the possibility that Al^{3+} or possibly $[AlOH]^{2+}$ could occupy Mg^{2+} sites of various kinds in cells. The suggestion arises that these sites could be phosphorylated tubulins, and there were some who believed that this could cause disease of the nervous system such as Alzheimer's disease (Sections 3.5 and 3.6), but recent evidence does not support this. The systems for uptake of Mg^{2+} are likely to be especially vulnerable to Ni^{2+} since the ion has the same simple geometry and size. However, to keep Mg^{2+} free from competition other cations in the cytoplasm had to be restricted to values often greatly below 10^{-6} M (see Fig. 2.8).

9.17 Lithium and magnesium

Figure 9.1 shows that there is a relationship in the periodic table between Li and Mg, termed the diagonal relationship, and a similar relationship between Na^+, Ca^{2+}, and Y^{3+}. The relationships are based on the fact that the elements above the line form small ions, radius less than 0.75 Å, while the ions of the elements below the line have radii greater than 0.75 Å. The result of this size difference is a markedly greater affinity by small ions for anions of high charge density, F^-, OH^-, and all phosphates $ROPO_3^{2-}$, than is shown by the larger cations and a considerable difference in kinetics also exists even in the exchange of water of hydration (Table 9.4). In biological systems we have seen how this has led to the close association of Mg^{2+} and phosphate metabolism. Lithium (beryllium and aluminium) are not essential elements, but they can interfere with biochemical processes *in vivo* by interacting with phosphate-containing molecules. One interesting possibility is the involvement of lithium with phosphate compounds. This can occur at two levels: (1) lithium could interact with the phosphate compounds of the Na^+/K^+ or Mg^{2+}/Ca^{2+} pumps; or (2) it could interact with the inositol phosphates. Lithium is, of course, a major psychotherapeutic drug, and its action may in part be based on these possibilities. Could Ln^{3+} compounds be useful here while interacting with Ca^{2+} sites?

9.18 Evolution and genetics of Mg^{2+}-proteins and networks

Undoubtedly many Mg^{2+}-proteins are constitutive and have been and are so in all cells, e.g. certain kinases. The number of these proteins has increased in a staggering manner as shown by genome analysis. Prokaryotes have largely histidine dependent phosphate transfer in their protein kinases, involving self phosphorylation at the histidine, which are unlike the threonine/serine or tyrosine kinases, which largely came later in eukaryotes, and phosphorylate

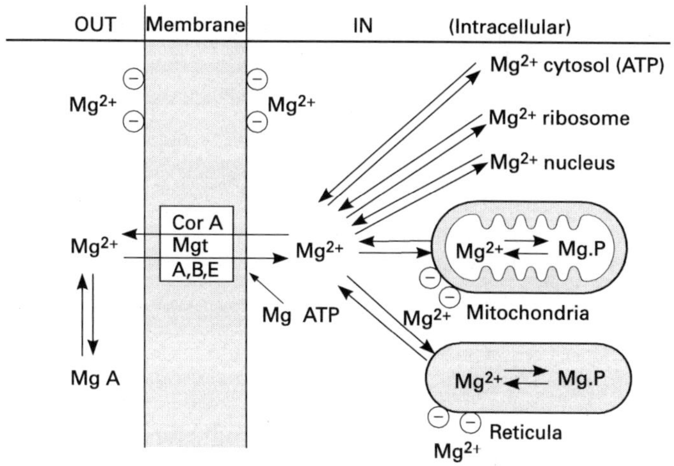

Fig. 9.17 The connection of Mg^{2+} to many proteins, P, enzymes and pumps, and with ATP all with binding constants of about $10^{3.5}$ M^{-1}.

other proteins, not themselves. We may consider all kinases as part of these three essentially different families. Prokaryotes, such as *E. coli*, have some 20 (histidine) kinases, yeast about 300 serine/threonine (ST) kinases, *C elegans* (a worm) some 500 (ST) kinases and the human genome probably has in excess of 2000. The tyrosine protein kinase increase from zero in *E. coli*, to 4 in yeast and 150 in *C. elegans*. The use of Mg^{2+} dependent kinase activity has therefore increased at least 20-fold in evolution. The involvement with other Mg^{2+} proteins has probably increased in a similar way. Meanwhile, the control of Mg^{2+} uptake is linked by feed-back to Mg^{2+} concentrations in all cells through the production of controlled uptake pumps (see Fig. 9.17 where the connection to nucleotides, RNA and DNA is also shown). It is probable that all the Mg^{2+} exchanges shown are in equilibrium with free Mg^{2+}.

The general evolution of these so-called P-type ATP-ases, that is, the pumps (P) for ions and substrates, has been studied in detail. Their conventional biological tree based upon sequences is shown in Fig. 9.18. There are few in *E. coli*, 16 in yeast and hundreds in man's cells, and while only some of them pump Mg^{2+} all use this cation in Mg^{2+} ATP.

We must be careful that we do not see the biological chemistry of Mg^{2+} separately from that of other elements. Wacker and Williams observed that Mg^{2+} concentrations in cells were linked with ATP levels and with Na^+ and K^+ concentrations. There is intense feedback between many ion levels dependent on Mg^{2+}. ATP. These relationships appear in the development of many tissues, but are best illustrated using the genome of *Salmonella typhimurium*.

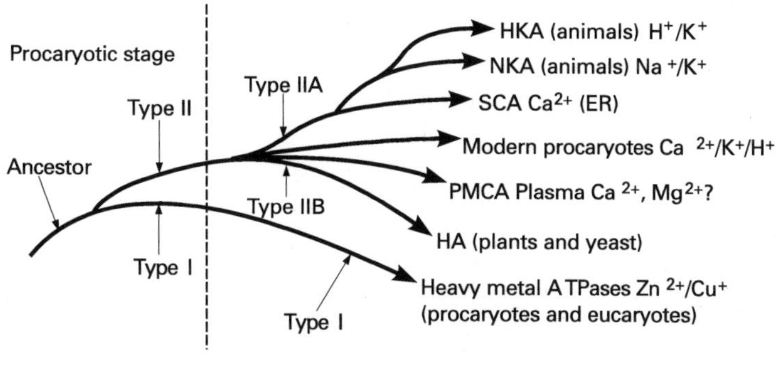

Fig. 9.18 A possible evolutionary connection between the P-type ATP-ase pumps. All are Mg^{2+} dependent. [After Møller J.V. *et al.* (1996) *Biochimica and Biophysica Acta*, **1286**, 1–51].

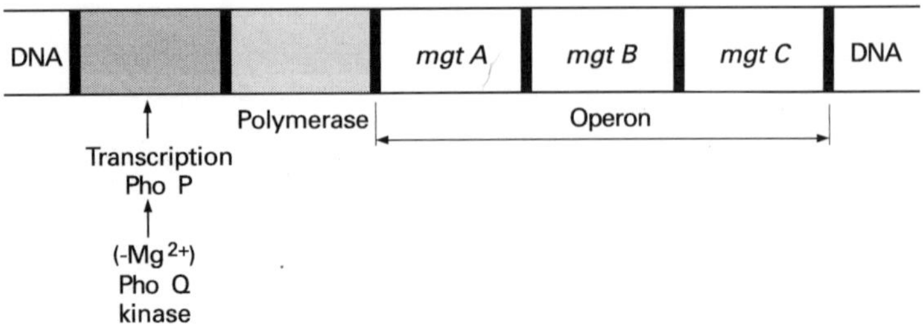

Fig. 9.19 The genes of Mg^{2+} uptake in *S. typhimurium*. The Pho system activates many other genes. The roles of the Mgt proteins, A, B, C, are in Mg^{2+} transport by pumping. [After the work of Moncrief M. B. and Maguire M. E. (1999) *Journal of Biological Inorganic Chemistry*, **4**, 523–7.]

Figure 9.19 shows the feed-back of Mg^{2+} levels to the uptake genes while Fig. 9.17 shows the connection through equilibria to many other parts of eukaryote cells.

9.19 Conclusion

The high concentration of magnesium ions in very many biological compartments is put to use in assisting the neutralization of many anionic centres in cells. This neutralization can allow binding together of anions, it can protect anions from hydrolysis, or it can be used to assist chemical reaction. The role of magnesium in stabilizing some structures and energizing others is to be found in all the types of biopolymer–DNA, RNA, proteins, polysaccharides, and lipids. The role extends to interaction with a multitude of small molecule anions including di- and triphosphates. The movement of Mg^{2+} from compartment to compartment seems to be less important than the maintenance of a level close to 1.0 mM almost everywhere in biological systems. The whole is, of course, an energized steady state and we must be aware that through small changes in Mg^{2+} concentrations in the cell cycle magnesium could have dramatic effects dependent on $[Mg^{2+}]^n$ and the variable concentrations of polyamines.

A point of great interest is that free Mg^{2+} is in equilibrium with nearly all its bound forms to substrates, proteins, and DNA/RNA. In order that they should all be functional, active, and sensitive to free Mg^{2+} concentration changes, all bindings must have similar binding constants in a given compartment. The value for Mg^{2+} is $\log K = 3$ to 4 almost everywhere. We shall see that parallel considerations apply to Ca^{2+}, Mn^{2+} and Fe^{2+} but not so obviously to Ni^{2+}, Cu^{+} and Zn^{2+}.

Given the central role of phosphate in all these polymers in metabolism, in bioenergetics, and in metabolic pathways, especially inside cells (DNA, RNA, protein, polysaccharide, and lipid synthesis), it is very difficult to escape the conclusion that cells evolved with magnesium as the essential partner for anions, especially phosphates, in biological solutions. Thus magnesium is and always was an element equal in importance to phosphorus in cells. Part of the problem in recognizing the role of the metal cations is that chemists and biochemists write compounds which are salts as if they were uncharged. Examples are DNA and RNA and frequently ATP, while the real species are $DNA^{n-}\,(M^{+})_n$, $RNA^{n-}\,(M)^{+}_{n}$, and $ATP^{4-}.\ Mg^{2+}$. Like phosphate and potassium the role of magnesium in living processes has evolved little since the early stages of life and it has few vital functions outside cells. Here is a striking difference from Na^{+}, Ca^{2+} and Cl^{-}, which were initially rejected from cells and so evolved in value late in evolution especially with the coming of multicellular systems. While stressing differences between Mg^{2+} and other ions in this chapter we shall show in Chapter 20 that there is cross-talk between them all.

A final word concerning the Mg^{2+} ion. The concentration of Mg^{2+} in the sea may well have always been as it is now; that is, greater than 10^{-2}M. However, all cells examined so far have a free internal Mg^{2+} concentration of about 10^{-3} M. It appears that all life, always based on phosphate as well as carbon chemistry, has, from conception, had this concentration of Mg^{2+} in the cytoplasm. Is this a *quantitative characteristic* of the cytoplasm metallome of all living systems?

Further reading

Birch, N.J. (ed.) (1993). *Magnesium and the cell*. Academic Press, New York.

Brautigam, C.A., Aschheim, K. and Steitz, T.A. (1999). Structural elucidation of the properties of an exonuclease. *Chemistry and Biology*, **6**, 901–8.

Brewer, J.M. (1981). The nature of enolase. *Critical Reviews of Biochemistry*, **11**, 109–54.

Carrell, H.L., Hoier, H. and Glusker, J.P. (1994). Modes of binding to D-xylose isomerase. *Acta Crystallographica*, **D50**, 113–23.

Cowan, J.A. (1992). Magnesium coordination chemistry. *Comments on Inorganic Chemistry*, **13**, 293–312.

Cowan, J.A. (ed.) (1995) *Biological chemistry of magnesium*. VCH, New York.

Feig, A.L., Scott, W. G., and Uhlenbeck, O.C. (1998). The hammerhead ribozyme cleavage reaction. *Science*, **279**, 81–4.

Graves, B.J. (1995). Integrin binding revealed. *Nature Structural Biology*, **2**, 181–5.

Hanna, R. and Doudna, J.A. (2000). Metal ions in ribozyme folding and catalysis. *Current Opinion in Chemical Biology*, **4**, 166–170.

Kennelly, P.J. and Potts, M. (1999). Life among the primitives: protein phosphatases in prokaryotes. *Frontiers in Bioscience*, **4**, 372–85.

Kim, E.E. and Wycokoff, H.W. (1991). Mechanism of alkaline phosphatase. *Journal of Molecular Biology*, **218**, 449–64.

Moncrief, M.B. and Maguire, M.E. (1999). Magnesium transport in prokaryotes. *Journal of Biological Inorganic Chemistry*, **4**, 523–7.

Saris, N.E., Mervaala, E., Karppanen, H., Khawaja, J.A. and Lewenstam, A. (2000). Magnesium: physiological, clinical and analytical aspects. *Clinical Chimica Acta*, **294**, 1–26.

Scott, W.G. and Klug, A. (1996). Ribozymes. *Trends in Biochemical Science*, **21**, 220–4.

Sigel, H. and Sigel, A. (1998) Compendium on magnesium and its role in biology, nutrition and physiology. *Metal ions in biological systems*, vol. 26. Marcel Dekker Inc., New York.

Tipton, P.A., Quinn, T.P., Peisach, J. and Cook, P.F. (1996). Role of divalent metal ions in the NAD: malic enzyme reaction. *Protein Science*, **5**, 1648–54.

Urbauer, J.L., Bradshaw, D.E. and Cleland, W.W. (1998). Kinetics and mechanism of the malate enzyme. *Biochemistry*, **37**, 18026–31.

Wei, C.-H., Chou, W.-Y. and Chang, G.-G. (1995). The metal coordination of liver malic enzyme. *Biochemistry*, **34**, 7949–54.

Notes This chapter should be read together with the chapter on phosphate, Chapter 18. Phosphates and magnesium have been closely connected throughout evolution and the increasing involvement of them can be seen through comparative genomics. Rubin G.M. *et al.* (2000) Comparative genomics of the eukaryotes. *Science*, **287**, 2204–15. Mg^{2+} is the essential catalyst associated with DNA and RNA e.g. in transcription see Klug, A. (2001) A marvellous machine for making messages. *Science*, **292**, 1844–6.

CHAPTER 10

Calcium: controls and triggers

10.1 Introduction

It is almost impossible to overemphasize the role of calcium in *advanced* biological systems (Fig. 10.1). Everywhere one turns in the study of controls over metabolic pathways there is a calcium-dependent step (Table 10.1). Again, outside cells calcium is involved in multiple structural and cross-linking functions which we refer to mainly in later sections of this chapter. To those unfamiliar with the deep involvement of calcium in metabolism we give just three examples of central controls: (1) in light harvesting and its coupling to photosynthesis of dioxygen; (2) in the control of dehydrogenases in oxidative phosphorylation; and (3) in a multitude of kinase reactions. While metabolic control is one role of calcium there are others.

Table 10.1 Calcium-controlled events in cells

Activity	Controlled events or systems
Photosynthesis	Dioxygen release
Oxidative phosphorylation	Dehydrogenases
Receptor responses	(a) Nerve synapse (b) IP_3-linked reactions
Contractile devices	(a) Muscle triggering (actomyosin) (b) Cell filament controls (tubulins)
Phosphorylation	Activation of kinases, e.g. in fertilization
Metabolism	Numerous enzymes inside cells
Membrane/filament organization	Annexin-like proteins modulate tension
Cell division	S-100 proteins, immune system
Cell death (apoptosis)	Internal proteases
Hormone/transmitter release	Homeostasis
Binding to membranes	C-2 domains of enzymes
Cross-linking	Outside cells
Enzyme-activation	Outside cells; in membranes

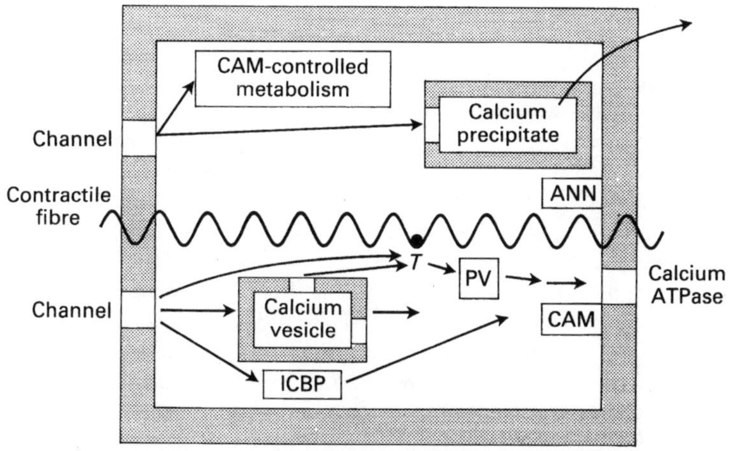

Fig. 10.1 A diagram to illustrate the intracellular activities which are controled by the calcium ion concentration. Included are calmodulin (CAM) control over metabolism, biomineral formation, tension (*T*) in filaments, and pumping (ATPase). ANN is the protein Annexin, which controls release of vesicle contents and contacts both membrane and filaments. The calcium transport is controlled by intestinal calcium binding protein, ICBP, and calcium buffering is limited by parvalbumin, PV.

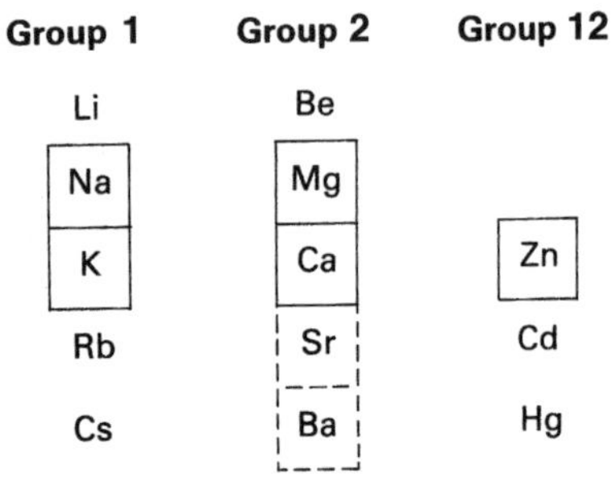

Some metal ions dominant in homeostasis (in boxes)

1. Calcium controls the mechanical stability of walls of some cells, of membranes of others, and of the tension in filaments in a whole variety of cells. Muscle contraction is stimulated by calcium as is exocytosis.
2. Calcium is intimately associated with fertilization, cell division, and hormonal activities.
3. Calcium salts, carbonates, phosphates and oxalates, dominate biominerals.
4. The extracellular matrix and its activities are calcium-dependent in multicellular organisms.

The difficulty in an attempt to summarize such a wealth of activities can be seen by looking at the flood of books and conference reports on calcium in organisms. The purpose of this chapter, therefore, has to be an attempt to give these detailed functional observations a rational basis at the risk of over-simplification. Only in the final chapter (Chapter 20) will this calcium biochemistry be pulled together with the biological chemistry of the other similar metals (see ASIDE) and non-metal elements which we believe to be involved in the homeostasis of acid–base cellular activity in a very general way. It will be made very apparent that there is a special relationship between calcium, magnesium, phosphate, and zinc, especially in hydrolytic reactions of advanced cells, and that this applies to plant, animal and fungal cells.

10.2 Free calcium ion levels

In Chapter 1 we discussed the levels of different metal ion concentrations in sea and fresh water. There is a huge variation in calcium concentration as anybody knows who attempts to wash clothes in hard as opposed to soft water never mind in the sea. Hardness of water is closely related to its calcium ion content. Despite this variation from, say, 10^{-6} to 10^{-2} M calcium in solutions outside organisms, the free calcium ion inside organisms is very, very, tightly controlled both inside cells and their compartments and in extracellular fluids. The free calcium in extracellular fluids is held close to 10^{-3} M and this allows a cellular system to manipulate the precipitation of carbonates (shells), phosphates (bones), and oxalates (plant structures) in a species- and organ-dependent manner by the way in which phosphate, bicarbonate, and oxalate concentrations as well as those of calcium itself and some special biopolymers are pumped into different zones. (Here again there are numerous books on this topic.) We stress the constancy and the absolute need for constancy of free extracellular calcium levels in complex organisms since there is a very narrow concentration range between desirable calcium for mineral precipitation in one location and the slightly lower desirable calcium levels for the maintenance of extracellular structures such as connective tissue at the other. There is little room for error, as we well know by observations on ageing populations who suffer from osteoporosis, calcifying arteries, and other malfunctions of calcium control.

In the cytoplasm of eukaryote cells a similar very, very careful control of free calcium levels is required. Here the level is near 10^{-8} M, but the authors believe that this level is not exactly the same in different cells types and that the differences are related to the biological functions of the cells. We shall see that those calcium-binding proteins which help to buffer calcium are present in different concentrations in different cells. Despite the fact that cytoplasmic free calcium can be measured relatively accurately it is difficult to get the *very accurate* data necessary about these low concentrations. This is especially the case since cells have shapes and, although it is assumed generally when these measurements are made that there is a homogeneous cytoplasm, there is good reason to believe today that not all cells are homogeneous in cytoplasmic calcium ion concentration due to the positioning of their pumps and channels (see Chapter 3). Calcium ions are then in a local steady state of flow. In passing note that, if free calcium in the cytoplasm reaches 10^{-5} M, a eukaryote cell, appears to commit suicide (apoptosis) by turning on calcium activated internal proteases, calpains. The situation in prokaryotes is very different and this will lead us to an intriguing possible role for calcium in the evolution of eukaryotes (see Section 10.22, and Chapter 20).

Now, there are further well-defined compartments in eukaryotic cells which lie behind membranes (Fig. 10.1): they are the organelles and vesicles described in Chapter 3. Figures 10.11 and 10.12 indicate how calcium is moved into many of them from the cytoplasm but we do not know accurately the steady-state free calcium level in these microvolumes. It is close to 10^{-3} M in the sarcoplasmic and other recticula. In general, vesicle space, as opposed to organelle space, is high in free calcium, but different vesicles may be widely disparate. Organelles, such as mitochondria, can remain viable with quite high internal calcium. (Organelles behave rather like prokaryotes, from which they originated, section 10.22.)

10.2.1 Calcium levels in pulsed cells

Eukaryote cells receive much of their information about the external environment from perturbation of their cytoplasmic membranes which allows a pulse of calcium to enter the cell. The pulse is quenched rapidly so that in $<10^{-2}$ s the cell returns to rest. The pulses may be located to extremely small areas of membranes and volumes of the cytoplasm, but they may act cooperatively under certain conditions. The detailed understanding of the individual events is slight, but the effect of the single pulse can be readily followed and interpreted. However, many cells receive series of pulses of the calcium ion rather than a single pulse. For example, the brain is constantly receiving messages about the outside world, and the heart and arteries beat regularly. When we examine the effect of fast trains of pulses it is clear that the effect of one pulse is such that the resting free calcium level cannot be fully relaxed before the next pulse is received. If we take the resting level of $[Ca^{2+}]$ in cells to be 10^{-8} M and that a single pulse raises the level in the cell at its maximum to 10^{-6} M then it may well be that a second pulse arises when the $[Ca^{2+}]$ has only fallen

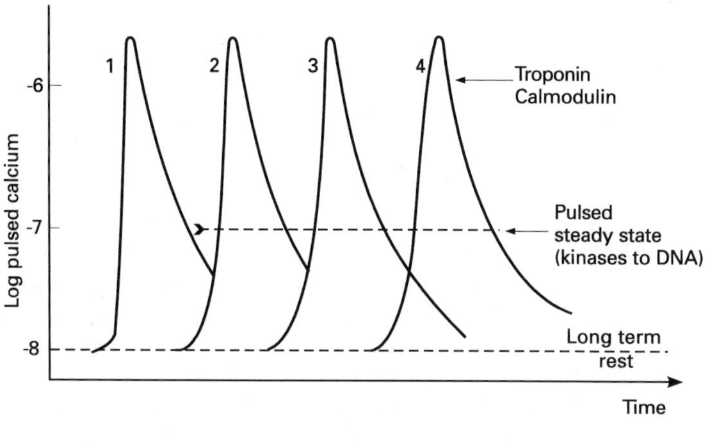

Fig. 10.2 The levels of free calcium in a cell following a train of fast pulses (1) to (4). The pulses must be fast enough so that there may never be complete recovery from a fast pulse before the next pulse is generated. Note the 'resting' level is raised from 10^{-8} M to 10^{-7} M by the train of pulses.

Table 10.2 Examples of Ca^{2+}-linked triggers

Example of triggering system	Effect
NO production linked to Fe proteins by calmodulin	Relaxation of slow muscles
Ca directly bound to fibre, Mg assisted	Tension of fast muscles
Kinase C linked to Mg, Ca, and Zn, then to transcription, metabolism, etc.	Protein phosphorylation (to gene expression)
Calcineurin (Ca) linked to Fe and Zn, then to immune system expression	Protein dephosphorylation (to gene expression)
Recoverin (Ca) linked to GTP.Mg action	Response to light (eye)
Dehydrogenases (Ca) linked to Fe/Cu, etc.	Mitochondria Oxidative phosphorylation
Water → NADPH + O_2 linked to Mn protein (Ca)	Chloroplast Oxygen release
Phosphorylase kinase activates glycogen use (Mg^{2+})	Messenger generation
Calmodulin (Ca) activates adenosylate cyclase (Mg)	Messenger generation
Neurocalcin (Ca) activates guanylate cyclase (Mg)	Nerve response
Gelsolin removes excess Ca	Actin control (cell movement)

to 10^{-7} M. If there is a constant series of pulses, the effective (resting) free [Ca^{2+}] in the cell remains at 10^{-7} M not 10^{-8} M (Fig. 10.2). Hence there can be at least two triggering processes following the initiation of pulses. The first involves a fast single pulse and calcium binding with a binding constant of 10^6 M^{-1} followed by complete relaxation, while the second effect can arise from a series of pulses with slower calcium binding and with a binding constant of 10^7 M^{-1}. Thus, a single pulse causes a skeletal muscle to be activated through the 10^6 M^{-1} binding constant of calcium to troponin followed by complete relaxation, while a train of pulses can lead to constant tension as in an ovarian cell activated on fertilization. Examples of calcium triggering are given in Table 10.2.

It is against this background that we have to follow the flow of calcium in the resting state and the ability of cells to be switched in activity by allowing pulses of calcium to enter the cytoplasm from outside or from vesicular systems. First, we observe the characteristics of the calcium ion which have allowed biological systems to utilize it in so many very important functions (see Table 10.1).

10.3 The calcium ion

Calcium is a large divalent cation (0.95 Å radius) and it is widely available since there is plenty of it in most natural waters and its common salts are not very insoluble. It is the intermediate strength of its binding in inorganic lattices and to organic molecules, as compared with the stronger binding of cations at the other end of transition metal series, such as zinc, which makes it so useful in control of conformational changes and then cell activities. However, switches of control (triggering) must be fast as well as of intermediate strength. We have analysed not only calcium binding and structure but also calcium kinetics in Chapters 4 and 9 and hence we can summarize our chemical knowledge under three headings.

1. *Calcium binding* is selective, see Aside, as a result of: (a) its ability to interact with neutral oxygen donors such as carbonyls and ethers, an interaction largely absent from magnesium chemistry in water, together with anions, hence also defeating competition from sodium; and (b) its ability to bind to a large number of centres at once—again an ability not shared by magnesium or any of the small cations such as zinc. These two factors give calcium a selective binding chemistry based on its charge-to-size ratio (see Tables 10.2 and 9.2) and, in addition, they make it an ideal cross-linking agent, as is seen in the structures of its compounds and complexes.
2. *Calcium structures* differ from those of the other available divalent element, magnesium; calcium is like sodium, potassium, strontium, barium, and the lanthanide cations in the high coordination numbers often observed in their compounds and complexes (Table 10.3 and 10.4). The geometry is not that of any simple regular symmetry and is controlled as much by the second co-ordination sphere, and solvation in solution, as by the first sphere. The contrast with magnesium (or aluminium) chemistry (Table 9.1) could not be greater. (Aluminium is mentioned here because of the acid rain problem, when aluminium may well compete for calcium sites but will not give the same structures.)

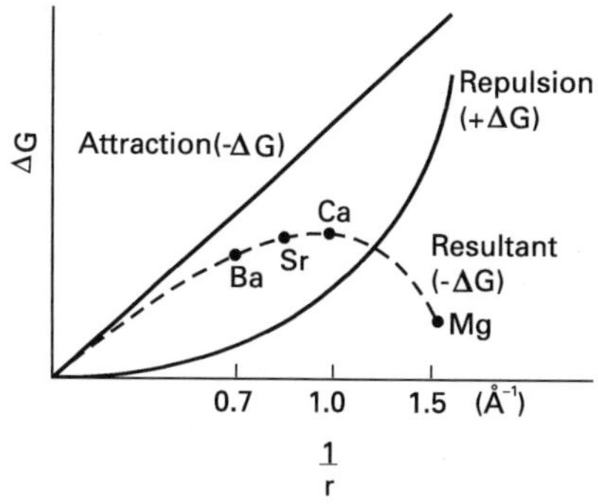

The figure shows how Ca^{2+} binding is favoured by ligand repulsion. *r* is the cation radius.

Table 10.3 The properties of some cations

Property	Na^+	K^+	Mg^{2+}	Ca^{2+}	Mn^{2+}	Ln^{3+}
Size (Å)	1.00	1.33	0.60	0.95	0.75	1.05 to 0.85
Coordination number	6(+)	8(±)	6	7(±)	6(+)	7(+)
Bond lengths	Irregular	Irregular	Regular	Irregular	Regular	Irregular
Bond angles	Irregular	Irregular	Regular	Irregular	(Regular)	Irregular
H_2O exchange(s)	10^{-10}	10^{-10}	10^{-6}	10^{-9}	10^{-7}	10^{-7}–10^{-9}

Mn^{2+} and Ln^{3+} are useful probes for Ca^{2+} activities. All these cations bind to oxygen centres almost exclusively, with the exception of Mn^{2+}. Na^+ and K^+ bind weakly to all donors except those in special cavities.

Table 10.4 Some typical calcium salt structures

Salt	Ca(II) coordination no.	Ca-O distances (nm)	
		Min	Max
$CaHPO_4 \cdot 2H_2O$	8	0.244	0.282
$Ca(H_2PO_4)_2 \cdot H_2O$	8	0.230	0.274
Ca 1,3-disphosphoryl-imidazole	6	0.226	0.236
	7	0.227	0.278
Ca dipicolinate $\cdot$ $3H_2O$	8	0.236	0.257
$CaNa(H_2PO_2)_3$	6	0.231	0.233
Ca tartrate $\cdot$ $4H_2O$	8	0.239	0.254
$Ca(C_5H_9O_7)_2 \cdot 2H_2O$	8	0.239	0.247

3. *Calcium kinetics* are also very different from those of magnesium in two respects. First, calcium exchanges water at close to the collisional diffusion limit of 10^{10} s^{-1} and its reaction on-rates are often diffusion-limited, whereas its off-rates are limited by binding strength. Second, within a complex the ligands on the calcium ion are *fluctional*. Magnesium (or aluminium) does not resemble calcium in either of these properties (see Section 4.6). In fact, in these respects calcium is again more like sodium and potassium and the lanthanide ions. It is the similar sizes of these ions, all around 1.0 Å in radius, which generates the similarities despite the differences in charge (Table 10.3).

With these features of calcium chemistry in mind we must turn to the properties of calcium-binding proteins, at first in general terms. The proteins are overwhelmingly from eukaryotes not prokaryotes.

10.3.1 Condensation equilibria

Now there is an intermediate condition between simple complex ion formation and precipitation. Consider a large polymer such as DNA with a distribution of negative charge all along its backbone. There is association with cations, (see Section 8.8) but it is observed that while the DNA remains in solution it can collapse, condense, into a very small volume at a critical concentration of the cation. It is not necessary for the cation to be bound directly to the polymer since it is held by the high surface potential generated by the fold. This is one of the risks of exposure of the cytoplasm of cells to high Ca^{2+} concentration. We believe that this is the mode of retention of Ca^{2+} by calsequestrin in the ER vesicles (Section 10.10.1). Here the protein surface has a very high concentration of carboxylate groups. Segments of it collapse, or rather now condense, but do not form precipitates, around some 40 Ca^{2+} ions, so that effectively they all have the same binding constant to the protein. Since the Ca^{2+} is in equilibrium with free ions it is readily released in quantity on exposure to solutions of low free Ca^{2+}.

10.4 Protein ligands for calcium

The major donor groups from proteins to the calcium ion are carbonyl and carboxylate centres. These groups can be put together in many ways, but three basic and very different constructions are known. The first involves a limited set of donors close together in the amino-acid sequence where the binding of calcium does not cross-link the protein through remote parts of the sequence so as to stabilize its fold grossly (Table 10.5). This is the mode of binding in calmodulin (Fig. 10.3) and in other trigger proteins from inside cells. It is likely that in the calcium-free state these calcium-binding regions are mobile and that there is considerable electrostatic repulsion between their large number of negatively charged centres. The second mode of binding involves groups distant in the amino-acid sequence and is largely dependent on a strongly predetermined fold which generates a rather immobile cavity for calcium. An example is phospholipase A_2 (Table 10.5 and Fig. 10.4 and 6.1). The striking differences between these two kinds of site relate to: (1) the effect of calcium on the structures; and (2) the rate of the on-reactions. The looser sequential binding site binds calcium very rapidly and adjusts the protein tertiary structure considerably (Section 10.7). The well-defined cavity type of binding site accepts calcium relatively slowly and there is virtually no conformational change. The third type of site has a longer series of anions on the surface often

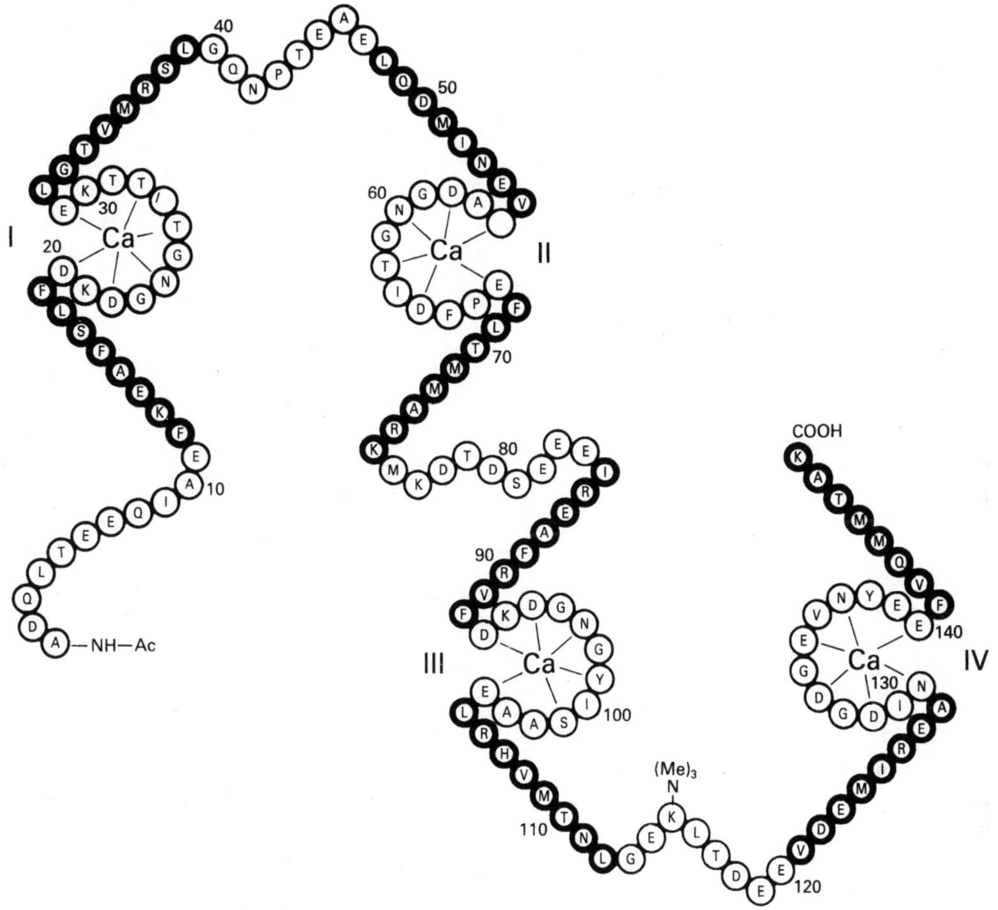

Fig. 10.3 An outline of the domains of calmodulin. The sequence of the protein is shown where heavy circles are helical stretches. Four calcium binding hands are represented: note acidic groups D and E. The helix–loop–helix motif forms the calcium-binding 'hand' as illustrated by the concave loop between a thumb and forefinger (see Fig. 6.2).

of a random polymer which come together through the sequence (or the fold) and amongst which the calcium can travel rapidly, for example, some phosphoproteins, the anionic glycoproteins of connective tissue, and the γ-carboxyglutamate (Gla) (see Section 10.5) containing peptides of prothrombin and the bone protein osteocalcin.

We can ask a series of questions about these sites. How do they avoid competition from Mg^{2+}? How is the calcium binding relayed into the structure? If magnesium should bind, would it have the same effect on the trigger? How does calcium diffuse in the cell to the sites? We must remember that the biological effect is related to calcium binding not only through binding constants, *K*, but also through limitations of rate based upon diffusion and the transmission of the binding through the structural change. It is therefore essential that we notice the character of the proteins involved. Finally we must ask where are the different calcium proteins within the cell and its various compartments (Chapter 3). The full implications of all these questions will not be addressed in this chapter and we shall return to their discussion in Chapter 20.

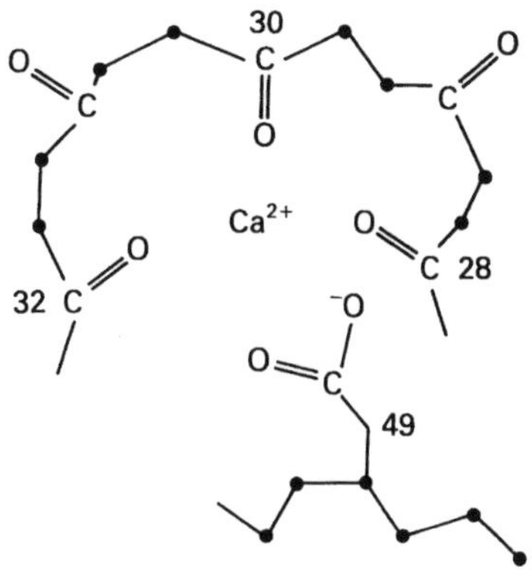

Fig. 10.4 The calcium-binding site of phospholipase A_2 (see Fig. 7.1). A relatively rigid hole is formed from the backbone carbonyls and the remote cross-linking carboxylates.

Table 10.5 Calcium-binding sites in some proteins

Extracellular	Sequence of ligands	Sequential?
Phospholipase A_2	Tyr-28, Glu-30, Gly-32, Asp-49, $2H_2O$	No
Staphylococcal nuclease	Asp-19, Asp-21, Asp-40, Thr-41, Glu-43, $1H_2O$	No
Trypsin	Glu-70, Asn-72, Glu-80, $2H_2O$	(Yes)
Thermolysin	Asp-138, Glu-177, Asp-185, Glu-187, Glu-190, $1H_2O$	No
	Asp-57, Asp-59, Glu-61, $3H_2O$	Yes
Intracellular		
Parvalbumin	Asp-51, Asp-53, Ser-55, Phe-57, Glu-59, Glu-62	Yes
	Asp-90, Asp-92, Asp-94, Lys-96, Glu-101, $1H_2O$	Yes
Calmodulin	Asp-20, Asp-22, Asn-24, Thr-26, Thr-28, Glu-31	Yes
	Asp-56, Asp-58, Asn-60, Thr-62, Asp-64, Glu-67	Yes
	Asp-93, Asp-95, Asn-97, Tyr-99, Ser-101, Glu-104	Yes
	Asn-129, Asp-131, Asp-133, Glu-135, Asn-137, Glu-140	Yes
Intestinal calcium-binding protein (calbindin)	Ala-15, Glu-17, Asp-19, Gln-22, Ser-24, Glu-27	Yes
	Asp-54, Asn-56, Asp-58, Glu-60, Ser-62, Glu-65	Yes
Troponin C	Asp-27, Asp-29, Gly-31, Asp-33, Ser-35, Glu-38	Yes
	Asp-63, Asp-65, Ser-67, Thr-19, Asp-71, Glu-74	Yes
	Asp-103, Asn-105, Asp-107, Tyr-109, Asp-111, Glu-114	Yes
	Asp-139, Asn-141, Asp-143, Arg-145, Asp-147, Glu-150	Yes

10.5 Magnesium/calcium competition

We know the structures of several magnesium-bound proteins, and we deduced in Chapter 9 that Mg^{2+} preferred sites of high negative charge density, e.g. ATP^{4-}, coordination number six, and/or of small cavity size. In this context we know that Mg^{2+} ions do not bind with any great strength to such calcium sites as that in phospholipase A_2 and this we attribute to the large size as well as the low charge of the cavity. However, Mg^{2+} will bind weakly to many of the calcium calmodulin sites, for example in an *intracellular* trigger protein, which are flexible and of higher charge (Fig. 10.2) with a constant close to 10^3 M^{-1}. We must suspect that, in the absence of calcium, Mg^{2+} ions (concentration near 10^{-3} M) will bind at least 10 per cent to some of these sites (and almost 100 per cent to others), since 10 per cent binding needs a binding constant of only 10^2. Even so, Mg^{2+} ions do not act so as to produce the calcium/calmodulin response since the conformational change in the calmodulin produced by Mg^{2+} is not the same as that produced by Ca^{2+}. Thus, while on the one hand magnesium may be excluded from calcium sites, on the other it may be an effective competitive inhibitor since the replacement of Ca^{2+} by Mg^{2+} does not lead to action.

A simpler example of the differences between the chemistry of Mg^{2+} and Ca^{2+} may be given by reference to the binding of the model malonate headgroup (see Aside) to the two ions. The Mg^{2+} compound, $Mg(C_3H_3O_3)_2 \cdot 2H_2O$, shows simple chelation to the malonate anion, the complex is octahedral, and the bond lengths and bond angles are regular. By way of contrast, the calcium compound $Ca(C_3H_2O_4) \cdot 2H_2O$ shows a co-ordination number of above six, the geometry, bond angles, and bond lengths are irregular, and cross-linking between anions occurs. These studies should be considered in the context of the differential binding effects of Ca^{2+} and Mg^{2+}, especially to the extracellular γ-carboxyglutamate (Gla)-containing proteins, such as prothrombin, since the function of calcium here includes cross-linking of the protein and a cell membrane while Mg^{2+} is inactive. The headgroup of Gla is malonate-like, see Aside. In cells the bonding ligand is usually such that calcium and magnesium chemistries are quite distinct; hence magnesium binding does not trigger calcium proteins.

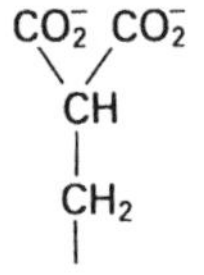

Malonate headgroup of γ-carboxyglutamate

Against the background of this summary of calcium ion chemistry in eukaryotes we shall divide the discussion of calcium function as follows. First, we describe the activity associated with calcium in the resting cell cytoplasm. This is followed by the analysis of triggering of cytoplasmic cell activities by the influx of a pulse of calcium from outside the cell. The recovery from the pulse as well as the entry of calcium requires calcium to pass through a membrane and this topic is the third in sequence. Next we look at the extracellular calcium and its steady state but we shall not concern ourselves overmuch with bones or shells. Finally, the calcium in vesicular systems and in organelles will be described. It must be understood that at rest the whole system is in almost perfect homeostasis so that there is always a dynamic exchange between all the calcium pools (see Chapter 20). This means that we must be aware at all times that the calcium ion is involved in cellular activities, currents, in the resting state, as it is in the extracellular compartments.

10.6 The resting state of calcium in cells

A problem that we face in the discussion of calcium biochemistry in relation to cellular activity is our lack of understanding of the cell at rest. Triggering to a new activity (see Section 10.7) is relatively well understood. At rest a cell is highly energized in all the ways described in Chapter 5 and, of course, is highly active in maintaining this state. This means not only that there is a metabolic rate at rest but that ionic gradients, units of structure, and tension in filaments are maintained in a highly regulated manner. Many of the activities depend on calcium levels and, while some new activities are produced by stimulants, many effects are just changes (up or down) of the resting actions. Two factors need particular attention. First, cells are not all alike and we would be wise to assume that their exact cytoplasmic steady-state levels of free calcium differ although they are all in the range 10^{-6}–10^{-8} M (Fig. 10.5). As far as calcium is concerned this means, for instance, that a slow posture muscle has a different free calcium concentration at rest, controlling its different metabolic rate, from that of a fast twitch muscle. It may well be that

Metal ion concentration

10^{-8} M	10^{-6} M	10^{-4}M	10^{-2} M
In-cell calcium	← Calcium pulse →	$[Mg^{2+}]$ →	← Extracellular calcium
Very few proteins bind calcium	Many in-cell proteins bind Ca^{2+}	Not observed for calcium	Many extracellular proteins bind Ca^{2+}
	$[Mg^{2+}]$ chromaffin granules		Na^+ and K^+ range

Fig. 10.5 The relative concentration of calcium and several other ions inside and outside cells, and the nature of the calcium ion pulse.

fast responses are characteristic of animals, e.g. motor muscles, and that plants are triggered more slowly, e.g. in the action of stomata. Even a twofold difference in the concentration of Ca^{2+} could be sufficient to alter a cell grossly since the effect of this change will be greatly amplified if the calcium dependence is proportional to $[Ca^{2+}]^n$ where n is large. (The resting cells are constantly receiving messages, of course, which keep the cell in the rest state.) Second, cells have shapes and in Chapter 3 we have seen that the arrangement of enzymes both along membranes and internal fibrils is related to cell shape. We know that the positionings of entry channels and exit pumps for calcium are also related to cell shape, so that a calcium current flows in particular ways in the cells and around them. The resting state is a state of constant calcium gradients everywhere inside and outside cells and must therefore be in constant flow. We return to this topic in Chapter 20. We describe next the activities of calcium when it is pulsed in and then out of the cytoplasm.

10.7 Calcium triggering: calmodulins

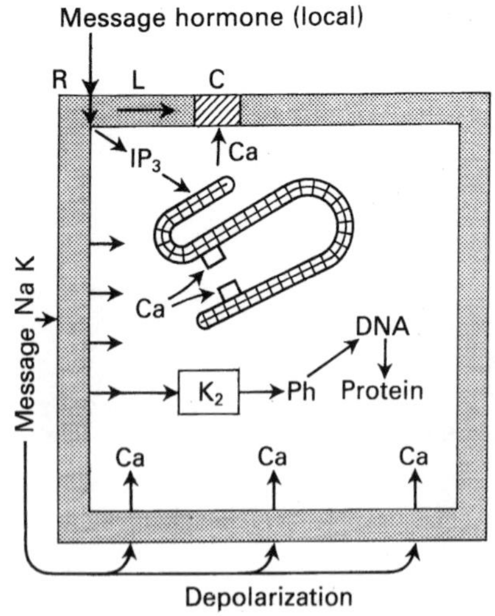

Fig. 10.6 There are two major message modes involving calcium. In the first, calcium enters the cell through depolarization of the whole cell membrane, Na^+/K^+ message. In local triggering a chemical message at a receptor, R, activates the breakdown of phospholipid, releasing IP_3, which then releases calcium from an internal store; finally, a kinase C is activated. Other kinases, K_2, are activated in the cell. Ph, protein phosphorylation.

A first message reaching a eukaryote cell can be of two kinds—electrical, e.g. a Na^+/K^+ ion nerve message, or chemical. In the first case the wave of depolarization produced at a cell membrane can be used generally to open appropriately located calcium channels which had been gated previously by the voltage across either the outer membrane or an extensive reticulum (Fig. 10.5). In the second case a chemical, a hormone or a transmitter, binds to a receptor protein on the outer surface of a cell and, by adjusting mechanically the calcium channel attached to the receptor, it opens the channel to calcium influx locally (Fig. 10.6). In both cases a calcium pulse enters the cell (Fig. 10.2), and this pulse, which is short-lived due to the recovery of the voltage or the removal of the hormone or transmitter, interacts with a whole series of related calcium-binding proteins. These are called troponin C, when bound into an actomyosin contractile fibre, or calmodulins, when bound to kinases or some other enzymes, or annexins, when associated with vesicular discharges, and so on. The effect of calcium is to cause a change in cell tension or cell chemistry (Fig. 10.1) and, simultaneously, to switch on recovery by outward pumping, especially stimulated by calcium activated energy production, part of which is used to reject the calcium via Ca^{2+}-ATPase pumps either to mitochondria or extracellular fluids. Fast muscle cells are triggered in this way.

Two summary points must be made. First, the calcium pulse can be localized by the locality of the stimulation and by the speed of recovery, e.g. in nerve termini. Second, as stated, if a new pulse is repeated regularly, the whole of the cell may adapt to a quite new steady state since the resting steady-state level of calcium must be affected (Section 10.2.1). If continued for some time the calcium may directly or indirectly, through changes in protein phosphorylation, generate altered gene expression. For example, a change in the rhythmic supply of calcium pulses can alter a fast muscle to a slow muscle. In other words, long repetitive calcium messages do not just interact with proteins but they strike indirectly to the level of RNA and DNA. The interaction through organic phosphates may be more subtle in that Ca^{2+} can stimulate the cyclization of nucleotides NTP, and then the cyclic-NMP can set off a series of reactions, once again reaching to DNA.

This triggering process concerns just the direct entry of calcium through the outer membrane or through a very extensive reticulum as in fast muscle. A somewhat more subtle process allows the calcium to enter the cell cytoplasm locally from the inner vesicles (Fig. 10.6). This process, used for example in smooth, not fast, muscle stimulation requires a message to be transmitted from the outer to the inner membranes. Routes for this excitation are via the release of inositol phosphates from the lipids of the outer membrane, or of cyclic ADP-ribose and or of nicotinamide adenosine diphosphate. Clearly, the process of inositol phosphate (say, IP_3) release, which we take as an example, followed by its diffusion to an inner membrane and then calcium release from vesicles e.g. the reticulum, is slower than direct release of calcium, but it is obvious that the space affected by the eventual release of calcium is more controlled since IP_3 activates only via its own receptors. In fact, the calcium release channels on the inner reticulum membranes are not exactly the same as those on the outer membrane (see Section 10.13). In some mechanisms the calcium back reacts with the zone of the outer membrane from which the inositol phosphate was released. Here, at kinase C, it meets the lipid L as shown in Fig. 10.6 which was released with the IP_3 and which acts as an in-membrane signal. The extremely limited zone then activated, usually a kinase C, can generate locally a combination of sequential events. The kinases C are phosphorylating agents for many proteins and many cellular events fall under their control. We shall return to such local effects in Chapter 20, where we shall see that other actions of calcium (1) at phospholipases also release fats in and from membranes, e.g. arachidonic acids, which lead forward to changes via prostaglandin signals, and (2) at phosphatases, generating transcription factors. Here again note that the removal process for calcium can be locally dependent on Na^+/Ca^{2+} exchanges of mitochondria as well as on the Ca^{2+}-ATPase of the reticulum from whence the calcium came (see Section 10.13).

There is a clear second distinction between the two kinds of message, electrical and chemical, general and local. The electrical stimulation, say in a fast muscle, is all over in 10^{-3} s, but chemical stimulation in a great variety of cells may well be much slower since there are many steps. Now, while there are proteins with the fastest response times to calcium, e.g. the calmodulins (troponins), there are also somewhat more sluggish calcium-binding proteins, e.g. the calbindins, S-100 and annexin families. Since the different proteins are distributed differentially in cell types, cells respond at very different rates and in different ways. The huge diversity of cellular responses to calcium make it difficult to give a coherent full account. Clearly, genes in differentiated cells of an organism are not equally active or activated by calcium signals. [This is an example which reinforces the need to know the proteome and the metallome in each compartment (see Chapter 20).].

Table 10.6 Some internal calcium-binding proteins

Protein	Response
Calmodulin	Kinases Phosphatases cAMP diesterases
Troponin C	Actomyosin ATPase
Gelsolin	Actin organization
Calpain	Protease
Parvalbumin	Calcium buffer?
Calbindin	Calcium transport?
S-100	Cell cycle? Inflammation
Annexins	Vesicle reactions and membrane attachments
C-2 domains	Enzymes

NB. Most of these proteins are largely helical and undergo helix/helix motions.

10.8 The calcium trigger proteins

The intra cellular calcium trigger proteins are listed in Table 10.6 and in this paragraph, for the sake of simplicity, we shall call all these proteins

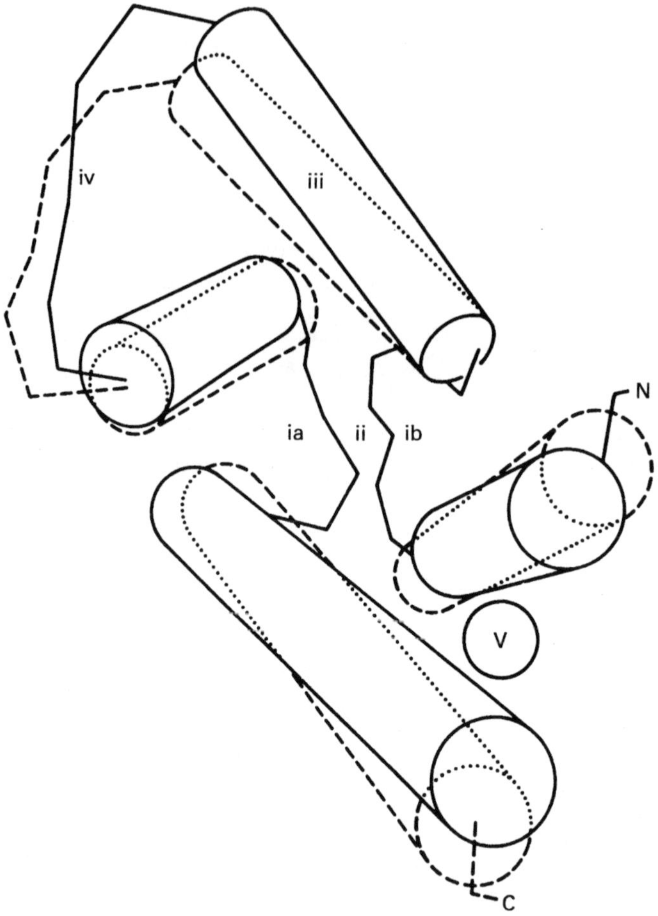

Fig. 10.7 The proposed helix/helix motion in the calmodulin family of trigger proteins on binding calcium. The calcium binds at (ia) and (ib) and the sites are connected by a short β-sheet (ii). The rolling of the helices reveals hydrophobic sites for binding drugs (v) (see Fig. 10.8), and changes the structure of linking units (iv). This type of helix/helix motion may well be the general mechanical action of membrane proteins, of DNA-binding proteins, and so on. The structure is after the work of R. Kretsinger (see also Fig. 6.3). The detailed workings of these proteins can be followed in the recent publications on calcium-binding proteins (see 'Further reading').

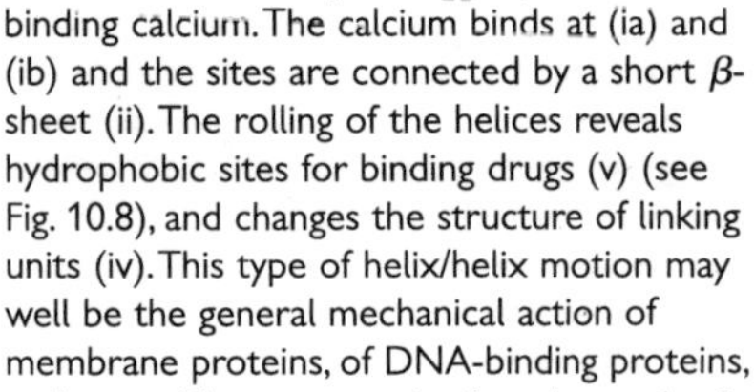

Trifluoperazine (TFP)

calmodulins. Here troponins C are included since they are all based on the same structural principles. The basic design of the calmodulin triggers is as follows. There are pairs of short sequences ia and ib (Fig. 10.7) which are called calcium-binding (EF) hands, placed back to back, and joined as a short two-strand β-sheet unit. This unit can be likened to the porphyrin (a sheet-like object) of haemoglobin. Leading from the unit, as in haemoglobin, are helices—two from each hand in the calcium-binding proteins (Figs 10.3 and 10.7). These helical stretches of sequence depend for their relative orientation and mutual disposition upon the exact bond angles within the β-sheet (iron porphyrin in the case of haemoglobin (Fig. 13.1) and on extraneous factors which bind at the linker regions of the helices or along the helices and which, through adjusting the helices, also adjust the sheet (porphyrin). The arrangement of the helices relative to one another is therefore not fixed. The changes in the linkers and exposure of the helices provide the connection to other proteins and the exposed helices are the sites of binding of non-competitive drugs (Fig. 10.7) such as trifluoperazine (TFP; see structure in margin). (Compare phosphoglycerate binding to haemoglobin.) However, given this outline of multiple helical stretches connected to several binding sites as well as to calcium hands, the choice of the binding sites for particular co-operating agents can be varied to give many combinations within calmodulin-like pro-

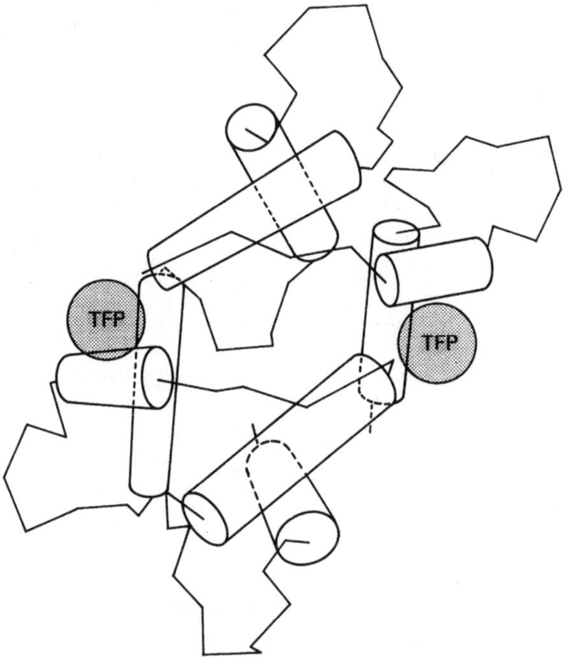

Fig. 10.8 The binding zone of a drug, trifluoperazine (TFP), to calmodulin so as to modulate triggering. Note that calmodulins have two four-helical bundles (four calcium sites) and binding to targets, after calcium binding, adjusts the two bundles relative to one another, see references to Bax *et al.* and Herzberg and James.

teins (Table 10.6). Calcium binding itself (compare the binding of O_2 to haemoglobin) causes a small change in the geometry of the β-sheet (compare changes of the haem) which twists and, in turn, rotates and moves laterally the helices (Fig. 10.9 and see Fig. 10.7). The calcium binding in calmodulins therefore interacts co-operatively with any object binding to the surfaces of any other part of the protein. In this way, not only can we see how long-range drive from calcium to a change in conformation of another protein (Fig. 10.10), e.g. an enzyme such as a kinase or the troponins which control contraction, can be managed via rotation and lateral motion of helices, but we also see how the drive can be modulated by any molecule (e.g. a drug) which binds to the surface of the helices (such as TFP) or by competitive binding agents at either end of the protein. A more complex unit can be made by doubling the number of calcium-binding sequences (Fig. 10.8 and Fig. 6.2).

It is possible that the activity of many mechanical devices in biological systems is based on the principles shown in Fig. 10.10. The exact way in which proteins like calmodulin activate kinases has been elaborated by detailed structural studies (see references).

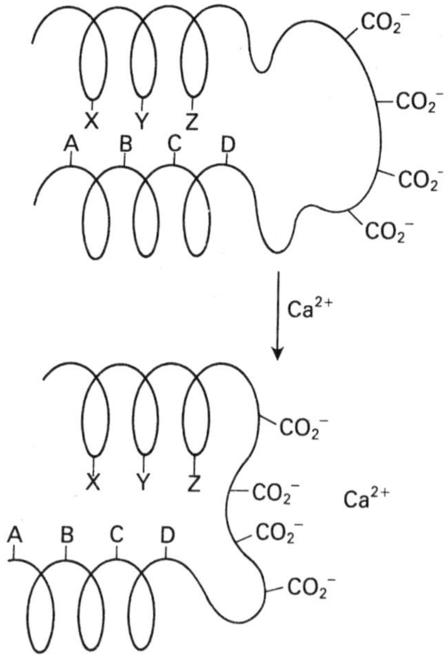

Fig. 10.9 A schematic representation of a calcium site in a subdomain (EF hand) of a calcium-binding protein. The donor groups, largely carboxylates (shown) and carbonyls, are in a sequence of a dozen amino acids. The site connects to helices which are adjusted relative to one another by the calcium binding. The rolling and the lateral motions of the helices are the essence of the triggers.

10.9 S-100 proteins

There is one group of calmodulin-like trigger proteins which does not act in quite the same way as those described, the so-called S-100 proteins, which were first found in brain tissue but are now known to be widely distributed. Their function is uncertain, but they are believed to be associated with calcium-dependent responses such as chemotactic actions, immune functions, and inflammatory responses. They frequently work as heterodimers. One unusual feature of the calcium binding is that one of the EF binding hands has very few anionic sidechains and resembles another group of proteins which are not triggers—the calbindins (Section 10.12). Another unusual feature of the

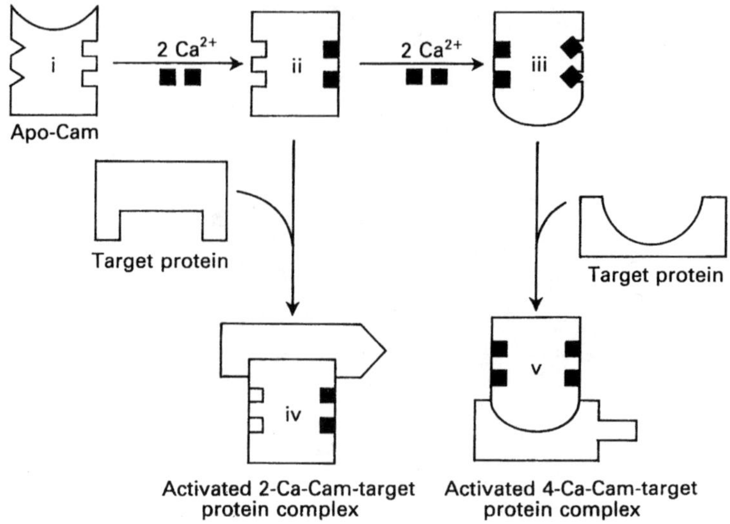

Fig. 10.10 A schematic representation of the sequence of binding steps where calmodulin. Cam, bound by calcium adjusts the conformation of a target protein. Two calcium levels of triggering for different targets are represented.

proteins is that the second of the two, the S-100B domain, binds zinc with a binding constant of around 10^6 M^{-1} or greater, which could be effective as a zinc-based trigger (see Chapter 11). In this case the levels of calcium and zinc may act co operatively in cells. The more usual co operativity is between calcium pulses and phosphorylation since the calcium pulse is often directed at kinases and phosphatases. The extensive activities of those S-100 proteins includes exchanges between cells, but these functions have not been fully described.

There are yet other kinds of these calmodulin-like proteins which undoubtedly enter into a huge variety of cell controls. Since their uptake and release is slow they can act as carrier in the cytoplasm (see calbindin in Section 10.12).

10.10 Other triggering modes: annexins and C-domains

We have just described the very similar characteristics of calmodulin and of the calcium trigger troponin C, of the fibrillar systems, e.g. in muscle contraction, and of the S-100 proteins which bind Ca^{2+} with a high affinity. There are at least two other types of intracellular triggering which show a calcium ion dependence. They are the vesicular release of chemicals by exocytosis, e.g. hormone release, and the alteration of cell shape on contact with a surface or with another cell. The calcium proteins associated with these membrane processes are usually of low Ca^{2+} affinity unless a charged, anionic phospholipid is present.

10.10.1 Vesicular contents and their release: exocytosis

There is a large group of calcium-binding proteins in cells, the annexins, which do not have quite the same amino-acid sequence in the calcium-binding sites as that found in the EF-hands of calmodulins and troponins. The sequence is similar in that it is has carbonyl groups and carboxylate sidechains, but it is shorter. When the proteins are isolated they have a much lower calcium-binding affinity, around 10^4 M^{-1}. The proteins are thought to

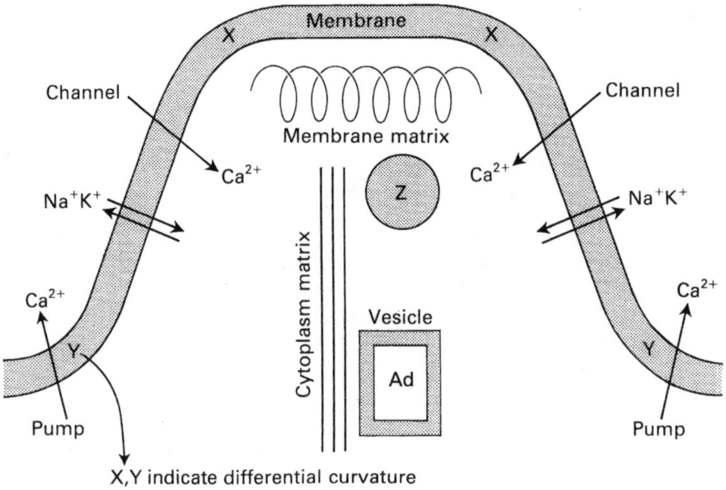

Fig. 10.11 The link between membrane, membrane-associated protein structures, calcium, and vesicle compartments. The diagram indicates that the whole is a unit of activity in that tension, curvature, positioning of pumps and channels, and movement of vesicles *all* depend on the activity of a cell in an interactive manner (see text). Cell shape is a steady-state property. Ad is adrenaline.

be bound to filamentous structures close to membranes (Fig. 10.1). It has also been observed that in the presence of anionic phospholipids the binding constant increases to around 10^6 to 10^7 M^{-1}. Clearly, only in these circumstances are they functional since the concentration of intracellular calcium on triggering only reaches 10^{-6} M. The function of these proteins is associated with the release of vesicular organic fast-acting transmitters or hormones, e.g. adrenalin. The following is no more than a suggestion as to how they may act.

Consider an arrangement of filaments extending to the cell membrane and which contains the negatively charged sequence of the calcium-binding protein Z near to the point of connection between proteins and membrane (Fig. 10.11) and perhaps in the membrane. Z clearly repels negatively charged phospholipids in the membrane so that there will be a distribution of these lipids in patches (Fig. 6.28). Let us further suppose that a vesicle having neutral phospholipids in its membrane, as is known to be the case for adrenalin vesicles, is held in not too distant a relationship to this membrane and that its position is decided partly by the pressure of the filaments, pushing it toward the membrane, but partly by an inability to reach the membrane due to the presence of proteins, including Z, and mutual incompatibility between the neutral vesicle lipids and the charged phospholipid of the outer membrane, Fig. 10.11. Addition of calcium, i.e. a triggering pulse, will alter the disposition of all the components in that calcium ions can cross-link Z to the anionic phospholipids due to its high binding constant when both are present, leaving more lowly charged lipids opposite the vesicle. The constraints on the position of the vesicle are removed and, moreover, the two membrane lipid layers, vesicle and cytoplasmic, are now compatible. Fusion follows, perhaps driven by calcium-controlled contractile factors in the filaments, and the contents of the vesicle are released. On removing the calcium by calcium pumps or by exchangers the former situation is restored and an empty vesicle is produced by endocytosis on the reversal of all the above steps. The vesicle can be sent back into the cell for recharging (see Chapter 3 and Fig. 3.5).

The preceding is no more than a proposal for vesicle discharge but it is given in order to show how calcium triggering could work in a rather different way from the triggering of muscle or of kinases. The action is now direct cross-linking by calcium, but it will generate conformational changes just the same. The calcium protein may well be designed in a way which is very similar to the design of calmodulin to allow conformation changes.

A further series of calcium binding units are the C-2 domains of certain enzymes such as the kinases-C and calpains. These domains assist internal attachment of the enzyme to the membrane after the cell has received a calcium message. Note that the kinases-C also contain zinc.

Once again it is important to observe the co-operative features of the calcium action. In all cases the proteins have several calcium sites so function depends on $[Ca^{2+}]^n$. In the case of annexins several phospholipid molecules interact with one protein molecule. Finally there is a selectivity in the construction of the individual helix-hand-helix unit which probably adjusts both the thermodynamics and kinetics of calcium binding.

10.10.2 Calcium, filaments, and cell shape

As we have stressed in Chapter 6, there appear to be several types of filaments in cells. There are: (1) semi-permanent multihelical structures such as actomyosins, which traverse the cell; (2) the multihelical proteins which lie under the membrane and are related to red blood cell spectrins; and (3) proteins related to properties of the nucleus as well as membrane such as tubulins. We have discussed the way in which calcium interacts with actomyosins and with filaments associated directly with the membrane. Much of this interaction can be visualized with the help of Fig. 10.11. The interaction with tubulin is less well defined and we have therefore described cation/tubulin effects under magnesium (Chapter 9) where it was stated that excess calcium opposed the polymerization of tubulin. It is thought that S-100 proteins, (Section 10.9) are closely involved with tubulins. The purpose of this paragraph is to pull together all these observations so that the dominant role of the Group 2 ions, Mg^{2+} and Ca^{2+}, in the control of the shapes of cells and their mechanical properties does not pass unremarked. The construction of filaments in cells seems to be Mg^{2+}/Ca^{2+}/phosphorylation-dependent and it appears likely that constant activity leads to the shape of cells as shown in Fig. 10.11. Remember, too, that the movement of vesicles along tubulins may be calcium dependent.

10.11 Calcium and protein phosphorylation

In Chapter 9 we have stressed the dependence of phosphate reactions upon the internal magnesium concentration: virtually all phosphate transfers, especially from nucleotide triphosphates, NTP, are Mg-dependent. This action is due to the association of magnesium with phosphate carrying small molecules at active sites. The role of calcium through calmodulin or calmodulin-like proteins is to switch, modulate, these enzyme actions, especially of the kinases. The calcium is not at the active site and works by remote conformational control. However, it does not affect all phosphate transfers but is mainly concerned with the initiation of a chain of signals.

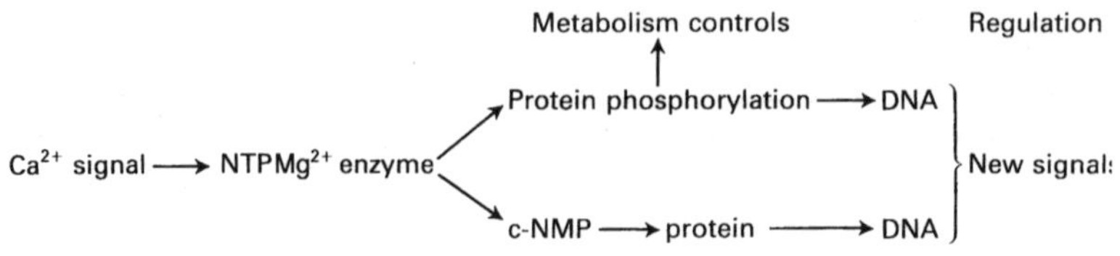

There is in this chain an increasingly sustained (in time) activity culminating in new protein synthesis. Earlier than this last action phosphorylation modulates the activities of all the proteins described in this chapter on calcium. Phosphorylation and dephosphorylation are intimately associated with calcium levels in an integrated system. We elaborate on the theme in Chapter 20 but the connection of calcium and phosphorylation with (1) homeostasis at rest, (2) rapid signals and relaxation, and (3) gross slower change in synthesis as in fertilization and development is a measure of their overwhelming importance in *eukaryotic* cells.

A particularly important phosphatase, calcineurin, is activated by calmodulin. The protein contains Fe^{2+} and Zn^{2+} at the active site and is heavily involved in immune responses since it acts, though indirectly, through dephosphorylation of transcription factors.

10.12 Calcium buffering and calcium transport in cells

So great is the need to control calcium levels in cells that any, even moderate, excesses after stimulation must be quickly reduced so that the cell is ready for action once more. Very fast muscles, which may be called upon quickly several times a minute, contain a protein called parvalbumin (Fig. 10.1 and Fig. 3.13), which is similar to a calmodulin but has an extra non-calcium binding segment, the presence of which seems to block action in the sense that it is not known to activate any metabolism nor to bind to any construct in the cell. Since it binds both Mg^{2+} and Ca^{2+} more strongly than calmodulin does, it can act as a delayed buffer in the way described in Chapter 9.

Equally important in some cells is the redistribution of calcium within them when calcium enters locally. We pointed out in Chapter 3 that the input and output systems for ions were located in different parts of certain cells (Fig. 3.16). For example, many epithelial cells see on one side a supply of food which includes calcium ions and on the other the circulation fluids of the organism which must have a maintained Ca^{2+} concentration. It has been observed that these cells in particular have very high levels of small calcium-binding proteins called calbindins (Fig. 10.1). (There are in fact varieties of this protein.) This location of the calbindin and is association with vitamin D activity, as well as its known binding affinity for calcium, $K_{Ca} \simeq 10^7\ M^{-1}$, suggest that its role is to transport calcium across cells from the calcium channels at one side of the epithelial cells to the calcium ATPases at the other side. This carrier transport action, due to 'metal chaperones', has been shown to be more effective than simple diffusion of Ca^{2+} ions at 10^{-7} M since the calcium calbindin association constant can be as high as 10^{-3} M. If the calbindin is correctly described, then the cellular system has gone to extraordinary lengths to manage calcium distribution within these and other cells, and indeed within the organism. It is certainly the case that no other potential role of calbindins has been found. The protein is deeply associated with the maintenance of the life of the whole organism via calcium levels since its synthesis is under the control of vitamin D, which in turn is controlled by sunlight, and this vitamin is known to control too the well-being of bones through calcium levels in the blood, (see Table 10.7). We note also that neuronal cells often contain high calbindin levels, but their function is uncertain, [see Williams (1992) in 'Further reading.'].

10.13 **Calcium currents: movement through membranes, channels, gates, and pumps**

In certain cells, e.g. Purkinje cells of the heart, the Na^+/K^+ current is replaced by a calcium ion current. The full significance of these currents as well as of proton currents around cell membranes will be analysed in Chapter 20. Here we illustrate first some features of calcium current using the skeletal muscle cell. In Fig. 10.12 the contact zone between the sarcoplasmic reticulum vesicular system and the muscle fibres is shown. The calcium flows constantly at a very low level through the channels in the region at the tip of the vesicular systems and is buffered in this reticulum region by the presence of the protein calsequestrin with binding constant, $K = 10^3\ M^{-1}$. Calcium levels ($<10^{-7}$ M) around the fibres are maintained by the pumps, ATPases, in a different part of the reticulum membrane. The organization is very like that in Fig. 10.11 and the whole of the activity is shape-related. A pulse from a nerve causes depolarization of the membrane (see Fig. 10.13) and release of calcium into the fibre region gives contraction and then recovery. The effect on the muscle fibres is shown in Fig. 10.13. Should the pulsing be maintained in sustained rhythmic sequences, then the system changes even its morphology since such activity changes the calcium levels in the cytoplasm and eventually affects DNA expression. This description of the muscle cell and its dependence on calcium levels may well be quite general to cells.

The movement of calcium into a cell is managed through calcium channels which, as stated earlier, may be electrically operated, voltage-gated, or opened by signal molecules which bind to receptors on the outside of membranes. Little is known about any of the voltage-gated channels for cations or indeed of the receptor-operated system except for sets of amino-acid sequences of their proteins. The sequences have been used to describe the channels as multihelical transmembrane proteins when the gating requires either charge

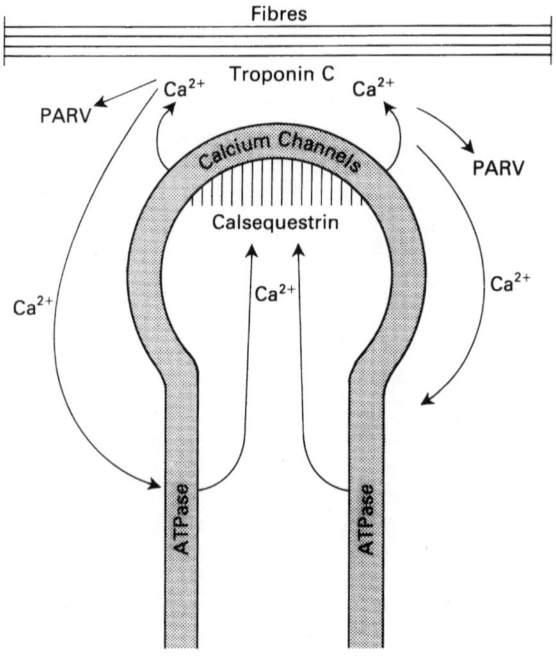

Fig. 10.12 The release system for calcium at the tip of a sarcoplasmic reticulum of a muscle cell. The calcium channels are placed very close to the contractile fibres and here calcium is stored in a calcium-binding protein, calsequestrin. The calcium is removed by parvalbumin, PARV, and then by pumps (ATPase). The whole organization has a weak current flow which can be stimulated by a nerve or chemical pulse.

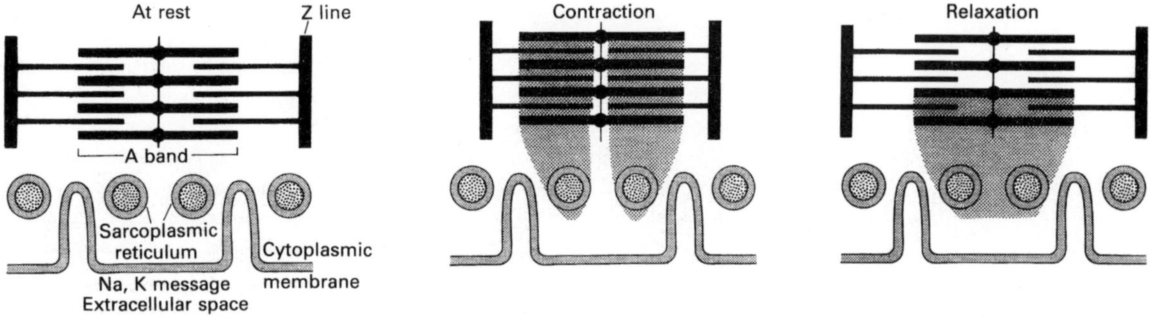

Fig. 10.13 The sliding filament model of muscle contraction. The calcium is shown as dots released from and returning to the sarcoplasmic reticulum (see Fig. 10.11). The contact between the filaments, actomyosin, which is activated by calcium . (see Fig. 10.1) is via a device in the myosin head which acts mechanically by helical movements like that of the calmodulins (see reference to muscle; see also Fig. 10.16).

rearrangement (voltage gates) or the realignment of the helices through a mechanical device to allow ion passage (chemical gating). We have described both modes of action in Chapter 8 under K^+ channels since the principles do not differ for any ion or molecule channel. Selectivity for calcium is achieved through the operation of structural factors outlined at the beginning of the chapter. A major group of drugs is the 'calcium blockers', for blood pressure control, one of which is shown in the margin.

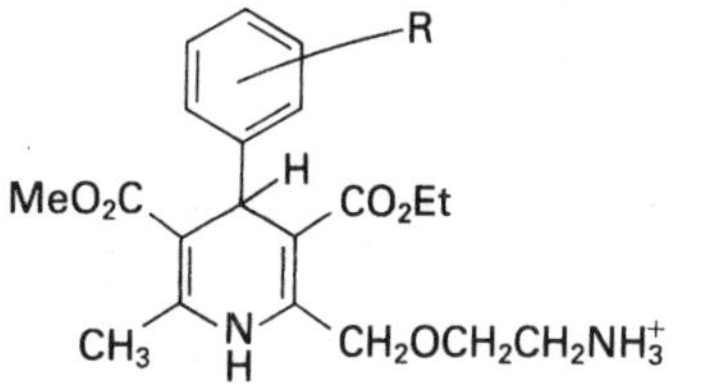

One series of dihydropyridine calcium channel blockers, e.g. nifedipine. R, a wide variety of substituents.

Calcium leaves the cell through pumping (or exchangers) which requires the input of energy to the pump in which the calcium ion sees a high-affinity site from one side of a membrane and a low-affinity site from the other. The metal is driven from the high- to the low-affinity state supposedly by movement of helical segments of the ATPases in the membrane. This enzyme is illustrated in Fig. 10.14 (compare the Na^+/K^+ and H^+ pumps) and the binding site for pumping is presumably in the membrane though other sites may exist. The pumping is slow, 100 times s^{-1}, hence the need for parvalbumin in very striated fast muscles (see Section 9.8.2). For the evolution of pumps, see Section 9.18. The Ca^2 exits to mitochondria, to the endoplasmic reticulum and to the extracellular fluids.

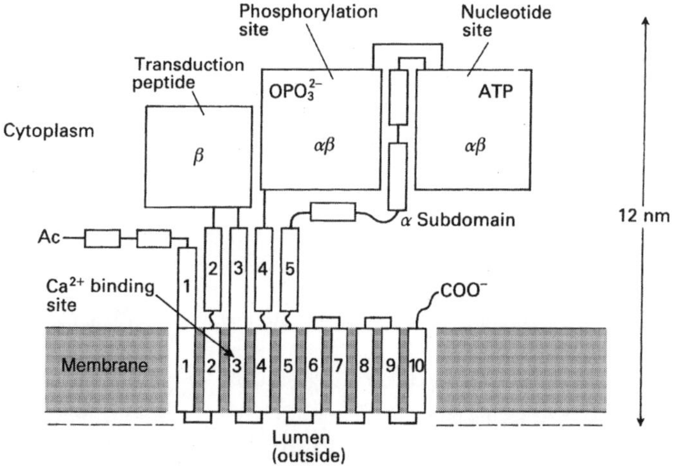

Fig. 10.14 A diagram of an ATPase metal-ion protein pump here for calcium. The ATPase segment is inside the cytoplasm. It undergoes conformational changes (see Fig. 6.9) and drives rearrangement of helices in the membrane where calcium binding is adjusted (after M. Green). A simplified picture has been given in Fig. 8.7 for the general case of ion pumps. Large motions occur between the $\alpha\beta/\beta$ subunits. The full structure is now known (see 'Further reading').

10.14 Calcium exchangers

There are alternative modes of moving Ca^{2+} in or out of cells by exchangers. In bacteria there are Ca^{2+}/H^+ exchangers even where the above ATPases may be absent. Presumably, these exchangers kept Ca^{2+} low in the primitive cytoplasm. The exchangers are usually of high capacity, but do not work to as low a Ca^{2+} level as the ATPases. They are also under feed-back control. It is particularly noticeable that the Ca^{2+} exchangers are confined to certain cell types and to organelles. Cells which have large amounts of Na^+/Ca^{2+} exchangers are, for example, those of the heart, which act relatively slowly. These exchangers are also present in organelles and may well be linked to the control of energy production by calcium both in mitochondria and chloroplasts (see Section 10.1).

The final movement of calcium and many other ions and molecules is through the gap junctions which join many cells keeping their metabolism in contact. The only reason for describing this junction here is that the gap junction is governed by calcium ion concentration. In the presence of 10^{-5} M calcium in any cell all gap junctions between it and neighbouring cells are closed. The cell is excommunicated and doomed.

10.15 Internal calcium-induced proteases: apoptosis

There is a further series of calcium-dependent proteins in cells which have both C-2 and calmodulin-like segments but which do not bind calcium very strongly. They are the calpains which are proteases when calcium is bound. Their function is not known but we can conceive a plausible use for what can only be a very damaging activity in a cell.

These proteins bind calcium with a binding constant of 10^5 M^{-1}. Now when the concentration of calcium entering a cell reaches 10^{-5} M a vast range of activities is switched on and a great number of otherwise dissociated units become cross-linked. The cell can be so seriously damaged that it may well be the case that to complete its degeneration, *apoptosis*, is a better solution than an attempt at recovery. A switch-on of proteases activated at 10^{-5} M calcium would bring about cell death. It is known that when calcium reaches such high levels the calcium also closes the gap junctions between cells so that death of one cell does not spread to others.

In Section 10.17 we shall describe a quite different series of calcium controls on *extracellular* proteases which are stored in vesicles in cells that are not exposed to internal calcium fluxes. While we know of these intracellular proteases there is no knowledge of calcium-dependent intracellular nucleases although nucleases are required to remove damaged DNA. As we shall see some extracellular nucleases are calcium dependent, while intracellularly they use Mg^{2+}.

10.16 General remarks concerning control systems in cells

In the light of the proceeding sections and the information concerning Na^+, K^+, Mg^{2+}, and Ca^{2+} ion actions let us look at the time sequences in the intra cellular control systems of biology (Fig. 10.15). We shall start from two assumptions.

1. The fastest movements possible for transfer of information are limited by diffusion of small ions or molecules. The usual number to quote is the

maximum rate constant of diffusion, $10^{10}\ s^{-1}$. An electrolytic current can not move charge faster than, say, 10–100 Å in 10^{-10} s.

2. To create an effect such as a specific switch of a protein to a new conformation will require some minimum of, say, 10 kcal, so as to avoid accidental thermal switching, which means that thermally the switch is from a ground state which in the absence of the 10-kcal energy input is some 10^7 times more likely to be inactive than active. This is a reasonable value, 10^3 being too small and 10^{10} unnecessarily large. The switch is a secondary structure change. The fastest possible triggering of a control is then given by the off-rate, i.e. relaxation

$$\frac{\text{Diffusion on-rate}}{\text{Off-rate}} = \frac{10^{10}}{\text{Off-rate}} = \boldsymbol{K} = 10^7$$

since it is only binding which can give energy to a chemical system. The off-rate is thus $10^3\ s^{-1}$ and biological systems operate at their fastest in the 10^{-3} s range. It follows that this is the speed of the nerve with a calcium-dependent protein booster system (Fig. 10.15) of muscle contraction, of hormone release, and so on. It is also the optimal rate of protein phosphorylation controlled by calmodulin. It is very important to observe that all calcium controlled switches have binding constants $K \simeq 10^7$ to allow enough energy to enter the system and to permit fast removal. Only calcium (or strontium, which is not available) ions can demonstrate this property.

If a more selective response is wanted and the system can afford time then there are many alternatives, including the following:

1. We can make coming together slower so that more precise fitting is achieved, i.e. $k_{on} < 10^{10}\ s^{-1}$.
2. We can arrange for better binding which will allow the use of rare species, zinc or most organic triggers, and a consequent slow off-rate.
3. It is possible to adjust the resting Ca^{2+} level of a cell by rapidly repeating input waves (Section 10.2). This is a separate way of generating a long-term message which can communicate with the DNA.

In Fig. 10.16 the local concentrations of Na^+, K^+, Ca^{2+} and of the fastest organic trigger, acetylcholine, are greater than 10^{-3} M. They all act very rapidly. Molecules such as sterols are insoluble in water and difficult to

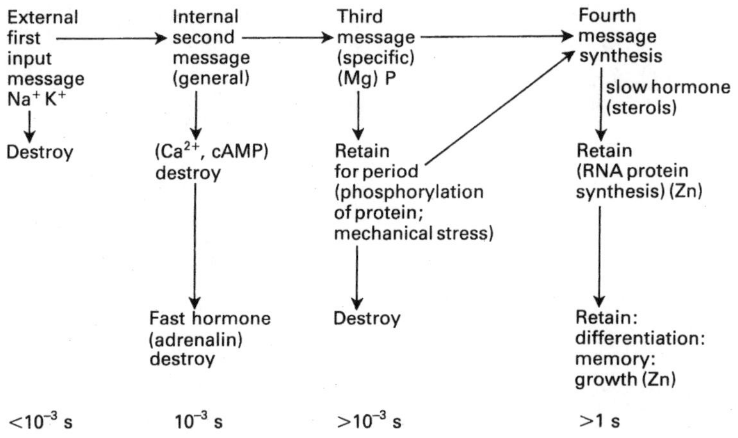

Fig. 10.15 The time sequences in biological systems and the metal ions involved.

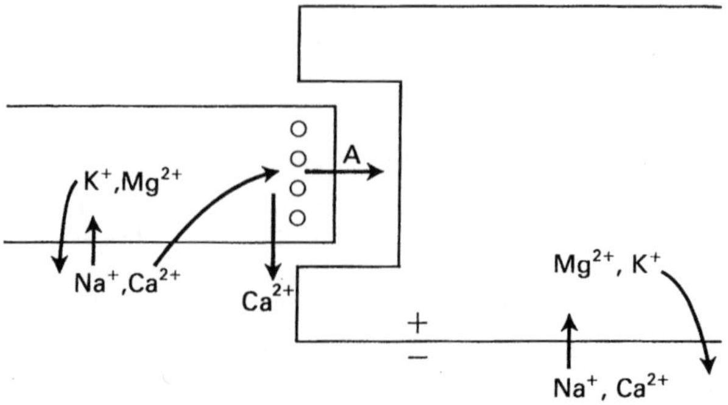

Fig. 10.16 The left-hand cell has a Na′/K′ gradient. Influx of Na′, a nerve message, causes eddy currents to flow from left to right until a Ca^{2+} channel is opened near A (acetylcholine or GABA). The calcium releases the contents of the vesicles, see Section 10.10, which now trigger the nerve message in the right-hand cell. There follows flow of Na′ and K′ across the membrane of this second cell.

transfer. They must have high binding and work as regulators not triggers. With metal ions such as Zn^{2+} they do not exceed 10^{-10} M in cells. Such ions as Zn^{2+} are buffered at that level. If they are to be useful in maintaining steady states then K for them has to be $\geqslant 10^{10}$, as for sterols, when their on/off controls cannot be faster than 1 s^{-1}. We begin to see how inorganic chemistry has been put to use in extremely sophisticated ways in living organisms. (Section 4.5).

10.17 Extracytoplasmic calcium in vesicles

In a multicellular organism it is essential that all the calcium activities, i.e. in the cell cytoplasm, in vesicles, in organelles, and in the extracellular fluids, are in communication with one another since all the calcium is locked into a vast network of controls. The effect of a calcium pulse into the cytoplasm is obviously deeply connected to the non-cytoplasmic calcium concentration in the vesicles, as we have mentioned already.

10.17.1 Vesicular calcium

The calcium which is pumped away from the cytoplasm is not all pumped to the circulating fluids. Much is pumped into vesicles including the reticula and some is pumped into organelles. The pumps are arranged in all the respective membranes, but there are also ion-exchanger systems which keep the balance between a variety of ions in different zones of the cell.

In reticula there are special proteins for calcium (Table 10.7), e.g. calsequestrin in the sarcoplasmic and endoplasmic (ER) reticula, which are positioned to assist local release or stimulation (see Fig. 10.12). In the Golgi apparatus and ER (Fig. 3.3), there is also a variety of calcium-binding proteins acting as *chaperones*, including calreticulum and calnexin. Particularly interesting is the management by these chaperones of the folding of iron and copper proteins for export. Zinc also binds to calreticulum.

Table 10.7 Calcium proteins in vesicles

Vesicle	Protein	Function
ER/Golgi	Calnexin	Chaperone
	Calreticulum	Chaperone (Fe, Cu proteins)
	Calsequestrin	Ca^{2+} store
ER or calcizomes (vesicles)	Calsequestrin	Ca^{2+} store

10.17.2 Organelle calcium

The functioning of organelles, mitochondria and chloroplasts, is equally dependent on calcium levels as is that of the cell cytoplasm but in a very different way (Table 10.1). In some ways this is not surprising since these organelles may well be of independent evolutionary origin. The peculiarities are that there are differences between mitochondria from plants and animals. *In vitro* animal mitochondria can pump in calcium even to the extent of precipitating calcium salts internally although this may never occur *in vivo*. Calcium regulates some of their enzymes, including some essential dehydrogenases. In plant mitochondria the uptake of calcium seems to be less. In chloroplasts the functions of calcium are not really known, although it is found that this ion and chloride anions have essential roles in early events of photosynthesis especially associated with dioxygen release. It would appear that the roles of calcium are much better understood in advanced cells than in organelles.

Table 10.8 Vitamin D controlled expression

Protein	Function
Calbindin	Ca^{2+} transport
Osteocalcin	Binding to bone
Carbonic anhydrase	Carbonate/CO_2 equilibration
Alkaline phosphatase	Phosphate ester hydrolysis
Collagen	Connective tissues
Metallothionein	Zinc 'buffer'?

NB. It may well be that thyroxine also links to a variety of calcium binding proteins and to connective tissue proteins, see also the role of retinoic acid.

10.18 Extracellular calcium in circulating fluids

In this chapter we have insisted that the level of extracellular calcium in the circulating fluids of higher animals is as exactingly controlled as intracellular calcium. The controls are exerted via epithelial cell pumps, through the kidney tubules, and through buffering by bone. There are two major organic hormonal controls based on the parathyroid hormone, the release of which is itself dependent on the extracellular calcium level and the vitamin D related controls which are stimulated by light and activate the uptake through the epithelial cells into the circulating fluid (Table 10.8).

This calcium interacts directly with a variety of biological polymers and with many biological minerals. The polymers are of great variety. Some are concerned with the matrices which hold cells together—connective tissue—and these polymers include polysaccharides as well as proteins. In plant tissue the polysaccharides greatly exceed the proteins, and note that a mixture of these two forms the cell walls of unicellular organisms—both eukaryotes and prokaryotes. Calcium cross-links many of these polymers in a well-balanced sol or gel state depending on the cells under discussion, Fig. 6.26. Binding is to carboxylate and phosphate sidechains as well as to neutral O-donor groups.

Some of the polymers involved with calcium are enzymes, Table 10.9. Probably in the majority of cases the calcium only helps to stabilize the fold of these proteins but in several cases it assists in activation. There are two outstanding examples—digestion and blood clotting. The triggering here is reminiscent of that of calpain, the intracellular protease.

Table 10.9 Some exported calcium enzymes

Protein	Function of Ca^{2+}
Chymotrypsin, etc.	Activation of protease
Thermolysin	Stabilization of protease
α-Amylase	Hydrolysis of polysaccharides
Phosphatases	Activation of hydrolysis
Nucleases	Splitting of DNA and RNA
Phytase	Hydrolysis of phytic acid

10.18.1 Calcium in digestion and blood clotting

In the fluids around cells, or in specialized cells in some organisms, it is necessary to have hydrolytic enzymes to degrade polymers into small fragments so that monomers can be pumped back into cells for synthesis and energy metabolism, Table 10.9. This applies to fats, polysaccharides, polynucleotides, and proteins. The degradation is carried out by a series of enzymes released from cells, by calcium input, often from vesicles where they are held in precursor or pro-enzyme forms. The need to keep enzymes in inactive states is also apparent in processes such as blood-clotting, antigen responses, and so on (Table 10.10). Enzymes are made available from pro-enzymes by various devices including binding to the calcium ion which may or may not affect catalysis directly. In the cases of enzymes such as the protease chymotrypsin

Table 10.10 Extracellular calcium proteins

Protein	Binding centre
EGF-domains of proteins (as in integrins and fibronectin)	(Hydroxy-aspartate) or two aspartates (blood-clotting)
C, S, Z, and VII, IX, and X Factors Osteocalcin (bone) Prothrombin	γ-Carboxy-glutamate plus aspartate or glutamate
Complement C_{1r} and C_{1s}	EGF-like?
Lectins	Carboxylates and carbonyls (cadherin)
Connective tissue	Many examples

NB. Most of these proteins are cross-linked by –S S– bridges and many contain β-sheet structures. Viral coat proteins often contain calcium

the function of calcium is to unwind a part of the tail of the enzyme which, in the pro-enzyme form, prevents reaction. It goes almost without saying that the release of this digestive enzyme from the pancreas when food is in the stomach is due to a calcium signal to that organ. The pancreas contains vesicles which hold the pro-enzyme in a harmless condition and in a solution of low calcium.

Now while this mechanism of altering the conformation of pro-enzymes is used in many digestive enzymes and also, in a somewhat different manner, in the blood-clotting cascade discussed later in this section, calcium has a different influence on some nucleases and phospholipases. The case of phospholipase A_2 has been particularly well studied.

Phospholipase A_2 has the fold illustrated in Fig. 6.1. The enzyme is without activity in the absence of calcium ions and in their presence it can digest phospholipids in aggregated states. The site of calcium binding is a hole in the protein surface close to the known attacking groups for hydrolysis. The function of calcium is presumed to be to hold the phosphate group of the substrate in a suitable orientation for attack by the active site histidine. The calcium is held in a completely different manner from that seen in calmodulin. The binding groups are mostly neutral oxygen donors held on the rigid framework of the protein (Fig. 10.3) and, to limit mobility, the protein has both a β-sheet construct and helices cross-linked by seven –S–S– bridges (Fig. 6.1). A more rigid structure could hardly have been constructed, again in marked contrast with the loose assembly of helices of calmodulin.

Enzymes of the same kind as phospholipases and external proteases are found in all digestive juices and in toxins, e.g. snake toxins. Such enzymes are also needed in fertilization. Much faster peptidases are derived from very different enzymes (see Chapter 11).

In blood-clotting there is again a new calcium trigger device (Figs 10.17 and 10.18). Here the extracellular calcium has a dual role with the membrane phospholipids rather like that with annexins in cells. Once injury to the organism has occurred, the calcium activates a set of proteases, through unfolding a series of multidomain proteins, and simultaneously binds domains locally to cell fragments and membranes, represented by PL (phospholipids) in Fig. 10.17. The effects are brought about by a very open binding site for calcium composed of a specially synthesized amino acid, γ-carboxyglutamic acid (Gla) (Table 6.2 and Fig. 10.18). By having several Gla units close together in sequence the effect of calcium binding is to change a very stretched out but random polymer segment into a highly organized one while leaving calcium ions (several) in part exposed so that they can bind phospholipid or other anionic headgroups of the damaged tissue. The multidomain protease prothrombin, for example, also readjusts its

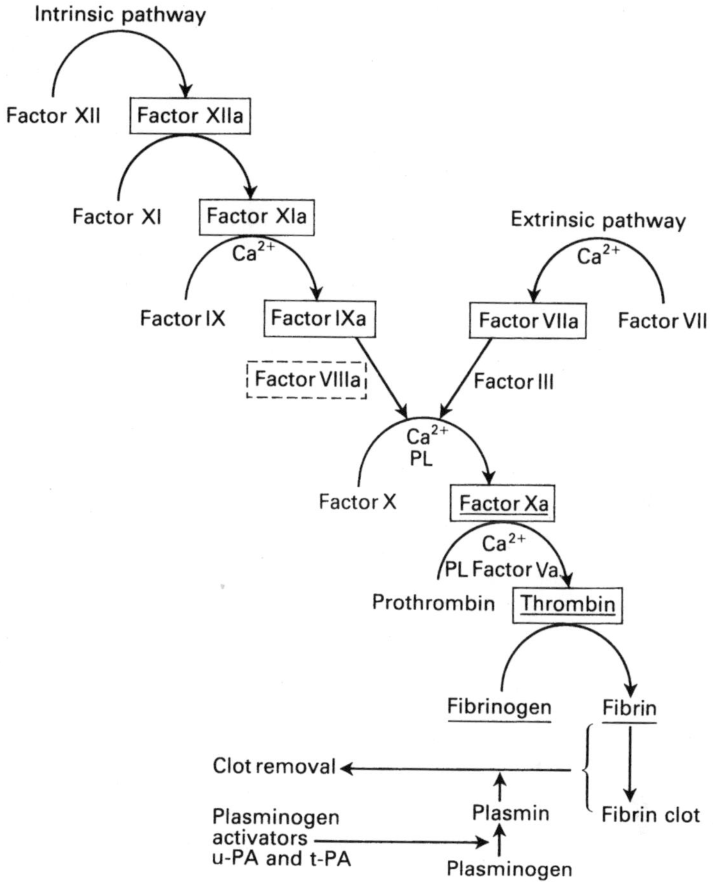

Fig. 10.17 The two blood-clotting cascades leading to the fibrin clot. A series of proteins (factors) in blood, described in detail in Fig. 10.18, becomes activated when tissue is damaged, releasing enzymes shown as square boxes. On release, the first enzyme acts on the second factor and so on. Each protein has its own receptors, C, G, K, F, for several additional activator molecules including calcium, Fig. 10.18. The last factor, fibrin, coagulates to form the clot. Removal of the clot is by a similar process (bottom left) and the two actions are always in a somewhat precarious balance.

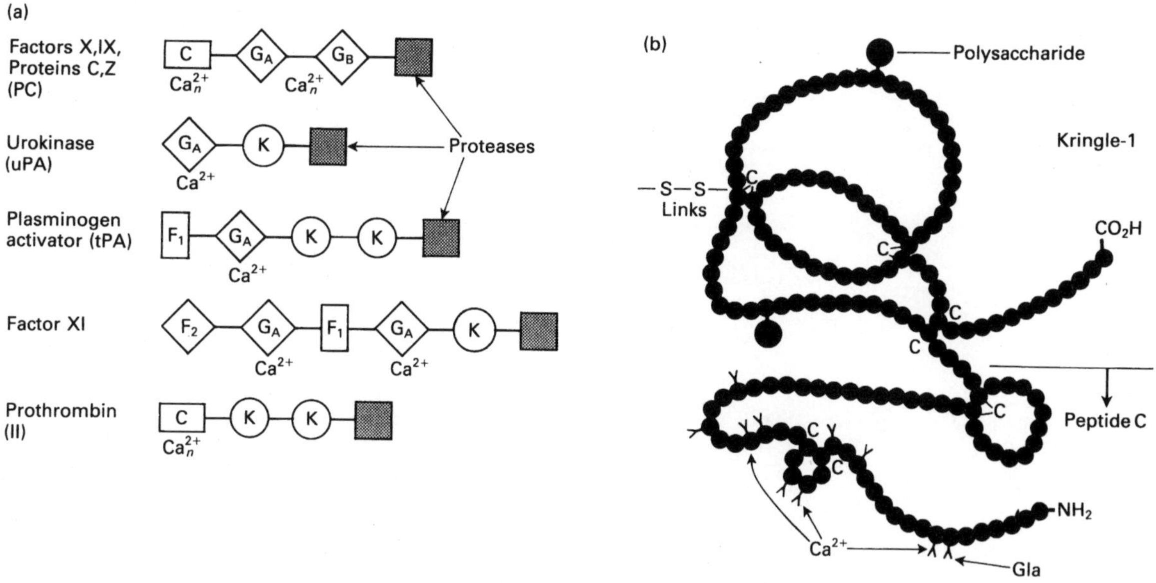

Fig. 10.18 (a) The complexity of the domain structures of some proteins in the clotting and clot-removal sequence of Fig. 10.17. The protein domains, peptide C, growth factor G, kringle K, and finger, F, control the proteases and in turn binding factors, such as calcium and phosphatidylserine (PS), control the domain protease interaction. (b) The sequence of the calcium-binding peptide, C, and the first kringle, K, of prothrombin. The Gla units (Section 10.7) are indicated. The –S–S– bridges in (b) are shown by C.

domain/domain interactions so that the hydrolysis necessary to make the blood clot can occur. The role of calcium is very like that in osteocalcin where Gla groups are also involved but where the exposed calcium can bind to a bone mineral rather than a damaged cell surface. The rates of these extracellular changes are all much slower than that of the intracellular calmodulin. Figure 10.18 includes several proteins from the blood-clotting system and Table 10.10 shows two from the complement components of the immune system.

10.18.2 Calcium in cell–cell connections

The extracellular matrices of higher cells have many functions partly controlled by the presence of calcium; hence, organization of cells requires calcium, see Figure 10.19. Some examples of such proteins are given in Table 10.10. The proteins have a modest affinity for calcium and their fold is

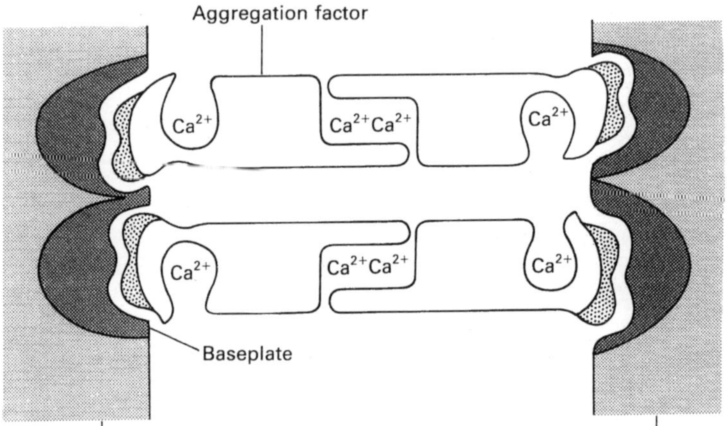

Fig. 10.19 The role of calcium in cell/cell interactions shown in schematic form.

Table 10.11 The main inorganic solids in biological systems

Cation	Anion	Formula	Crystal	Occurrence	Function
Calcium	Carbonate	$CaCO_3$	Calcite Aragonite	Widespread in animals and plants	Exoskeleton Gravity sensor, Ca store
			Vaterite		Eye lens
	Phosphate	$Ca_{10}(PO_4)_6(OH)_2$	Hydroxyapatite	Shells, some bacteria, bones, and teeth	Skeletal, Ca store (piezo-electric)
	Oxalate	$Ca(COO)_2 \cdot H_2O$ $Ca(COO)_2 \cdot 2H_2O$	Whewellite Weddellite	Insect eggs, vertebrate stones, abundant in plants	Deterrent Cytoskeleton Ca store
	Sulphate	$CaSO_4 \cdot 2H_2O$	Gypsum	Coelenterate statocysts Plants	Gravity sensor S store, Ca store
	(Silicate)			(Phytoliths)	
Iron	Oxide	Fe_3O_4	Magnetite	Bacteria, animals, teeth	Magnetic device
	Hydroxide	$Fe(OH)_3$	Ferritin	Widespread	Iron store
Silicon	Oxide Hydroxide	SiO_2 $SiO_n(OH)_{4-2n}$	Amorphous (opaline)	Sponges, protozoa, abundant in plants	Skeletal Deterrent
Magnesium	Carbonate	$MgCO_3$	Magnesite	Reef corals	Skeletal

decided by calcium-binding. Here calcium could be termed a structure forming element. The proteins are sometimes part of permanent extracellular filaments, e.g. cadherin, or can be induced, e.g. antibodies. They often contain series of relatively simple motifs, e.g. EGF-domains, see Aside. Well-known proteins in this group are the integrins, cadherins, fibrillins, and fibronectins. Here Mg^{2+} and Ca^{2+} act together to stabilize cell–cell structures (Fig. 10.19).

Cells may also communicate directly through gap junctions. These junctions are controlled by the levels of free calcium (see Section 10.14).

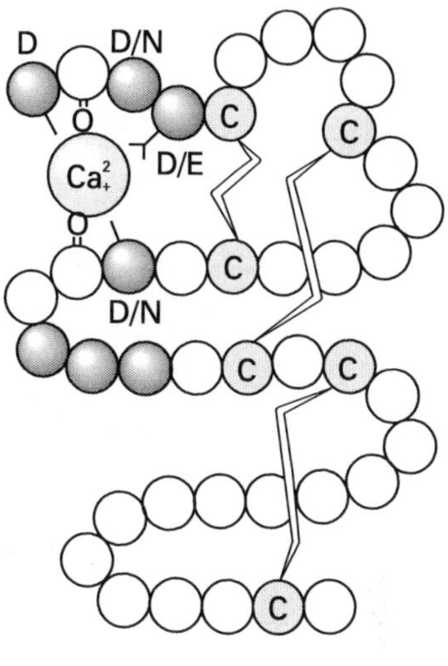

Schematic illustration of the EGF-like domain. There may be as many as 50 in fibrillins.

10.19 Calcium proteins of biominerals

Finally, calcium proteins are deeply involved with the laying down of biominerals—bones, shells, and teeth (Table 10.11). The proteins involved here are heavily glycosylated and phosphorylated in many cases but the exact function of these proteins is not known. What is known is that they have virtually random-coil structures in the absence of calcium. A typical small protein of this kind is osteocalcin which carries Gla (γ-carboxyglutamic acid centres.) The fold is shown in Fig. 10.20. There is a surface from which calcium ions exchange easily and presumably move readily over the great number of anionic centres. This protein binds directly to bone, but there are other anionic surfaces amongst the bone and teeth proteins which do not bind directly and may act to control the diffusion of calcium or as a reservoir of calcium somewhat like calsequestrin. Bone is, of course, a huge extracellular calcium buffer. It is thought that the soluble proteins such as osteocalcin and casein (in milk) inhibit precipitation while the matrix calcium-binding proteins initiate crystallization.

10.20 Calcium biominerals

There is clearly a relationship in evolution between the development of mineralization and the development of organisms. An obvious point is that load-bearing structures are most easily devised in a growing tissue system using minerals. The hard organic exoskeletons of insects are disadvantageous

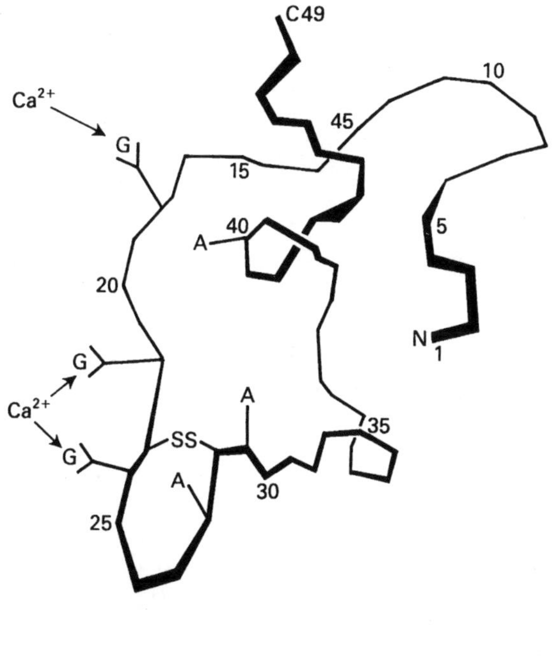

Fig. 10.20 The outline structure of osteocalcin. All the Gla units with malonate headgroups, formula I in Section 10.7 (shown as G) are to one side so that the folded protein binds many calcium atoms in an almost planar array. A is aspartate. (After work of A. Atkinson.)

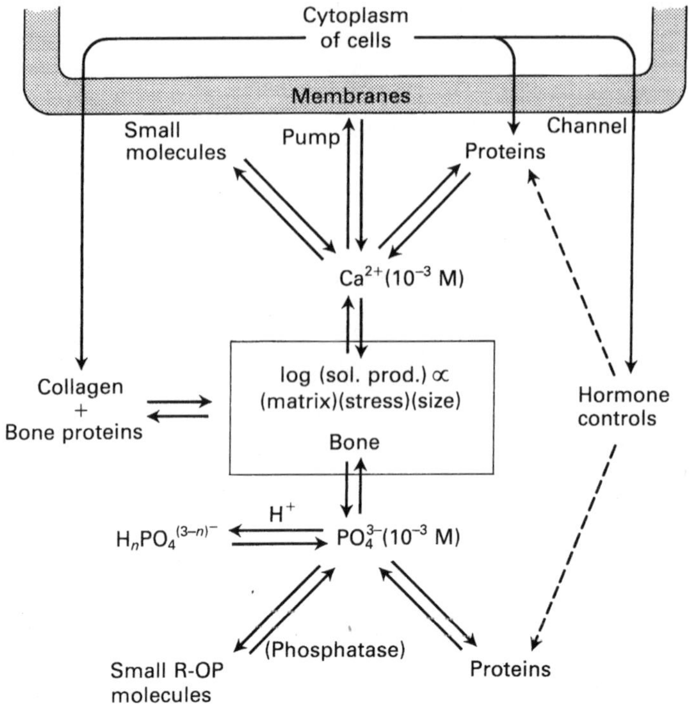

Fig. 10.21 The homeostasis of a multicellular organism containing bones can be shown to be sensitive to continuous *external* effects. Consider a local pressure stress. The stress affects the solubility product (sol. prod.) of the bone matrix. This matrix helps to stabilize the connective tissue. Thus, an extracellular pressure interaction develops, much as it does in the inside of a cell when the cell makes contact with a surface. Note that calcium effects are closely linked to those of pH and phosphate, again as in the interior of cells. The connection between inside and outside calcium is made via many means. Note that the solubility product is not absolutely fixed but varies between very narrow limits with pH.

since they restrict growth. The connection between minerals and growth is shown in Table 10.11 and Fig. 10.21. Here we must look on the mineral deposits from the point of view of solution chemistry. The minerals, especially calcium and iron salts, are in tight steady-state relationship with the extracellular (calcium) or intracellular (iron) free ion levels through the solubility products (see Chapters 1 and 2).

The insolubility of calcium salts arises from its doubly charged state, Ca^{2+}, and its large size which more nearly matches that of inorganic and organic anions. As a general rule, given equal concentrations of Mg^{2+} and Ca^{2+}, the calcium salt will precipitate with all common highly charged anions, e.g. SO_4^{2-}, HPO_4^{2-}, CO_3^{2-}, and organic anions such as dicarboxylates. The common biological minerals, with silica, are calcium carbonate, phosphate, and oxalate. (Table 10.11). The solubility products are such that at about pH = 7 (lower for oxalate) the salts precipitate with a calcium concentration of about 5×10^{-3} M and a given concentration of any of the anions HCO_3^-, HPO_4^{2-}, or oxalate^{2-} in the range of $1–10 \times 10^{-3}$ M. This statement is made with little precision since the precipitation reactions are sensitive to pH, temperature, pressure, general salt concentrations, and specific effects from the presence of especially Mg^{2+}, F^-, OH^-, and so on. Thus, we have to relate the relationship of precipitation to the particular conditions in an organism and in an environment, e.g. shellfish in the sea. (In krill it is probable that the coat is protected by CaF_2!) However, it is very clear that the precipitation is very carefully controlled. In this chapter the implication is that at the site of precipitation the solubility product is only just exceeded. The uptake of calcium and the related anions in the case of HPO_4^{2-} and oxalate^{2-} must be extremely carefully balanced with the deposition. There can be but one conclusion: the calcium diet is controlled such that the concentration level of this ion is held by feedback controls at the level of the solubility product of the precipitated salt

$$[Ca^{2+}]\,[X^{2-}] \simeq \text{solubility product}$$

From the point of view of inorganic chemistry this statement is extremely peculiar since it implies very considerable risks from unwanted deposition and very strong limitations on the flexibility of much other chemistry. On the other hand, such a balanced chemostat in calcium and its partner anion (and presumably pH) has great biological advantages in that it sets the conditions in a very firm manner for the evolution of organic reactions and structures. The highest forms of life we know have evolved further and further toward an extraordinary set of conditions which generate almost complete dominance over the organism by fixed levels of free calcium. This is not an accident but an inevitable result of a search for the best and most sensitive devices for controlled growth (Chapter 20), for controlled response, and for interaction with external events. The fact that the control is associated with calcium is dependent upon the availability of this element and the solubility of its salts, and it is very difficult to see an alternative solution for a general external chemostat given the abundances of the elements (Chapter 1). The choice of partner anion seems to have evolved (with fluctuations) toward carbonate (shells) and then to phosphate (bone). In the first case the external environment is largely responsible for the HCO_3^- level or the CO_2 level so that the organism is bound into the environmental chemostat. In the case of calcium phosphates (bone) there is a body fluid-dependent upon the required internal interactive control between free calcium and free phosphate but independent of the external environment. They are then the most finely tuned extracellular fluids in organisms. The link to phosphates also connects to the inside of the cell and here we must also mention Mg^{2+} which is an important minor part of bone. The discovery of bone by evolution (possibly on change from the sea to fresh water) was undoubtedly an irreplaceable benefit for the development of animal organisms (see Chapter 20).

Two other features of this chemistry must not escape notice. First, the dominant precipitates, $CaCO_3$ and $Ca_2(OH)PO_4$, contain the elements carbon and phosphorus, respectively, with all this holds for the organic chemistry of life. Second, both precipitates buffer the pH, so that there is feed-back between all possible pairings of Ca^{2+} to HPO_4^{2-} or HCO_3^- and to pH. We stress that solubility products are sensitive to temperature and pressure and, if the chemostat is to be maintained, then it must be a physicostat as well. It may be that in plants calcium oxalate performs this role at a lower pH. While pressure control rests largely on contact with the environment, temperature control internally is imperative and is found in higher animals with bones. Surely it is this chemostat feature which allows a very great sensitivity to have

Table 10.12 Calcium light isotope enrichment

Tissue	$^{40}Ca/Ca^{44}$ enrichment*
$CaCO_3$ of foraminiferae formed directly from sea water	1.00†
$CaCO_3$ of coccoliths formed by internal precipitation of calcium in vesicles	2.15†
$Ca_2(OH)PO_4$ of deer bones. Note passage of calcium through many steps in plants and animals	2.50‡

*Enrichment of lighter isotope in thousandths.

Data from †P. Zhu and J.D. MacDougall, *Geochim. Cosmochim. Acta*, **62**, 1691–8, 1998; and ‡J. Skulan, D.J. DePaolo, and T.L. Owens, *Geochim. Cosmochim. Acta*, **61**, 2505–10, 1997.

NB. The method of isotope fractionation is also applicable to a range of heavy metal isotopes, e.g. Mg, Cu, Fe, and Zn, allowing their incorporation into proteins to be analysed also (O'nions, Zhu, and Williams, to be published).

evolved in biological computers, brains, which so strongly depend on inorganic chemistry.

Recently it has become possible to follow the kinetics of calcium incorporation in minerals using isotope fractionation. As is usual, lighter isotopes pass over barriers more easily than heavier ones. In the case of calcium this means that the $^{40}Ca/^{44}Ca$ ratio increases on passage through membranes. Table 10.12 shows that the light isotope enrichment is greatest in those tissues which are made by passing calcium through the greatest number of steps. In principle therefore we can follow the age of tissues from the calcium isotope ratios in fossils much as is done with carbon dating.

On looking again through this chapter the reader is asked to keep in mind the fundamental importance of the sparingly soluble calcium salts to the whole of the network of calcium controls, which were described in previous sections.

10.21 Intra- and extracytoplasmic calcium balances

We now wish to bring into focus further features of calcium controls, both of homeostasis and of the linking between the different calcium pools. We begin with a description of a common feature, the sensitivity of these controls.

10.21.1 Multiple sites for calcium

We have concentrated attention on the individual sites of calcium binding but in many proteins several atoms of calcium bind cooperatively. Examples are troponin C, calmodulins, osteocalcin, and prothrombin. As explained in Chapter 2, multiple binding, which is co-operative, gives an added intensity to effects of change of calcium similar to dioxygen binding to haemoglobin. The extent of co-operativity increases from the monomeric sites to multi cooperative sites of crystals.

The use of multiple binding sites on the surfaces of the extracellular calcium proteins is somewhat strange since several of the sites must be saturated. Their binding constants exceed 10^3 in the presence of 10^{-3} M calcium. Good examples are provided by the Gla-containing proteins of the blood-clotting system and by osteocalcin. A possible explanation is that it is only on binding the last one or two out of some four or five calcium ions that the active form of the protein is created. We may then write

$$\underset{unfolded}{\text{Protein}} + n\text{Ca}^{2+} \underset{K \geqslant 10^3\ \text{M}^{-1}}{\rightleftharpoons} \underset{fold(1)}{[\text{Protein}(\text{Ca}^{2+})_n]} \overset{+\,m\text{Ca}^{2+}}{\underset{K < 10^3\ \text{M}^{-1}}{\rightleftharpoons}} \underset{fold(2)}{[\text{Protein}(\text{Ca}_{n+m})]}$$

In fold (2) all of the calcium atoms match a surface of either a phospholipid membrane or a crystal. This mode of triggering is very dependent on free calcium and clearly resembles a very sensitive buffer.

10.21.2 Cell shape

The steady-state, dynamic shape of a cell depends on the interaction of the outer membranes with both the *internal* matrix and the *external* connective tissue and its tension. Calcium and magnesium ions are deeply involved with both and they require metabolic energy constantly to maintain gradients. We can therefore anticipate that the future will show that there is a close involvement in the maintenance of a steady state of flow, now inside and outside cells, of these two ions and of sodium and potassium currents with cell shape. Of course, these flows are monitored by hormonal (and morphogen?) organic chemicals too, but in the context of this book we can be forgiven for thinking that the complicated organic hormone/inorganic messenger relationships of present-day life are latecomers in evolution compared with the initial system using filaments made from proteins and glycoproteins and inorganic ions only.

10.21.3 Cell morphogenesis

Most of the classical descriptions of calcium biochemistry concentrate upon triggering. We have insisted that the steady state of free calcium at rest is of equal importance in homeostasis. The homeostatic condition can only be maintained if any pulse of calcium entering a cell is followed rapidly by calcium expulsion and provided that no calcium-dependent message affects the cell's DNA and its translation activity. The opposite is required in morphogenesis, when the expression of DNA through to proteins must be altered. We do not know how this is achieved but it has to be the case that in some way the level of calcium following the initial pulse is sustained after, say, fertilization. It has been suggested that the calcium is released from vesicular stores in the egg. Does it then interact directly with DNA or is this process dependent on phosphorylation? How are external as well as internal proteins changed in concert and how is morphogenesis connected to these changes?

10.22 The calcium network today

A major development in evolution, probably prior to the development of all but very low levels of dioxygen, was the beginning of the involvement of calcium described in this chapter. Dioxygen at low levels seems to have been needed for the production of cholesterol and hence the development of the new membranes found in eukaryotes and not in prokaryotes. Before this initial rise of dioxygen, calcium ions only strengthened external membranes or formed crudely constructed external precipitates as seen today in algal coral reefs, while calcium in the cell cytoplasm was kept below 10^{-5} M. The next development, which we can follow in fossils, was in single cell *eukaryotes* as seen in the *foraminiferae* (dating back to more than 10^9 years ago), which have well-ordered crystalline externally formed shells, and in *coccoliths*, which form shells from units made in internal vesicles in fossil deposits. It is also notable that features in the cytoplasm common to many present-day unicellular protozoa, depend on calcium. Examples are the cilia beating of *paramecium*, the stalk contraction of *vorticellia*, and cytoplasmic streaming in *physarium*. As we have seen higher organisms, both plants and animals, have an increasing dependence on calcium concentrations through evolution. Again yeasts, which are fungi and separated from both bacteria and the antecedents of plants and animals at least 1–2 billion years ago, have well-developed calcium control systems. In all the early examples, calcium gradients are associated with slow development, or even constant activity, so that it would appear that the stress of today's scientific investigation on fast triggering by calcium in higher animals is a study of a more recent innovation. Calcium concentration changes remain an essential feature of differential growth, too, starting from an egg, and only later (see next chapters) did they become used in triggering. Growth is, of course, a dominant feature of all multicellular eukaryotes. While the importance of calcium in early single cell eukaryotes is well established, it concerns relatively few processes and proteins (see below).

Undoubtedly, the first multicellular organisms were plants. Hence, it is to plant life we must turn to appreciate how the functions of calcium developed after those in eukaryote single cells. Immediately we see that large calcium gradients are maintained across the cytoplasmic membrane, with fixed cytoplasmic concentration, but the extracellular fluids are not particularly well controlled. The endoplasmic reticulum and the vacuole of plant cells quite generally are high in calcium too. The novel functional value of calcium lies now in the relationship through its concentration in the cytoplasm to the level of the newly acquired set of organic messengers which go between cells—the plant hormones and control molecules such as phytochromes. It is these molecules which control the path of differentiation during growth. Thus, calcium became a major player in linking the outside yearly environmental changes with the internal metabolism and differentiation into organs during the plant life-cycle:

Environmental signal $\rightarrow$	Hormonal changes $\rightarrow$	Calcium message changes $\rightarrow$	Differentiation

A very interesting example is the involvement of the steady-state calcium concentration with the exposure to light and even light of different wavelengths. It is thought that the changes in calcium alter the organization of filamentous constructs such as actomyosin and tubulin so allowing chloroplast reorganization and streaming, and so changing development. Calcium is also required in chloroplasts to permit dioxygen generation. Additionally, the new calcium levels act to adjust phosphorylation, via kinases, and then to alter gene expression. A very similar response in plant roots is to the earth's gravitational field. Here the calcium current circulating at the root tip is affected by gravity so that the internal calcium levels change and then affect the downward growth of the root.* Calcium is again not to be seen as a trigger for immediate fast response but as the intermediary in the slow responses of development during growth and opposite a changing environment. (Of course, some fast responses such as the closing of a fly-trap plant or other reactions of sensitive plants are related to calcium, too). It may be that calcium circulates constantly through the cytoplasm of the plant cells, maintaining homeostasis in different ways according to the environment. Such changes in flow adjust the calcium input to the cytoplasm and the new stationary level of this ion then is relayed *via* phosphorylation to bring about cellular development. It is becoming apparent that calcium is involved in a vast number of steps in cell homeostatic states and differentiation by connecting the external conditions to plant cell development. Because of the complexity of changes linked to this process we must use network diagrams to illustrate its role in assisting cell homeostasis as well as triggering (see Figs 10.22 and 10.23).

In animals, as we have stated already, the functional importance of calcium developed to an even greater extent, Calcium became the communicating messenger between extracellular events and intracellular activity, including now fast nerve and muscle triggering. One additional feature of animal life is then fast physical response, movement. Thus calcium is required in light-sensitive devices (vision), in hearing, and in all nervous and muscle physical activity. In animal mitochondria calcium controls the supply of reducing equivalents for oxidative phosphorylation. Taking together all the

* (This might be tested by experiments on growth in the absence of gravity, e.g. in space vehicles.)

different uses of calcium in all the differentiated cells of animals with the uses in plants and single cell eukaryotes (see Fig. 10.22) we can observe how the use of calcium has developed by noting the numbers of calcium binding proteins of different kinds during evolution (Table 20.18).

If we consider the internal (cytoplasmic) proteins first we find that in single cell eukaryotes such as the fungus yeast (but very probably not in primitive prokaryotes) there are a few Ca^{2+}-binding helical proteins some of which are just calmodulins, but there is already a calcium channel as well as Ca^{2+}-ATPase pumps, and one or two calcium dependent kinases, phosphatases and phospholipases. On going to multicellular organisms of greater and greater sophistication, the number of calcium-dependent proteins increases. Hence we find that whereas few if any of these proteins exist in prokaryotes such as *E. coli* there are some ten to twenty in yeast and seventy in the worm *C. elegans*. Multicellular plants have an intermediate number of these proteins. The number in the human genome possibly exceeds several hundred. New features were necessary in the nervous communication network of the worm, obviously not present in yeast, and it has been found that there are also some five voltage regulated Ca^{2+} channels. There are many more, perhaps 50, in higher animals. Other intracellular proteins such as annexins, C-2 domains, and S-100 appear to be absent from unicellular eukaryotes such as yeast but are found increasingly in higher animals.

Turning to extracellular proteins, then *E. coli* and yeast have a few calcium dependent extracellular proteins and enzymes, but in the worm these increase and are augmented by calcium dependent endothelial growth factors, EGF domains. *C. elegans* has fifteen *notch* genes containing EGF domains. As extracellular fluids developed further new binding motifs appeared. A late development was the γ-carboxyl-glutamate found in many extracellular proteins of higher animals, for example in bone and in blood-clotting factors.

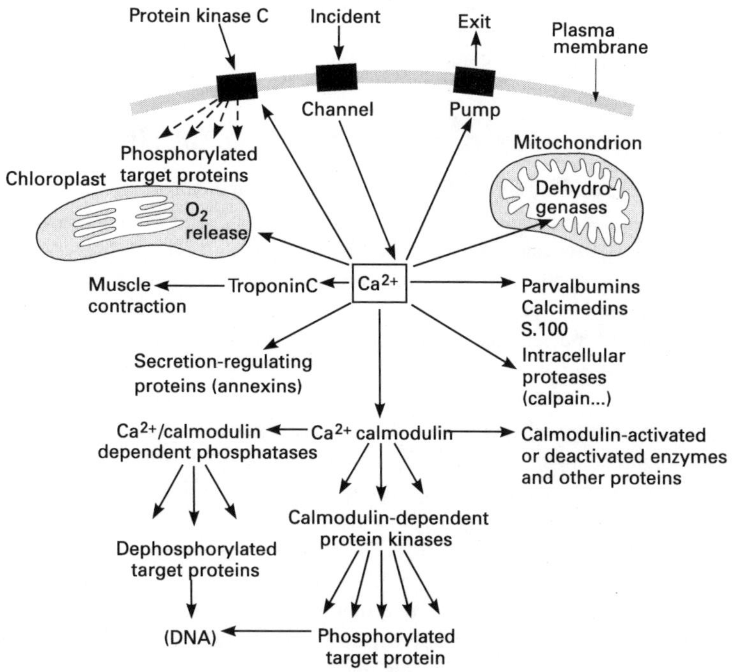

Fig. 10.22 Condensed view of the interaction of Ca^{2+} with the intracellular proteins. The Ca^{2+} can come from outside or from vesicles, e.g. the endoplasmic recticulum, to which the Ca^{2+} is returned. Signalling to the nucleus is shown as (DNA).

It is controlled by vitamin K, a coenzyme which requires dioxygen for its formation. The range of calcium uses are seen to extend in evolution from membrane protection, through simple signalling to complicated signalling (resulting in muscle, nerve and brain), to protein protection outside cells and finally to internal skeleton formation, and cell death, apoptosis.

Now we need to note also the links of the developing use of calcium with the ever-increasing use of organic signal molecules outside cells as well as with the link to Na^+/K^+ signals along membranes. We cannot list here all the transmitters and hormones in higher animals (and plants), but the number certainly exceeds 100. Finally, the calcium levels themselves are under feed-back control, which includes the control over synthesis of the calcium binding proteins at DNA level (see Chapter 20).

10.23 The genetic controls of calcium-binding proteins

The link between the internal calcium, its binding proteins and overall genetic regulation is probably via protein phosphorylation and dephosphorylation (kinases and phosphatases) and through cyclic nucleotides, though there are one or two calcium-binding proteins which act directly on DNA, e.g. the protein DREM. These proteins act as transcription factors. Basic production may be constitutive, but much of the synthesis is tightly linked in a feed-back circuit with the free calcium levels (see Section 6.13). Cell differentiation is also linked to the circulating hormones which affect calcium concentrations. The extracellular calcium proteins are linked more directly to these hormones.

Some of these calcium-binding proteins are expressed through Zn-finger promoters which bind the vitamin D steroids (Table 10.8). They are known to include one or two of the α-helical intracellular calcium proteins such as calbindin, and the epithelial calcium channel, as well as bone proteins such as osteocalcin. The transcriptional control over such internal proteins as calmodulin and annexin is not fully explored yet.

10.24 Summary of calcium biological chemistry

A striking feature of multicellular organization is now apparent. Within any compartment there is probably close to equilibrium distribution of calcium ions in a given resting or active site but this equilibrium is opposite very different concentrations of calcium ions and very different protein assemblies. It is not just that the state of the cytoplasm has one calcium concentration and the outside of the cell has another. Each type of organelle and each type of vesicle has its particular set of calcium-binding proteins and has a given free calcium concentration. The calcium concentration also equilibrates with activity switches, with filamentous tension, and connections between cells. All these features are specific in each compartment and the compartments are connected by the gates and pumps for calcium. A low level of calcium current then flows across all membranes, but since the gates and pumps are open to modulation at receptors the current can be boosted and then the cell network is either changed, leading to differentiation (by a sustained series of pulses), or for a very short period tension and metabolism are increased before fast relaxation occurs. Clearly, these effects and the resting states are not independent of other element and organic molecule activities. We have noted, for example, that both activity and tension changes due to Ca^{2+} are linked to reactions of $ATP{\cdot}Mg^{2+}$ and to changes in $Na^+/K^+/Cl^-$ ion flow. We shall not pursue this analysis of the connections of the network of calcium flows to that of other elements yet, but we shall return to it in Chapter 20. Here we cannot stress too strongly that calcium became the central element connecting together hundreds of cellular activities, in different compartments, see Table 10.13 and Fig. 10.23 for examples, starting from with the evolution of the special membranes and the compartments in eukaryotic cells and then undergoing a multifold increase in use with each step forward of complexity in multicellular organisms up to man.

Table 10.13 Some biological functions of calcium

Function	System or effect
Structural unit	Bone, shells, teeth, cell walls
Triggering	Muscle, synapse, hormone release
Control	Morphogenesis, fertilization, sporulation
Catalysis	Control of digestive enzymes (proteases), blood clotting
Current flow	Heart, barnacle muscle, general around cells(?)
Cell lysis	Protease trigger
Sensors (gravity)	Otoliths
Cross-linking	Calcium proteins, cell fusion

We see that whereas free calcium was kept at $<10^{-5}$ M in primitive prokaryotes, its value in the cytoplasm of all eukaryote cells fell to a value $<10^{-7}$ M. This appears to be a quantitative characteristic of the metallone of the cytoplasm of all modern organisms. Moreover, the extracellular and vesicular calcium has settled at around 10^{-3} M. We shall place these data in the context of the metallome of cells in Chapter 20.

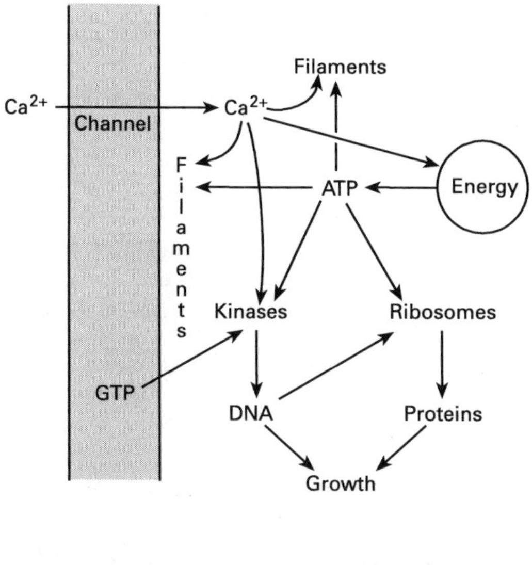

Fig. 10.23 The effect of Ca^{2+} on some actions of phosphates. Note also Ca^{2+}. + ATP enzymes, adjusts c-AM.

Further reading

Summaries of calcium functions are to be found in the first three references.

Carafoli, E. and Krebs, J. (eds) (2000). *Calcium homeostasis, Topics in biological inorganic chemistry*, Vol. 3. Springer-Verlag, Berlin.

Carafoli, E. and Klee, C. (eds) (1998). *Calcium as a cell regulator*. Oxford University Press, New York.

Celio, M.R., Pauls, T., and Schwaller, B. (eds) (1996). *Guidebook to the calcium-binding proteins*. Oxford University Press, Oxford.

Allen, T.J.A., Noble, D., and Reuter, H. (eds) (1989). *Sodium–calcium exchange*, Oxford University Press, Oxford.

Axelrod, J. (1990). Phospholipase A_2. *Biochemical Society Transactions*, **18**, 503–7.

Baum, G. and Trewavas, A.J. (2000). Shedding light on the use of light in plants. *Biochemist*, June, 25–9.

Bax, A., Tjandra, N., Crivici, A., Ikura, M., Structures of calcium-binding proteins in Carafoli and Klee (see reference above) pp. 152–70.

Berridge, M.J. and Irvine, R.F. (1989). Inositol phosphates and cell signalling. *Nature*, **341**, 197–205.

Berridge, M.J., Lipp, P., and Bootman, M.D. (2000). The calcium entry pas-de-deux. *Science*, **287**, 1604–5 (and H.-T. Ma *et al. Science*, **287**, 1647–51).

Coghlan, A. (1998). Sensitive flower. *New Scientist*, 26 Sept., pp. 24–8.

Goldbeter, A., Dupont, G., and Berridge, M.J. (1990). Calcium oscillations and protein phosphorylation, *Proceedings of the National Academy of Sciences USA*, **87**, 1461–5.

Heizmann, C.W. and Hunziker, W. (1991). Intracellular calcium-binding proteins. *Trends in the Biochemical Sciences*, **16**, 98–103.

Herzberg, O. and James, M.N.G. (1988). Crystal structure of troponin C. *Journal of Molecular Biology*, **203**, 761–79.

Huber, R., Schneider, Mayr, I., Romisch, J., and Paquer, E.-P. (1990). Annexin calcium-binding sites. *FEBS Letters*, **275**, 15–21.

Humphries, M. J. (2000). Integrin structure. *Biochemical Society Transactions*, **28**, 311–40.

Kater, S.B., Mattson, M.P., Cohen, C., and Connor, J. (1988). Calcium regulation of the neural growth cone. *Trends in Neurological Science*, **11**, 315–21.

McAinsh, M.R., Gray, J.E., Hetherington, A.M., Lockie, C.P., and Ng, C. (2000). Ca^{2+} signalling in stomatal guard cells. *Biochemical Society Transactions*, **28**, 476–81.

Nager, B., Overduin, M., and Rini, J.M. (1996). Structural basis of calcium induced cadherin changes. *Nature*, **380**, 360–4.

Petrobon, D., Di Virgilio, F., and Pozzan, T. (1990). Calcium homeostasis. *European Journal of Biochemistry*, **193**, 599–622.

Pochet, R. (ed.) (2000). *Calcium: The molecular basis of calcium action*. Kluwer Academic Publishers, Dordrecht.

Ricardi, D. (1999). Cell surface, Ca^{2+}-sensing receptors, *Cell Calcium*, **26**, 77–83.

Rini, J.M. (1995). Lectin structure. *Annual Review of Biophysics and Biomolecular Structure*, **24**, 551–77.

Smith, R.J. (1995). Calcium and bacteria. *Advances in Microbial Physiology*, **37**, 83–133.

Squire, J.M. (ed.) (1990). *Molecular mechanism of muscle contraction*. Macmillan, London.

Toyoshima, C., Nakasako, M., Nomura, H., and Ogawa, H. (2000). Ca^{2+}-ATPase—high resolution structure. *Nature*, **405**, 647–52.

Trewavas, A.J. and Knight, M. R. (1994). Calcium functions in plants. *Plant Molecular Biology*, **26**, 1329–35.

Williams, R.J.P. (1970). The biochemistry of sodium potassium, magnesium and calcium. *Quarterly Reviews of the Chemical Society, London*, **24**, 331–65.

Williams, R.J.P. (1992). Calcium fluxes in cells. *Cell Calcium*, **13**, 273–5.

Note In this chapter we have stressed the link between the Ca^{2+} message and phosphates in kinases and phosphatases as well as with c-NMP activities. These are linkages to very primitive signals in the cytoplasm. There are also connections between Ca^{2+} flows and more recent response systems such as photo-receptor and NO-release, see *Biochimica Biophysica Acta* (2000) **1498**, 81–284 which is devoted to Ca^{2+} biochemistry.

CHAPTER 11

Zinc: Lewis acid catalysis and regulation

11.1 Introduction to Lewis acids

At the beginning of each chapter on a given metal in living organisms we should have in mind the question as to why and how it has been selected for its particular tasks. In the present case the question itself has three parts: (1) why use zinc rather than another metal ion? (2) why use a metal ion at all since zinc is only an acid catalyst comparable to the proton? and (3) how has zinc been combined in order to use its properties optimally? As we stressed in Chapter 6, there are certain advantageous properties of metal ions both in their binding strengths and their rates of exchange of ligands which make them a potential source of chemically and hence biologically interesting selective properties. Zinc, Zn^{2+}, is the case in point here. Its most obvious distinction is its highly concentrated charge since it is a small ion (radius 0.65 Å). In themselves, however, an electrostatic charge of two and small size give quite modest binding to anions such as carboxylate and phosphate even where several anions occur together (Table 2.2). This is due to the competition from water of hydration (see Chapter 2). Furthermore, the electrostatic binding by zinc is a property shared almost equally with Mg^{2+}, and to a lesser degree Ca^{2+}, as well as with other metal cations such as Cu^{2+} and Ni^{2+}, but such charge concentration is not known amongst organic ions. Thus, although this electrostatic part of its chemistry is a feature of certain metal ions, it is not peculiar to zinc. The second characteristic of metal ions is their high affinity for electrons as shown by the energy required to get from the gas atom M to M^{2+}, i.e. their ionization potentials (Fig. 2.5). The ionization potential is a rough guide to Lewis acid strength and here zinc differs from Mg^{2+} and Ca^{2+} as well as, quite obviously, from organic cations, since its electron affinity is much higher. While it is therefore a strong *Lewis* acid, see Aside on p 316, it remains similar to copper and nickel in this respect and in all model properties described so far (see Fig. 2.8). All these three ions can then bind strongly to donors such as thiolates and amines, which is a property of strong Lewis acids not shown by magnesium or calcium. Zinc does differ from copper and nickel in yet other respects, however; for example, it does not show variable valence (oxidation state change) and could therefore be preferred in some biological compartments for this reason alone since redox activity introduces the risk of free radical reactions. On the other hand Zn^{2+} is still quite like Cd^{2+}, and Cd^{2+} is known to be able to substitute for Zn^{2+} in many zinc enzymes. It might have been thought that cells would have made some use of cadmium, but cadmium is a relatively rare element and zinc is much more available (Chapter 1). We see, therefore, that

Mg	Mn	Fe	Co	Ni	Cu	<u>Zn</u>

Divalent ions

Al	Cr	Mn	Fe	Co

Trivalent ions

Lewis acids available to biological systems

zinc is not quite unique, though it is highly selective as a cation; it has a special set of structural and thermodynamic properties (Table 11.1) and is widely available today, and these properties are further enhanced in value when rates of reaction are considered.

In Section 4.3 we showed that when zinc was forced to take up a tetrahedral or 5-coordinate geometry with one vacant site, it was a particularly effective acid and its acid function was kept available, see constrained (entatic) state. [In passing, note that zinc was not really available in the early sulphide environment and we shall see that its use has increased vastly during the course of evolution.]

Reactions require reorganization of atoms. Fast reactions will be more easily carried out by metal ions which, while polarizing groups, can take up and release molecules from their coordination spheres rapidly (Chapter 4). We can write K, the binding constant, as a ratio of rate constants.

$$K = \frac{k_{\text{on}}}{k_{\text{off}}}.$$

The values for both k_{on} and k_{off} for zinc reactions usually greatly exceed those of nickel (see Fig. 4.3) while K is roughly the same. Moreover, the rearrangements of groups on the surface of zinc without release, i.e. fluctional reactions, are usually fast unlike those of nickel. The differences arise from the fact that in a field of ligands the zinc ion is not effectively polarized so that its ligand field stabilization energy is zero (Section 2.5). As a result, zinc, but not nickel, passes readily from one symmetry of its surrounds to another without exchange and it exchanges ligands quickly. Such factors favour a higher catalytic activity since they allow rapid atom flow including substrate binding, changes in the bound intermediates, and product release to, on, or from the metal ion, in enzymes. If we use I_2, the ionization potential to the doubly charged state, i.e. the polarizing power, multiplied by k_{on} (i.e. the rate constant for water release from the co-ordination sphere) as a guide to acid catalysis, zinc looks like a better catalyst than almost any other divalent cation readily available to organisms except copper, which is objectionable on redox grounds. It may even be that, if we examine more highly charged relatively available cations such as Al^{3+}, Fe^{3+}, and Co^{3+}, (see Aside) zinc (Zn^{2+}) would still turn out to be the best catalyst in neutral aqueous media since in this medium the higher valent ions are hydrolysed and lose their functional strength by binding to hydroxide ions. The higher valent ions also exchange water much more slowly and are less fluctional.

The need for zinc in catalytic sites of enzymes may not even still be thought to be obvious since many enzymes which hydrolyse esters or amides, the major substrates of zinc enzymes, do not require metals. As already mentioned, Zn^{2+} has highly localized charge and high electron affinity and is therefore a very effective attacking group in a sense quite different from that of any organic centre or even bound protons (NB pH = 7.0). Looking at the substrates it attacks (in enzymes) we see first that they are often *small molecules* such as H_2O, CO_2, and C_2H_5OH, and that another major target is amino- or carboxyl- *terminal* peptide links generally. A clear possibility, therefore, is that zinc enzymes can be used to attack substrates in a more or less chemically non-selective but physically powerful constrained site. Enzymes without such powerful attacking centres as zinc can attack only large substrates (with very high selectivity) by using many points of attachment. The active site groups are here relatively poor attacking groups *per se*, and attack in these non-metal enzymes is based on very well regulated orientation

Table 11.1 Why zinc is so valuable in biology relative to other metals

Availability > nickel, cadmium, iron, copper, etc. but not to primitive life
Strongly retained > manganese, iron (II), magnesium
Fast ligand exchange > nickel, magnesium
No redox reaction; contrast copper, iron, manganese
Flexible coordination geometry > nickel, magnesium
Good Lewis acid (only copper amongst divalent ions is better)
Supplies hard base, hydroxide (only Cu(II), Fe(III), Mn(III) are better)
Common in some cells and may well be a link between controls and regulation

and strain in the substrate generated by the precision of binding. Zinc enzymes would then be reserved (1) for attack on small substrates which cannot be handled by organic groups in a multipoint attachment, and (2) for less specific but very rapid attack when they could work even with relatively weak binding of substrate. Lack of selectivity, which would usefully go with powerful attacking groups, is of obvious advantage in digestion, for example. One case is the zinc nucleases. Speed (also at the price of selectivity) could also be advantageous in hormone turnover, and we notice that zinc is particularly associated with peptide hormone release from propeptides and peptide hormone destruction.

Metal ions need not be used just in catalysis since they can be triggers of systems or play a structural role. We know that triggering, like catalysis, requires good binding with selective recognition, but also reasonably rapid on and off reactions since there is a need both to stimulate a biological system and to relax it in controlled ways quickly. Amongst the divalent cations zinc, like calcium, has the required properties for a trigger or control ion since it binds well and exchanges rapidly (Section 11.5). The parallel role of calcium in catalysis and triggering activity was described in the previous chapter. Naturally, it is used in different circumstances from those in which calcium is used, as the calcium ion binds to O-donor ligands only but zinc binds largely to S- and N-donors in biological systems; moreover, zinc is at a much lower free ion concentration than calcium. Since zinc binds much more strongly than calcium its exchange from bound sites in proteins is that much slower, hence it is not so much a trigger or transmitter as a *regulator* (or a hormone).

The similarities (and differences) between zinc and calcium lead to further parallel (not identical) activities in biology where both metal ions are of extreme importance in structural cross-linking, but once again different chemical specificity is involved. It may be that the fluctional properties of both these metal ions are useful in all these functions.

Finally, since the zinc ion is a dissociable but indestructible unit and can therefore be used as a homeostatic control between many reaction systems, it could be employed simultaneously as a catalyst and as a control. We must ask if there is some hidden cross-talk between different uses of this metal ion. To answer this question, we shall have to distinguish sites which exchange relatively quickly from those which are more or less permanent.

11.2 Zinc in biological space

Before making statements about particular functions a few general remarks about the three roles of zinc (acid catalyst, control ion, and structural ion) and their spatial distribution in cellular systems are worth careful note. At the risk of overgeneralization we make the following statements:

1. Zinc is the most common trace metal ion in the cytoplasm (including the nucleus) of advanced aerobic cells, but is much less common than some group 1 and 2 metal ions. Iron is more common in membranes, organelles and primitive cells, manganese in some vesicles or organelles, and copper is as common outside the cytoplasm and outside cells. Nickel, cobalt, and all other metal ions are not common anywhere in eukaryotes.
2. Zinc enzymes in cells are involved in a great variety of reactions but relatively few are in the cytoplasm; most occur in vesicles or in extracellular space.
3. Zinc digestive enzymes are especially active outside cells or in vesicles.

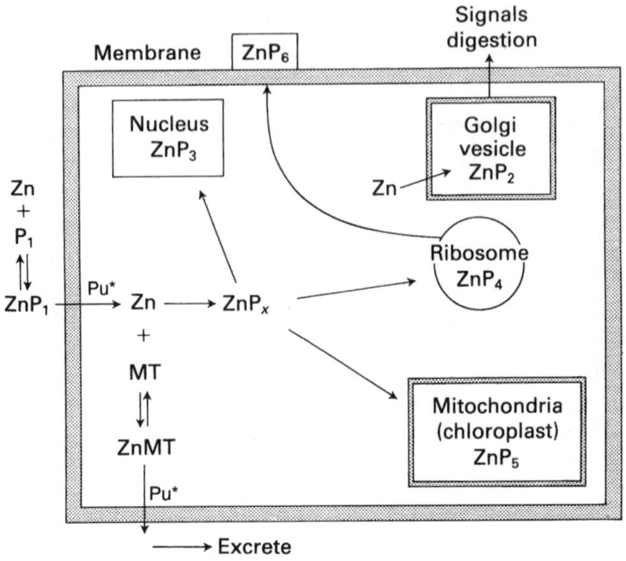

Fig. 11.1 The variety of roles played by zinc inside and outside cells. P, protein; MT, metallothionein; Pu*, pump.

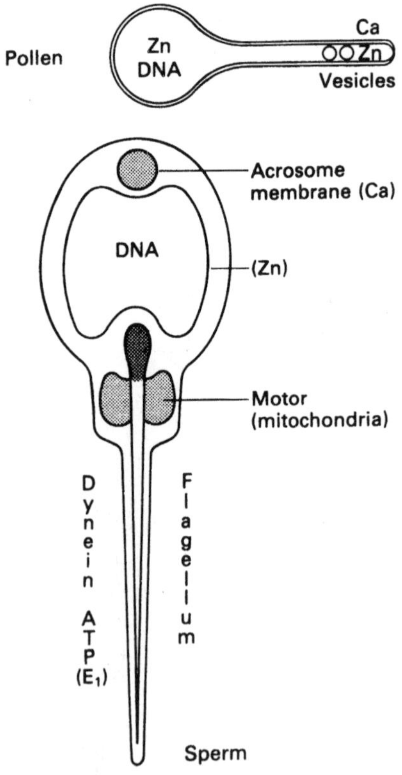

Fig. 11.2 The compartments of pollen and of sperm showing the uneven distribution of zinc and calcium. Note the localization of the ATP(ase), E_1, to drive the cell movement.

4. Zinc is rarely intimately associated with a membrane.
5. The structural role of zinc extends from stabilisation of filamentous, keratin-like structures to the organisation of chromosomes, largely inside cells.
6. The extensive use of zinc in DNA regulatory functions is confined to eukaryotes.

The peculiarities of the chemistry of zinc, summarized in Table 11.1, are then put to particular use in particular parts of biological space (Fig. 11.1), which differ greatly from those used by iron or copper, for example. A particularly fascinating example is given by the male gamete, (Fig. 11.2). We turn first to the availability of zinc and the ways it can be incorporated into biological cells, and to the general kinds of proteins with which it is associated.

Note again that zinc was not very available in the primitive sea, as we have explained in Section 1.4, and we shall return to the effect of changing zinc availability in evolution later (Section 11.13).

11.3 Availability and concentration of free Zn^{2+} ions

Biological systems are energy-consuming. Energy economy is therefore important and this implies that cells will use available elements rather than those which have similar properties but are much less available (Table 11.2). Zinc is relatively available today even in sea water and is amongst the most available of the trace elements. It is more available than copper, iron, cobalt, or cadmium. Considering its potential value, it is not surprising that it is relatively common in biological systems, about as common as iron, but there is an additional feature. The relative concentration of free zinc ions within biological cells varies from about 10^{-11} M in the cytoplasm of many cells to about 10^{-3} M in some vesicles. (Calcium has a similar localized concentration in cytoplasm and vesicles.) Clearly, a mechanism must be available for moving the element into selected compartments by pumping and the elevated concentrations then allow triggering as with calcium. These pumps require selective binding like all the other activities associated with zinc described in Section 11.2 In extracellular fluids, zinc may well be bound to a level as low or lower than the intracellular value 10^{-11} M, but since it can be released from vesicles the level may be much higher locally for a short time. In fact, zinc concentrations are high in a variety of vesicles of the male reproductive tract and of nerve terminals of the forebrain, where in mammals they have evolved into a complex interactive network that interconnects most of the cerebral cortices and limbic structures in particular in 'mossy fibres' of the hippocampus, (see Frederickson in 'Further reading'). It is not known why zinc should be present in the synaptic vesicles, but one interesting possibility is that its release under stimulus assists the production of 'long-term' memory in the form of growth of the nerve. Zinc seems to interact with the glutamate (a transmitter) receptor. We consider another possibility in Section 11.14. A great deal of work is required to define fully the way in which zinc is distributed (see Section 11.8). A similar comment concerns the transport and circulation of zinc in the blood of higher organisms where an albumin together with histidine appear to be involved.

We turn next to the types of chemical neighbours with which zinc is associated in organisms and we start with proteins since a great deal is now known about zinc proteins from crystallographic and chemical studies.

Table 11.2 Logarithmic concentration* of elements in sea water (mol dm^{-3})

Na	−0.3	Ca	−2.0	Ni	−8.1
Mg	−1.3	Sc	−11.0	Cu	−8.4
Al	−7.7	Ti	−8.0	Zn	−8.1
Si	−4.0	V	−7.5	Mo	−7.0
P	−5.7	Cr	−8.4	Sr	−5.0
S	−1.6	Mn	−9.3	Cd	−9.0
Cl	−0.3	Fe	−9.0		
K	−2.0	Co	−10.7		

Data from Bruland, K.W. (1983). Trace elements in sea water. In *Chemical oceanography*, Vol. 8 (ed. J.P. Riley and R. Chester), Chapter 45. Academic Press, London.

*These are not the free ion levels but the total concentrations (see Chapter 1).

11.4 Types of protein associated with zinc

In Chapter 6 we described three very different extreme types of protein: (1) the extended β-pleated sheet, (2) sets of α-helices; and (3) almost random structures, in relation to their relative mobility and therefore their functional potential (Table 11.3). There are, of course, varieties of proteins which have combinations of β-sheets and α-helices in them. Inspection suggests that, even in these proteins, the $\beta\alpha$ alternating structures are relatively rigid, for example in domains, while where there are patches of structure containing only helices these patches are internally mobile. We observed in Chapter 10 that the trigger proteins, e.g. those which bind calcium, are almost totally helical and mobile, and that this is also true of the *allosteric* oxygen-carrier proteins which use iron or copper. On the other hand, the electron transfer proteins which contain copper, the blue proteins, or simple iron, are based on immobile β-sheets (Section 6.8). Of course, chemical cross-linking also reduces mobility. Cytochrome *c*, another electron transfer protein, although it

Table 11.3 Major fold characteristics of some zinc proteins

Helical	β-sheet	Mixed*	Random
Insulin, Zn†	Carbonic anhydrase, Zn	Carboxypeptidase,† Zn	Metallothionein‡, Cu, Zn, Cd
S-100, Zn, Ca	Superoxide dismutase†, Zn, Cu	Aspartate transcarbamylase, Zn	σ-factor IIIA, and Zn-fingers generally‡
Phospholipase C†, Zn		Alcohol dehydrogenase, Zn β-lactamase, Zn	

*The active site is always closely associated with the β-sheet.

†Cross-linked by the metal ion or -S-S-bridges.

‡Fold only on binding metal ions.

is helical, is cross-linked three times and is also of low internal mobility where it is cross-linked. Such electron transfer proteins do not require gross mobility and, in fact, that would be disadvantageous (see Section 4.7). Turning our attention to enzymes generally, it is usually true that active sites are held in proteins close to or on β-sheet-like structures and we may suppose that the relative rigidity of these structures helps to control specificity of binding and the strength and directionality of attack by acid–base or redox centres. The stability of the fold also prevents exchange of metal ions from such a protein. We look therefore at zinc proteins and enzymes in this context.

11.4.1 Zinc enzymes

Crystal structures are available for many *enzymes* which require zinc at the active site (Table 11.3). It is immediately clear that with one or two exceptions they are based on extensive β-sheets and we presume that it is the strength of the fold which generates the peculiar and particular active-site geometries of the metal ion (Fig. 11.3) since these proteins fold in the absence of the zinc. We have described the imposition of a special geometry at the metal by the strength of the fold as an entatic, constrained condition of the metal geometry (Section 6.8) and we shall return to this feature in Section 11.7.3 since probes allow the metal binding in the protein to be inspected. An enzyme in which zinc is not at the active site is superoxide dismutase, another β-sheet enzyme, where copper is the active metal ion. Here both the binding of the zinc in a β-sheet fold and the β-sheet itself now stabilize the unusual rhombic geometry around the copper ion, whereas the zinc geometry is a conventional tetrahedron. Zinc is but an additional stabilizing factor and when we look at copper biochemistry (Chapter 15) we shall see why it was necessary to build such a stable structure.

Now, the active-site attacking groups of enzymes for the catalysis of larger substrates, zinc liganded water and nearby organic sidechains in the present case, must retain some local mobility since, unlike the active sites of electron transfer proteins, considerable relaxation of the local structure is needed to allow for more rapid on/off reactions of substrates and products and for the coordinates of the whole system, attacking groups plus substrates, intermediates and products, to *flow* along the reaction pathway easily (see Section 4.3). The zinc and substrate bindings are both to sidechains of proteins, all of which are somewhat flexible since they have β-CH_2-sections to their sidechains, which are very different from the more constrained environments of metal ions in Fe_4S_4, cytochromes, vitamin B_{12} or F-430 enzymes. The flexibility at the open-sided zinc enzyme site will be illustrated by an example later. A number of extracellular enzymes with similar functions have

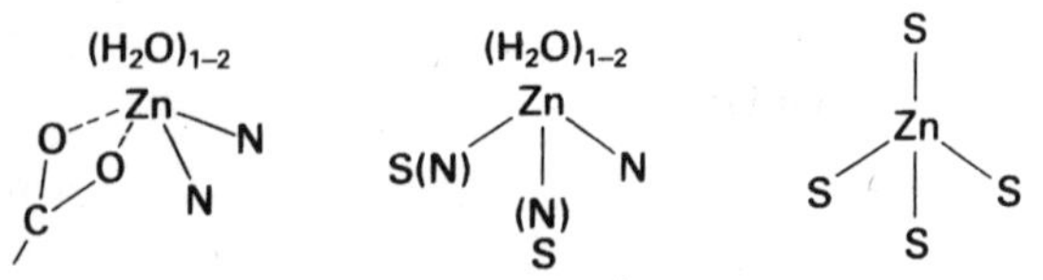

Fig. 11.3 The immediate donor atoms in some zinc binding sites where S is a thiolate of cysteine and N is an imidazole donor; contrast calcium and magnesium. The coordination geometry is generated by the relatively rigid protein. There are a few proteins in which zinc is bound by three oxygen donors from protein sidechains (see Table 11.4).

Table 11.4 Zinc enzymes

Class	Coordination	Spatial distribution
Carboxypeptidases(1)	H E H	Vesicles, extracellular
Aminopeptidases(1)	H E H	Vesicular, extracellular
Alkaline phosphatases(2)	D S D H	Extracellular + Mg^{2+}
Nucleases(2)	W H D H H D	Extracellular
Collagenase(1)	H H X,[(C) or (H_2O)]	Extracellular
Phospholipase C(3)	H H E (H_2O) H W D (H_2O) H H D D	Extracellular (Intracellular ?)
β-Lactamases(2)	H H D H H	Extracellular
Glyoxalases II(2)	H H D H H	Extracellular
Aryl sulphatases(2)	H H D H H	Extracellular
Phosphohydrolases(2)	H H D D	Extracellular
Leucine amino peptidase (2)	K D D D E	Cytoplasmic
Calcineurin(1)	D N H H (H_2O)	Cytoplasmic
Aspartate transcarbamylase(1)	C C C C	Cytoplasmic
RNA/DNA Polymerase(1)	C C H H	Cytoplasmic
Carbonic anhydrase(1)	H H H (H_2O)	Cytoplasmic

H = histidine, C = cysteine; D = aspartate; E = glutamate; S = serine; W = tryptophan; K = lysine. ( ) = number of zinc atoms at active site.

[L. Aravind and E.V. Koonin, *Trends in Biochemical Sciences* (1998) **23**, 469–72 and references therein.]

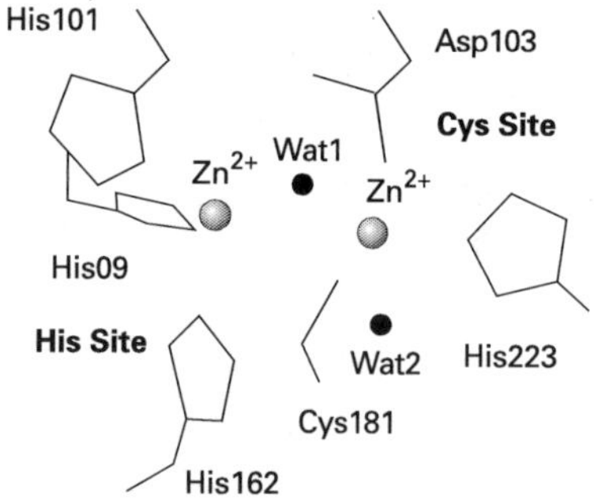

The active site of a β-lactamase

two or more zinc ions at the active site (Table 11.4, and see Aside). The advantages of such a combination of two or more acidic attacking groups has been described in Section 4.3. The vast majority of these enzymes are extracellular and they function both in digestion and in matrix adjustment.

There are a few cytoplasmic zinc enzymes which differ from the above enzymes. In one or two (Table 11.4) the active site has O-donor ligands only, and in one case while the ligands are not unusual N- and O-donors the protein is totally helical, see phospholipase C, (Section 11.7.1). One or two enzymes have N/S ligands and bind Zn^{2+} very tightly.

Turning to other kinds of *enzymes* in which zinc is bound but not at the active site there are crystallographic data for a few (Fig. 11.3). In these proteins the metal acts as a cross-link between thiolate groups, usually four, or a combination of thiolates and histidines to make a total of four ligands. This is a double or knotted cross-link and is found in cytoplasmic proteins, e.g. alcohol dehydrogenase. The cross-link could be, in part, a replacement for the –S–S– or calcium bridges of extracellular proteins since inside a cell most –S–S– bridges are unstable to reduction and calcium cross-links are unstable because in the cytoplasm of cells there is very little calcium. In this case zinc binding (cross-linking) controls the amount of the protein in the cell by stabilizing it in bound form so that it is not hydrolysed. In the cross-links of alcohol dehydrogenase and aspartate transcarbamylase zinc has a geometry which is now that of a conventional tetrahedron.

Even in the most primitive as well as the most advanced organisms zinc is also involved not just in hydrolysis but also to some degree in syntheses in the cytoplasm. Typical of the few examples known is the presence of zinc in reverse transcriptase and in RNA polymerase (see Table 11.7). Once again notice that these enzymes have high rates of reaction but may not be very discriminating with regard to the nucleotide bases. As stated in Section 11.1, zinc is most valuable where speed is most important, but its aggressivity as a Lewis acid implies that care must be taken in the selection of targets or else a general attack must be anticipated on peptide, phosphodiester, or polysaccharide bonds. While we make these observations we must be aware that the nucleotide and amino-acid polymerases must be of high fidelity and that this is managed by units other than the zinc enzymes in the ribosomes.

Another example in synthesis is the zinc enzyme in plants which has replaced the requirement of these organisms for vitamin B_{12} (Co) for methyl transfer by a zinc/thiolate combination (see Matthews in 'Further reading').

11.4.2 Zinc proteins other than enzymes

So far, all the zinc binding we have examined in enzymes is strong, with binding constants $>10^{11}M^{-1}$. However, in insulin complexes, where the zinc and insulin are retained in a *vesicle*, the complexes are weak and the concentration of free zinc must be high. The zinc is cross-linked to this *helical* protein in two different but conventional ways, one each in two different forms of the protein, through histidine sidechains. In the two forms the helices are slightly differently disposed relative to one another and the zinc site geometries are closely related to an octahedron. There is a remote link here to calcium- and zinc-binding to calmodulins or S-100 proteins where the metal ions bind to and modulate the properties of a helical protein (see Fig. 10.6). We do not know why insulin should have a secondary structure which is inherently mobile but a cross-linked interior. Notice particularly that conventional zinc binding geometry, α-helical structure, and non-enzymic function occur together. There is an exception to this generalization in that phospholipase C is a helical protein. However, it is strongly cross-linked by *several* zinc ions and mobility must again be constrained.

Metallothionein is, in the absence of zinc, a random coil protein and only on binding zinc, cadmium, or copper does it form a series of turns so as to hold the metal ions in two types of cluster of M_4 and M_3 units, respectively, formed by regular tetrahedral, thiolate, donors. The structure of metallothionein bound by different metals, zinc, copper, and cadmium, is not the same but the essential core structure is fixed. While it is reasonable to assume that all the other proteins described so far are very highly zinc-selective, this is not true of metallothionein which is in effect a scavenger of heavy metal ions and the protein structure clearly adapts itself to the particular metal ion. There is no question of a truly constrained, entatic state here, the zinc binding constant is lower, 10^{11} M^{-1}, and exchange can be quite fast. The suggestion is that it is a buffer and the value of such a common scavenger will be examined again, especially in the discussions of the interactive role of zinc and copper in organisms and the protection of organisms from pollution caused by some metal ions. Note that while metallothionein is a eukaryote protein a somewhat similar protein is found in some aerobic prokaryotes.

11.4.3 Proteins for zinc distribution

Apart from metallothionein which may transport zinc in the cytoplasm there are now known to be zinc pumps which contain the membrane located sequence cysteine-proline-x, and are driven by ATP. They are often called CPx-type ATPases and will be described further under copper (Section 15.6). There are other special zinc proteins, labelled ZnT, probably pumps, asso-

Table 11.5 Zinc fingers

Zinc-finger type	(a) DNA binding proteins
(C_3H)	Phage protein, P-53
(C_4)	Steroid receptor, LIM domains
(C_2H_2)	TFIIIA, Zia R transcription factors
Multi Zn (C_6, C_3HC_5)	Fungal fingers
	(b) Enzymes or ancillary proteins
(C_3H)	Protein kinase C
(C_2H_2)	RNA (DNA ?) polymerases, DNA-primase
(C_4)	Aspartate transcarbamylase
(C_4)	Alcohol dehydrogenase
(C_3H)	Retroviral capsid protein

C = cysteine; H = histidine

Notes All zinc fingers in group (a) have random coil apoproteins with binding constants of about 10^{11} M^{-1} while the apoproteins of group (b) may have folded apoproteins and binding constants > 10^{12}. [See Klug, A. (1999) *J. Molec. Biology*, **293**, 215–8 and references therein for the different 'zinc-finger' types.]

ciated with membranes which contain many histidine residues (see Section 11.13.2). There are further zinc transporters in membranes with labels such as DCT 1 in neurons, ZIP in plants and ZRT in yeast cells, but they have been less investigated. As stated above, zinc in extracellular fluids is transported to these membrane transporters by albumin and maybe by histidine. It is not critical for selectivity that different metals use the same binding sites if the stereochemistry induced is very different, e.g. for Ni, Cu, and Zn, since receptors can then differentiate between them. The metal ions, here specifically zinc alone, can then be taken up by or rejected from cells by the pumps into extracellular space or vesicles. It is not known to what degree zinc is transported inside the cytoplasm and to what degree it diffuses as free zinc.

11.4.4 Zinc fingers

We have deliberately left until last a further extremely important group of zinc proteins which have no enzymatic activity, the zinc-finger proteins (Table 11.5), which have only been found in eukaryotes. Most of these zinc proteins bind to DNA promoter sites, but others are part of critical enzymes and appear to have a structural or a regulatory role.

The structure of the most common repeated zinc-finger sequence is shown in Fig. 11.4. The zinc is in repeating loops of protein of some 30 amino acids (Fig. 11.5). In the protein (free from DNA) the fingers are not well structured relative to one another and the zinc, even in the bound state, is rather exposed and exchanges relatively easily. Other zinc fingers are of somewhat different structure (see Table 11.5). Unlike the enzymes, in which one or at most two or three zinc ions are involved, the zinc fingers can occur in multiples of up to 30. Their binding constants are also close to $10^{11}M^{-1}$. The zinc-free protein lacks structure and so zinc binding is accompanied by a large change. We return to their probable function in Section 11.9.

As stressed in previous chapters, it is important to observe whether the ligand centres of the protein for metal binding occur close together in the sequence or not. The distribution of these two different constructs between calcium trigger and calcium enzyme sites is very clear, see Table 10.5. The

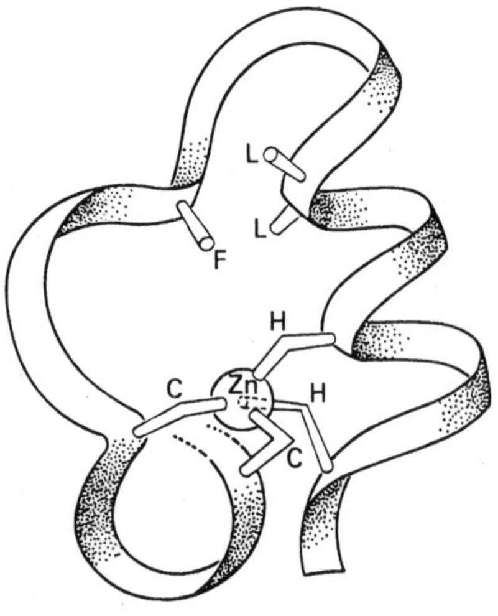

Fig. 11.4 The structure of one kind of zinc finger as determined by NMR methods. Here certain amino-acid residues are shown: L, leucine; F, phenylalanine; H, histidine; and C, cysteine. (After R. E. Klevit and P. E. Wright.)

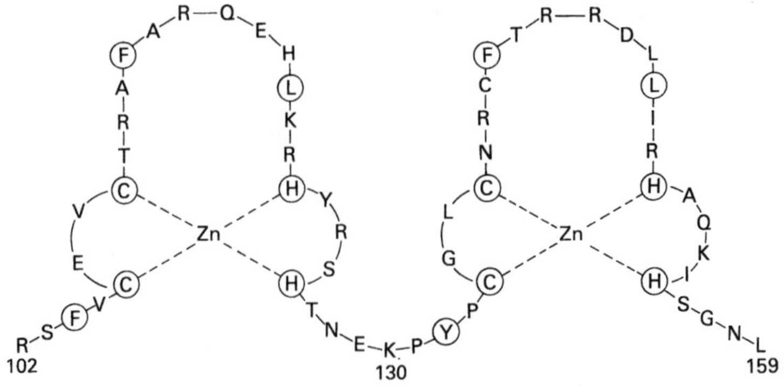

Fig. 11.5 An example of a usual sequence around zinc sites in two zinc fingers. The circled amino acids are conserved and are either ligands or part of the hydrophobic core.

situation is much the same for zinc. The DNA binding zinc finger is composed of a run of residues closely in sequence and we must expect that zinc removal will be fast and alters the fold at least locally, as is required of a control. The connections of the metal in enzymes, by way of contrast, are usually to residues at large distances apart in the sequence and so are related to total fold stability. We shall see that, much as for calcium, the two different types of protein allow very different zinc exchange rates.

Table 11.6 Exchange rates of zinc proteins

Protein	Ligands	pH	Exchange rate $t_{1/2}$(h)
Carboxypeptidase	N_2O_3	8,0	>24
Alkaline phosphatase			
Site A	N_2O_3	8.0	84
Site B	N_1O_3	8.0	24
Gene 32 protein	S_3N	7.8	1–2
Gal4	S_4	7.8	1–2
	S_4	7.8	1
	S_4	7.8	0.75
Metallothionein	S_4	7.0	~1

This table is from unpublished work of Professor J.E. Coleman, Yale University, New Haven, Connecticut, USA, and is presented with his permission. The exchange rates were measured in the presence of cadmium ions. We have also had helpful discussions with Professor J. Kaegi of the University of Zurich, Switzerland.

11.5 Zinc exchange rates

The exchange rates of metal ions from protein complexes are not an intrinsic feature of their metal ion binding strengths, i.e. just to the ligand groups, but are dependent on protein fold energies; this has already been observed for calcium. In the case of zinc it is well known that many zinc enzymes can be extracted from cells without significant loss of the metal. It is also found that zinc, incorporated in these β-sheet–helix structure, does not exchange over days (Table 11.6). The zinc in the zinc fingers, metallothionein, and in a few cytoplasmic enzymes (Section 11.5) exchanges over a period of the order of 1.0 h; thus, here zinc has the same exchange time as any strongly bound organic reagent such as a sterol hormone. Exchange rates of this order are suitable as regulators over periods of the order of days, but cannot be functional in rapid triggering. Exchange can occur directly between ligands called 'metal chaperones', see Section 10.12, so that a metal ion need not have a large dissociation rate to be distributed in a cell.

Now that we have examined the general types of zinc proteins we must return to the sites themselves and their activities.

11.6 The number and selectivity of ligands to zinc

The proteins in Table 11.4 nearly all bind zinc through at least three groups, two of which are imidazole and/or cysteines (Fig. 11.3). In some cases, four such groups are used when the zinc acts as a cross-link and is blocked from interaction with any small-molecule substrate. In many enzymes, zinc can have as few as two imidazoles when it is also bound to a carboxylate, e.g. in carboxypeptidase, or it can have three such donors, e.g. in carbonic anhydrase. Table 11.4. The fourth (and fifth) positions are then occupied by water molecules which become involved in the catalytic act, or, at least, are readily replaced allowing access of substrate to the zinc. The zinc site is then organized in several ways: (1) type of ligands; (2) angular distribution of ligands around the zinc; (3) bond distances; (4) leaving groups; and (5) the next nearest neighbours and surrounds of the metal. The last term includes the whole of the active site groove in which the zinc sits. We should be able to rationalize all the peculiarities of these different parts of the zinc sites in terms of the reactions which they carry out. An example, carbonic anhydrase, is given in Section 11.7.1.

Against the background of the preceding structural and chemical characterization of zinc sites we must examine the sites of other metal ions in proteins in an attempt to appreciate selectivity. We observed in Chapter 2 that, if we used a naïve approach based just on stability constant data and free competition amongst cations, it was very difficult to see how the selection of zinc in a given site is so specific—an estimate suggests that zinc sites contain less than 1 per cent of any competing element. We repeat that most competing metal ions are excluded on the basis of low accumulation (Ni), of lower affinity for N(S) co-ordination (Mg), by a dislike for tetrahedral coordination (Cu), and, finally, by a deliberate segregation of other elements (Cu, Mn) away from the cytoplasm or the special vesicles into which zinc is pumped. Again the production of the apo-forms of zinc proteins, like those of other metals seems to be restricted by the available zinc. Only a combination of such factors can have made it possible for biological systems to select zinc into the above sites. For example, other proteins use the same four groups (four RS^-) to bind Fe (in rubredoxin). Again, the site of zinc in an almost octahedral hole bound only by histidines in glyoxalase is a unique situation for zinc in an enzyme so far, but a very parallel site binds iron in the photosynthetic reaction centre complex and possibly in lipoxygenases. To understand this competition we must know the redox state of competing ions, their concentration locally and so on, before the effects of all the factors mentioned in Chapters 2, 3, and 5 can be appreciated. All such matters are considerations of effective stability constants (Section 2.4). However, we must also remember too that metal ions can be kinetically inserted into sites (Section 2.12).

11.7 Zinc as a catalytic group in enzymes

11.7.1 Zinc as a Lewis acid

A zinc ion is a good electron acceptor, or Lewis acid, in inorganic and organic chemistry. As such, it polarizes groups to which it binds and either increases the attacking power of a bound base or increases the probability of attack on the bound group by acting as an acid. An example of the first case is bound water which becomes an effective bound OH^- at some pH, usually above 9.0 for model complexes,

$$Zn^{2+}(H_2O) \longrightarrow Zn^{2+}(OH^-) + H^+,$$

which means that zinc as an acid can generate a very effective localized hydroxide ion (base) which can attack other molecules. This is the proposed function of zinc in several enzymes including carbonic anhydrase where the OH^- attacks CO_2,

$$O{=}C{=}O + OH^- \longrightarrow HCO_3^-,$$

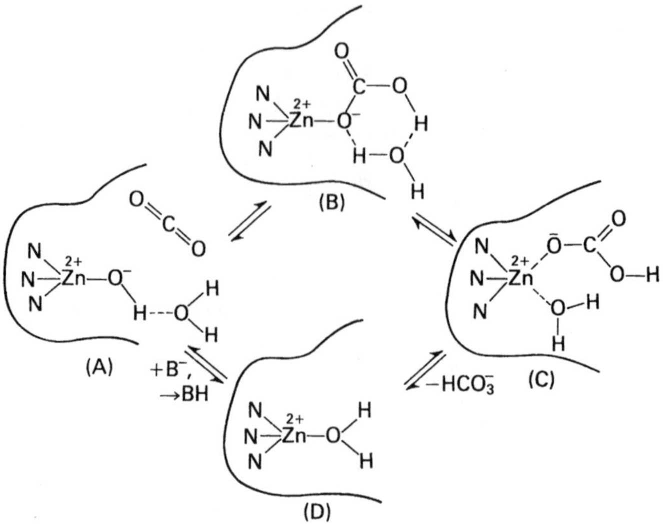

Fig. 11.6 A proposal for the pathway of formation of bicarbonate from carbon dioxide using carbonic anhydrase as catalyst. The function of a base, B, is not clear.

We take carbonic anhydrase as an example of a typical zinc enzyme. Water is a substrate in this reaction, and in carbonic anhydrase the zinc ligands, all neutral imidazoles, and the protein structure are 'designed' to make the zinc of $Zn(H_2O)$ as acid as possible so that the ionization $H_2O \rightarrow OH^-$ occurs at pH = 7. In Section 4.3 we have shown why the development of Lewis acidity is possible in a sterically constrained entatic zinc site. In addition, the movement of the substrate atoms along the path of intermediates is aided by the rapid known switching of coordination geometry of the zinc between four and five coordination which reduces barriers between steps (Fig. 11.6). Furthermore, the groove which leads in the enzyme from the surface to the active site is designed as a proton conductor by the incorporation along it of several imidazoles. It is the combination of the geometry, bond angle and bond length, the ligand groups, the site and mobility, and the immediate surrounds, effectively the solvent, of the zinc complex, together with the more remote proton conduction groups which makes the enzyme so powerful. The overall structure is typical of an enzyme in that it is largely a β-sheet (Fig. 11.7). We begin to see why a very large molecule might be required to carry out even such a simple reaction with speed and selectivity.

An example of a second use of zinc is in a direct hydrolysis of a zinc ester or amide substrate complex where the zinc acts as an attacking acid on this second substrate while making water attack it as well (see Section 4.3).

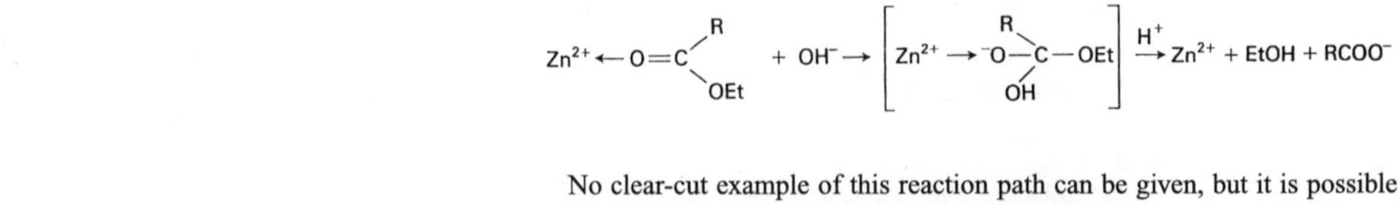

No clear-cut example of this reaction path can be given, but it is possible that such a path is present in carboxypeptidases. (Note again how firmly the zinc is bound in these enzymes.)

An alternative to this pathway is again the activation of the water by the zinc and by a neighbouring carboxylate group so that OH^- is readily produced in the presence of substrates. This OH^- then attacks the amide bond.

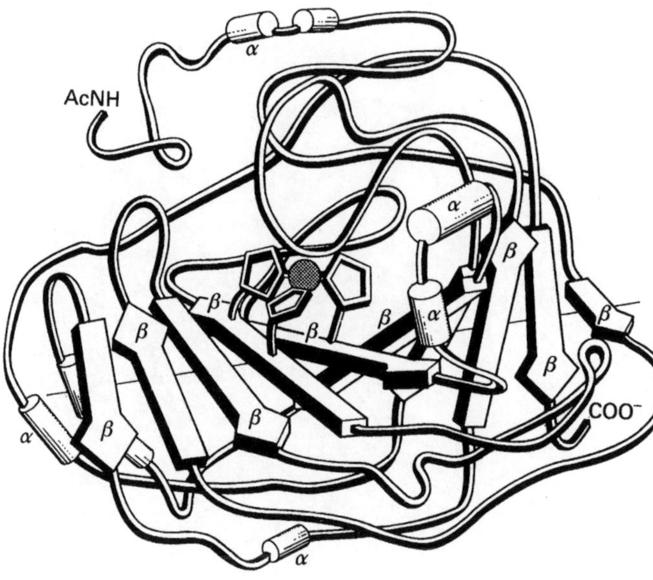

Fig. 11.7 Secondary and tertiary structure of human carbonic anhydrase C. The cylinders represent the α-helices and the arrows the strands of the β-structures. The zinc atom together with its histidine ligands is also shown connected to the β-sheet. [After Lindskog, S. *et al.* (1971). In *The enzymes*, Vol. 5 (ed. P. Boyer), Chapter 21. Academic Press, New York.]

Chemists have clearly developed models of some of these metal ion catalysed hydrolyses and group transfers using inert trivalent Co(III) metal ion centres, see Aside (A). These centres substitute very slowly and therefore allow analysis of single steps of reaction while they show rates for these single steps of the same order of magnitude as those of the enzymes themselves. Of course, in this chemistry there is no flow of substrates since the products cannot leave the cobalt(III). Now, the rate of steps in these model reactions can be enhanced by the presence of more than one metal ion in the synthesized complexes. When one turns to the zinc enzymes handling phosphate esters and some other substrates, one observes that, indeed, in several of them there is more than one metal ion as the model chemistry indicator is more effective section 4. The usual metal ions are the divalent cations Mg^{2+} and Zn^{2+} (Fig. 11.8). Examples of this type of enzyme are given in Table 11.4. The question arises as to how these divalent ions can be as effective as the trivalent cobalt(III) even in the single steps of reaction. Consider their different ability to generate a bound hydroxide ion as earlier in this section. The difference in hydrolysis constants for $M(III)(H_2O)_6$ and $M(II)(H_2O)_6$ is greater than a factor of 10^6 for cobalt (III) against Zn(II) or about 10^{10} for cobalt(III) against magnesium(II) so it is probable that the magnesium ion in an enzyme will act only to orient the phosphate-containing substrate. The zinc is then the enzyme attacking group and we know from the studies of carbonic anhydrase that this metal in an enzyme active site bound to three imidazoles is specially activated to $Zn\text{-}OH^-$ at a pH not far from 7.0. The hydrolysis constant of zinc in the enzyme is some 10^3 greater than for free zinc, but, clearly, the enzyme site provides still extra catalytic factors. It is intriguing that here model chemistry is giving much pause for thought about the exact mechanism of enzymatic metal ion catalysis. The discussion applies equally to those *acid* phosphatases which contain not zinc and/or magnesium but two atoms of either iron or manganese.

(A)

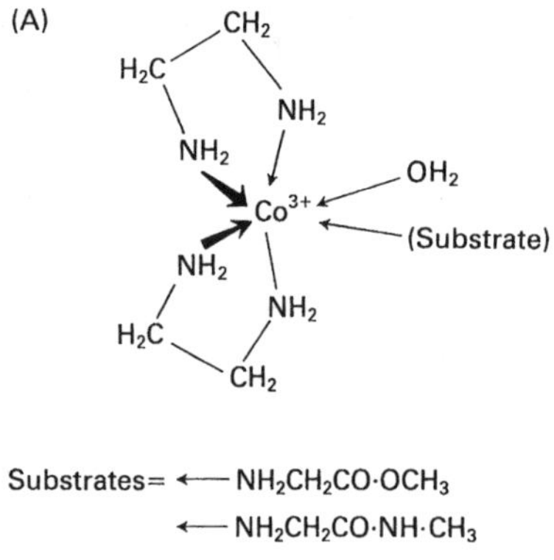

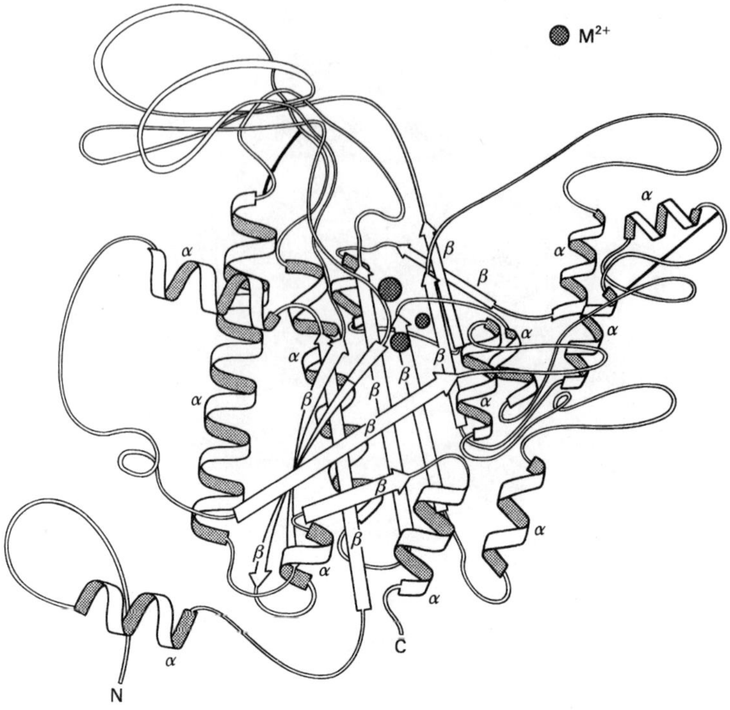

Fig. 11.8 The structure of alkaline phosphatase showing the grouping together of three metal ions in the active site. Notice the β-sheet. (After J. Coleman and H. W. Wyckoff.)

11.7.2 Zinc and redox reactions

Zinc is also found in several enzymes which are involved in redox reactions. A typical example is alcohol dehydrogenase

$$NAD^+ + C_2H_5OH \longrightarrow NADH + CH_3CHO + H^+.$$

The metal, as a Lewis acid, assists in the activation of the alcohol to make hydride removal easier. The acceptor in the enzymes is usually NAD (NADP). A particularly important example of these redox reactions is an enzyme in the shikimate pathway where the hydride removal step uses the zinc-dependent enzyme dehydroquinate synthase. The importance of the pathway is that it leads to the synthesis of the three aromatic amino acids phenylalanine, tyrosine, and tryptophan. The enzyme, like several other zinc enzymes, is in the cytoplasm. The question arises as to whether the zinc has more than a catalytic role in this particular enzyme. The parallel with the use of zinc in nucleic acid synthesis, that is in aspartate transcarbamylase, is clear in that zinc is again linked to some of the most fundamental steps in biological synthesis of major polymers. Finally, zinc appears in the enzyme levulinic dehydratase, which controls porphyrin synthesis and hence a vast range of redox reactions. As for magnesium and calcium, the involvement of zinc in so many different pathways leads us to ask how extensive is its role in controls in organisms.

11.7.3 The constrained (entatic) state and probes

In all the above enzymes zinc can be replaced *in vitro* by other divalent metal ions (Fig. 4.2). Outstandingly, cobalt(II) is the best replacement for activity despite the fact that the expected order of Lewis acid catalytic power is

$$Cu > Zn \leq Ni > Co > Fe > Mn.$$

(Note that this switching of metal ions at the active site of an enzyme is a parallel inorganic study to site-specific mutagenesis of organic protein sidechains). The implication is clear: the enzyme site is often designed for a site symmetry which is not octahedral and both direct structure determination and spectroscopic probe analysis show this. In several cases the site is of irregular symmetry, four or five coordinate, which is readily attained by Zn^{2+} or Co^{2+}. As stated in Section 11.4, the enzyme imposes a geometry on the attacking divalent metal ion to optimize attacking strength and minimize barriers for reaction. Similar comments have been made earlier concerning electron-transfer enzymes. The strain is in the free energy of binding, of course, and not just in the bond energy (See section 6.8). Note the different problem where in helical proteins the strain forces motion in the helices.

11.8 Summary of zinc proteins

It is clear that all the bound zinc, much like the bound calcium, cannot be considered to be in one class of functional value. Thus, one class of calcium was seen to be required in long-lasting cross-linking functions, while another class was in triggering. Function here depended on rate of exchange which was related to binding constants. Consider the case of zinc. There would appear to be at least *six* classes of zinc as judged by binding constants and exchange.

(i) In *extracellular enzymes* the zinc binding constant ($> 10^{12}$ M^{-1}) is often such that zinc is only lost on destruction of the protein. Some of these abundant enzymes are present in all phylla, including prokaryotes where there are fewer (see Section 11.10). Frequently they are functional in hydrolytic reactions and are outstanding in digestion and in matrix-manipulation. Some similar zinc enzymes are used in the cytoplasm, e.g. carbonic anhydrase. The binding is by two or more histidines and/or one carboxylate plus a water molecule. It is questionable if these extracellular enzymes, usually based on β-sheets, are primitive. Zinc was not very available in early periods of life.

(ii) Usually in intracellular *enzymes* of all phylla zinc is bound strongly to three or four N/S donors, $K > 10^{12}$ M^{-1}, and does not exchange (see Table 11.7). These enzymes too are based on β-sheets. It is possible that zinc is inserted into such enzymes by zinc selective pumps and carriers. However, there are some enzymes in which zinc is bound by O-donors from which zinc will exchange with free zinc. Zinc here can be used and probably was so used initially in synthesis. We note that, in one way curiously, zinc is not a major element in hydrolytic catalysis in the cytoplasm. Rather it is Mg^{2+} and now and then Mn^{2+} or Co^{2+}, or even Fe^{2+} in prokaryotes. Maybe the lack of zinc enzymes in the cytoplasm, apart from its association with polymerases, arises from the low availability of zinc in the early environment.

(iii) In zinc fingers, parts of *transcription factors* and some *control features* of enzymes only found in the eukaryotes, zinc is bound less tightly, $K = 10^{11}$ M^{-1}. We shall consider that it exchanges possibly with free zinc and definitely with metallothionein sufficiently rapidly to give 'free zinc' a *regulatory role* on a slow time-scale. All zinc fingers belong to a coordinated set. These proteins are random coils in the absence of zinc.

Table 11.7 Zinc enzymes in synthesis

Synthetic step	Zinc enzymes
RNA	RNA polymerase
DNA	Reverse transcriptase
Viral synthesis	Terminal dNT transferase
Transfer RNA	tRNA synthetase
Essential amino acids	Dehydroquinate synthase
Essential nucleotides	Aspartate transcarbamylase
DNA-repair	Excinucleases

(iv) In some *vesicles* zinc is weakly bound, $K < 10^6\ M^{-1}$, and exchanges very rapidly. The zinc can act as a *messenger* when discharged. We note the presence of zinc in the male reproductive tract and in the brain, especially in the forebrain and the mossy fibres. Here free zinc *controls* certain activities while its value in the human insulin vesicles is unsure.

(v) There are *ATP-ase pumps and other transport proteins* for zinc in membranes. On the inside of the membrane facing the cytoplasm the binding constant may be quite high, say $10^{+11}\ M^{-1}$, so as to accept zinc from zinc carriers. Outside the cytoplasm in vesicles it must be much lower, but on the outside of cells it may be high again, as stated in (i).

(vi) The *transport of zinc by albumins* in extracellular fluids of advanced organisms must have a zinc binding constant $> 10^{+10}\ M^{-1}$. In cells there are also some transport proteins, called 'metal chaperones'.

In the next sections we shall look at the functional value of these proteins separately, beginning with the regulatory role of zinc.

11.9 Regulatory and control roles of zinc

As stated in Section 11.4, an extensive group of zinc proteins is now known in eukaryotes which bind to promoter regions of DNA (Table 11.5). These zinc finger proteins are connected to hormones deeply associated with developmental process, e.g. thyroxine and many steroid hormones. Lack of zinc has long been associated with various problems of growth and healing of damaged tissues. At present (2001) the number of zinc finger transcription factors exceeds several thousand in man and maybe more than 10 per cent of eukaryotic DNA expression is regulated by them. The role of zinc in life subsequent to its primitive stages cannot be over-stated. Mutations in zinc fingers in several cases, e.g. the protein p.53, are known to give rise to cancer.

As is the case for calcium, the *regulatory* role of zinc can be made to depend on a high power of the free zinc ion concentration by the binding of several metal ions to one protein. This type of cooperativity is seen in the transcription zinc finger σ-factors, e.g. IIIA (see Table 11.3), where as many as 10 zinc atoms may occur (Fig. 11.5). The equilibrium reaction must now be written in a series of steps up to n where maximum cooperativity is n

$$A(Zn^{2+})n \rightleftharpoons B + nZn^{2+}$$

Alternatively, each zinc complex $A(Zn^{2+})$, $A(Zn^{2+})_3$, etc. could have a separate role to play so that the gearing of reactions in a cell could be regulated by zinc concentration in a smooth fashion. It is the authors' opinion that these roles of zinc, in which the zinc bound states can be active or inhibitory in regulation, will be found in many more cases than we know at present, especially associated with fertilization and growth.

The high dependence on Zn^{2+} concentration can also be of value in homeostasis provided the zinc ion is effectively 'buffered'. The most sensitive homeostatic mechanism is based on a high power, n, of the free metal ion concentration, here $[Zn^{2+}]^n$, here we note again the clusters of metallothionein. However, we have to note that the same or a similar protein seemingly maintains the homeostasis of copper (and cadmium). Let us suppose that the random coil protein metallothionein folds differently according to the

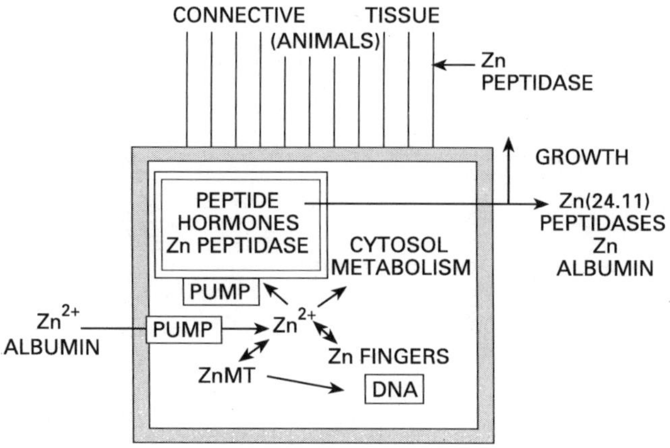

Fig. 11.9 The relationship of zinc activities to cellular structure. MT, metallothionein.

metals to which it binds; there can then be two sets of receptors for copper and zinc metallothioneins as well as one for the apoprotein on DNA, but this means that the metabolisms of the two metals have become inextricably linked.

Now, if the zinc homeostasis is to be maintained, then the zinc entering and leaving a cell must be managed much as this is so for calcium. The requirement is for pumps with feed-back control. Again, as previously stated, to maintain the occupation of zinc fingers and metallothionein the zinc pump on the cytoplasmic side of the cell must have a binding constant very closely similar to those of these proteins, $\sim 10^{11}\ M^{-1}$. In this way we can see that there is a resting state of slow zinc flow in the cell when zinc is connected in a network (see Fig. 11.9). An approximately 10^{-11} M zinc concentration may well be a characteristic of the free metallome of eukaryotes. The network is probably an integral part of the endocrine system.

A final role of zinc in a message transmission is the release of free zinc from vesicles especially in the brain and in reproduction. Free zinc in the brain is stored in high concentration with the transmitter glutamate. It is of uncertain value but is reminiscent of free zinc stored with insulin and released with it. Whether zinc has a growth-promoting role opposite certain synapses is still not clear.

11.10 The export of zinc enzymes: digestion and peptide messages

Apart from the transport of zinc itself, completed zinc enzymes can be moved in or out of cells. The extracellular enzymes can be generated in two ways: first, by direct export of protein across the outer membrane when, if they are zinc enzymes, they find and fold around zinc in the extracellular fluids; second, by synthesis in vesicles. Here the protein is secreted into a space where zinc is also secreted. Both modes are common in zinc biochemistry and we should ask why both are used.

It is particularly noticeable that zinc proteases are stored in vesicles in such systems as the pancreas (digestive carboxy- and aminopeptidases) and in venoms (Table 11.8). The proteases are extremely important as matrix cutting devices in growth (see Chapter 2) and hence in the development of cancer (Table 11.8). It must not be forgotten that zinc is also present in other extra-

Table 11.8 Zinc in degradation processes

Degradative process	Zinc enzyme
Pancreatic juice action	Carboxypeptidase
Venom haemorrhagic action	Zn proteases
Extracellular digestion (yeast)	Zn aminopeptidase
Extracellular digestion (bacteria)	Thermolysin
Breakdown of DNA and RNA	3′Nucleotidase
Peptide hormones	Peptidases 24.11
Connective tissue degradation	See Table 11.9

Table 11.9 Zinc and external filamentous structures

Filamentous structure	Role of zinc
Collagens	Zinc collagenases
Proteoglycans	Zinc stromelysin
Denatured collagens	Zinc gelatinase
Keratins	Zinc cross-links

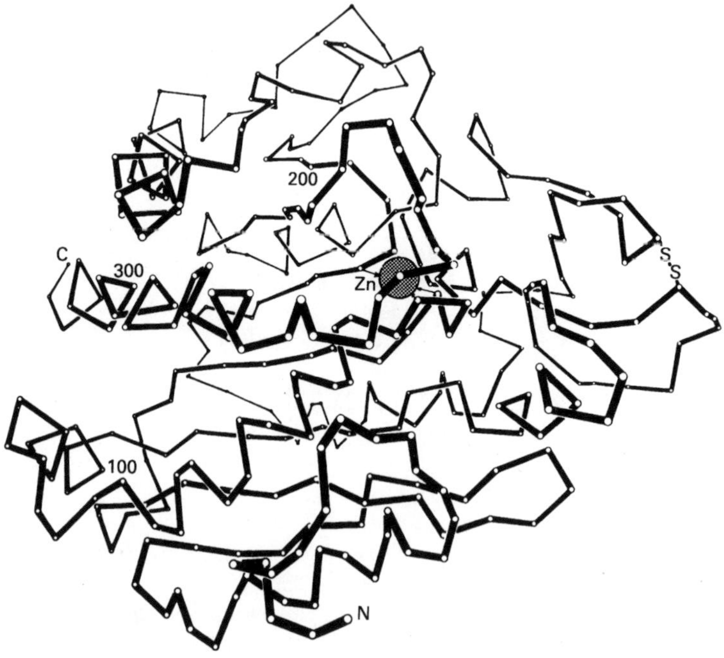

Fig. 11.10 The fold of carboxypeptidase showing the buried metal ion connected to a β-sheet. (After W. Lipscomb.)

cellular hydrolytic enzymes, e.g. phospholipases, nucleases, and saccharases, all active in digestion. (In passing observe the β-sheet structures of these enzymes; Fig. 11.10.)

11.11 Zinc enzymes and peptide hormones

There is a requirement for the activity of many peptide triggers and hormones such that both their release and their removal must be very rapid. Consider the usual reaction sequence for a peptide hormone, where E is an enzyme,

$$\text{Stored precursor} \xrightarrow{E_1} \text{Hormone (transmitter)} \xrightarrow{E_2} \text{Local hydrolysis of hormone (relaxation).}$$

The familiar parallel is with stored acetylcholine or adrenalin and *local* rapid degradation after action by esterases and amine oxidases (Cu-proteins). The rapidity of catalysis by E_1 and E_2 and their location are equally important. Zinc enzymes are at their best in such situations and the known distribution of

Table 11.10 Zinc proteins related to peptide hormonal action

Hormone	Mode of association with hormones
Insulin	Zinc associated with hormone storage
Angiotensin	Zinc in antiotensin-converting enzyme
Enkephalin	Zinc in enzyme enkephalinase
Several peptides	General hydrolysis by endopeptidase 24.11
Neurotensin	Degradation by zinc-dependent enzyme
Other peptides	Amino and carboxypeptidases (Zn) at synapses

zinc in certain parts of organisms may well be linked to this function. Certainly the zinc enzyme peptidase 24.11 and its close relatives are associated directly with the destruction of enkephalins and other peptide hormones (Table 11.10).

11.12 Other functions of zinc outside cells

11.12.1 Cross-linking and hardening of extracellular matrices

There is growing evidence that zinc is used to cross-link phenolic materials in the strengthening of teeth of insects and lower animals. These teeth or mandibles are made from chitins. The presence of high zinc in the teeth has not been definitely associated with phenolate sidechains, but this remains highly probable. The movement of the zinc down hollow central tubes of the teeth toward the cutting edges is thought to occur by the transfer of vesicles high in this metal.

11.12.2 Zinc in solid-state devices

Zinc also occurs in some simple crystalline materials within some organisms. The most obvious are the deposits of zinc phosphates in what may be storage vesicles of such animals as snails. The real purpose of the vesicles is unknown, but they could act as depositories of unwanted excess zinc until it is rejected by exocytosis. Another certainly useful zinc deposit is crystalline zinc cysteinate in the tapetum of night-seeing animals. The crystals act as reflectors giving added sensitivity to the eye. It seems likely that these crystalline deposits were also laid down in vesicles.

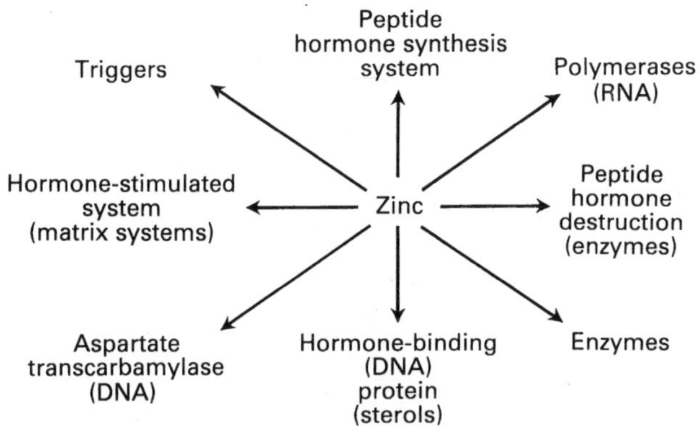

Fig. 11.11 The extensive connections of zinc to multicellular controls and regulations. Zinc links to sterol-like hormones, to peptide hormones, to the external matrix, and to a vast list of enzyme and control activities in cells.

11.13 Zinc genetics

We have described the zinc network of interactions and controls over concentration and enzyme activities in Section 11.9. and see Fig. 11.11. We need now to relate these observations to the genome. The relationship to genome is in two parts: the expression of proteins relating to the control over zinc concentrations and the expression of the production of zinc-binding proteins generally. We are no longer concerned about the expression of a wide range of proteins due to transcription factors which contain zinc — the zinc fingers of which there are very, very many (Sections 11.4.4 and 11.9). We turn first to the control of zinc uptake in prokaryotes.

11.13.1 FUR and ZUR

Bacteria have two uptake (u) regulatory (r) proteins for iron (Fe) and zinc (Zn); FUR and ZUR, respectively. In fact, FUR has a zinc binding site and ZUR has an iron binding site. The proteins are transcription factors of similar sequence and presumably structure. The two binding sites exchange with free zinc (site A) and free iron (site B) selectively. The selection is achieved by the chemical differences in binding site, strengths of binding and the limitations on free metal ions by buffering and/or pumping, as well as limitations on proteins production by regulation of DNA.

The site for zinc is tetrahedral, using two thiolate ligands and probably one histidine N donor, giving a binding constant of about 10^{11} M^{-1}. The iron (Fe^{2+}) site (see Section 12.16) is octahedral, with one or two histidine and one or two carboxylates. Its binding constant for Fe^{2+} is around 10^7 M^{-1}. In the presence of limiting concentrations of Zn^{2+} (about 10^{-11} M) and Fe^{2+} (about 10^{-7} M) the two sites will be occupied by zinc and iron separately.

ZUR is required for the zinc-specific repression of two operons, *yciC* and *ycdH*, which appear to encode a low affinity and a high affinity zinc carriers (pumps) and are implicated in the uptake of this metal (see Helmann and Gaballa in 'Further reading').

More is becoming known about the expression of zinc uptake systems in higher organisms. The zinc 'buffer' protein metallothionein may well be related to its eukaryotic transcription factors paralleling the system found for copper. In yeast there is a transcription factor ZAP 1, which controls expression of zinc carriers/pumps, and itself has five zinc finger domains, indicative of a feed-back relationship with transferable zinc.

11.14 Zinc and evolution

Zinc was not readily available to primitive organisms due to its precipitation as a sulphide. Hence the most primitive use of zinc, probably inside cells, needed very strong binding sites, e.g. in polymerases. It is very likely that a few zinc proteases and peptidases were first employed somewhat later by prokaryotes simply for digestive functions. This meant that zinc proteases, with strongly bound zinc had to be exported. As zinc became more available, single cell eukaryotes could refine this function by holding the proteases in vesicles, releasing them when there was information that external food was available. Later still, when there was even more environmental zinc, multicellular organisms turned around this use of hydrolysis by zinc enzymes in that zinc proteases and peptidases could be placed in vesicles to digest precursors of peptide messengers and could also be held on the surface of cells to destroy these hormonal peptides in the circulating fluids and to break up connective tissue when needed. Thus, zinc not only remained part of digestion but became an essential part of connective tissue and of peptide message removal and sometimes of its production, and thence of growth control. These hydrolytic enzymes, usually associated with extra-cellular activities, cannot be part of *internal* cellular homeostasis. Once again as zinc levels increased with the oxidation of sulphide a second marked development in evolution was the appearance of zinc fingers and related transcription factors (Table 11.11). The number of zinc-finger related proteins increases from zero, *E. coli*, to 150 in yeast (they are largely of certain types), to 550 in the worm *C. elegans*, to over 2000 in humans. They represent a change in function of N/S sites and are probably connected to zinc exchange and thence regulation of a wide variety of eukaryotic growth functions. Similarly, the special zinc dependent metallo-proteases increase from 1–2 in yeast to about 100 in the worm to a very much larger number in humans (see Table 11.11) for their association with peptide hormone metabolism in evolution.

The use of zinc as a free ion released from certain vesicles is a very recent development, connected with bisexual activity and the brain.

11.15 **Summary: is zinc today a master hormone?**

Inspection of zinc activities

1. in transcription factors, zinc fingers; related to sterol, thyroid and other hormones;
2. in the core proteins of DNA, i.e. the division processes;
3. in RNA synthetase and reverse transcriptase;
4. in cytoplasmic enzymes, some extremely critical for the synthesis even of the monomers of proteins, DNA and RNA, and porphyrins, e.g. haem, vitamin B_{12} and chlorophyll;
5. in many enzymes for the degradation of external peptides (hormones), which are stored in vesicles, together with (1) in general control of growth and structures (collagen, see Section 11.10.)
6. in its storage (free ?) in critical zones of the body, e.g. the brain and the male reproductive tract;
7. in digestion;

leads the possibility that zinc is not just the major extracellular acid catalyst of increasing use from primitive times but that the internal homeostasis of zinc is finely tuned so that zinc is one of the dominant *cytoplasmic* growth hormones. Rapid cell division will then depend upon the presence of stored zinc and its transfer. This suggestion that zinc is effectively a hormone is no more than an hypothesis, but the importance of zinc, which equals iron in concentration in living systems, must not be missed because it is so hard to follow by analysis. We see again the all-embracing role of this metal when we turn to *extracellular* activities closely connected to growth. Perhaps in early evolution and still in prokaryotes iron and manganese played the main metal ion role of communicating directly with DNA (Chapters 12 and 20), but in the evolution of the eukaryotes this role has been largely taken over by zinc in the cytoplasm. The importance of zinc then increased again as development of complex multicellular systems evolved which were dependent on hormones such as peptides, thyroxine and sterols. Is all this chemistry linked to the

Table 11.11 Communication networks

Simple current carriers	**Triggers (second messengers)**	**Hormones**
Na^+, K^+	Ca^{2+} (Mg^{2+})	Zn^{2+}, Fe^{2+}, Mn^{2+}
Acetylcholine	cAMP (cGMP)	Steroids

Other control elements are not usually classified in the above manner, e.g. Fe^{2+}, Mn^{2+}, ATP, ADP, since they serve many other functions.

increased concentration of dioxygen some one billion years ago which released zinc from zinc sulphide? As stressed earlier, zinc is not redox-active and yet has particular value in organisms developed after a dioxygen atmosphere was generated. This change is due to the relative solubilities of its oxide and sulphide (Fig. 1.12) which allowed the metallome and the proteome to develop using zinc. We return to these problems in Chapter 20.

In conclusion, while in primitive organisms zinc probably occupied only fixed non-exchangeable sites in enzymes both inside and to some degree outside the cell due to the strong fold of the apoproteins, and this is still a major feature of zinc biochemistry, in more advanced cells three new features have emerged. First, this early use of zinc enzymes has been vastly extended so as to manage extracellular filamentous structures (matrix metallo enzymes), extracellular messengers and many digestive processes. Second, zinc has become incorporated into a vast coordinate system for control of DNA expression during development. Third, free zinc is an essential messenger in its own right in some compartments of the brain and the reproductive tract. In this discussion we must never forget that the production of zinc-binding proteins is in feed-back relationship to the concentration of free zinc so that there is never such an excess of apo-proteins that they could bind adventitiously to other metal ions (see Chapter 20). The metallome and proteome are linked but the metallome is also connected directly to the environmental availability. (Note that in keeping with the essential nature of zinc for all development, zinc deficiency causes stunted growth and inability to go through puberty.)

Once again we must see these developments in relationship to limited expression systems in different cell types and in feed-back relationship with free cytoplasmic zinc levels. Expression of growth through binding to zinc finger proteins is under the control of factors other than zinc, especially of the steroid family, vitamin A related retinoic acid, thyroid hormones and so on, which require iron for synthesis. Thus, these transcription factors and proteins

Table 11.12 Evolution of zinc functions

Binding groups	**Early Zn functions**	**Late added Zn functions**
N/O	Digestive enzymes (possibly a few)	Matrix metallo-enzymes and hormonal peptide hydrolysis*
N/S	RNA polymerases Substrate enzymes	Zinc transcription factors Developmental control proteins (Metallothionein ?)
O/O ?		Free zinc in brain and reproductive tract
N/N	Carbonic anhydrase	Some proteases

*See Coates *et al.* in references.

are not just zinc dependent.

It is difficult to avoid the conclusion that in diagrams of survival strength against composition (see Fig. 7.9) the switch in the use of non-redox elements in evolution, including K^+, Na^+, Ca^{2+}, and Zn^{2+}, was integral to the changes of organisms. Many of the properties which higher organisms evolved could not have been developed without the changes in use of these elements: Ca^{2+} (earliest) in chemical message transfer connected to organic ester phosphates, Zn^{2+} (intermediate) as a Lewis acid giving novelty to hydrolytic enzymes and transcription factors, allowing multicellular growth (Table 11.12) (see also copper), and K^+/Na^+ (last) in electrolytic messages (see Section 8.10) permitting the evolution of nerves and brain. The changes in the free metallome of the cytoplasm were never very large, but in the extracellular and vesicular fluids it developed very considerably with the environment changes. (The change of free zinc in the cytoplasm may be only in the range 10^{-15} M to 10^{-11} M in the progression from the anaerobic sulphide medium to that of today. The free metallome of vesicles and in extracellular space could be from below 10^{-11} M up to 10^{-5} M). The bound forms of elements, seen easily in the total proteome, developed in all compartments. In the following chapter we shall examine parallel evolutionary developments of redox active metals (Chapter 12 to 17), and then of non-metals (Chapters 18 and 19) before we look at the integrated advance of them all (Chapter 20).

Note. Probes of metal sites by isomorphous substitution and by indicator dyes

(a) *By isomorphous substitution*

It has proved particularly easy to substitute zinc in protein (enzyme) sites by other metal ions. In particular, Co(II), Mn(II), and Cd(II) have proved rewarding in exchange studies. The detailed spectroscopy of the active sites of the zinc enzymes made possible by these substitutions and the kinetic analysis that has been carried out with them has revealed the very subtle nature of the structure and dynamics of active sites. This work, which precedes site-specific mutagenesis of organic groups but is parallel to it and more amenable to interpretation, is often overlooked by biochemists who do not follow the intricacies of the analysis available through the spectroscopy of metal ions. The potential as reporter groups is of another order from that provided by organic centres. In this book little of this work is recorded, and many deductions depend on the depth of work given in volumes devoted to this type of analysis. Table 11.13 gives some possibilities of probe studies, not only for zinc but also for many other metal ion substitutions.

Table 11.13 Some probe (isomorphous) substitutions

Natural metal	Probe substitution
K^+	Tl^+, Rb^+
Mg^{2+}	Mn^{2+}, Cr^{3+}, $[M(NH_3)_6]^{3+}$ where M = Co, Cr
Ca^{2+}	Mn^{2+}, Lanthanides (M^{3+})
Zn^{2+}	Co^{2+}, (Mn^{2+})

(b) *By indicator dyes*

An alternative approach to an examination of the role of zinc is the use of zinc-selective dyes such as Zinquin, compare Quin-2 for free calcium detection. Zinquin has a binding constant of 10^8 M^{-1}. Use of it has shown that in many cells zinc is concentrated in local vesicular zones, which is not yet fully understood. Much zinc is required during development so that storage is essential. A zinc-rich protein in eggs is vitellogenin.

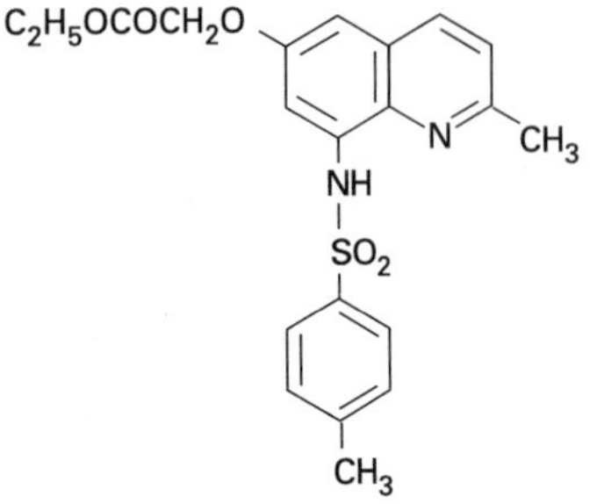

Zinquin

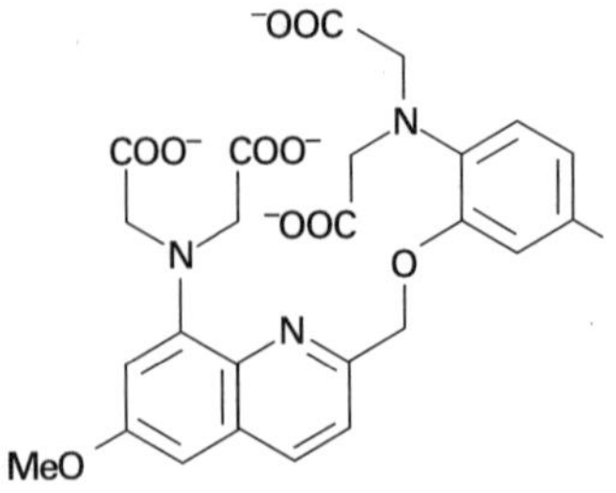

Quin–2

Acknowledgements. We are grateful to Prof. Y. Mély (Strasbourg), Prof. R. Palmiter (Washington), and Prof. N.J. Robinson (Newcastle) for considerable assistance with this chapter.

Further reading

Berg, J. and Shi, Y. (1996). The galvanization of biology: a growing appreciation of the role of zinc. *Science*, **271**, 1081–5

Christianson, D.W. and Cox, J.D. (1999). Catalysis by metal-activated hydroxide in zinc and manganese metalloenzymes. *Annual Review of Biochemistry*, **68**, 33–57

Coates, D., Siviter, R., and Isaac, R.E. (2000). Exploring genomes for neuropeptide and peptidase function. *Biochemical Society Transactions*, **28**, 464–9

Coleman, J.E. (1998). Zinc enzymes. *Current Opinion in Chemistry and Biology*, **2**, 222–34

Coleman, J.E. (1992). Zinc proteins: enzymes, storage proteins, transcription factors and replication proteins. *Annual Review of Biochemistry*, **61**, 897–946

Colvin, R.A., Davis, N., Nipper, R.W., and Carter, P.A. (2000). Zinc transport in the brain: routes of zinc influx and efflux in neurons. *Journal of Nutrition*, **130**, 1484 S–7S

Endo, T., Fujii, T., Sato, K., Taniguchi, N., and Fujii, J. (2000) A pivotal role of Zn-binding residues in the function of the copper chaperone for SOD. *Biochemical Biophysical Research Communications,* **276**, 999–1004.

Frederickson, C.J., Suk, S.W., Silva, D., Frederickson, C.J., and Thompson, R. B. (2000). Importance of zinc in the central nervous system: the zinc containing neuron. *Journal of Nutrition*, **130** (5 S Suppl), 1471 S–1483 S

Gaballa A. and Helmann, J.D. (1998). Identification of a zinc-specific metalloregulatory protein, Zur, controlling zinc transport operons in *Bacillus subtilis*. *Journal of Bacteriology*, **180**, 5815–22

Gonzalez T. and Robert-Baudouy, J. (1996), Bacterial amino-peptidases. *FEMS Microbiological Reviews*, **18**, 319–44

Grotz, N., Fox, T., Connolly, E., Park, W., Guerinot, M.L., and Eide, D. (1998). Identification of a family of zinc transporter genes from Arabidopsis that respond to zinc deficiency. *Proceedings of the Natural Academy of Science USA*., **95**, 7220–4

Klug, A. (1999). Zinc finger peptides for the regulation of gene expression. *Journal of Molecular Biology*, **293**, 215–18

Lindskog, S. (1997). Structure and mechanism of carbonic anhydrase. *Pharmacology and Therapeutics*, **74**, 1–20

Lipscomb, W.L. and Sträter, N. (1996). Recent advances in zinc enzymology. *Chemical Reviews*, **96**, 2375–2433

Maret, W. (2000). The function of zinc metallothionein: a link between cellular zinc and redox state. *Journal of Nutrition*, **128**, 667–70

Mély, Y., De Rocquigny, H., Morellet, N., Roques, B.P. and Gérard, D. (1996). Zinc binding to the HIV-1 nucleocapsid protein. *Biochemistry*, **35**, 5175–82

Matthews, R.G. and Goulding, C.W. (1997). Enzyme catalysed methyl transfers to thiols: the role of zinc. *Current Opinion in Chemical Biology*, **1**, 332–9

McMahon, R.J. and Cousins, R.J. (1998). Mammalian zinc transporters. *Journal of Nutrition*, **128**, 667–70

Palmiter, R.D., Cole T.B., Quaife, S.D. and Findley S.D. (1996). ZnT-3, a putative transporter of zinc. *Proceedings of the National Academy of Science*, U.S.A. **93**, 14934–14939

Patzer, S.I. and Hantke, K. (2000). The zinc responsive regulator, *Zur. Journal of Biological Chemistry*, **275**, 24321–32

Radisky, D., and Kaplan, J. (1999). Regulation of transition metal transport across the yeast plasma membrane. *Journal of Biological Chemistry*, **274**, 4481–4

Rauser, W.E. (1999). Structural and functions of metal chelators produced by plants: the case of organic acids, aminoacids, phytin and metallothioneins. *Cell Biochemistry and Biophysics*, **31**, 19–48

Roesijadi, G. (2000). Metal transfer as a mechanism for metallothionein-mediated metal detoxification, *Cellular and Molecular Biology*, **46**(2), 353–405

Sandstead, H.H., Frederickson, C.J., and Penland, J.G. (2000). History of zinc as related to brain function. *Journal of Nutrition*, **130** (2S Supplement), 4965–5023

Smith, K.S. and Ferry, J.G. (1998). Prokaryotic carbonic anhydrase. *FEMS Microbiology Reviews*, **24**, 335–66.

Takatsuji, H. (1999). Zinc-finger proteins: the chemical zinc-finger energies in plant sciences. *Plant Molecular Biology*, **39**, 1073–8

Vallee, B.L. and Falchuk, K.H. (1993). The biochemical basis of zinc physiology. *Physical Reviews*, **73**, 79–118

Vallee, B.L. and Williams, R.J.P. (1968). Probes for metal ions in enzymes. *Chemistry in Britain*, **4**, 397–406

Vasak, M. and Hasler, D.W. (2000). Metallothioneins: new functional and structural insights. *Current Opinion in Chemical Biology*, **4**, 177–83

Zhao, H. and Eide, D.J. (1996). The yeast *zrt 1* gene encodes the 3i-c-transport protein of a high affinity uptake system induced by zinc limitation. *Biochemistry*, **93**, 2454–8

Zhao, H, and Eide, D.J. (1997). ZAP1, a metalloregulatory protein involved in zinc-responsive transcriptional regulation in *S. ceresace. Molecular and Cellular Biology*, **17**, 5044–52

Notes (1) The control of the zinc concentrations in cell compartments and that of its proteins, the free and total metallomes, is under intensive study. It appears to be confirmed that intracytoplasmic free zinc is held close to 10^{-11} M. The most recent work can be found by reference to the Abstracts of a meeting of The Society for Experimental Biology (2001) Section C.1. Metals and Cells, held at Canterbury, UK.

(2) It must not escape notice that zinc enzymes are very common metalloproteases affecting a wide variety of extracellular responses such as growth of tumours and Na^+ levels (hypertension) through angiotensin release. Many metalloproteases are released via Ca^{2+} message. Here Zn^{2+}. Ca^{2+} and Na^+ are linked.

CHAPTER 12

Non-haem iron: redox reactions and controls

12.1 General introduction to transition metals in biology

In the introductory chapters we stressed that the different mineral elements had different biological functions and that these were based on a selection corresponding to potential for activity. This is the course evolution had to find, but much thought is necessary to see the complications which inevitably arise since many elements are capable of more than one function (see Chapter 7). The problems of understanding have already increased as we have moved from the chemistries of Na^+/K^+ to those of Mg^{2+} (note phosphate chemistry, CO_2 fixation and chlorophyll), to those of calcium (note fast resting state controls and triggering), to those of zinc (note catalysis and slower control). However, all these activities relate to an increasing Lewis acid strength and a decreasing availability from group 1 to group 2 elements and then zinc. Moreover, in evolution we have little reason to suppose that the relative availability of these elements except for zinc, changed grossly from the anaerobic to the aerobic eras or even from one biological niche to another. When we leave this group of elements and turn to the transition metals (A) we meet quite new problems since: (1) the elements have increased functional capacity in that while they remain good Lewis acids they can be both π-electron donors and redox-active; (2) their availability has undergone remarkable changes during evolution (Chapter 1); (3) biological systems have synthesized selected organic partners for a considerable fraction of each of the elements, e.g. Fe (haems), Co (corrin, B_{12}), Ni (F-430), and Mo (a pterin), (remember Mg (chlorin)), while leaving much of the element in a freer form; and (4) functional properties, e.g. redox potentials, are now very sensitive to ligand coordination, including its geometry. There is then complexity to be met which is made even less simple by biological diversity during evolution. We have seen that advanced multicellular organisms developed special systems of control based on calcium and zinc with the proteins calmodulin and zinc fingers, respectively, and there arose a major difference between prokaryotes and eukaryotes. This distinction, although not so immediately obvious in the biological chemistry of transition metals, is equally important for them; for example, the redox chemistry of eukaryotes is largely removed into compartments separated from the cytoplasm. There are further new problems for us to appreciate, however, since the chemistry of the transition metal ions is not just specialized according to different functions, or between prokaryotes and eukaryotes, but has evolved differently in different higher organisms. There was a primitive separation between organisms heavily dependent on iron, nickel, and cobalt, the archaebacteria, and other prokaryotes where the dominance of iron is more obvious. It is also clear that copper is of little significance in most of these organisms relative to its multitude of roles in multicellular eukaryotes while in these eukaryotes the role of nickel and cobalt is further diminished. Finally, the dependence on vanadium is highly specific amongst certain eukaryotes. Element role in evolution is obviously specialized and has clearly changed with biological niche and with time for the transition elements as well as for zinc and calcium. Understanding rests here as elsewhere upon an analysis of the proteins associated with the redox-active metal ions. In Tables 12.1 and 12.2 we divide them into two extreme groups which separate α-helical, more mobile structures, from β-sheet or cross-linked, immobile structures for reasons which have been mentioned earlier (see Chapter 6)

V	(Cr)	Mn	Fe	Co	Ni	Cu
	Mo					
	W					

(A) Transition metals

Table 12.1 Less rigid structures in redox proteins

Redox protein	Structural feature	Function
Cytochromes *c*	Helical (α) two thioether bridges	Electron transfer
Cytochromes *c′*	Helical (α) one thioether bridge	Allosteric change on binding NO/CO
Cytochromes *b* (membrane)	Helical (α)	Electron transfer, proton pumping/coupling
Cytochrome oxidases (*b* or *a*, a_3)	Helical (α)	Electron transfer O_2 reduction proton pump/coupling
Cytochrome P-450	Helical (α) small β-sheet	Hydroxylation
Iron binding protein of ribonucleotide reductase	Helical (α)	Changes conformation on redox reaction
O_2-carrier proteins	Helical (α)	Allosteric

Another problem, of which we must be constantly aware, is the need for selectivity of binding so that each element can be used optimally and be placed correctly in the cell. We saw that zinc could be bound quite separately from calcium, magnesium, sodium, and potassium using the fact that the product [L][Zn]K, where L is a ligand and K the binding constant for the ligand, would generally exceed that of [L][M]K for the other four metals provided (1) the binding site was based on N/S donors or (2) if based on certain O-donors, which was rare, the site was of tetrahedral geometry. Now, the preferred site of binding for metal ions such as Mn^{2+}, Fe^{2+}, Co^{2+}, Ni^{2+}, and Cu^{2+} is much more similar to that for Zn^{2+}. The affinity order and the discrimination within that order is increasingly for N/S over N/O or O/O ligands in the series (Fig. 2.5)

$$\mathrm{Mn < Fe < Co < Ni < Cu > Zn}$$

Table 12.2 Highly immobile tertiary structures in redox proteins

Redox protein	Structural feature	Function
Blue copper proteins	One β-sheet domain	Electron transfer
Copper oxidases	Three β-sheet domains	O_2 oxidation to free radicals
Peroxidases		
HRP	-S-S- cross-linked helices	H_2O_2 free radical reaction
CcP	β-sheet domain plus helices	
Cytochrome b_5	β-sheet domain plus helices	Electron transfer
Superoxide dismutases		
Cu/Zn	β-barrel	Superoxide dismutation
Mn/Fe	$\alpha\beta$-structure	
Flavodoxin	β-sheet	Electron and hydride transfer
Ferredoxins	β-sheet	Electron transfer
Rubredoxin	β-sheet	Electron transfer

We expect and find that Mn, the largest and weakest Lewis acid is only accepted by mainly O-donor centres, usually with but one or two N-donors, while copper and zinc are more frequently bound by N/S centres. Since there is so little nickel (removed to vesicles in aerobes) and cobalt (mostly bound in vitamin B_{12}) the real problem in making [L][M]K very selective lies between copper and zinc and iron and manganese, respectively.

One way of removing the problem of copper competition is to bind it as Cu^{+}, the cuprous ion, and apart from the carrier proteins in the fluids outside cells this is the option used in the cytoplasm of cells. Of course, Cu^{2+} is always favoured over Zn^{2+} by greater strength of binding, especially in a tetragonal field of ligating atoms as seen in external ligands such as albumins. The Cu/Zn separation problem is then easily solved provided there is a reducing medium and relatively little excess of copper apo-proteins to which zinc could bind. The separation of them from Fe and Mn can also use a valence state switch to Fe^{3+} and Mn^{3+}. It is frequently observed that these two metals are transported in the oxidation state III and can act in these states in very similar enzymes (Table 12.3). The separation between them through thermodynamic factors is then difficult. However, it is clear that Fe^{3+} can be held by thiolates or in iron clusters with sulphide, but this is not the case for Mn^{3+}. Thus, while both Mn^{3+} and Fe^{3+} appear in M–O–M clusters and also together with phenolate donors (tyrosines), which can serve the same function, only Fe^{3+} appears in M_n/S_n clusters where there is now often a mixed Fe^{3+}/Fe^{2+} complex.

Competition for binding sites between transitional metal ions

The separation of single ion lower oxidation states, Mn^{2+}, Fe^{2+} and Zn^{2+} rests with

1. The number of N-donors which increases from one or two for Mn^{2+} to two or three for Fe^{2+} and Zn^{2+}.
2. The cavity size: $Mn^{2+} > Fe^{2+} > Zn^{2+}$.
3. The geometric demands: Fe^{2+} and Mn^{2+} octahedral, Zn^{2+} tetrahedral or five coordinate.

We can also consider that *in vivo* the production of a protein for one of the metals is limited by feed-back control so that in each case only the selected ML is formed, and finally that selective pumping will remove an ion, especially Mn^{2+} (Golgi) or Zn^{2+} (vesicles), from competition with Fe^{2+} in the cytoplasm. Different organisms, species, behave differently; for example, quite generally plants are heavily laden with Mn^{2+} compared with animals (see Chapter 20).

We shall give a specific example of selectivity in Section 12.9. An important feature of many of the sites for Mn^{2+}, Fe^{2+}, and to a lesser degree Zn^{2+} is that they allow relatively fast exchange. Thus, catalysis is linked to control even to the production of their own individual proteins by metal-binding transcription factors. All the sites for a given exchangeable metal must work in concert depending on uniformity of their binding constants.

Given this multitude of problems in dissecting biological selection we have decided to start the description of the redox role of the elements with the one transition metal element certainly common to all life—iron. As we stressed in Chapter 1, there is no way life could have existed without the introduction of a redox-active metal to carry out functions such as the activation of small molecules for redox chemical change and for the transport of electrons. It is these functions which allow the essential energy-capture reactions related to redox processes to be connected to the equally essential acid–base functions connected to polymerization of phosphate in ATP. The connection is made at the non-metal level via the chemistry of H^{+}, X-H, and H^{-} (see Chapters 1 and

5), but only through the assistance of metal-ion redox catalysis using overwhelmingly iron, at first in simple complexes and then in haem.

Starting from iron we shall progress to the biological chemistries of manganese, copper, and molybdenum, plus vanadium, tungsten and chromium, and then to those of cobalt and nickel. We shall be in a position, at this stage, to look at the comparative roles of non-metals. Here we shall find that, corresponding to the functional division of metals Na^+/K^+: Mg^{2+}/Ca^{2+}: Zn^{2+} from the redox-active transition metals, there are the divisions of the non- metals Cl^-, silica, and phosphate (RPO_4^{2-}) from S, Se, and I, where we again deal with the acid–base chemistry of the first three (Chapter 18), separately from redox chemistry of the second three (Chapter 19). This leaves aside the major 'organic chemistry' of metabolism, synthesis and degradation, based on H, C, N, and O, which is well described in standard biological texts and will receive only passing reference here. (The chemistry of biological minerals in which the metals and non-metals combine has a chapter of its own (Chapter 20) in the first edition of their book). At this stage we shall be left with the way in which all this chemistry is brought together in some harmonious fashion to form an active organism. Chapter 20 speculates on this problem.

12.2 Introduction to iron biological chemistry

The involvement of iron in organisms is complicated, perhaps more complicated than that of any element other than the more common H, C, N, and O. It is not just involved in oxidation–reduction catalysis and bioenergetics but is connected to many control systems and to several acid-base reactions (Table 12.3). In an effort to simplify the description of its biochemistry we shall divide the material into:

1. uptake and storage of iron;
2. iron/sulphur proteins;
3. additional iron proteins using simple N/O co-ordinating ligands;
4. feedback controls from uptake to enzymes;
5. haem proteins (which will be described separately in Chapter 13 since they are so widely distributed and important).

Table 12.3 Some iron proteins and enzymes: localization and function

Protein or enzyme	Localization	Function
Haemoglobin (Fe)	Cytoplasm	Transport of O_2
Transferrin (Fe, Mn)	Outside cells (plasm)	Transport of Fe and Mn(?)
Ferritin (Fe)	Cytoplasm	Storage of Fe
Rubredoxin, ferredoxin (Fe)	Cytoplasm (prokaryotes) and membranes	Electron transfer catalysis, and sensing O_2, NO
Hydroxylases (Fe)	Cytoplasm and membranes	Hydroxylation
Peroxidases (Fe)	Vesicles and outside cells	Utilization of H_2O_2
Acid phosphatases (Fe, Mn)	Vesicles and outside cells	Hydrolysis of phosphate groups
Superoxide dismutases	Mitochondria (prokaryotes)	Dismutation of O_2 (Fe, Mn)
FUR, FNR	Cytoplasm	Gene regulation
Calcineurin	Cytoplasm	Phosphatase

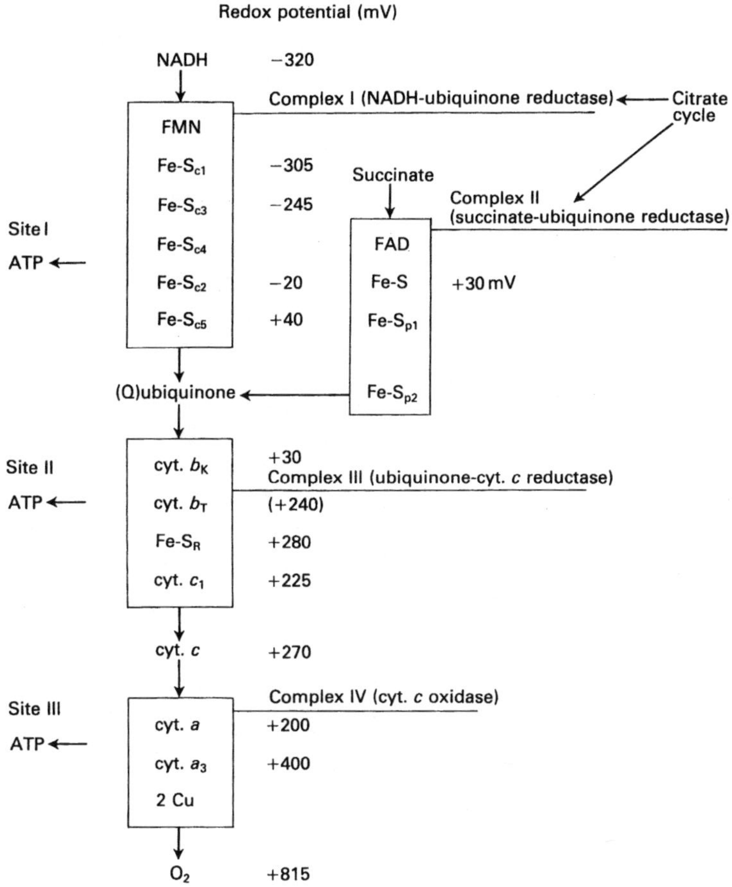

Fig. 12.1 The energy-transducing respiratory chain of mitochondria. Note the Fe/S proteins and their redox potentials. See also Photosystem I and the reaction centre complex and its Fe/S proteins (Section 5.3).

As a consequence of this division, largely in terms of iron co-ordination chemistry, functions of similar kinds but not always in similar locations will appear under different headings. For example, electron transfer chains involve (2) and (5), oxidases can be found from (3) and (5) as can dioxygen carriers, while controls (4) are linked to (2) and (3) via free iron but to (5) via haem iron. To remind the reader of the way in which some of the different iron proteins are involved in one organization, here energy capture, we show in Fig. 12.1 the essence of the electron transfer chain of mitochondria. Note that complex I and II are the most primitive.

The first problem we meet in the biological chemistry of iron is its low availability today, generated by the upsurge of dioxygen (see Chapter 1). Due to this fact, all living organisms in aerobic zones had to develop complicated uptake and control mechanisms to remain viable since in the presence of dioxygen at pH = 7 iron is precipitated as $Fe(OH)_3$.

12.3 **Iron uptake**

The basic problem of iron incorporation is shown in Fig. 12.2. There is a multitude of steps in which iron is moved from place to place in a living organism. Starting with aerobic prokaryotes, the simplest involves the use of small chelating agents which are synthesized by many prokaryotes and used to scav-

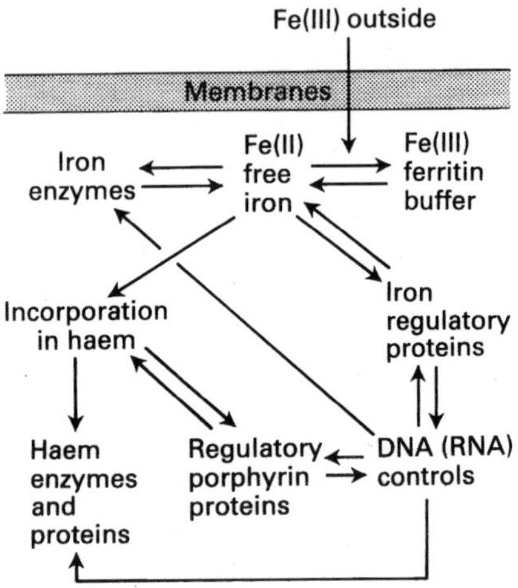

Fig. 12.2 The uptake of iron and its subsequent controlled distribution and metabolism.

enge the external media for iron (Fig. 12.3; see Braun and Killmann in 'Further reading'). The design of these reagents is such that they are soluble in water when not bound by iron, but on binding they turn inside out and become able to pass through lipid membranes. In more complex systems the iron chelate is taken up by a receptor and channelled to or through the inner membrane by the TON system. A very similar final transfer mechanism exists for the uptake of cobalt in vitamin B_{12}. Parallel mechanisms are available in the membranes of mitochondria, but here the iron is taken up as its citrate from the cytoplasm. (In plants iron may also pass up the stem associated with citrate.) The iron is released to the cytoplasm of the prokaryote by a variety of methods, including reduction to iron(II) and destruction of the carrier. The metal can be kept as iron complexed with protein, and is then distributed to many compartments even to the extent that some iron oxidase proteins are extruded, e.g. from soil bacteria, giving oxidases to break down lignins. Alternatively, iron can be incorporated into porphyrin (Fig. 12.2) and used as discussed in Chapter 13. The excess iron is held in a number of storage forms, the best known of which is ferritin, but the name ferritin covers a number of protein/iron stores which differ considerably.

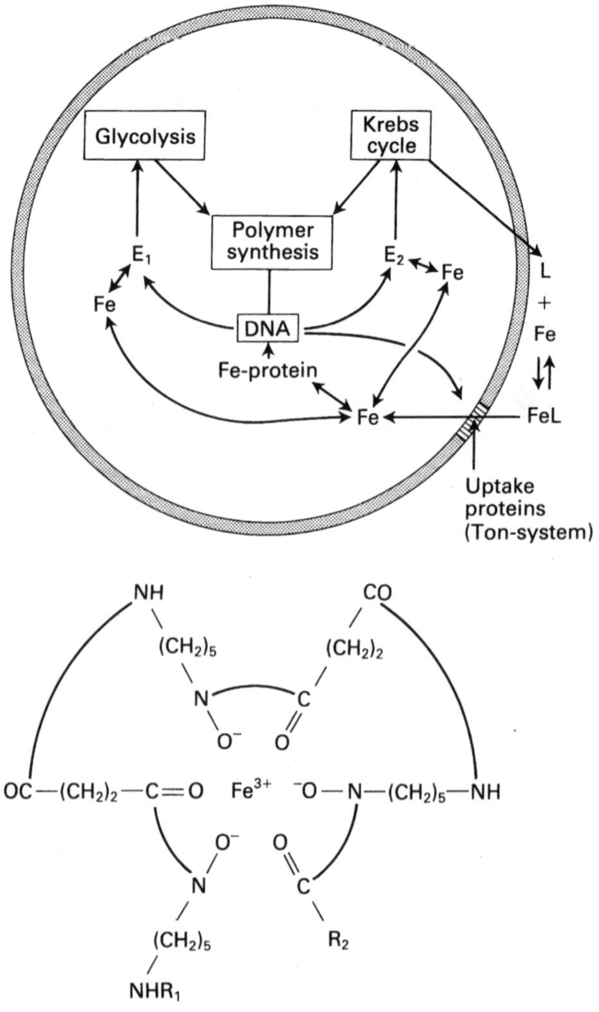

Fig. 12.3 A schematic representation of iron metabolism in bacteria such as *Escherichia coli*. Note the feedback loops. E, enzyme; L, Ferrioxamine B; R_1, H; R_2, CH_3, shown below.

The uptake of iron is complicated further in higher animal organisms where proteins handle the metal in body fluids (Fig. 12.4). The proteins called trasferrins (Fig. 12.5) are recognized, when bound to iron, by special receptors on the outer membrane of cells and are them engulfed in minimicrosomes (vesicles) (Fig. 12.4). The vesicles transport the transferrins to a lysosome where the metal is released. Part of the mechanism depends on the fact that carbonate is bound initially with the iron to transferrin at pH = 7 and the acidity of the lysosome destroys this complex. The iron is passed to the cytoplasm by an Fe^{2+} transport protein, NRAMP-2, which has a binding constant of around 10^7 M^{-1}. In the cytoplasm, at a more alkaline pH, series of non-haem iron proteins and Fe(III) complexes form with other carriers, and some iron is subsequently pumped into mitochondria or stored as ferritin. In organelles it can form then a variety of thiolate proteins or it can be handled in a chelatase converting the iron to haem iron. The iron in haem is inserted into various proteins, not only in mitochondria and chloroplasts, but finds its way to a number of vesicles, e.g. peroxisomes and cell membranes, or to cellular dioxygen carrier proteins in circulating fluids. We return to the intercommunication network between the various haem and non-haem proteins after we have described each of the classes of proteins.

In the early days of evolution the problem of iron uptake may not have existed since in the anaerobic sulphide-rich media there could well have been 10^{-7} M in Fe(II) and contained small clusters of Fe(II)/Fe(III) sulphides. The uptake principle could then have been simple incorporation of the iron, possibly as sulphide complexes, and cells would have had direct access to both electron transfer and hydrogenation catalysis. We need not speculate further except to note that to this day the incorporation of iron sulphide complexes is still an essential feature of all biological systems. The way in which these

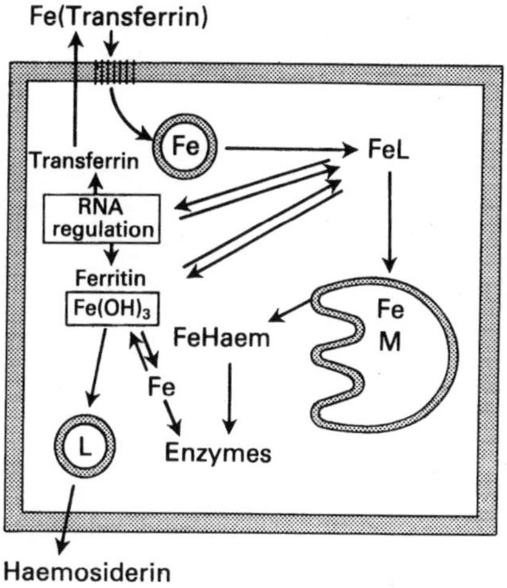

Fig. 12.4 A scheme corresponding to that of Fig. 12.3 for the uptake of iron into an advanced eukaryote cell. Note: (1) the feedback control; (2) the role of the mitochondria, M; (3) the danger of excess iron leading to haemosiderin deposits. FeL = citrate complex; L = rejection vesicle.

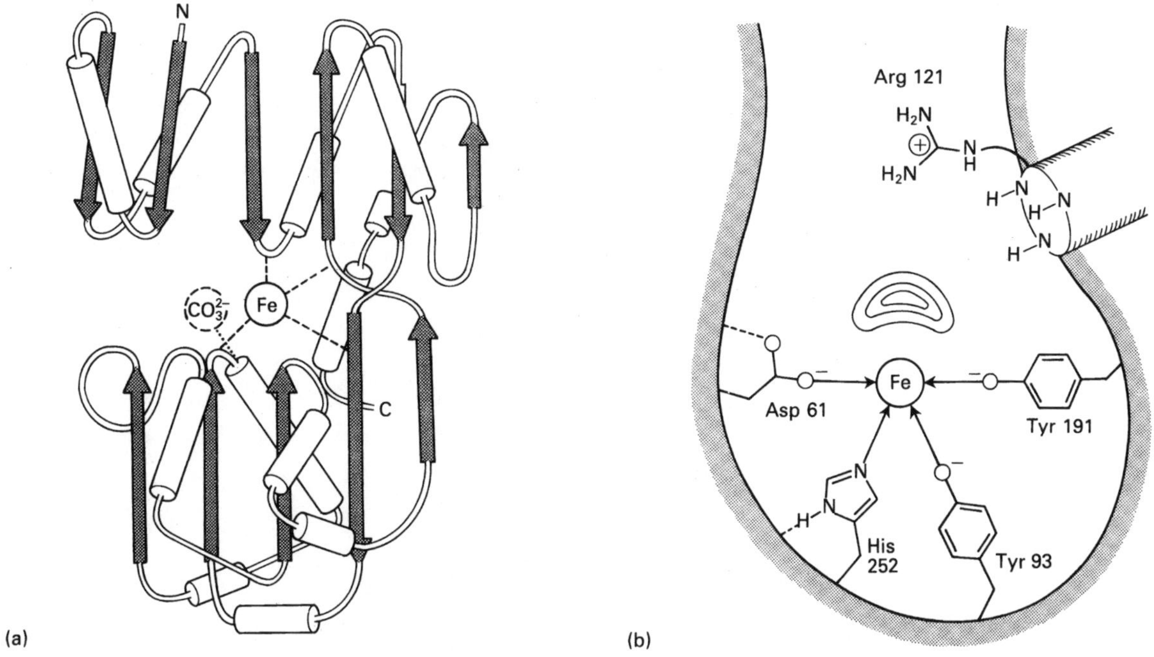

Fig. 12.5 (a) The outline fold of transferrin, a hinged protein. (b) Details of the binding site for iron; the open side interacts with bicarbonate. (After E. N. Baker and T. Blundell.) A protein similar to (a) is used to transport SO_4^{2-}, SeO_4^{2-}, and MoO_4^{2-}.

sulphides are formed in the presence of dioxygen is described in Chapter 19. Finally, the free Fe in the cytoplasm of all cells seems to have been held constant throughout evolution at 10^{-7} M.

12.4 The non-haem iron proteins

The series of these proteins include very distinctive classes. The most prominant are the iron sulphur proteins which are taken to include mononuclear rubredoxins, and the various multinuclear ferredoxins, 2Fe/2S, 3Fe/4S, 4Fe/4S, and some higher clusters, in which all iron atoms are bound either to thiolate or sulphur atoms with only one or two exceptions (Fig. 12.6). The exceptional cases where one coordinating atom is not sulphur will be discussed after the main group has been described. The second well known series is based on the oxo-bridged unit, Fe–O–Fe, first seen in haemerythrin, the dioxygen carrier of certain worms. The final group is a set of mononuclear iron proteins where the ligands are nitrogen and oxygen donors. The functions of these three classes of proteins are very different. The structures of these complexes are not just those expected from model chemistry since the protein constrains the geometry while manipulating redox potential or exchange of metal ions.

12.5 The iron/sulphur proteins

The structures of many of these small proteins are known (Fig. 12.7). To a first approximation, in the vast majority of them each iron is in a tetrahedron of RS^- alone (rubredoxins) or together with S^{2-} donors (ferredoxins). The fold of these small proteins is baed on β-strands forming small β-sheets (Fig. 12.7); the iron coordination sphere is usually saturated and is held in a stable condition by this fold. The cluster is held in a protective groove in the protein (see Section 6.7). The exact stereochemistry even in rubredoxins (several classes of this type of protein are now known) is clearly dependent on the exact protein fold. The proteins are variously soluble in water or membrane phases and in this stable form in either phase they appear overwhelmingly to have a simple single function: they are part of electron-transfer wires often in

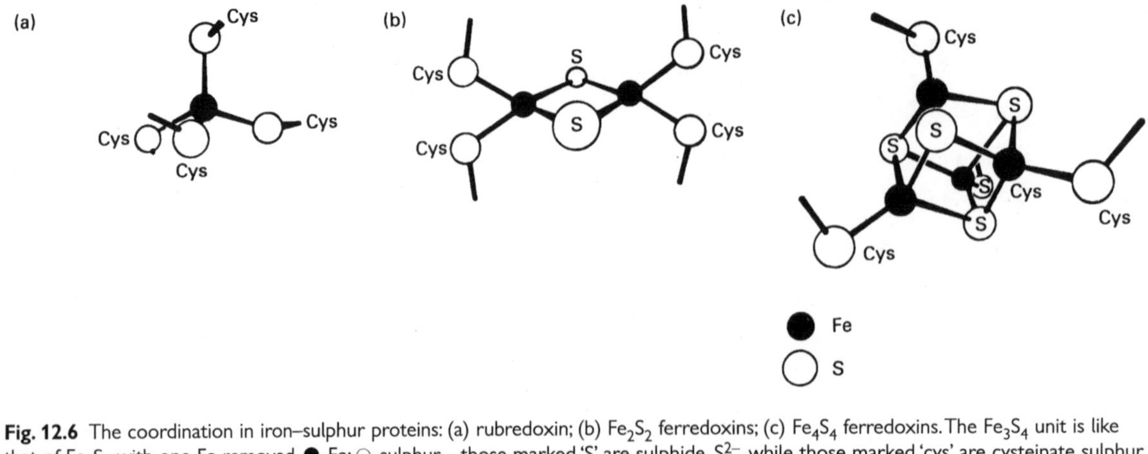

Fig. 12.6 The coordination in iron–sulphur proteins: (a) rubredoxin; (b) Fe_2S_2 ferredoxins; (c) Fe_4S_4 ferredoxins. The Fe_3S_4 unit is like that of Fe_4S_4 with one Fe removed. ●, Fe; ○, sulphur—those marked 'S' are sulphide, S^{2-}, while those marked 'cys' are cysteinate sulphur.

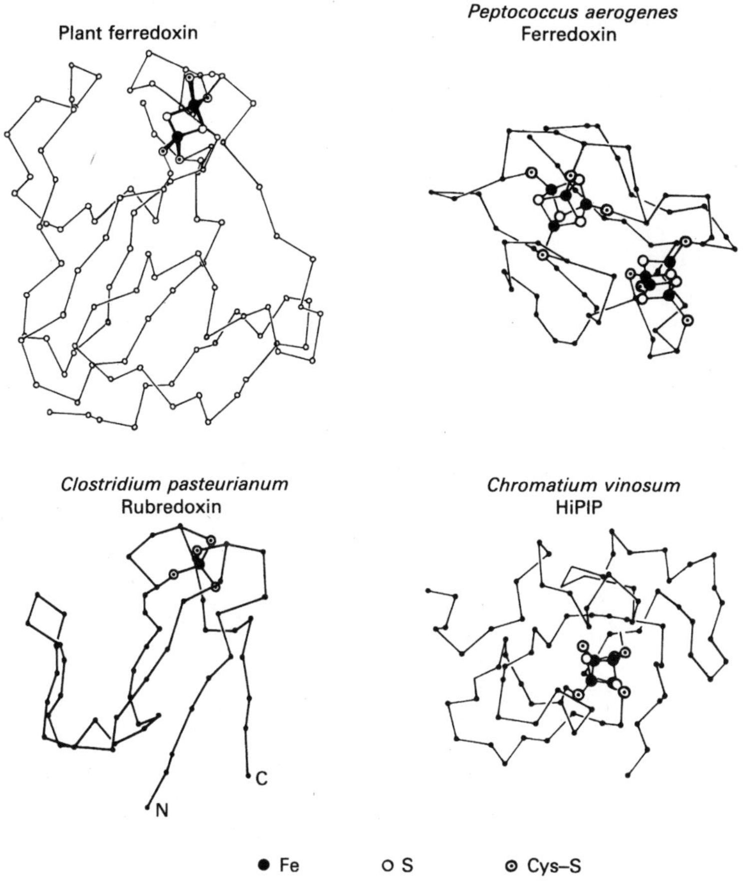

Fig. 12.7 The folds of some typical iron sulphur proteins. Note the β-sheet constructions. (After the work of many authors including L. Jensen, K. Fukuyama, and R. Stroud.)

domains of complex enzymes, (Fig. 12.1). (Exceptions are described later.) The redox switch of high-spin Fe(II) to high-spin Fe(III) involves only a small change in bond length due to the fact that it occurs in an interactive sulphide cluster. There is little relaxation energy change (see Section 4.7) so the sites are ideal for function. Conformational strain induced by the protein fold, which is undoubtedly observed in some of the Fe/S proteins, is not necessary for the basic function but may be used to control redox potentials or Fe exchange as may the general surrounds of the iron centre.

For the most part the iron–sulphur proteins have very negative redox potentials (Table 12.4) while covering the range from 0.0 to –0.5 V. Their frequent occurrence in the early steps of oxidation-reduction of organic molecules, e.g. NADH (i.e. bound H_2) and succinate (Fig. 12.1) suggests that they belong to the most primitive parts of energy-capture machinery (see Chapter 5). At an even earlier stage in evolution they may have been involved in direct hydrogen transfer reactions from dihydrogen and in photosystems. This function brings us to the exceptional Fe/S proteins which have substrates other than the electron, e.g. dihydrogen (see Section 12.6).

As mentioned, some of the Fe/S clusters as well as some rubredoxins are rather different from the majority. Three exceptional features are found (Table 12.5). The first is the high redox potential, +300 mV, of the cluster in one protein which is part of the mitochondrial and chloroplast redox chain. The protein is called HIPIP (high potential iron protein). This protein is on the

Table 12.4 Midpoint potentials* of examples of different iron-sulphur cluster types

Cluster	Protein	E_m (mV)
[Fe]	*Clostridium pasteurianum*: rubredoxin	–58
[Fe]	*Clostridium thermoaceticum*:	
	Rubredoxin 1	–27
	Rubredoxin 2	–20
[Fe]	*Desulphovibrio gigas*: desulphoredoxin	–35
[2Fe-2S]	Spinach ferredoxin	–420
[2Fe-2S]	*Spirulina maxima*: ferredoxin II	–310
[2Fe-2S]	*Nostoc* strain MAC: ferredoxin II	–455
[2Fe-2S]	Adrenal ferredoxin	–270
[3Fe-4S]	*Azotobacter vinelandii*: ferredoxin I	–424
[3Fe-4S]	*Desulphovibrio gigas*: ferredoxin II	–130
$[4Fe\text{-}4S]^{1+/2+}$	*Clostridium pasteurianum*: 8Fe ferredoxin	–403
$[4Fe\text{-}4S]^{1+/2+}$	*Bacillus stearothermophilus*	–280
$[4Fe\text{-}4S]^{1+/2+}$	*Desulphovibrio gigas*: ferredoxin I	–455
$[4Fe\text{-}4S]^{2+/3+}$(HIPIP)	*Chromatium vinosum*	356
$[4Fe\text{-}4S]^{2+/3+}$(HIPIP)	*Paracoccus* sp	282
Clusters in complex electron-transfer systems		
[2Fe-2S]	Xanthine oxidase (bovine milk)	–300
[2Fe-2S]	Rieske centre	
	Bovine heart	260
	Spinach chloroplasts	290
[4Fe-4S]	Photosystem I (spinach chloroplasts)	
	Centre A	–550
	Centre B	–590
	Centre X	–705
	Succinate dehydrogenase (bovine heart mitochondrial complex II)	
[2Fe-2S]	Centre 1	0
	Centre 2	–260
[4Fe-4S] or [3Fe-4S]	Centre 3	70

*Potentials are versus the standard hydrogen electrode.

Table 12.5 Unusual Fe/S centres in electron transfer systems

Protein	Fe/S	Unusual features
Nitrogenase	$P\text{-}Fe_4S_4$	Spin states
High potential iron protein	Fe_4S_4	Nitrogen coordination
Aconitase*	Fe_4S_4	Loses Fe, binds ligands
D. gigas protein	Fe_4S_4	Loses Fe
Sulphite (nitrite) reductase	Fe_4S_4	Fe/S unit bound to haem
Hydrogenase	$H\text{-}Fe_4S_4$	Unusual EPR, binds H_2
Ribonucleotide reductase	Fe_2S_2	Glycine radical

*At least six dehydratases are known.

oxidizing side of the quinone oxidases of energy capture devices and is, therefore, at a higher potential than the membrane cytochrome *b* (Fig. 12.1). One of the ligands is a histidine nitrogen. It is not easy to think of any role of such a protein which could not be performed by a haem protein, but see below.

The second group of unusual Fe/S proteins are enzymes as opposed to electron-transfer proteins. The functions are now much more complicated since the acid–base properties of the iron atoms become important. The iron must be to some degree exposed, and some of it can dissociate making a potential link to controls. Some similar proteins form the third group, for example the FNR proteins which are major transcriptional regulators. For a more complete account see Beinert *et al.* in references.

12.5.1 Electron-transfer chains

Figure 12.1 shows the degree of involvement of the Fe/S proteins in mitochondrial electron-transfer associated with oxidation of a range of substrates by dioxygen eventually (see Fig. 5.4). Probably, in primitive anaerobic systems such chains finished before complexes III and IV with donation of electrons to sulphur. There would then be no cytochromes involved. Complex I (see Aside) can pump protons apart from the production of proton gradients by oxidation (see also complexes III and IV). The number of Fe/S clusters in this and other groups of electron-transfer proteins is very large, more than 120 different types, and all of them sit on membranes or in the cytoplasm. Some are concerned with essential reactions of small molecules such as H_2, CO, NO, and N_2. Very many are associated with prokarytes or mitochondria of eukaryotes.

Some of the units shown in Fig. 12.1 are closely similar in several isolated enzymes listed in Table 12.4. We described the complex III Fe/S cluster in Section 5.1 and see Section 13.6. In the light of these observations the enormous importance of Fe/S systems in the most primitive as well as the most advanced forms of life is very clear, (see 'Further reading').

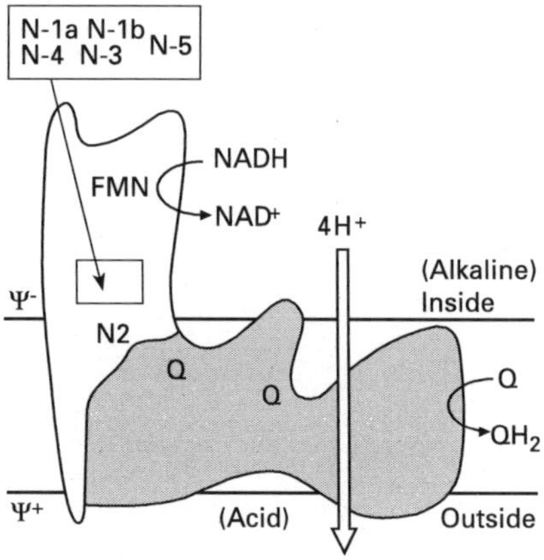

The probable structure of complex I of the electron-transfer chain. N-*x* are the Fe/S centres, Q are quinones, FMN is flavin mononucleotide, Ψ is membrane potential. A very similar array of Fe/S proteins is found in some hydrogenases and in pyruvate oxido-reductase. Note the pumping of protons.

12.5.2 Succinate dehydrogenase: particle II of mitochondria

The structure of particle II, succinate dehydrogenase, of the mitochondria chain has not been determined but a very closely similar $[Fe/S]_n$ protein, fumarate reductase, is of known structure. The protein, which may work as a dimer, has three subunits A, B, and C. A contains the reactive site flavin (FAD), subunit B has three iron–sulphur centres some 10 Å apart and subunit C has two *b*-cytochromes and lies in the membrane. Electron transfer is no problem since the haems are not more than ~ 15 Å from their electron donors. Of greater interest in this chapter are the three Fe/S clusters which are one of each kind Fe_2S_2, Fe_3S_4, and Fe_4S_4. The significance of this finding is unknown. Finally, the chain of centres from flavin, the beginning of the chain, to the receptor from the second haem, b, the end of the chain, is almost linear and covers a distance of 70–80 Å: a considerable molecular wire (see Lancaster *et al.* in 'Further reading').

Some primitive metabolic paths linked to non-haem iron

Metabolite	Comment
Dihydrogen	Hydrogenases need either Ni and Fe, or just Fe
Dinitrogen	Nitrogenases need Fe
Citrate cycle	Several enzymes need Fe, see aconitase
Fat synthesis	Enzymes require Fe
Nitriles	Hydratases, Fe (Co)

12.6 Fe/S centres active as enzymes

It has become clear recently that the variety of Fe/S centres which bind to substrates rather than transfer electrons is quite considerable [see Aside and Beinert (2000) in 'Further reading']. Table 12.5 refers to some of these, but there are also at least six dehydratases. At the same time, physical studies indicate that these centres are quite different from those simply involved in electron transfer. The peculiarity of the sites rests not just in the type of protein ligation to the iron atoms but also in the site geometry, in the number of iron atoms in a cluster, and in the exposure of at least one iron atom to exchange with iron in the medium. The types of coordination vary and include one

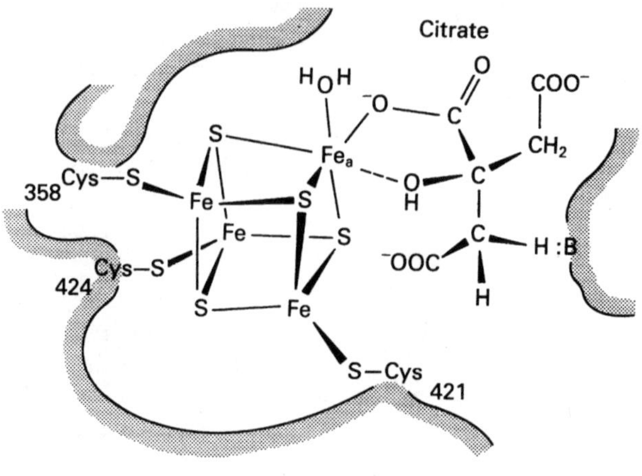

Fig. 12.8 The proposed active site of aconitase showing substrate bound to one iron. (After H. Beinert.)

bound histidine and one bound carboxylate from the protein. When the substrate binds, the single exposed iron of, for example, an Fe_4S_4 cluster, can become six coordinate and the centre probably generates a series of trapped free radical intermediates. Two examples stand out—dihydrogen-reacting and aconitase enzymes for which there is a structure available. In aconitase the iron centre acts directly on the substrate, aconitic acid (Fig. 12.8), which is involved in the citric acid cycle (Fig. 5.10), allowing the creation of a short-lived free radical. The reactive cluster is Fe_4S_4, which is able to lose an iron atom, but the consequently formed Fe_3S_4 cluster is not active. The suggestion follows that the switch from active to inactive forms of the enzyme by loss of iron is a *control* on the citric acid cycle related to iron availability. Other enzymes in this class are the ferredoxins III from *Clostridia*, which are thought to be associated with reactions of dihydrogen and which, due to exchange of iron, control these reactions. It is probable that the association/dissociation of one iron atom of a cluster, often bound to an aspartate, is involved in the ferroxidans bacteria which utilize the redox reaction of Fe^{3+} of the environment as an energy source. In conclusion it would appear that the effect of the protein fold on the Fe/S centres in these enzymes is much stronger than in the simple electron transfer proteins, so that the idea of constrained states with easy transition between intermediates, as in zinc and copper enzymes, becomes plausible (see entatic state).

In passing, we note that the association of redox centres with free radical chemistry and especially with rearrangement of molecular skeletons mooted here for the Fe/S enzymes is common to vitamin B_{12}, some copper enzymes, and the Fe-O-Fe centres described in Section 12.10.

The observation of freely dissociable iron from clusters and its possible connection with control also leads forward to a general discussion of control of cellular metabolism based on the levels of iron in a cell (see Chapter 20).

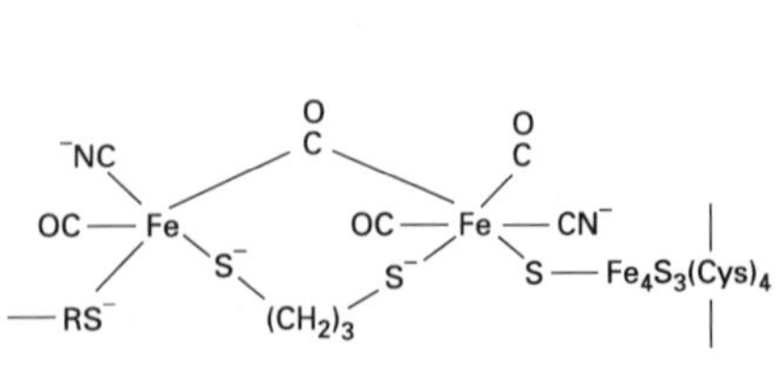

An hydrogenase active site.

12.6.1 The peculiarities of hydrogenases

There is a remarkable active site in hydrogenases of both the all Fe and the NiFe kinds. The coordination of the iron in these hydrogenases is not mostly to ligands from the protein but to CO, CN^-, and a propane-dithiolate, (see Aside). The reasons for the structure are not understood, though CN^- is an obvious site for the uptake or release of H^+. The binding site here is not an Fe/S cluster, but the active complex is bound to an Fe_4S_4 unit, (see Nicolet *et al.* in 'Further reading'). (The only parallel known is with sulphite reductase where the Fe_4S_4 unit is linked to haem iron.) This enzyme also raises the problem of proton diffusion in proteins, here of H^+ to or from the active site, which has not yet been understood here or within the cytochrome oxidase chain. There is evidence that H^+ can be taken up by Fe/S clusters.

12.7 Location of Fe/S proteins

The Fe/S proteins are rarely found in extracellular space. Thus, in bacteria the proteins are usually in the cell inner membrane or in the cytoplasm and less often in the periplasmic space, and this is also true for the organelles, mitochondria and chloroplasts. In eukaryotes most of the Fe/S proteins are in mitochondria. It would appear to be the specific function of copper (blue) proteins and cytochromes *c* and not of Fe/S proteins to carry out electron transfer in the external aqueous phases especially at the high redox potential of aerobic life. The Fe/S proteins are not very resistant to oxidation by dioxygen. Exceptionally, however, there are a few Fe/S complex proteins in the extracellular fluids of higher organisms. Amongst these proteins are the xanthine (aldehyde) oxidases and it is not clear why such a centre has been used when apparently a more stable group could have been found. In early evolution very little electron transfer chemistry could have been necessary outside a cell, but this changed with development concomitant with the advent of dioxygen (see Chapters 15 and 17).

The membrane Fe/S proteins of photosystem I can be related to dehydrogenases of oxidative phosphorylation, particle I, or to very low potential reactions of dihydrogen (hydrogenases), both of which involve NADH, and with reactions of dinitrogen (nitrogenases) which also release dihydrogen. (In Chapters 3 to 5 we referred to the NADH and NADPH coenzymes as forms of bound H_2.) The oxidation of these coenzymes is directly connected to membrane Fe/S proteins or to flavins, which themselves transfer electrons to Fe/S at a potential of around –0.5 V. Very many of these systems are membrane bound.

There is an extremely important group of non-membrane Fe/S prokaryote transcription factors known as FNR (section 12.5) which are associated with DNA and therefore, with the related proteins in eukaryotes, they are in the nucleus. The Fe/S clusters have dissociable Fe/S units and control a range of expressed proteins including those of the anaerobic/aerobic switch (see also Sections 12.17 and 12.18).

12.8 Why are there Fe_nS_n clusters in cells

The first question is 'Why do such clusters form at all?' General chemistry leaves us with a good understanding of the selective uptake of Fe(III) into these highly anionic sulphide sites (Section 2.5.7). Sulphide binding brings the Fe(II)/Fe(III) redox potential into the range below 0.0 volts required in anaerobes. It is also reasonable that in a reducing (H_2S) atmosphere inorganic clusters of Fe(III) with Fe(II) and sulphides will form. The studies of Beinert and Holm (see 'Further reading') have been very revealing as to the nature of this non-biological chemistry which resembles anaerobic chemistry. The problem for the chemist is to appreciate how such species can be formed today in the interior of cells which are exposed to dioxygen. Indeed, it is not always possible even to extract the ferredoxins without damage from dioxygen. This is a general problem, which concerns also sulphide clusters containing other metals, e.g. Mo, V, Ni.

The fundamental value of Fe/S proteins in the redox equilibria of cells

Table 12.6 Primitive Fe/S proteins (enzymes)

Protein	Cofactors
Glutamate synthase	flavin
Fumarate dehydrogenase	see particle I NADH dehydrogenase (flavoprotein)
Pyruvate ferredoxin oxido-reductase	thiamin dephosphate
Succinate reductase	see particle II. FAD-enzyme
Hydrogenases	(Ni) CO/CN
Nitrogenases	Mo/V
Dehydratases	flavin
Aconitase	none
FNR	none

If we believe that biological systems have developed in an understandable evolutionary sense, the question must be asked also as to why Fe_nS_n clusters are found in virtually all biological systems (see Table 12.6). This question can be put at two levels: first, what is the particular value of clusters; and, second, if they have such a value why was an Fe_nS_n cluster chosen rather than some other metal cluster?

Now, apart from functioning as an electron transfer agent and as an Fe-exchanger, the cluster of M_nS_n, where M is now any transition metal of variable valence, must have come into ready equilibrium with two redox reactions if it was to become biologically useful. The first is the reaction with protons and dihydrogen (the dihydrogen could be bound as NADH)

$$M_nS_n^{p+} + H_2 \longrightarrow M_nS_n^{(p+-2)} + 2H^+,$$

and the second is the reaction with sulphide

$$M_nS_n^{p+} + S^{2-} \longrightarrow M_nS_n^{(p+-2)} + S\downarrow.$$

These two reactions are connected through H_2S formation. Given a suitable organization and energy source they could be utilized taking hydride, originally from a carbon source, or H_2, and in place of the dioxygen used today in mitochondrial reactions there would be a sulphur acceptor when the reaction system would run between –0.5 and 0.0 V. It may well have been a primitive necessity to carry out these reactions and they are found in some extremophile archaea, and it may also have been that the catalysis of the reactions was present before life. The question could then be answered as to why Fe_nS_n was chosen in particular in early evolution for this function. It may well be that of all such M_nS_n clusters, assuming clusters are the best catalysts for the above reaction sequence, the iron complexes were the most valuable and available. Notice that iron sulphides readily give persulphides, pyrites, which is a step on the way to polymerized sulphur. In the earliest forms of life one or another inorganic reaction system Fe/S/H_2 may have formed the basis of energy, life-giving processes themselves. Wächterhauser (see 'Further reading') has stressed this approach and gone on to examine how the reactions from pyrite using iron and sulphur could be used to drive carbon assimilation.

We note again that there are no other simple M_nS_n clusters in organisms, except perhaps one copper complex (Section 15.3). There are a few other biologically active small clusters of iron (Fe_2O) and manganese (Mn_2O and possibly Mn_4O_4 in the dioxygen carriers and oxidases), but none is based on sulphur chemistry. While there are no other M_nS_n, *mixed* clusters based on

Fe_nS_n units, e.g. Fe_6MoS_8 or Fe_6VS_8, exist in nitrogenases, We discussed the chemistry which allows these clusters to exist in Chapter 3.

12.8.1 Genetics of Fe_n/S_n assembly in bacteria: genetic structure

The Fe_nS_n unit is found in more than 120 different types of proteins in *bacteria*. The essential nature of especially the Fe_2S_2 proteins is revealed in that they are required even for the synthesis of leucine. The iron is imported into the cell from the siderophores or other bound sources (aerobes) or directly (anaerobes). The sulphur comes from cysteine desulphurase as shown in Fig. 19.3 or directly from H_2S. The initial steps in Fe_nS_n formation follow from the formation of an Fe_2S_2 protein. The gene structure for the synthesis of this protein is complicated. It appears that there is an iron–sulphur assembly protein, ISC-A, which controls the insertion of the Fe_2S_2 unit into its carrier NIF-U with the help of one or two other proteins, NIF-S and NIF-V, where NIF stands for nitrogen fixation. One of the necessary genes in the same operon is the cysteine synthesis gene, *cys*-E1, since the sulphide of Fe/S proteins comes from organic sulphur. There are also further proteins in the operon for the distribution of Fe_2S_2 and the formation of larger clusters Fe_nS_n. There are promoter regions ahead of the operon which may be of the kind IRP (Iron regulatory protein) or FNR (formate/nitrate reductase). Some operons also contain NO or O_2-sensors for production of proteins which function in the protection of Fe_nS_n units. Note that the same series of proteins appears in nitrogen fixation as in the general operon for Fe_n/S_n synthesis.

The Fe_nS_n proteins are made in eukaryotes from Fe_2S_2 clusters synthesised in mitochondria, remnants of prokaryotes. There is a specific carrier for Fe_2S_2 in the cytoplasm of the organelle sometimes called a YAH-1 protein. The Fe_2S_2 unit is probably exported from the organelle to the eukaryote cytoplasm through an ABC type ATPase pump, ATM-1p. Note that such ABC ATPase pumps are very common in prokaryotes and eukaryotes and *E. coli* has as many as fifty.

12.9 The organization and selectivity of ferredoxins

The ferredoxins introduce a problem in metal/protein chemistry which we have put aside so far—the fact that they are incorporated into organized assemblies. The self-assembly is here essential to function since it controls spatial ordering and the electron transfer distance (Chapter 4). The selective, almost specific, recognition of surfaces by proteins is often a difficult problem for us to solve since the surface residues are very mobile and it must be remembered that some mobility in the organization may be essential for function (Chapter 6). Moreover, the combination of several protein units must be within a *particular* membrane. Here the location of protein synthesis may be important since several of the Fe/S proteins of the electron-transfer chains of mitochondria and thylakoids are coded in mitochondrial and chloroplast DNA and not in cytoplasmic DNA. In fact, there is at least one mitochondrially coded peptide or protein in each of the three major particles of the cytochrome chain and one in the mitochondrial ATP synthetase (see Fig. 12.1). In the development of a complex cell it may be important to restrict the site of synthesis of a given protein since this may decide the membrane into which it and then other proteins are incorporated. Proteins may also have 'leader' sequences which indicate where a protein should be positioned.

There is a further interest in the synthesis of Fe/S proteins. When an Fe/S protein is made there does not appear to be much excess in the cell of either iron, in an Fe/S cluster or not, of sulphide or of its apoprotein. Here we return to the remarkable selectivity of combination of metals and proteins which is greatly enhanced if the concentration of specific partners, chosen metal or metal complex and chosen protein, is matched by feedback control of the concentration of both via links to DNA. The implication is that the synthesis, and perhaps the destruction of the proteins, is related to the free Fe or Fe complex concentration, and that the uptake of Fe or Fe complex is related to the availability of protein for metal incorporation. The introduction of sulphide into clusters which was easy in the primitive world became extremely difficult in aerobes and sulphide is made in aerobes from sulphate and organic sulphur (see Chapter 19) and is once again gene regulated.

A disturbing feature of this selective uptake using thiolates for binding to the proteins, mentioned in the previous chapter, is that the zinc finger sites look exactly like the Fe(III) sites of rubredoxins (and ferredoxins), yet the two must not be confused. Now, there are no zinc fingers in prokaryotes, and in eukaryotes much Fe/S chemistry is confined to mitochondria and chloroplasts. The suggestion is clear that the coming of dioxygen with release of Zn^{2+}, brought about a radical reorganization in the ways in which Fe/S proteins were positioned and the way in which (RS^-) coordination chemistry was used. Possibly at this time FeO clusters appeared also. As we discussed in Chapter 2, the selection in these proteins may not be at the level of the donor atoms in the coordination sphere, but through selection of cavity size by second and third coordination sphere sidechains due to the fold of the protein, and of pumping into compartments.

12.9.1 Competition for Fe/S cluster sites by other metals

We turn next to part of the selective formation of Fe_4S_4 units in cells. There is no doubt that the formation of this core unit in solution is specific for iron due to the binding of sulphide, S^{2-}, by Fe^{3+} ions. The very fine model chemistry, (see reference to Beinert *et al.* in 'Further reading') has established this. There is no other M^{3+} ion available in cells. However, the final step, like the steps from the Fe_2S_2 to the Fe_3S_4 unit, is in a protein:

$$Fe_3S_4 + Fe^{2+} \rightleftarrows Fe_4S_4 \text{ Metal ion}$$

and is open to competition from other M^{2+} ions, e.g. Zn^{2+} (see Fig. 12.9). That Fe^{2+} and not Zn^{2+} binds to the core follows from the removal of Zn^{2+} by stronger and more selective N, S-donor ligands, the removal of excess Zn^{2+} by pumps of high affinity, and the limited formation of the Zn^{2+} binding proteins so that they cannot sequester Fe^{2+} (see Fig. 12.10). This is an excellent example of control over synthesis by conditional or effective stability constants (see Section 2.4).

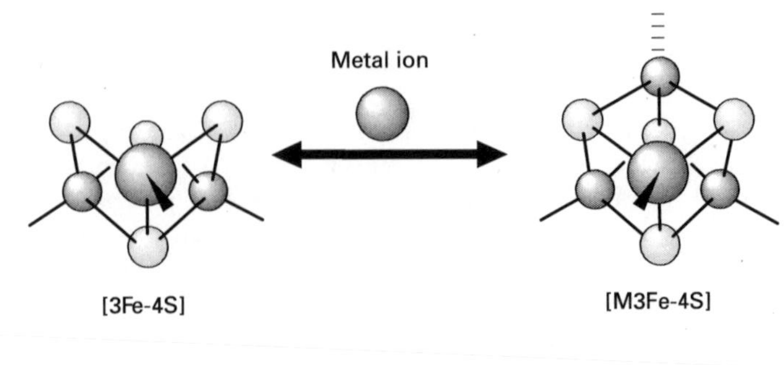

Fig. 12.9 Addition of a metal ion to a [3Fe-4S] cuboidal cluster to yield a [M3Fe-4S] cubane, see ☰ on right.

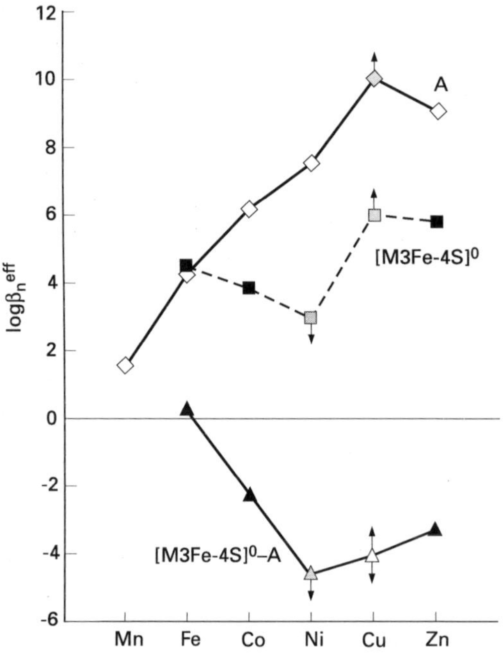

Fig. 12.10 Graph showing the overall binding constants β_n for various M^{2+} with the ligand $^{-}SCH_2CO_2^{-}$ (n = 2) yielding $[M(SCH_2CO_2)_2]^{2-}$ (A), and with the $[3Fe\text{-}4S]^0$ core (n = 1) to yield $[M3Fe\text{-}4S]^{2+}$ as measured in *Da* Fd III. The data for $[M(SCH_2CO_2)_2]^{2-}$ have been re-scaled to give effective β_n values at pH 7 by subtracting an increment of 6 log units to compensate for the high pKs (10) of the –SH groups in the two ligands required to form the complex. The lower plot shows the difference in β_n for $[M3Fe\text{-}4S]^0$ —(A), revealing the advantage afforded to Fe^{2+} by incorporation in an [4Fe-4S] cluster. Arrows indicate uncertainties in values where only a lower or upper limit can be given. The reaction of Cu^{2+} with $^{-}SCH_2CO_2^{-}$ produces Cu^{+} and disulphide, so a lower limit for the Cu^{2+} value has been estimated on the basis of Irving–Williams trends for other ligands.

12.10 The Fe-O-Fe cluster

It was thought at one time that the Fe_2O cluster was confined to certain dioxygen carriers, the haemerythrins. This seemingly strange cytoplasmic carrier is pictured in Fig. 12.11. It is assumed to bind dioxygen as the peroxide anion O_2^{2-}, since the Fe(II) of the deoxygenated state is converted to the oxo-bridge Fe(III) condition on uptake of O_2. The protein itself has a four multihelical bundle fold. The expectation then is that such multisubunit haemerythrins can be cooperative since the oxidation state of the iron would be expected to adjust the relative disposition of the helices as there is a radius contraction of >0.1 Å going from high-spin Fe(II) to high-spin Fe(III). In fact, some haemerythrins are cooperative, but the cooperativity is not as high as in haemoglobins. Examination of the structure shows that the Fe-O-Fe unit cross-links the helices and must constrain motion to some degree. Even so the proteins illustrate again the idea that multihelical proteins are mechanically adaptable in contrast with the Fe_nS_n clusters in β-sheet proteins (Tables 12.1 and 12.2).

The Fe–O–Fe centre in other proteins (Table 12.7) is easily recognized by the peculiar high intensity of the spin-forbidden d-d absorption bands in the visible region of the spectrum, as well as by use of Mössbauer spectroscopy. It was these peculiarities which led scientists to describe one of the iron ribonucleotide reductases, the second to be found in the Fe–O–Fe series of proteins, in terms of this bridged oxo structure (Fig. 12.11). However, this protein is an enzyme and contains an even more unusual feature—a tyrosine free radical which remains stable within the protein all through extraction procedures. It would seem that the purpose of the iron centre is to assist in the formation of this radical. The starting point for such an assertion is that the reaction

Ribose → deoxyribose

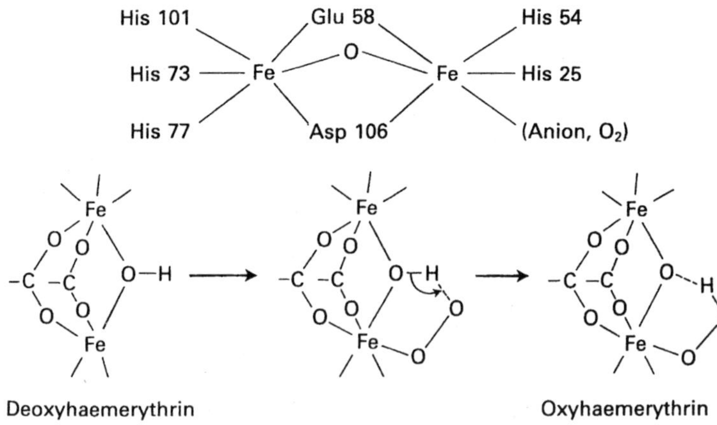

Fig. 12.11 The structures of the haemerythrin-active site. (After L. Jensen.)

is a reaction requiring first a hydrogen atom abstraction, a radical intermediate. The two-electron reduction then goes via effective $OH^{\bullet}$ abstraction and removal. (See the similar redox and rearrangement reactions in coenzyme B_{12} catalysis in Chapter 16 and note that the enzyme with coenzyme B_{12} must have come earlier in evolution and perhaps is being replaced by the iron-dependent ribonucleotide reductase.) Thus, it is believed that the tyrosine radical acts as a hydrogen-abstracting agent prior to reduction of the ribose. Notice the extreme example of an unusual energy state of a group, here tyrosine, in a protein, which parallels the formation of constrained states of metal ions (see Table 6.12). There are two other forms of ribonucleotide reductases from more primitive sources: one has an Fe_2S_2 centre and a glycine radical, the other a B_{12} centre and a Co(II) 'radical'.

There is another very powerful and unusual enzyme in this Fe–O–Fe series—methane mono-oxygenase. There are no known cofactors in this enzyme other than the Fe–O–Fe centre. Once again we notice that the reaction is probably a hydrogen atom abstraction

$$CH_4 \rightarrow CH_3^{\bullet} + H^{\bullet}.$$

(The easiest way in which to break a C–H bond at a saturated carbon is the radical path, Section 4.12.) The suggestion then arises that the reaction

Table 12.7 Fe–O–Fe sites in proteins

Protein	Function
Haemerythrin	O_2-carrier
Ribonucleotide reductase	Ribose → deoxyribose
Methane mono-oxygenase	Methane → methanol
Rubrerythrin	Unknown (Part of NADH peroxidase)
Purple acid phosphatase	Hydrolysis of phosphates Fe(III).Zn(II)
Stearoyl Δ^9 desaturase	Removal of 2H from stearate → arachidonic acid
Toluene mono-oxygenase	Hydroxylation
Alkene epoxidase	Epoxide of double bonds
Phenol hydroxylase	Creation of quinone

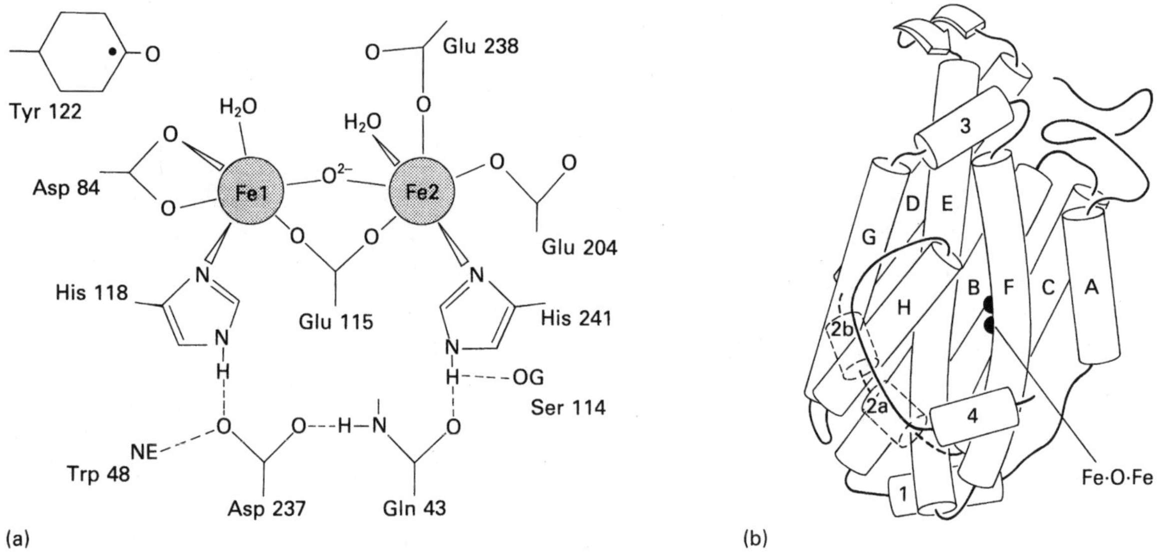

Fig. 12.12 The structure of the active site of ribonucleotide reductase (a). The protein is helical and its structure is shown at (b). (After Norland *et al.*, see Further reading.)

requires a free radical in the protein, when the initial step parallels that of the cytochrome P-450 reactions which include ω-hydroxylation of long chain fats (see Chapter 13) and the reactions of certain other monoxidases, e.g. copper galactose oxidase. There is the distinct possibility that the P-450 catalysts are a recent development of great value especially in eukaryotes, though they occur in prokaryotes, while the non-haem radical catalysts are usually found only in prokaryotes and vesicles of eukaryotes.

A very instructive feature of the Fe–O–Fe enzymes is that the domain which includes the Fe–O–Fe unit is helical and presumably adjusts conformation as this domain and the tyrosine free radical are formed. Here as in other helical 'enzymes' a required order of reaction is found which is compatible with a series of conformation changes (Section 6.4).

12.11 **Mononuclear non-haem/non-Fe/S iron and oxidative enzymes**

The examples of this third class of enzyme (Table 12.8) include those which act outside cells, e.g. those expressed by soil bacteria, and do not involve a cycling of oxidation states. It is the substrate which is attacked by molecular oxygen. The function of the iron is to activate the substrate, S, for ease of O_2 attack

$$\text{Fe(III)S} \longrightarrow [\text{Fe(II)S}^{\bullet}] + O_2 \longrightarrow \text{S-}O_2 + \text{Fe(III)}.$$

where $[\text{Fe(II)S}^{\bullet}]$ represents a partial charge-transfer complex. A typical reaction is the addition of dioxygen across an activated benzene ring. The enzyme is maintained in the Fe(III) state so that there is no loss of iron.

The iron mono-oxygenases in cells which contain a single iron atom usually go through the Fe(II), Fe(III), and higher oxidation states such as FeO(IV). The mechanism of oxidation is due to insertion of oxygen atoms derived from dioxygen. A typical case is the reaction of the penicillin ring

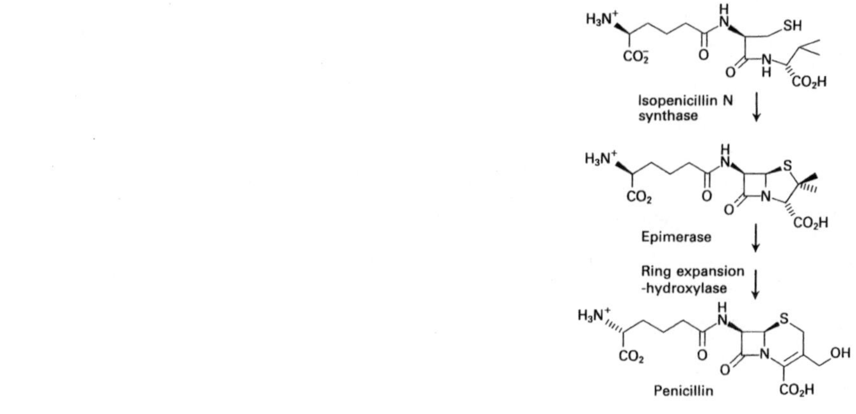

Reaction scheme A

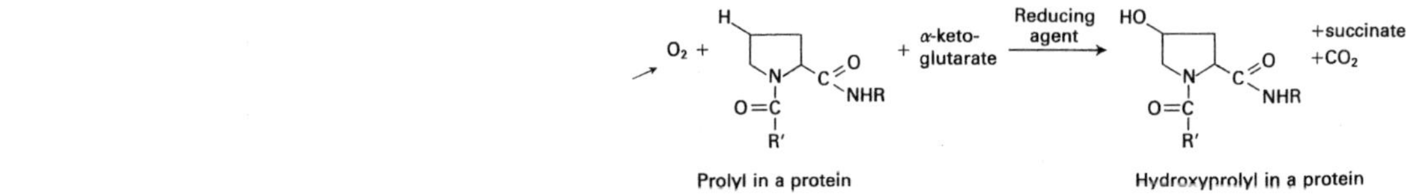

Reaction scheme B

Table 12.8 Some mononuclear non-haem/non Fe/S oxygenases

Protein	Reaction
Lipoxygenase	Produces lipid peroxides Fe(II)
2-Oxoglutarate dependent oxygenases	Collagen, plant hormone, antibiotic syntheses Fe(II)
Prostaglandin synthetase	Arachidonic acid → prostaglandin
Protocatechuate dioxygenase	Ring cleavage Fe(III)
Catechol dioxygenase (Extradiol)	Opens aromatic ring Fe(III)
Phthalate dioxygenase	Ring opening Fe(II)
Ferulate hydroxylase	Hydroxylation of lignins
Lysyl oxidases	Precursor of collagen
Prolyl oxidases	Precursor of collagen
Penicillin cyclases	Cyclization
Rubrerythrins*	Detoxify from H_2O_2 or O_2

*Also contains a rubredoxin site.

cyclase, reaction scheme A, and many such enzymes are found in prokaryotes, but there are parallels with eukaryote enzymes such as the connective tissue lysyl and prolyl hydroxylases, reaction B, used in collagen maturation See also the synthesis of hormones and prostaglandins in Fig. 12.13. Another interesting feature is that the oxidation is linked to the use of α-ketoglutarate as a hydrogen donor. This substrate connects the metabolism back to the citric acid cycle and we are reminded of the presence of isocitrate in the enzyme nitrogenase, Chapter 17. Are these controls of C/H/O/N chemistry related to Fe^{2+}?

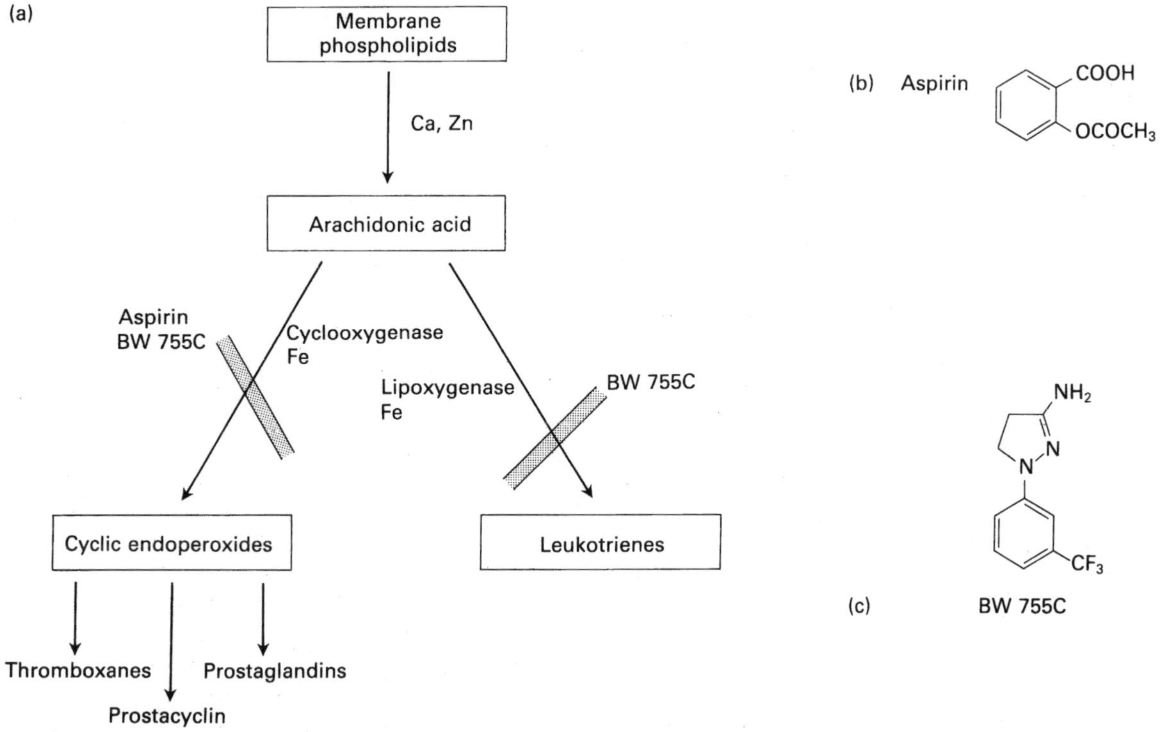

Fig. 12.13 (a) The pathway of prostaglandin synthesis from arachidonic acid (see reaction scheme B), produced from membrane phospholipids. Prostaglandin production can be blocked by (b) aspirin and (c) BW 755C, which inhibit the cyclo-oxygenase system. BW 755C also inhibits lipoxygenase and blocks the production of leukotrienes.

Once again the fact that the iron in the resting state of the enzyme, Fe(II), may (freely) dissociate connects the activity of the enzymes of this kind in prokaryotes to the levels of free iron in the cell according to the equilibrium

$$\underset{\text{Active}}{\text{Fe}\bullet\text{E}} \rightleftarrows \underset{\text{Inactive}}{\text{E}} + \underset{\text{Control}}{\text{Fe}^{2+}}$$

Now, while it is clearly permissible for the lower organisms to have such enzymes with freely dissociable iron in their cytoplasm (see Table 12.5), this is no longer desirable in the eukaryotes and especially in multicellular organisms. Free Fe(II) and Mn(II) or, for that matter, Cu(I), Co(II) and many other free transition metal ions, easily activate molecular dioxygen as

$$O_2 \rightarrow O_2^{\bullet} \rightarrow H_2O_2 \rightarrow OH^{\bullet} (\rightarrow H_2O).$$

The three central compounds in this series are powerful oxidizing agents and are known to be mutagenic and carcinogenic; hence the risk of adventitious dioxygen reactions had to be contained. It is a general observation that the reactions of the above intermediate steps of reduction of dioxygen are handled with great precautions in eukaryotes, for example:

1. Many of the reactions involving iron are in organelles or vesicles, e.g. mitochondria, peroxisomes, reticula, or vesicles, or are two electron reactions.
2. Many of the reactions outside cells use copper proteins which do not release copper easily.

3. Many iron enzymes are held in the Fe(III) state so as to protect them from loss of iron even in vesicles.
4. The iron oxidases in the cytoplasm are often haem proteins which do not release iron, e.g. cytochrome P-450, and/or are on membranes or in microsomal fractions.
5. Some of the enzymes have more than one metal to ensure that intermediates between O_2 and H_2O are not released, e.g. cytochrome oxidase.

[An interesting enzyme which is conveniently placed in this group is the superoxide dismutase of prokaryotes, mitochondria, and chloroplasts; this enzyme is replaced by a copper/zinc superoxide dismutase in the cytoplasm of all eukaryotes. Here we meet another Fe(III) enzyme which has a very closely related Mn(III) form. The character of all three superoxide dismutases (see Chapter 15 on copper proteins) is that the redox potential of the M(I)/M(II), Cu, or M(II)/M(III), Fe, or Mn, is placed approximately between those of the $O_2/O^{\cdot-}_2$ and the $O^{\cdot-}_2/H_2O_2$ couples. The superoxide dismutases are a clear example of proteins in which there is a switch away from a risk-carrying iron (or manganese) enzyme (prokaryotes) to a safe copper enzyme (eukaryotes)].

12.11.1 Unusual iron enzymes

In this group of non-haem, non-sulphur–iron enzymes there are two or three which have quite distinctive properties or functions. All are intracellular. Outstanding is the hydrolytic enzyme deformylase. This iron enzyme, which is only found in prokaryotes, removes the formyl of formylmethionine, the amino-acid required for the initiation of all prokaryote protein synthesis. The removal of terminal methionine itself is catalysed by a different, possibly another iron (or cobalt) enzyme. The use of iron here suggests that the final necessary steps in protein synthesis in prokaryotes and eukaryotes (?) is under the control of free ferrous ions.

Another unusual intracellular enzyme is nitrile hydratase, which is a low-spin Fe(III) enzyme when bound to substrates. It has a strange coordination sphere of three peptide nitrogens and two thiolates. There is a corresponding low-spin Co(III) protein.

Finally, there is the calmodulin moderated phosphatase calcineurin, which is influential in the immune system. This enzyme also contains zinc in a bi-metal ion pair. Again the iron may be a control related to free iron.

12.12 Iron and secondary metabolism

The substrates of many of the above enzymes and their products are of considerable interest in secondary metabolism (Table 12.8 and Fig. 12.13), which produces compounds outside the major required metabolic paths. These secondary metabolites are usually made by oxidative reaction so that either hydroxyl or halogen atoms are introduced into organic molecules. Halogenation will be described in Chapter 19. Both haem and non-haem enzymes are active in halogenation and hydroxylation. In prokaryotes the enzymes seem to be merely a part of the catalytic armory for the production or removal of offensive chemicals. An obvious example is penicillin, but many halogenated compounds and sterols fall in this class. In multicellular eukaryotes it appears that evolution has adapted these compounds so that they become part of the messenger or hormone systems. Another use is in the synthesis of the progenitors of the cross-linking agents of extracellular matrices such as hydroxyproline and hydroxylysine.

There is a peculiarity about the production of some hydroxylated compounds especially in plants. The wide range of alkaloids, which appear to have little value today, may well have a defence role, and perhaps they are waiting to become messengers or hormones. Note that plants make much greater use of hydrogen peroxide than of dioxygen, but the reactions of this oxidant are usually controlled by haem enzymes.

12.13 Binding ligands in non-haem/non-Fe/S proteins

As we have seen there is a large series of iron enzymes (non-haem/non-Fe/S) which cover a variety of functions. The binding site, whether in the Fe–O–Fe or the single Fe series of proteins, is to either several histidines or histidine and carboxylate ligands and the geometry approaches octahedral (Figs 12.11 and 12.12 and Table 12.9). The disposition of the ligands and probably the bond lengths are such that these sites do not favour either zinc or copper relative to other protein (N/S) centres of lower coordination number which remove these other metal ions by binding them more strongly. The iron sites do not form very stable complexes, especially with the reduced metal state, so that exchange takes place, certainly in prokaryotes

$$\text{Fe(II)}\cdot\text{E} \rightleftharpoons \text{Fe(II)}+\text{E}.$$

This allows a control over reaction systems which is dependent on the equilibrated free Fe(II) concentration. We shall describe the way in which this concentration connects to iron-controlled reactions generally in Section 12.17. The secondary structures of these enzymes are based upon helices, but structures of some of the proteins, e.g. superoxide dismutase, include considerable amounts of β-sheet, as is common amongst Cu and Zn enzymes generally.

12.14 Extracellular iron as an acid catalyst

The requirements for catalysts in relatively acidic media ($pH < 5$) which prevent or reduce the stability of the binding of divalent ions to proteins may well be the cause of the development of a few M(III)-dependent hydrolytic enzymes in the acidic ($pH = 4$ to 5) zones of some lysosomes (Fig. 12.4). The major proteins are either Fe(III)- or Mn(III)-dependent. As in a number of other enzymes which require Fe(III) it would appear that Mn(III) is an acceptable substitute. Some of the enzymes have a pair of metal ions. The advantages of a site with more than one metal ion in acid-base catalysis have been discussed in Section 4.3, but there the metal ions were quite far apart in the enzyme. Perhaps the use of the M.O.M. cluster is just to stabilize the M(III) state. There are also some Fe(III) enzymes which activate substrates to attack by O_2 outside cells, e.g. the dioxygenases of soil bacteria.

12.15 Summary of non-haem iron (non-Fe/S enzyme) functions

We see in this chapter that non-haem non Fe/S iron is often involved in redox chemistry and that much of this chemistry involves free radicals. The free

Table 12.9 Some binding sites for simple iron proteins†

Protein	**Ligands**
(a) Sites for intracellular Fe(II) cations	
Diphtheria toxin repressor (DtxR)	His, His, Glu (octahedral)
Ferric uptake repressor (FUR)	His, His, Glu (probable)
Dioxygenase	His, His, Glu (2/3 H_2O?)
Isopenicillin *N*-synthetase*	His, His, Glu $3H_2O$ (octahedral)
Calcineurin (intracellular)	His, Asp, Asp (2/3 H_2O)
Superoxide dismutase	His, His, Glu (trigonal)
Lipoxygenase	His, His, His, His, Asp (H_2O)
Photoreaction centre	His, His, His, His, ($2H_2O$)
(b) Sites for extracellular Fe(III) proteins†	
Transferrin (extracellular)	Tyr, Tyr, His, Asp, CO_3^{2-}
Protocatechuate-3-4-dioxygenase (extracellular)	Tyr, Tyr, His, His, H_2O
Phosphatases (extracellular)	(Two metals)
(c) Sites for Fe–O–Fe proteins†	
Haemerythrin	(1) 3His, Glu, Asp. (2) His, Glu, Asp
Ribonucleotide reductase	(1) His, Asp, Glu. (2) His, Glu, Glu
Methane Mono-oxygenase	(1) His, Glu, Glu. (2) His, Glu, Glu

*It is thought that the family of α-ketoglutarate/ascorbate dependent Fe(II) hydroxylases are related to this structure.

† Note that manganese can often substitute for iron in very many of these sites.

radical may be just an electron-transfer intermediate, but it could be part of the ground state of an enzyme or it might be involved in the reactions of substrates. (Free radical reactions will be apparent again in cobalt and copper biological chemistry.) Some of these enzymes are listed in Table 12.9 where we include the mononuclear iron site of the photoreaction centre. (This site is of uncertain function, but its structure is known to be like the others in that iron is in a rough octahedron coordination sphere including four histidines. The site in lipoxygenases (Table 12.9) may be similar). Undoubtedly a critical feature of all this iron chemistry is the compartment it is in and we shall keep this in mind, especially in Chapter 20. The locality of Fe/S proteins in association with low redox potential chemistry (H_2, NADH, H_2S, aconitase) differs clearly from that of high redox potential chemistry (O_2) where the Fe or Fe_2O is bound by imidazole and carboxylate. Haem chemistry will highlight this point again (see Chapter 13). A further essential point is the *control* over iron uptake and use.

12.16 Iron buffering and carriers

When describing the biological chemistry of sodium, potassium, calcium, and zinc we observed different ways of keeping ion concentrations constant. The first three ions are managed, at least in part, by pumping mechanisms with feedback controls which operate against leaks. In the case of calcium and zinc

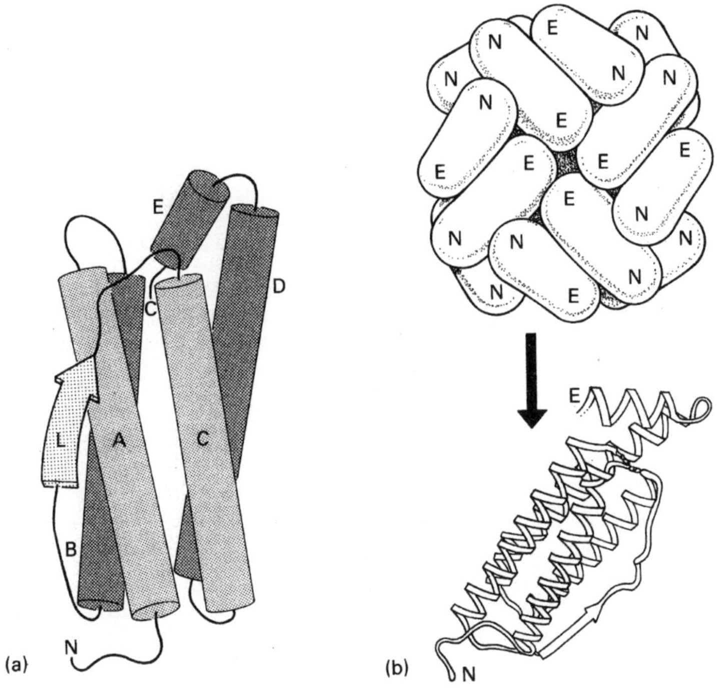

Fig. 12.14 (a) Schematic representation of a ferritin subunit. Four long helices, A, B, C, and D, form a bundle with the shorter fifth helix, E, lying at an acute angle to the bundle. There is only one short region of intermolecular anti-parallel β-sheet in the loop L. Some of these proteins, perhaps all, bind haem (Chapter 13). The way they pack together is shown in (b). (After P. Harrison.)

there are, in addition, binding proteins which bind more than one metal ion and so form buffering agents sensitive to $[M]^n$ where n may be greater than three. Calcium is also buffered by precipitates such as bone and shell. Iron buffering is managed by a small precipitates phase in a protein capsule, ferritin (Fig. 12.14), which is a four-helical-bundle protein (12 copies) sometimes containing haem *in vivo* and which very probably is an allosteric protein. The protein is not the same in bacteria and eukaryotes and neither is the precipitate. We turn to the complexities of such forms of storage in Chapter 20 on the homeostasis of cells. As stated in Section 12.8 in evolution Fe/S clusters probably controlled (buffered) cells on the early sulphide-rich earth.

We have described iron carriers outside cells in Section 12.3. Inside cells iron is distributed partly as free Fe^{2+} and partly by carrier proteins for units such as Fe_2S_2 apart from in coenzymes. We shall contrast this behaviour with that of copper in Chapter 15.

12.17 **Iron controls of metabolism**

Looking at the synthesis of intermediates either for scavenging for iron or for the synthesis of porphyrin (Fig. 12.15), we see that they are derived from the citric acid cycle. In this cycle there are key enzymes which can be controlled by the levels of free iron since the iron dissociates from them with some ease. An example in the citric acid cycle is aconitase, and in the pathway from the product of glycolysis, pyruvate, to the citric acid cycle, it is pyruvate decarboxylase. Both enzymes use Fe_4S_4 clusters. Of course, there is also indirect control of haem synthesis since the insertion of iron in porphyrins uses a chelatase enzyme. At the fundamental level of nitrogen fixation there is the gene product NIF U which also dissociates iron readily.

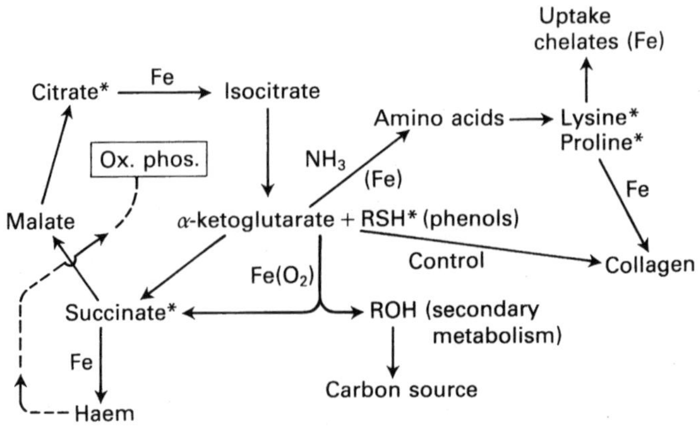

Fig. 12.15 The link between the citrate cycle and iron controls. It is suggested that iron levels are essential for switching between the different metabolic paths required for uptake and energy generation and for balancing synthesis and degradation.

Apart from control over basic metabolism common to prokaryotes and eukaryotes, much of which is locked into organelles in the eukaryotes, there are a variety of secondary metabolic pathways in aerobes which depend on hydroxylation. In fact, most secondary metabolites have an oxidative step requiring an iron enzyme. Any list of such products would include a wide range of hormones including sterols, thyroxine, and adrenalin. Lack of iron is critical to the levels of all these compounds not only at the cellular but at the multicellular level. Distribution, as well as uptake and incorporation, is essential. Finally, iron enzymes are linked not only to the metabolism inside cells but also to extracellular products such as hydroxyproline for connective tissue. Connective tissue is based on *extracellular* collagen, but the requirement for 2-oxoglutarate as well as iron in the synthesis of hydroxyproline, a part of collagen, once again links control back to the intracellular citric acid cycle. All these enzymes dissociate iron readily and are therefore linked by equilibrium control to transcription factors (see Section 12.18). All of them must have a very similar binding constant to Fe^{2+} of around 10^7 M^{-1}. Now this gives a value of free Fe^{2+} in the cytoplasm of cells of the 10^{-7} M. Is this value of 10^{-7} M iron a quantitative characteristic of the metallome of the cytoplasm of all life? Notice that it is not connected to haem-iron exchange.

Given the complexities of control and its link with several elements as well as iron we shall leave further discussion until Chapter 20.

12.18 **Iron regulation relationship to genes and evolution**

Because the role of iron is so central to redox metabolism, it is probable that there never was life without it. (There are very few exceptions known today which use Mn, as is the case of the Lyme disease pathogen *Borrelia burgdofori*. In some theories of early evolution (see Section 12.8) iron takes part in the catalysis of extremely primitive reactions which were in existence even before they were incorporated into living systems). At first the iron–sulphur–hydrogen reactions may have 'buffered' life's chemistry in tune with the environment. The activity may well have been reductive as in a reversed steps of the metabolic pathway which became the citric acid cycle (Section 5.6). Be this as it may, today iron is required absolutely. In its absence oxidative systems must be shut off so that essential polymers and monomers are preserved while scavenger molecules are sent out to find this metal. The implication is clear. Iron

Table 12.10 Some iron regulatory proteins (IRP) at DNA (RNA)

Protein (IRP)	Function
FUR	Controls iron uptake and several metabolic reactions (prokaryotes)
FNR	Controls aerobic/anaerobic switch in prokaryotes
'Aconitase'	Controls iron uptake in eukaryotes
IRP.ALA	Controls haem synthesis via amino levulinic acid (NB Zn^{2+})
TFII B	Transcription factor of RNA polymerases in Archaea
SOX R	Activation of superoxide or nitric oxide related genes in eukaryotes
DtxR/Ide R	Control of toxins, Fe import, enzymes

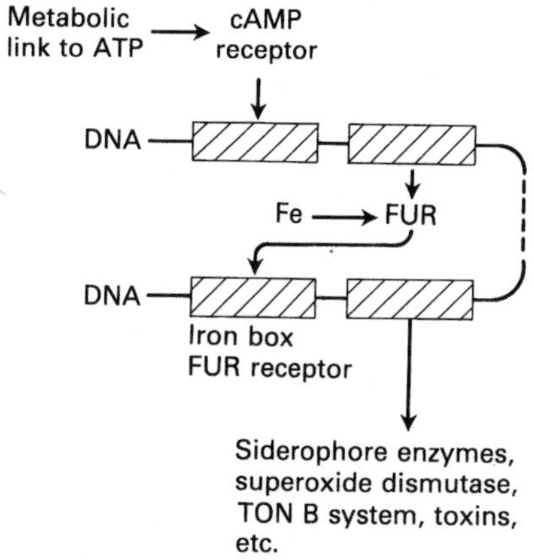

Fig. 12.16 The protein FUR regulates the iron-box gene so as to produce the range of proteins shown. FUR itself is produced under the control of a gene regulated by cAMP. Thus phosphate and iron are connected through the genes and in metabolism.

concentration must be connected to DNA-regulated synthesis of its own scavenging chelating agents and to a considerable number of other pathways. Starting from this premiss the iron (Fe) uptake (U) regulatory (R) protein (FUR) was looked for and found in aerobic prokaryotes by J.B. Nielands. It binds both iron and DNA and it is presumed that the way in which it alters its own conformation and that of DNA when iron is bound is such as to prevent the synthesis of enzymes for the production of scavenger chelates (siderophores such as ferrioxamins) (see Fig. 12.16) for a generalized genetic scheme, also applicable to DtxR/Ide R. In eukaryotes a parallel scheme uses the aconitase like Fe/S centre (see Section 12.5), which controls the production of transferrin, ferritin, and a number of other uptake proteins. Here a scheme parallel to that of Fig. 12.17 is required.

Now, these are mechanisms in aerobes and cannot be primitive. Notice that they demand very high affinity binding of Fe^{3+}, $K \geq 10^{17}\ M^{-1}$, outside the cell, either by siderophores or by transferrin, and that production of these scavengers is related to an internal Fe^{2+}, gene-linked, concentration of 10^{-7} M. In the sulphide atmosphere at the time of the beginning of life external Fe^{2+} was probably close to this value of 10^{-7} M and uptake mechanisms were less necessary. However, control was required and this may well have been managed by the $Fe_4S_4 \rightleftarrows Fe_3S_4 + Fe^{2+}$ equilibrium of the FNR protein, discovered as *F*ormate/*N*itrate *R*eductase but now known to act as a general transcription factor for many iron redox enzymes and haem synthesis in anaerobic bacteria. It also activates the anaerobic/aerobic switch in facultative bacteria. The parallel proteins in eukaryotes are shown in Table 12.10. Note that Sox R regulates oxidative stress. Such general redox gene controls must be coupled to the major acid–base metabolism of cells. FNR apparently works together with the ARC (arginine catabolism) kinases which then links iron and phosphate controls (see also Fig. 12.16). Such a very tightly knitted scheme of genes linking phosphates (acid–base) and iron (redox) is essential for all life, connecting gene regulation and metabolic control levels (Fig. 12.18). The 10^{-7} M Fe^{2+} of the cytoplasm is one central feature of the metallome.

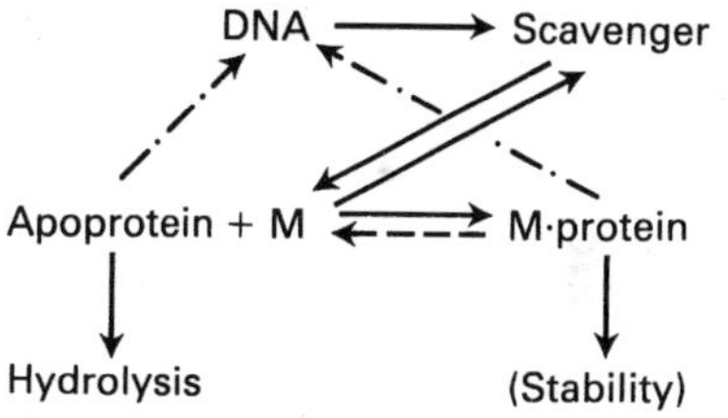

Fig. 12.17 As well as feed-back regulation of DNA as a source of synthesis of proteins there is a controlled degradative system for protein hydrolysis in all cells. The ratios of protein to metal are then carefully controlled. In the figure full arrows include fast reactions, broken lines show slow reversibility, and dash-dot lines refer to regulators of expression.

Note. The discovery of Fe/S centres

As a side-issue it is interesting to review briefly how the Fe/S centres were discovered. The spectroscopy of the Fe/S proteins, especially their EPR and Mössbauer spectra, generated a very interesting discussion in the 1960–70

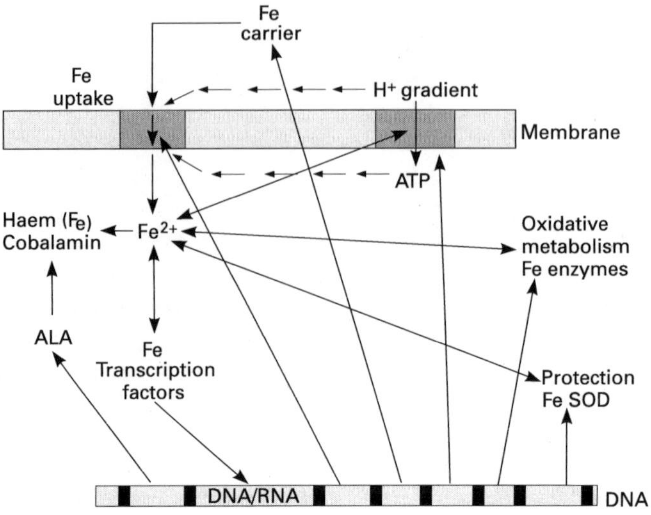

Fig. 12.18 The wide range of types of gene product which must all be related to free Fe^{2+} controls through the DNA.

period. This book is not concerned with such analyses but here they provide an interesting insight into the difficulty faced by chemists in the examination of biological molecules. When the EPR signals were first observed physical scientists refused to believe that these proteins contained iron. It was thought and firmly stated that no metal centre with five or more d electrons could have an EPR signal with $g < 2.0$. There are very good theoretical arguments for this if the iron is in an isolated site and no iron models had shown such signals. Proof of the origin of the signal as being due to a coupling of spins on *two* iron atoms was provided when two groups of scientists were brought together by one of us (R. J. P. W.) to analyse the EPR spectra. When a satisfactory theory of the spectra based on spin-coupling between two or more iron atoms had been established the biologists believed that the condition of the iron atoms was unusual. Again dispute arose in which chemists who could synthesize related small-molecule clusters argued that there is nothing particularly unusual about them. As the number of Fe/S clusters increased, however (see Table 12.4), it became clear that there was a wide variety of them with different EPR signals and very different redox potentials. Slowly it has become obvious that some clusters of this kind are held in particular energy (spectroscopic) states by the fold of the protein to which they are attached. The general position is that which is maintained throughout this book: each metal site in a protein can only be understood thoroughly with reference to the total protein which holds it and gives it its function. Parallels but not copies can be studied in small molecule models.

By way of comparison models using UV/visible spectroscopy led the way to the discovery of Fe–O–Fe centres in haemerythrin and ribonucleotide reductases (see Section 12.10).

Further reading

Agar, J.N., Yuvaniyama, P., Jack, R.F., Cash, V.L., Archer, A.D., Dean, D.R., and Johnson, M.K. (2000). Modular organisation and identification of a mononuclear iron-binding site within the NIF U protein. *Journal of Biological Inorganic Chemistry*, **5**, 167–77.

Al-Karadaghi, S., Hansson, M., Nikonov, S., Jonsson, B., and Hederstedt, L. (1997). Crystal structure of ferrochelatase. *Structure*, **5**, 1501–10.

Andrews, N.C. (1999). The iron transporter DMT 1. *International Journal of Biochemistry and Cell Biology*, **31**, 991–4.

Arakaki, A.K., Ceccarelli, E.A., and Carrillo, N. (1997). Plant type ferredoxin-NADP$^+$ reductases: a basal structure and a multiplicity of functions. *FASEB Journal*, **11**, 133–40.

Averill, B.A. (1996). Dissimilatory nitrite and nitrite oxide reductases. *Chemical Reviews*, **96**, 2951–64.

Bagg, A. and Nielands, J.B. (1987). Ferric uptake regulation protein acts as a repressor employing iron(II) as a cofactor to bind the operator of an iron transport operon in *E. coli*. *Biochemistry*, **26**, 5471–7.

Baltscheffsky, H. ed. (1966). *Origin and evolution of biological energy conversion*. VCH, New York.

Bauer, C.E., Elsen, S., and Bird, T.H. (1999). Mechanisms for redox control of gene expression. *Annual Review of Biochemistry*, **53**, 495–523.

Beinert, H., Holm, R.H., and Munck, E. (1997). Iron–sulfur clusters: nature's modular, multipurpose structures. *Science*, **277**, 653–9.

Beinert, H. (2000). Iron-sulfur proteins: ancient structures still full of surprises. *Journal of Biological Inorganic Chemistry*, **5**, 2–15.

Braun, V. and Killmann, H. (1999). Bacterial solutions to the iron supply problem. *Trends in Biochemical Sciences*, **24**, 104–9.

Chasteen, N.D. (1998). Ferritin uptake, storage and release of iron. *Metal Ions in Biological Systems*, **35**, 435–77, (see Vol. 35 of this journal for other papers on iron homeostasis and iron storage).

Coulter, E.D., Shevri, N.V., and Kurtz Jr, D.M. (1999). NADH peroxidase activity of rubrerythrin. *Biochemical and Biophysical Research Communications*, **255**, 317–23.

Dickman, S.R. (1977). Ribonucleotide reduction and the possible role of cobalamin in evolution. *Journal of Molecular Evolution*, **10**, 251–60.

Green, J., Trafeser, M., Six, S., Unden, G., and Guest, J.R. (1991). Iron-binding transcriptional regulator, FNR. *Proceedings of the Royal Society of Biological Sciences*, **244**, 137–44.

Gruenheid, S., Canonne-Hergauz, F., Gauthier, S., Hockam, D.J., Grenstein, S., and Gros, P. (1999). The iron transport protein, NRAMP2. *Journal of Experimental Medicine*, **189**, 831–41.

Harrison, P.M., Andrews, S.C., Ford, G.C., Smith, J.M.A., Treffrey, A., and White, J. L. (1987). Ferritin: an ion sequestering molecule from microbes to man. In *Iron transport in microbes, plants and animals* (eds G. Winkelman, D. van der Helm, and J.B. Nielands), pp. 445–75. VCH Verlagsgesellschaft, Weinheim.

Lancaster, C.R.D., Kröger, A., Auer, M., and Michel, H. (1999). Structure of fumarate reductase. *Nature*, **402**, 377–85.

Lill, R., Diekert, K., Kaut, A., Lange, H., Pelzer, W., Prohl, C., and Kispal, G. (1999). The essential role of mitochondria in the biogenesis of cellular iron-sulfur proteins. *Journal of Biological Chemistry*, **380**, 1157–66.

Nicolet, Y., Lemon, B.J., Fontecilla-Camps, J.C., and Peters J.W. (2000). A novel FeS cluster in hydrogenases *Trends in Biochemical Sciences*, **25**, 138–42.

Ponka, P., Schulman, H.M., and Woodworth, R.C. (eds) (1990). *Iron transport and storage*. CRC Press, Boca Raton, Florida.

Rolfs, A. and Hediger, M.A. (1999). Metal ions transport in mammals: structure, function and pathological implications. *Journal of Physiology*, **518.1**, 1–12.

Shergill, J.K. and Mason, J.R. (eds) (1996). Non-haem iron and oxygen. *Biochemical Society Transactions*, **25**, 65–95.

Spiro, S. and Guest, J.R. (1990). FNR and its role in oxygen-regulated gene expression in *E. coli*. *FEMS Microbiological Reviews*, **6**, 399–428.

Stephens, P.S., Jollie, D.R., and Warshal, A. (1996). Protein control of redox potentials of iron-sulfur proteins. *Chemical Reviews*, **96**, 2491–514.

Vollack, K.U., Hartig, E., Korner, H., and Zumpf, W.G. (1999). Multiple transcription factors of the FNR family. *Molecular Microbiology*, **31**, 1681–94.

Wachterhäuser, D. (1988). Origin of life and iron sulfides. *Microbiology Reviews*, **52**, 482–6.

Williams, R.J.P. (1982). Iron and manganese in controls. *Federation of European Biochemical Societies Letters*, **140**, 3–10.

Note: The Ton B system for uptake of iron siderophores and molecules such as vitamin B_{12} is now known to be dependent on a series of β-barrel proteins extending through the periplasmic space and through both the inner and outer membranes. Crystal structures of the proteins such as FHU A have been determined (Ferguson, A.D., Hofmann, E., Coulton, J.W., Diederichs, K., and Welte, W. (1998) *Science* **282**, 2215–2220). Energy coupling to the Ton B system is described by Larson, R.A., Thomas, M.G., and Postle, K. (1999) *Molecular Microbiology*, **31**, 1809–1824.

A novel iron centre has been found in superoxide reductase. The iron has four histidines and one cysteine coordinated in a rough octahedron (Yeh, A.P., Hu, Y., Jenney, F.E., Adams, M.W.W. and Rees, D.C. (2000) *Biochemistry* **39**, 2499–2504).

The structures of several iron regulatory repressors are now known, Pohl, E., Holmes, R.K. and Hol W.G.J. (1999) *Journal Molecular Biology* **292**, 653–657. They bind 2 atoms of Fe(II) with octahedral coordination, one site possibly using a cysteine and methionine but all other ligands are N/O donors. For further examples see Abstracts C.1, Metals and Cells, the Society of Experimental Biology (2001), Canterbury, UK.

CHAPTER 13

Haem iron: coupled redox reactions

13.1 Iron in porphyrins

Our knowledge of the biochemistry of iron porphyrins is probably more extensive than that of any other metal-factor unit: literally hundreds of proteins from different sources have been examined. The functional value of this cofactor is immediately seen by the diversity of its uses within biology (Table 13.1) and by the amounts of its proteins in a wide variety of cell types. The availability of the proteins has led to very detailed studies of them by chemists and biologists, but this book is not concerned with the detailed physical chemistry of the haem or its proteins. We shall concentrate here on appreciation of the functional value of haem proteins and perhaps the best place to start is with an examination of some features of the isolated haem units.

13.2 Properties of isolated haem units

To a first approximation the porphyrin ring is a di-anionic chelating agent which provides a relatively rigid hole for a cation, close to but smaller than the diameter of high-spin Fe^{2+} or Fe^{3+}, with four pyrrole nitrogens as its ligands [Fig. 13.1 (a), (b)]. As stated in Chapter 2, the fact that iron and not some other metal is almost invariably associated with the porphyrin rings in biology is not due to a thermodynamic bias of metal iron equilibrium constants: starting from high-spin aqueous ions the binding constants to porphyrin must follow the Irving–Williams series of Fig. 2.8. It is known that of all the metals in this series only Fe^{2+} is incorporated *in vivo* but all the other elements have been put into the ring by chemical syntheses. There are closely related rings in organisms which do not contain iron but do contain magnesium, cobalt, or nickel (Fig. 2.13). Iron is associated with porphyrin by a biologically evolved kinetic pathway—an energized synthesis—just as cobalt is associated with corrin, magnesium with chlorin, and nickel with F-430, to the exclusion of the other metal ions in each case. The synthesis of all these rings has a common route via the uroporphyrins, see section 13.12, and we must look later at the precursors in the synthetic pathway seeking for any feedback controls of interest. The synthesis of iron porphyrin is partly controlled by free, Fe^{2+}, iron levels at the insertion step in mitochondria. The very fact that evolution has produced this compound in such quantity speaks to its functional value as a chemical.

The iron porphyrin complex has some immediately interesting physical and chemical properties where it differs from simple iron.

1. It is limited in water solubility and, especially after screening of the iron by two extra axial ligands [Fig. 13.1(a)], readily enters hydrophobic solvents.

Table 13.1 Modes of incorporation of iron-protoporphyrin in different proteins to generate different functions in different cells, organelles, and membranes

Metalloprotein*	Iron† coordination number	Links	Electronic 'states'‡	Oxidation states
O_2 carrier				
Haemoglobin	5	His (H_2O)	HS/LS (O_2)	II
Myoglobin	5	His (H_2O)	HS/LS (O_2)	II
NO binding				
G protein	5 ?	His/?	HS/LS (NO)	II
CO binding				
Cytochrome *c′*	5	His/Met	HS/LS (CO)	II-III
O_2 activator				
Oxidases	5?	His (H_2O)	HS/LS (O_2)	II-VI
Cytochrome a_3	5?	His (H_2O)	HS/LS (O_2)	II-VI
P-450	5/6	Cys (H_2O)	HS/LS (O_2)	II-VI
Peroxidases (O_2)	5	His (H_2O)	HS/LS (O_2)	II-VI
H_2O_2 activator				
Peroxidases (H_2O_2)	5	His (H_2O)	HS	III-V
Catalases (H_2O_2)	5	Tyr (H_2O)	HS	III-V
Electron transfer				
Cytochromes *a*	6	His (2)	LS	II-III
Cytochromes *b*	6	His (2)	LS	II-III
Cytochromes *c*	6	His/Met	LS	II-III
Cytochrome *f*	6	His/Tyr (N)	LS	II-III
Reductases				
Nitrite reductase to NH_3	5 1	His/Met Lys	LS HS(NO_2^-)	II-III
Hydroxylamine reductase	6	His/His	LS	II-III
Nitrite reductase to NO (2 haems)	6 6(5)	His/His → His/Met His/Tyr → His/H_2O	LS HS/LS	II-III
Sulphite reductase	(6)	Fe_4S_4	LS	II-III
Sulphur to H_2S	6	His/Met	LS	II-III

*All these proteins are helical to a large degree.

†Many of the five-coordinate sites could have H_2O bound.

‡HS, high-spin; LS, low-spin.

2. It is easily oxidized by molecular dioxygen to higher iron oxidation states; then there follows a slow attack on the ring structure. There is, in fact, an enzyme, haem oxygenase, which assists degradation of free haem to eliminate it from those parts of an organism where it would be a dangerous oxidation catalyst.
3. The organic substituents around the outside of the ring are not fixed (contrast vitamin B_{12}) and a variety of porphyrins are chosen for different functions.

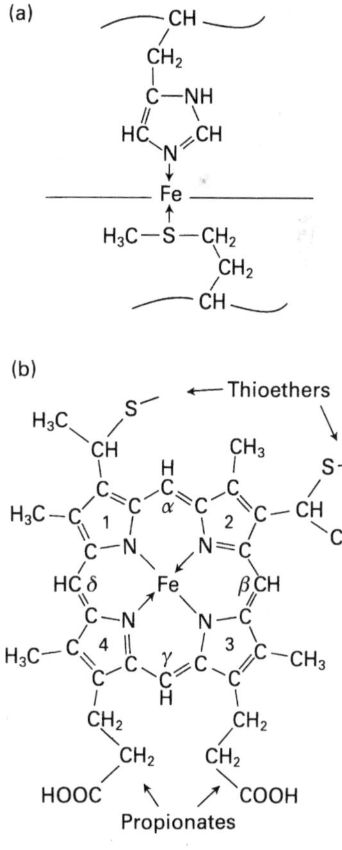

Fig. 13.1 (a) The axial ligands, methionine and histidine, to the iron porphyrin in cytochrome c. (b) The formulae of protoporphyrin bound here through thioether links as in cytochrome c.

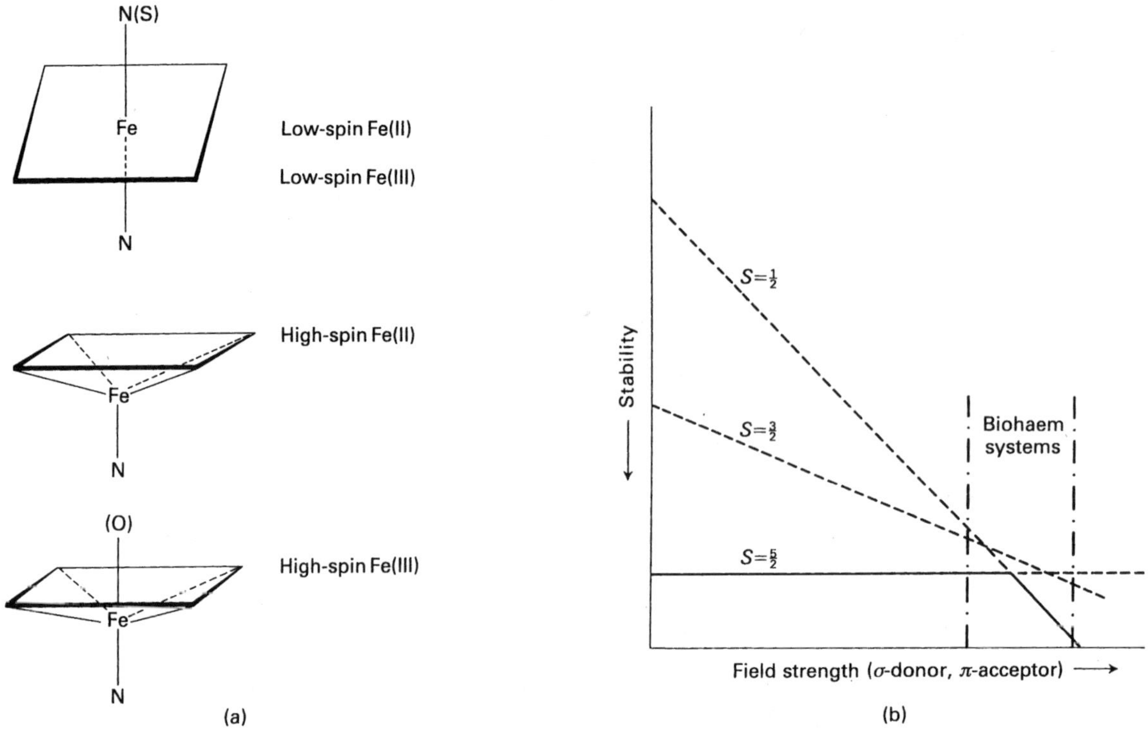

Fig. 13.2 (a) A generalized picture of the variety of high- and low-spin iron porphyrin structures. Note that bond lengths, bond angles, and the conformation of porphyrin are involved. (b) The manner in which the stabilities of the spin states (*S*) of Fe(III) haem varies with the field strength of the fifth and sixth ligands. A similar diagram can be drawn up for the Fe(II) ion.

4. The high-spin iron in the ring, whether in the Fe(II) or the Fe(III) state, is quite close in energy to the low-spin state, and indeed to intermediate spin states (Fig. 13.2), and this spin-state equilibrium balance is one of the delicate controls which the porphyrin ring together with the axial ligands to the iron and the proteins hold over the metal. The low-spin state of iron has special chemical properties.
5. While the ligand, porphyrin, is obviously a good four-centred σ-donor, it is a negatively charged $>N^-$ donor from two centres, and is also an effective π-acceptor. The energy level diagram of the iron-porphyrin unit makes it clear that the metal and the ligand are a strongly cooperative unit. In the low-spin state Fe(II) is made into a good π-donor, i.e. a π-base, which strongly assists binding of O_2, CO, and NO.
6. Porphyrin compounds, like chlorins and phaeophytins (Chapter 9), can be oxidized to useful free radicals. The metal and the ligand oxidation states can be treated separately to some degree depending, of course, on the symmetry of the orbitals in which the unpaired electron(s) sit. NB High-spin iron is always a weak (multi) free radical.
7. Some of the ring substituents are reactive; for example, the vinyl groups can suffer addition to give thioethers, e.g. in cytochromes *c* (Fig. 13.1(b)), and various oxidations or reductions are possible to give haems *a* and rare iron chlorins.

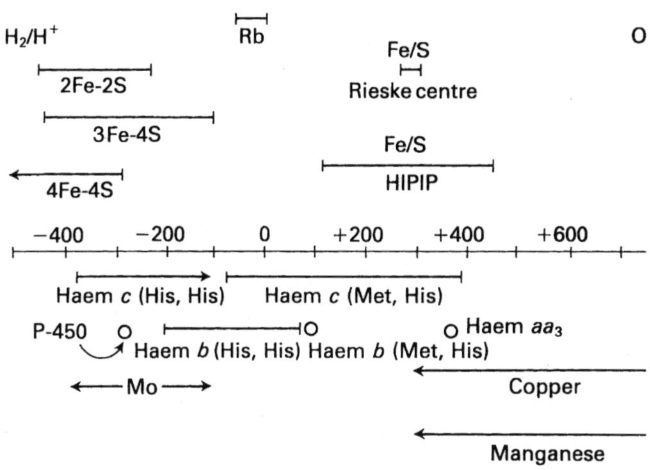

Fig. 13.3 The redox ranges of haem proteins and a comparison with the redox potentials of other metal centres. The haem group in all cases is in a cytochrome. Note Fe(IV), Rb, rubredoxin; HIPIP, high potential iron protein.

8. The hydrophobic nature of the porphyrin surface makes for strong interaction with hydrophobic small molecules as well as with hydrophobic protein surfaces.
9. Iron does not dissociate from haem to any degree, so that haem iron is almost a new 'element' distinct from iron.
10. The redox potentials of the iron porphyrins are readily adjusted by the proteins (see Fig. 13.3).
11. Haem can act as a cell regulator, quite separately from free iron through dissociation from its proteins. Binding constants in some proteins are around 10^{10}.

Given the complexity behind these properties it pays us to simplify the description of haem binding in proteins.

13.3 Classification of haem proteins by iron properties

Table 13.1 classifies the haem proteins by function. (A different approach is to classify them by the chemical bonding.) In the table, in addition to functions, we give spin and oxidation states for the metallo proteins and also classify them as containing five-or six-coordinate iron. The next valuable descriptive feature is the ligation from the protein sidechains. If there are two axial ligands making a rough octahedron around the iron the two donors are almost invariably one imidazole (histidine) plus either a second imidazole or a thioether of methionine, but cases of lysine, carboxylate, or phenolate binding are known. In the methionine complex the haem is usually covalently bound through thioether links via its vinyl groups to the protein. This construction is a feature of cytochromes *c* (Figs 13.1 and 13.4), but there are variations of this structural pattern with but one thioether link. cytochromes *c*′. Such is the complexity of this cytochrome *c* group of proteins, most if not all of which are extracellular, that there are books which deal with them alone (see 'Further reading'). The next group of proteins, with two ligand histidines, can have the porphyrin bound covalently as above or, more usually, not bound to the protein. The two series of proteins correspond roughly to a further group of cytochromes *c*, that is, those covalently linked into the protein, and to the

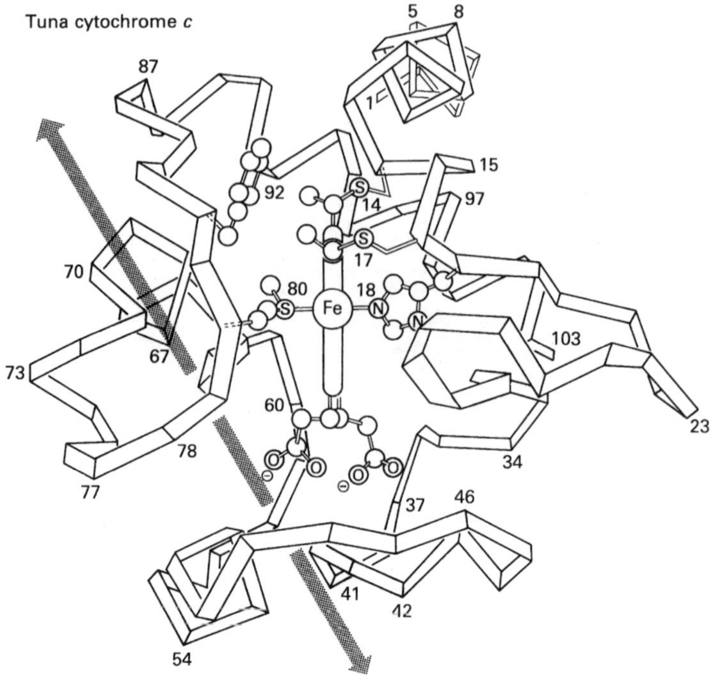

Fig. 13.4 An outline, somewhat impressionistic, picture of cytochrome *c*. The haem and the ligands are in the centre and the haem is covalently linked to the sequence. Note the many helices. The arrow shows a hypothetical protein-coupled track for protons due to known H-bond breaks at Gly-41, Thr 78, Met 80, Tyr 67, Lys 78, and a re-adjustment within helices 60–69 and 70–77 with redox state changes (see Fig. 13.8).

cytochromes *b*. Again there are occasional variations from these general statements since it is known that a further few cytochromes with two axially co-ordinated protein sidechains are bound by a lysine $-NH_2$ group and a histidine, cytochrome *f* (see also ferritins in Fig. 12.13). Finally, within this pattern are cytochromes which are based on oxidized ring sidechains, cytochromes *a*, and which have two histidines coordinated. Almost all the cytochromes *a*, *b*, and *c* are in low-spin states. We turn to their protein structures later, but the haem in them is usually well buried in a hydrophobic slot formed by several strands, often of helices (Fig. 13.4). To a first approximation they are electron transfer proteins and the noticeable feature of them is that they cover a redox potential range Fe(II)/Fe(III), from –0.3 to + 0.4 V (Fig. 13.3), providing a redox range reaching to much higher potentials than the Fe/S proteins. It may well be that the development of the higher potential proteins followed the coming of dioxygen during evolution.

The second group of haem proteins are five-coordinate in the Fe(II) state and may or may not have one water molecule in the sixth position when in the Fe(III) state (Table 13.1). The iron is largely high-spin in both conditions and is usually bound by histidine as its fifth ligand, but there are a few cases of thiolate, phenolate, or even carboxylate binding. Usually the proteins do not bind covalently to the haem, but the haem lies in a hydrophobic pocket. There are variations in the haem substituents in, for example, spiro-haem of sulphite oxidase and haem a_3 of cytochrome oxidase. The outstanding feature of these proteins is that they are 'designed' to transport or bind dioxygen (sometimes NO also) or to activate such molecules as hydrogen peroxide, superoxide, nitrite, sulphite, or dioxygen rather than to be electron transfer proteins. Their names give their functions, for example, dioxygenases and dioxygen transfer proteins use the sixth co-ordination position as a binding centre for dioxygen. In the course of many of their reactions as oxidases the iron cycles through oxidation states (II) to (VI), and the porphyrins often pass through a radical state. The iron also cycles through changes of spin state and, with these changes and changes of oxidation state, the porphyrin ring alters its ruffling. Then, unlike most of the first group of haem proteins, the redox and binding reactions are obviously coupled to changes in the haem and its ligands, and to some degree can be linked to cooperative changes in the protein. The degree of coupling is an essential part of different functions. Coupling is frequently associated with allosteric effects, e.g. in the uptake of dioxygen by haemoglobin. We shall need to examine the structures of the associated proteins in order to appreciate why and where coupling within haem proteins occurs since this is one of the the most fascinating features of haem biochemistry.

13.4 Classification of haem proteins by secondary structure

We have now divided the haem proteins generally by function and by structural coordination chemistry, but it is the protein that gives each of them its specific functional character. Of course, this applies to the second substrate they bind, i.e. in addition to O_2 or H_2O_2, but first and importantly we need to look at the protein fold. It appears that, quite unlike Fe_n/S_n proteins where the active site is usually multiply cross-linked to thiolate sidechains of the protein and the strands of the protein form β-sheets, i.e. cross-linked H-bonded structures, the essence of haem proteins is a loose hold on the iron through perhaps only one coordination group, a relatively loose hold even on the porphyrins, and a secondary structure around the haem of helices or loose strands. Helices, unless they are cross-linked by –S–S–bonds or joined to β-sheets, are intrinsically mobile relative to one another (see Chapter 6). It would therefore seem that helices plus haem form an excellent structural motif for the development of many *mechanical* functions. Our next step in this analysis is to see in which haem proteins the helices are cross-linked and in which they are not, so that we can know where evolution has clamped the haem to prevent cooperative motion of haem and protein during reactions. It is the development of cooperative interaction between iron and many proteins through the use of the porphyrin frame that has led us to use *coupled reactions* in the title for this chapter.

Table 13.2 classifies the secondary structures of haem proteins. The first group includes the *c*-type cytochromes which, as already stated, are structures which are covalently linked to the protein. The proteins themselves do not fold without the haem, but once the fold is completed and the thioether cross-links have formed the cytochromes *c* are the most stable and have the least

Table 13.2 Secondary structures of haem proteins

Secondary structures	Examples
Helical but cross-linked by sulphur links	Cytochrome *c,c'*, peroxidase
Helical plus β-sheet	Cytochrome *c* peroxidase, cytochrome b_5, catalase
Helical	Membrane cytochromes *b*, *a*, a_3, *o*, cytochrome P-450 (largely), haemoglobin, myoglobin

NB. Note the difference from the Fe/S proteins where the Fe/S is locked in a β-sheet. Haem is usually surrounded by a more mobile framework of α-helices or loose strands.

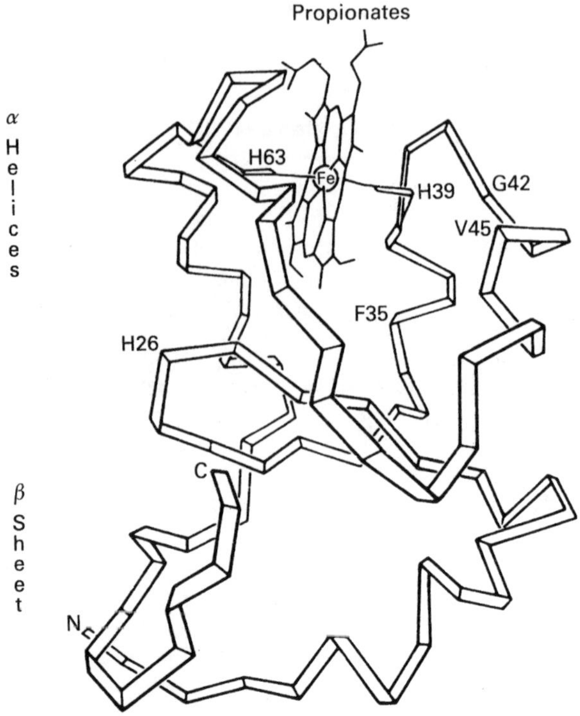

Fig. 13.5 An outline structure of cytochrome b_5. There is a β-sheet holding the helices which in turn hold the haem in a relatively rigid protein with two histidine, H, ligands. (After F.S. Matthews.)

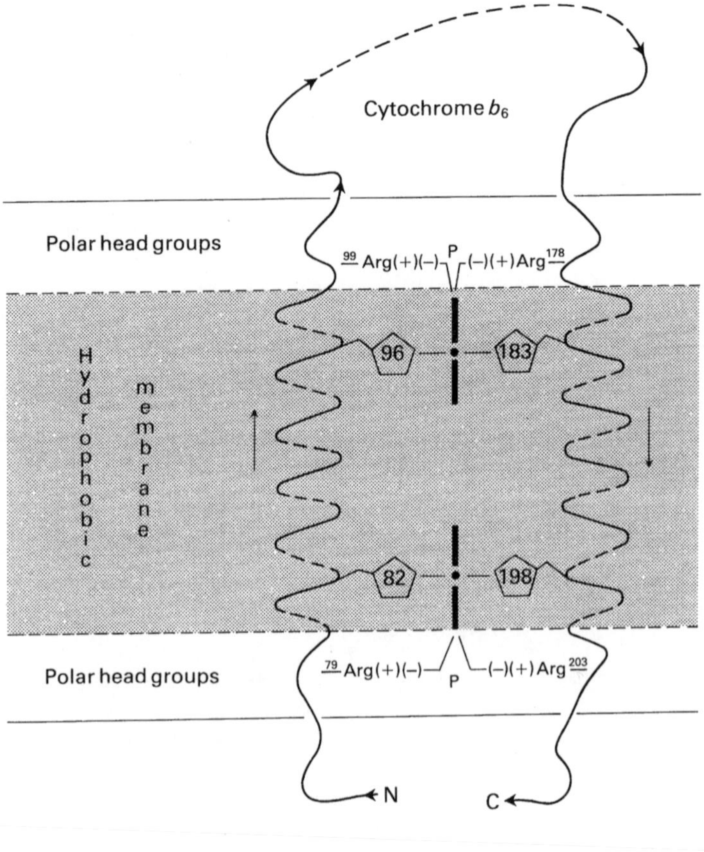

Fig. 13.6 The critical section of the cytochromes b_6 of the energy-capture membranes. It is considered that the haem units are held as shown by histidines protected by helices of which two are shown. The structure is not rigid. The numbers are of the sequence. Note the buried propionate carboxylates (P) and compare cytochrome *c* while contrasting cytochrome b_5.

internally mobile sidechains of all proteins studied in depth (Fig. 6.7). They are invariably simple electron transfer proteins in which electron gain or loss is accompanied by little coupling to changes in protein conformation (but see Section 13.6.3). The haem cannot exchange from these proteins and very generally they are placed outside cells where dissociation would be a problem.

The other two groups are also based, at least in part, on helices around the haem. There is one group in which the haem is wrapped around by helices, usually four, but where these helices are linked to a β-sheet (Fig. 13.5). These proteins, cytochromes b_5 for example, are simple uncoupled electron transfer proteins but operating in a somewhat different way, in a lower redox potential range, and in a different cell compartment, from cytochromes *c*. The structure of the third group is similar around the haem, but there is no β-sheet section. These proteins may also take part in electron transfer and in binding of and in the reaction with small molecules such as dioxygen, but their reaction, even electron uptake or release, is connected to a protein conformation change which may be linked to energy capture devices or to the relay of information. Some of these proteins are found in the energy-capture membranes (Fig. 13.6). The protein of cytochrome *a* is of this kind also but the haem is different and it has a high redox potential.

The group also includes the dioxygen carrier proteins; here the helical structure (Fig. 6.3) is more complicated than in the cytochromes *b* and the haem is open-sided. The helical structure leaves a narrow groove to the iron.

The cooperative helix–helix movements linked above to electron-transfer are also similarly linked here, but more importantly they are linked to dioxygen uptake and release by several haem groups in a multihaem protein, leading to the allosteric equilibria of haemoglobins. Notice that this linkage is much reduced in the dioxygen storage protein, myoglobin.

Some members of the first and second groups of proteins are also based on five-coordinate haem, but now either the helices are cross-linked by –S–S– bridges, as in peroxidases and oxidases, some of which use superoxide, or they are attached to a β-sheet, as in cytochrome *c* peroxidase. These proteins have second substrates as well as dioxygen or hydroperoxide, do not have mobile folds, are usually found outside cells or in vesicles, and are freely diffusing molecules but uncoupled systems. The distinction between coupled and uncoupled proteins is strongly stressed in Table 13.1 and 13.2.

A further point of interest here is the degree to which the haem is buried. In electron transfer proteins the haem lies just buried so that the iron is about 10 Å from the surface. In proteins which pick up dioxygen or hydrogen peroxide and where a small organic substrate has to be oxidized, e.g. peroxidases, the haem is perhaps somewhat deeper in the pocket and in these the substrate usually sits in a groove close to the entrance to the haem. In catalases, which only destroy H_2O_2 and must not undergo electron transfer reactions, the haem is buried some 20 Å into the interior. In all proteins the haem is displaced away from the centre of the protein. Once again, we see how the protein conditions the metal ion, both directly in its coordination geometry and indirectly at greater distances so that functional value is optimal. When the haem is buried, either its propionate sidechains are buried in the protein (cytochrome *c* and membrane cytochromes) or they are exposed to water (cytochrome b_5 and haemoglobin). The significance of these arrangements will become clear later.

Finally, there is a group of haem proteins in which the helical structure dominates but where the haem coordination to the protein fluctuates in the reaction cycle. In one subsection of haem proteins, the iron ligands are a thiolate from a cysteine sidechain and one water. These are the cytochrome P-450 enzymes (see Chapter 6). They are known to be largely based on helices but to have some short β-strands and a small β-sheet not too far from the haem. The second substrates are bound here and are very inert molecules, often simple hydrocarbons. There is a compulsory order of addition of substrates and reaction pathway, which demands a cooperative series of protein rearrangements (Section 6.5). In both these latter respects the enzymes are very different from simple peroxidases and oxidases and resemble cytochrome a_3, although they differ from cytochrome a_3 in their location in a cell. Cytochrome a_3, which is in a membrane, probably uses a similar mechanical action of helices in order to act as a proton pump. The haem protein which binds NO and modulates guanidine cyclase enzymes is likely to be of this kind also. A final example is nitrite reductase in which the ligand to one of the cd_1 haems, d_1, changes from tyrosine to open-sided, and the ligand to the other, *c*, switches from a histidine to a methionine during the activation of the enzyme (see 'Further reading').

Whereas we have seen that the haem does not dissociate from the cross-linked proteins and enzymes such as cytochrome *c* and peroxidases, dissociation from several of the other proteins, e.g. cytochromes *b*, is much easier. As a consequence it is possible for free haem at about 10^{-10} M to be a control molecule acting much like free iron (Section 13.11). Overall we see that, as for free metal ions, haem functions are constrained by each protein.

13.5 Where are haem proteins in cells?

Organisation of haem proteins

The low-levels of haem proteins in some cells and even organs is reflected in the fact that they are almost white (contrast red, haemoglobin-containing cells with white blood cells) but the distribution is often due to the different content of compartments in cells. The conclusion is obvious that cells with a high oxidative capacity, which have lots of mitochondria, or cells which carry oxygen will appear darkly coloured. In plant tissue a parallel contrast exists between green, chlorophyll-containing tissues, and white tissues, e.g. between leaves and roots. Since different haem proteins are widely distributed in compartments it is easier to start this description of the topographical distribution of haem proteins with a statement of the compartments in which there is no red colour. Haem proteins are very rare outside cells of higher animal organisms, i.e. in circulating fluids though they are found outside plant and bacterial cells, e.g. peroxidases. They are also virtually absent from the nucleus, from lysosomes, from the Golgi apparatus, and from the interior of reticula. The compartments where they are present in greatest amounts are the membranes of aerobic prokaryotes and their periplasmic space, the membranes of organelles, and the aqueous phase of peroxisomes (Table 13.3). In prokaryotes and in mitochondrial and chloroplast membranes they form the basic electron transfer functions at high redox potential (cytochromes *a*, c_1, *f*, and some cytochromes *b*; Table 13.1). This function extends to the periplasmic space of prokaryotes and organelles where there are several low and high potential cytochrome *c* proteins. There are also some cytochrome *b* electron transfer proteins anchored to the cytoplasmic membrane. In strict contrast, the haem catalases and peroxidases in eukaryotic cells are in peroxisomes while in prokaryotes they are again in the periplasmic space. The haem oxidases may be associated with a variety of membranes in eukaryotes and prokaryotes. For

Table 13.3 Haem protein organization in eukaryotes

Haem protein	Compartment
Haemoglobin	Specialized red blood cells
Myoglobin	Cytoplasm of red muscles
Peroxidases	Peroxisomes
Catalases*	Peroxisomes
Cytochromes (*c*)†	Mitochondrial membrane or periplasm
Cytochrome P-450*	Microsomal membrane
Cytochrome b_5	Microsomal membrane
Cytochromes *a*, *b*	Mitochondrial or thylakoid membranes

*Haem enzyme that, in prokaryotes, is free in the cytoplasm.

†Haem enzyme that, in prokaryotes, is on the outside of membranes.

example, the P-450 cytochromes belong to the eukaryote reticula membrane, but these same enzymes are in the cytoplasm of prokaryotes. The low level of haem proteins in the cytoplasm of eukaryotes is most striking and yet the cell with the most haem in it, the red blood cell, has all its haem protein, haemoglobin, in the cytoplasm. Now this is a most peculiar cell since in many higher species it has lost its nucleus. It is almost as if haem protein reactions and the nucleus were best kept separate. Perhaps we should understand this as a further removal of eukaryotic DNA from any danger due to side-reactions of an oxidative kind. Haemoglobin is, in principle, only a dioxygen carrier, but there is constant risk of some superoxide production together with Fe(III) haemoglobin. The blue appearance of blood under some conditions is due to this oxidation and the haemoglobin is protected from oxidative reactions in many red blood cells by the presence of a strong soluble reductase using cytochrome b_5 as an electron carrier.

Finally, we should not forget the possible restriction of location based on synthesis. The cytochrome *b* membrane protein and one subunit of cytochrome oxidase are synthesized via mitochondrial DNA. We can only ask: is it this localization of synthesis which keeps these two proteins within the mitochondrial membrane? It will be interesting to see which proteins are synthesized by the chloroplast DNA. We have also noted that during the course of some haem-protein syntheses the protein itself or the haem and the protein are cross-linked. These proteins are external to cell or organelle cytoplasm or membranes.

13.6 **Haem protein functions I: electron transfer**

13.6.1 Simple electron-transfer proteins

We now turn to the more detailed examination of the functions of haem enzymes. The most obvious function consists of electron-transfer reactions often at redox potentials higher than 0.0 V (Fig. 13.3 and Table 13.4). The least complicated examples are the cytochromes *c* of the periplasmic space of bacteria and organelles, the cytochromes c_1 and *f* on the outer surface of the same membranes, and perhaps the cytochrome *a* of cytochrome oxidase. In each case electron-transfer is not complicated to a marked functional degree

Table 13.4 Effects of proteins on haem electron-transfer sites

Control over structure of coordination sphere affects the relaxation energy (constrained, 'entatic' states)
Control over structure, first- and second-coordination spheres, and solvent exposure, affects redox potential and relaxation
Degree of burial affects the distance of electron transfer
Degree of burial protects from adventitious small-molecule reaction
Surface charge and surface structure affect the recognition between proteins
Internal mobility affects relaxation energy
Internal mobility affects long-range coupling possibilities whether this mobility refers to single proteins, pairs of proteins, or particles
Coupling can be to energized proton diffusion paths (pumps) in membranes

by coupling of this reaction to other changes. The structures of these electron-transfer proteins are governed by the features listed in Tables 13.1, 13.2, and 13.4, which apply equally to the simple electron-transfer proteins based on Fe/S and Cu (see Chapters 12 and 15 and the discussion of the theory of electron-transfer in Chapter 4). We shall assume that the 'design' of these electron-transfer proteins concerns mainly protection of the centre from accidental attack and the creation of a useful distance between centres to give a wire-like construction—say 10–20 Å. There is no certain knowledge of organic group pathways in the proteins (see 'Further reading').

13.6.2 Multihaem electron-transfer proteins

There are several proteins which contain more than one haem (see Fig. 13.7). We leave to one side here the membrane haem proteins cytochrome b_6 and cytochrome aa_3, which provide a path for electron-transfer across or into membranes, since these proteins are coupled and are part of energy-

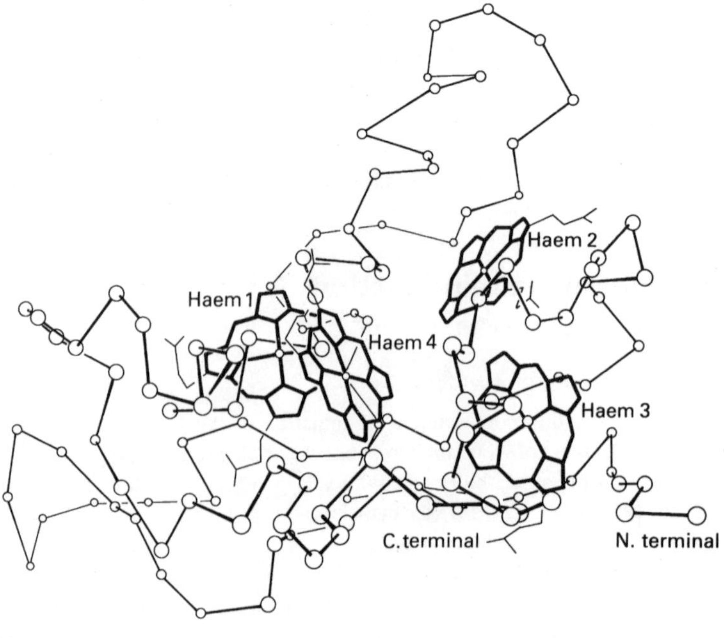

Fig. 13.7 The outline structure of a cytochrome c_3. (After R. Haser.)

transducing devices, as described in Section 13.6.3. Others appears to act merely as a supply of electrons to coupled systems as in the photosystem I of bacteria (Fig. 5.3) and in sulphite and nitrite reductases (Table 13.1). In some sulphur bacteria there are water-soluble multi-haem cytochromes, c_3, c_7, and so on which contain from three to eight haem units. They are usually found in the periplasm of the bacteria. They are of unknown function but one hypothesis is that they are electron storage devices rather like condensers in man's electronic circuits. They then hold *n* electrons ready to be released on contact with a reactant. These *c*-type proteins have a low redox potential and can act between the very low potentials of the particle I and particle II of mitochondria, that is between NADH or even H_2 (–0.5 V) and quinones (0.0 V). They could scavenge for electrons much as cytochrome *c* scavenges between particles II and IV. Another possibility is that, through small conformational changes, they have a control function to which we turn next.

13.6.3 Electron-transfer/proton transfer coupling: models

A major consideration in bioenergetics is the coupling of electron and proton flow in the mitochondrial and chloroplast membranes (see Chapter 5). There are electron-transfer haem proteins involved, cytochromes aa_3 and cytochromes *b*, and it is commonly thought that the redox reactions and some binding reactions, e.g. of O_2, are also coupled to proton movements by the readjustments within their helical frameworks. In this way ideas on proton transfer in energy coupling are obviously linked to ion pumping by reversible ATPases (see Chapter 8). The structure of cytochrome oxidase is now known (Section 13.9).

13.6.4 The cytochrome bc_1 complex

In different chapters we have described in outline the structures of the electron-transfer systems (1) of chloroplasts (Chapter 5), (2) of particle I and of particle II, succinate dehydrogenase, of mitochondria (Chapter 12) and of cytochrome oxidase, particle IV of mitochondria (this chapter). There remains particle III to be examined, the bc_1 complex of mitochondria. In essence this protein carries out a Q-cycle electron transfer which involves (Fig. 5.4) three haem-iron centres; two (b) in a membrane and one (c_1) cytochromes outside it, and one Fe/S Rieske protein also outside the membrane. The electrons can cycle in the membrane using the quinones, Q, and the haems (b), or leave the membrane via the c_1 and Fe/S protein (see Section 5.3). The detailed structures are described by Crofts *et al.* and Xia *et al.* in the references. In themselves the electron transfer steps appear normal but the enzyme is made so that the membrane section carrying haems (b) and the section outside the membrane (haem c_1, Fe/S) are able to move relative to one another by some 25 Å making and breaking the electron transfer in a switch-like activity. The full functional significance of these motions is discussed in the references. The electron transfer is coupled to proton movements across membranes and a parallel system is found in photosynthetic energy capture devices (Fig. 5.3). All the structures of these large electron-transfer proteins are extremely difficult to obtain at high enough resolution (see Section 13.9) to follow proton switches, and working models using smaller proteins are the most we can devise as yet. One such working model which we describe next follows from very detailed studies of cytochrome *c* (Fig. 13.4), which shows minor coupling of redox reaction to the protein matrix and H-bond rearrangements. We give this example in detail to illustrate the subtle nature of bio-energy coupling (see also bacteriorhodopsin, Section 5.3).

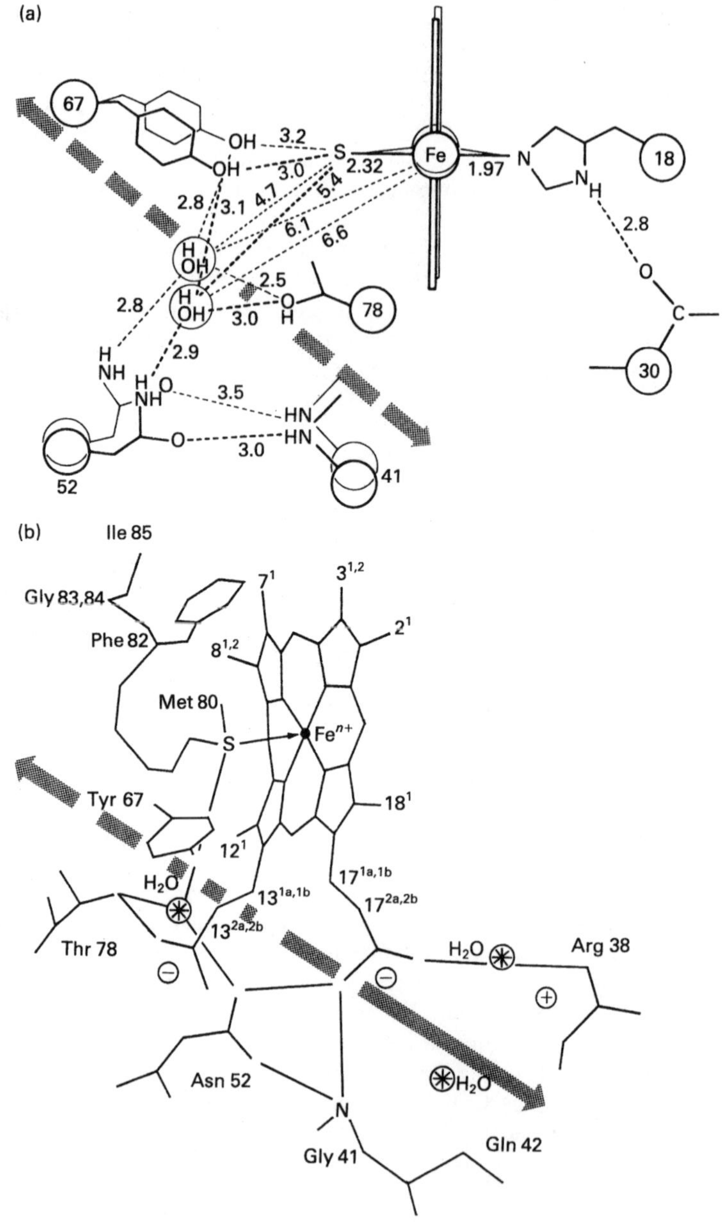

Fig. 13.8 (a) A crystallographic study showing the major changes in structure of cytochrome *c* on redox reaction which lie between residue 40 and 80. Note the internal water molecule and its movement (after T. Takano and R. E. Dickerson). The heavier lines are for the reduced state. The grey arrow is the line of H-bond breaks. (b) A diagram using crystallographic studies showing main region of changes seen by NMR (in solution) of the structure of cytochrome *c* on change of redox state. Taken with Fig. 13.8(a) the suggestion is that the electrostatic energy change from Fe^{2+} to Fe^{3+} causes a shift of the propionates and a series of hydrogen bond rearrangements (arrow) occurs in the region shown (see Fig. 13.4).

As explained above, cytochrome *c* has the following structural features and redox state dependent changes [Figs 13.4 and 13.8(a), (b)].

1. A haem iron bound to a methionine sulphur (Met-80) and a histidine (His-18) where the numbers refer to the position in the sequence.
2. Two long helices from residues 1–12 and 87–102. The first helix is rigidly fixed in space by groups immediately following it, cysteine 14 and 17, which form covalent thioether bridges to the vinyl group of the protoporhyrin haem. The second helix is locked into this first helix.

3. Following residue 18 there are series of turns on one side of the haem running to residue 38 which is an arginine holding the buried propionic acid sidechain of the haem (Fig. 13.4). Effectively all these parts of the protein are relatively rigid, i.e. from residues 1 to 38 and from 87 to 102.
4. The sequence from 38 on contains a set of rather poorly defined helices from residues 50 to 55, 60 to 69, and 70 to 76 before it runs along the haem through Met-80 to the terminal helix, 87–102.
5. The haem propionates are buried in the protein and associated with an internal hydrogen-bond network which even includes water molecules (contrast haemoglobin and cytochrome b_5).
6. NMR studies using ^{15}N resonance shifts show that amino-acids far from the active site are affected by redox state change. In particular, it is the set of loose helices and the hydrogen-bond network between residues 38 and 86 which is of interest here. On the electron-transfer redox reaction, Fe(II) to Fe(III), there is not just some alteration of the ligand, Met-80, and the strand from residues 79 to 85 but there is a rearrangement of the hydrogen bond network in which some three or four H-bonds are broken and some made (see Fig. 13.8, or details). At the same time the accessibility of this region of the molecule to protons from the solvent is considerably changed. Thus we observe on a small scale features which are demanded in a proton/redox pump: (1) the redox-coupled alteration of proton-binding accompanying helix readjustment, and (2) a change of access of protons to these regions of the protein which is equivalent to a gating and is essential for redox-controlled proton flow. (In Chapter 4 proton flow is described in terms of rotational diffusion requiring the making and breaking of H-bonds); (3) the presence of anionic proton binding groups deep in the membrane; here we note particularly the propionates, which take part in the H-bond network and which are responsive to the redox state change of the iron; (4) the H-bond network to water molecules is considerably changed. Increasingly it is being found that there are networks of H_2O inside these proteins (see Sections 5.4 and 5.5). All these features are probably important in membrane bound proteins.

13.6.5 The membrane electron-transfer proteins: sidechain proton coupling

A particularly intriguing feature in the possible connection between redox reactions and proton energization and gating (pumping) is the potential value of these *buried* haem propionate sidechains in membrane-bound proteins. The protonation and deprotonation of such carboxylic acid centres as well as their connection into H-bond networks are adjusted by the redox and binding states of the iron in several cytochromes (see above) much as the reactions of haemoglobin are coupled to protonation states elsewhere. It may well be that these or other buried carboxylate groups associated with redox reactions (e.g. with haem or quinones (Fig. 13.9) or of light-absorbing centres) or other energizing reactions, such as those of ATP-dependent pumps, are essential parts of the energization of protons in membranes and that, through changes in helical structures, movements of these protons are gated. Thus, electrostatic energies of buried charges could become extremely valuable in coupling. The appearance of the haem unit in the association with helical structure appears to have a very deep significance. Figure 13.9 shows the structures of the *b* type

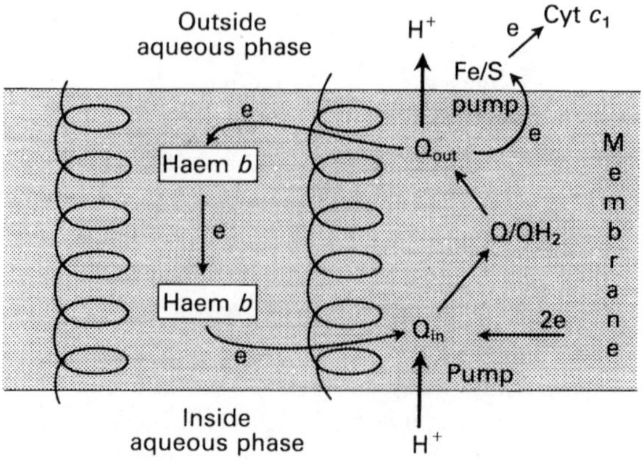

Fig. 13.9 The electron-transfer events in the cytochrome *b* of particle III of mitochondria and thylakoids (Fig. 11.1), are coupled to the uptake and releases of protons by the quinone Q at its redox sites. The coupling probably uses the buried propionate sidechains of the haem and other acid anion centres of the proteins. QH_2, carries the protons over the longest distance.

membrane proteins which have buried propionates and quinone (Q) coenzymes which are often close to buried carboxylate protein sidechains that change protonation state with redox reactions. These structures clearly have the potential to act as pumps or gates coupled to electron flow. We return to this topic after examining the uptake of small molecules by haem proteins, when we also discuss alternative models of coupling.

13.7 The surfaces of haem proteins

We interrupt the discussion of haem-protein function with an important diversion about their surfaces. The discussion applies to all other kinds of proteins, but most information is available concerning cytochrome *c* (Fig. 13.10). The haem proteins are of three different kinds with respect to organized multiprotein systems:

1. isolated proteins, catalase and peroxidase;
2. carrier proteins which bind to several sites, cytochromes *c*;
3. proteins bound in complexes, e.g. cytochromes *a* and *b* of mitochondrial complexes III and IV. These proteins are at best only laterally mobile in their membranes.

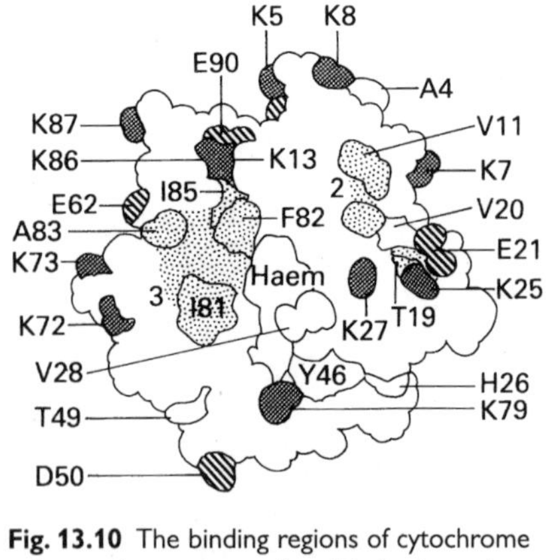

Fig. 13.10 The binding regions of cytochrome *c* due to the presence of positively charged sidechains are shown in black. There is a low level of selectivity of binding. Negatively charged sidechains are cross-hatched; hydrophobic groups are stippled.

It is useful to discuss these observations here since amongst electron-transfer proteins the difference between these three classes greatly affects functional value. The description of their surfaces, which generates site location and the degree of restriction on diffusion, has to be analysed in terms of the charged and hydrophobic groups on the surfaces and the matching of surfaces to one another (see Chapter 6).

The simplest situation was discussed in Chapter 5 where we analysed electron transfer in the reaction centre of photosystems and noted that the organization was tightly maintained (see also Figs 13.9 and 13.15). Now at the other extreme there are proteins such as cytochrome *c* which need to diffuse between many partners. They scavenge the intermembrane spaces for electrons and must be able to recognize several target proteins, and to bind to them and to dissociate from them rapidly. As explained in Chapter 6, the best device for the first group is a lock-and-key or a dye-and-mould fit and here hydrophobic interactions may well be very valuable. The best device for the second is a hand-in-glove matching using electrostatic surfaces and a more mobile underlying frame may also be of value. Perhaps these requirements explain the charged surface (Fig. 13.10) and the internal mobility of one side of cytochrome *c* near its binding surface. It is an adjustable protein. Finally, peroxidases have even more hydrophilic surfaces due to glycosylation so that they are totally protected from binding (see Chapter 6).

13.8 Haem-protein functions II: storage and transport

13.8.1 Dioxygen: storage, transport, and signalling

The study of haemoglobins and myoglobins has been as intensive as that of almost any other proteins. The following features are outstanding.

1. Dioxygen binding alters the haem iron spin states, the ruffling of the haem plane, and the iron-ligand bonds [Fig. 13.11 (a)].
2. The changes in the iron porphyrin and its ligands on dioxygen binding are relayed to changes in each protein unit to different degrees. Internally there are sidechain and mainchain helix adjustments [Fig. 13.11(a), (b)].
3. In multi-subunit haemoglobins, see Fig. 6.3, the conformational changes are transmitted from one subunit to another [Fig. 13.11(c)] giving allosteric, cooperative, dioxygen-uptake curves, see Banerjee *et al.* in Further Reading.
4. The binding of protons and several small molecules adjusts the binding constants and the cooperativity through alterations of the helical structures.
5. Amino-acid changes are species or even family specific, of course, and give a wide range of affinity and cooperativity differences leading frequently to advantages in particular biological niches but also to diseased conditions, e.g. sickle cells.

The analysis of haemoglobins has thus led to major development in our knowledge, but for details the reader must go to specialist reviews on this protein (see 'Further reading').

Finally, there is a control link via a dioxygen-binding proteins, probably another haem protein, to the synthesis of the globin proteins themselves. The body adjusts such synthesis even at different altitudes. Thus, dioxygen is regulatory through its binding to haem in proteins other than haemoglobin.

There are other haem proteins of unknown function which behave in a somewhat similar way on binding to ligands. In particular, the cytochromes *c′* of bacteria are allosteric, four-helical bundle, haem proteins which alter conformation on binding CO, NO, or CN^-. It is not known if they act so as to control metabolism in the presence of such poisons. There is the distinct possibility that control was the primary function of haem proteins in evolution before their use in electron transfer, which could always have been done at low potential by Fe/S proteins, or before their handling of reactions of dioxygen. The value of signalling the presence of poisons, e.g. NO, CO, or even O_2 cannot be overestimated and in evolution the neutralization of environmental poisons may have always been the first step before making use of them. An example is provided perhaps by nitric oxide.

13.8.2 Nitric oxide and haem receptors

Nitric oxide, NO, is an intermediate in the reduction of nitrate or nitrite to ammonia, and until recently it has been considered as a dangerous poison.

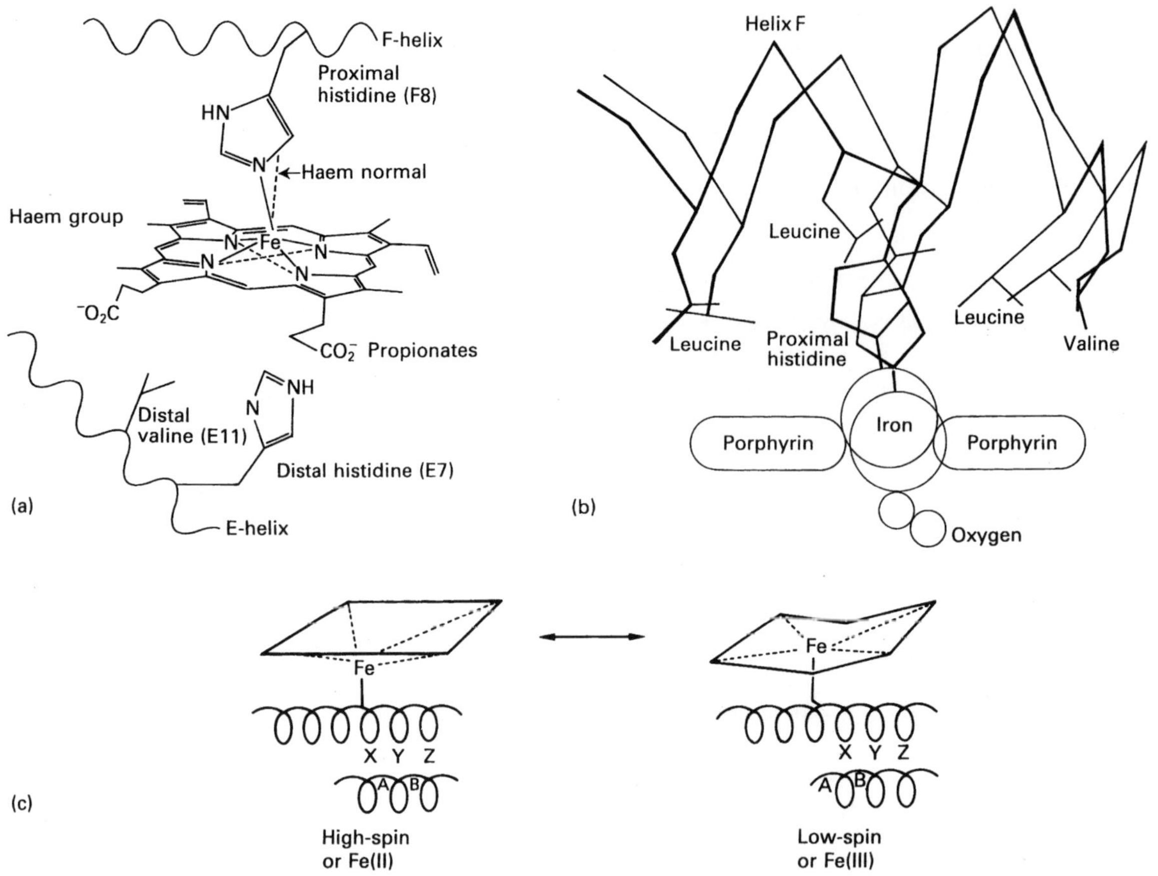

Fig. 13.11 (a) A detailed picture of the O_2-binding sites of myoglobin or haemoglobin. (b) The particular changes of the protein parts of haemoglobin on binding O_2. (After Pertuz, Baldwin, and Clothia.) (c) A sketch of the way in which helices slide (schematic) on change of spin state with O_2 binding in haemoglobin. (Note the relative positions of AB and XYZ.) This causes allosteric cooperativity.

Surprisingly, NO is also found to be produced *in vivo* by oxidation of l-arginine and acts as a neurotransmitter, a regulator of blood pressure in cellular defence mechanisms and as an endogenous trigger of smooth muscle relaxation when it is produced in nearby endothelial cells. The generation of NO from arginine involves a family of three distinct calmodulin dependent NO synthases (NOS), stimulated by Ca^{2+} invasion of cells. It appears that the highest values of NO are to be found in neurons. The action of NO is on a haem requiring GTP-cGMP, enzyme, where NO generates a conformational change reminiscent of haemoglobin and cytochrome c′ reactions (see Section 13.8.1 and Figs 13.12 and 13.13). The importance of NO interactions with haem proteins is today equal to that of the O_2 binding by haem-containing dioxygen carriers.

13.8.3 Cytochrome *c* and cyanide

Once again we can use cytochrome *c* to probe the effects of ligand binding to haem proteins just as it was used to follow the effects of redox state change. Cyanide is a powerful poison for O_2-binding centres much as is CO. Cyanide binding to cytochrome c requires; (1) a groove-opening reaction; (2) a ligand

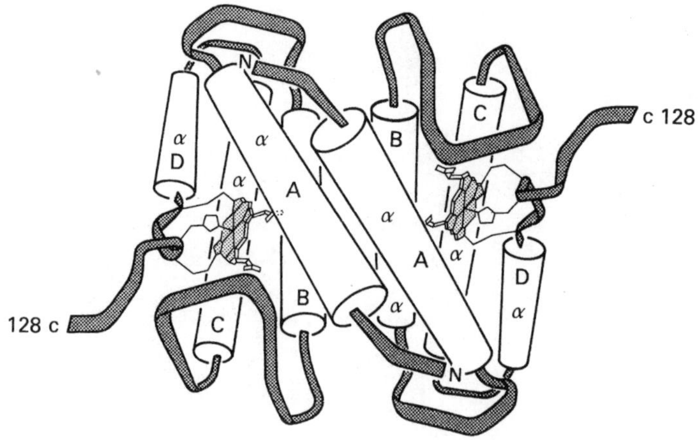

Fig. 13.12 The α-helical fold of cytochrome *c*′, shown as a dimer. The iron is five-coordinate. (After P. Weber.)

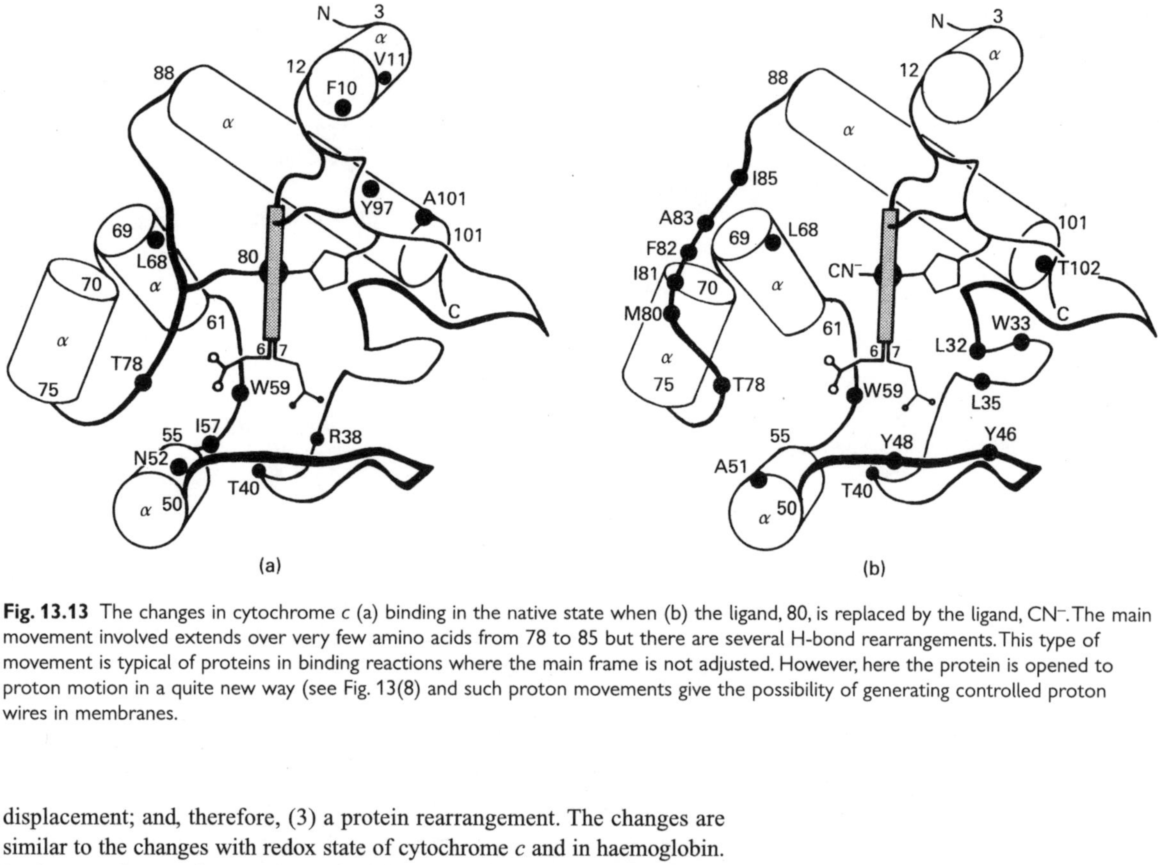

Fig. 13.13 The changes in cytochrome *c* (a) binding in the native state when (b) the ligand, 80, is replaced by the ligand, CN^-. The main movement involved extends over very few amino acids from 78 to 85 but there are several H-bond rearrangements. This type of movement is typical of proteins in binding reactions where the main frame is not adjusted. However, here the protein is opened to proton motion in a quite new way (see Fig. 13(8) and such proton movements give the possibility of generating controlled proton wires in membranes.

displacement; and, therefore, (3) a protein rearrangement. The changes are similar to the changes with redox state of cytochrome *c* and in haemoglobin. In the cytochrome the reaction, which has quite a high activation energy, involves several hydrogen-bond breaks, as described for the redox state change, and minor adjustments in the whole sector of the protein from residues 40 to 85 in the sequence (see Fig. 13.13), but most atoms move little more than 1.0 Å except, of course, for the haem ligand at 80.

Table 13.5 Haem monoxygenases and dioxygenases

Haem protein	Functions
Monoxygenases	
Cytochrome P-450	Hydroxylation of a large range of inert organic molecules: detoxification
Dioxygenases	
Aromatic ring oxygenase	Bacterial ring cleavage reactions.
Haem oxygenase	Degradation of haem outside cells?
Indole amine dioxygenase (tryptophan)	Attack on viruses? Uses superoxide? Cleavage of indole ring.

13.9 Haem-protein functions III: oxidases and dioxygenases

There are in fact relatively few *simple* haem monoxygenases or dioxygenases and many of these are of bacterial origin (Table 13.5). Where possible, oxidation of organic molecules is carried out by *complex* oxidases starting from hydrogen abstraction so as to keep the energy of the reaction with dioxygen controlled in the cytochrome chain of mitochondria. Of course, this is not a possible route for the oxidation of very resistant molecules such as simple aliphatics, e.g. alkyl chains, terpenes, and aromatics like benzene and naphthalene compounds. Here very aggressive attack is needed—see cytochrome P-450 in Section 13.9.2. These enzymes are associated with a great variety of reactions (Fig. 13.14).

We tackle the different oxidases in the following order:

1. cytochrome oxidase, so as to link immediately to electron transfer proteins;

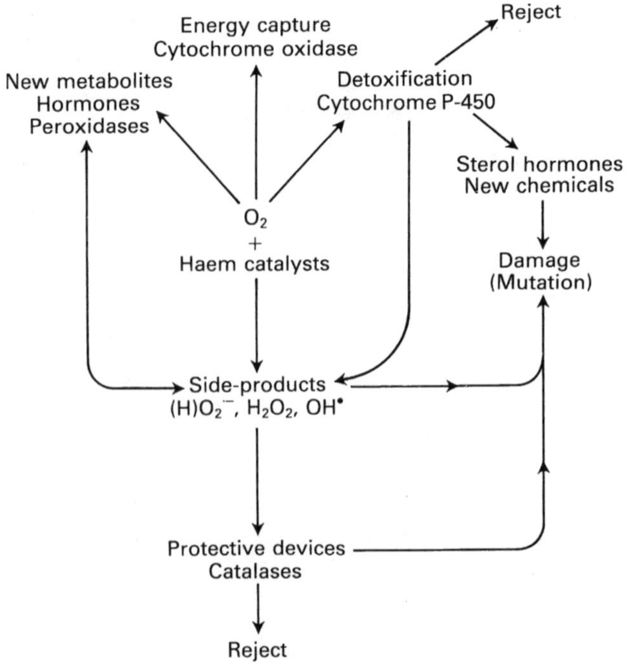

Fig. 13.14 Reactions of dioxygen which are generated by haem catalysis, summarizing the utilization of haem enzymes in cellular chemical reactions.

2. cytochrome P-450, a typecial monoxygenase;
3. peroxidase and catalases;
4. oxidases for organic substrates.

In many cases these reactions are used in the degradation of poisonous organic molecules, but some of the products are extremely valuable as hormones and we find that both oxidases (cytochrome P-450) and peroxidases are employed to produce them (see Fig. 13.14).

13.9.1 Cytochrome oxidase

This is a membrane protein and has in mitochondria seven metal ions which are thought to be essential, including one zinc and one magnesium which control the fold. (There are many similar energy conserving oxidases in different organisms.) The five active metal ions are three copper and two haem iron atoms (Figs 13.15 and 13.16). Two coppers, Cu_A, and one haem *a* are largely involved in electron transfer functions, while one copper, Cu_B, and one haem, haem a_3, carry the O_2 molecules through its four steps of reduction to $2H_2O$. While so doing they neutralize four protons and also pump four additional protons across the membrane in which the enzyme sits (Fig. 13.15). The states of the metal ions have been the subject of extraordinarily intensive study, but, while we may start from a state $Cu_B(I)$ Fe^{2+}: a_3 the interactions with dioxygen are so very intimate that they are hard to follow and there is variation depending on small factors, e.g. pH (see Fig. 13.17). The function of the unit is to capture the energy between the redox potential of O_2 (+0.8 V) and that of cytochrome *c* (+0.25 V) as a proton gradient. This is probably achieved, in parallel with the reactions of cytochrome P-450 (see Section 6.4), by a cyclic series of conformational changes of helices here roughly arranged perpendicular to the membrane. This description, flimsy as it is, is sufficient for the purpose of this chapter since it gives the essence of the value of haem Fe (a_3) and Cu_B with their flexibility inside proteins in contrast to the rather rigid Fe_4S_4 and some copper units The obvious parallels between this description and that of the mechanical movements of dioxygen carrier proteins should not be missed. An hypothesis as to the helical motions that could be involved in

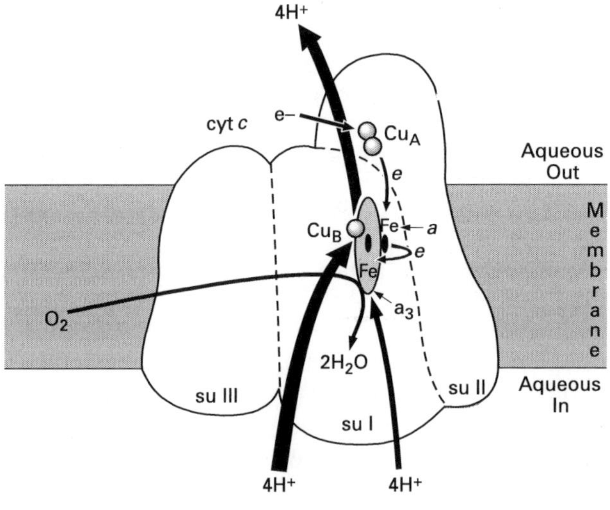

Fig. 13.15 A schematic diagram of cytochrome oxidase. (After A. Iwata *et al.* see references). su = subunit.

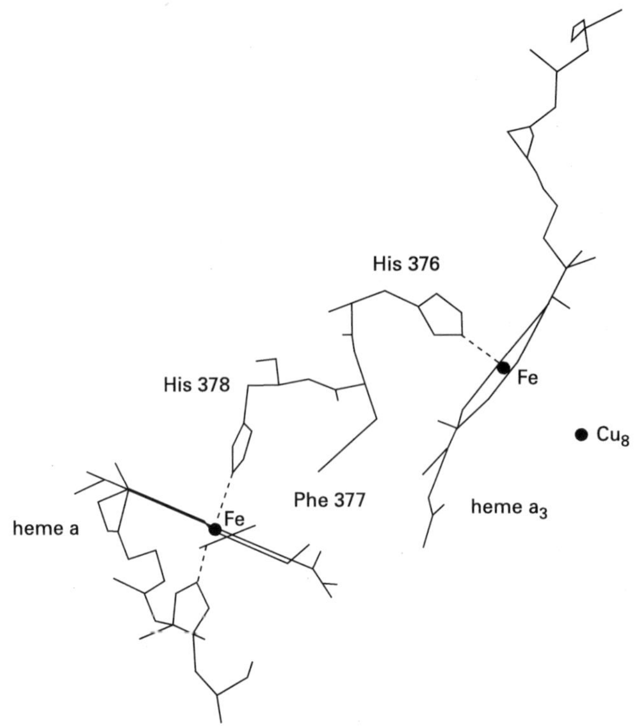

Fig. 13.16 The relationship of haem a and haem a_3 to the Cu of cytochrome oxidase. (After S. Yoshikawa, *et al.*, see reference.)

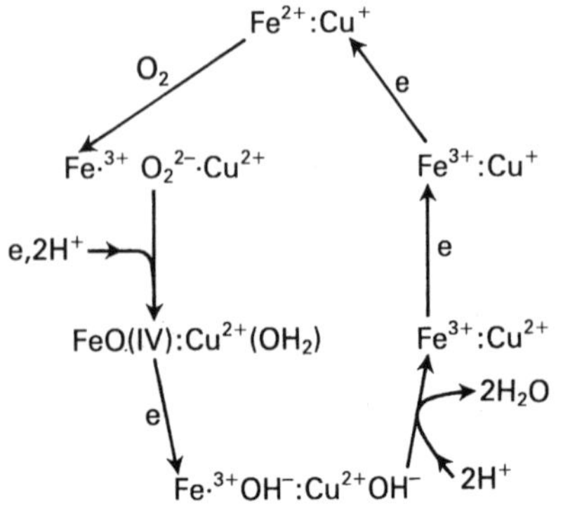

Fig. 13.17 The intermediate steps of the reaction of cytochrome oxidase.

proton pumping and gating has been given in Section 8.5 while describing ATP/proton pumps.

The proton movements in cytochrome oxidase are shown in outline in Fig. 13.15. It is here that the detailed knowledge of features of cytochrome *c* and its redox reactions becomes of value. In Fig. 13.8 we showed the way in which H-bond breaking was linked to the changes in the states of the iron in cytochrome *c*. This can happen either on ligand binding (O_2, CN^-) or on electron transfer. It is likely that in the case of cytochrome oxidase all changes of condition of the iron and Cu_B will be coupled to small protein movements and H-bond breaking.

A particularly interesting model requires movement of one of the histidine ligands of Cu_B using it as a proton carrying shuttle (see 'Further reading'), but it is dangerous to break down proton movements too closely into binding associated with particular reaction steps of electron transfer.

Before leaving the description of cytochrome oxidase it is worthwhile to draw attention to certain parallels in the descriptions of other haem enzymes and a wide variety of enzymes which use mechanical devices. First, in Chapter 9 we described the kinases in terms of hinge-bending using α-helical zones and movements of domains to limit reactions. We shall see that a very similar description applies to the P-450 cytochromes. Second, we analysed the coupling of the ATP hydrolysis to pumping of ions including protons, or the reverse in ATP-synthases. The mechanism was one where energy was applied to sets of α-helices in membranes so that the location of (anionic) groups would be relocated by applied energy gating and/or pumping of ions (Fig. 8.6). The same description has just been used for cytochrome oxidase where gating and pumping of protons is required. (Note that here the proton pumping is secondary to the stoichiometric reaction of protons and dioxygen.) Now in each of these cases there are parallels with the α-helical motions of calcium triggers (Chapter 10) which can also be connected to kinases, and with dioxygen carrier proteins (Chapters 12 and 13). In fact, there are also haem kinases dependent upon O_2-binding and NO-receptors links in G-proteins (GTP hydrolysis).

In the description of non-haem iron enzymes we observed that there were a number of oxidases. Similar oxidases are not so common in eukaryote as opposed to prokaryote cells. It may be the case that in evolution haem is taking over these oxygenase functions. It remains the case that all such activity is largely confined to compartments outside the cytoplasm.

13.9.2 Cytochrome P-450

We have described the character of compulsory-order enzyme reactions in Chapter 4. In P-450 oxidases the sequence of reactions is (1) binding of organic substrate RH; (2) iron reduction to high-spin Fe(II); (3) O_2-binding to give low-spin Fe(II) or as superoxide bound to low-spin Fe(III); (4) reduction of Fe(II).O_2 to FeO; and (5) attack of O on RH to give ROH, Fig. 13.18, probably using radical intermediates. There is a series of conformational switches which generate this pathway (Fig. 13.18). They are allowed by the protein which is very largely helical (see Section 6.4). The P-450 enzymes are concerned with animal hormone metabolism as well as detoxification (Fig. 13.14).

13.9.3 Peroxidases and catalases

The peroxidases are intimately concerned with plant hormone metabolism, e.g. the removal of indole acetic acid, and with the production of the animal hormone thyroxine. In the plants they also help lignification and destruction of lignins in protection and growth. In effect, they provide ways of producing free radical polymerization and depolymerization outside the cell (compare

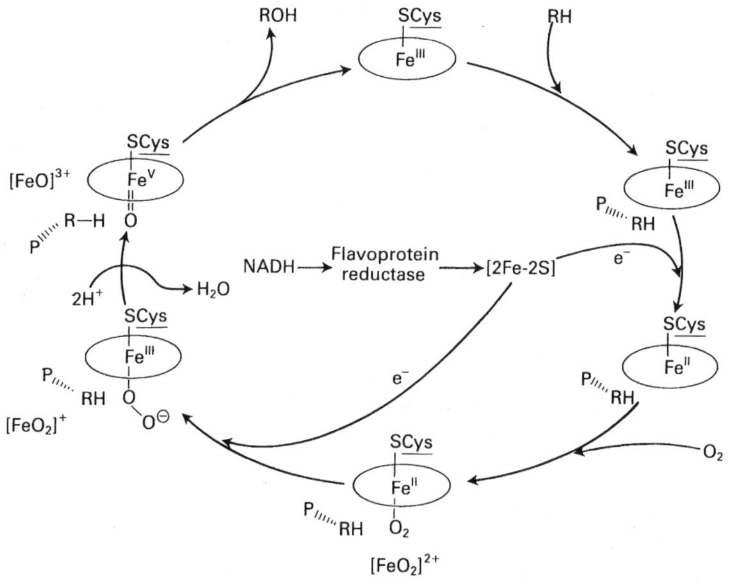

Fig. 13.18 The pathway of the reactions of cytochrome P-450.

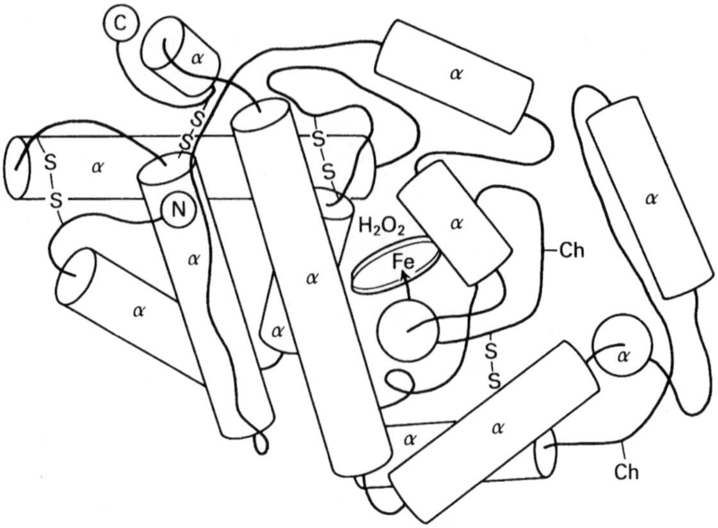

Fig. 13.19 The active site of horseradish peroxidase is buried in the middle of the protein and is accessible only to small molecules such as H_2O_2, O_2, and CN^-. The second substrate binds near the surface. Note the -S-S- bridges and the glycosylation, Ch. (After Y. Morita.)

copper oxidases such as laccases and collagen cross-linking enzymes). The reactions are often of low selectivity and both phenols and indoles are attacked. The reaction involves long-range electron transfer from the substrate to the oxidized iron porphyrin (Fig. 13.19), but there is no suggestion of coupling to protein conformational changes since the proteins are heavily cross-linked. An outline reaction mechanism is given in Fig. 13.20.

The catalases, like the peroxidases, are often in vesicles—in this case the peroxisomes—since many of the oxidative reactions in these enclosed zones generate the dangerous chemical H_2O_2. The structure of catalase is such that the haem is hidden in a deep pocket, presumably to prevent the escape of even more reactive intermediates such as $OH^{\cdot}$ and $O_2^{\cdot-}$ As a consequence, catalases are only effective against small primary peroxides and, of course, they can only act in aqueous solutions. Cells have two other methods of disposing of peroxides—using vitamin E in membranes and a selenium enzyme in the cytoplasm (see Chapter 19).

13.10 **Substrates of haem enzymes and secondary metabolism**

The oxidative chemistry of haem iron and indeed of iron generally is closely linked to secondary metabolism, i.e. not the major synthetic and degrada-

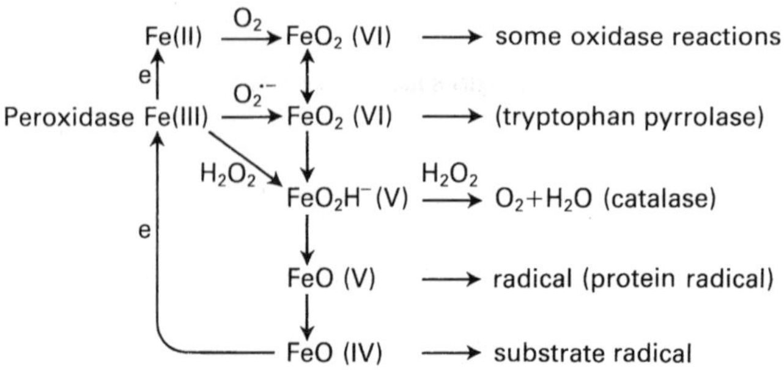

Fig. 13.20 The intermediates in peroxidase reactions.

tive routes of biology but side-products. The products of this metabolism early in their evolution may have been merely useful poisons, e.g. hydroxylated saturated ring compounds and halogen compounds. Some are listed in Table 13.6. Further development would require that protection from these poisons evolved, hence, the same metabolites probably came to be destroyed by mechanisms similar to those that were used to produce them. Much activity of unicellular systems especially in connection with oxidative secondary metabolism appears to be summarized in this way (see Section 12.12).

In multicellular organisms messengers were required. These messengers are hormones and they look very like the poisons of an earlier evolutionary period. In particular, enzymes such as haem oxidases and peroxidases are involved both in syntheses and degradation of hormones. This is a new role for iron quite separate from electron-transfer and the primary metabolism of C/H/O compounds to give energy. What we lack at the moment is a detailed connection from the production and destruction of hormones to their use in the cell machinery. It appears to be the case that the hormones have to be present in an organism in well-controlled steady-state quantities. Their concentrations must therefore be under feed-back control and we shall ask later to what extent this feed-back involves levels of metals such as haem iron.

The most extensive utilization of secondary metabolism based on haem-enzymes appears to have evolved in plants. Series of peroxidase enzymes, sometimes coupled to manganese reactions, e.g. in ligninases, are found in extracellular plant fluids, but there are no corresponding peroxidases in animals where corresponding oxidation requires O_2-using copper enzymes. Again peroxidases are found to be extremely important in plant defense mechanisms as well as in the production of important hormones. Plants evolved considerably before animals and maybe they have used the more available H_2O_2 before O_2 was sufficiently concentrated in the atmosphere. For example, ascorbate oxidase is frequently a haem-enzyme in plants but a copper protein in animals.

Table 13.6 Haem enzymes and hormones

Haem enzyme	Hormones (synthesis or removal)
Cytochrome P-450	Sterols Ecolysome (plants)?
Peroxidase	Thyroxine Indole acetate (plants)
NO synthase	Nitric oxide

13.10.1 Haem and higher oxidation states of nitrogen and sulphur: reductive reactions

The relative functions of haem and Fe/S proteins can be appreciated in the schemes

$$NO_3^- \xrightarrow{\text{Mo}} \begin{matrix} NO \\ NO_2^- \end{matrix} \xrightarrow{\text{haem}} \begin{matrix} N_2O \\ N_2 \end{matrix} \xrightarrow[\text{Fe/S}]{\text{Mo}} NH_3$$

$$SO_4^{2-} \xrightarrow{\text{Mo}} SO_3^{2-} \xrightarrow[\text{Fe/S}]{\text{haem}} H_2S \xrightarrow{\text{Fe/S}} S_n$$

As stated before, Fe/S catalyses are concerned with reactions of low redox potential while Fe/haem enzymes, apart from electron transfer reactions, are more used in high potential reactions. This leaves the differential use of haem and single Fe or Fe.O.Fe centres in high potential redox reactions. The obvious disadvantage of haem is that it has only one coordination position for binding to reagents, usually O_2 or H_2O_2 but in the above schemes NO_2^-, SO_3^{2-} or NO. The second disadvantage is the ease with which haem iron is blocked by CO or CN^-. This reflects in part the value of haem iron in that it readily goes low-spin. The greatest value, however, of the simple Fe centres must be in the use of coupling to ancillary reactants such as α-ketoglurate, using two or three substrate binding positions and only three or four protein donor ligands to iron, and its exchange with free Fe^{2+} at 10^{-7} M giving direct connection through the free metallome to the genome.

13.11 Haem in controls

Haem is turned over quite rapidly in cells. It is synthesized through insertion of iron into protoporphyrin by ferrochelatase in mitochondria, which is one of the enzymes which binds Fe^{2+} in moderately fast exchange. Excess protohaem is destroyed by a haem dioxygenase. There is then a steady state of haem which only in part is related to the level of free iron.

The haem binding to some proteins is made irreversible by cross-linking to the protein, e.g. in cytochromes *c*, *c′*, and *f*. These proteins are possibly all extracellular and are not controlled by free haem. In cells, the haem proteins are based on hydrophobic binding in pockets, e.g. in cytochromes *b*, myoglobin, haemoglobin, and cytochrome P-450. (Presumably cytochrome *b*-type proteins are the most primitive since they are found in all bioenergetic chains with redox potentials around 0.0 volts). The binding constants for the free haem to the apoproteins are estimated to be around 10^{10} M^{-1} giving a free haem concentration in the cytoplasm of 10^{-10} M. Thus, haem in its dissociative reactions is not very different from Mn^{2+}, Fe^{2+}, and most like Zn^{2+}. It is this that allows it to be used in control (see Section 13.11.1).

Free haem controls, in what is presumed to be an allosteric manner, the activity of a number of proteins through haem-response motifs (HRM). The proteins include a kinase, which phosphorylates the translation initiators, EIF2α and NF–E2, and the enzymes, 5-aminolevulinic acid synthase (ALAs), haem lyase and oxygenase-2. Interestingly, bacterial ferritin is a haem-binding protein (the ferritin haem linkage is quite peculiar, being formed in bacteria from two methionine ligands) and it is known that the haem redox reactions are important in iron release from bacterial ferritin. The release of iron from eukaryotic ferritin may also depend on haem binding to the protein. If this is the case, it must dissociate from the protein readily, as it does from *b*-type cytochromes, myoglobin, and haemoglobin, and all these proteins are then linked to a haem pool as well as to iron release from ferritin. The controlled release of haem from ferritin may depend on the iron load of the ferritin since this, a helical-bundle protein, has a stability (and even a redox potential) which is strongly dependent on the load of iron.

13.12 The synthesis of haem and its genes

The control over the concentrations of haem in a cell does not just rest with the level or iron and the synthesis of a protein since the porphyrin itself has to be prepared. This problem exists for the ring systems of corrin (Co), F-430 (Ni), and for chlorin (Mg). We are not concerned here with the niceties of the stereochemical problems in making such a ring, but, with the nature of the precursor, δ-aminolevulinic acid. This molecule comes directly from a combination of glycine and succinate. Notice that succinate and glycine (from acetate) are linked to the citric acid cycle, so that switching off this cycle in mitochondria stops haem synthesis as well as iron scavenging (Chapter 12). As mentioned in Section 13.11, the synthesis of haem from iron and porphyrins occurs in the mitochondria where the citric acid cycle occurs, indicating again that very much of the iron chemistry in all organisms is locked into the redox activity of primitive (prokaryote) cells or related cell compartments. Strangely, vitamin B_{12} is also important in the synthesis of haem, hence B_{12} probably predated haem.

Haem itself, iron-protoporphyrin IX, is synthesized from δ-aminolevulinic acid (Ala) by condensation to give an intermediate porphyrin which is then suitably modified and iron is encapsulated. Regulation of haem synthesis is by inhibitory feedback, Ⓕ, of haem at the transcriptional level on the production of the enzyme glutamyl -t-RNA reductase which yields Ala, or by direct inhibition by haem of this enzyme.

Glutamate ⟶ δ -aminolevulinic acid ⟶ Uroporphyrin III

Uroporphyrin III ⟶ Precorrin ↗ Vitamin B_{12} ; ↘ Haem d_1

Protoporphyrin IX ⟶ Haems *c*,*o*,*a*

DNA	*Haem a* gene	*Haem d* gene	*Haem e,f,n,g* genes	*Haem h* gene
	Glutamine t-RNA Reductase	Uroporphyrin III Synthetase	Protoporphyrin Synthesis	Ferrochelatase

Haem genetics: production and transcription factors

Each box may represent more than one gene, so that there can be a total of nine genes in the chain of synthesis of a fully substituted haem. In *E. coli* the corresponding apoproteins to make cytochromes also are under transcription factors FNR and ARC (see Section 12.18). Since FNR is related to the aerobic/anaerobic switch and itself is dependent upon free Fe^{2+}, the whole oxidative phosphorylation (cytochrome) chain is dependent upon Fe^{2+} availability and dioxygen. In particular, cytochrome oxidase is also dependent on dioxygen for the reactions which produce haem *a* and a_3.

In some organisms the haem genes are closely contiguous, but in others they are far apart. It is interesting that it is the final product, haem, which inhibits the first step.

13.12.1 Genes for haem-containing proteins

The genes which give rise to the proteins for haem synthesis described above are known under various letterings. Genes for cytochrome proteins are known by the initial letter *c* followed by additional letters and numbers to distinguish different genes in a family. Genes expressing simple electron proteins, e.g. cytochrome *c*, are products of the expression of genes labelled *cyc-1* while the proteins are labelled CYC (capital letters, not italics); more complex genes which express enzymes such as cytochrome P-450 are labelled *cyp-1*. Cytochrome oxidase genes are labelled *cyo* followed by the letters *A, B, C,* etc. Genes expressing units of cytochrome oxidase, which are not really cytochromes, are labelled *cox-1*, and the numbering may extend to *cox-13*. Many of these genes are regulated by haem-binding proteins which act as transcription factors and they often have labels which refer to the genes they control. Other genes, which also act as a haem transcriptional control sensors, are of the O_2 (redox state) of a cell environment, opposite nitrogen fixation gene expression, and are called *fix-1* genes (Table 13.7). The haem sensor gene for detecting carbon monoxide and acting as a transcriptional regular for carbon monoxide metabolizing proteins is *cooa-1*, but more difficult to

Table 13.7 Haem-dependent receptors

Protein	Sensed agent	Activated System
FIX-L (ORF-1)*	O_2	Kinase leading to oxidase expression
Guanylate cyclase*	NO (not O_2)	GTP → c-GMP
COOA	CO (not O_2)	CO dehydrogenase Hydrogenase

*Activation depends upon spin-state change.

NB. Guanylate cyclase is the enzyme which responds to NO as a messenger in higher organisms.

recognize is the gene for the haem-containing NO sensor labelled *sgc-1*. Since expression of soluble protein guanylate cyclase (GC) is under its control, this enzyme is in a response system to the NO messenger. Like many gene expressions of extremely important proteins, these genes are connected also to other transcriptional genes here related to non-haem iron containing proteins and their effectors. One example is the *crp/fnr* family of redox control genes where *crp* stands for the genes of the cyclic nucleotide response system and *fnr* for the genre of the fumarate/nitrate reductase, here an Fe_n/S_n protein described in Chapter 12. Another example is the family of *hap-1* genes, after Huntington associated proteins. HAP-1 (a protein) is a transcriptional activator of cytochromes c_1, b_2, of catalase and of certain proteins for the repression of anaerobic genes in yeast. The *hap-1* gene has a zinc-finger binding region coupling haem systems to zinc in a quite different way from the involvement in the enzyme, ALAs. Yet another is the family of *gre-1* genes, standing for glucocorticoid response element, genes, which are under the general control of steroids (Table 13.7). As shown above there is, of course, a whole series of 10–20 genes for haem synthesis and again multiple genes for the transcription factors which control their expression, probably including zinc-finger domains in modern organisms.

The importance of this brief outline of the genes for expression of haem containing proteins and the genes for control of that expression is the appreciation of their internal complexity but also of the linking of haem protein expression to the concentrations of effectors other than free iron (see Chapter 12) such as zinc in GRE transcription (see Chapter 10) and energy status represented among other factors by cyclic nucleotide (c-AMP and c-GMP) receptors. It is the complex networking of biological chemistry which is so apparent in these gene constructs (see Chapter 20). While in each chapter we concentrate on a particular element or element complex (here haem) we note that genetic controls do not work under such restrictions but cross-coordinate activities of different elements or element complexes. We have already seen this in the dependence of $Na^+/K^+/Ca^{2+}$ on Zn-transcription factors (GRE), and between free Fe/cAMP/Zn and phosphate transcription factors, and here we see cross-dependence between free haem and several of these other element controls. The increase in networking of metabolic paths is a reflection of the progressive increase in complexity of living organisms all along the evolutionary process, characterized by additions to the previously extant chemistry. Note that the haem-proteins were added on very early in prokaryote development (see Chapter 20).

13.12.2 Haem and the metallome

While we have insisted in earlier chapters that a full understanding of the chemical system underlying a living organism cannot be limited to a catalogue of its genome (DNA) but requires also a detailed description of its proteome and its metallome, we must now add to the metallome of free elements the concentration of haem in a cell, since it is independently affected by environmental factors such as O_2 pressure. It is this and other small units, especially some coenzymes, in a cell which relate directly to the environment. We have noted also the response of plant chlorophyll production to light and we must not be surprised if there is a wider connection between cofactors and coenzymes to available elements (e.g. C, O, N, H) and energy. In this book, which concentrates on the inorganic elements, we shall tacitly include, for simplicity, all the small organic molecules connected to the environment under the heading of the metallome but we shall hardly refer to them. The free haem in the cytoplasm is settled at a value of around 10^{-10} M even in rather primitive cells and was connected to CO, NO, CN^-, and later to O_2.

13.13 Summary of haem-iron functions

It is worth putting some of the characteristics of haem proteins together, which provide the basis for their functional value.

1. The porphyrin traps the iron, so that there is no equilibrium with free iron.
2. The porphyrin unit, like the chlorin unit, fits together very well with helical proteins. The hydrophobic character of the porphyrin ring means that it does not dissociate easily from its proteins.
3. Helices together with porphyrin unit generate a cooperative coupled unit suitable for mechanical energy transfer. This is expressed in allosteric systems, in preferred order reactions, and in oxidative phosphorylation couplings.
4. The porphyrin unit assists long-range electron-transfer by developing low-spin states of iron so that relaxation energies are small (Chapter 4).
5. The iron in the porphyrin complex can be left with one open position for small substrates. The low-spin Fe(II) then acts as a π-donor. Links to haem are not just for enzyme reactions but for carrier (O_2) and transmitter (NO) functions.
6. The iron and the porphyrin readily yield higher oxidation states of very high redox potential (Section 13.9). The fruit of this activity is a very large part of secondary metabolism involving dioxygen and protection, as well as energy capture.
7. Haem acts as a feed-forward control on many non-haem non-iron enzymes linking haem levels to various metabolic paths.
8. Haem binds to transcription factors regulating, by feed-back, haem synthesis and also the synthesis of many proteins involved in (3) to (6) above.

There is clearly a great advantage in the use of haem over that of simple iron or iron in Fe/S or Fe_2O centres in that it can be held so as to minimalize adventitious reactions especially with reagents on the oxidizing side of 0.0 volts. This removal of iron deep into proteins also allowed its very selective use at high redox potential (see Section 13.9). Again cross-linking allowed haem to be used outside cells. We note that both cobalt and nickel are also removed to a large degree into large chelates when they are used in the cytoplasm, while manganese is removed into vesicles and copper is exported. The chelating agents are σ-donor, π-acceptor anions in all three cases (Fe, Co, Ni) and bring oxidation states and spin states of these oxidation states close together which is very useful for reactivity.

Further reading

Allen, J.W.A., Watmough, N.J., and Ferguson, S.J. (2000). A switch in heme ligation in cytochrome cd_1. *Nature Structural Biology*, **7**, 885–8.

Aono, S., and Nakajima, H. (1999). Structure and function of CooA. *Coordination Chemistry Reviews*, **190**, 267–82.

Babcock, G. T., and Wikström, M. (1992). Structure and function of cytochrome oxidase. *Nature*, **356**, 301–8.

Banerjee, R., Alpert, Y., Leterrier, F. and Williams, R.J.P. (1969). Spectra of the isolated chains of human hemoglobin. Discussion of chain mediated heme-heme interaction. *Biochemistry*, **8**, 2862–8.

Beinert, H. (1995). Crystals and structures of cytochrome *c* oxidases—the end of an arduous road. *Chem. Biol.* **2**, 781–5.

Ciba Foundation Symposium 180: (1994). *The biosynthesis of the tetrapyrrole pigments*. John Wiley.

Cooper, C.E. (1999). Nitric oxide and iron proteins. *Biochimica Biophysica Acta*, **1411**, 290–309.

Coutinho, I.B., Turner, D.L., LeGall, J., and Xavier, A.V. (1992). The haem core of tetrahaem cytochrome c_3. *European Journal of Biochemistry*, **290**, 329–33.

Crofts, A.R., Hong, S., Zhang, Z., and Berry, E.A. (1999) Physicochemical aspects of the movement of the iron/sulphur protein in the be_1 complex. *Biochemistry*, **38**, 15827–39.

Einsle, O., Messerschmidt, A., Stach, P., Bourenkov, G.P., Bartunik, H.D., Huber, R., and Kroneck, P. M. (1999). The structure of nitrite reductase. *Nature*, **400**, 476–9.

Gruez, A., Pignol, D., Zeghouf, M., Coves, J., Fontecave, M., Ferrer, J.L., and Fontecilla-Camps, J. C. (2000). Four crystal structures of sulphite reductase. *Journal of Molecular Biology*, **299**, 199–212.

Hasemann, C.A., Kurumbail, R.G., Boddupalli, S.S., Peterson, J.A., and Deifenhofer, J. (1995). Structure and function of cytochromes P450: a cooperative analysis of three crystal structures. *Structure*, **3**, 41–62.

Isaac, I.F., and Dawson, J.H. (1999). Haem-containing peroxidases. *Essays in Biochemistry*, **34**, 51–69.

Iwata, S., Ostermeier, C., Ludwig, B., and Michel, H. (1995). Structure at 2.8 Å resolution of cytochrome c oxidase from *P. denitrificans Nature*, **376**, 660–9.

Jackson, J.D. and Richardson, D. (eds) (1998). The diversity of bacterial redox proteins. *Biochemical Society Transactions*, **26**, 385–427.

Michel, H., Behr, J., Harrenga, A., and Kanut, A. (1998). Cytochrome c oxidase structure and spectroscopy. *Annual Review of Biophysics and Biomolecular Structure*, **27**, 329–56.

Moore, G.R., and Pettigrew, G.W. (1987). *Cytochrome c: biological aspects*. Springer-Verlag, Berlin.

Moore, G.R., and Pettigrew, G.W. (1990). *Cytochrome c: structural and physico-chemical aspects*. Springer-Verlag, Berlin.

Morimoto, A., Tanaka, M., Takahashi, K., Ishimori, K., Hori, H., and Morishima, I. (1998). Reaction mechanisms of horse peroxidases. *Journal of Biological Chemistry*, **273**, 14753–14760.

Musser, S.M., and Chan, S.I. (1998). Evolution of the cytochrome c oxidase proton pump. *Journal of Evolution*, **46**, 508–20.

Omura, J. (1999). Forty years of cytochrome P450. *Biochemical and Biophysical Research Communications*, **266**, 690–8.

Ponka, P. (1999). Cell biology of heme. *American Journal of Medical Science*, **318**, 241–56.

Poulos, T.L. (1985). Heme enzyme crystal structures. In *Advances in inorganic biochemistry*, Vol. 7, 1–36. (eds G.L. Eichorn, and L.G. Marzilli). Elsevier, Amsterdam.

Putnam, C.D., Arvai, A.S., Bourne, Y., and Tainer, J.A. (2000). Active and inhibited human catalase structures: ligand and NADPH binding and catalytic mechanism. *Journal of Molecular Biology*, **296**, 295–309.

Rodgers, K.R., Lukat-Rodgers, G.S., and Barron, J.A. (1996). Structural basis for ligand discrimination and response initiation in the heme-based oxygen sensor FixL. *Biochemistry*, **35**, 9530–48.

Ruckpaul, K., and Rein, H. (eds) (1990). *Frontiers in biotransformations*: cytochrome P-450. Vol. 1–7. Taylor & Francs, London.

Schnuller, D.J., Wilks, A., Ortiz de Montellano, P.R., and Poulos, T.L. (1999). Crystal structure of human heme oxygenase-1. *Nature Structural Biology*, **6**, 860–7.

Snyder, S.H., Jaffrey, S.R., and Zakhary, R. (1998). Nitric oxides and carbon monoxide: parallel roles as neural messengers. *Brain Research Reviews*, **26**, 167–75.

Sono, M., Roach, M.P., Coulter, E.D., and Dawson, J.H. (1996). Heme-containing oxygenases. *Chemical Reviews*, **96**, 2841–88.

Wikström, M. (2000). Mechanisms of proton translocation by cytochrome *c* oxidase. *Biochimica Biophysica Acta*, **1458**, 188–98.

Williams, R. J. P. (1999). Electron-transfer and proton-coupling in proteins. *Journal of Solid State Chemistry*, **145**, 488–95.

Xia, D., Yu, C.-A., Kim, H., Xia, J.-Z., Kachurin, A.M., Zhang, L., Yu, L., and Deisenhafer, J. (1997). The structure of the bc_1 complex, *Science*, **277**, 60–6.

Yoshikawa, S., Shinzawa-Itoh, K., and Tsukihara, T. (1998). Crystal structure of bovine heart cytochrome c oxidase at 28 Å resolution. *Journal of Bioenergetics and Biomembranes*, **30**, 7–14.

Zaslevski, D., and Gennis, R.B. (2000). Proton pumping by cytochrome c oxidase: progress, problems and postulates. *Biochemical and Biophysical Acta*, **1458**, 164–79.

CHAPTER 14

Manganese: dioxygen evolution and glycosylation

14.1 Introduction

The importance of manganese in living systems, though known to be very considerable, especially in plants, remains poorly explored in many areas. The manganese biochemistry which presently attracts greatest interest is that involved in O_2 release by photosynthesis system II. There are two other well-known enzymes which contain manganese and also evolve O_2, the superoxide dismutase of prokaryotes and of the organelles of eukaryotes and the catalase of lactobacilli. Besides these cases, manganese biochemistry largely results from the association of the metal with a few enzymes and proteins, some of which are listed in Table 14.1. Looking at this list there would appear to be no general rhyme or reason for the way in which manganese is used, despite the fact that it has a particularly unique chemistry. In this chapter we summarize this chemistry before we turn to an attempted rationalization of the uses of the element in comparison with the uses of other transition metal elements of somewhat similar chemical potential. The first functional values of manganese is as a Lewis acid (compare magnesium, calcium, and zinc, (M^{2+}), and the second, in higher

V	Cr	Mn	Fe	Co	Ni	Cu

The first transition metal series

oxidation states, is as an oxidation catalyst (compare iron and copper). It will be shown that compartmental separation of manganese is very closely connected to its particular functional value in all cases. It may be that manganese is in fact largely removed from the cytoplasm of aerobes, as is copper, to protect the DNA from oxidative attack.

14.2 Manganese chemistry

14.2.1 Oxidation states

The most obvious features of the chemistry of manganese in simple aqueous solution are the stability of Mn^{2+} in acid solution and the greater stability of MnO_2 in alkaline solution and in the presence of oxygen. This chemistry is brought out in an oxidation state diagram (Fig. 1.7) which shows that oxidation states higher than Mn(IV), i.e. than MnO_2, are extremely powerful oxidizing agents in all conditions and are therefore unlikely to be of much

Table 14.1 Manganese enzymes and proteins

Protein	Valence	Comments on binding site	Function
Concanavalin A	II	Glu, Asp, Asp, His, H_2O, H_2O	Saccharide binding (extracellular)
Galactosyl transferase	II	Mn specific *in vivo*	Transfer of galactosyl group (Golgi)
Arginase	II	4 Mn atoms per enzyme molecule (Mn dimers)	Hydrolyses, arginine to ornithine and urea (cytoplasm.)
Glycosylaminase	II	S–H group near active site	Hydrolyses, guanidino- acetate to glycine and urea (mitochondria, cytoplasm)
Phosphoenolpyruvate carboxykinase	II	Binds through O. (Fe completes, not Mg)	C–C carboxylase (mitochondria)
Pyruvate carboxylase	II	4 Mn atoms per enzyme molecule	C–C carboxylating synthetase (mitochondria)
Proline dipeptidase	II	O (and S) ligands proposed	Hydrolyses C-terminal proline dipeptides (ubiquitous)
Isopropylmalate synthase	II	S–H group near site	C–C oxo-acid lyase (mitochondria)
Mn-ribonucleotide reductase	II	Probably N, O ligands (see iron enzyme)	Reductive elimination of 2-hydroxyl from ribonucleotide (prokaryote)
Mn-dioxygenase	II	Low-symmetry Mn(II) centre	Catalyses extradiol cleavage of catechols (prokaryote)
Acid phosphatase	III	His residues essential	Hydrolyses orthophosphoric monoester (prokaryote)
Mn-catalase	III	–	Redox enzyme; acts on H_2O_2 (prokaryote)
Mn-peroxidase (ligninase)	II/III	–	Redox enzyme as above, degrades lignin (extracellular)
Mn-superoxide dismutase	III	His, His, Asp, His, H_2O	Redox enzyme; acts on superoxide radical (mitochondria, chloroplasts, prokaryotes)
MnP_2O_7	III	O-ligands	Acts in place of SOD in lactobacillus
Photosystem II	II/IV	Assumed typical N/O donor site	Dioxygen release in plants (chloroplasts)
Imidazole glycerol phosphate dehydratase	II?		Histidine synthesis
Glutamate synthetase	II	N/O donor site	Glutamine synthesis (mitochondria)
Isocitrate dehydrogenase	II?		Citric acid cycle (mitochondria)
Oxalate oxidase	II	N/O site	Oxalate + O_2 gives CO_2 H_2O_2. uptake of Mn
NRAMP-2	II	N/O site	Uptake of Mn

concern in biological systems. We shall deal here only with the chemistry of Mn(II), and Mn(III), and Mn(IV).

In detail, the oxidation state diagram (Fig. 14.1) shows how different is the chemistry of manganese from that of, for example, iron and molybdenum. It also differs greatly from that of the other major element among oxidation catalysts, i.e. copper. In each case there are three obvious oxidation states to discuss: II, III, and IV for Fe and Mn; I, II, and III for Cu; and IV, V, and VI for Mo. In the cases of Fe and Cu it is the middle oxidation state, Fe(III) and Cu(II), respectively, which is the most stable under mild oxidizing conditions and at neutral pH, while in the case of manganeses the middle oxidation state, Mn (III), readily disproportionates to Mn(II) and Mn(IV) see Fig 14.1; molybdenum differs again in that all its three states are of comparable stability. It is these differences in redox potential which evolution has used to the utmost advantage, given the additional features of the different chemistries of the

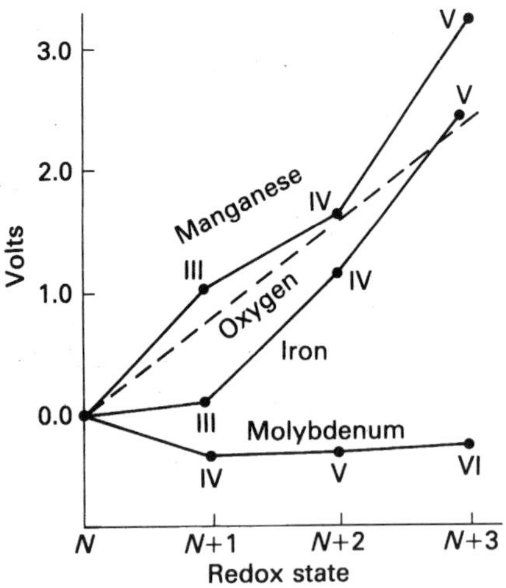

Fig. 14.1 An oxidation state diagram for Mo, Fe, and Mn, where *N* is oxidation state II for Mn and Fe and III for Mo, versus the normal hydrogen potential (= −0.42 V) at pH = 7. The slope of the lines joining two points gives their redox potential. Note the instability of Mn(III) and Mn(V).

most common oxidation states, Mo(VI), Mn(II), Fe(II) and Cu(I), in solution at pH = 7 and at a redox potential of about 0.0 V. It is clear to all chemists that Cu(I) is useful in similar ways to Fe(II), e.g. both bind O_2 and CO, but both Mn and Mo in all their oxidation states are very different indeed and act as O-atom donors. We turn next to a description of the chemistry of the individual Mn oxidation states before looking at their biochemistries.

14.2.2. Structures of Mn(II) complexes with organic ligands

When we describe the chemistry of a divalent ion, the foremost observation is the ionic radius. The radius of Mn(II), 0.75 Å, lies between that of Mg(II) and Ca(II). The half-filled, $3d^5$, shell of the ion makes it behave as a spherically polarizable ion, with no crystal-field stabilization energy; hence we expect that in the structure of its complexes it will behave in a manner between those of Mg and Ca. This is generally true; Mn(II) complexes are usually six-coordinate and octahedral, but there are some examples of irregular seven-coordinate complexes, e.g. with ethylenediaminetetracetate, EDTA, while Mg(II) complexes are almost invariably six-coordinate and octahedral and Ca(II) complexes are usually of irregular coordination with seven or more ligands. The ease of transition between the six- and seven-coordinate states presumably accounts for the lability of Mn(II) complexes compared with those of Mg(II) and in this respect Mn(II) resembles Ca(II). We must not be surprised, then, when we find that in some biochemical systems Mn(II) can substitute for Ca(II), while in others it can substitute for Mg(II) (isomorphous replacement). This substitution makes Mn(II) a useful probe of the activities of the other two ions since it is easy to follow the Mn(II) ($3d^5$) ion using either its EPR signal or its enhancement effect on the relaxation times of the NMR signals of nuclei in molecules in its vicinity. Mn(II) can therefore be used to map biological space and has been much used in this way in many substitution studies of enzymes, e.g. metallo-enzymes of Zn(II), Mg(II), and Ca(II), and in NMR imaging.

14.2.3. Manganese(II) equilibria and biochemistry

The main difference between Mn(II) and Mg(II) chemistry lies in the somewhat higher affinity of the manganese ion for nitrogen or just possibly sulphur ligands (Fig. 2.8). If we consider first the affinity for simple purely oxygen donors, then Mn(II) parallels Mg(II) with but small differences, e.g. with carboxylate or phosphate donors, and this is also true of the solubility of its salts with oxyanions. As is well known, both of these cations have a somewhat higher affinity for small simple anions than has Ca, but Ca is noticeably more stable in association with a ligand carrying a larger number of donor groups (compare EGTA with EDTA in Fig. 14.2). The difference arises through the larger size of Ca(II) which also permits it to form relatively stronger lattices and so gives more insoluble salts. The following related facts can now be put together.

1. The concentration of Mg(II) in biological systems is usually about 10^{-3} M in all compartments with the possible exception of some vesicles; that of Ca(II) is 10^{-3} M in extracellular fluids but 10^{-7} M inside cells, while the Mn(II) concentration is close to 10^{-7} M almost everywhere inside or outside cells. (Especially in eukaryotic cells cytoplasm Mn(II) concentration may be even lower than 10^{-8} M, but is much higher than this in vesicles.)
2. Of the three cations Ca(II) forms the most insoluble salts.

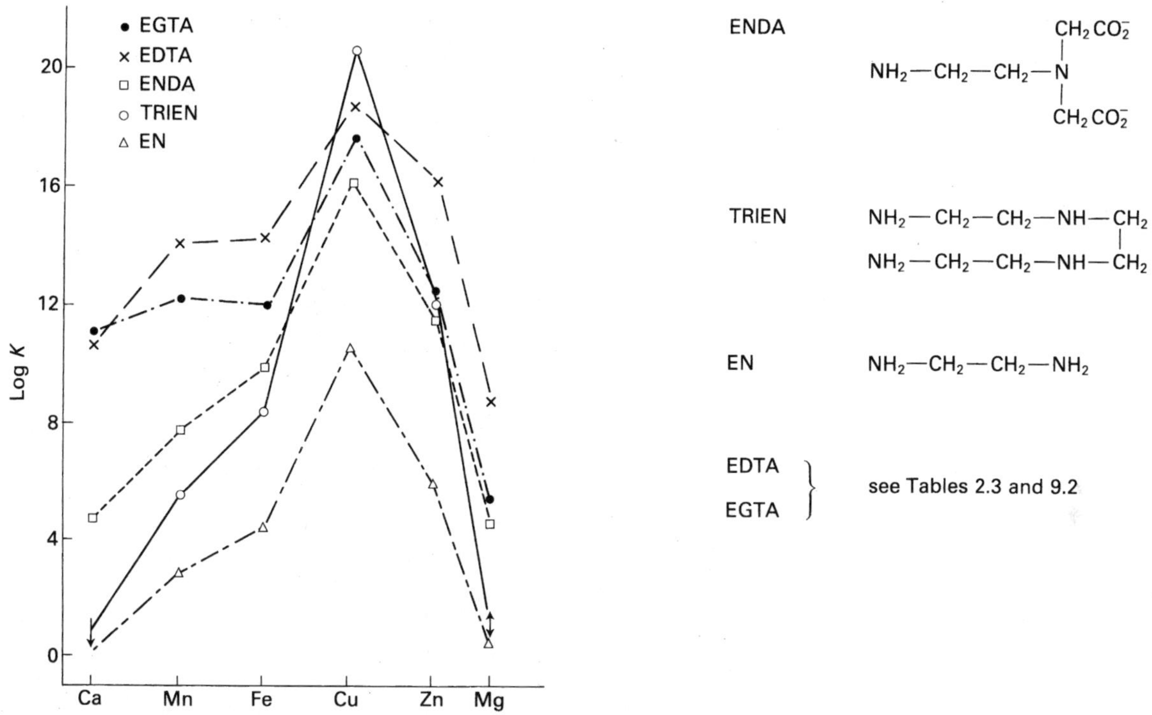

Fig. 14.2 The stability constants, log *K*, of some complexes of the elements in their divalent M(II) states compared with Mn(II). The symbols refer to ethylenediamine, EN; triethylenetetramine, TRIEN; ethylenediamine diacetate, ENDA; ethylenediamine tetraacetate, EDTA; and ethyleneglycol-diamine tetracetate, EGTA. Note how the increase in the number of O-donor groups stabilizes Mn(II) and Ca(II) relative to other elements, even relative to Fe(II).

3. Mg(II) and Mn(II) have roughly equal binding strengths to simple ligands based on oxygen donors, e.g. polyphosphates, and Ca binds of only somewhat more weakly.
4. Calcium alone binds relatively more strongly to larger numbers of donor centres, especially those of lower charge, e.g. in EGTA and in calmodulins.

It is then easy to understand the following biological observations concerning purely the O-donor centre chemistry of the three cations.

1. The insoluble salts of biology are largely Ca salts—in animal bones and shells and in plants, especially calcium oxalates; Mn is found only in a few solid systems, mostly oxalates or in excretion nodules (see Section 14.8).
2. Small-molecule complex ions are found in the cytoplasm of cells as Mg(II) complexes, e.g. Mg(II) ATP, i.e. a phosphate complex.
3. Calcium can be and is used as the major signalling trigger ion of biology; this value arises from its large concentration gradient across the cell membranes.

It is seen that manganese is generally excluded from simple oxygen-donor chemistry; its low availability relative to magnesium and calcium (see Section 14.4) makes it less valuable in biological chemistry than either. The role of manganese, like that of other trace elements, is then in special situations, with specially designed protein ligands which usually have one or more nitrogen donor groups.

Table 14.2 Radii and ionization potentials (*IP*)* of some elements including manganese

	Ca	Mg	Mn	Fe	Co	Ni	Cu	Zn
Radius M(II) (Å)	0.99	0.65	0.78	0.72	0.69	0.65	0.65	0.65
IP M→M(II) (eV)	11.87	15.04	15.64	16.18	17.06	18.17	20.29	17.96

*The ionization potential, $M \rightarrow M^{2+}$, is the same as the electron affinity of the reverse reaction $M^{2+} \rightarrow M$ and is a good measure of the acceptor power of the ion, M^{2+}. Similar tables can be given for M(III). etc. The order of 'softness' to *IP* (see Table 2.8).

Now, when we turn to those more favourable circumstances for Mn(II) as opposed to Mg(II) and Ca(II), i.e. binding to nitrogen donors, there is a different competition to be faced by the Mn(II) ion, which is that from the other trace elements, especially Zn(II) and Cu(II) (see Fig. 2.8). Simple inspection shows that these two elements dominate the competition for the binding to N/S donors through the absence of Ni(II) and Co(II), and due to the very low external availability of Fe(II) which is easily oxidized to Fe(III). The strength of the competition faced by Mn(II) from Zn(II) and Cu(II) can be seen in the slopes of the Irving–Williams series of stability constants (Figs 2.8 and 14.2). Clearly, if there is a limited supply of centres based only on nitrogen/sulphur donors, they will be overwhelmingly occupied by Zn or Cu; this expectation is borne out in a number of proteins. The divalent Mn(II) is left to be sequestered by some donor system of intermediate strength between that of all O-donors (Mg, Ca) and all N/S donors (Zn, Cu). Even here we must look for situations peculiarly suited to Mn(II) and its somewhat larger size since Zn(II), especially, is relatively available. As already mentioned, Mn(II) has the peculiarity that it is not only an ion of intermediate electron acceptor strength but it also has an intermediate size between Ca(II) and Mg(II) while Zn(II) and Cu(II) are of the same radius as Mg(II) (Table 14.2). The study of complex ion formation constants has shown that framework ligands can be designed very selectively for ions of a given size so that a multidonor ligand made up from O-donors and N-donors, giving a total of six or at most seven donors and such that is not able to collapse around very small ions, is then ideal for Mn(II) once Zn(II) and Cu(II) free ion concentrations have been suitably reduced by N/S donors. Illustration of the small difference in stability constants between Mn(II) and Zn(II) complexes with ligands of this N/O type and which occupy considerable space around a cations is given in Fig. 14.2 (EGTA). We must not be surprised then at that coordination chemistry of Mn(II) as seen in cells. Figure 14.3 gives an example. It may well be that the few Mn(II) enzymes in the cytoplasm are also favourably selected by size factors even relative to Fe(II), see Fig. 14.2.

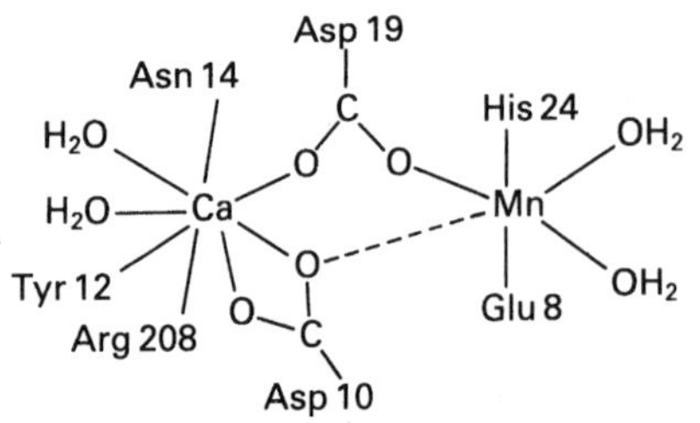

Fig. 14.3 The formula of the binding site of concanavalin A showing the difference between a Ca site with three carbonyl ligands and a Mn site with one nitrogen ligand, apart from carboxylates.

14.2.4 The kinetics of Mn(II) complexes

As stated in Chapter 4, Mn(II) exchanges water and ligands rapidly; this has some advantages in controls since, given the relatively weak binding constants of the metal in proteins, it is readily separated from a protein partner, but it also means that it is lost from proteins during many extraction procedures. Generally, Mn(II) does not form metalloproteins but exchangeable metal complexes with proteins, much as Mg(II) and Ca(II) do, in marked contrast with Cu(II) and, to a lesser degree, with Zn(II). Here Fe(II), which can also exchange quickly and forms relatively weak complexes, is similar to Mn(II). The lability of many Mg(II) enzymes and the fact that Mn(II) can replace Mg(II) at the active site has led to the unlikely suggestion that some Mg(II) enzymes are, in fact, Mn(II) enzymes. (In the cytoplasm of cells [Mg] exceeds that of [Mn] by at least a factor of 10^4, so here Mg(II) is presumably the active cation, but in the Golgi the situation may well be rather different.)

Before passing on, the reader should note that we have so far looked at Mn(II) as a competitor ion with Mg(II), Ca(II), Ni(II), Cu(II), Zn(II), and Fe(II), for ligands. Biological systems, are of great compartmental complexity, so that isolation of particular biological spaces behind membranes can be used to separate the functions of one ion from others (see Chapter 3). We have already mentioned functional separation of Ca, which is pumped out of all cells allowing its use as a trigger. Consequently we shall see that, while the above logic is well-founded on thermodynamic principles, kinetic control (pumping) can affect some of the conclusions. The evidence available suggests that Mn(II) is pumped into vesicles where its competitive strength is enhanced, e.g. in the Golgi apparatus.

We turn to the chemistry of manganese in higher oxidation states.

14.3 Monomeric Mn(III) and Mn(IV) chemistry

14.3.1 Structures

The higher oxidation states of manganese

The Mn(III) ion ($3d^4$) forms complexes with a tetragonal symmetry having four short and two longer bonds in the general case and in this respect it resembles Cu(II). It is the only trivalent ion of biological consequence which falls in this structural class and therefore it has the potential to be introduced into selected sites in proteins; in this respect it differs from Fe(III) which, being a $3d^5$ cation, enters octahedral or tetrahedral sites readily [see also Mn(II)]. These trivalent ions are small, of course, with radii of about 0.6 Å, and the exchange of simple ligands, e.g. H_2O, is relatively fast as for most Jahn–Teller ions, e.g. Cu(II).

The Mn(IV) ion ($3d^3$) resembles the Cr^{3+} ion in its electron configuration and hence it should seek octahedral coordination; however, the very strong polarizing power of M^{4+} ions leads to close association in water with one oxygen dianion, giving the oxocation MnO^{2+} whose preferred stereochemistry is that of a very distorted octahedron. The inorganic chemist will note the parallels with VO^{2+} (Chapter 17).

14.3.2 Thermodynamics

The high charge on the M(III) ions leads them to bind strongly to anions, but their small size leads them to prefer the smallest of anions such as RO^- and F^- rather than RS^-; this is true of Mn(III) and Fe(III). In addition, RS^- is open to oxidation and Mn(III) is a very powerful oxidizing agent. In this chapter we shall assume that high-valent Mn chemistry does not involve sulphur coordination; there are as yet no known exceptions in organisms. There is, therefore, some difference from Fe(III) chemistry, which is probably due to the oxidizing power of Mn(III), but we shall stress the overall similarities between these two elements at first. A problem with M(III)ions at pH=7 is hydrolysis, which grossly limits the free ion concentrations and, especially in the case of biological structures, it is often unclear if a bound 'water' molecule is in fact an hydroxide anion, or even if the metal ion forms a $[M–O–M]^{4+}$ complex using two M(III) ions (see Chapter 2).

Mn(IV) shows this hydrolysis to a much more marked degree and it exists generally as MnO^{2+}. The chemistry of this oxidation state in water is not well known, but we can use the chemistry of VO^{2+} and UO_2^{2+} as a guide. The available data show that MO^{2+} complexes behave in the stability of their coordination compounds very much like M^{3+} complexes, i.e. while one coordination position is totally screened by the oxo group the others act as if the residual electrostatic positive charge on the cation were three not two. It has then a strong affinity for small anions, e.g. F^- and RO^-; it too readily forms bridged hydroxy/oxo complexes and once again the free ion concentration is limited.

The relative stability of the Mn(III) complexes when compared with Mn(II) and Mn(IV) complexes is the factor which controls the redox potentials relative to the *aqueous* couple values (Fig. 14.1). It is generally true that the Mn(III) ions form much more stable complexes than the Mn(II) ions with almost every kind of ligand. On the other hand, since little difference is expected between the stability constants of the complexes of MnO^{2+} (Mn(IV) and Mn(III), since MnO^{2+} behaves like an M^{3+} cation, there will remain a tendency for Mn(III) to disproportionate to Mn(II) and Mn(IV) species in multinuclear assemblies, no matter what the coordination centre. Here Mn is very different from Fe where chemistry drives Fe(II) plus Fe(IV) towards 2Fe(III) even in complex ions, i.e. in the opposite direction; in a similar manner, Cu(III) and Cu(I) give the stable intermediate state Cu(II). We can only expect to see stable Mn(III) compounds in organisms or elsewhere in isolated mononuclear units (see superoxide dismutase (Section 14.7) where it is bound in a rather unexpected manner. The use of manganese to make dioxygen and of iron and copper to reduce it becomes clearer.

14.3.3 Kinetics of Mn(III) and Mn(IV) reactions

The kinetic aspects of primary interest in the Mn(III) and Mn(IV) states are:

1. The increased stability of complexes relative to Mn(II) leads to slower dissociation and there is not much loss of ions of this kind from metalloenzyme sites. Mn(III) ions, therefore, separate with their proteins.
2. The relatively fast loss or exchange of OH or O from these species is of primary concern since these reactions can be one-electron ($OH^{\cdot}$) or two-electron transfers (O); these reactions, which concern, O_2, H_2O_2, and O_2^- and their oxidation/chemistry, have been described in detail in Sections 13.9 and 13.10.

The energetics of the oxidation states is very important in controlling the reaction cycles of manganese. We observe too that Mn(III) readily generates organic free radicals.

14.3.4 Manganese cluster chemistry

In oxidation states III or IV, Mn, like Fe(III) and Cr(III), readily forms clusters related to simple hydroxide or oxide chemistry. The three elements can also form clusters in mixed oxidation states. The exact formulation of many such complexes is now known from X-ray diffraction studies; it is found that oxo groups can bridge two or three Mn atoms in many types of cluster (Fig. 14.4), including cubane and adamantane forms. Interest in the Mn clusters, focuses, of course, on the O_2-generating proteins of chloroplasts. However, in these biological clusters the special geometry enforced by the protein may well lead to peculiarities, as is well known in some Fe_4S_4, Fe_3S_4, and Fe_2S_2 clusters in proteins. (Note that some Fe_nO_n clusters are also known in cells, e.g. in ferritin and haemerythrin-like proteins.)

Throughout this account of Mn chemistry there is always the running thought of the parallel chemistry of Fe. How can organisms select for Mn in the higher as well as the lower oxidation states and then in clusters? It would

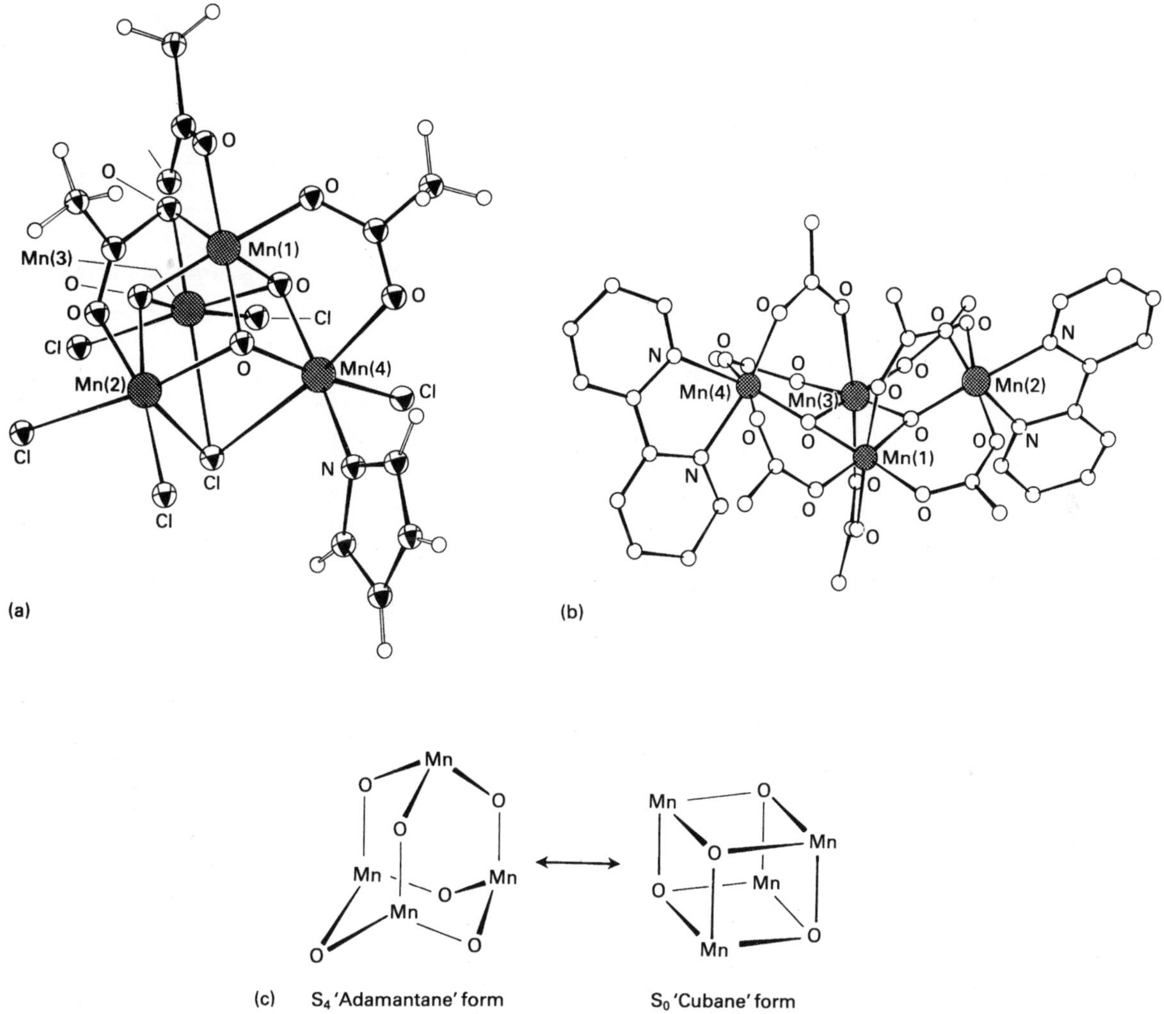

Fig. 14.4 Two complexes made by Christou and co-workers (see 'Further reading') to model the manganese mixed-valence cluster used in dioxygen formation. (a) $[Mn_4O_3Cl_6\ (acetate)_3\ (imidazole)]^{2-}$ and (b) $[Mn_4O_2\ (acetate)_7\ (bipyridyl)_2]^+$. (c) Some other proposed structures of the manganese cluster involved in dioxygen release in biology (see Fig. 14.6).

appear that the answer must be through the preparation of *all* the Mn complexes through the Mn(II) state since Mn(II) and Fe(II) are far from equally available and not so similar as Mn(III) and Fe(III). In fact, it is suspected that Mn(III) and Fe(III) are so similar in binding and kinetics that some biological systems hardly distinguish between them, e.g. in superoxide dismustases, in acid phosphatases, in ribonucleotide reductases, and in the uptake and transfer of both metal ions across membranes using transferrin and its receptors. It is an important peculiarity of the cluster biochemistry of Fe and Mn that this confusion does not arise. There can be little doubt that it is the disproportionation of Mn(III) that removes it from such clusters to give Mn(II).Mn(IV) and so allows the build-up of oxo-species MnO(IV).

The structures in Fig. 14.4 and those of the Fe_nS_n clusters in Fig. 12.6 illustrate some similarity but we must remember there is a difference of one volt in redox potential and of $>2 \times 10^9$ years in their evolution in biology.

14.4 The biological chemistry of manganese

From the preceding considerations the biochemistry of Mn ought to be dominated by: (1) the chemistry of Mn(II), which should be intermediate between that of Mg(II) and Ca(II) and Zn(II); and (2) the powerful oxidizing action of the stable higher oxidation state, Mn(IV). This expectation is only partially fulfilled and we shall have to try to understand in addition why considerable Mn(III) chemistry is observed in cells.

14.4.1 The occurrence and availability to organisms of manganese

Manganese is not a rare element and is about as common as Fe or Zn (Fig. 1.5). At pH = 7 many of its salts are quite soluble so that at equilibrium the free Mn(II) concentration is relatively high in biological systems (10^{-7} M), even in the presence of oxygen and manganese oxide precipitates. For the same reason, in the very early periods of life, Mn(II) must have always been available since manganese sulphides are the most soluble of all transition metal sulphides from Mn(II) to Zn(II) (Fig. 1.12). Availability of Mn(II) at about 10^{-7} M is not a problem. The presence of Mn^{2+} in the cytoplasm of unicellular *anaerobes* does not present a hazard for DNA, but in aerobes it is reduced in the cytoplasm.

14.4.2 The uptake of manganese

Assuming a reasonable concentration of this trace element as Mn(II) in the sea or in soil solution, then, initially, Mn(II) will be taken up in a similar manner to Ca(II) and Mg(II) and it is undoubtedly transferred in this oxidation state in the more reducing environment of biological systems, to various organs, cells, and vesicles. It is known to be concentrated in plant cell vacuoles and in chloroplasts, for example. This uneven distribution means, of course, that it must have an energized, pumped distribution.

In the bloodstream of animals the Mn(II) can be oxidized on binding to transferrin and then, as Mn(III) transferrin, it may be transported into their cells, only to revert to Mn(II) on release. It is then again pumped into animal cell vesicles and organelles.

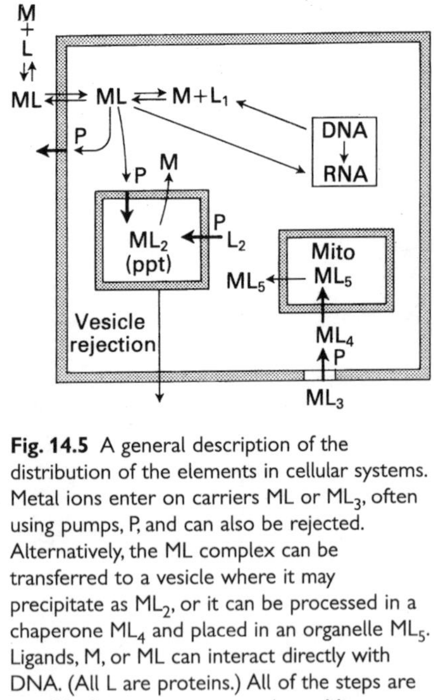

Fig. 14.5 A general description of the distribution of the elements in cellular systems. Metal ions enter on carriers ML or ML_3, often using pumps, P, and can also be rejected. Alternatively, the ML complex can be transferred to a vesicle where it may precipitate as ML_2, or it can be processed in a chaperone ML_4 and placed in an organelle ML_5. Ligands, M, or ML can interact directly with DNA. (All L are proteins.) All of the steps are possible for manganese complexes. Mito, mitochondria.

14.4.3 Pumping of manganese

As stressed in Section 2.12, ions are frequently handled by pumping mechanisms of a considerable degree of selectivity. Without such specialized pumps we could not understand the differential distribution of metals within biological species, organs, cells, organelles, and vacuoles. Only if we accept such kinetic control of free ion distribution can we systematize not only the special compartments but also the particular way in which chemical partners of the metal ions are 'selected' in cells. Manganese is pumped out of the cytoplasm of all eukaryotic cells (Fig. 14.5) and into a large variety of organelle and reticula (vesicle) compartments (Table 14.3). The pump is, in all probability, very similar to that for calcium (Section 10.13). The transfer operation of such pumps is as stated not based on thermodynamic competition for sites since it will depend upon selected recognition of the stereochemistry of the compound to be transferred by the accepting compound in the pump (see Chapter 3). We remind the reader that, since the process is coupled to energy. it can be a one-way selection. Consider again that

$$\text{All M bind to } L_1 \text{ to give } ML_1,$$
$$\text{but only } M_1 \text{ allows } M_1L_1 \xrightarrow{P_i} M_1L_2 \text{ (transfer).}$$

This means that the binding of L_1 is thermodynamically controlled, but the transfer of a particular M_1 to L_2 from ML_1 is the only M transfer allowed since the protein, P_1, recognizes the shape of M_1L_1 only. The principles are exactly the same as those discussed in the placing of magnesium in chlorophyll.

14.4.4 The distribution of Mn in biological systems

Mn is known to be a mutagenic agent and it is clear that, since its binding to proteins is relatively weak, it could not be allowed in high concentration in the cytoplasm of aerobic eukaryotic cells. The problem of mutation, so serious a threat to eukaryotic DNA, is not such a problem for prokaryotes since mutation can be helpful in generating variation of such species. Mitochondria and chloroplasts are organelles which can be looked upon as descended from prokaryotic cells. It is noteworthy that these organelles, as well as the cytoplasm of prokaryotes, accumulate Mn to a much greater degree than does the cytoplasm of eukaryotes. The eukaryotic cell also pumps manganese into almost any vesicular partition of the cell rather than leave it in the cytoplasm. Some of the vesicles known to have high manganese are listed in Table 14.3; in such vesicles Mn(II) does not face such strong competition from Mg(II) or Fe(II) which are retained in the cytoplasm of all cells. The enzymes then formed are discussed below. Particular examples are some Mn-dependent peroxidases as shown in Table 14.1, and they often are found outside cells.

Manganese poisoning may well arise when an excess of Mn is forced into a vesicle where it has damaging effects; for example, chromaffin granules of animals, into which Ca is deposited, can receive considerable Mn (if it is made available) which is then menacing since it can affect the adrenalin of the vesicles through oxidation reactions. To some degree Mn is also rejected to the outside of cells, but this rejection is much less notable than that of Cu or Ca; it could well be just a detoxifying mechanism. We next look at the value of this energized pumping of Mn for control of certain cellular activities in local compartments.

Table 14.3 Manganese-containing vesicles

Vesicle	Known Mn protein
Lysosome (acid)	Acid phosphatase (Mn(III))
Chloroplast	Superoxide dismutase, enzymes
Mitochondria	Superoxide dismutase, enzymes
Golgi apparatus	Glycosyl-transferases
Thylakoid (acid)	Dioxygen generation
Vacuole (acid)	(Mn(II) is free)

See also Table 14.1

14.5 The production of dioxygen

The description of the active site for the generation of dioxygen from water (see photosystem II in Section 5.2) is limited by the absence of a structure;

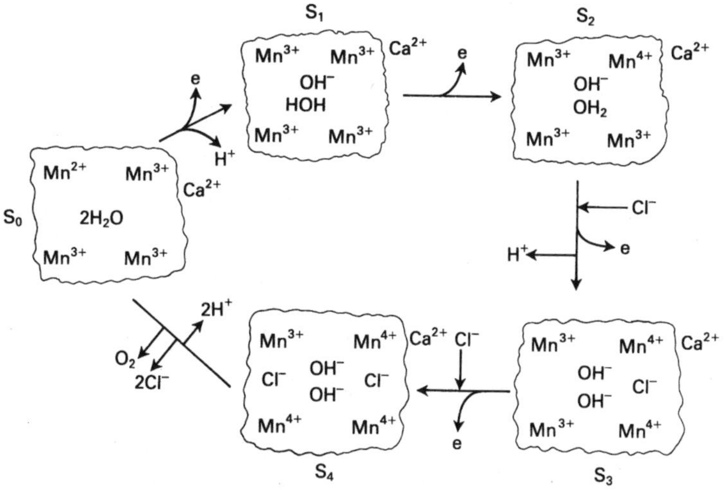

Fig. 14.6 A chemical scheme for O_2 production at a four-manganese cluster. The S-states are the five observable spectroscopic conditions. Note the involvement of Ca^{2+} and Cl^-.

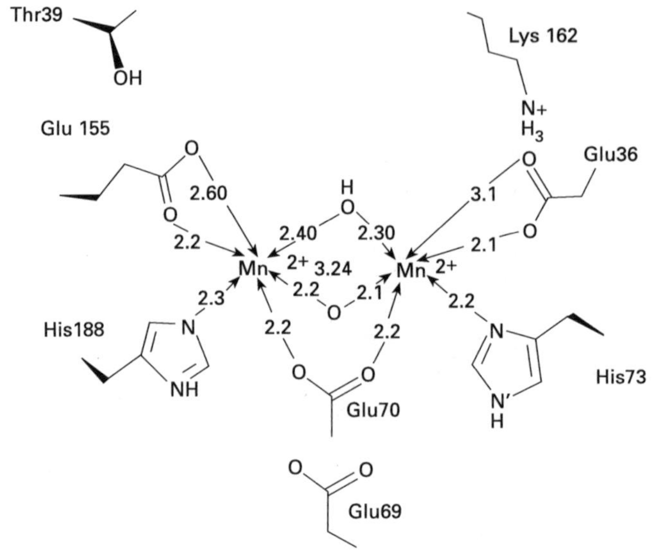

Fig. 14.7 The metal center of di-Mn catalase in T. thermophilus. (After V.V. Barynin, see Further Reading.)

what is thought to occur is shown in a scheme in Fig. 14.6. There are four manganese atoms involved and there are major influences of calcium and chloride, which must be looked upon as control functions. There is also a redox active tyrosine, hydrogen-bonded to a histidine, which is not represented. The exact way in which the four manganese atoms work together is not resolved, but the step(s) in the reaction, one electron oxidation at a time, can be readily followed by a variety of spectroscopic techniques. It is now accepted that the process involves hydrogen atom-abstraction by the tyrosyl radical from water molecules bonded to the manganese atoms, which undergo redox cycling, (see Fig. 14.6) until O_2 is released.

A very interesting model system for the O_2-evolving centre has been devised by Limberg *et al.* (see '*Further* reading'). The manganese ions, two, are bound by terpyridyl (Section 4.3) a ligand with a constrained coordination geometry. The metal ion has to cycle through a variety of oxidation states and the release of oxygen occurs when one oxygen atom bonded to a Mn(V) reacts with another oxygen atom of a water molecule. Another interesting reference for the O_2-generating complex is that seen in a special catalase (Fig. 14.7). The reaction here is $2H_2O_2 \rightarrow O_2 + 2H_2O$. The manganese ions are in a bridged pair and undergo redox cycling acting to reduce and oxidize H_2O_2 and liberating dioxygen. Notice how each manganese is bound by one imidazole and five oxygen donors in close to octahedral geometry. It is highly likely that all manganese enzymes have a coordination geometry rather similar to this, which is also similar to the iron centre of haemerythrin (Fig. 12.9; see T.S. Stammler *et. al.* in 'Further reading'). The need for Ca^{2+} and Cl^- in the generation of dioxygen suggests that the whole unit of the dioxygen release by $(Mn)_4$ is open to conformational change and control, and it is tempting to expect a Mn-binding helical bundle like that of haemerythrin. The linking between Ca^{2+}, Cl^-, and Mn^{2+}, all of which are exported from the cytoplasm in eukaryotes, is a very intriguing aspect of homeostasis.

A further control feature may depend on the stability of the Mn(II) binding; there is the distinct possibility that Mn(II) can dissociated from the protein complex and in this dissociation could lie a control by manganese availability over the use of light energy. Photosystem II generation of O_2 is secondary to the production of reduced CO_2 and of ATP in biological growth and it is, therefore, necessary to keep the locations of the two types of reaction separate and to have controls allowing different pathways to dominate in different conditions. Note that the large concentrations of Mn^{2+} are remote from DNA.

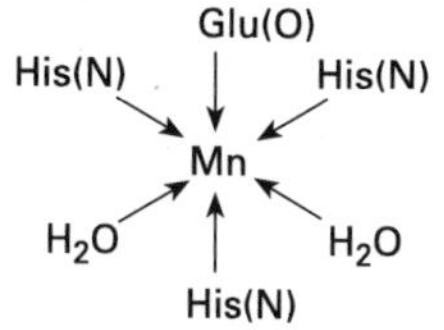

Manganese in superoxide dismutase and also in oxalate oxidase.

14.6 **Manganese and peroxide metabolism**

As shown in Table 14.1, and referred to in Sections 14.3 and 14.5, Mn is found in certain peroxidases, catalases, and superoxide dismutases (Fig. 14.8 and Aside). However, these uses are much more dominant in plants than in animals. The peculiarity then is that Mn, like Fe, is useful in all the reactions of oxygen except, unlike Fe, the direct use of dioxygen itself. The Mn peroxidases and catalases are found in several extracellular or vesicular systems. The first is associated with the strengthening of cross-linking in plant lignins—a function which can also be served by the haem peroxidases and Cu oxidases released from vesicles. It is likely that the switch to copper took place later in evolution.

The coordination sphere of loosely held manganese in the plant manganese/haem containing lignin peroxidases has been found to be six oxygen atoms in a fairly regular octahedron with four donors from protein sidechains and two water molecules, see Aside. This site is of low anionic charge and is not likely to be open to strong binding by Mn^{3+} ions. Since the enzyme is extracellular Mn^{2+} could, in principle, face competition from other transition metals, e.g. Fe^{2+} and Zn^{2+}. However, all are much less available externally (Fe^{2+} due to oxidation, Zn^{2+} due to stronger binding to other ligands).

The exchange rate of Mn^{2+} from this site in ligninase is fast. Moreover, Mn^{2+} goes to Mn^{3+} upon peroxide attack at the haem (the Mn^{2+} is closely

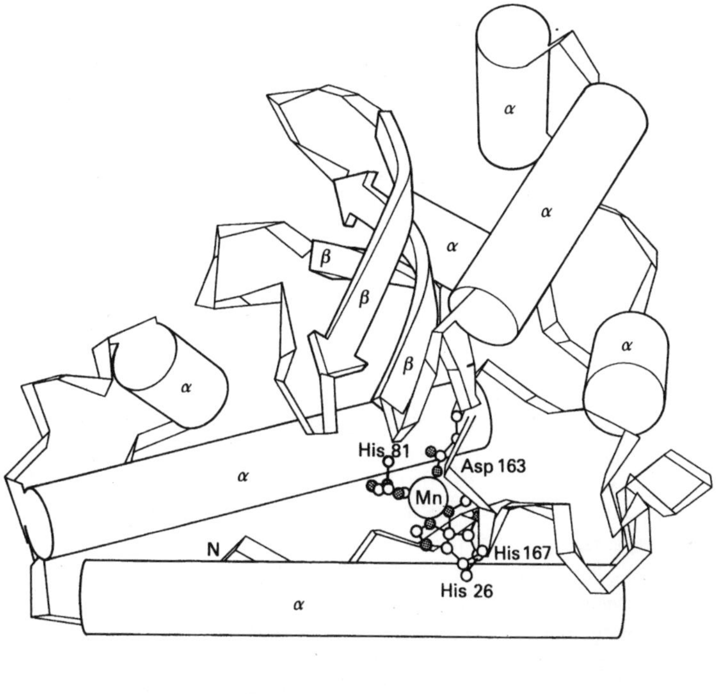

Fig. 14.8 A view of manganese superoxide dismutase showing the α-helices and β-sheet and the octahedral Mn(III) coordination. (After C.C.F. Blake and M Ludwig.)

linked to the haem propionate side-chain) and Mn^{3+} then dissociates. Mn^{3+} is probably the leaving group product while Mn^{2+} is the substrate. The Mn^{3+} attacks organic phenols of lignins and so on to produce radicals which cross-link the extracellular matrix. The association of manganese redox chemistry with free radicals is also found in the O_2-release reaction centre of plant photosynthesis which utilizes tyrosine radicals, as stated above. It is relatively easy to match the high redox potentials of organic sidechains, e.g. phenolate, with that of the Mn^{2+}/Mn^{3+} couple.

Now, while we can explain why Mn and not Fe is used in dioxygen production and why Fe or Cu, but not Mn, are used in dioxygen transport, in pathways for energy production, and in many oxygenases, this leaves no reason for the use of Mn(III) in either catalases, superoxide dismutases, ribonucleotide reductases or in peroxidases rather than Fe(III). This problem comes again in the choice of Mn(III) or Fe(III) in acid phosphatases (see also the comment on Mn(III) transferrin in Section 14.4.2). The simplest explanation is the higher availability of manganese as Mn(II) than Fe(II) outside cells and in the vesicles which hold these enzymes. Once incorporated into a protein site of largely O-donor ligands the metal is then oxidized, but disproportionation is prevented by isolation of the Mn(III) in a monomeric complex. We are suggesting that the preference for Mn(III) over Fe(III) may be almost an accident of availability in some enzymes. At the same time manganese is preferred in plants more than in animals. Do not forget that the animal cells can circulate carrying risks with them if peroxide reactions occur.

14.7 Manganese and hydrolytic reactions

Nitrogen is lost by three different routes in different organisms. In many cases, e.g. plants, loss is of ammonia to external water or perhaps even the air, but this method of removal is not possible for organisms such as birds and mammals whose body is made to prevent surface loss of water solutions. The second mode of removal is by the synthesis of urea which is lost after it has been concentrated in the kidney. Urea is not poisonous to the degree that ammonia is. This method is common in mammalian metabolism. The third method is by the synthesis of uric acid in the form of a pellet which is then rejected. This last process is common in birds. Huge deposits of guanine are found on some Pacific islands from bird droppings and used to be a source of nitrogen fertilizers. A major interest in this book is the use of different elements in such reactions. The catalysts are peculiar in their metal content. Arginase, which is used in the urea cycle and is an enzyme common to many forms of life from prokaryotes to man, is a dinuclear manganese (or sometimes zinc?) enzyme. Very few hydrolytic enzymes use manganese in this way in the cytoplasm and in the Mn(II) state. Urease is another enzyme in the urea cycle that is not found in man but is common in plants which try to retain nitrogen as NH_3. It is no less puzzling in that it is a hydrolytic enzyme based on nickel.

Another strange manganese hydrolytic enzyme is the prolidase for splitting proline peptides. It also occurs in the cytoplasm. There may be a few other such enzymes in the cytoplasm, but the vast majority are exported (see Table 14.1). Remember, however, that a correct homeostasis of manganese is critical for plants, so that plant-life must have genetic controls over free Mn^{2+} uptake and circulation through and in cells. Excess manganese causes brown spots (MnO_2 deposits) in leaves and is dangerous in man, especiallyin the central nervous system, see Graham *et al.* (1958) in 'Further reading'. Hence there must be manganese regulatory proteins in organisms, some of which function equally with Fe^{2+}.

A very important function of manganese enzymes in vesicles is in glycosylation. In fact the association of manganese with sugar chemistry extends to the synthesis of glycogen and to its use in some plant lectins, e.g. concanavallin A. The general sense, unproven, is that manganese has a very important regulatory role, especially in plants, (See section 14.9).

14.8 Manganese precipitates

It has been observed that in certain plant tissues, e.g. the base of the stinging hair of the common nettle, there is a cyclic appearance during development of the hair cell of a very large deposit of manganese (Fig. 14.9 and Table 14.4). There is also excess phosphate present and we wonder if this manganese system is being used to protect oxalate while Ca oxalate is laid down in the hair. The manganese would then be removing the products of the glyoxalate cycle, such as O_2^- and H_2O_2 which are dangerous for oxalate; this cycle also occurs in peroxisomes where Mn is in high concentration. The store, which is removed at maturity, is possibly $Mn_2P_2O_7$, a known mineral in a vesicle of a lactobacillus that is thought to act as a superoxide dismutase in just this way.

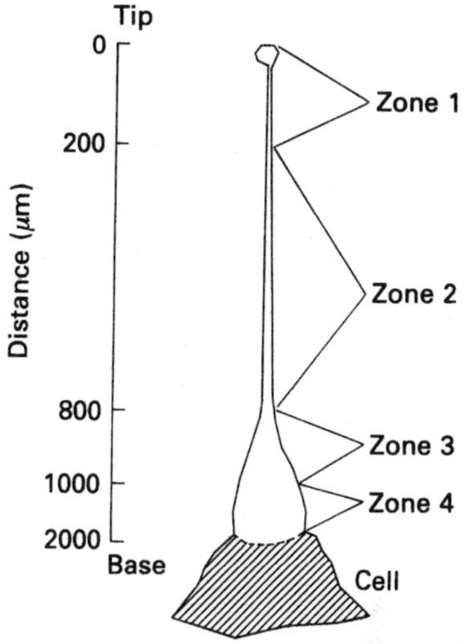

Fig. 14.9 Diagramatic representation of a stinging hair of a simple nettle showing various zones. Elemental maps of immature and mature stinging hairs, Emergence I and II, were produced using proton-induced X-ray emission combined with a scanning proton microprobe. Elements determined are Ca, Si, P, S, K, and Mn, in the zones of the figure as given in Table 14.4.

Table 14.4 Analytical data extracted from analyses performed at different points along the length of each hair (Fig. 14.9)

Distance from tip (μm)	Concentration (μg/cm²) Ca	Si	P	S	K	Mn
Emergence I						
At tip	14	171	6	0.3	20	–
100	82	139	14	0.9	58	–
200	91	56	11	1.7	128	–
300	121	27	9	0.9	86	–
400	127	14	7	0.9	67	–
600	149	17	6	1.4	87	–
800	167	20	6	0.8	57	–
1000	242	84	10	0.9	34	–
1200 (at base)	110	21	60	3	124	42
Emergence II						
At tip	8	190	2	0.05	1	–
100	206	117	2	0.6	5	–
200	248	71	3	0.8	6	–
300	252	49	4	0.5	6.5	–
400	294	21	2.5	0.8	7	–
600	304	16	1.2	0.4	5	–
800	285	17	2.5	0.5	5	–
1000	246	62	3	0.7	5	–
1200 (at base)	82	5	2	0.7	30	–

Emergence I is an earlier stage of development than emergence II.

Manganese-depositing bacteria have been known for many years; specifically *Gallionella, Sphaerotilius, Leptothrix*, and *Chonothrix* have been shown to deposit manganese oxide extracellularly. Extensive work has been carried out on control of the deposition process (allied with the extracellular deposition of Fe hydroxide) and on the physiological aspects of the deposits in these systems. It can be that up-and-down movement in the water column, from oxic to anoxic zones, allows the organisms to capture energy using a cycle of changes

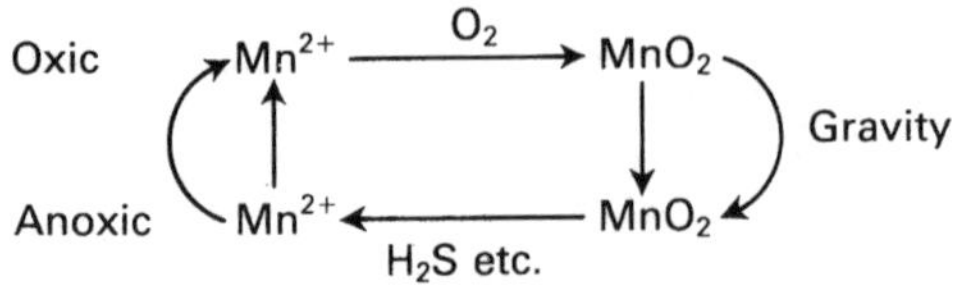

Furthermore, intracellular granules of manganese salts have been observed within the common garden snail; these granules frequently contain a wide variety of other environmentally available metals and therefore have been interpreted as part of a cellular detoxification process.

14.9 Manganese control systems and genetics

There is now the clear knowledge that DNA expression is controlled in part by Zn proteins, the so-called zinc fingers, especially in eukaryotic cells (see Chapter 11). The ways in which Fe (and Mn) are built into biological systems and compartments led to the earlier prediction that, especially for the prokaryote systems, there would be regulatory genes controlling the relationship between uptake and growth. The similarities between the chemistry of Mn(II) and Fe(II) lead us to suppose that the control systems connected to Mn could be related to the FUR system, but the best parallel is the analogy with the NRAMP 2/DCT-1 system (see Section 12.3).

The basis for the idea that such a system exists is the link between Mn and photosynthesis, and the handling of sugar moieties so produced in some biological compartments, see Aside. The complexity of control over such complicated interactive systems demands gene regulation. However, as with Zn so with Mn, the full ramifications of the use of this metal have not been uncovered. Zinc, we can say, is in some way like a general growth factor in eukaryotic systems; magnesium is intimately linked to phosphate transfer. Can we pin-point a central biochemical function for Mn from which we can seek its regulatory systems or is its use too generalized? Mn is connected closely to: (1) organelle functions, e.g. in thylakoids and mitochondria (see Table 14.1); and (2) the handling of glycosylation, which is a peculiarity of the Golgi apparatus. It would appear that in plants in particular we must seek a manganese pump protein which is regulated by the same gene systems as either that for the synthesis of photosystem II (dioxygen production) or that for the production of glycosylases. Table 14.1 includes some other enzymes which could be involved in metabolic control at the organelle level, mitochondria or chloroplasts. Particularly interesting in the context of the comments in Section 12.16, on iron control, is the requirement for manganese in some of the enzymes of amino-acid synthesis and some in the citric acid cycle; the picture is then where manganese controls the production of (E), the protein of manganese enzymes, but manganese excess switches off Mn(II)

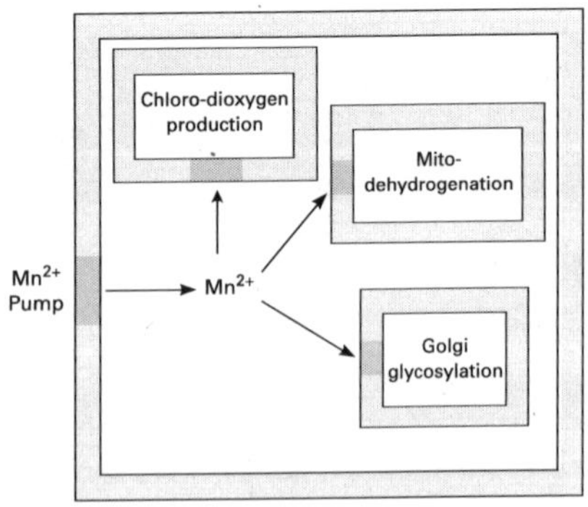

The distribution of Mn^{2+} in eukaryotes.

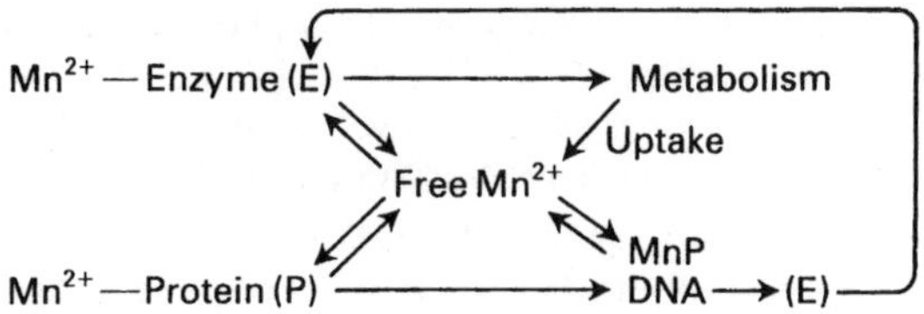

uptake, and manganese deficiency switches it on via free E. It would appear that such a regulatory system has been found controlling the production of manganese peroxidases (ligninase) in *Phanerochaete chrysosporium*.

Two genes, *smf 1* and *smf 2*, which were indentified in yeast, appear to encode manganese transporters. They are homologous of the *nramp 2* genes in mammalian cells which encode intestinal and endosomal Fe(II) transporters. The same occurs for *smf 1* and *smf 2*, but the regulation of the transcription of these genes does not appear to be affected by either Fe or Mn. The activity of the SMF proteins, however, is regulated at post-translational level by the product of the *bsd 2* gene, an endoplasmic recticulum membrane protein which responds to metal concentrations. BSD 2 directs SMF 1 to vacuoles for degradation when the concentration of the metals is high.

14.10 The evolution of manganese functions

This section is purely speculative. Manganese, in the sulphide solutions that may have dominated early life, was relatively more available than any other transition metal ion since it does not have a very insoluble sulphide but see tungsten, Chapter 17). On the other hand it is a considerably better Lewis acid than either magnesium or calcium which were the two other divalent ions of easy availability. We might well concede then that an early use for Mn(II) would be as a useful acid catalyst for easy reactions such as those of phosphatases. In such reactions we know that two metal ions are better than one, e.g. as seen in many acid and alkali phosphatases to this day. Thus manganese (II) clusters may have evolved.

At some stage life developed the ability to use light linked to the discharge of electrons so as to generate reducing equivalents, but it was simultaneously necessary to have a source of reducing equivalents on the opposite side of the membrane in which light capture occurs (Fig. 5.1). In the presence of sulphide the reaction $H_2S \rightarrow S\downarrow + 2H^+ + 2e$ is an ideal source, but it requires metal ions to catalyse the reaction (see Section 17.10). The preferred metal ion must go readily through the cycle.

$2M^{2+} + h\nu \rightarrow 2M^{3+} + 2e$ and then $2e \rightarrow$ reduction system

$2M^{3+} + H_2S \rightarrow 2M^{2+} + S\downarrow + 2H^+$ $2H^+ \rightarrow$ gradient energy (ATP) production.

Now manganese is peculiarly suitable for such a reaction as Mn^{3+} sulphides are open to ready reduction with sulphur release and metal dissociation. The reaction is best carried out in a vesicle so as to trap the sulphur produced which can then be rejected. Note that the vesicle would also be acid. It would pay to put manganese in the vesicles and to couple the whole reaction to ATP production (see Chapter 5).

Given that this chemistry is all speculative, the further consideration that, given the very high redox potential of Mn(IV), the above system would come to generate dioxygen from water using a Mn(IV) dimer may be acceptable. Moreover, dioxygen rejects itself by diffusion and the greater proton gradient generated in the acid vesicle can be used to make more ATP (see Chapter 5).

As a consequence of a separation of the dioxygen-generating system into plants the question arises as to whether manganese has a higher level of control in plants than in animals. It is certainly present in much larger quantities. The link between photogeneration of dioxygen and the incorporation of reducing equivalents in carbon assimilation followed by the extensive use of more polysaccharides by plants is suggestive but as yet we know little more than could be used to suggest that there may well be a general network of links in metabolism in plants based on free Mn^{2+}. Especially intriguing are the further connections to deoxyribonucleotides (Table 14.1) and amino acids via the shikimic acid pathway and we notice the ability of Mn(III) to act as a replacement for Fe(III) in a few other important reactions. Thus manganese biochemistry underwent dramatic changes with dioxygen introduction and we note how much of this new chemistry is put outside the cytoplasm into vesicles (compare calcium and copper). It is possible that manganese importance in gene controls in higher organisms is only to be found in multicellular plants. As stated, other very important reactions outside the cell cytoplasm are the oxidation of lignin by white rot mycetes, which utilize free Mn^{2+} ions, and glycosylation. It has been shown that the production of the enzymes for these reactions is probably controlled by a manganese protein, i.e. it is regulated by free Mn^{2+} in the cell. Another interesting plant Mn-enzyme is oxalate oxidase, and again oxalate is common in plants but not in animals.

14.11 **Summary**

The biochemistry of manganese in all oxidation states is dominated by transfer to selected compartments and by its association with N/O, largely O, donor systems. These allow the use of Mn(II), and even Mn(III), as weak acid catalysts and the cycling of oxidation states (II), (III), and (IV) in reactions of especially H_2O, O_2, H_2O_2, and O_2^-. The very special association of manganese in the higher oxidation states is with the release of O_2 from within chloroplasts. Mn(II) appears to have an additional association with glycosylation in Golgi derived vesicles and Mn(III) is found in mitochondria and chloroplasts in such enzymes as superoxide dismutase as well as in lysosomes in acid phosphatases. Finally, there is the suspicion that there are regulatory proteins which bind manganese. The connection of all this special manganese biological chemistry with that of other elements will be described in Chapter 20, where we analyse the metallome of both animals and plants. The concentration of free manganese in cell cytoplasm is $<10^{-7}$ M, probably less than that of free Fe^{2+} due to outward pumping of Mn^{2+} (see Section 20.7).

Further reading

There are several articles on manganese in *Metal ions in biological systems* (2000) **37**, ed. Siegel H, and Siegel, A.

Antonyuk, S.V., Melik-Adamyan, V.R., Popov, A.N., Lamzin, V.S., Hempstead, P.D., Harrison, P.M., Artymyuk, P.J., and Barynin, V.V. (2000). Three-dimensional structure of the enzyme dimanganese catalase. *Crystallographic Reports*, **45**, 105–16.

Banci, L., Gofi-Baffrin, S., and Tien, M. (1999). Lignin and Mn-peroxidase catalysed oxidation of lignin. *Biochemistry*, **38**, 3205–10.

Christou, G. and Vincent, J.B. (1987). A molecular double-pivot mechanism for water oxidation. *Inorganica Chimica Acta*, **136**, L42–3.

Debus, R.J. (2001). Amino-acid residues that modulate the properties of tyrosine Y(Z) and the manganese cluster in the water oxidising complex of photosystem II. *Biochimica Biophysica Acta*. **1503**, 184–6.

Ghiorse, W.C. (1984). Biology of Fe- and Mn-depositing bacteria. *Annual Reviews of Biochemistry*, **38**, 515–50.

Graham, R.D., Hannam, R.J., and Uren, N.C. (eds) (1988). *Manganese in soils and plants*. Kluwer Academic Publishers, Dordrecht.

Hughes, N.P., Perry, C.C., Williams, R.J.P., Watt, F., and Grime, G.W. (1988). A scanning proton microprobe study of stinging emergences from the leaf to the common stinging nettle *Urtica dioica L. Nuclear Instruments and Methods*, **B30**, 373–7.

Kehres, D.G., Zaharik, M.L., Finlay, B.B., and Maguire, M.E. (2000). The bacterial NRAMP proteins are manganese transporters. *Molecular Microbiology*, **36**, 1085–1100.

Khangulor, S.V., Barynin, V.V., Voerodskaya, N.V., and Grebenko, A.I. (1990). Binuclear cluster in manganese catalase. *Biochimica et Biophysica Acta*, **1020**, 305–10.

Limburg, J., Vrettos, J.S., Liable-Sands, L.M., Rheingold, A.L., Crabtree, R.H., and Brudvig, G.W. (1999). A functional model for 0–0 bond formation. *Science*, **283**, 1524–7.

Paidhungat, M. and Garrett, S. (1998). CDC1 and vacuole coordinately regulated Mn^{2+}. *Genetics*, **148**, 1787–98.

Radisky, D. and Kaplan, J. (1999). Regulation of transition metal transport. *Journal of Biological Chemistry*, **274**, 4481–4.

Requena, L. and Bornemann, S. (1999). Oxalate oxidase. *Biochemistry Journal*, **343**, 185–90.

Stemmler, T.L., Sossong, T.M., Goldstein, J.I., Ash, D.E., Elgren, T.E., Kurtz, D.M., and Penner-Hahn, J.E. (1997). EXAFS comparison of the dimanganese core structures of manganese and manganese substituted proteins. *Biochemistry*, **36**, 9847–58.

Sundaramoorthy, M., Kishi, K., Gold, M.H., and Poulos, T.L. (1994). The crystal structure of manganese peroxidase. *Journal of Biological Chemistry*, **269**, 32759–67.

Vijayan, M. and Chandra, N. (1999). Lectins. *Current Opinion Structural Biology*, **9**, 707–14.

Westphal, K.L., Tommos, C., Cukier, R.I., and Babcock, G.T. (2000). Concerted hydrogen-atom abstraction in phatosynthetic water oxidation. *Current Opinion Plant Biology*, **3**, 236–42.

Williams R.J.P. (1982). Free manganese and iron cations can act as intracellular controls. *Federation of European Biochemical Societies Letters*, **140**, 1–10.

Yocum, C.F. and Pecoraro, V.L. (1999). Recent advances in understanding the biological chemistry of manganese. *Current Opinion Chemistry and Biology*, **3**, 182–7.

Notes: (1) A major problem which involves the levels of free manganese, as well as those of free iron, zinc, and copper is oxidative stress in the cytoplasm. This is an important reason for keeping all these free metal ions in the metallome below 10^{-7} M. However it is not associated just with one metal ion (or non-metal such as sulphur or selenium) but is a general problem faced by the cytoplasmic anaerobic (primitive) metabolism of all cells in the modern oxygen atmosphere.

(2) A good outline structure of photosystem II is now available showing the Mn_4 unit of dioxygen release. The four Mn atoms are close together in an arrangement not far from a tetrahedron. Zouri, A., Witt, H.-T., Kern, J., Fromme, P., Krauss, N., Saenger, W., and Orth, P. (2001). Crystal structure of photosystem II at 3.8 Å resolution. *Nature*, **409**, 739–47; also described in *Science*, (2001), **292**, 447–8.

CHAPTER 15

Copper: extracytoplasmic oxidases and matrix formation

Ni	Cu	Zn
Pd	Ag	Cd
Pt	Au	Hg

The Ni, Cu, and Zn groups of the periodic table

15.1 Introduction

There are five distinctive characteristics of the biochemistry of copper, four of which are chemical:

1. Cu(II) is the most effective available divalent ion for binding to organic molecules (Fig. 2.8). Since this is in fair measure due to its electron affinity, it is also the best attacking acid available at around pH = 7, but it rarely used as such in organisms.
2. Cu(I) is the most effective available monovalent ion for binding to organic molecules. Again this is due to its high electron affinity (see Aside on page 419).
3. The almost equal tendency of Cu(I) and Cu(II) to form complexes with many organic ligands, notwithstanding their very different sizes and stereochemistries, makes the Cu(I)/Cu(II) couple particularly useful and adaptable in the range of redox potentials from about + 0.2 to + 0.8 mV, i.e. on the oxidizing side of the biological scale (Fig. 13.3).
4. Cu(I) can act as a π-donor and can bind such ligands as O_2 and CO.
5. The fifth factor, which is non-chemical, is that only one or two of the many copper proteins are found in the cytoplasm of cells.

A most striking aspect is the strong correlation of copper enzymes with the reactions of small non-metal compounds which only became available in quantity when dioxygen entered the atmosphere (Table 15.1). We then may make the conjecture that biological systems did not use copper extensively before the advent of an oxidizing atmosphere based on dioxygen. Now this event coincides with the development towards multicellular organisms and we shall therefore seek special roles for copper in that development. One of particular note is that a copper(I) protein is the ethylene receptor. Ethylene is a major plant hormone and ethylene itself is an oxidative product like many hormones.

Table 15.1 Some substrates for copper enzymes

Enzymes	Substrate
Simple oxidases	O_2 plus phenols, amines, ascorbate, ferrous ions, etc.
Cytochrome oxidase	O_2 plus cytochromes *c*, *a*, and a_3
Nitrogen oxide enzymes	NO, NO_2^-, N_2O + oxidizing or reducing agents
Specialist oxidases	O_2 + proteins or polysaccharides for cross-linking

Table 15.2 Some copper proteins and enzymes: localization and function

Protien or enzyme	Localization	Function
Cytochrome oxidase	External face of mitochondrial membrane	Reduction of O_2 to H_2O
Laccase, Tyrosinase	Extracellular	Oxidation of phenols (reduction of O_2 to H_2O)
Caeruloplasmin	Extracellular (blood plasma)	Oxidation of Fe(II) to Fe(III)? (reduction of O_2 to H_2O)
Haemocyanin	Extracellular (blood plasma)	Transport of O_2
Lysine oxidase	Extracellular	'Cross-linking' of collagen (reduction of O_2)
Ascorbate oxidase	Extracellular	Oxidation of ascorbate (reduction of O_2 to H_2O)
Galactose oxidase	Extracellular	Oxidation of primary alcohols to aldehydes in sugars (reduction of O_2 to H_2O_2)
Amine oxidase	Extracellular	Removal of amines and diamines
Blue proteins	Membranes (high potential) thylakoid vesicles	Electron-transfer (Many different kinds)
Superoxide dismutase	Cytosol	Superoxide dismutation (eukaryotes)
Nitrite reductase	Extracellular	Reduction of NO_2^- to NO
Nitrous oxide reductase	Extracellular	Reduction of N_2O to N_2
Metallothionein	Cytosol	Copper(I) storage
ACE-1 (MAC)	Cytosol	Transcription factor
Dopamine mono-oxygenase	Vesicular	Hydroxylation of Dopa
Coproporphyrin decarboxylase	Extracellular	Production of Proto-porphyrin IX
Ethylene receptor	Membrane	Hormone signalling
Methane oxidase	Membrane	Oxidation to methanol
Terminal glycine oxidases	Vesicular	Produciotn of signal peptides
CP-x type ATP-ase	Membrane	Copper pump
ATx-1 (Lys 7)	Cytosol	Copper transfer

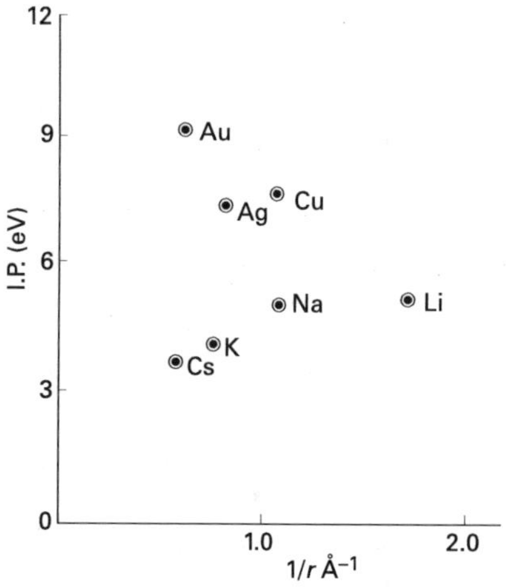

The relationship between ionization potential (I.P.) and radius (r) for some monovalent cations.

15.2 Copper and electron-transfer

Perhaps the most common use of copper is in electron-transfer, especially associated with oxidative enzymes and energy capture. These electron-transfer systems are often made from small 'blue' proteins, Table 15.2, that ferry electrons to reaction centres and some of them exist as freely mobile small proteins. None of these blue copper proteins is found in the cytoplasm, though some are bound to the external surface of cytoplasmic membranes. In Chapter 6 it was shown that the construction of all these copper sites is constrained, entatic, i.e.

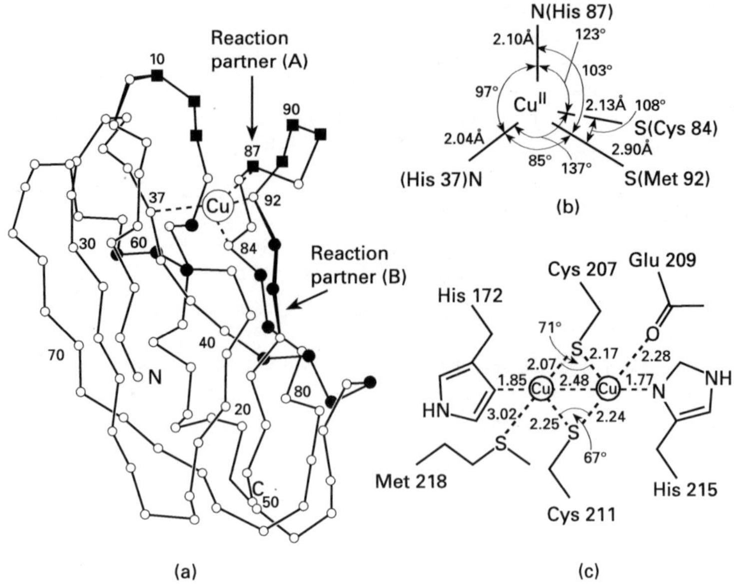

Fig. 15.1 The active site of plastocyanin and azurin (after the works of H. Freeman and L. Jensen). (a) The β-barrel construction, the ligands, and the substrate binding sites at A and B for two different proteins. (b) Details of the active site. The blue, single copper site has been classified as a type I site. When it appears in a dimeric or two copper centre it is sometimes called Cu_A [Fig. 15.1(c)]. The list of other types of copper site is given in Tables 15.2 and 15.3.

with unusual but somewhat variable geometries from protein to protein and held in those geometries by a β-barrel fold of the apoprotein. The rigidity of the geometries (Fig. 15.1) is then easily recognized as one that has evolved with a ligand set 'designed' to assist electron-transfer. In energy transduction devices where it is necessary to pass several electrons to the dioxygen-binding centre, e.g. in cytochrome oxidase and in nitrite reductase, the site, Cu_A, has two such copper atoms in a coupled unit [see Fig. 15.1(c)]. In many enzymes other copper sites are open-sided, differently designed to accept and activate small molecules such as O_2. These various cases are illustrated in Table 15.2 where we also note the localization of these copper proteins. Some of these reactions use high redox potentials which require that the site and its geometry are designed to give both fast reactions and high potentials (Table 15.3). The distortion of the copper site toward tetrahedral geometry has just this effect since it stabilizes Cu(I) relatively to Cu(II); in fact, it would appear, from all our knowledge of structure and function of the blue copper sites, that their proteins are rigidly designed as a match for Cu(I) and a mismatch for Cu(II). In Chapter 6 it was pointed out that, upon removal of copper from the protein, the structure of the protein and the coordination site remained intact. Thus the very robust character of the protein allows it, like cytochrome *c*, to be used as a freely mobile scavenger of electrons. It is very probable that the selection of copper in certain sites is based on the binding of Cu(I) which is the only singly charged cation which is a good Lewis acid and is available to living organisms.

15.3 Copper and dioxygen

15.3.1 Haemocyanin

There is a complex copper protein in a few species which carries dioxygen in their blood. This colourless protein, haemocyanin, has two closely associated copper(I) atoms which bind dioxygen reversibly to give a purple $[Cu(II)]_2.O_2^{2-}$ complex. The ligands to the copper are three histidines and the

Table 15.3 Redox potentials (pH = 7.0) and constrained states of copper in proteins

Copper protein	State of copper
Blue protein (type (I)) Azurin, Laccase, Caeruloplasmin many Oxidases	Distorted tetrahedron or trigonal. Thiolate/3 Histidine. E 300–800 mV. Electron-transfer. β-sheet
Superoxide dismutase (Type II)	Distorted tetragonal, Open-side 4 Histidine. E 280 mV Superoxide reactions. β-sheet
Hydroxylase	
Dopamine β-hydroxylase	(a) 3-coordinate (open) Met/His/His. β-sheet. E > 500 mV.
Peptide glycine α-hydroxylase	(b) 3-coordinate (open), 3His E (?) β-sheet
Lysine oxidase (?)	O_2-activation via H_2O_2, E (?) Substrate activation. β-sheet
Galactose oxidase	5-coordinate (tetragonal prism, open); 2 His, 2 Tyr. β-sheet. E (?)
Laccase (Type II) Caeruloplasmin, many Oxidases	Distorted tetragonal 3 Histidines. E > 500 mV. O_2-reaction to H_2O). β-sheet
Copper ATP-ase Carrier proteins (ATx-1)	Linear, 2 Thiolates. E > 500 mV. Copper(I) Pump. β-sheet/α-helix
Laccase (Type III) Caeruloplasmin, many Oxidases	Distorted Tetragonal, 3 Histidines. E 400 mV? O_2-reactions to H_2O. β-sheet
Cytochrome oxidase	Distorted Trigonal coordination
Copper(A) pair	as in Type I E ≈ 400 mV Electron-transfer. β-sheet
Metallothionein (type) (ACE-1)	Tetrahedral(?). Thiolate clusters. (Random coil) in apo form E > 500 mV Buffer and transcription factor
Haemocyanin (type)	Tetrahedral. 3 Histidines. E > 700 mV(?) O_2 carrier Helical
Nitrous oxide reductase	4 Cu(I) in tetrahedral cluster 7 Histidines, one S^{2-} (?) β-sheet, E < 500 mV.

Note that the redox potentials are higher than that of the copper aqueous ion system at pH = 7.0 (+170 mV) which means that Cu(I) is bound with greater strength than Cu(II).

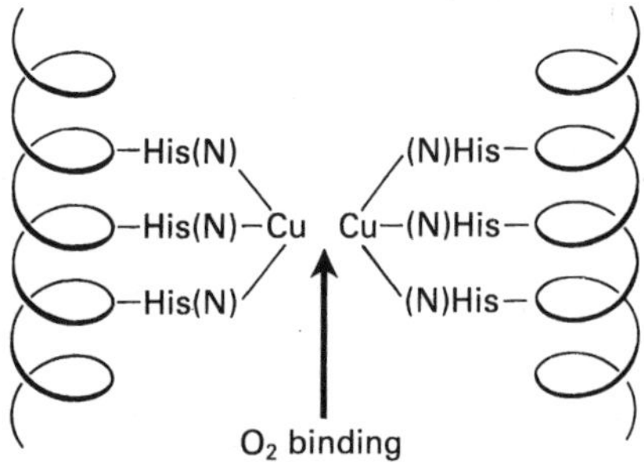

Fig. 15.2 The dioxygen binding site of haemocyanin drawn schematically to call attention to the α-helical nature of the local structure.

Table 15.4 A comparison of the three dioxygen carrier proteins

Larger ions	Smaller ions	O_2 transport protein
Cu(I)	Cu(II)	Haemocyanin
Fe(II) (high-spin)	Fe(II) (low-spin)	Haemoglobin
Fe(II) (high-spin)	Fe(III) (high-spin)	Haemerythrin

Cu(I) ion is stabilized in roughly open-sided tetrahedral geometry. This coordination is also found in a number of copper enzymes as already seen in Table 15.3. The region around the copper atom in haemocyanin is, however, based on protein helices (unlike the blue copper proteins) (Fig. 15.2) which are adjustable on uptake of ligands. The metal changes size on binding dioxygen and the binding of several O_2 molecules is cooperative as in iron-containing haemoglobin or some haemerythrins. It is interesting to compare the size changes of the ions in the three examples of Table 15.4. The use of helices to generate cooperativity via the ion size contraction on dioxygen

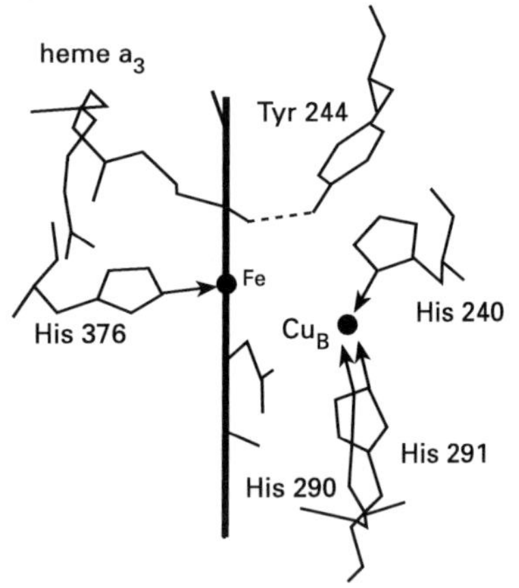

Fig. 15.3 The Cu_B site with haem a_3 (Fe) of cytochrome oxidase. (After Yoshikawa *et al.*, see reference in Chapter 13).

binding is common to all these dioxygen carriers. The haemocyanin protein itself is an extracellular protein which appears to be synthesized in a special way restricted to some mollusks and arthropods (but note that arthropods are the largest group in the entire animal kingdom, 85 per cent of all animal species). Clearly, a large intake of copper is required and the protein synthesis is in the hepatopancreas. The protein is put out into the circulating bloodstream as a large aggregate of monomers and in this it is quite unlike haemoglobin and haemerythrin which are found in cells. Unlike them, too, haemocyanin does not require a reductase to be kept in the Cu(I) state since Cu(II) haemocyanin is very difficult to make and, quite possibly, haemocyanin has a high redox potential which prevents auto-oxidation. Haemocyanin is stabilized further by binding with calcium ions, which cross-link the protein structure.

15.3.2 A comment on copper of cytochrome oxidase

Cytochrome oxidase is another protein (enzyme) which binds to dioxygen but now it uses dioxygen in reactions (see Section 13.9). It has three copper atoms, widely separated in space and, when we examine them, we observe that there is in a β-sheet domain a pair of copper atoms, Cu_A, which look individually somewhat like the blue copper discussed in Section 15.2, [see Fig. 15.1(c)] and which are presumably for electron transfer, while the third copper, Cu_B, is directly associated with a haem group, cytochrome a_3, at the dioxygen uptake site (Fig. 13.15 and see Fig. 15.3) that resembles the copper of haemocyanin both in binding and in that this part of the protein is helical. The dioxygen reaction site, then, is a combination of one haem iron of the haemoglobin kind, here the cytochrome a_3, and one copper, Cu_B, of the haemocyanin kind, bound by three histidines, and the surrounding protein is a typical α-helical bundle membrane protein (Fig. 13.15). As we have just seen in O_2-carrier proteins, such an α-helical construction is a very useful one for converting chemical energy to conformational change. In these two coupled electron transfer centres of cytochrome oxidases we suppose that the coupling goes from the reduction of dioxygen to the mechanical movement of helices to the transfer of protons uphill across the membrane: the system is a proton pump (see Fig. 5.5). Note that this part of the energy capture of the mitochondrial and bacterial membranes is the only one which depends on copper. Similarly, the only part of the photocapture organization of the thylakoid in chloroplasts which depends on copper is in photosystem II, which leads to the production of dioxygen. In these thylakoids only a simple soluble copper blue protein is involved and it appears that this is a late evolutionary replacement, now for a cytochrome, in an electron transfer pathway. In both cases the copper system is external even to the cytoplasm of the organelle or prokaryote in that it is either in a special vesicle (a thylakoid) or locked in a membrane on the periplasmic surface. The potentials of all these copper proteins are around or above +0.30 V and, obviously, these copper-utilizing systems developed late in evolution.

Recently, it has been found that there is quite a variety of membrane oxidases which resemble cytochrome oxidase. The analysis of these oxidases, all of which may be proton pumps, has shown that Cu_B is present in most but not all of them and that its partner, cytochrome a, may be replaced by a cytochrome b. What we take to be a very early example, a quinone oxidase, has no copper but iron in its place. The details of cytochrome oxidases and the way in which they can act as proton pumps have been described in Chapter 13.

15.3.3 Copper and extracellular oxidases

The major extracellular oxidases to which this paragraph refers are given in Table 15.2. They usually consist of a combination of 'blue' copper (Type I), pairs of copper atoms (Type III), and extra atoms of a type of copper (Type II) associated with the two Type III copper atoms (Fig. 15.5). Enzymes of this kind include laccase, ascorbate oxidase, caeruloplasmin, and nitrite reductase. (The blood coagulation factors V and VIII appear to have similar copper centres). These enzymes usually have an organic substrate site which is thought to be bound some 10 Å away from the Type I site, so that there is the electron-transfer chain

$$\text{Substrate} \xrightarrow{e} \text{Type I} \xrightarrow{e} \{\text{Type II} + (\text{Type III})_2\} \xrightarrow{e} O_2.$$

(It is possible, at the high redox potentials used, that electron transfer is through some of the peptide bond constructs and not through space as discussed in Chapter 4). The substrate leaves as a free radical. The dioxygen remains bound and is reduced in steps to $2H_2O$, i.e. the oxidases are designed to retain O_2, $O_2^{\cdot-}$, O_2^{2-} (i.e. $H_2O + O$), and $O^{\cdot-}$ before giving the second H_2O. The structures of two such oxidases are known (Fig. 15.5), and it is possible to imagine that the three copper atoms (two Type III and one Type II) can act as an interactive system. The reaction steps of the metal ion oxidation states are not very different from that of cytochrome oxidase or, in the reverse direction, from that of the release of dioxygen from the $(Mn)_4$ centre of photosystem II (Chapter 14). Here we suppose that any pair of the three coppers could be involved in the initial uptake of O_2 since the inhibitor azide (N_3^-), which can be compared with $O_2^{\cdot-}$, or O_2^{2-}, is known to bridge between the Type II copper and the Type III copper atoms.

Although the coordination chemistry at the active site for the non-blue Type III copper is again somewhat like that in haemocyanin in that histidines are involved in the binding of copper, in one striking respect the oxidases differ absolutely from the O_2-carriers and cytochrome oxidase. They are built from several β-sheet domains, like most enzymes (Table 15.5 and Fig. 15.5), and they are then constrained so that they do not undergo conformational changes. These oxidases are 'simple' enzymes, not mechanical transduction devices, and this allows them to have more selective surfaces for different substrates. Note that they are all based on similar series of β-sheet domains put together in different ways (Fig. 15.4).

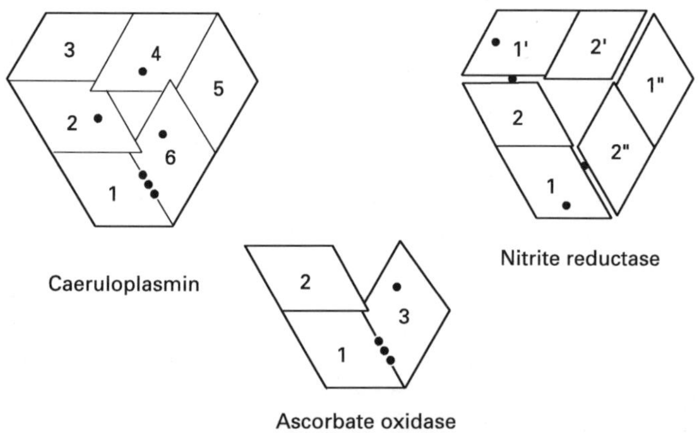

Fig. 15.4 The combinations of rhombic-like β-sheet protein domains which make up a variety of copper enzymes.

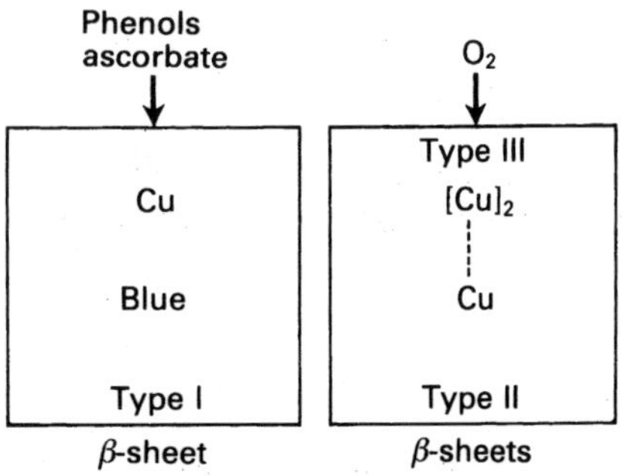

Fig. 15.5 The scheme of binding of many copper oxidases where each domain is a β-sheet, often a β-barrel. Contrast cytochrome oxidase (Chapter 13). (See Huber (1989) in Further reading.)

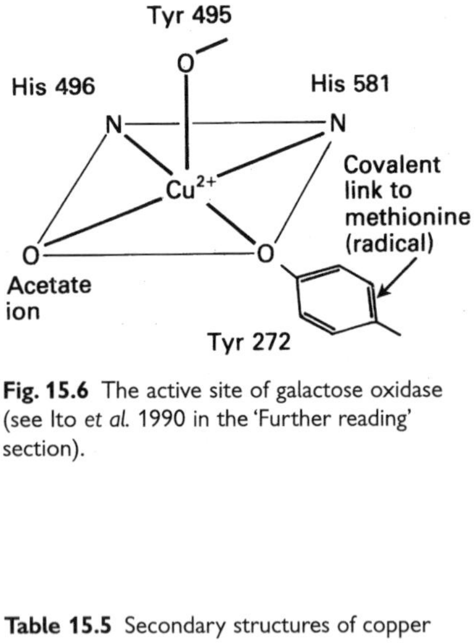

Fig. 15.6 The active site of galactose oxidase (see Ito *et al.* 1990 in the 'Further reading' section).

Table 15.5 Secondary structures of copper enzymes

Enzymes	**Secondary structure**
Electron transfer proteins	Single domain β-barrel
Ascorbate oxidase	Three β-sheet domains
Galactose oxidase	Three β-sheet domains
Nitrite reductase	Two β-sheet domains
Cytochrome oxidase	Cu_A domain. β-sheet: Cu_B domain, helical (?)
Haemocyanin	α-helical at active site

There are one or two copper oxidases which do not fall into this group as judged by both structures, as far as they are known, and spectroscopic data. One example is galactose oxidase (Fig. 15.6), which has but one copper with a tyrosine ligand that becomes a tyrosine radical during the reaction cycle. The reason for the use of a radical in many sugar reactions is not clear, but we have seen this use of a tyrosine radical before, namely in the photoreaction-centre chemistry (Chapters 5 and 14) and we shall meet it again in the PQQ copper oxidases, in riboside reductases and in some other reactions and rearrangements using either Fe–O–Fe or vitamin B_{12} centres (cobalt). These are again examples of constrained (entatic) groups, now of organic sidechains of proteins.

A feature of this copper *enzyme* which is common to most other copper enzymes (see Section 15.5.2) and which has been apparent in the description of the magnesium, calcium, zinc, non-haem iron, manganese *enzymes*, and the previous copper oxidases is the presence of a β-sheet secondary structure at or near to the active site. We stress again that this structure is not present in calcium trigger *proteins* nor at haem active sites which have α-helical constructs.

Another point of interest is the occasional binding of copper to a phenolate which is an indication of the very strong Lewis acid character of Cu(II), compare Fe(III) and Mn(III). This type of binding is not known with other divalent ions in living systems. Finally, note that the association of copper here and elsewhere with high potential *radical* reactions makes it unsuitable for use in the cytoplasm but extremely valuable in external fluids (Section 15.4).

The copper hydroxylases of Table 15.3 belong to yet another class of *vesicular* (extracellular) copper proteins in which copper is bound to two histidines and a methionine. The copper is then open sided and ready to bind dioxygen. There are other oxidases in Table 15.2 which are not yet well described, so that the types of copper centre may be of much greater complexity than those shown in Tables 15.3 and 15.5 (see Section 15.3.5).

15.3.4 Substrates of copper oxidases

The small substrates for copper enzymes are listed in Table 15.1. Leaving them aside, the enzymes oxidize ascorbate, some amino-acids, phenols, some amines, some sugars and perhaps ferrous ion. These substrates are often produced as (organic) free radicals which can diffuse in the aqueous media outside cells. We can imagine that the objective is protection since the oxidation of phenols or indoles in this way leads to free radical polymerization observed in strengthening skins of insects and of plants, i.e. in keratinization of insect cuticle and browning, lignification of cut plant surfaces, and in the pigmentation of animals. [Some of these reactions may have been catalysed by haem proteins earlier in evolution]. Again, the production of ascorbate free radicals may be best seen as the necessary introduction of a feeble free radical in extracellular fluids to scavenge for other dangerous free radicals. In this same sense, caeruloplasmin may remove Fe(II) from the blood, which could otherwise generate free radicals from dioxygen. Amine oxidases, also referred to in Table 15.2, are important since random oxidation of them would produce nitrosamines and other dangerously reactive entities; the amine oxidases help to remove excess transmitters or hormones such as adrenaline, which also requires copper for its production in the vesicles of the adrenal gland.

Reactions with larger substrates of special copper enzymes include the final cross-linking oxidation of collagen and elastin, and in other animals of chitin. Collagen is produced in the cytoplasm but processed in (Golgi) vesicles by

Table 15.6 Zinc and copper enzymes of the external matrix

Enzyme	Metal	Activity Hydrolysis of
Elastase	Zn	Elastins
Collagenase	Zn	Collagens
Stromelysin	Zn	Proteoglycans
Gelatinase	Zn	Gelatins
Protease 24.11	Zn	Peptide hormones
		Oxidation
Lysine oxidase	Cu	Cross-links collagen and elastin
Laccase	Cu	Cross-links via phenols (plants)
Phenol oxidase	Cu	Cross-links via phenols (plants)
Amine oxidase	Cu	Removes amine hormones
Glycine oxidase	Cu	Amidated peptide hormones

N.B. This table is used again – Table 20.15

iron oxidases before being released to the extracellular medium; in this medium it is cross-linked by a copper oxidase. The degree of cross-linking is controlled in different parts of an organism, so that the mechanical properties of much connective tissue are dependent on copper; for this reason, copper deficiencies introduce connective tissue weaknesses. The interactive role of copper and zinc enzymes in the management of connective tissue (Table 15.6), is particular noteworthy since the production of these enzymes must have been a critical step in the evolution of multicellular life. An interesting addition to the known substrates of copper oxidases is the synthesis of amidated peptides in multicellular organisms cross-linked by collagens. Here the initial peptide has a terminal glycine at the carboxyl terminus but the copper enzyme oxidises it away to the amide

$$-CO-NH-CH_2-CO_2H + O_2 \rightarrow -CO-NH_2 + CO_2 + H_2O$$

The reaction occurs in vesicles where the products, which are signalling peptides, are stored. A typical example is enkephalin.

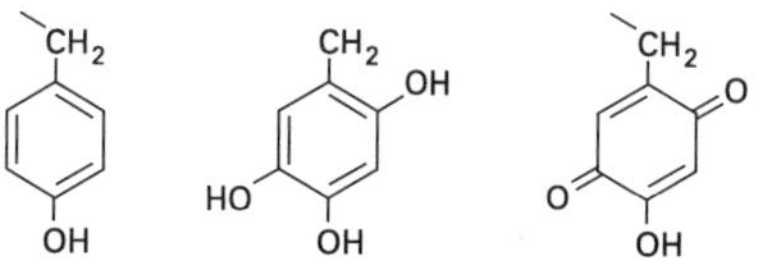

Tyrosine Trihydroxy phenylalanine Quinone (TPQ)

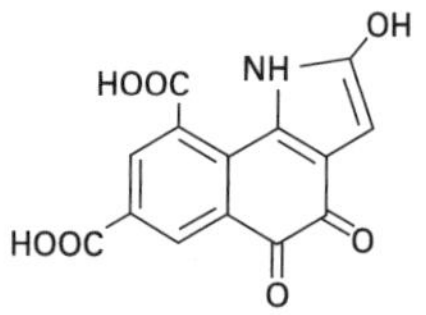

Pyrroloquinoline quinone (PQQ). There is some confusion in the literature as to whether PQQ exists.

15.3.5 Copper proteins and coenzyme PQQ (pyrroloquinoline quinone)

Some copper enzymes have an in-built dihydroxylphenol quinone or a similar dihydroxyaromatic compound as a redox partner (see Aside). Known examples are plasma amine oxidase, diamine or histamine oxidase, choline dehydrogenase, dopamine β-hydroxylase, dopa decarboxylase and the enzyme for deamination of lysine to give desmosin cross-links. The unit, the modified amino acid or coenzyme PQQ, which is to be found only in enzymes outside cells or inside vesicles, goes reversibly to an orthoquinone (see Aside). A possible reason for this curious reactive group is that in order to be part of an effective 2-electron reaction, copper, a one-electron centre, needs either a second metal ion or such an organic group as a second one-electron radical partner. There is a strong suggestion of a general link between this especially modified amino-acid or PQQ (see Aside), ascorbate, and copper in the control of connective tissue (collagen formation) through the requirement for all of these three cofactors in lysyl and prolyl hydroxylation.

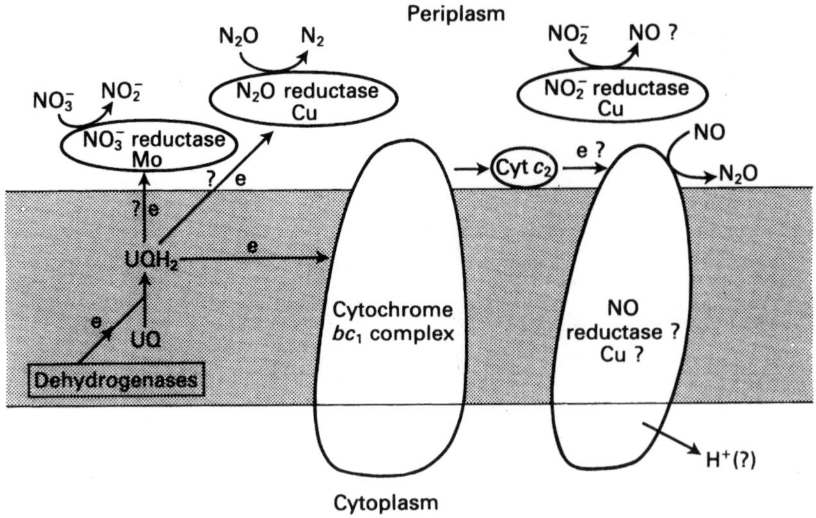

Fig. 15.7 A proposal for the involvement of the stages of nitrate reduction by cells. Nitrate reductase is a molybdenum enzyme while nitrite and nitrous oxide reductases use copper or haem iron. The reactions are largely in the periplasmic space. For a review of these enzymes see Moura, I and Moura, J.J.G. (2001). Structural aspects of denitrifying enzymes. *Current Opinion in Chemical Biology*, **5**, 168–75.

15.4 Copper enzymes and nitrogen oxides

Amongst chemicals which are today quite readily available due to the dioxygen in the air are nitrate and its reduction products, NO_2^-, NO, and N_2O. In a general way these are dangerous chemicals for multicellular organisms, e.g. nitrates must not be high in drinking water, but they are possible sources of energy (see Fig. 1.8). In fact the denitrifying bacteria utilize this possibility and some copper enzymes assist in the reduction of the lower oxides, but not of nitrate which has to be handled first in an atom transfer reaction by molybdenum (see Chapter 17). The reactions of the lower oxides of nitrogen, like the reactions of N_2 itself, are largely confined to bacteria. Their metallo-enzymes, not all of which are copper enzymes, are usually either in the periplasmic space or in the membranes of the bacteria (Fig. 15.7). We shall draw attention to parallel molybdenum enzymes in Chapter 17. One of these enzymes, nitrous oxide reductase, has a Cu_A site [Fig. 15.1(c)]. It may well be that once again some of these reactions were carried out by iron (haem) enzymes early in evolution (see Section 13.10).

15.5 Superoxide dismutase

15.5.1 The location

We have already stated that most copper proteins are outside eukaryotic cells or in the periplasmic space of prokaryotes. Others are in vesicles (Table 15.2). We view the contents of a vesicle as being close to extracellular space since vesicles do not contain reductive reaction pathways and, frequently, they equilibrate their internal fluid with extracellular fluids by endocytosis. As stated in Table 15.2, we find copper blue proteins in the inner solution of thylakoids which are vesicular systems of chloroplasts. We also find copper in the chromaffin granule vesicles of the adrenal gland where a copper enzyme is involved in the last step of oxidation to produce adrenalin. Thus copper, like iron, can act both in synthetic and oxidative clearance steps associated with important hormones. There is, however, an exception to this general placing of copper proteins since virtually all eukaryotic cells have one copper *enzyme* in the cytoplasm, superoxide dismutase, apart from the inactive copper proteins concerned with carrier, buffer, and transcription functions (see Section 15.6). The latter are maintained in the Cu(I) state.

15.5.2 The reactions of superoxide dismutases

The involvement of copper in superoxide dismutase in the cytoplasm could be thought of as an anomaly especially since the prokaryotes do not need such an enzyme to remove superoxide. It is known that bacteria, algae, mitochondria, and chloroplasts manage to eliminate superoxide with iron or managanese enzymes. Thus, if we are to maintain the point that copper entered biology at a late stage, probably when dioxygen accumulated in the atmosphere, we must ask why was it necessary to supersede the prokaryote Mn and Fe superoxide dismutases. We can well believe that the first superoxide dismutases in the

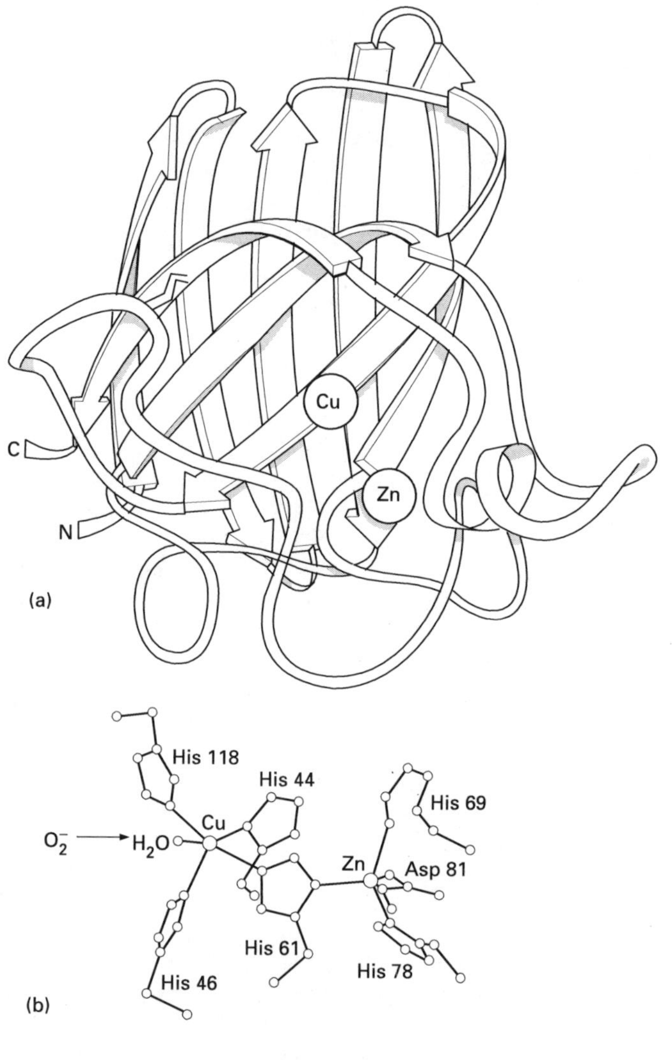

Fig. 15.8 (a) The structure of copper superoxide dismutase which, like most copper proteins, is a β-sheet. (Note the inclusion of zinc.) (b) Details of the metal binding sites.

bacteria would be iron or manganese enzymes since their synthesis would utilize these available ions; no copper was available. The reason for dispensing with these enzymes late in evolution could well be that these Mn or Fe enzymes are not very stable and readily lose Fe(II) or Mn(II). They would, therefore, be dangerous to DNA in the cell cytoplasm of eukaryotes and especially multicellular organisms once O_2 appeared. Since both Cu(I) and Cu(II) bind very strongly, the novel dismutase enzyme of eukaryotes does not give a measurable concentration of free metal ion in the cytoplasm. The structure shown in Fig. 15.8 is clearly designed to protect the copper ion from release. The copper superoxide dismutase is a strong β-barrel and, apart from binding copper via *four* histidine sidechains, the enzyme active-site residues are cross-linked into other parts of the structure by zinc bridges. It does appear as if a special dismutase was required in the cytoplasm of multicellular organisms, which can protect DNA from $O_2^{\cdot-}$ and its products but at the same time would not allow a high free metal ion concentration. The copper of the enzyme is now in a rigid site, more closely demanded by Cu(II), so that Cu(I) is more constrained, entatic. It is also essential that the reaction is very fast, and since it involves electron transfer but at relatively low redox potential, $E \simeq 0.3$ volts, the structure is a very fine 'design' preventing metal ion relaxation during reaction. Finally note again the collaborative role of zinc with copper in this enzyme.

15.6 The transport and homeostasis of copper

15.6.1 The buffer and carrier proteins

High levels of copper in living systems do not appear frequently since copper minerals are rather insoluble. Nutritional deficiency is more frequent due to the fact that copper can be lost by precipitation in the gut of many grazing animals living in molybdenum rich soil areas, for example Cu_2MoS_4. The deficiency leads to connective tissue weakness, as one might expect (see Section 15.3.4). Once copper is digested by animals, its further uptake is controlled by an albumin and by histidine circulating in the blood which deliver the copper to the liver. The binding site of albumin utilizes four nitrogen donor centres, one of which is an ionized peptide. Here again organisms utilize the very strong Lewis acid properties of Cu(II) to selectively carry copper in a now orthodox square coplanar site (see Section 2.4). The redox potential is now very low ($E^\circ < 0.2$ volts). The liver is the source of many of the copper oxidases circulating in blood. These proteins are exported presumably through the Golgi and it may well be that copper is incorporated directly in the vesicles; if excess copper is delivered, then there are various powerful chelating peptides and proteins for its removal.

As stated in Section 15.5.2, copper is maintained at very low levels in the cytoplasm of all cells; the excess of copper is buffered by special proteins (not enzymes) in eukaryotes, metallothioneins (MT), which bind eight Cu(I) atoms in a thiolate complex and are common in animals, and also by *phytochelatins*, which occur mostly in plants. The copper, as copper(I), is bound by thiolates in clusters but some of it is rather exposed. The exact structure of these MT complexes is now known, and it is worth noting that it is different from those of the zinc and cadmium metallothioneins; this allows bound copper to be recognized as Cu.MT and transported to or even exocytosed through vesicles,

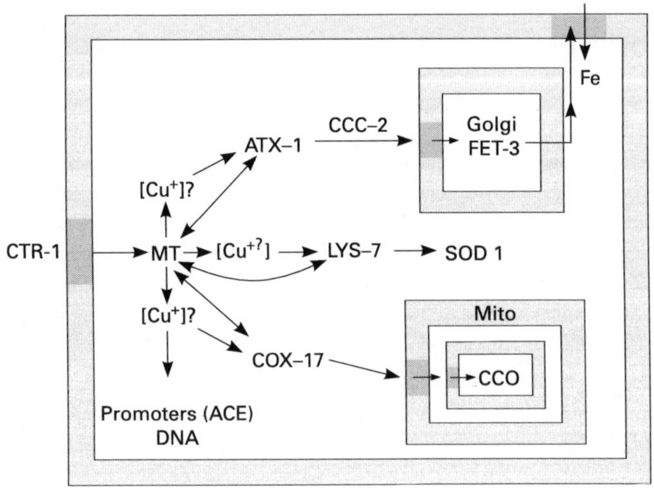

Fig. 15.9 The copper transport system of proteins in yeast. CTR copper transport protein (pump); COX-17, part of cytochrome oxidase system; ACE transcription factor; Lys-7, transporter to superoxide dismutase; CCC, copper pump to Golgi; ATX, carrier to Golgi; FET-3, copper protein exported to plasma membrane to assist iron uptake, compare caeruloplasmin; MT, metallothionein.

while zinc metallothionein remains as a zinc buffer in the cytoplasm. Metallothionein protein syntheses are strongly regulated and it is likely that their concentrations are always just in excess of the free Cu, Zn, and any Cd in a cell. Homeostasis is built here around feedback regulation of the synthesis of metallothioneins for these three metal ions (see Section 15.6.4). Note that the binding constant of metallothionein to copper(I) is very high so that free $[Cu^+]$ does not exceed 10^{-15}M in cells. Free copper(I) is then only a useful notional parameter and most copper is transferred by carriers, not as free ions even in cells.

This very low level at which free copper can possibly equilibrate with proteins in cells leads us to consider how the handling of the metal in movement across cells and through membranes is managed. The problem has been seen before in the transport of calcium (Section 10.12) where the metal ion was handled by the use of pumps and carrier proteins. Today a series of such protein pumps and carriers for copper are known which have confusingly different labels in different species (see Fig. 15.9 for the example of yeast). We need to understand first how the copper can be moved from the outside of the cell by a pump into the cytoplasm and then into a final site by using a carrier in the cytoplasm to transfer the metal to an enzyme, or to a vesicle or an organelle membrane, there to be pumped further before being inserted in a final protein, now outside the cytoplasm. In a vesicle, for example the Golgi, it is known that copper, like iron, is incorporated into its proteins with the help of 'chaperones', here calcium binding proteins, (Section 10.2). From the vesicles the formed copper proteins may be exported to the extracellular space. We clearly need to understand the nature of the pumps and carriers especially since certain copper-related neurodegenerative diseases, Menkes and Wilson, have been tracked to mutations in the pumps and may be there are others due to mutations in carriers. Recent studies have also implicated copper in the pathogenesis of prion-mediated encephalopaties (see 'Further reading').

15.6.2 The copper pumps

If the supply of copper proteins, mainly extracellular or in vesicles, is to be managed then the copper must be transported across membranes and the production of the various copper proteins must be synchronized with its build-up. A copper pump protein is now known which has resemblance to the calcium ATP-ase (outward) pump, but the copper pump uses two linear thiol groups in place of carboxylates to bind the metal in the external pump regions, as well as a cysteine/proline (CP) region in the membrane, (see Fig. 10.13). The pumps are not very different from those for Cd and Zn but are selective against these ions while they may pump Ag and Hg. It is probable that the copper carrier proteins are very similar to the external domains of the CPx type copper ATP-ase (see Section 15.6.3) and that they carry copper bound in between two thiolate groups. In many organisms there are pumps both for the input and output of copper. External copper is reduced at the membrane by the iron proteins FRE 1 and FRE 2 of yeasts.

15.6.3 The copper carriers

Copper pumps carriers and enzymes and their genetics

The way in which cells handle copper so as to produce the essential copper enzymes is very dependent upon the type of cell under discussion. Consider bacteria first. Anaerobic bacteria do not appear to use copper. Moreover, it does not appear that there is an essential function for copper in the aerobic bacterial cell cytoplasm. However, the production of the copper proteins of *aerobic bacteria*, all of which are either in the periplasm or on the outside of the plasma membrane, must be dependent on appropriate modes of transport. We already mentioned that the two proteins for handling copper in the membrane are two CPx-ATP-ase pumps—one moving the Cu^{+} ion inwards, the other rejecting it. In fact, the copper is handled in a more complicated way at an earlier stage of transfer between the outer membrane and the inner membrane by a transport protein in the periplasm. This may well belong to the TON system of transporters we have mentioned in Section 12.3.

The difficulty of controlling copper is increased in all *single-cell* eukaryotes since the metal ion has to be distributed to organelles (again in their protected space such as the periplasm or the interior of thylakoids), and to other spaces outside the cytoplasm, e.g. the Golgi or other vesicles. To maintain homeostasis of copper the element has several complex transport proteins. In these aerobic eukaryotes there is now a copper protein in the cytoplasm, superoxide dismutase, and it again has a special carrier for delivery of copper to it. Fig. 15.9 indicates the complexity of the system for copper transfer in yeast from an initial uptake to metallothionein, MT, and then to three different transport, so-called (mistakenly) 'chaperone' proteins, ATx-1, LYS-7, and COX-17. The destination of the copper from these three supply routes is shown. Here we note in passing that all these systems have parallel genetically coded proteins in man. The role of metallothionein is pictured as being not just a buffer but a protein reagent for distribution of the metal, see the corresponding calcium and zinc proteins, but it may have further roles. Other points of note are that FET-3 of yeast is a plasma membrane copper-dependent protein which controls iron uptake. As described above, we do not know the exact ways in which copper is transferred from protein to protein. The proteins CTR-1, ATx-1, LYS-7 like the copper pumps all contain the sequence Met-X-Cys-X-X-Cys, to which Cu^{+} is undoubtedly bound on the surface of the protein. Now, all these proteins are produced only under instruction from the DNA, hence further proteins sensitive to copper are required to bind to promoter regions of DNA, (see Section 15.6.4).

In *multicellular* organisms the homeostasis of copper increases in difficulty since different organs and even different cells in different organs require copper in different amounts. Moreover, copper is concerned with the reactions in circulating extracellular fluids. As stated, in some animals, e.g. octopus, copper proteins in the extracellular fluid transport dioxygen bound in haemocyanin and must have copper delivered to them. Other notable features are the extra involvement of copper proteins in the production of hormones or transmitters such as adrenaline or amidated peptides, and the use of copper enzymes in the cross-linking of extracellular matrices. Of course, all these functions require further transport proteins and new promoter regions of DNA for their production and their activation by copper. A further difference from unicellular yeast is that the copper protein equivalent to FET-3 (for bringing iron across the yeast plasma membrane) is the extracellular 'carrier' caeruloplasmin of multicellular organisms, which is a copper oxidase for converting Fe^{2+} to Fe^{3+} before incorporation in transferrin but which may also transport copper. We shall turn to what is known about the DNA (gene) structures relating to copper proteins, after an analysis of exchange rates.

15.6.4 Copper exchange rates

There are three reasons for supposing that, for the most part, there is little signalling due to free copper or copper *enzymes*. First, the copper exchange rates between enzymes and free copper are very slow due to the strength of Cu(I) and Cu(II) binding to ligands. A binding constant at pH = 7 of about 10^{15} M^{-1} is to be expected in many of the enzymes we have described. Second, copper in enzymes is largely extracellular and it is difficult to imagine how these copper enzymes would send back messages to the DNA. Third, copper is usually bound deeply inside enzymes and these proteins are made from β-sheets, hence exchange is inhibited, both in the forward and reverse directions, by the strength of the protein fold.

At least four groups of proteins bind copper differently from these final sites in *enzymes*. They are the buffers, carriers, pumps and transcription factors, mentioned above. All these proteins have copper somewhat exposed in thiolate complexes and they may well have somewhat weaker binding constants though they are still close to 10^{15} M^{-1}. The exchange of copper between these proteins is found to be quite rapid. There is reason to suppose that the mechanism of diffusion of copper between them all is not by exchange with free copper but, as shown in Fig. 15.9, by direct exchange, so that copper behaves as a covalently attached atom like H or NADH. The final step of incorporation into an enzyme is then by insertion.

Recently, methods have been devised which may help us to follow the kinetics of copper and other metal ion incorporations. The method is based on the isotope effect on exchange where the light isotope is always preferential transferred. The method is well-known in H, C, N, O, and even S chemistry, but it has only become possible for heavier elements recently due to advances in mass spectrometry. So far the method has yielded quantitative observations which have yet to be given detailed interpretation (see O'nions *et al.* in 'Further reading').

15.6.5 Copper genetics

The major controls over copper metabolism as seen above rest with six processes: (1) reduction of Cu(II) in extracellular fluids to Cu(I) for uptake;

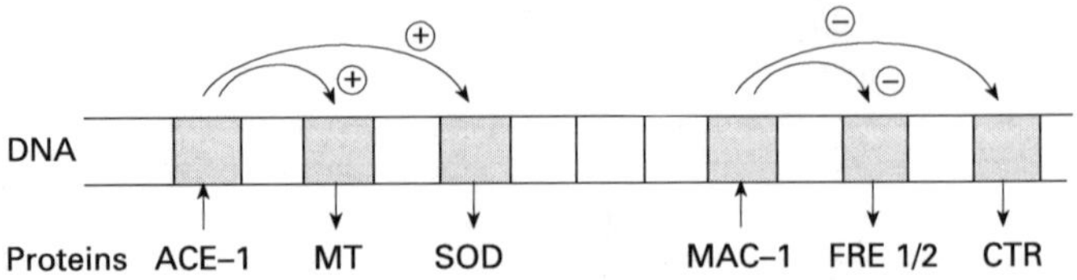

Fig. 15.10 A scheme for the structure of copper regulated genes in yeast. ⊕ Increase in expression, ⊖ Decrease in expression.

(2) uptake or export pumps for copper to or from the cytoplasm, from or to extracellular fluids and vesicles; (3) buffering or storage of copper in metallothionein; (4) transport of copper by 'chaperones'; (5) incorporation of Cu(I) into the relevant proteins by transfer; (6) export of copper proteins. The genes for the synthesis of the corresponding proteins are all known to be dependent upon the availability of copper (probably bound) in the cytoplasm which is recognized by the state of copper(I) binding transcription factors. The proteins concerned are shown in Fig. 15.9, that allows us to draw a scheme of copper-regulated copper-protein expression for yeast, but which is thought to be similar to that for higher plants and animals (Fig. 15.10). In the figure, change in copper in the cytoplasm, leads to change in activity of ACE and MAC transcription factors. ACE activity leads to increase in metallothionein (for buffering) and of superoxide dismutase (for reducing possible oxidative damage through Cu/O_2 reactions), while MAC activity leads to a decrease in the Cu^{2+}-containing-reductase (FRE) and of the copper uptake (pumps and transporters represented by (CTR). In the context of this book we again draw particular attention to the fact that FRE, controlled by copper transcription factors, is an iron uptake protein, while the transcription factors ACE and MAC, as well as binding four coppers in a cluster, contain a zinc-binding site. Thus Cu, Fe, and Zn activities are again seen to be linked. Plants and yeasts have an alternative mode of sequestering excess copper: *in vacuoles*. They synthetize a polypeptide, not directly a gene product, called phytochelatin (see above) which is transported into the vesicle together with copper (pumped) where they are stored as a rather complicated Cu_n (phytochelatin)$_m$(S^{2-})$_p$ product. The genetic control of this metabolism is not known.

15.7 The overall functions of copper

The curiously intriguing feature of copper biochemistry is the way in which the reactions are separated from parallel oxidative functions controlled by iron (haem) or manganese enzymes. The explanation, we believe, is that copper was not used until the coincidental appearance of substantial quantities of dioxygen. The scheme of events would be:

1. early life in the presence of extremely insoluble copper sulphides (Fig. 1.11); no copper proteins;
2. increasing amounts of dioxygen in the atmosphere;
3. increase in concentration of copper(II) resulting from the oxidation of copper sulphides and the coincidental loss of iron as $Fe(OH)_3$ (see Fig. 1.11);
4. protection against 'poisonous' copper by sequestration (MT) and export to outside or into vesicles.

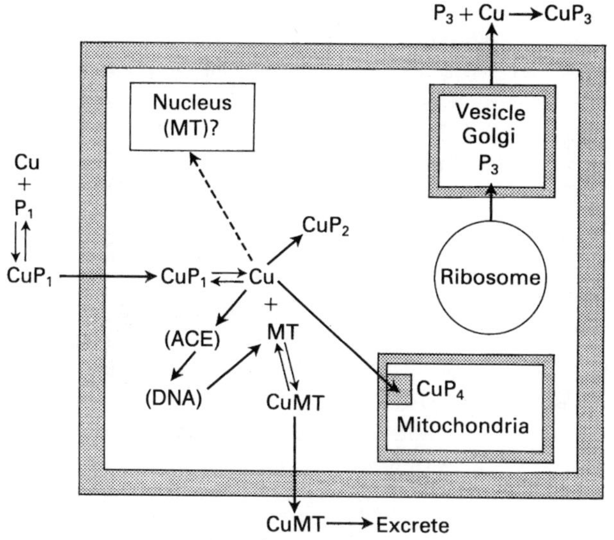

Fig. 15.11 A general scheme for the handling of copper in cells and their compartments. MT is metallothionein.

5. use of copper in such energy-capture systems as cytochrome oxidase on the outside of membranes;
6. internal protection by copper both in cells (superoxide dismutase) and in the bloodstream from adventitious free radicals;
7. increase in free radical attack, due to copper, on the extracellular matrix of unicellular organisms to produce cross-linked proteins;
8. evolutionary development of cross-linked connective tissue, e.g. collagen and chitins via free radical polymerization using copper enzymes (leading to multicellular organisms).

The further use of copper in multicellular organisms could then follow:

9. protective activity and as a receptor for ethylene in plants;
10. the development of hormonal messengers, e.g. adrenaline and amidated peptides in animals

The complexity of the uses of copper is illustrated in Fig. 15.11. If all this is true, available copper became the new essential extra-cytoplasmic element for multicellular development in life. It followed the use of iron and haem iron since once it was available and controllable, it was easier to employ than iron or haem *outside cells* even at high potential close to the O_2/H_2O value of +0.8 volts. It is curious to note that this development probably arose from the introduction into the environment of an extremely 'poisonous' substance (dioxygen) followed by another one (copper) and that adaptation to their presence led to such striking evolution as that of multicellular life. Once again we find the curious evolutionary sequence, see calcium and zinc uses,

$$\text{poison} \rightarrow \text{protection} \rightarrow \text{signalling} \rightarrow \text{use}$$

The levels of free copper in the cytoplasm after the atmosphere contained dioxygen then became fixed at a nominal 10^{-15} M. Did this value then become a quantitative characteristic of all aerobic organisms? Is it a characteristic of their metallome? In what way did the genome adapt (or was it pushed) by the environment?

Further reading

Averill, B.A. (1966). Dissimilatory nitrite and nitric oxide reductases. *Chemical Reviews*, **96**, 2951–64.

Bertini, I., Hartmann, H.-J., Klein, T., Gaohua, L., Luchinat, C., and Weser, U. (2000). Structure of Cu_7 metallothionein. *European Journal of Biochemistry*, **267**, 1008–18.

Brown, K., Djinoric-Carugo, K., Holtia, T., Cabrito, I., Saraste, M., Moura, J. J. G., Moura, I., Tegoric, M., and Cambillon, C. (2000). Revisiting the CuZ cluster of Nitrous Oxide Reductase. *Journal of Biological Chemistry*, **275**, 41133–36.

Cooper, C. (2001). Metals and disease: copper on the brain. *The biochemist*, June 2001, 11–13.

Eipper, B.A., Stoffers, L.A., and Mains, R.E. (1992). Oxidative production of some hormonal signals. *Annual Review of Neuroscience*, **15**, 57–85.

Ferguson, S.J. (1987). Denitrification *Trends in Biochemical Science*, **12**, 354–8.

Forbes, J.A., Hsi, G., and Cox, D.W. (1999). Copper binding domains of copper pumps. *Journal of Biological Chemistry*, **274**, 12408–13.

Gaykema, W.P.J., Volbeda, A., and Hol, W.G.J. (1986). Structure of haemocyanin. *Journal of Molecular Biology*, **187**, 255–75.

Gray, H.B., Malmström, B.G., and Williams, R.J.P. (2000). Copper coordination in blue proteins. *Journal Biological Inorganic Chemistry*, **5**, 551–559.

Guss, J.M. and Freeman, H. (1983). The blue copper electron transfer proteins. *Journal of Molecular Biology*, **169**, 521–63.

Harris, E.D. (2000) Cellular copper transport and metabolism. *Annual Reviews of Nutrition*, **20**. 291–310.

Harris, E.D., Qian, Y., and Reddy, M.C. (1998). Genes regulating copper metabolism. *Molecular Cell Biochemistry*, **188**, 57–62.

Hu, S., Furst, P., and Hamer, D. (1990). DNA and copper binding functions of ACE 1. *The New Biologist*, **2**, 544–55.

Huber, R. (1989). Copper clusters in oxidases. *Angewandte Chemie (International Edition)*, **28**, 848–69.

Huffman, D.L., and O'Halloran, T.V. (2000). Energetics of copper trafficking. *Journal Biological Chemistry*, **275**, 18611–18614.

Ito, N., Phillips, S.E.V., McPherson, M.J., Keen, J.N., Stevens, C., Yadav, K. D.S., and Knowles, P.F. (1991). Galactose oxidase. *Nature*, **350**, 87–90.

Jimenez-Cervantes, C., Garcia-Borron, J.C., Valverde, P., Solano, F., and Lozano, J.A. (1993). Tyrosinase enzymes in melanocytes. *European Journal of Biochemistry*, **217**, 541–56.

Larsson, S. (2000). Energy saving electronic pathways in proteins. *Journal Biological Inorganic Chemistry*, **5**, 560–564.

Malmström, B.G. and Leckner, R. (1998), The chemical biology of copper. *Current Opinion in Chemical Biology*, **2**, 286–92.

O'nions, R.K., Zhu, X., and Williams, R.J.P. (2001) to be published.

Rasmussen, T., Berks, B.C., Sanders-Lochr, J., Dorley, D.M., Zumft, W.G. and Thomson, A.J. (2000) The nature of copper clusters in nitrous oxide reductase *Biochemisitry 39*, 12753–56.

Rauser, W.E. (1990). Phytochelatins. *Annual Reviews of Biochemistry*, **59**, 61–86.

Richards, M.P. (1989). Copper regulatory proteins. *Journal of Nutrition*, **11**, 1062–70.

Radisky, D., and Kaplan, J. (1999). Regulation of transition metal transport across the yeast plasma membrane. *Journal of Biological Chemistry*, **274**, 4481–4.

Ryde, U., Olsson, M.H.M., Roos, B. O., De Kerpel, J.O.A., and Pierlot, K. (2000). The role of strain in blue copper proteins. *Journal of Biological Inorganic Chemistry*, **5**, 565–74.

Sarkar, B. (2000). Copper in diseases. *Journal of Inorganic Biochemistry*, **79**, 187–91.

Stites, T.E., Mitchell, A.E., and Rucker, R.B. (2000). Physiological importance of quinone cofactors. *Journal of Nutrition*, **130**, 719–27.

Solomen, E.I., Sundaran, V.M., and Mcahonkin, T.E. (1996). Multi copper oxidases and oxygenases. *Chemical Reviews*, **96**, 2563–603.

Weser, U. (1997) Biological chemistry of copper in *Biological Chemistry* (ed) Trautwein, A.X. Wiley-VCH. Weinheim, Germany.

Waggoner, D.J., Bartnikes, T.B., and Gitlin, J.D. (1999). The role of copper in neurodegenerative diseases. *Neurological Diseases*, **6**, 221–30.

Williams, R.J.P. (1995). Energised (entatic) groups in metalloproteins. *European Journal of Biochemistry*, **234**, 363–81.

Winge, D.R., Gradon, J.A., Posewitz, M.C., Martins, L.J., Jensen, L.T., and Simon J. R. (1997). Sensors that mediate copper-gene activities. *Journal of Biological Inorganic Chemistry*, **2**, 2–10.

Note: Through the work of N. J. Robinson (Newcastle, UK) the evolution or at least the involvement of copper in prokaryote photosynthetic organisms can be followed from very primitive organisms up to plants and animals. It appears that many life forms have both uptake and rejection pumps for copper and have regulatory proteins for its transport and buffering. While copper metallome levels may well have increased from the earliest anaerobes ($\leq 10^{-20}$ M free copper) to the most recent organisms ($\leq 10^{-15}$ M) the basic control rests in the chemistry of thiolate binding to copper which exceeds greatly that to any other available metal ion, see Chapter 2. However detoxification of very rare metal ions such as Hg, Cd and Ag undoubtedly uses very similar genetically controlled thiolate proteins. See Abstracts C.1 Metals and Cells, of the Society for Experimental Biology (2001), Meeting held at Canterbury, UK.

CHAPTER 16

Nickel and cobalt: remnants of early life?

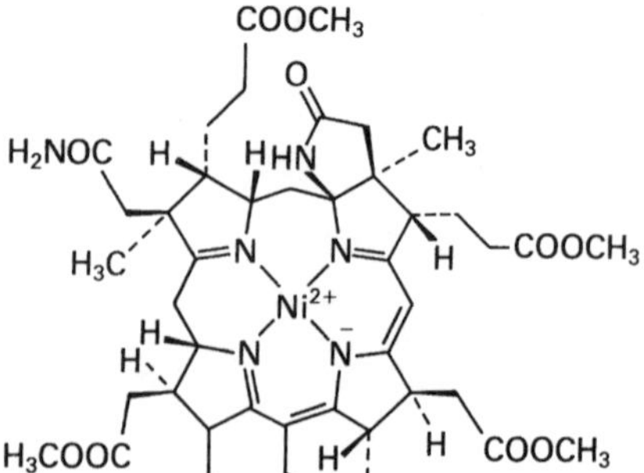

F-430

Fig. 16.1 The structure of the nickel-containing coenzyme, F-430.

16.1 Introduction

We start this chapter with the remark that the functions of these two elements appear to belong to very primitive systems indeed. In some sense it looks as if the elements were at their most useful when metabolism was based on such chemicals as CH_4, CO, and H_2, but that after the advent of dioxygen their value diminished. Thus, the concentration of nickel and cobalt in some anaerobic bacteria can be quite high. The concentration of both in human serum is below 1 μg/L compared with 100 μg/L for zinc, iron and copper.

In higher organisms the activity of *nickel* is apparently confined today to one curious enzyme, urease, although the symbiotic anaerobic bacteria of these higher organisms still use nickel in several reactions. In vertebrates there is no known direct use of nickel based on their own genes. Free anaerobic bacteria, especially methanogens, have also kept the nickel hydrogenases and other nickel enzymes, but these methanogens belong to a special domain of life, the archaebacteria, most of which have been relegated to anaerobic niches within the geochemical world. They are thought to be of very primitive ancestry and also use uniquely special cofactors including the F-430 ring chelate of nickel (Fig. 16.1) as well as a special sulphydryl coenzyme-M (see Chapter 19).

The well-recognized activity of *cobalt* is almost totally confined to functions of vitamin B_{12} (Fig. 16.2). There are also two or three known cases of the occurrence of cobalt porphyrins in bacteria. Most other cases of cobalt-activated systems have been identified in a somewhat unsatisfactory way by adding back cobalt to extracted systems and we shall not describe them further (but see Chapter 11 on zinc enzymes) although there are a few apparently well-defined non-B_{12} cobalt enzymes which do need attention. Even the use of vitamin B_{12} in living organisms may also be on the way out, in an evolutionary sense, since the transfer of methyl groups and the rearrangement reactions by radical mechanisms, which depend on B_{12} in some organisms, can be done by methionyl CoA and zinc, iron- or manganese-utilizing enzymes, respectively, in other organisms, and especially in plants. The iron and manganese enzymes are dependent on dioxygen, but see primitive ribonucleotide reductase, Section 12.10. Vitamin B_{12} is also required in archaebacteria and this association of nickel and cobalt with early anaerobic organisms suggests immediately that their organometallic chemistry may have had a considerable importance in early life-forms. It is possible too that B_{12} and F-430 functions predated iron porphyrins.

Whereas in haem biochemistry there are a variety of chemical modifications of the haem itself through such reactions as formation of thioethers or oxidation, the rings of vitamin B_{12} and, as far as we know, that of F-430, are fixed.

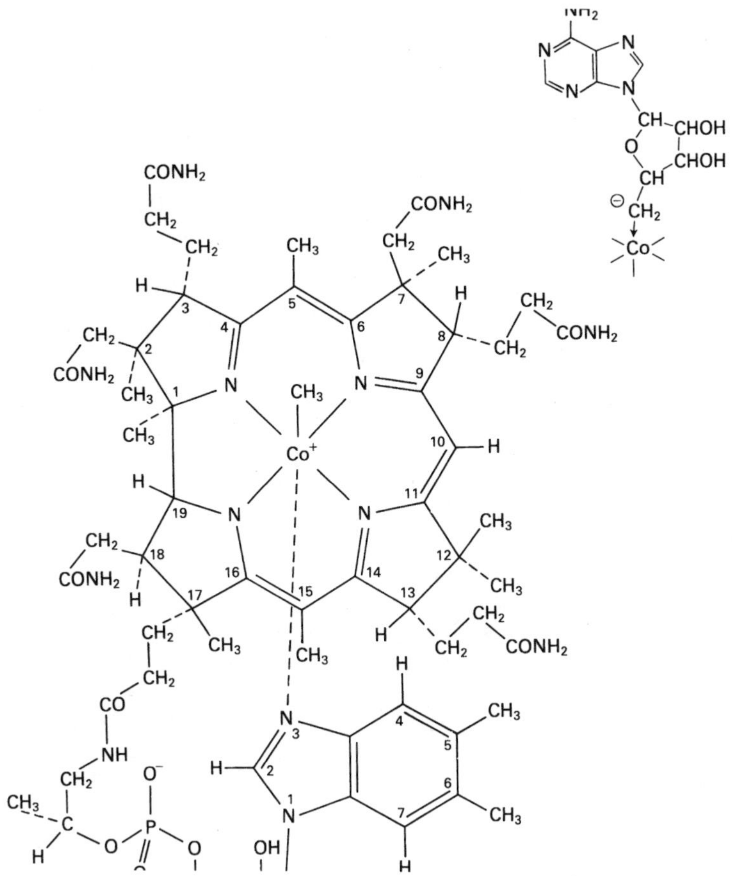

Fig. 16.2 The structure of the methylated coenzyme from vitamin B_{12} which contains cobalt. The B_{12} coenzyme for many other reactions has a cobalt-carbon bond to the 5′-CH_2^+ of adenosyl, see inset.

Variation is introduced in the B_{12} series by changes in the nature of the bound base, benzimidazole. The cobalt of B_{12} is sometimes bound by sidechains of proteins, histidines, as is the case for the metals of haem and chlorophyll. The situation is less clear in the case of the nickel of F-430. It is quite possible that much of the diversity of haem chemistry is of recent evolutionary origin, following the advent of dioxygen, e.g. cytochromes *a*, and that it was relatively simpler in the distant past (see Chapter 13).

Good reasons for the tendency to eliminate nickel and cobalt from biology today are that neither seems to have a very special chemistry in acid/base catalysis which could not be replaced by zinc, which has become available more recently and has chemical advantages (Chapter 11), and their redox functions, in a dioxygen-dominated atmosphere, would seem to be replaceable by the more accessible iron, copper, or manganese. The source of carbon today is virtually confined to CO_2, so that O_2 must be lost in synthesis of carbon compounds, in direct contrast with the necessary loss of H_2 when CH_4 was a carbon source. It is noteworthy that nickel in bacteria is mostly used in H_2 reactions. Again, cobalt is a relatively rare element and today nickel is not easily available from silicate rocks, e.g. soils, since it is highly stabilized in the octahedral holes of silicate lattices. However, nickel is available from the sea and we must suppose that it is unwanted by organisms since its exchange kinetics are too slow (Chapter 4) and it could be a threat to many Mg^{2+} reactions since both Ni^{2+} and Mg^{2+} have very similar structural demands. Because of the extensive primitive use of nickel and cobalt we describe this redox biochemistry first following a brief introduction to the chemistry of these two elements. Later in this chapter we return to the few uses of these elements in hydrolytic reactions.

16.2 The chemistry of cobalt and nickel

Iron, cobalt, and nickel have two features in common which separate them from earlier and later elements in the periodic table. All three elements give a variety of oxidation states and a variety of spin states and all have more than five 3d electrons in their lower oxidation states, i.e. they are electron-rich. While stability of oxidation state (III) falls along the series Fe to Cu, the stability of oxidation state (I) increases, while oxidation state (II) is dominant through all this series of elements. Through a combination of increasing electron affinity and decreasing size in any one oxidation state the balance of bound ligands shifts from oxygen donors to nitrogen donors, while sulphur donors retain a reasonable and increasing binding power relative to oxygen donors throughout the series (Chapter 2). This general picture is complicated by the ligand field stabilization energies of the different metals in different oxidation states (Chapter 2) which allow stability to particular electronic states of the different ions to appear and also connect particular stereochemistries with these states. Table 16.1 lists some features. The elements nickel and cobalt can clearly be described as special, in that they are not only generally electron-rich, especially in lower oxidation states and in low-spin states, but in that some of their 3d electrons are forced into exposed σ-(or π-) orbitals in these low-spin states by the preferred symmetry of their complexes. In a conventional tetragonal field it is usual for a d^8 low-spin ion, Co(I) or N(II), to have a very exposed doubly occupied d_{z^2} orbital, so that it is a donor rather like RS^-, i.e. it is a very strong Lewis base. The tetragonal low-spin d^7 ion, Co(II) or Ni(III), is a reactive free radical, able to give or take an odd electron, also in the exposed d_{z^2} orbital, to or from other molecules. In this respect d^7, low-spin ions are to be

Mn Fe	Co Ni (Cu)	(Cu) Zn
3d shell reactivity in π-orbitals	3d shell reactivity in σ-orbitals	3d shell reactivity almost removed
O_2,NO	H_2,CH_3^-	

Table 16.1 Occurence of ligand field states

		High-spin Octahedral	Tetrahedral	Low-spin Octahedral	Tetragonal
Iron(II)	d^6	Usual	Less common(S)*	Quite stable(N)*	Rare
Iron(III)	d^5	Usual	Common(S)	Less stable(N)	Rare
Cobalt(I)	d^8	?	?	Rare	Usual
Cobalt(II)	d^7	Common	Common	Rare	Usual
Cobalt(III)	d^6	(Rare)	Rare	Usual	Rare
Nickel(I)	d^9	(Tetragonal)	Rare	Rare	Usual
Nickel(II)	d^8	Dominant	Rare	Rare	Usual
Nickel(III)	d^7	Common(?)	Common(?)	Rare	Usual
Cooper(II)	d^9	(Tetragonal)	Rare	–	–

*S and N refer to the nature of the donor atoms.

compared with σ-organic free radicals. This exposure of d electrons in σ-orbitals is only possible in low-spin states after d^6, Fe(II) in the first transition metal series, since the division of orbitals in an octahedral field protects at most six 3d electrons in the lower set of t_{2g} orbitals of Fig. 4.8. These orbitals have π- and not σ-character. We see that low-spin nickel and cobalt ions have a very special potential reactivity associated with occupied σ-orbitals as well as having π-donor capability and lower symmetry (tetragonal) than is usually seen for low-spin d^6, Fe(II). The lower symmetry and coordination number expose an active site (Figs 16.1 and 16.2). On the other hand it is difficult to see why a biological organism should select high-spin cobalt or nickel ions in acid/base catalysis in preference to zinc ions in a case of equal availability. First we look at the reason for incorporating these metal ions in ring chelates (Figs 16.1 and 16.2). Tetragonal ligands have been developed from porphyrin for Fe, Co, and Ni based on such ring structures, but the evolved rings are not the same for the three metal ions (Section 3.12). For iron, the porphyrin ring (together with imidazole donors, histidines of proteins) is highly conjugated and symmetrical to protect to a marked degree *all* the out-of-plane 3d electrons of Fe(II) (six 3d) or Fe(III) (five 3d) in π-orbitals. We have already described how the fifth and sixth coordinating positions can be used to modulate the properties of the iron in these porphyrin complexes especially between high and low-spin states (Chapter 12). Where one site is open, as in oxidases, the switch to low-spin d^6, especially for iron, gives a high affinity to the binding of dioxygen. For cobalt, the ligand ring (corrin) size is smaller (Fig. 16.2) giving a smaller hole to stabilize Co(III) by σ-bonding, and now in B_{12} the remaining fifth and sixth coordination positions are not necessarily from the protein. This makes the coenzymes from vitamin B_{12} into apparently stable octahedral *low-spin* d^6 complexes. However, two other factors have changed from iron porphyrin chemistry: (1) in the structure, the fifth coordination position is occupied either by a non-protein base, usually a benzimidazole, or a protein histidine, and (2) the sixth coordination position can be occupied by a saturated σ-carbon donor to make a metal–carbon bond. The unit becomes a very stable and transferable coenzyme and B_{12} derivatives are vitamins. The chemistry with a carbon-metal bond is quite unusual for elements normally forming rather polar bonds in water and, in effect, it provides a relatively weak bond which can break in two ways:

$$Co^{III}\text{–}CR_3^- \rightarrow Co(I) + {}^+CR_3,$$
$$Co^{III}\text{–}CR_3^- \rightarrow Co(II) + {}^{\bullet}CR_3.$$

Hence, cobalt-carbon chemistry in low-spin complexes is associated with easy development of carbonium ion and carbon-radical intermediates involving saturated carbon chemistry and tetragonal d^8 or d^7 oxidation states. The combination of cobalt with the B_{12} ring system and carbon chemistry is a masterpiece of synthetic judgement produced by evolution. Yet it is being replaced.

The ligand for nickel is F-430, which is no longer a planar ring due to its heavier saturation (Fig. 16.1). This ligand can be used to provide open space mainly to one side of the nickel ion. In this complex the low-spin d^8 ion Ni(II) (compare Co(I)) can accept a CH_3^+ group to form a Ni-CH_3 bond. If we look upon this state as a Ni(IV)-CH_3^- compound, a $3d^6$ complex, we see that it can go on to form an octahedral unit. It is now well suited to combining with CO in the absence of dioxygen, much as is the open-sided form of Fe(II) porphyrin ($3d^6$) in haemoglobin. A reaction typical of organometallic chemistry could then follow because this metal ion has two *adjacent* sites not blocked by the ring chelate, in contrast to porphyrin chemistry. CH_3CO^+ is a leaving group for general synthesis. (The exact pathway may not perhaps be of this kind since we

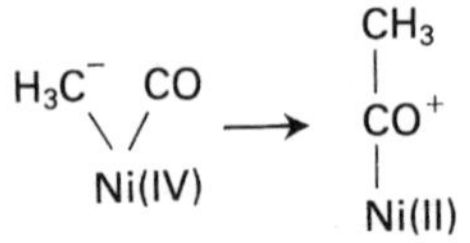

(16.1)

know that in methanogens the first carbon to enter must come from CO_2 and it must be reduced to CO before it can be activated in the organometallic synthesis (16.1), see Fig. 16.3. Insofar as this chemistry is correctly described, it is possible to see how both cobalt and nickel systems were delicately developed to *activate very simple* C_1 carbon compounds. C_1 syntheses are now unusual and most biochemistry starts from C_2, except for photosynthesis where the C_1 unit is incorporated into sugars directly. A further use of the nickel in the F-430 ring system appears to be as a desulphurization catalyst.

In conclusion, we can see how evolution has prepared ring systems so that special chemistry could be carried out with very particular and admirably useful metal ions in particular spin states and geometries (Table 16.1). The elements selected and discussed in this chapter, Ni and Co, have special value opposite σ-bond organometallic chemistry, i.e. M-H and M-CH_2R. The switch to high redox potential reactions and other factors, see Chapter 20, that dioxygen brought about removed the absolute need for these low potential reactions which vitamin B_{12} and F-430 achieved. We have described the reactions of haem proteins in Chapter 13, but much of that chemistry came later in historical time. While noting the sophistication of this small-molecule chemistry, we must also note that it developed within proteins. Cooperativity between protein and metal ion chemistry is demanded by evolutionary pressure so that the special reactivities at the metal centres, already highly developed by the ring ligand, were further enhanced by proteins. The imposed structures generate constrained 'entatic' conditions in the coordination sphere, which are illus-

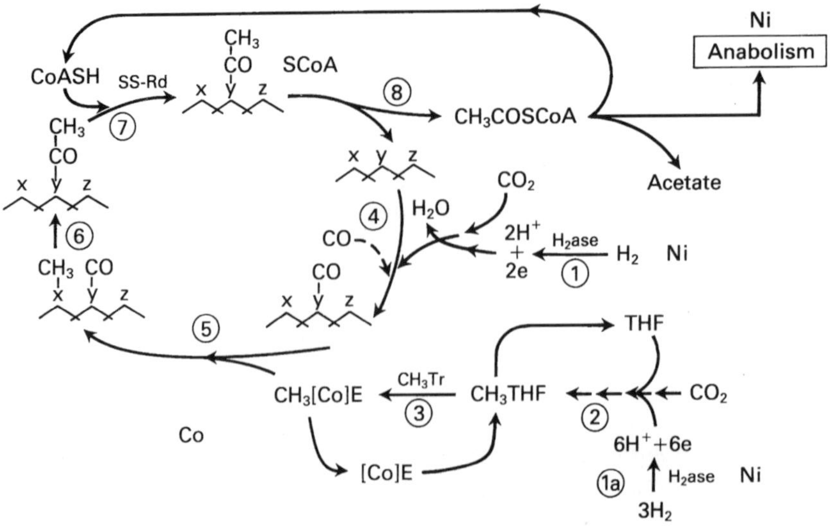

Fig. 16.3 The reaction pathway of carbon fragments in archaebacteria. x, y, and z are catalytic sites on a framework; THF, tetrahydrofolate. Note the involvements of Ni and Co (B_{12}). (After the work of Wolfe and Thauer.)

trated in Section 16.4. Again, selective sites for substrates evolved, which do not rely on binding to the rings. We note particularly the significance of the ring propionate in energy capture (Section 13.9) using haem. We look first at some further nickel enzymes which were required to activate another early reactant, dihydrogen, and which are still used by anaerobes.

16.3 **Hydrogenases**

The ability to react with dihydrogen, H_2, or to produce dihydrogen from water is well known in chemistry. The reactions are catalysed by a variety of compounds containing at least the following metals which are available to organisms: Fe, Co, Ni, Cu. It is also easy today for man to make hydrogenation catalysts from rarer metals, such as Ir, Rh, Pt, and Ag. Clearly, the best metal ions are from the electron-rich ends of the transition metal series in the periodic table, but if we assume that copper was not available to life in its earliest period and that cobalt was more of a rare element than iron or nickel, then these last two elements, as sulphur complexes or clusters, could well have been the preferred catalysts in biology for these reasons alone. Since the reactions are redox reactions and involve the divalent state, a considerable advantage from the use of nickel is that it dissociates less readily from a protein. The easiest reaction to envisage for the activation or release of a molecule of dihydrogen is hetereogeneous dissociation, which can be written reversibly

$$M^+ - X^- + H_2 \rightleftharpoons M^+H^- + H^+X^- \quad (16.2)$$

This is the way the reaction of copper(I) or silver acetate with dihydrogen is written and the mechanism involves the loss of one ligand from the coordination sphere. Now, it is known that the hydrogenases which involve iron alone are derived from very unusual Fe_4S_4 clusters. No matter what the value and disadvantages of iron, however, one system which evolved for hydrogenases uses the more strongly bound, non-exchanging nickel. The structure of the nickel/iron hydrogenase is now known. The active site is described in Section 12.6.1. These enzymes must be very primitive and some are known which contain an atom of selenium in place of one sulphur. The use of the extraordinary ligands CO and CN does remind one of organometallic chemistry. Some other special reaction of nickel enzymes are included in Table 16.2 for comparison.

Table 16.2 Nickel-containing enzymes: localization and function

Enzyme	Localization	Function
Urease	Extracellular (Vesicle)	Hydrolysis of urea
Co-dehydrogenase (+ B_{12} (cobalt) active site)	Membrane	Catalysis of the reaction $4H_2 + 2CO_2 = CH_3COOH + 2H_2O$ (acetogenic bacteria)
Factor F-430	Membrane	Catalysis of the reaction $4H_2 + CO_2 = CH_4 + 2H_2O$ (methanogenic bacteria)
Bacterial hydrogenases (several)	Membrane	Catalysis of the reaction $H_2 = 2H^+ + 2e$
Superoxide dismutase	Cytoplasm (rare)	Catalysis $O_2^{\cdot-} \rightarrow H_2O_2 + O_2$

NB Nickel enzymes may well be involved in desulphurization too.

16.4 The reactions of vitamin B_{12}

Table 16.3 lists some of the enzymes which contain vitamin B_{12}, either in the simple form or as a coenzyme. Only one, methyl malonyl CoA mutase, occurs in mammals and most others are confined to bacteria. There are no known such cobalt enzymes in plants. (We shall describe the simple cobalt proteins in Section 16.6.2). There are four groups of important reactions:

1. the reduction of ribose to deoxyribose;
2. the rearrangement of diols and similar molecules;
3. the rearrangement of malonyl to succinyl;
4. the transfer of methyl groups.

An immediate observation is that all these reactions require the approach of quite big substrates to the cobalt since even in methyl transfer the transfer is to methionyl CoA. Therefore, one side of the vitamin B_{12} ring system must be very open during reaction and we know that this is the side of the carbon σ-bond (Fig. 16.2). The benzimidazole base remains attached in some enzymes, but in others it is replaced by a histidine of the apoprotein. The histidine nitrogen appears to form a long Co–N bond and it has been suggested that this is another example of the constrained (entatic) condition at active sites (see 'Further reading'). The protein groove containing the cobalt ring must also be able to open very considerably as in cytochrome P-450 but not in haemoglobin (Fig. 16.4). Such a system could only remain secure if it did not react with dioxygen or if it was protected from dioxygen. It is known that vitamin B_{12} Co(II) and Co(I) complexes separated from enzymes are vulnerable to dioxygen and so we conclude that the B_{12} enzymes must react today by preferred order mechanisms, i.e. in the order:

blocked centre; binding of substrate to open the centre; reaction; close centre

as products leave, see Mancia *et al.* (1999) in Further Reading. The coenzymes are known to have an α-helical protein which is mobile and undergoes conformational change.

The exact nature of the cobalt intermediate has been established as Co(II) in reactions (1), (2), and (3), but is Co(I) in reaction (4). The use of a radical

Table 16.3 Some examples of enzymes requiring B_{12} coenzymes: localization and function

Enzyme	Localization (in bacteria)	Function
Methionine synthetase	Cytosol	Conversion of homocysteine into methionine
Methylmalonyl CoA-mutase	Cytosol	Conversion of methylmalonyl-Co-A into succinyl-CoA
Glutamate mutase	Cytosol	Conversion of L-glutamate into L-threo-β-methylaspartate
Diol dehydratase	Cytosol	Conversion of diols into aldehydes
Lysine mutase	Cytosol	Conversion of D-lysine into 2,5-diaminohexanoic acid
Ethanolamine deaminase	Cytosol	Cleavage of amino group in $NH_2CH_2CH_2OH \rightarrow CH_3CHO + NH_3$
Ribonucleotide reductase	Cytosol	Reduction of ribonucleotides to the corresponding deoxyribonucleotides

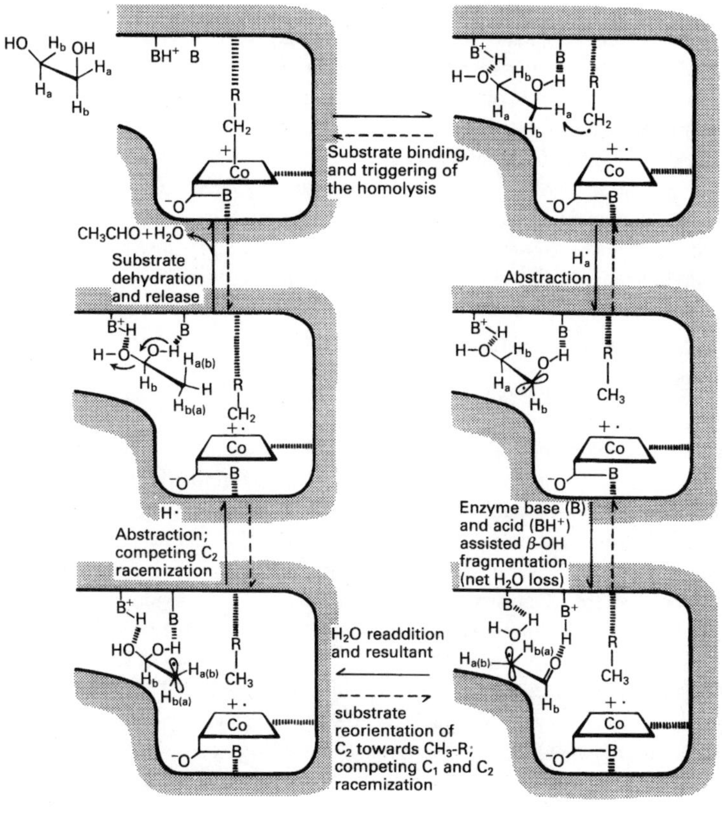

Fig. 16.4 A plausible reaction pathway for the diol-dehydratase reaction catalysed by coenzyme B_{12}.

pathway follows from the homolytic splitting of Co(III) bond to the carbanion of adenosine (Fig. 16.2). The diradical formed has as its primary reaction centre the carbon radical of a sugar, which allows H-atom transfer to and from substrate while rearrangement occurs (Fig. 16.4). The mechanism has been thoroughly analysed and the Co-C bond breaking in the enzyme requires a small amount of energy (10 kcal) to be compared with 20 kcal in free solution. Such is the power of the protein.

In methyl transfer it is Co(I) which is the acceptor/donor of CH_3^+. This is a curious reaction since the redox potentials of cobalt in the vitamin indicate that Co(I) is extremely difficult to obtain in water, but there appears to be a starter reaction involving Co(II) in the enzyme reaction sequence. Interestingly, Co(I) in these complexes does not form a Co-H bond with H^+ until a pH of around zero (contrast Ni and Fe hydrogenases). The discrimination in favour of CH_3^+ against H^+ may be due to the very large size of the d_{z^2} orbital.

Of very special concern is the reduction of ribose to deoxyribose, catalysed by ribonucleotide reductase, which is considered to go via the radical pathway. In eukaryotes and many bacteria the enzyme is not of this kind and uses iron (see section 12.10). In the earliest form this was an Fe/S protein. There is a certain bewilderment in the thought that there was an early requirement for a metal complex as complicated as vitamin B_{12} in such a central reaction; perhaps we can accept that the earliest life was based on RNA, but it looks almost impossible to conceive how a switch to DNA took place if this switch needed vitamin B_{12}. It is therefore probable that vitamin B_{12} had evolved much earlier to assist organometallic chemistry in organisms.

The still later change for this reaction to an iron or manganese enzyme with an M_2O structure, a dependence of seemingly greater simplicity (Section 12.10), clearly requires a very high redox potential, i.e. of dioxygen, to generate tyrosine free radicals. Cobalt allows free radical chemistry at very low redox potentials. Something of extreme importance for the understanding of the early development of living system is still missing. As stated there is known an iron–sulphur form of ribonucleotide reductase which generates a glycine radical and perhaps this was the really primordial form.

Methylation by B_{12}-enzymes may easily go astray and is the cause, through methyl mercury, for example of poisonings in fish and man.

16.5 The organometallic chemistry of life

Before we discuss the relevance of organometallic chemistry to early life we must examine the nature of the probable ligands. Of course this organometallic chemistry could not be that of non-aqueous phases using such ligands as phosphines or cyclopentadienes which are so common in man's organometallic chemistry. The obvious electron-rich ligands in water are compounds of sulphur, sulphide, and thiolates. Today organometallic chemists avoid this chemistry advisedly since it is very reactive; organisms could not avoid it, but it could protect this chemistry in proteins. Now, as well as providing water-soluble electron-rich compounds, small colloidal sulphide particles, M_nS_n, and anions such as MoS_4^{2-}, the early earth provided large sources of H_2S and various sulphide surfaces all far out of equilibrium in the sea. The source of the unstable minerals and chemicals was and still is volcanoes and ocean vents.

The reactions and the development of these surfaces and particles as non-living *autocatalytic systems* in reactions with CO_2, H_2, CH_4, and so on, could have been the point at which the organic chemistry of life began. It could then have generated just random polymers, at first associated only with templates of sulphide or clay particles; the next step to ordered polymer production we all would recognize as genuinely living organometallic chemistry. Much research on the reaction of sulphide complexes, colloids, and surfaces would appear to be a priority to an understanding of these possibilities. Some residual chemistry of this kind could rest in the hydrogenases and CO chemistry (Ni), nitrogenases (Mo, V, Fe), and ferredoxins (Fe). Note that the sulphide chemistry could have been energized from the very nature of its sources. Here we wish to include the low-spin cobalt(III) nitrile hydratases, see also the Fe(III) enzyme (Section 12.6) and note the ligands which includes three peptide nitrogens, three thiols and can bind cyanide.

The development of porphyrin-related systems for Co, Ni, Fe and Mg could have followed from the relatively easy synthesis of this very special macrocycle starting from metal cyanides. A scheme has been proposed by Eschenmoser (see 'Further reading').

16.6 Hydrolytic catalysis by nickel and cobalt

Today it is recognized that nickel and cobalt are not just used in redox reactions but have occasional value in hydrolytic activity in special organisms. One or two uses are more general such as nickel in urease and perhaps cobalt in methionine removal from the termini of newly synthesized proteins. This raises two questions: (1) is there special value in the use of these elements? and (2) how is their selection managed? We examine their uses before turning to uptake and incorporation. A list of the enzymes of interest is given in

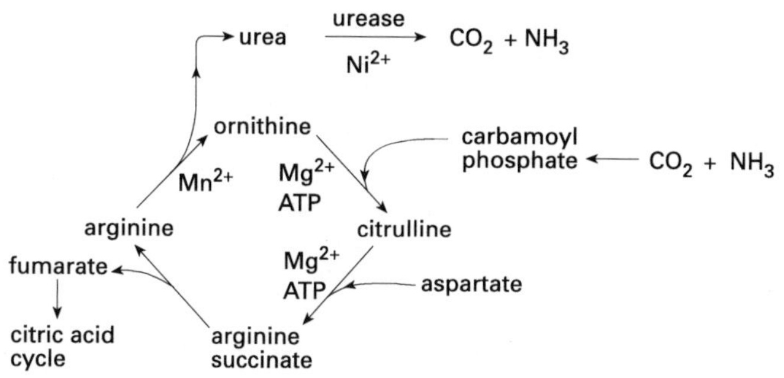

Fig. 16.5 The urea cycle showing the involvement of Mn^{2+} and Mg^{2+} with the side-reaction of Ni^{2+} in urease.

Tables 16.2 and 16.4. The most studied is nickel in urease which we describe first. It is part of the urea cycle which has a peculiar set of metal-ion dependencies (see Fig. 16.5).

16.6.1 **Urease: nickel in vacuoles**

The disappearance of nickel chemistry from higher organisms can be accounted for partly by the strength with which nickel is held within oxide rocks in octahedral holes, partly by the kinetic risk which it generated since the style of redox chemistry it promotes is very vulnerable to dioxygen, and partly by its ability to compete with Mg^{2+}. There is quite curiously residual nickel in urease, where it acts in the animal (not the vertebrates) and plant kingdoms as an acid catalyst for the particular pathway of decomposition of urea so as to generate ammonia. The reaction is

$$H_2N \cdot CO \cdot NH_2 \xrightarrow{H_2O} H_2N \cdot CO_2^- \cdot NH_4^+ \xrightarrow{H_3O^+} 2NH_4^+ + HCO_3^-.$$

Two nickel ions are used in the enzyme in some concerted fashion. The structure of the active site is shown in Fig. 16.6. It is a peculiar geometry again revealing the use of constrained (entatic) states. We must suppose that some special advantage accrues from the use of nickel rather than zinc, but it is hard to see what that advantage is. Perhaps it was that nickel was more available than zinc early in evolution. Notice that the nickel enzyme is in a vesicle and, generally, nickel taken up by plants is deposited in vacuoles. Modern

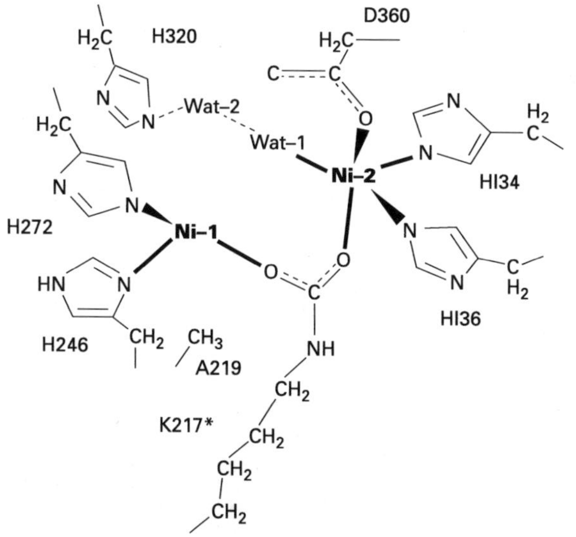

Fig. 16.6 The structure of the active site of urease, (see Jabri *et al.* in 'Further reading').

organisms reject nickel, but at the same time the urea cycle is very important for many of them, requiring nickel to keep ammonia balance.

This urea cycle carries out the net reaction

$$NH_4^+ + HCO_3^- + 2H_2O + \text{aspartate} + 3\ \text{ATP}$$
$$\downarrow$$
$$\text{Urea} + \text{fumarate} + \text{AMP} + \text{P}\cdot\text{P} + 2\text{P} + 2\text{ADP}$$

We are not concerned in this book with the intermediate steps but we observe that this is one route for the elimination of excess nitrogen ($NH^+{}_4$) as urea. Excess ammonia is very poisonous and different species have different modes for its elimination. Man, amongst many land animals, uses urea formation in this way. Now urease itself does the reverse in that it generates eventually ammonia from urea. While this could be advantageous where nitrogen is in short supply, e.g. in bacteria and plants, it seems pointless in land animals. The mechanism of hydrolysis via ammonium carbamate may be pointing to a quite different function for the enzyme but this is not known. Finally, we note again that the use of this enzyme with two *nickel* atoms at the active site is indicative of a very special use of bound nickel. Why is it locked away in vesicles? A similar problem arises with glyoxalase I.

A further peculiarity of this nickel enzyme is that the incorporation of nickel apparently requires some five proteins as judged by its gene structure (Section 16.8). This kinetic control over synthesis is another example of the lengths to which organisms go in order to develop special enzymes and to keep 'dangerous' metals away from the cytoplasm.

16.6.2 Cobalt enzymes not using B_{12}

The most interesting use of simple cobalt (see Table 16.4) appears to be in methionine amino peptidase which occurs in the cytoplasm of many

Table 16.4 Cobalt proteins* (excluding B_{12}-dependent)

Proteins	Function and comment
Methionine aminopeptidase	Removal of terminal methionine Co(II) dimer (bacteria to animals).
Prolidase	Hydrolysis adjacent to proline Co(II) dimer.
Glucose isomerase	Unusual: in actinomycetes. Other M(II) cations are active.
Methylmalonyl-CoA carboxytransferase	Bacteria [Co(II) + Zn(II)].
Aldehyde decarbonylase	Algae (not higher plants?).
Lysine 2,3 amino-mutase	(Bacteria); Zn(II), Fe_n/S_n also.
Adenylate kinase	Sulphur bacteria, Zn(II)
Bromoperoxidase	Bromination (bacteria); V or Fe in other organisms (e.g. algae).
Nitrile hydratase	–CN bond hydration (bacteria) Co(III). Sometimes Fe(III) enzyme.
Cobalt transporter COT 1, GRR1, HOXN	Transfer of Co(II) (bacteria and yeast) Note activity with Zn(II) and Ni(II)
Cobalt porphyrin protein	Electron-transfer? (bacteria).

*NB Note the absence of plant proteins and the rare appearance in animals. All proteins are in the cytoplasm. It is not always clear that cobalt is in fact the *in vivo* catalyst.

organisms. The active site has two cobalt atoms and the binding groups are sidechain protein N/O donors. There is no obvious reason why this site does not select zinc since many zinc sites in enzymes have an identical set of ligands. One possibility is that these and other cobalt (and nickel) enzymes are remnants of a period when zinc was of low availability. To avoid zinc competition today the cobalt (or zinc) must then be carried to a site of insertion in the enzyme so that it is never exchanged with free aquated ions, see next section. The same descriptions and rationalisations may apply to other enzymes of Table 16.4. In passing one reason for retaining cobalt in methionine peptidase could relate to the use of cobalt (B_{12}) in methionine synthase.

16.7 **Nickel and cobalt uptake**

The uptake of nickel into bacteria and plants must be mediated by some energized system. There is the strong suggestion that the pump is, in fact, the magnesium pump. The hydrate of nickel (or of cobalt) is very similar in all respects to that of magnesium. Of course, this coincidence has its dangers since magnesium is absolutely essential and nickel must be selectively removed, e.g. by plants into vacuoles. Risks to the health of humans from nickel poisoning is not inconsiderable. The way in which nickel and cobalt are so selectively incorporated in their special ring systems is not totally understood, but we have described potential procedures in Chapter 2.

The carrier for nickel uptake in plants is known to be histidine and it is thought that the subsequent handling of nickel by pumps and carriers is by histidine sidechains of proteins, Fig. 16.7. Competition with zinc is avoided since zinc is preferentially handled by thiolate donors and here selection against zinc may well arise through the different stereochemical demands of the two elements: zinc tetrahedral and nickel octahedral. Inside the cell it may well be that nickel (and both cobalt and zinc) are handled by carriers which accept them directly from pumps and give them directly to enzyme sites.

16.7.1 **Genetic controls over nickel and cobalt uptake**

The genetic structures controlling expression of cobalt and nickel proteins have been studied in several bacteria, (Section 16.8). It is commonly observed that the genes related to proteins of one metal are clustered together so that

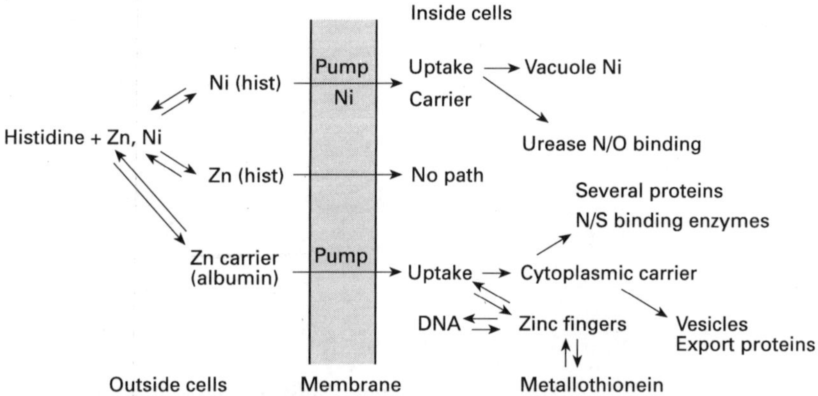

Fig. 16.7 The uptake systems and their connections to proteins for Ni^{2+} and Zn^{2+} in a modern cell.

regulation is of the genetic complex. The usual structure, where expressed proteins are in brackets, is

promoters. polymerase. [pump, carriers, enzymes]

In bacteria there may be ten or so genes in total for nickel or cobalt separately. Knowledge of the chemistry suggests that the system is switched on (or off) by the presence (absence) of the enzyme substrate or by the presence (absence) of the metal ion concerned. Inevitably there is amongst the pumps a nucleotide triphosphate requiring (energizing) protein. The common ground between nickel and cobalt systems hides the need for selective chemistry. This is obvious in the requirement for nickel in F-430 but cobalt in B_{12} exclusively. The two ring systems have the same precursor compounds and these compounds also serve to make porphyrins (Fe) and chlorins (Mg). Since little free ligand of any kind or indeed much free approteins for the elements are observed there has to be shut-down mechanisms at the points of departure of the synthetic routes based on feed-back from metal ions and the substrates of their enzymes. Obviously we are far short of a complete appreciation of these genetic regulations.

16.8 Cobalt and nickel genes in *E. coli*

Cobalt and nickel uptakes in *E. coli* are most needed under anaerobic conditions since they are especially valuable in the metabolism of small molecules and fragments such as H_2, CO and -CH_3. Interestingly there is induced in *E. coli* both a nickel uptake and a cobalamin synthesis series of reactions in anaerobic but not in aerobic conditions. One regulatory gene is *fnr* which is associated with expression of formate nitrate-reductase (FNR) and hydrogenase genes which are part of a coupled enzyme system. The FNR transcription factor is an iron/sulphur protein which is one of the switches used for the change from aerobic to anaerobic life (see Chapter 12). In anaerobic conditions Fe^{2+} is increased so that [Fe^{2+}] is the master switch. Note that formate dehydrogenase is a molybdo-seleno-protein while hydrogenase is a nickel–iron protein.

The nickel transport system has five genes *nik ABCDE* and there is an additional gene for the insertion of nickel into dehydrogenases. The NIK ABCDE proteins are presumed to include an ATP-ase pump which has a series of histidines in it. It is notable that nickel binding to histidines appears to be of great selectivity. In archaebacteria there has to be a series of gene products for the formation of the F-430 organic ring system. Much more is known of the synthesis of the vitamin B_{12} ring for cobalt where there are some fifteen genes, labelled *cob* in a sequence. The expression of these genes is under control of the usual kinase protein phosphorylation genes *arc A* and *arc B*. The cobalt transport system is uncertain but both cobalt and nickel are transported to some degree by the magnesium uptake protein COR.

The simple combination of Fe^{2+} and phosphate switches, FNR and ARC, here in Co and Ni uptake is very common and links redox activity to phosphate metabolism and then to a vast number of biological regulatory and control systems.

16.9 Conclusion

The environment in which life began could be called electron-rich; it had an atmosphere of reducing gases and much sulphide was available at the surface.

Living organisms in such an atmosphere needed special catalysts to handle compounds such as CH_4, H_2, CO, and H_2S. Cobalt and nickel centres are very valuable in these circumstances since they have reactive sulphides and their complexes could develop to give both π and σ-donors in low-spin open-sided systems. The advent of dioxygen changed the environment to one which was relatively electron-poor, containing O_2, N_2, and CO_2. This change demanded a switch in catalysts to metals such as manganese, iron, and copper. Such is the depth of involvement of the trace elements, the metallome, in life. In the 'hidden-world' of archaebacteria, major catalysts remain Ni- and Co-based (F-430 and corrin, for example), but today in higher organisms these catalysts are lost or trivial in extent when compared with those of iron and copper oxidases. In large part nickel is now rejected, cobalt as a simple ion is not even taken up but became part of a *vitamin*. Where nickel is used in higher organisms it is in vesicles. At all times the free cobalt or nickel in the cytoplasm, in its free metallome, may not have exceeded 10^{-10} M.

Further reading

D'souza, V.M., Bennet, B., Copik, A.J., and Holz, R. C. (2000). Divalent metal binding properties of the methionyl aminopeptidase from *E. coli*. *Biochemistry*, **39**, 3817–26.

Eschenmoser, A. (1994). B_{12}: reminiscences and afterthoughts. *Ciba Foundation Symposium*, **180**, 309–29.

Eitinger, T. and Mandrand-Berthelot, M.A. (2000). Nickel transport systems in microorganisms. *Archives of Microbiology*, **173**, 1–9.

Ferry, J.G. (1999). Enzymology of one-carbon metabolism in methanogenic pathways. *FEMS Microbiology Reviews*, **23**, 13–38.

Fontecilla-Camps, J.-C. and Ragsdale, S. W. (1999). Nickel-iron-sulphur active sites: hydrogenase and CO-dehydrogenase, In R. Cammack and A. G. Sykes (eds). *Advances in Inorganic Chemistry*, Vol. 47 283–332, Academic Press, San Diego.

Kobayashi, M. and Shimizu, S. (1999). Cobalt proteins. *European Journal of Biochemistry*, **261(1)**, 1–9.

Krämer, V., Cotter-Howells, J.D., Charnock, J.M., Baker, A.J.M., and Smith, J.A.C. (1996). Free-histidine as a metal chelator in plants that accumulate nickel. *Nature*, **379**, 635–8.

Jabri, E., Carr, M.B., Hausinger, R.P., and Karplus, P.A. (1995). The crystal structure of urease from *K. aerogenes*. *Science*, **268**, 998–1004.

Lowther, W.T. and Matthews, B.W. (2000). Structure and function of the methionine aminopeptidase. *Biochimica and Biophysica Acta*, **1477**, 157–67.

Ludwig, M.L. and Matthews, R.G. (1997). Structure-based perspectives on B_{12}-dependent enzymes. *Annual Review of Biochemistry*, **66**, 269–313.

Mancia, F., Smith, G.A., and Evans, P.R. (1999). The crystal structure of complexes of methylmalonyl-CoA reductase. *Biochemistry*, **38**. 7999–8005.

Marsh, E.N. (1999). Coenzyme B_{12}-dependent enzymes. *Essays in Biochemistry*, **34**, 139–54.

Ragsdale, S.W. (1998). Nickel biochemistry. *Current Opinion in Chemical Biology*, **2**, 208–15.

Ragsdale, S.W. and Thauer, R. (2000). One-carbon chemistry: CO_2, CO, CH_4, formate. 1-Reductive Chemistry. In J. S. Valentine, I. Bertini, and H. Gray (eds), *Biological inorganic chemistry: structure and reactivity*, Univ. Science Books.

Roderick, S.L. and Matthews, B.W. (1993). Cobalt-dependent methionine amino-peptidase. *Biochemistry*, **32**, 3907–12.

Thauer, R.K. and Bonacker, L.G. (1994). Biosynthethesis of coenzyme F–430. *Ciba Foundation Symposium*, **180**, 210–22.

Thomä, N.H. and Leadley, P.F. (1998). Mechanistic and structural studies in methyl-CoA-mutase. *Biochemical Society Transactions*, **26**, 293–8.

Volbeda, A., Charon, M.-H., Pires, C., Hatchikian, E.C., Frey, M., and Fontecilla-Camps, J.C. (1995). Crystal structure of the nickel-iron hydrogenase. *Nature*, **373**, 580–7.

See also:

Vitamin B_{12} and B_{12}-proteins (1998). B. Krantler (ed.), Wiley-VCH, Weinheim.

CHAPTER 17

Molybdenum, tungsten, vanadium and chromium

17.1 Introduction

In this chapter we shall describe the biological chemistry of the metals molybdenum, tungsten, and vanadium, and mention the possible involvement in organisms of chromium. There is some stress in the chapter on the relationship of Group 6 elements (see Aside) with the non-metal elements of Group 16, i.e. with sulphur and selenium. Thus, the elements are often found covalently bonded in cofactors and not as simple metal ions, and in Chapter 19 we shall see that phosphate and sulphate are also often held in such cofactors or coenzymes. There is a close parallel between molybdenum chemical properties and those of tungsten and vanadium, and it has been found that in one biological cofactor, FeMoco, Mo can be replaced by V (or Fe) while in the other, Moco, Mo can be replaced by W. We give first some background information about the availabilities of molybdenum, tungsten, vanadium, and chromium.

17.2 The availabilities of molybdenum, tungsten, vanadium, and chromium

The availability of molybdenum from the sea today is relatively high (see Fig. 17.1), but its distribution on land is extremely patchy. In a small country such as Great Britain there are areas where soils generate an agricultural problem due to a deficiency of molybdenum and other areas where molybdenum is in excess. Perhaps the greatest problem arises from the excess molybdenum in soils since it causes copper deficiency in animals grazing on such lands. It would appear that the difficulty arises from the metabolism of the anaerobic bacteria in the rumen of these animals. One of the reactions of molybdenum on entering the anaerobic first digestive system of the ruminant is the conversion of the molybdate into thiomolybdate by the bacteria of the stomach. Unfortunately. MoS_2^{4-} is a very good scavenging precipitation reagent for copper and so the copper content of the diet is reduced almost proportionately with the excess of molybdenum. The animals then suffer from weakened connective tissue since copper enzymes are required to cross-link collagen external to the animal cells (Chapter 15). In other geographical areas there is molybdenum deficiency which frequently relates to the difficulties of incorporation of molybdenum from molybdate of soils (see Chapter 2). While vanadium is yet more available (Fig. 17.1), tungsten is less so by a factor of ten, but is still as available as copper and more available than iron and manganese. Chromium is as available as tungsten and copper. The use of these elements, V and W, in what we take to be primitive life may relate to the

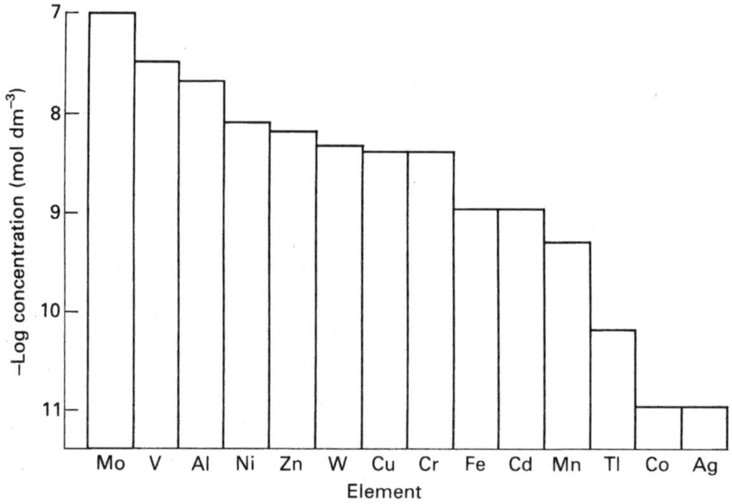

Fig. 17.1 The concentrations of some elements in sea water. Note that these are total concentrations and do not refer to free ion species.

lower availability of molybdenum early in earth's history (see below). Today molybdenum is the most important element of this group.

To get molybdenum chemistry in its proper biological perspective let us turn next to a brief survey of molybdenum in enzymes.

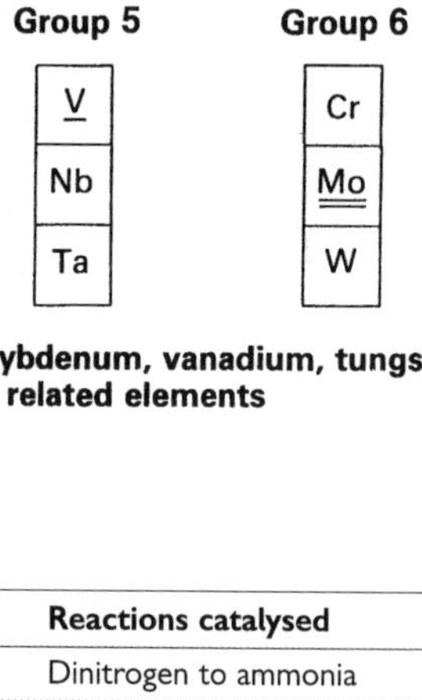

Molybdenum, vanadium, tungsten and related elements

17.3 The molybdenum enzymes: a first overview

Table 17.1 lists many of the known molybdenum enzymes and their cofactors. The enzymes are functionally quite intriguing; in effect, while nitrogenase catalyses the six electron reduction of N_2 to NH_3, others, e.g., xanthine oxidase, aldehyde oxidase, sulphate reductase, nitrate reductase, and fumarate dehydrogenase, catalyse an oxygen atom transfer or two-electron reaction, i.e.

Table 17.1 The major molybdenum enzymes

Enzyme type	Enzyme families	Examples	Reactions catalysed
Multinuclear M centre	Nitrogenases (Mo, V, Fe)	Mo-nitrogenase	Dinitrogen to ammonia
Mononuclear (pterin bonded)	Oxido-reductases	Xanthine oxidase	Purine or pyrimidine catabolism Aldehyde to acid
	I. Xanthine oxidase family (hydroxylases) (15–20 members)	Aldehyde oxidase Formate dehydroferase	 Formate to CO_2
	II. Sulphite oxidase family (eukariotic oxotransferases) (2–3 members)	Sulphite oxidase Plant nitrite reductase (assimilatory)	Sulphite to sulphate Nitrate to nitrite
	III. DMSO-reductase family (bacterial oxotransferases) (2 pyranopterins bonded to Mo) (8–10 members)	DMSO reductase Nitrate reduction, dissimilatory; terminal respiratory oxidase	DMSO to DMS Nitrate to nitrite
Others		Pyridoxal oxidase Xanthine dehydrogenase Pyrogallol transhydrolase	

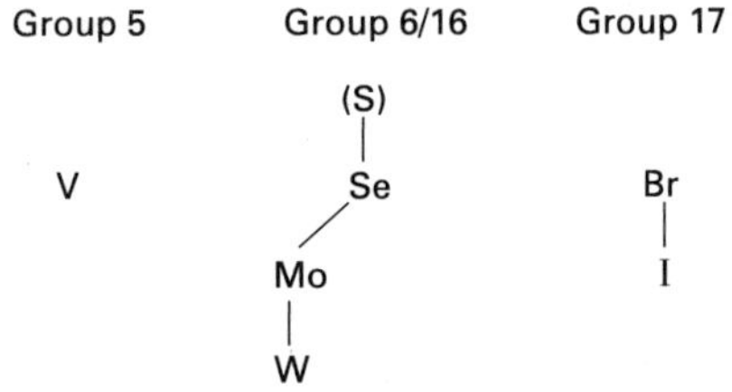

Fig. 17.2 Atom centres from which oxygen atoms are relatively readily transferred.

$$X + O \rightarrow XO$$
$$\text{or } XH + H_2O \rightarrow XOH + 2H^+ + 2e$$

(Note that molybdenum is essential for nitrogen fixation by two routes). Some of these enzymes are primitive and are present in bacteria and even archaebacteria. Immediately we see that these reactions avoid free radical redox steps and a good reason is apparent for the use of molybdenum. This element is one of the first metals in the periodic table to undergo relatively easily, low redox potential $M = O$ bond breakage to $M + O$, i.e. a 'two-electron leaving-group reaction'. (We know this from the exchange rates of O from M.) In fact, this is a typical reaction of a heavy non-metal oxide (note also tungsten) and it is well known in the chemistry of, for example selenium (SeO_2 oxidations) and iodate (see Fig. 17.2 and Chapter 19), but while selenium catalyses such reactions *in the cell*, molybdenum carries out these oxygen atom transfer reactions mostly *outside* the cell cytoplasm, for example in the periplasmic space of bacteria (Fig. 17.3). [Note in passing that selenium, molybdenum and tungsten belong to the same group (VI) as sulphur in the old short form of the periodic table.] Thus, the outside of the cell has both a one-electron series of copper catalysts (Chapter 15) and a two-electron molybdenum series (with other metals, e.g. tungsten and vanadium). We can contrast also the chemistry of iron (and copper) which can also carry out $M = O$ bond cleavage but their reactions are at high potential, easily involve one electron reactions, and were only possible after the increase of dioxygen in the atmosphere.

Nitrogenase, a type of enzyme different from all others, is found in the cell cytoplasm mostly of some prokaryotes. The complexity of the enzyme is startling, but it is fairly clear that each molybdenum atom acts at a separate catalytic centre in the FeMoco cofactor shown in Fig. 17.5(a). The question is: why molybdenum here? It is not to do just with the formation of $Mo–N_2$ complexes since this property is common to many elements and, therefore, the underlying reason has to be this ability plus that of being able to go between oxidation states readily at low potential, e.g. from Mo(III) to Mo(VI) (Fig. 1.7). Here lies a second property to add to the leaving group strength of molybdenum—the small change in redox potential from Mo(III)/Mo(IV) up

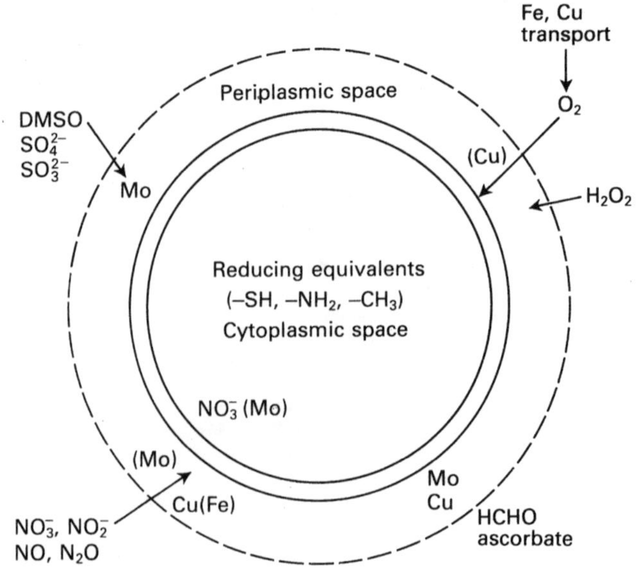

Fig. 17.3 The distribution in space of the major molybdenum (and copper) enzymes except nitrogenase. Note the types of substrate.

to Mo(V)/Mo(VI) and the low values of these potentials, close to zero, (see Fig. 14.1). We shall see that this is valuable when we consider all the enzyme reactions of molybdenum. In passing notice the similar range and trend of redox potentials of tungsten, and also of vanadium from V(II) to V(V) which, however, is not so narrow and is more oxidising.

All the other molybdenum-containing enzymes are very large and complex and it is notable that they have multiple units of other cofactors besides molybdenum, namely Fe/S clusters in them all and flavin in all except nitrogenase which requires MgATP. In fact, a separate flavo-protein feeds electrons to nitrogenase.

The reasons for some of these complications are obvious, e.g. a series of necessary electron transfer and storage sites, the Fe/S centres, for many-electron reactions. The involvement of MgATP in an ATP-splitting reaction in nitrogenase is not so clear, but must mean that the chemical energy of more than one ATP → ADP + P, yielding 10–15 kcal mol^{-1}, is, of necessity, being transferred first to mechanical and then to reductive energy, so changing the redox potential of an hydrogen atom or of an electron transfer centre. We must attempt to explain how such a unit functions and eventually relays its reductive power to the nitrogen molecule presumably associated with molybdenum.

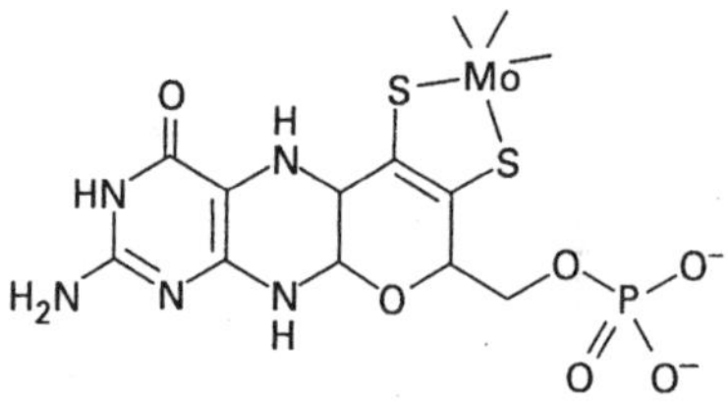

M.Molybdopterin cofactor – Moco

Fig. 17.4 The formula of one of the mononuclear molybdenum cofactors, Moco. Others have a nucleotide phosphate extension, (see 'Further reading').

In the cases of the O-transfer reactions that occur at molybdenum centres coupling is also via Fe/S centres, but now include flavin. (Once again we need to understand the reasons for the train of electron transfer centres.) The molybdenum *in these enzymes* is also bound by special organic pyranopterin cofactors, and need not be held directly by the sidechains of proteins (Fig. 17.4). The pyranopterin cofactor actually is a dithiolene complex.

There are no stability data available but it may be that the special feature of the dithiolene is the bite of the five-membered ring which selects very small high-valent metal ions.

17.4 **Biological chemistry of molybdenum**

Now that we have seen the particular value of molybdenum we shall take a more detailed look at its biological chemistry.

17.4.1 **Uptake of molybdate**

A particular problem is posed by the *initial* uptake of molybdenum in today's aerobic conditions which must involve the electrostatic binding of MoO_4^{2-}. This anion resembles HPO_4^{2-} and SO_4^{2-} in geometry and is intermediate in H-binding ability as judged by pK_a values. It is, therefore, very difficult to see at first sight how competition with SO_4^{2-}, which is highly concentrated in the sea and even in inland water, is avoided. In fact, biological constructs, especially proteins, are extremely selective in binding on the basis of differences in ion size (see Aside).In Chapters 8 and 9 we have seen how large selectivity factors are generated for the individual elements of Groups 1 and 2 of the periodic table using ligand crowding or its absence to create large or small selective cavities. The same principles apply to anions (see Section 2.15), but now the binding is through positively charged amino-acid sidechains such as lysine, histidine, and arginine and their H-bonding capabilities. The structure of the molybdate (MOD A) and sulphate external carrier proteins are now known, see Section 19.2.5, and it has been clearly shown that binding selectivity in these two very similar protein folds arises from very similar ligands but different cavity sizes. Once

The radii of some anions (Å)

anion	radius	anion	radius
CO_3^{2-}	1.85	SO_4^{2-}	2.30
NO_3^-	1.89	SeO_4^{2-}	2.43
PO_4^{3-}	2.38	MoO_4^{2-}	2.54
CrO_4^{2-}	2.30	IO_4^{2-}	2.49
VO_4^{3-}	2.35		

molybdenum or tungsten or vanadium enter a cell then selection can be based on redox potential as well as size differences.

17.4.2 Oxidation states and redox potentials in enzymes

When discussing enzymes which are involved in catalysis of redox reactions we need to look carefully at the oxidation states of the metal in the enzyme as far as they are known. Any redox potential in the proteins is related to the relative binding constants of the appropriate two different states of the element, and to the intrinsic redox potentials of the free metal ion species in water, which we obtain from the oxidation state diagrams. The approximate oxidation state diagram of molybdenum relative to H^+/H_2 and at pH = 7 where the redox potential for the H^+/H_2 couple is –0.41 V, is shown in Figs 1.7 and 14.1. The relative stability (redox potential) between any two states is given by the slope of the line from any point to any other point. Since the stability constants of different oxidation states are not too dissimilar, we do not expect complex formation to have much influence over the redox potentials of hydrolysed ions such as MoO_2^{2+}, MoO_2^+ and MoO^{2+} and, in fact, the redox potentials fall in the range 0.0 ± 0.3 V (see Table 17.2). This means that the redox potentials for molybdenum enzymes are such that this element can carry out reversible redox changes near 0.0 V, such as mild oxidations and reductions by similar two-electron reactions. Examples are SO_4^{2-}/SO_3^{2-} (–0.05 V), -COOH/-CHO (+ 0.05 V) and reductions of high potential anions such as NO_3^-/NO_2^- (+ 0.7 V). Now, these are all two-electron or oxygen-atom transfer reactions and, while we have shown here that the thermodynamic potentials make molybdenum a possible redox catalyst, we need to see also how it functions in overcoming kinetic barriers of such reactions.

17.4.3 The types and rates of molybdenum reactions

Molybdenum is involved in three very different classes of reaction in living organisms. The first is *substitution* where we include addition through the expansion of the co-ordination sphere since it is generally believed that

17.2 Redox potentials in molybdenum enzymes (pH 7 and H^+/H_2 = – 410 mV)

Enzyme	Redox states	Potential (mV)
Aldehyde oxidase	Mo(VI)/Mo(V)	–360
	Mo(V)/Mo(IV)	–360
Sulphite oxidase	Mo(VI)/Mo(V)	+40
	Mo(V)/Mo(IV)	–163
Nitrate reductase	Mo(VI)/Mo(V)	+180
	Mo(V)/Mo(IV)	+220
Xanthine oxidase	Mo(VI)/Mo(V)	–350
	Mo(V)/Mo(IV)	–350
Formate dehydrogenase	Mo(VI)/Mo(V)	–330
	Mo(V)/Mo(IV)	–450
Flavins	$Fl^- \rightarrow Fl$	–300
	$Fl \rightarrow Fl^+$	–300
Fe/S	Cluster reaction	–300
Cytochromes *b*	Fe(III) → Fe(II)	–100

substitution for this element occurs through a higher coordination number intermediate. For example,

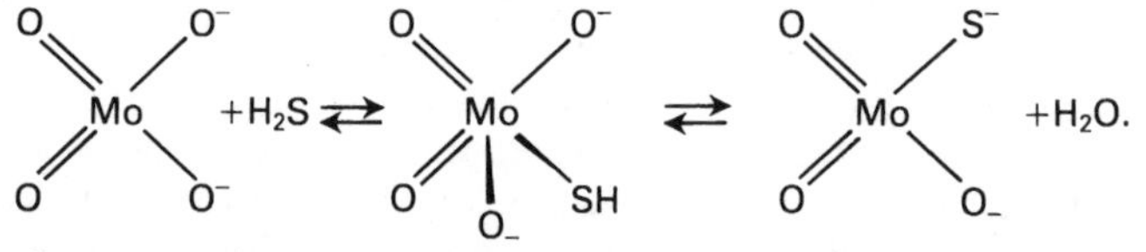

These exchanges are easier in the higher (six and seven) coordination complexes. Such S_N2 mechanisms (see Chapter 4) are common amongst heavy elements. Second, molybdenum may also allow relatively easy *atom transfer* in reactions such as

$$MoO_4^{2-} + SO_3^{2-} \rightleftarrows MoO_3^{2-} + SO_4^{2-}$$

but these reactions may be solvent-assisted and are not always simple reactions of MoO_4^{2-}, that is, the species are always in relatively fast equilibrium with higher coordination species. It is a feature of molybdenum chemistry that the cationic states are very flexible between different coordination numbers.

The third reaction is *electron transfer* as also is observed in the molybdenum blues and bronzes. The reactions are relatively fast indicating that one-electron redox reactions of molybdenum which require little rearrangement of the coordination sphere could be used in biological systems.

17.4.4 Oxygen-atom transfer reactions

The reactions which we wish to understand (Table 17.1) in more detail are, for example

$$MoO(VI) + NO_2^- \rightleftarrows Mo(IV) + NO_3^-.$$

There is a parallel set of sulphur reactions, since sulphur transfer is also known in molybdenum enzymes:

$$MoS(VI) + CN^- \rightleftarrows Mo(IV) + SCN^-.$$

The understanding we need is of the possible intermediates along the pathways since, as stated, we know that bond-breaking is relatively easy for molybdenum, a heavy atom, and that its redox potential changes are close to those of the substrates. Note that instead of O or S it is possible to transfer OH or SH (see Table 17.1).

The probable intermediates of some of the parallel enzyme reactions are known through EPR studies. For example, in aldehyde oxidases we can write

$$MoOS + RCHO \rightarrow Mo(SH)OCOR;$$
$$Mo(SH)OCOR \rightarrow MoS + RCOOH.$$

Here there is no change in the coordination number of the molybdenum. The scheme indicates that the molybdenum oxene group is a good attacking group on the positive carbon atom of the carbonyl. Equally this implies that probably the oxidation of nitrites and sulphites goes via direct binding of nitrite and sulphite to an oxygen of MoO(VI).

17.5 The structure and function of the molybdenum enzymes

There are now very detailed atomic level structures of several of the mononuclear molybdenum enzymes. In this book we cannot describe structure at the atomic level nor can we examine the detailed analysis of the kinetics in terms of the structures. For such information we refer the reader to the primary and review literature [see 'Further reading', especially Hille (1996)]. As shown in Table 17.1 the pyranopterin-bound molybdenum enzymes are best divided structurally and functionally into three classes. The classes have major features in common, and it is important to note that the five-coordination of the molybdenum in virtually all these enzymes is very unusual, has not been copied in model syntheses, and is apparently *constrained* to match reaction requirements. The function of the pyranopterin moiety is not known.

The reoxidation of Mo(IV) does not go via direct attack of molecular oxygen in the molybdenum oxidases. It is found that the inserted oxygen on molybdenum comes from water, not from dioxygen. In fact, the oxidation goes by electron transfer from molybdenum to dioxygen via an ancillary flavin, which does react with O_2. As in cytochrome P-450, there is, in addition, an Fe/S electron carrier site in the protein to keep the reaction centres apart. The comparison between these two very different oxidases reveals the selectivity of living organisms and the use of two different metals in two different parts of space for the oxidation of very different substrates but using comparable electron transfer chains. There is another parallel of unknown significance: both the molybdenum and the iron which take part in O-atom transfer reactions are often bound to thiolate groups.

Molybdenum and the fixation of nitrogen

17.5.1 Nitrogenase and its reaction centre: FeMoco

The structure of nitrogenase is now known in considerable detail though the sites of binding of N_2 and H_2 remain elusive (see Howard and Rees in 'Further reading'). Central features of the structure will be described here but the atomic details are again left for the reader to inspect in the primary literature. There are three subunits in the dimeric protein $(\alpha\beta)_2\gamma_2$ [Fig. 15 (b)]. The major interest in the $\alpha\beta$ subunit lies in the FeMoco units ($Fe_7Mo_1S_6$, the iron/molybdenum coenzyme, two in the whole enzyme), which is the catalytic site (a vanadium and an all-iron enzyme are also known), and in the Fe_8S_8 P-clusters which resemble two Fe_4S_4 units bridged by two cysteines. The enzyme also includes a dimeric iron protein, γ, which has an Fe_4S_4 cube and two MgATP sites. FeMoco is shown in Fig. 17.5 (a) to have a homocitrate molecule bound to the molybdenum, the function of which is unknown. The units are placed in the whole structure shown schematically in Fig. 17.5 (b). All the sites are constrained away from simple model structure expectations. The bound Mg.ATP, in the γ subunits, on hydrolysis causes a conformation change in the structure forcing electron transfer from the Fe_4S_4 centre to each P-centre of the MoFe protein. The P-centre then carries electrons to the FeMoco for reduction of N_2. It is not understood why several ATP molecules must react to give ADP in order for N_2 to be reduced, nor it is known why H_2 production has to occur simultaneously with N_2 reduction. The source of reducing equivalents is NADH which acts upon a required flavoprotein, itself the donor of electrons to the Fe_4S_4 cube. A detailed scheme of the kinetics of reaction has been given (see Burgess and Lowe in 'Further reading').

Now it is believed that the molybdenum is the site of dinitrogen reaction but that one of the necessary steps in the activation of the molybdenum is prior formation of a hydride, Mo–H_3. In the reduction of nitrogen it is observed that one mole of dihydrogen is released. Under optimal conditions the reaction is $N_2 + 8H^+ + 8e + 16\ MgATP \rightarrow 2NH_3 + H_2 + 16\ MgADP + 16P_i$. (It is also thought that the vanadium and pure iron nitrogenases are less effective in

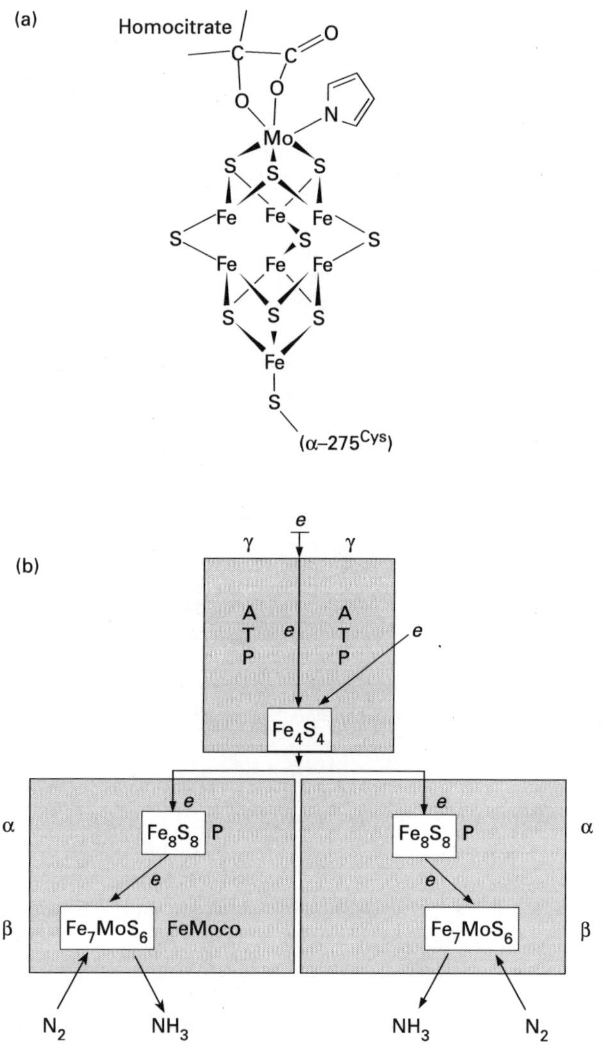

Fig. 17.5 (a) The formula of FeMoco. (b) The essence of the structure of nitrogenase. The Fe protein is in two subunits. The path for electrons is shown but it requires ATP hydrolysis and conformation changes to be activated.

nitrogen reduction because they release larger quantities of hydrogen). We may therefore write, schematically, the following catalytical cycle

$$\mathrm{Mo} \xrightarrow[e]{H^+} \mathrm{MoH} \xrightarrow[e]{H^+} \mathrm{MoH_2} \xrightarrow[e]{H^+} \mathrm{MoH_3{*}} \xrightarrow{N_2} \mathrm{MoHN_2} + \mathrm{H_2}\uparrow$$

$$\mathrm{H^+} \downarrow \mathrm{e}$$

$$\mathrm{MoH_2N_2}\ (\mathrm{Mo = N{-}NH_2})$$

$$\mathrm{H^+} \downarrow \mathrm{e}$$

$$\underset{+\,\mathrm{NH_3}}{\mathrm{Mo}} \xleftarrow[e]{H^+} \mathrm{MoNH_2} \xleftarrow[e]{H^+} \underset{+\,\mathrm{NH_3}}{\mathrm{MoNH}} \xleftarrow[e]{H^+} \mathrm{MoH_3N_2}\ (\mathrm{Mo \equiv NH_3})$$

The overall reaction is

$$N_2 + 8H^+ + 8e \rightarrow 2NH_3 + H_2$$

The evolution of hydrogen may be one way of keeping the enzyme free from dioxygen interference. The further reduction of N_2 is seen as a series of switches of oxidation state of molybdenum, due to the introduction of electrons, before $H^{\cdot}$ is introduced to the N_2.

17.5.2 Molybdenum exchange and possible control functions

Molybdenum outside (and here and there inside) cells in the form of the pyrano-pterin complexes exchanges from proteins quite readily, but molybdenum itself does not dissociate. It is not known if the pyrano-pterin-Mo complex has any control function outside (or inside) cells.

In nitrogenase there is the possibility that the molybdenum is exposed in the Fe_7MoS_6 cluster. As stated, it is associated with the homocitrate. The very presence of this organic moiety in the enzyme is reminiscent of the structure of aconitase which has an iron sulphur cluster that binds exchangeable citrate. Maybe homocitrate is a dissociable control unit but observe that one iron atom of aconitase is easily lost from its cluster and may control the citrate cycle (Chapter 12). It could be that either the molybdenum or the iron of Fe_7MoS_6 is also a dissociable control atom, but this is not known.

17.6 Molybdenum genetics

The expression of the molybdenum using proteins in bacteria has been well studied. It is known that several proteins are regulated by the same promoter(s) and that they lie sequentially along the genome. The genes involved in the uptake of molybdenum are referred to by the label *mod* and the expressed proteins by the label MOD. To simplify we then have the genes MOD A is an

Promoter	Polymase	*mod* A	*mod* B	*mod* C	*mod* D

external scavenger or carrier for MoO_4^{2-}; MOD B is an uptake enzyme, possibly an ATP-requiring pump; Mod C is an internal distributing carrier; and MOD D is a transcription factor which switches on this uptake system when no molybdate is bound to it. In addition to this set of genes, elsewhere on the genome there are the *fix* genes, some of which are for producing FeMoco, which are part of the nitrogenase family of genes shown in Fig. 17.6. The *fix* genes express a haem-containing protein which is a transcription factor for

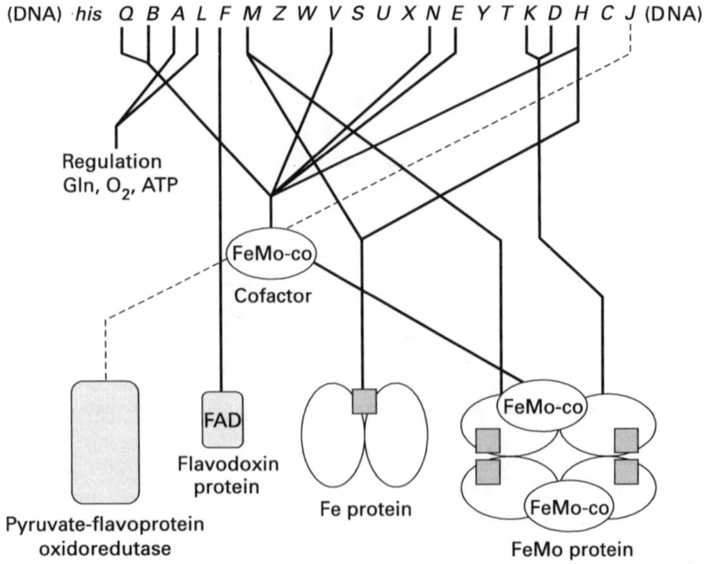

Fig. 17.6 The genes required for the synthesis of nitrogenase. *his* is the histidine gene.

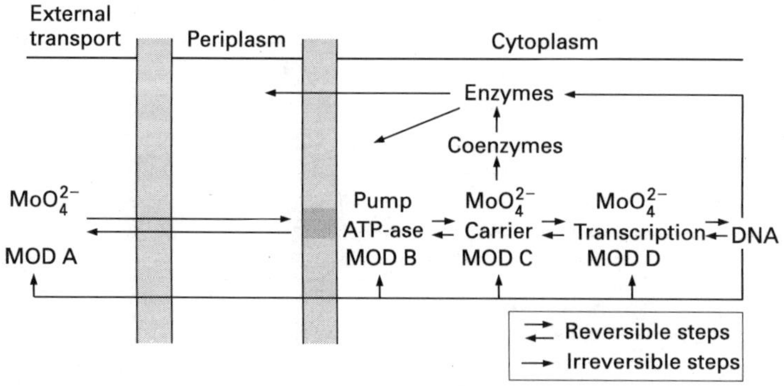

Fig. 17.7 The uptake and incorporation of molybdenum into its various sites under the control of the gene products labelled MOD.

switching off nitrogenase genes in the presence of dioxygen. There are also the genes for the production of the molybdenum/pterin coenzyme, Moco. The genes for these and other molybdenum using proteins are under the regulation of substrates. Nitrate reductases are to be found under the *nap* and *nar* gene labels and sulphite oxidases under the gene labels *sor*.

For a more comprehensive discussion see Hille in 'Further reading' and references therein.

17.7 Molybdenum: conclusions

We conclude the first part of this chapter with the most intriguing problem. There is no denying that molybdenum is central to four reactions without which biological systems would fail

$N_2 + 3H_2 \rightarrow 2NH_3$,
$NO_3^- \rightarrow NO_2^- + (O)$,
$SO_4^{2-} \rightarrow SO_3^{2-} + (O)$,
$HCOOH \rightarrow HCHO + (O)$

The first three are essential for biology's source of nitrogen and sulphur and the fourth is central to carbon metabolism. There is here a conundrum for the chemist and the biologist or indeed for anybody interested in the nature of living systems. Is the requirement for molybdenum an absolute necessity or could the chemistry have been done in some other way and is what we observe an evolutionary accident? No matter how we answer this question we know that such a need, for molybdenum, is reflected back to the genetic apparatus (Fig. 17.7). The essential nature of molybdenum, although it has switched role in evolution from a part of carbon redox chemistry to a component of nitrogenase, to a catalyst for handling sulphate and nitrate, is persistent. We have to consider that it is one of very few elements which can carry out many electron reactions at low potentials, e.g. O-atom transfer, in which regard it resembles the *slower-reacting* non-metals sulphur (and selenium). However, it is quite probable that molybdenum was not so available during the most primitive periods of life and that some of the earliest of these required functions were performed by tungsten or vanadium. We turn to these elements in biological systems.

17.8 Tungsten biological chemistry

Tungsten as a biological metal is a most unlikely choice; there are indeed very few examples of selection for functional purposes of elements of atomic

Table 17.3 Tungsten enzymes

Enzymes	Sources	Functions
1st group (2 pterins non-nucleotide form per W)		
Aldehyde ferredoxin-oxireductase (AOR)	*Pyrococcus furiosus, Pyrococcus strain ES-4, Thermococcus strain ES-1*	Reduction of non-activated carboxylates to aldehydes (and reverse reactions)
Formaldehyde ferredoxin-oxireductase (FOR)	*Pyrococcus furiosus, Thermococcus litoralis*	
Glyceraldehyde ferredoxin-oxireductase (GAPOr)	*Pyrococcus furiosus*	
Carboxylic-acid reductases (also aldehyde oxidation) (CAR)	*Clostridium thermoaceticum* (2 forms), *Clostridium formicoaceticum*	
Aldehyde dehydrogenase (ADH)	*Desulfovibrium gigas*	
2nd group (1 pterin + nucleotide form + 1 Se per W)		
NADP-dependent formate dehydrogenase (FDH)	*Clostridium thermoaceticum, Clostridium pasteurianum*	Reduction of CO_2 to formate; CO_2 fixation in anaerobes
Formyl methanofuran dehydrogenase (FMDH) *Methanobacterium wolfei*	*Methanobacterium thermoacetotrophicum,*	Conversion of CO_2 to CH_4 (first step)
3rd group		
Acetylene hydratase (AH)	*Pelobacter acetylenicus*	Hydration of acetylene to acetaldehyde

number higher than 35 (see Chapter 1). The outstanding exceptions are molybdenum and iodine and neither of them is generally required by all living species. Other rare examples are tantalum and niobium, which seem to be accumulated by certain tunicates instead of vanadium (see Section 17.9.2), as well as strontium and barium used as sulfates in the skeleton of some acantharians and radiolarians. None of these last four metals is involved in any general metabolic mechanism.

However, since 1990, over a dozen tungsten enzymes from archaea and bacteria have been purified and characterized. The genes for some of these enzymes have been cloned and sequenced, and in one case the enzyme, an aldehyde ferredoxin oxireductase, has been extracted from the hyperthermophilic archaea *Pyrococcus furiosus* and the corresponding crystal structure was determined by X-ray diffraction methods to 2.3 Å resolution (curiously, even before any of the molybdenum enzymes). Thus, a biological role for tungsten has been demonstrated for several prokaryotes, but until now (2001), no requirement for a tungsten enzyme has been observed in eukaryotes.

Interestingly, it was also observed that for almost all the known tungsteno-enzymes there is an analogous molybdo-enzyme in the same or in a very similar organism. The only exceptions are the enzymes isolated from hyperthermophilic archaea which have an absolute requirement for tungsten. This element, therefore, not only has a biological role but is essential in a large class of very primitive organisms.

The known tungsten enzymes can be divided broadly in three main categories on the basis of the substrate on which they act (Table 17.3). In all the enzymes, tungsten is at the active site bound either to *two* dithiolene groups of *two* pyranopterin moieties, (Group I), (Fig. 17.8), resembling the DMSO reductase family of molybdenum enzymes (Table 17.1), or to one dithiolene group of one pyranopterin moiety plus a selenium atom from a selenocysteine sidechain (Group II). In the active site of the aldehyde ferredoxin oxireductase (AOR) the two pterin containing molecules are bridged by a magnesium ion bound to the phosphate groups. The holoenzyme is a homodimer and the two sub-units each containing one tungsten and one Fe_4S_4 cluster are bridged by a mononuclear Fe ion coordinated to sidechains of a histidine and a glutamate residue from each sub-unit. This iron may act as a control. Enzymes other than the aldehyde oxireductases (AOR), formate and glyceraldehyde-3-phosphate ferredoxin oxireductases (FOR and GAPOR), have different accessory elements, including zinc. The requirement for W is not replaceable by Mo in these enzymes.

The general reaction catalysed by the first group of tungsten enzymes (AOR, FOR, GAPOR, CAR and ADH) can be written schematically

$$RCHO + H_2O \rightleftharpoons RCOO^- + 3H^+ + 2e.$$

These reactions are similar to those catalysed by some molybdenum enzymes, (see Table 17.1), and are effectively the reverse of the atom-donor reaction (two electron-reduction) of oxotransferases, but they occur at lower redox potentials.

Now, the second group of tungsten enzymes includes two types of dehydrogenases: NADP-dependent formate dehydrogenases (FDH) in acetogenes, and formyl methanofuran dehydrogenases (FMDH) in methanogenes, (Table 17.3). FDH catalyses the conversion of CO_2 to formate using NADPH as the physiological electron donor.

$$CO_2 + NADPH \rightleftharpoons HCOO^- + NADP^+$$

or

$$CO_2 + H^+ + 2e \rightleftharpoons HCOO^-$$

The second enzyme, FMDH, catalyses the first step of the conversion of CO_2 to methane in methanogens,

$$CO_2 + 8H^+ + 8e \rightleftharpoons CH_4 + 2H_2O$$

$$\text{via} \quad CO_2 + MFR^+ + H^+ + 2e \rightleftharpoons CHO\text{-}MFR + H_2O;$$

that is, methanofuran MFR, is the other substrate, but the physiological electron-donor is not known.

The last group of tungsten enzymes (Group III), contains only one type of enzyme, acetylene hydratase (AH), which, as the name indicates, catalyses the reaction

$$HC \equiv CH + H_2O \rightleftharpoons CH_3CHO$$

This enzyme resembles the aldehyde oxi-reductases (AOR) in cofactor content, but the reaction that it catalyses is now a hydration.

We stop here our overview of the tungsten biochemistry. Despite the number of papers published the discussion still remains very much at the speculative level for lack of necessary data. We shall return to the discussion of why tungsten is found in some primitive organisms in Section 17.10.

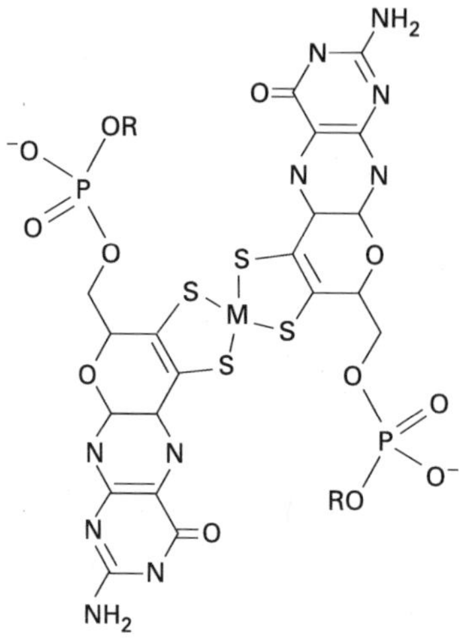

Fig. 17.8 The tungsten pyranopterin cofactor, where R can be H, AMP, GMP, HMP

17.9 Vanadium chemistry and biochemistry

17.9.1 The chemistry of vanadium in biology

Vanadium, like molybdenum, has two very different catalytic activities associated with it and two very different biological functions. Both elements have a rich oxygen chemistry associated with the oxyanion and oxycation species, VO_4^{2-} (MoO_4^{2-}), VO^{2+} and VO_2^+, (MoO^{2+}, MoO_2^+, and MoO_2^{2+}). These species are oxidizing agents which can react by one-electron steps, electron transfer, or by two electron steps when V (Mo) act as O-atom donors, but whereas molybdate is a mild oxidant vanadate is relatively powerful, $E \approx +0.5$ V at pH = 7.

The second class of chemistry of these two metals belongs to sulfur-rich reducing conditions, where vanadium, like molybdenum, can form sulfur-containing anionic centres, e.g. VS_4^{3-} (MoS_4^{2-}); sulfur-containing cationic centres, e.g. VS^{2+} and $VSSH^+$ (compare $MoOS^{2+}$); and iron-sulfur clusters Fe_7MS_6, associated in biology with nitrogen fixation. It is quite possible that in the primitive earth, when these reducing conditions were prevalent, vanadium and tungsten, not molybdenum, (and iron) had considerable catalytic activity (it may even have been that life became a possibility because such catalysts existed), but vanadium and tungsten may well have been more available since their sulfides are more soluble than that of molybdenum. (Note again the table of group separations in analytical chemistry.) Iron was then available as Fe^{2+} and Fe^{2+}/Fe^{3+} sulfide clusters and vanadium as V^{3+} and V(IV) sulfur complexes, e.g. VS^{2+}, $V(S)SH^+$, and, perhaps, $VS(S_2)_2^{2-}$.

17.9.2 The biological forms and functions of vanadium

The known forms of vanadium in living organisms, see Fraústo da Silva in Further reading, are given in Table 17.4. It is probable that only the vanadium-iron-sulphide cluster FeVco (similar to FeMoco) is primitive and all the other forms appeared later as the level of atmospheric oxygen increased; note that only the FeVco protein is inside cells and that other vanadium compounds are probably in vesicles or outside cells. This is certainly true for the vanadium-containing haloperoxidases occurring in some brown and red algae which contain V(V) bonded to oxygen and nitrogen donors. The accumulation by tunicates and *Amanita* toadstools may have started at an earlier stage, with atmospheric oxygen still at a rather low level. Note that in the tunicates vanadium is known to be stored mostly as V(III) (or VO^{2+} in some types) in the vacuoles (vanadophores) of specialized 'signet' blood cells (generally referred to as vanadocytes) together with SO_4^{2-} in stoichiometric proportions and at low pH (~2), and this may have been determined by the need to protect the metal ion from the oxidizing environment. The highest values of V were found in the blood cells of *Ascidia gemmata*, 247.2 mM, which corresponds to a 10^7

Table 17.4 Known forms of vanadium in living organisms

Form	Occurrence	Function
Free as V(III) or combined with a special pigment—tunichrome—or with low M.W. protein	Some ascideans (tunicates), e.g. *Ascidia gemmata, Phallusya mamillata, Ciona intestinalis*, etc. Also present in the worm *Pseudopotamilla occelata polychaeta*	Unknown; probably protection (synthesis of tunic), or, alternatively, an electron and proton sink
V(IV) bis-complex of *N*-hydroxyimino-di--*α*-propionate (amavadine)	Some *Amanita* toadstools, e.g. *Amanita muscaria, Amanita velatipes*	Peroxidase and catalase activity; probably part of protection system against injury of tissues; compare fungal tyrosinases
V(V) bonded to N,O-ligands	Some brown and red marine alagae, e.g. *Fucus vesiculosis, Ascophyllum nodosum, Ceramium rubrum*, etc, and terrestrial fungi, e.g. *Curvularia inequalis*	Bromo- and iodo-peroxidases (chloroperoxidase in terrestial fungi), probably defensive
Fe, V, S cluster	Azotobacteria, e.g. *Azotobacter vinelandii, Azotobacter chroococcum*	Cofactor of nitrogenase; fixation of dinitrogen

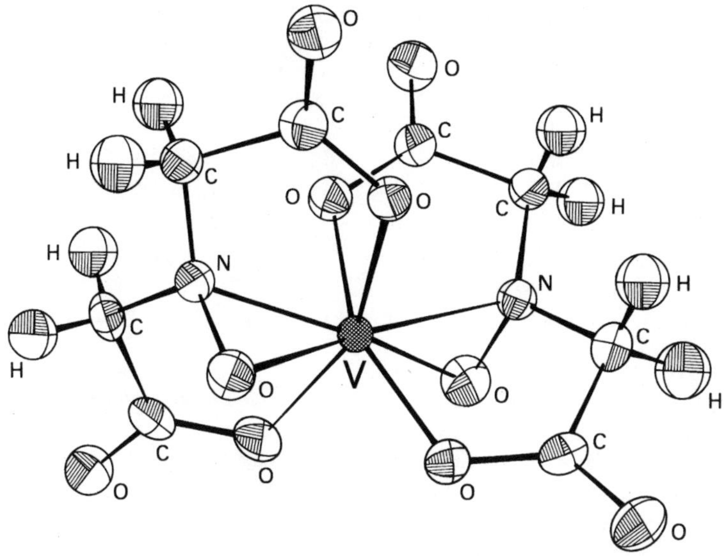

Fig. 17.9 The structure of the vanadium complex of (S,S) –2, 2′ N(hydroxy-imino) di-α-propionates. Amavadine, has methyl substituents on the CH_2 carbon of this model complex.

concentration ratio relative to the sea water. Nowadays, the animals capture vanadium as vanadate from sea water and incorporate it in the vanadophores through the phosphate channel profiting from the analogy between $H_2VO_4^-$ and $H_2PO_4^-$. Once inside the vacuole, vanadates are reduced to the cationic forms V^{3+} and VO^{2+} possibly by a polyphenolic pigment, tunichrome, with which the cations may combine, but this has not been unequivocally demonstrated since *in vitro* the reaction goes only to V(IV). In any case the metal is then retained and kept in a reduced free or combined form, perhaps even associated with a protein (see Michibata in 'Further reading').

In the toadstools we find another curious device that biology has used, this time an apparently simple ligand, (S,S)-2,2′-N(hydroxy-imino)di- α-propionate, which forms an extraordinary stable 2:1 complex with V(IV), (not VO^{2+}!) (see Fig. 17.9) and is therefore able to extract the metal from soils in which it is scarce relative to its potential competitors, e.g. Fe^{3+} and Cu^{2+}. Here we need to remember the stability constants of transition metal complexes which indicate that extraction of VO^{2+} would not have been easy due to strong competition by other ions. (We are able to discuss VO^{2+}, as such, binding since this ion is not hydrolysed like MoO_2^{2+}). The overall stability constant, log $\beta_2 \simeq 23$, of this complex (named amavadine) is enough to remove the oxene group to give a complex of the naked V^{4+} ion, actually an eight-co-ordinated species in which even the N-hydroxyl groups of the ligand are ionized.

This V(IV) complex may undergo a reversible one-electron oxidation at about $E \approx 0.8$ V to the corresponding V(V) complex, but it is quite stable in aqueous solution at pH ≈ 7 and does not need to be stored where it is protected from oxygen.

It has been found that amavadine can act as a mediator in the oxidation with H_2O_2 of certain biological thiolic compounds containing also carboxylic or ester groups, such as cysteine or glutathione, to the corresponding disulphides, thus acting as a kind of peroxidase (a Michaelis–Menten type of mechanism was demonstrated for an electrochemically driven similar reaction). However, in the absence of substrate, H_2O_2 is decomposed into $O_2 + H_2O$, when amavadine acts as a catalase (see Matoso and Fraústo da Silva, in 'Further reading').

This double catalytic role of amavadine can be represented schematically as

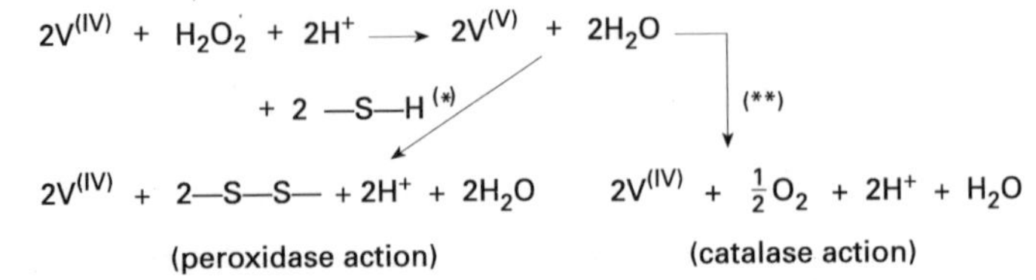

(*)In presence of adequate biological thiols
(**)In absence of biological thiols

Scheme 1 – The double role of amavadine (the ligand is omitted for the sake of simplicity)

It therefore looks as if amavadine might be a kind of primitive protective chemical, able to use H_2O_2 for self-regeneration of severed tissues or against foreign pathogens and predators, and decomposing it when not necessary. In this it reminds us of other electron-transfer catalysts acting at fairly high potential, such as those found in some blue copper oxidases, having dioxygen at the terminal oxidizing end. Such copper oxidases exist in other fungi and have protective functions, e.g. fungal tyrosinases. Again, amavadine does not react directly with dioxygen, so that there must be other cofactors and, since it is involved in one-electron transfer steps, radicals are likely to be formed. We do not speculate overmuch by saying that amavadine is likely to be formed external to the cell cytoplasm in order to be kept away from DNA.

The final case to consider is that of the vanadium haloperoxidases (see Tables 17.7 and 17.7).

While in some terrestrial fungi, e.g. *Curvularia inequalis*, the enzyme is a chloroperoxidase, i.e. able to catalyse the oxidation of Cl^- (as well as, of course, Br^- and I^-) by H_2O_2 in the marine organisms all the enzymes studied are bromoperoxidases or just iodoperoxidases.

The haloperoxidase catalysed reaction is

$$H_2O_2 + X^- + H^+ \rightarrow H_2O + HOX \xrightarrow{RH} R - X + H_2O$$

where X represents the halide, Cl^-, Br^-, or I^-, and RH the organic substrate. Recently the X-ray diffraction structures of a chloroperoxidase (CPO) and of its peroxide derivative were solved, (see Messerschmidt *et al.* in 'Further reading'). The active site lies in a pocket, where vanadium is bound as hydrogen vanadate (V), coordinated by four non-protein oxygen atoms and one nitrogen atom from a histidine sidechain (His 494) in a trigonal bipyramidal fashion with three oxygens in the equatorial plane linked by hydrogen bonds.

In the peroxide derivative the peroxide group is bound to the vanadium in η-fashion after release of the apical oxygen of the vanadate group; note the coordination changes in this reaction. The selectivity for vanadium over other metal ions in this site must be due to kinetically controlled insertion. The X-ray diffraction structure of a bromoperoxidase (BPO) has also been solved (see Weyand *et al.* in 'Further reading') which allows a direct comparison of the two types of enzyme, CPO (terrestrial fungus) and BPO (marine algae). This shows that the fold of the proteins near the active site and the architecture of the V-bonding centre are very similar.

A further point of interest concerning the vanadium haloperoxidases is the fact that the amino-acid residues contributing to the active-sites of these enzymes are conserved within three families of acid *phosphatases* (see Hemrika *et al.* and Plass in 'Further reading') which suggests that the active

Table 17.5 Haloperoxidases

Protein (halide)		Source	Metal
Chloroperoxidase (I^-, Br^-, Cl^-)	Myeloperoxidase	Neutrophilic granulocytes	Iron chlorin, Iron porphyrin
	Chloroperoxidase	*Cadariomyces fumano*	Iron porphyrin
	Chloroperoxidase	*Pseudomonas pyrrocinia*	Iron porphyrin
	Chloroperoxidase	*Curvaloria inequalis*	Vanadium (V)
Bromoperoxidase (I^-, Br^-)	Lactoperoxidase, Eosinophil peroxidase	Milk, saliva, tears, granulocytes	Iron porphyrin
	Bromoperoxidase	Some brown and red marine algae	Vanadium (V)
Iodoperoxidase (I^-)	Horseradish peroxidase	Horseradish roots	Iron porphyrin
	Thyroid peroxidase	Thyroid gland	Iron porphyrin
	Iodoperoxidase	Some brown and red marine algae	Vanadium (V)

sites of all these enzymes are very similar and that an acid phosphatase is likely to be the origin of them all. Vanadate and phosphate go through very similar coordination chemistry cycles and are very similar in size (see Section 17.4.1).

17.9.3 Vanadium and iron

From the information given in the previous sections, summarized in Tables 17.4 and 17.5, we conclude that in its location and functions vanadium does, in fact, have a chemistry related to that of iron, especially Fe^{3+}, as well as to molybdenum. VO^{2+} and Fe^{3+} are bonded in living organisms by similar chelating agents (catechols, oximes, hydroxamates, porphyrins, sulphur clusters, etc. (see Table 2.3). Both metals form nitrogenases, although their catalytic activities involve the liberation of much more hydrogen from water than in the case of the molybdenum enzyme which appears to have replaced the others in the present atmospheric conditions.

Another parallel is with the use of iron in peroxidases (Table 17.5) although the chemistry here appears to be different. Peroxidation using iron goes through a highly oxidizing FeO(IV) species and then by direct formation of XO^- or XO^+ where X is a halogen, while the most likely route for vanadium is via known peroxo complexes $VO_2^+(O_2)$, giving XO^-, the halogenating agent, and VO_2^+, i.e. it is the peroxide, not the metal, that is involved in the reaction, Fig. 17.10. Once again this kind of 'dangerous' chemistry must take place outside the cytoplasm, either in vesicles or in the external fluids of higher cells.

As to the other cases in Table 17.4 it is also known that one order of ascidians which does not accumulate vanadium (the Stolidobranchia) does accumulate iron (in ferrocytes) with apparently the same function, whatever it may be—sink for electrons and protons, defense against predators, tunic synthesis or some other. As we have seen, accumulation takes place in cell vacuoles, i.e. external to the cytoplasm.

In the case of the *Amanita* toadstools the analogy is perhaps more with today's copper oxidases (e.g. the fungal tyrosinases) than with iron enzymes,

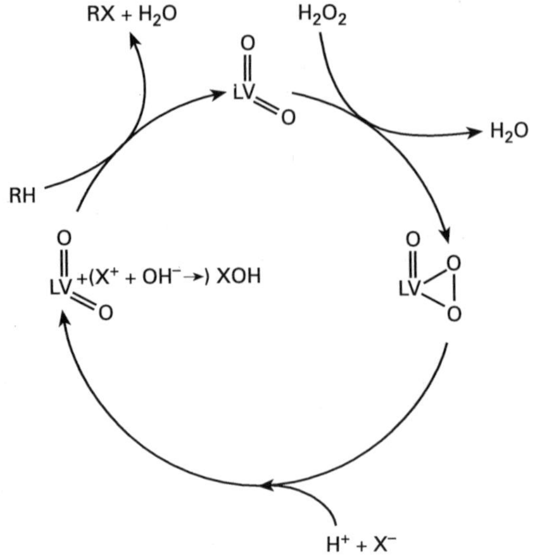

Fig. 17.10 A possible (simplified) mechanism of oxidation of halogens and halofunction of substrates by vanadium peroxidases.

although there are iron cofactors with functions similar to that of amavadine, i.e. electron-transfer catalysts, usually functional at lower potentials, and a catalase function for the vanadium compound has been reported.

17.10 Molybdenum, vanadium, and tungsten in evolution

It is observed that today molybdenum is largely associated with the reactions of NO_3^-, SO_4^{2-}, aldehydes, and N_2 in biological systems. Of these there is good reason to suppose that the first two were rare chemicals in the period before the appearance of at least some dioxygen in the atmosphere. We can make the (risky) conclusion that molybdenum only became incorporated in biological systems in the form used to attack these substrates, about 2×10^9 years ago, life having existed at least twice as long (Fig. 1.11). The appearance of this second form of molybdenum in life would then correspond to the known change of the chemistry of the sea at that time. In the earlier long period of the evolution of life in a reducing atmosphere this metal may have been present as sparingly soluble MoS_2 and perhaps some soluble Mo/S anions (thio-complexes), generally in oxidation states lower than six. Molybdate would not have been formed and the very low concentration of molybdenum in the sea would be due to these soluble sulphur complexes only.

To some extent the same is probably true for vanadium and tungsten, but the sulphides of these metals are more soluble (notice that they are not precipitated together with that of molybdenum in the classical analytical group separation method using H_2S). Especially vanadium may have been present not only as solid V_3S_2 and VS_4 and soluble sulphur complexes but also as VS^{2+}, $VSSH^+$, or even soluble simple and mixed metal/sulphur clusters like those of iron. As the result of the progressive oxygenation of the atmosphere, molybdenum, vanadium, and tungsten were oxidized and appear today in sea water as soluble molybdate, vanadate, and tungstate. The concentrations of these species in the sea are shown in Fig. 17.1, together with those of several other biological and non-biological trace metals.

It is striking to see that, although molybdenum and vanadium are now even more abundant than many of the more common metals, e.g. iron, zinc, or copper, only molybdenum has a significant biological role; vanadium and tungsten are restricted to a few primitive or last remnants of primitive species and do not seem to be required by many modern living organisms. Vanadate may even be objectionable since it can poison ATPases working as a phosphate mimic.

It is also interesting to note that both molybdenum and tungsten (but not vanadium), although generally available as oxygenated anions, revert to Mo-S and W-S compounds in biological systems, whereas the major species of vanadium today are nitrogen- and oxygen-ligated forms of V^{3+}, V(IV), and V(V). There are two reasons for this difference:

1. Higher oxidation states of vanadium are more oxidizing than those of molybdenum or tungsten, so that they oxidize thiolate groups.
2. The sulphur chemistry of vanadium is less stable than that of molybdenum and tungsten, i.e., to put it differently, vanadium has a more 'a' (hard) character, whereas molybdenum and tungsten have a more 'b' (soft) character.

This does not mean that vanadium cannot form any biological sulphur compounds or complexes *inside* cells, and vanadium nitrogenase, which may predate molybdenum nitrogenase, is an example.

Here, however, there is a quandary concerning evolution since we do not know whether nitrogen fixation was required early in life. Indeed, the early available forms of nitrogen are in doubt. For example, did the early atmosphere and aquatic environment have sufficient NH_3 or HCN? If the answer is affirmative then there was no need of a nitrogenase. The evidence for an early requirement of this enzyme is indirect: on one hand, nitrogenases are only found in prokaryotes and we may feel that they must have been present from a very early stage in life's history; on the other hand, both the molybdenum and vanadium nitrogenase active centres are quite different from those in any other known molybdenum and vanadium enzymes and are inside cells, whereas the others are outside cells. Finally, the high contents of vanadium in certain oils believed to be of remote biogenic origin, e.g. some Peruvian and Venezuelan oils, show that this metal must have been widely present in the ancestors of present-day algae, fungi, and higher plants, perhaps in the giant ferns that once seemed to have populated the tropical forests. The requirement for this metal points to an essential function and nitrogen fixation or CO_2 reduction in photosynthetic processes are likely candidates.

We may add still another speculative note to the above comments on what we consider to be one of the most primitive metals used in biology. While commenting on the possible evolution of the functions of manganese (Chapter 14) we considered that, when life developed the ability to use light, a source of reducing equivalents was necessary on one side of the membrane, where light capture occurs, and hydrogen sulphide could be an ideal choice due to the reaction $H_2S \rightarrow S^0 + 2H^+ + 2e$, provided that a metal catalyst was available. We have shown that Mn(II) was a possibility, but vanadium (nickel or iron) could also be used to drive this or a similar reaction, leading, for example, to the precipitation of $V(S_2)_2$ (and FeS_2). This could be the reason for the accumulation of vanadium by ancient photosynthetic species, accounting for the unusually high contents of this metal in some oils, particularly Venezuelan and Peruvian, and also for the occurrence of this metal as the sulphide VS_4 (patronite) in the ores which were once the most important mineral source of vanadium (the Ragra's mine in Peru, curiously in the same tropical region).

Finally, we must discuss the question of the preferential choice of tungsten by hyperthermophylic archaebacteria, the fact that many bacteria can utilize both W and Mo, and the finding that the majority of life species today use only Mo in the relevant enzymes.

First, we note that the thermal stability of W(VI) complexes is higher than that of Mo(VI) complexes due to stronger π-bonding, and this may mean that only the former are stable enough at high temperature.

Then, W complexes systems have much lower redox potentials and can operate at quite low potentials, e.g. –430 mV, needed, for example, for formate/CO_2 conversions. This may require stabilization of W(IV), explaining why two pterin coordination is required for tuning the redox potential to the required level. Last, these centres must be protected against O_2, hence application is restricted to obligate anaerobes (or a protective mechanism must be operating).

A further observation is that the kinetics of W complexes are generally slower than those of Mo complexes, but higher temperatures can ensure adequate catalytic rates. All these observations explain why W enzymes are absolutely required by hyperthermophiles, and all other organisms synthesize both Mo and W isoenzymes. In our opinion, W-containing enzymes predated the Mo-containing enzymes.

In conclusion we return to the role of these elements in the metallome during evolution. We note that they perform highly selected functions which have developed and changed with time especially through the slow coming of dioxygen. In very advanced species tungsten and vanadium have disappeared, leaving only molybdenum; compare the decrease of importance of nickel and cobalt compounds with the rise of those of copper. Now it is also seen that the three elements are associated with the chemistry of especially Fe/S and to a different degree P, Se, Mg, Zn, and halogens. We return to these themes in Chapter 20 where we consider the cooperative activity of elements in the metallome.

17.11 The biological chemistry of chromium

This element has the peculiarities that it has virtually no available oxidation state except Cr(III) and in this state it has little affinity for sulphide and is heavily hydrolysed. It is a d^3 cation and, as such, usually has octahedral complexes which show slow ligand exchange. The verdict of biology appears to be that this element is of little interest in either redox or acid/base properties. If anything, it appears to be a poison (compare Al(III) and Ni(II)), to be avoided by biological systems generally. Little plant life grows on Ni/Cr minerals. All these comments are made knowing that Cr(III) has now and then been reported to be essential in some function or another in higher animals and it was even considered to be a component of a so-called 'glucose tolerance factor', GTF, isolated from brewer's yeast and pork kidney powder, essential to maintain animal glucose levels in mammals. These reports have persisted for 50 years, but only recently has evidence appeared that Cr(III) may indeed have a biological role. In effect, it seems now probable that Cr(III) is linked in insulin signalling. The association of Cr(III) with a low molecular weight (LMW) peptide of <10,000 daltons, dubbed *chromodulin*, has been demonstrated. Four Cr(III) ions bind per monomer and the peptide binds chromium at least as strongly as apotransferrin. The LMw(Cr) complex is found to stimulate a kinase activity which is dependent upon the insulin receptor, but a full explanation of this effect has not yet been given (see J. B. Vincent in 'Further reading').

Further reading

Molybdenum

Burgess, B.K. and Lowe, D.L. (1996). Mechanism of molybdenum nitrogenase. *Chemical Reviews*, **96**, 2983–3011.

Eady, R.R. (1996). Structure–function relationships of alternative nitrogenases. *Chemical Reviews*, **96**, 3013–30.

Heffron, K., Léger, C., Rothery, R.A., Weiner, J.H., and Armstrong, F.A. (2001). Optimal potential window for catalysis by dimethyl sulphoxide reductase and the role of Mo(V) in the reaction. *Biochemistry*, **40**, 3117–26.

Hille, R. (1999). Molybdenum enzymes, *Essays in Biochemistry*, **34**, 125–37.

Holm, R.H. (1990). The biologically relevant oxygen atom transfer of molybdenum. *Coordination Chemistry Reviews*, **100**, 183–222.

Howard, J.B. and Rees, D.C. (1996). Structural basis of biological nitrogen fixation. *Chemical Reviews*, **96**, 2965–82.

Kappock, T.J. and Caradonna, J.P. (1996). Pterin-dependent aminoacid hydroxylase. *Chemical Reviews*, **96**, 2659–756.

Kisker, C., Schindelin, H., and Rees, D.C. (1997). Molybdenum-cofactor-containing enzymes: structure and mechanism. *Annual Reviews of Biochemistry*, **66**, 233–67.

McMaster, J. and Enemark, J.H. (1998). The active sites of molybdenum and tungsten enzymes. *Current Opinion in Chemical Biology*, **2**, 201–7.

Rajagopalan, K. V. (1988). Molybdenum: an essential trace element. *Advances in Inorganic Chemistry*, **35**, 103–77.

Romão, M.J., Knablein, J., Huber, R., and Moura, J.J. (1997). Structure and function of molybdopterin containing enzymes. *Progress in Biophysics and Molecular Biology*, **68**, 121–44.

Tungsten

Hagen, W.R., and Avendsen, A.I. (1998). The bioinorganic chemistry of tungsten. *Structure and Bonding*, **90**, 161–92.

Johnson, M.K., Rees, D.C., and Adams, M.W.W. (1996). Tungstoenzymes. *Chemical Reviews*, **96**, 2817–39.

Kletzin, A. and Adams, M.W.W. (1996). Tungsten in biological systems. *FEMS Microbiological Reviews*, **18**, 5–63.

Vanadium

Fraústo da Silva, J.J.R. (1989). Vanadium in toadstools. *Chemical Speciation and Biovailability*, **1**, 139–50.

Hemrika, W., Renirie, R., Dekker, H., and Wever, R. (1998). Vanadate containing haloperoxidases and acid phosphatases: the conserved active site. In: Tracey, A.S. and Crans, D.G. (eds). *Vanadium compounds—chemistry, biochemistry and therapeutic applications*. American Chemical Society Symposium Series 711, Chapter 16, pp. 216–27.

Hemrika, W., Renirie, R., and Wever, R. (2000). Glucose 6-Phosphatase and vanadium peroxidases. In. *Membrane structure in disease and drug therapy*, ed. Guido Zimmer. Marcel Dekker, New York.

Isupov, M.N., Dalby, A.R., Brindley, A.A., Izumi, Y., Tanabe, T., Murshudov, G.N., and Littlechild, J.A. (2000). Crystal structure of bromoperoxidase. *Journal of Molecular Biology*, **299**, 1035–49.

Matoso, C.M.M., Pombeiro, A.J.L., Fraústo da Silva, J.J.R., Guedes da Silva, M.F.C., Silva, J.A.L., Baptista-Ferreira, J.L., and Pinho-Almeida, F. (1998). A possible role for amavadine in some *Amanita* fungi. In: Tracey, A.S. and Crans, D.C. (eds.) *Vanadium compounds—chemistry, biochemistry and therapeutic applications*. American Chemical Society Symposium Series 711, Chapter 18, pp. 241–7.

Messerschmidt, A., Prad, L., and Wever, R. (1998). Chloroperoxidase from *Curvularia inequalis*: X-ray structure of native and peroxide forms reveal vanadium chemistry in vanadium haloperoxidases. In: Tracey, A.S. and Crans, D.C. (eds) *Vanadium compounds—chemistry, biochemistry and therapeutic applications*. American Chemical Society Symposium Series 711, Chapter 14, pp. 186–201.

Michibata, H., Uyama, T., and Kanamori, K. (1998). The accumulation mechanism of vanadium by ascidians. In Tracey, A.S. and Crans D.C. (eds). *Vanadium compounds—chemistry, biochemistry and therapeutic applications*. American Chemical Society Symposium Series 711, Chapter 19, pp. 248–58.

Plass, W. (1999). Phosphate and vanadate in biological systems: chemical relatives or more? *Angewandte Chemie, International Edition*, **38**, 9009–12.

Renirie, R., Hemrika, W., and Weser, R. (2000). Peroxidase and phosphatase activity of active-site mutants of vanadium chloroperoxidase. *Journal of Biological Chemistry*, **275**, 11650–7.

Weyand, M., Hecht, H., Kiess, M., Liaud, M., Vilter, H., and Schomburg, D. (1999). X-ray structure determination of a vanadium dependent haloperoxidase from *A. nodosum* at 2.0Å resolution. *Journal of Molecular Biology*, **923**, 5995–611.

See also *Journal of Inorganic Biochemistry*, (2000), May 30, **80** (1–2), a special issue devoted to vanadium biochemistry.

Chromium

Mertz, W. (1998). Interaction of chromium with insulin: a progress report. *Nutrition Reviews*, **56**, 174–7.

Vincent, J.B. (1999). Mechanisms of chromium action: low molecular weight chromium binding substance. *Journal of the American College of Nutrition*, **18**, 6–12.

Vincent, J.B. (2000). The biochemistry of chromium. *Journal of Nutrition*, **130**, 715–18.

Phosphate, silica, and chloride: acid-base non-metals

18.1 Introduction to the non-metals

It is not warranted to give in this book anything but passing reference to the non-metal chemistry of H, C, N, and O since these four elements form the central part of organic biochemistry and there are excellent volumes dealing with this subject. At the same time, it is pertinent to give a summary table (Table 18.1) of the involvement of metal ions with virtually all the small molecules which are the available precursors of biological organic chemistry. In fact, major features of the chemistry of hydrogen appear in Chapters 1 and 5, of that of carbon and nitrogen in Chapter 1, toward the end of Chapter 2, and in Chapter 7, and of that of oxygen in chapters on transition metal chemistry in organisms (Chapters 12–17). In each case we have tried to link the biological chemistry of these non-metals to that of other inorganic elements and we have noted in particular how they are handled as elements by specific co-enzymes. In this and the following chapter a somewhat more detailed account is given of the non-metal elements, P, Si, S, and Se, concentrating upon the mainly anionic chemistry corresponding to that of the cationic chemistry of metals. The anionic chemistry of chloride has been briefly described in Chapter 8 but we shall add some more detail in this chapter. Only when this anionic chemistry has been described can we go forward to the description of the way in which metals and non-metals are used together in cell systems: (1) in forming inorganic minerals in organisms; (2) in a complementary fashion to control homeostasis. In the this chapter on non-metals we shall describe the chemistry which is associated with acid/base that is, primarily, the chemistry of phosphorus and silicon but not redox chemistry, which is described in Chapter 19. This division follows that of the metals.

Group 13	Group 14	Group 15	Group 16	Group 17
B	(C)	(N)	(O)	F
(Al)	Si	P	S	Cl
	Ge	As	Se	Br
				I

The non-metals

Table 18.1 Selective metal-involvement with elementary reactions of non-metals

Non-metal	Metal involvement
H_2	Ni, Fe
CH_4	Fe, Cu
CH_3^-	Co, Zn
CO_2	Zn
CO	Ni, Fe
$O_2, O_2^{·-}$	Mn, Fe, Cu
H_2O_2	V, Mn, Fe, Se
N_2	Mo, V, Fe
NO	Fe
NO_3^-	Mo
NO_2^-, N_2O	Fe, Cu
SO_4^{2-}	Mo
SO_3^{2-}	Fe
$HPO_4^{2-}, P_2O_7^{4-}$	Mg, Zn, Fe, Mn
Cl^-, Br^-	V, Fe
I^-	V, Fe, Se
NTP^{4-}	Mg

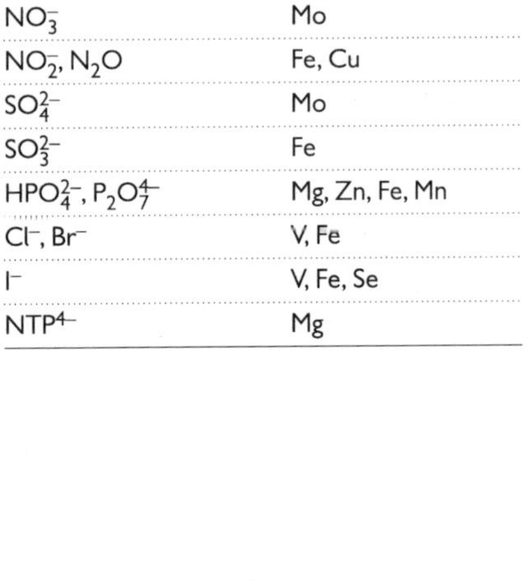

Fig. 18.1 A diagram of the uptake routes of phosphate for incorporation and silica ($Si(OH)_4$) for rejection and biomineral formation. Chloride and carboxylates are the other major anions.

18.2 Phosphate biochemistry

Apart from the major 'biological' elements C, H, O, and N, phosphorus and sulphur are central to all forms of life. In the present chapter we are especially concerned with phosphorus, which occurs always in oxidation state V as phosphate, free or combined. There is no life without phosphate, but the most impressive fact is that, directly or indirectly, this chemical species plays a decisive role in the majority of the metabolic steps of all organisms, from the most primitive to the most developed. It forms the linking units of the sugar esters of DNA and RNAs, it is present in the energy storing nucleotide triphosphates, e.g. ATP and GTP, in coenzymes such as FMN, NADH, and NADPH, in the phospholipids of membranes, in carriers such as glucose phosphate, in signalling molecules such as cAMP and inositol triphosphate, IP_3, and in biominerals such as bone. It can also be used as an agent for the modification (phosphorylation) of proteins which then undergo allosteric changes of conformation and act as triggers or signalling devices to promote chemical reactions. In the following sub-sections we summarize some relevant aspects of these various roles of phosphate (see Figs 18.1 and 18.2).

18.2.1 The chemistry of phosphate

Free phosphate is present as the anion HPO_4^{2-} inside cells and in circulating fluids at a concentration of about 10^{-3} M. It is at considerably lower concentration in the environment and, therefore, it has to be pumped into biological systems across cell membranes (Fig. 18.1).

Phosphate is tribasic and two of its protonation steps are those of a weak acid ($pK_a = 13.0$ and 6.5) while the third corresponds to $pK_a = 1.0$ which means that phosphate always carries an average charge of above one at pH = 7. The organic diesters and esters of phosphate also always carry a negative charge of one or more since they have pK_a values of 1.0, and 6.5, respectively. None of these anions bind appreciably to simple cations such as Mg^{2+} and Ca^{2+}, but all can bind appreciably to positively charged protein surfaces and to M^{3+} ions.

Metapolyphosphates, linear P_n such as pyrophosphate, P_2, or organic esters XP_n or XP_nY behave similarly with respect to their proton affinities. Where there is a terminal phosphate its pK_a values are about 1.0 and 6.5, while all the chain diester phosphates have a pK_a of ≃1.0, so that they are fully deprotonated in physiological conditions. All phosphate is therefore negatively charged and all molecules with a terminal phosphate are always partially protonated—the biochemistry of phosphate is inevitably linked to that of the proton. In addition, the binding to Mg^{2+} is now of great significance since, as shown in Chapter 9, Mg^{2+} binds all pyro- and higher metaphosphates, P_n. The involvement of especially kinases and phosphatases with Mg^{2+} and to some extent other metal ions is such that we have dealt with many of these enzymes in early chapters, especially Chapter 9.

The stable form of phosphate in water at pH = 7 is as free inorganic phosphate, $H_2PO_4^-$ and HPO_4^{2-}, and all the esters and anhydrides of phosphate are unstable, i.e. they are energized. The degree of energization is small for simple phosphate esters, but each pyrophosphate bond requires more than 10 kcal for its synthesis. The appearance of phosphate compounds in cells is then a sign of their limited kinetic stability, which implies that they will be formed or hydrolysed at selected places and at selected times where the

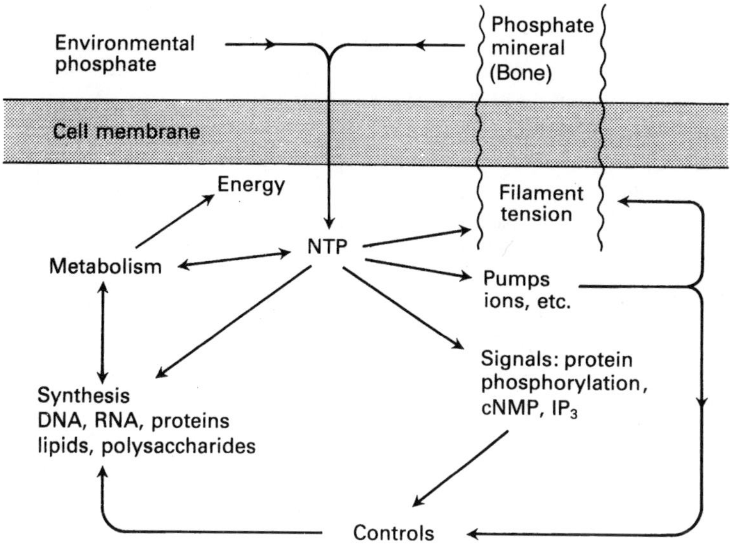

Fig. 18.2 The flow of phosphate into a cell and some of the connections it makes.

specific catalysts for the reactions of a given organic phosphate are to be found, e.g. glycolysis in the cytoplasm and glycosylation in the Golgi (see Chapter 5). In fact, as we have stressed, phosphorylated substrates are extremely common in biological systems.

The basis of the kinetic lability of compounds of phosphate, like those of silicate but unlike many carbon compounds, is that they react relatively quickly with attacking reagents due to the ease of phosphorus in going to the five-coordinate intermediate state:

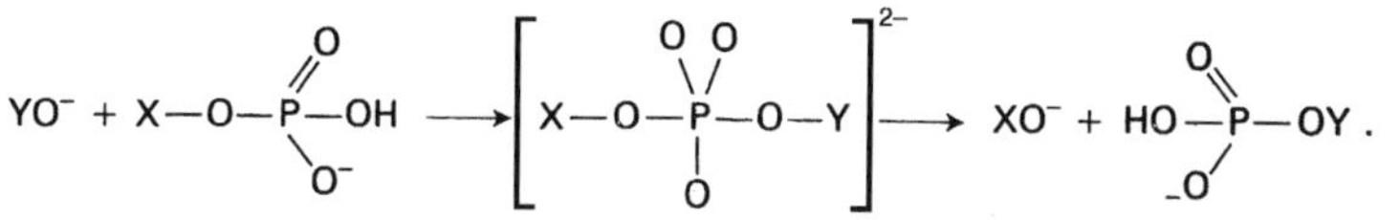

The intermediate state of a five-co-ordinate phosphorus is readily mimicked by stable vanadium complexes, which are then able to block the phosphate transfer sites of some enzymes, e.g. Na^+/K^+ ATPases. We turn next to the biological chemistry of phosphate.

18.3 **The forms and energies of bound phosphate**

Phosphate is a linking unit in sugar esters of DNA(RNA) but, as we have emphasized in Chapter 6, it remains a 'deliberate' centre of mobility and bond angle variation. DNA(RNA) must be able to change conformation easily and must be relatively open to hydrolysis: that is the essence of any polymer which carries rapidly reproducible instructions and which has to undergo correction of errors. Of course, phosphate compounds are especially vulnerable when their charge is neutralized by cations.

Phosphate is also found in phospholipids both as a diester and a monoester, especially in membranes. In IP_3 lipids (see Section 18.3.4) the phosphate bond is again a source of weakness (through hydrolysis) in these lipids, and a centre of mobility. Membranes are fluctuating fluid structures in shape and local composition. Phosphate, as a diester phosphate in lipids, as in DNA and RNA, has the advantage of deriving from a strong acid ($pK_a \sim 1$) and its interactions with H^+ are weak. In isolation its interactions with Mg^{2+} and Ca^{2+} are also weak, but diester phosphates in membranes and DNA(RNA) are in fact large one- and two-dimensional polyelectrolytes, respectively, which, although not neutralized by direct bonding to simple cations, hold counterions in the Gouy–Chapman layer, which help to stabilise these structures. The phosphate monoesters are very different in function (see Ca^{2+} triggering of membrane fusion in Chapter 10) since they bind directly to cations and are then held in the co-ordination sphere (see Chapter 9). There is competition on all these surfaces between the inorganic action binding, that of Mg^{2+} in cells and both Mg^{2+} and Ca^{2+} outside cells, and polyamines or positively charged proteins. The competition is of extreme importance in many associations of phosphate-carrying surfaces, e.g. DNA, RNA, membranes, polysaccharides, and even bone (see Section 18.3.7), and it may be involved in the initiation of cell division (Chapter 9).

Now, while the monoester phosphates are largely concerned with carrying small molecules in synthetic or degradative metabolic paths or in the construction of polymers and membranes, the pyrophosphates are concerned with the transfer of chemical energy and signalling related to conformational switches and only here and there to transfer of chemicals, e.g. glycosylation.

18.3.1 The energy of pyrophosphates and the resting state of cells

We have explained that all pyrophosphates, e.g. nucleotide pyrophosphates, NDP and NTP, are partially protonated at pH = 7. The binding constant of NDP and NTP, i.e. diphosphates and triphosphates, to magnesium ions is about $10^{3.5}$ and, since the free (Mg^{2+}) is about 10^{-3} mM in cells, there is an equilibrium such that there is always some free Mg^{2+}, some free NTP^{4-} or NDP^{3-}, and much $(NTP{\cdot}Mg)^{2-}$ and $(NDPMg)^{-}$ in solution, the triphosphate concentration being about 10^{-3} M and the diphosphate concentration about 10^{-4} M. Thus the energy-producing reaction paths, such as glycolysis, are regulated by the H^+, Mg^{2+}, NDP^{3-}, and NTP^{4-} levels. The link between H^+, Mg^{2+}, and NTP^{4-} is seen to be very immediate and through direct binding, and is indeed inevitable, so that, from the time in evolution when pyrophosphate became the distribution mode for chemical energy, H^+ and Mg^{2+} limited that use. Given the nature of evolution these interactions would obviously be turned to advantage and this advantage is enhanced by the way in which other cations, e.g. Ca^{2+} and adrenalin, modulate the reactions.

The maintenance of the levels of NTP in cells, e.g. all nucleotide phosphates, is through the synthesis of ATP. Even in a cell at rest there has to be constant synthesis since biological systems are not stable. [In what follows, and generally in this book and in the literature, we write NTPs in this form although these compounds are charged and largely bound by Mg^{2+}. Hence we have described kinases together with phosphatases in Chapter 9.] ATP is slowly hydrolysed in water itself and many other compounds are more or less unstable. Again tension in the cell and hence cell shape is maintained and constantly regenerated using hydrolysis of ATP and GTP (Chapter 5). Oxidative damage has to be repaired continuously, poisons are rejected, nerves send continuous messages, and so on and so on. Each process demands use of ATP. The supply of ATP is described in Chapter 5 and the way in which it is held constant is through feedback control of the enzymes of glycolysis or of the energy-transducing membranes. Section 9.8 describes the control in a muscle cell but the control is exerted everywhere. Phosphate is also found in other small-molecule highly energized forms in cells including mixed anhydrides, e.g. acyl phosphates, creatine phosphates, etc., but we have no space to include these reactions here. These componds are of higher energy than simple phosphate esters.

18.3.2 **Phosphate controls**

We should perhaps have analysed the biochemistry of phosphate immediately after that of magnesium, calcium, and zinc because we consider that these four elements have a significance in the control of eukaryote cytoplasmic cellular acid/base metabolism which sets them apart (Fig. 18.3). In one sense (organic) ionised phosphate behaves as if it were negatively charged calcium, as we shall show in this section, just as chloride behaves as negatively charged sodium or potassium in conductance. If this is a correct view, then we imply that the reasons for the association of the above four elements with control of hydrolytic systems in eukaryotic cells, and virtually only these four elements amongst all others, are that they are sufficiently abundant, bind with sufficient strength, have suitable (fast) kinetic parameters associated with their reactions, and do not undergo redox chemistry, so that together they act as basic non-oxidative homeostatic controls. Their concentrations of cations are manipulated (energized) so that, in fact, they are always in differentially balanced flow both between stable free ions and bound states as well as in gradients across membranes. Their rest and triggered states can then be governed by pumped gradients and opening of channels. The condition of cytoplasmic phosphate compounds is different but parallel. The reference state for phosphate in *the cytoplasm of a cell* is not the same as in the above cationic chemistry since phosphate *bond* chemistry, not its cross-membrane partition, is energized to such an extent that any cross-membrane gradient energization, though present, is not of comparable consequence (compare Mg^{2+}). (When an NTP is highly concentrated in vesicles, however, it, like other organic molecules such as adrenaline, can be used as a messenger.)

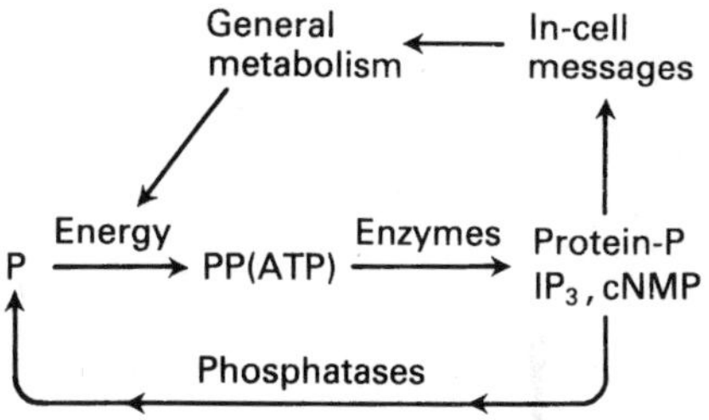

Fig. 18.3 Feedback loops related to phosphate controls. All connections to calcium are excluded.

Because of this degree of bond energization, we shall take as a reference state ATP concentration, [ATP], and not inorganic phosphate concentration. It is [ATP] which is held constant in cells and it is binding of phosphate in organic molecules relative to binding in ATP which represents the transfer potential of phosphates. (This corresponds to taking the *extracellular* Ca^{2+} concentration as its reference state for intracellular reactions.) The free energy ΔG of the phosphate in ATP is some 10.0 kcal (10^7 concentration units) away from that of free phosphate and ADP in cells. (Note that internal Ca^{2+} is energized by some 10^4 concentration units away from external calcium.) Thus, phosphate can be said to have a 'kinetically restrained' real binding constant ($>10^5$) to serine in proteins *relative to ATP* in regulating reactions such as

$$\text{ATP} + \text{Protein-Serine} \longrightarrow \text{Protein Serine-P} + \text{ADP}$$

and in reactions which give cyclic phosphate esters, e.g. cAMP, which are used as internal triggers, much as calcium injection is used. Other proteins undergo tyrosine phosphorylation, and in bacteria, as opposed to eukaryotes, histidine is also phosphorylated. The protein phosphate is an ester (and so is cAMP), which is still somewhat energized with respect to free phosphate, and a final part of the trigger or control is its hydrolysis by phosphatases (see 'Further reading') often using Mg, Zn, or Ca. Given that the base level from

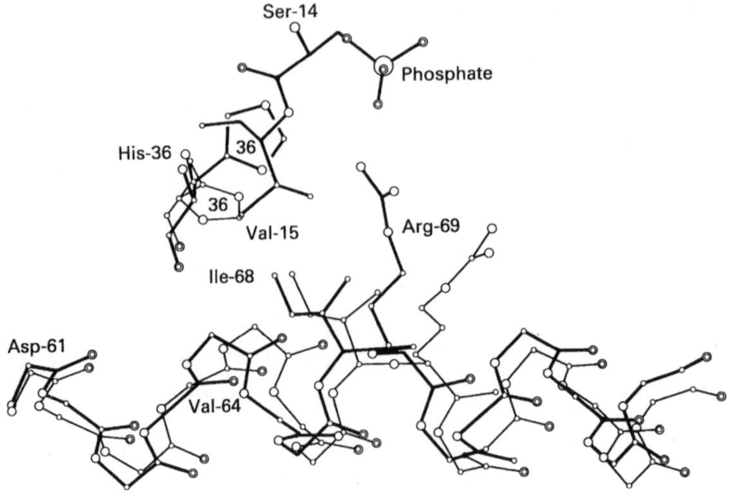

Fig. 18.4 The effect of phosphorylation of Ser-14 on a helical portion of phosphorylase. The inactive state, phosphorylase b, shown in thin lines, is adjusted by phosphorylation to give the active state. NB The effect of phosphorylation is transmitted to the active site by the adjustment of helices (see Chapters 7, 10, and 13). (After the work of L. Johnson.)

which phosphate reactions should be seen is ATP (i.e. a particular pyrophosphate), we note again that the kinetics of phosphate reactions are slow enough (they involve 'covalent' binding) so that ATP reactions are controlled in time and space by catalysts and not by reversible association/dissociation reactions as with metal ion complexes (Fig. 18.2), which are often located at membrane channel openings and closings. All the components of the phosphate trigger are in the cytoplasm and do not need to be divided by membranes because of these kinetic controls. The effect of binding phosphate to all allosteric proteins seems to be through the adjustable nature of helices (Fig. 18.4); compare the effects in calcium and haemoglobin triggers.

18.3.3 **Phosphate messengers**

Cyclic nucleotides, cNMP

Now although the primary source of *chemical* energy is adenosine triphosphate, ATP, there are other nucleotide triphosphates (NTP), e.g. of guanine, cytidine, and uridine, and also yet other pyrophosphate compounds for the distribution of saccharides and isopentyl units. All are synthesized from ATP and we have described their distribution in cellular vesicles in Chapter 3. All of these nucleotide triphosphates can be used in signalling, when they give cNMP at membrane surfaces, to be used in controls and in the generation of mechanical forces, for example in the construction of tubulin in the cytoplasm of eukaryote cells.

The main *second messengers*, cAMP and cGMP, so-called since they are generated by primary messengers such as hormones, are monovalent anions and are, therefore, related to chloride (Cl^-) just as the monovalent organic cations derived from the organic derivatives of ammonium, e.g. acetylcholine and adrenalin, are related to Na^+ and K^+. The skill of biological evolution is clear in using the simplest inorganic ions, Na^+, K^+, and Cl^-, and relating their message-carrying capacity subsequently to the much more selective actions of a much larger number of organic derivatives of ammonia and phosphate. The use of secondary messengers is exemplified in Fig. 18.5, where the role of magnesium and calcium in the whole of phosphate and message chemistry is largely put to one side (see Chapters 9 and 10). Now, yet other phosphates are

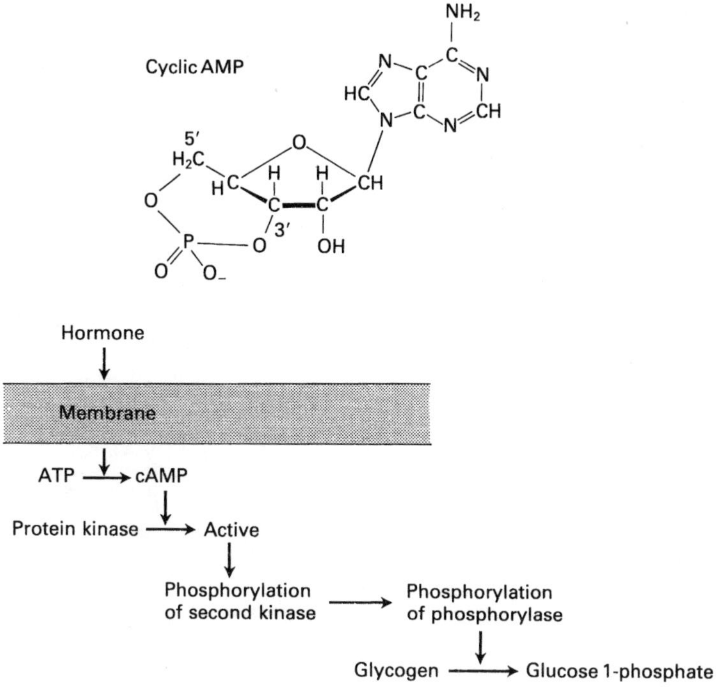

Fig. 18.5 Cyclic adenosine monophosphate, cAMP, and a typical set of reactions activated by it.

used in signalling through hydrolysis reactions since all phosphate esters are energized though much less so than pyrophosphates.

Finally, note that c-NMP formation and hydrolysis are normally very closely related to Mg^{2+} and Ca^{2+} levels.

Inositol phosphates

The inositol phosphates are one such further group of messenger organic phosphates which can either be in a membrane-bound form or free in solution. Their release is then different in kind from the formation of cAMP from ATP which pervades the cell (Fig. 18.6). IP_3 will be released locally if, through membrane composition or curvature, it is locally concentrated (in lipids), and if the stimulant is *local*. The release generates two extra signals: (1) free IP_3 which liberates calcium locally from nearby internal stores in the reticulae, Fig. 10.6; and (2) free lipids locally in the cytoplasmic membrane. The subsequent stages of transmission to enzymatic activity are complex but involve GTP-dependent membrane proteins and kinases C which cause protein phosphorylations. The complexity of this set of signals allows a high degree of chemical selectivity and, by its locality, of physical selectivity in space and time. Again the release mechanism is through energized (but now weakly energized) chemical bonds by specific enzymes, here phospholipases which often require metal ions (Mg, Ca, Zn). Part of the selectivity of cell response to stimuli is due to the fact that the IP_3 does not interact with the same calcium pools as does the cAMP response. The metal dependence of the phospholipases and the subsequent metabolism of lipids lead to complicated connections between signalling systems (see Chapter 20). Here too enzymes are required to hydrolyse the IP_3.

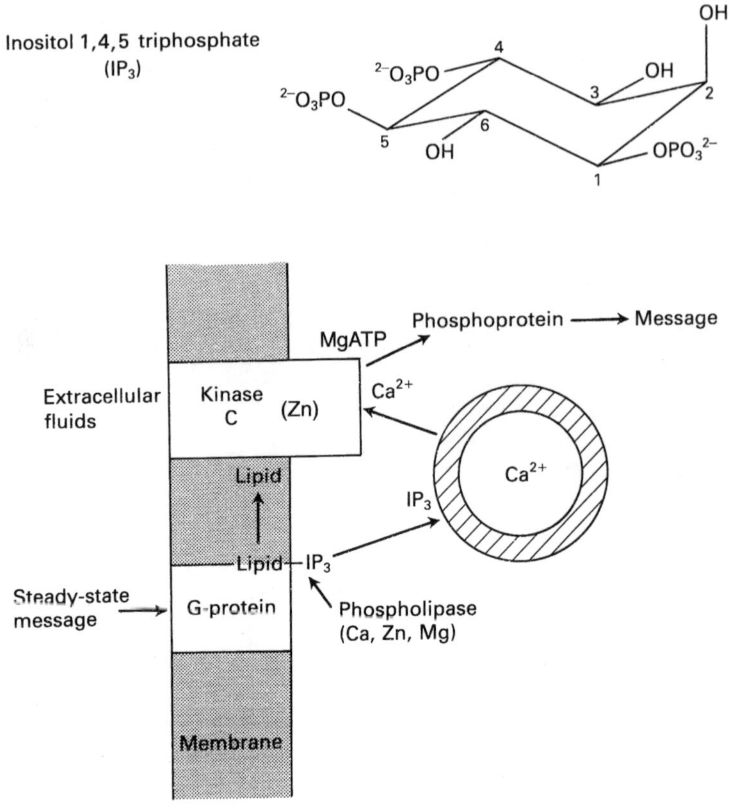

Fig. 18.6 Inositol triphosphate and some of its effects after release from phospholipid inside a cell.

Adenosine triphosphate (ATP)

It is now known that ATP, which can be released from storage vesicles to the extracellular fluids, is often used as a messenger *between* cells as well as in cells. In these vesicles it may well be stored with organic bases such as adrenaline and with calcium rather than magnesium. Such external signalling is also seen occasionally with c-AMP in some special organisms which are normally unicellular but associate under stressed conditions, e.g. slime moulds.

18.3.4 **Phosphate switches: protein phosphorylation, dephosphorylation and time hierarchy**

When a signal, usually a chemical signal, causes a change in level of cNMP or of protein phosphorylation then this change must be slower and more prolonged than that due to Ca^{2+} invasion. While the calcium switch is a series of fast ionic reactions, the phosphate switch is covalent compound formation and can only be relaxed by enzymic activity. Phosphate esters are easily hydrolysed, however, so that on/off switches can be in the 1 second to 1 hour range, but not in the millisecond range. The fastest phosphate switches are those due to anhydride formation in ion, sodium and calcium, pumps

$$\text{Protein-CO}_2^- + \text{ATP}^{4-} \longrightarrow \text{Protein-CO}\cdot\text{P}^{2-} + \text{ADP}^{3-}$$
$$\text{Protein-CO}\cdot\text{P}^{2-} + \text{H}_2\text{O} \longrightarrow \text{Protein-CO}_2^- + \text{P}^{2-} + \text{H}^+.$$

Since these reactions are slowish, the relaxation of really fast muscles uses as an intermediate an ionic buffer, parvalbumin, for calcium (Chapter 10) that is faster than the slower ATP-dependent pumps.

Some reactions of phosphate which are yet slower use ester formation,

$$\text{Protein-OH} + \text{ATP}^{4-} \longrightarrow \text{Protein-OP}^{2-} + \text{ADP}^{3-},$$
$$\text{Protein-OP}^{2-} + H_2O \longrightarrow \text{Protein-OH} + \text{HP}^{2-}$$

where OH is a serine, threonine, or tyrosine centre in eukaryotes and histidine in prokaryotes. There is, in fact, a hierarchy of kinetic stability within protein phosphate compounds, as there is in carbon compounds, but since phosphorus is always in the same oxidation state the speed variation is due to the variation in the carbon compound to which the phosphorus is bound. This hierarchy has to be seen against the background of kinetics in the periodic table which is

Na^+, K^+, Cl^- diffusion: very fast
Ca^{2+} ionic reactions: fast, close to equilibrium
IP_3, ATP highly polar covalent reactions: intermediate rates
Zn^{2+} binding close to equilibrium: tightly bound ionic reactions, slow rates
C-C non-polar reactions: very slow, not close to equilibrium.

Phosphate in a switch device introduces, therefore, a medium-term or even a relatively long-term chemically covalent memory trace, while the calcium (electrostatic or ionic) switch is as close to instantaneous as biology can manage and by itself leaves no trace. Carbon reactions are a permanent record or long-term memory as shown by changes of state, e.g. growth patterns.

18.3.5 Phosphates and gene transcription

We are only beginning to understand the complexities of regulation of gene transcription (DNA to RNA) and of translation (RNA to proteins). Here we refer to only the simplest control over gene transcription—direct action of a regulator gene via protein binding. The bound proteins may be in one of two conformational states, but in one such state they aid transcription (positive feedback) or they prevent transcription (negative feed-back), (see Section 5.8). The conformational switch is needed in the control over the production of messenger RNA and, therefore, of protein product. In this chapter we are especially interested in phosphate. It is already known that very many important enzymes which have their synthesis controlled at the gene level by proteins have their binding conformation affected by phosphorylation or by association with cAMP. Much of protein synthesis is then under the control of phosphate metabolism. It is known, in fact, that there are two classes of such interactions:

Protein phosphates are a dominant part of gene transcription

1. one which modulates protein production up and down and helps to maintain a cell in homeostatic harmony;
2. one which causes the cell to be irretrievably sent along a new path to a particular differentiation.

While there is some difficulty with an absolute separation of these two concepts in chemical terms it is clear that while (1) lies behind homeostasis, (2) is associated with morphogenesis. The idea of homeostasis is connected to that of maintaining and/or controlling concentrations in a dynamic steady state and we must therefore look at ways in which phosphate concentrations can be buffered. In the cytoplasm the buffering relates to the [ATP]/[ADP][P], the phosphorylation potential, but outside the cell phosphate concentration is managed by hormones and messengers sometimes linked to precipitates, e.g. bone.

18.3.6 Phosphate minerals

Phosphate salts are the basis of bone, where $Ca_2(OH)PO_4$ is a simple way to represent the apatite mineral. Calcium phosphates are also deposited by some lower organisms. Other phosphate nodules occur in a variety of worms. Phosphate is also found in ferritin, especially bacterial ferritin which begins to look like $Fe_4O(PO_4)_2$ rather than a derivative of ferrihydrite. It is quite possible that bacterial ferritin was an early phosphate buffer. In these cases we must not assume that the solid phosphates seen in organisms are just protections, structures, or even stores of phosphate; they act as controls as well. In Chapter 20 we shall look again at the controls over phosphate in body fluids. We shall not examine phosphate biological minerals here but we direct attention to Chapter 10, while noting that they are Ca^{2+} and OH^- buffers too.

18.3.7 Phosphate uptake and transport

The understanding of the processes of transport of phosphate in bacteria, plants, and animals progressed considerably in the last few years. For example, it has been shown that in *E. coli* there exists a phosphate transport system (Pit) that promotes the uptake of inorganic phosphate (P_i) in the presence of divalent cations, such as Mg^{2+}, Ca^{2+}, and Mn^{2+}, which form soluble metal phosphates $MHPO_4$. The metal phosphate is the transported solute in stoichiometric symport with H^+. Uptake is inhibited at low internal pH.

In plants, genes encoding phosphate (and sulphate) transporters have been identified in several species. The expressed transporter proteins are predicted to have 12-membrane spanning domains and function as $H^+/H_2PO_4^-$ co-transporters. Both high-affinity and low-affinity types have been identified Regulation at the transcription level appears to depend on the nutrient status of the plant (see Smith *et al.*, in 'Further reading').

In animals, recent studies concentrated on the sodium-dependent renal proximal tubular reabsorption of inorganic phosphate which plays a major role in the maintenance of P_i homeostasis. Two types of Na^+/P_i transporters, types I and II, have been identified in the renal cortex of various species, and a third, type III, transporter appears to be expressed also in a variety of other tissues. All of them are regulated by feedback from the P_i concentration, and the hypothesis of existence of a novel P_i-regulating hormone, 'phosphatonin', has been advanced (see Takeda *et al.* in 'Further reading'.).

18.4 Summary of phosphate functions

We have not attempted a comprehensive cover of phosphate biochemistry and here we give just an outline of its enormous significance which can only be compared with that of C, H, N, O, S, and Cl among non-metals and Na, K, Mg, Fe, Zn, and Ca among metal elements. These are the elements which allow polymer formation and control of the medium, water, in which all the reactions occur. Once again, we must stress, however, that phosphate, as far as we know, has been central to biological group transfer and control signalling as well as to structures of DNA(RNA) and membranes since the beginning of life. Note that its role is not in redox but in acid–base reactions. The other non-metal of comparable importance in these respects, transfer and signalling, is sulphur, but it can engage in redox reactions also. As stated at the beginning of this chapter, P then is to be placed with Na, K, Mg, Ca, and Zn, while S is to be placed with most non-metals and metals such as Fe. Of course, the acid–base systems and redox elements must cooperate in the reductive/condensation chemistry of cells. This is revealed in crosstalk between the two groups in compounds but also in the value of, for example, P (phosphate) and Fe in transcription factors. In bacteria such as *E. coli* it is found that genes such as *arc* (for kinases, P) and *fnr* (for Fe) act cooperatively in controls over metabolism (see 'Further reading'). As life has developed the number of phosphate-dependent processes has increased, for example in the number of kinases, which starting from tens in bacteria has increased to thousands in man.

18.5 Chloride channels pumps and exchangers

The requirement for channels and pumps to allow movement of chloride is equal to that for sodium and potassium (Chapter 8). Chloride channels have been well characterized and measurements of permeability ratios for common biological anions are shown in Table 18.2. The obvious point to make is that these channels have the properties of those anion sites which give rise to the lyotropic series and therefore have somewhat hydrophobic non-hydrogen-bonding character (Section 2.15.1). It was essential that they should not accept carboxylates, sulphate, or phosphates. However, anion balances require the export of HCO_3^- and there is a chloride/bicarbonate exchanger in many cells (Fig. 18.7). This exchanger is controlled by both cAMP and Ca^{2+} ions so that anion balances are tightly connected to other cell activities, including cation balances, and also to energy (ATP) status.

The Cl^- pump in some early Cl^--tolerant prokaryotes (archaea) is called halorhodopsin and it is obviously related to light-driven bacteriorhodopsin which pumps protons (Section 5.3). It is selective against other anions. The Cl^- channel in human cells keeps the cells anionic balance. It has been found that certain mutations of the gene which encodes the Cl^- channel protein give rise to the disease cystic fibrosis. The structure of the chloride pump is now known, see Further Reading.

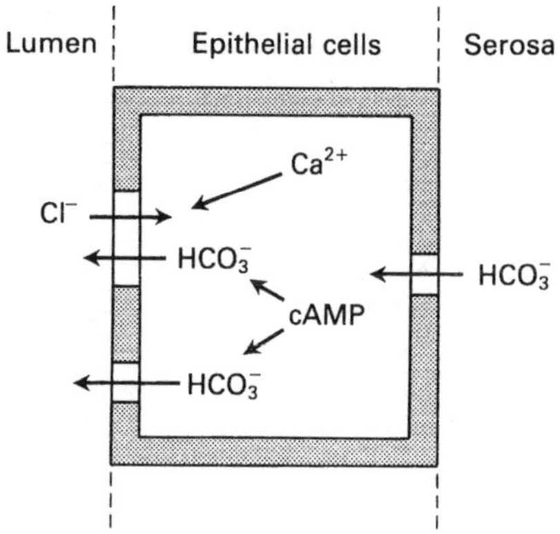

Fig. 18.7 A diagram showing bicarbonate secretion coupled to chloride balance across the duodenal epithelium. Note the involvements of cAMP, Ca^{2+}, and H^+ in controls.

Table 18.2 Chloride channels: relative permeability

Anion	Radius (Å)	Permeability (aprox.)
Cl^-	1.80	1.0
Br^-	1.95	1.0
I^-	2.15	1.0
NO_3^-	1.90	0.5
ClO_4^-	2.40	0.5
Carboxylates	>2.00	Very low
Phosphates	>2.40	Very low
Sulphate	2.30	Very low
Hydroxide	1.40	Very low

From Edwards, C. (1982) *Neuroscience*, **1**, 1335–66. The radii are derived from thermochemical data.

18.6 Anion balance: chloride, phosphate, sulphate, and carboxylates

In Chapter 8 we described the basic functions of Na^+ and K^+ in osmotic and electrolytic balance. Amongst anions these roles are taken on mainly by chloride ions which, like sodium ions, were exported from cells very early in evolution. A parallel use has developed in that some cells can use chloride as a current carrier (see the next section) and in yet other cells, e.g. of halobacteria, there are even light-driven chloride pumps. Among divalent ions the role of phosphate is somewhat similar to that of calcium but it is not driven out of cells; it is used in internal signalling. On the contrary, sulphate is either reduced or is driven out of cells and is not used in signalling. Of course, phosphate and carboxylate compounds are accumulated to give the total anion pool. We repeat here that most important biological compounds, including polymers, are negatively charged at the physiological pH, i.e. they are *anions*.

Before we treat those non-metals which appear to have a less essential role in organisms, Si, B, Br, and I, we must note that the concentrations of the anions Cl^- and HPO_4^{2-} as well as some phosphate compounds, such as ATP, have a closely fixed value in cells much like the concentrations of Na^+, K^+, Mg^{2+}. In this respect we may include them within the metallome of a cell. Thus the concentrations in the cytoplasm of most cells (Cl^- (10^{-3} M), HPO_4^{2-} [$10^{-3.5}$ M), ATP^{4-} (10^{-3} M)] are a characteristic of life. These values, like those of the metal ions are not directly related to the genome or the proteome but are linked to the availability of elements and the energy supply of the environment. Similar comments apply to the concentrations of $Si(OH)_4$, $B(OH)_4^-$ and other halides in the species where they are essential (see the next sections). Note that Cl^-, $Si(OH)_4$ and HCO_3^- concentrations in the sea may not have changed greatly since Earth formed but that life has reduced the HPO_4^{2-} content and O_2 has reduced that of HS^- and HSe^-.

18.7 **Silicon biochemistry**

18.7.1 **Occurrence and physical nature of silicon**

Silicon in the form of compounds of $Si(OH)_4$ is an extremely abundant element in minerals and soils, but its role in living systems has been poorly examined since it is *required* only as a trace element in most higher animals, although there is as much silicon present in man as there is magnesium. By contrast, in plant life, especially grasses, and in many unicellular organisms, such as diatoms, silicon is a major element. The turnover of silica by these forms of life is only one order of magnitude less than that of carbon. Huge deposits of SiO_2 on the floors of the oceans and in many deserts are made from tiny silicon shells, many of which belong to the radiolaria family, Fig. 18.8. Silicon as SiO_2, nH_2O, forms a major building unit of many amorphous hard structures even in lower animals. It is thought to be associated with polysaccharides whether it is found in bulk in plants or as traces in higher animals where it has

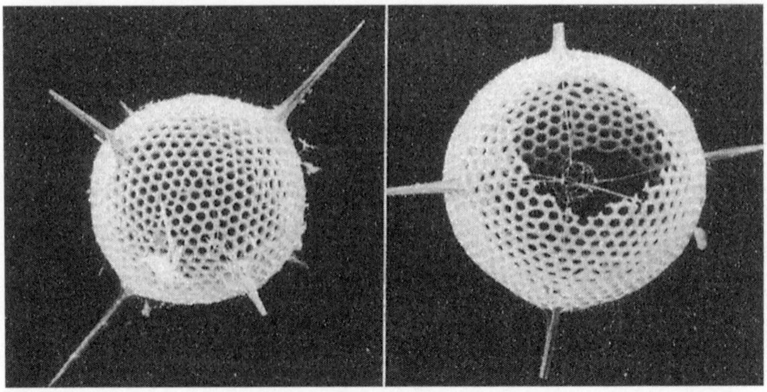

Fig. 18.8 The fine tracery of the silica shell of a diatom (×4100).

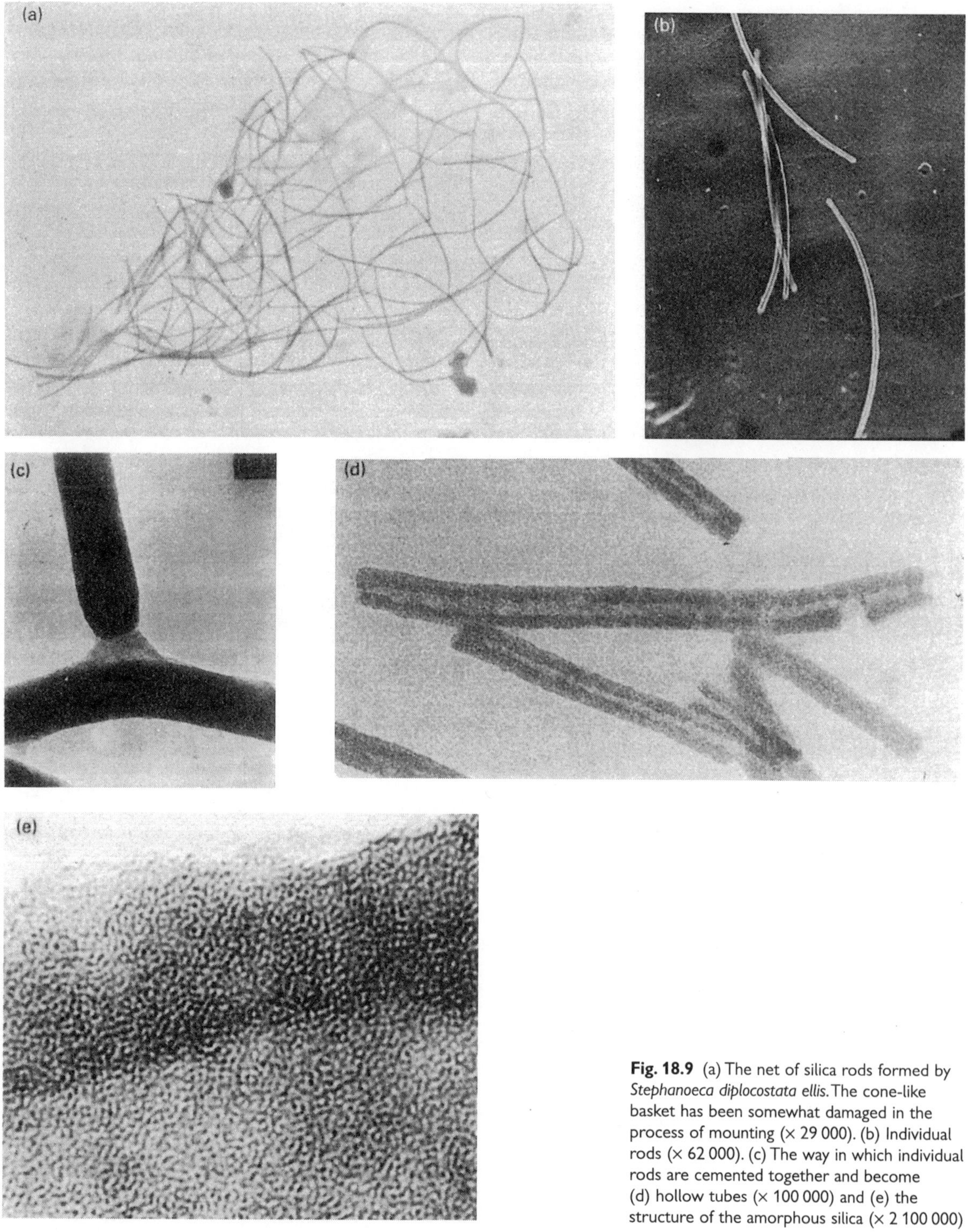

Fig. 18.9 (a) The net of silica rods formed by *Stephanoeca diplocostata ellis*. The cone-like basket has been somewhat damaged in the process of mounting (× 29 000). (b) Individual rods (× 62 000). (c) The way in which individual rods are cemented together and become (d) hollow tubes (× 100 000) and (e) the structure of the amorphous silica (× 2 100 000)

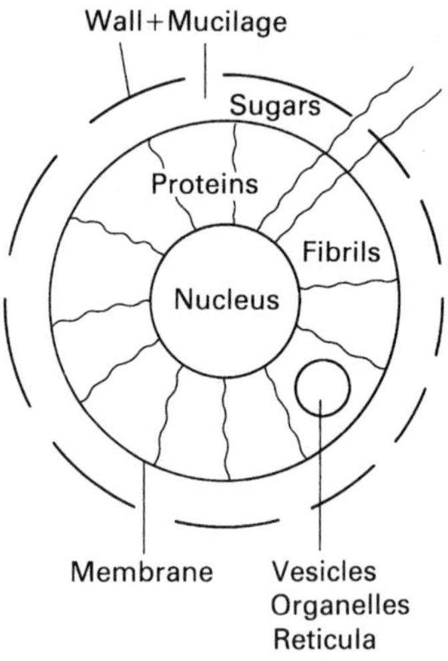

Fig. 18.10 The organization of vesicles by filaments so that eventually silica structures can be exported into the cell wall.

a role in bone formation, in connective tissue, i.e. in glycoprotein stabilization, and perhaps elsewhere. The much wider association with plant life probably arises through the much higher concentration of saccharides there, but it has been recently reported that there are *proteins* involved in silicic acid transport. Before looking at the chemistry it is worthwhile examining the physical forms in which silica appears since the lessons of silica in organisms may be more for materials scientists than for the conventional academic chemist.

Some pictures of the forms which silica takes in shells of diatoms and the baskets of certain protozoa, e.g. *Stephanoeca diplocostata ellis*, are shown in Figs 18.9(a) and 18.10(a). The architecture is almost beyond belief. The forms of silica in higher plants are simpler, although now and then they support large-scale needles or hairs of well-defined structure as in nettles (see Fig. 14.9). These constructions are all found external to cells and have been made in biological moulds. The moulds are either internal vesicles (in the cells) where their forms are produced using tensions developed on the vesicle containers by the internal filaments of the cell (Fig. 18.10) or they are extra-cellular polysaccharide containers. The purpose of the decoration on the structures is unknown, but it is hard to believe that all these intricate patterns have a simple structural function. Here, however, we must be careful since an amateur eye glancing at fan-tracery in the ceiling of a building readily sees that it is beautiful but cannot perceive the functional significance of the whole nor of its cross-bridging. Often elaboration of struts in structures is to do with the weight effectiveness of the overall construction in relation to the load it has to bear. We do not know the stresses and strains which must be met by silica structures, remembering all the time that the structure will have to permit feeding of the organism through the shell. We note here that the complexity of solid forms is clearly related in some way to the underlying stress in filaments belonging to cell membranes (Fig. 18.10) much as bone structure is managed by tension.

The silica rods of the protozoa, *Stephanoeca diplocostata ellis* (Fig. 18.9(b)), are particularly intriguing. It appears that these protozoa are capable of making and counting roughly the number of rods produced, say up to 150 ± 5 (Fig. 18.9(a)), and that a certain small number of the rods are of a thickness different from the larger number. The rods are ejected from the cell in what looks like an oriented bundle and are sorted later so as to make a basket lying on an outer film. The rods are apparently joined together by a cement (Fig. 18.9(c)) such that the structure is relatively rigid and can be used as a kind of 'lobster pot' to catch bacteria. The rods are of hydrated SiO_2 (see Section 18.7.2) and, at first sight curiously, they dissolve from the inside (Fig. 18.9(d)). Before discussing the chemistry involved we must be aware that the loss of material from inside a rod to give a hollow tube is not disadvantageous to the mechanical properties; most scaffolding in the building industry is made from hollow tubes. It seems probable that the flow of silica from the inside of the rods to the rod joints allows rod-rod joints to be cemented by silica (Fig. 18.9(c)) without noticeable weakening of the rods themselves. The whole structure is then made in accord with well recognized principles of mechanics.

We have given this somewhat detailed picture of one or two mineral frameworks here in order to show where the importance of silica lies within biology. There is evidence that the building of silica frameworks is directly controlled from the DNA by silicon receptor systems in such organisms as diatoms, (see Perry in 'Further reading'). Certainly silica is associated with some very early forms of life, for example in cherts found in Africa.

18.7.2 The chemistry of silica in biology

The silicon in biological systems is derived from $Si(OH)_4$ condensation reactions and there is no known reduction or organic chemistry involved. All the silica in organisms which has been examined in detail by high-resolution electron microscopy is amorphous down to atomic levels, i.e. to less than 10Å [Fig. 18.9(e)]. This means that there is no order in the silica and it resembles a rather dense, somewhat dried silica gel [compare Fig. 18.12(a) and (b)]. Nuclear magnetic resonance specroscopy shows that the silica is an ill-defined product of empirical formula $SiO_n(OH)_{4-2n}$ where n goes from 0 to 2. This agrees with its infra-red spectrum. The variable degree of hydration and the lack of structure allows the surface to redissolve readily. The surface can be stabilized by hydrogen bonding to organic matrices well-known outside biology, so that the structures of Figs 18.8 and 18.9(a) are probably based on H-bonded organic templates, see Fig. 18.11. This explains the dissolution of the rods of Fig. 18.9(c) from the inside which is the last part of the structure to be deposited and may well have no organic support.

While the above mechanism for the manufacture of the delicate silica forms may be reasonably clear, the detail of the movement of the silica in living systems is starting to be elucidated (see 'Further reading'). At one time it was thought that silica was laid down in plants simply by transpiration at the leaf surface. This is clearly unsatisfactory as a general way of handling silica in organisms since some silicon constructions are known in algae which live in water. Again the concentration of silica as $Si(OH)_4$ in the sea is quite far (tenfold too dilute) from the level required to precipitate amorphous silica. (The concentration may be closer to the solubility limits of certain silicates since sea water contains much calcium and magnesium.) Since many organisms in the sea can precipitate hydrated silica it is apparent that they must accumulate $Si(OH)_4$. Again the silica content of fresh water in soils is usually low. All plants transpire, yet only some plants, especially grasses, deposit silica. It follows that many plants must have a special uptake mechanism.

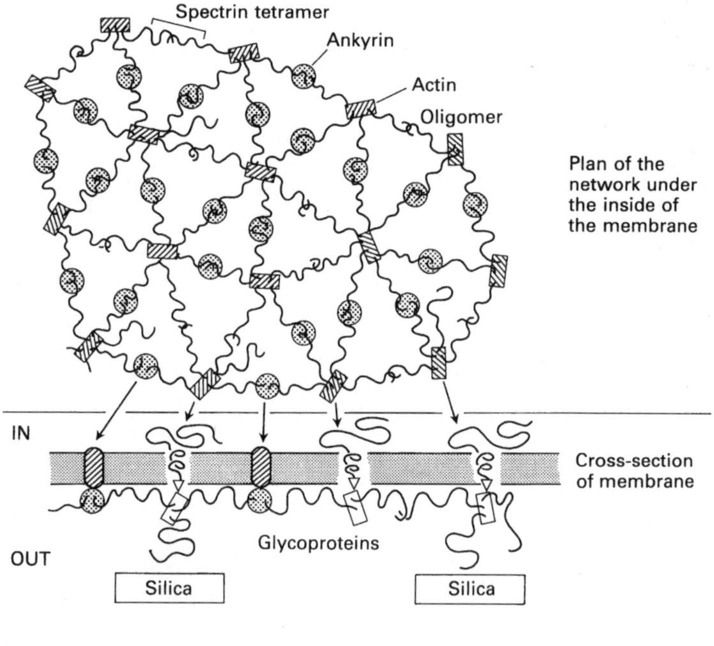

Fig. 18.11 A schematic representation of the network of crossed filaments which may well occur under many cell membranes. This network with protruding proteins through the membrane could well be the basis of patterns of silica such as that seen in Fig. 18.9(a).

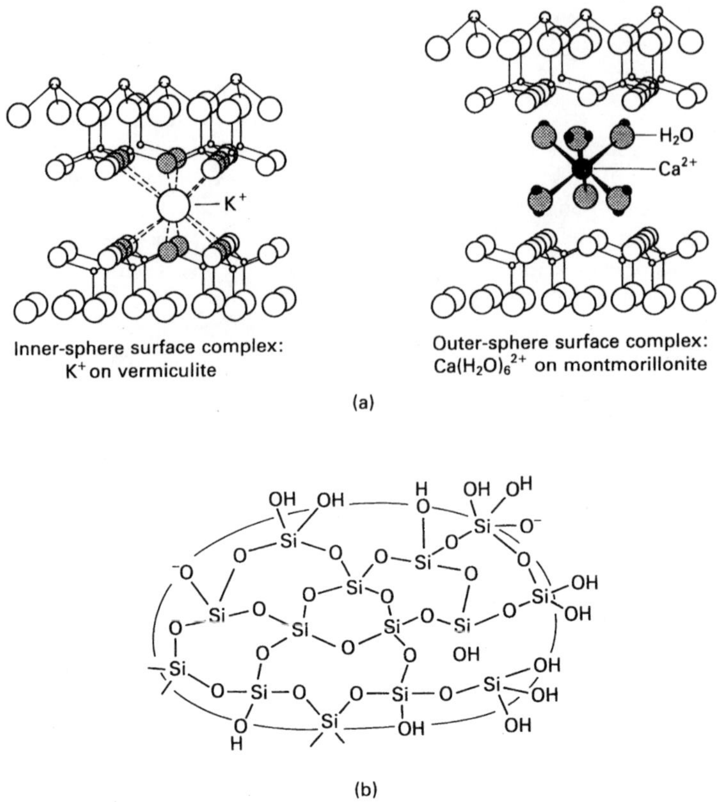

Fig. 18.12 (a) A characteristic pair of silicate layer structures which are crystalline. (b) The structure of the amorphous opaline silica (hydrated) of biological structures as shown by NMR and infra-red spectroscopies (see Fig. 18.9(e).

18.7.3 Uptake and transport of silicon

As stated above, silicon has been recognized as an essential element for many biological systems, from the tiny diatoms to higher plants and animals. In these, silicon was found to have structural and metabolic functions in connective tissues, especially in bone and cartilage, contributing to the strength of collagen and being involved in glycosoaminoglycan formation. It also appears to be a precursor of bone formation in osteogenic cells, in whose mitochondria silicon is particularly high. There even appears to exist a definitive relationship between silicon levels in connective tissues and ageing. The importance of silicon is clearer in plants and especially in diatoms and siliceous sponges.

The transport of silicon is intimately associated to the synthesis of the silicified cell wall of diatoms. In a recent study (see Hildebrand *et al.* in 'Further reading'), the isolation and sequence determination of five silicon transporter (*sit*) genes from the diatom *Cylindrotheca fusiformis* has been reported. The corresponding expressed SIT amino-acid sequences are induced prior and during cell wall synthesis. Several cysteine residues in their putative transmembrane domains are highly conserved, which may account for the sensitivity of silicon uptake to thiol-blocking agents. Summary accounts of recent findings on the gene family of silicon transport and of silicon metabolism in diatoms and sponges have been given by Hildebrand *et al.* and by Tacke (see 'Further reading').

It is sometimes suggested that silica could replace phosphorus as the central element in condensation reactions in biology. Three factors militate against this: (1) Si–O–Si bonds are of low energy content; (2) Si–O–Si systems do not form kinetically stable organic small units in water; and (3) Si–O–Si systems readily form cross-linked solids, not linear polymers. The reader is advised to look back at Fig. 6.1 and note how chemistry changes across the second short period of the periodic table. Organisms know this chemistry! One final peculiar difference between biology and geology is that there are no aluminosilicates in cells except perhaps those formed in pathological conditions such as Alzheimer's disease.

It is found that Al does accumulate with silica in the needle leaves of fir trees, but it is uncertain that there is association in a particular compound. Silica acts frequently as a protection against aluminium, for example in fish.

18.8 Boron in biology

A requirement for boron has not been definitely demonstrated for animals, but this element, associated with pectin, appears to be essential for vascular plants as a component of cell walls, and to be required for the growth of pollen tubes. The best documented case is that of the condensation of boric acid with pectic polysaccharides of plant walls. In this respect, the status of boron in plants is similar to that of the status of silica in both plants and animals—both are connected to the stabilization of connective tissues. Some presumably naturally occurring boron complexes have been isolated, as is the case of boromycin and aplasmomycin (see Sections 2.14 and 2.15.3). More recently, a boron-containing antibiotic has been isolated from a culture broth of a myxobacterium. The boron binding region of this antibiotic (tartrolon B) is identical to that of boromycin (see Irschik *et al.*, in 'Further reading').

Boron is unusual in that its anion is $B(OH)_4^-$ and like $B(OH)_3$ it condenses readily to polyborates. Boron itself is a strong neutron capture reagent which has value both in nuclear reactors and in medical practice, see Gahbauer *et al.* in Further Reading.

18.9 Halides and other non-metal trace elements

Little is known about the distributions of other non-metal elements. We shall therefore make short notes only upon them

1. **Halides.** Here we exclude the use of the simple chloride *anion* (see Section 18.6). Bromine and iodine appear to be activated (1) in vesicles to form covalent organic compounds external to eukaryotic cells, e.g. thyroxine, (2) to be used external to prokaryotes as protective poisons, and (3) to give protection to algae using vanadium halo-peroxidases. Here we may include covalent chlorine, as in the *antibiotic* chloromycetin. Man has made use of these toxic properties of covalent halogen carbon compounds for practical use, e.g. as insecticides, but they have been progressively banned due to their potential deleterious effect in higher animals, including man himself. As this is oxidative organic halogen substitution we return to it in Chapter 19.
2. Fluoride as a simple anion is extremely low in biological systems with an odd exception: its high concentration in krill is not explained. While it is useful in biominerals, e.g. in teeth, it is a dangerous competitor for OH^- in enzymic reactions. Shark's teeth are also peculiarly high in fluoride for no known reason.

3. The appearance of some arsenic compounds is not yet studied in enough detail for us to appreciate the functional value. Of course, it is a toxic chemical and this aspect has been studied in detail, but we have not included a discussion of it or other elements which also act as poisons and pollutants. These aspects were considered in Chapter 22 of the first edition of this book.

Further reading

No specific references to the main part of the chapter on phosphorus are required since in all biochemistry textbooks the properties of all the individual compounds described here can be found and their value appreciated in detail. The purpose of this chapter is to pull together the data on especially phosphorus and silicon. Nevertheless, some publications on phosphorus, silicon and boron may be useful, and a few recent papers are pertinent. The description of many minerals found in organisms is given in Chapter 20 of the first edition of this book.

Phosphate

Drezner, M.K. (2000). PHEX gene and hypophosphatemia. *Kidney International*, **57**, 9–18.

Kennelly, P.J. and Potts, M. (1999). Life among the primitives: protein phosphatases. *Frontiers in Bioscience*, **4**, 372–85.

Raghotama, K.G. (2000). Phosphate transport and signalling. *Current Opinion in Plant Biology*, **3**, 182–7

Smith, F.W., Rae, A. L., and Hawkesford, M. J. (2000). Molecular mechanisms of phosphate and sulphate transport in plants. *Biochimica Biophysica Acta* **1465**, 236–45.

Takeda, E., Taketami, Y., Morita, K., and Miyamoto, K. (1999). Sodium-dependent phosphate co-transporters. *International Journal of Cell Biology*, **31**, 377–81.

van Veen, H.W., Abee, T., Korstee, G.J., Konings, W.N., and Zehnder, A.T. (1994). Translocation of metal phosphate via the phosphate Inorganic transport system of *E. coli*. *Biochemistry*, **33**, 1766–70.

Silicon

Hildebrand, M., Volcani, B.E., Grassmann, W., and Schroder, J.I. (1997). A gene family of silicon transporters. *Nature*, **385**, 688–9.

Hildebrand, M., Dahlin, K., and Volcani, B.E. (1998). Characterisation of a silicon transporter gene family in *C. fusiformis*: sequences, expression analysis and identification of homologues in other diatoms. *Molecular Gen. Genetics*, **260**, 480–6.

Perry, C.C. (2000). Biosilicification: the role of the organic matrix. *Journal of Biological Inorganic Chemistry*, **5**, 537–50.

Tacke, R. (1999). Milestones in the biochemistry of silicon from basic research to biotechnology applications. *Angewandte Chemie, International Edition in English*, **38**, 3015–18.

Boron

Benderdour, M., Bui-Van, T., Dicko, A., and Belleville F. (1998). *In vivo* and *in vitro* effects of boron and boronated compounds. *Journal of Trace Elements in Medicine and Biology*, **12**, 2–7.

Irschik, H., Schummer, D., Gerth, K., Hofle, G., and Reichenbach, H. (1995). New boron-containing antibiotics. *Journal of Antibiotics (Tokyo)*, **48**, 26–30.

Gahbauer, R., Gupta, N., Blue, T., Goodman, J., Bath, R., Grecula, J., Soloway, A.H., Sauerwein, W., and Wanbersie, A. (1998). Boron neutron capture therapy: principles and potential. *Recent Results in Cancer Research*, **150**, 183–209.

Nakamura, H., Iitaka, Y., Kitahara, T., Okazaki, T., and Okami, Y. (1997). Structure of aplasmomycin. *Journal of Antibiotics. (Tokyo)*, **30**, 714–19.

Chloride

Gereneser, G.A. (1996). The chloride pump: a Cl^- translocating P-type ATP-ase. *Critical Reviews in Biochemistry and Molecular Biology*, **31**, 303–37.

Kolbe, M., Besir, H., Essen, L.V., Oesterhelt, D. (2000). Structure of the light-driven chloride pump halorhodopsin at 1.8 Å resolution. *Science*, **288**, 1390–6.

Spudish, J.L. (2000). A chloride pump at atomic resolution. *Science*, **288**, 1359–64.

Arsenic

Ma, L.Q., Komar, K.H., Tu, C., Zhang, W., Cai, Y., and Kennelley, E.D. (2000). A fern that accumulates arsenic. *Nature*, **405**, 579.

Sulphur, selenium, and halogens: redox non-metals

19.1 Introduction

Knowledge of the early atmosphere of the planet and of the mineral elements that were present gives rise to the intuitive feeling that sulphur must have been a very important element from the very beginning of life especially at the low redox potentials which pervaded the anaerobic initial states. Hydrogen sulphide in water may have been present at above millimolar levels and anyone familiar with the properties of transition metal sulphides, such as vanadium, iron, molybdenum, nickel, and cobalt, knows that they are good semiconductors and have reactive surfaces which make excellent catalysts. Even without biological interference there must have been very considerable organic chemical activity in the first 10^9 years of the cool earth. Anyone familiar with energy-capture devices in biology also knows that the Fe_nS_n complexes are uniquely important even today in many low redox potential reactions (Chapter 12) and we have had occasion to refer to nickel, molybdenum, and even vanadium and tungsten sulphides, in Chapters 16 and 17, in extremely important reactions such as those of CH_4, N_2, and H_2. In this chapter we do not wish to delve into these, origin-of-life, matters since the possible interaction of mineral sulphide chemistry and early biology can only be based on speculative discussion as yet. We do wish to summarize the involvement of especially the element sulphur in a number of important functions which have persisted throughout biological development as far as we know them (Table 19.1). However, it is not possible to give more than a glimpse of this

Table 19.1 Some key functions of sulphur in biological systems

Function	Example
Thiolate as a base	Thiol-proteases
Thiolate as a ligand	Metal-ion binding
Sulphide as a ligand	Metal-ion binding in clusters (Fe_n S_n)
Sulphide as a redox group	Lipoic acid
Thiolate in group transfer	Transfer of acetyl (CoA)
–S–S– as a cross-link	Many extracellular enzymes
$-S^-$ in controls	Thioredoxin, glutathione
$-S-CH_3$ ethers (methyl transfer)	Coenzyme (methionyl CoA), coenzyme M
$-S-CH_3$ ether as a ligand and cross-link	Cytochrome *c*
Coenzyme catalysis	Biotin, thiamin, lipoic acid
Inhibitors (drugs)	Penicillin

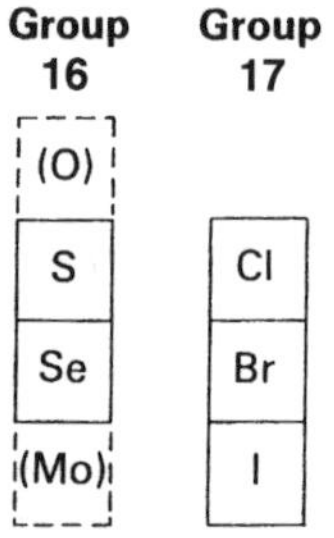

Redox non-metals in periodic table

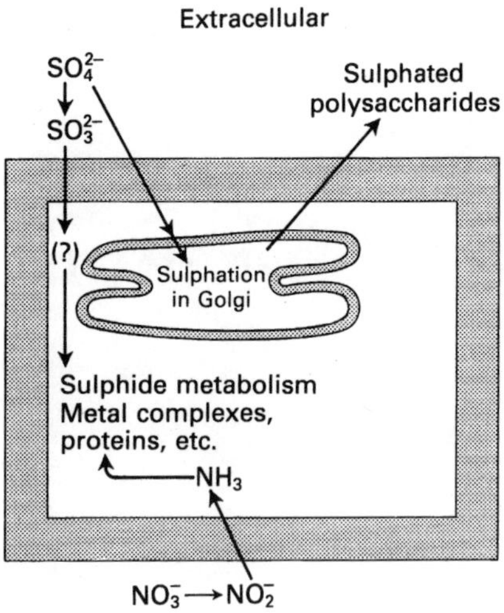

Fig. 19.1 The handling of sulphate to give both sulphated polysaccharides and the many forms of reduced sulphur requires extracellular molybdenum enzymes and a variety of intracellular processes.

chemistry in this book since it is so extensive. The related biochemistry of selenium and the halogens will be given in more detail.

19.2 Sulphur biochemistry

19.2.1 Oxidation states and redox reactions

Today sulphur enters many cells largely as sulphate, as opposed to the sulphide available to anaerobes, and it is processed in two very different ways (Fig. 19.1). In the first it is condensed with ATP in an energetically uphill reaction to produce APS (Figs 19.2 and 19.3). This compound, formed in the Golgi apparatus, is then used to sulphate polysaccharides and some proteins and thereafter sulphate becomes a useful bound anion especially in the open matrices of connective tissue. Although we shall not describe its chemistry further, since it interacts weakly with most cations, the importance of sulphated anionic polymers, which form open networks and are not cross-linked by calcium in multicellular organisms, cannot be overstressed (Chapter 6).

The second pathway involves successive reduction to SO_3^{2-} and then to sulphur and sulphide (Fig. 19.3). For this reduction biology uses a molybdenum enzyme in the first step and subsequently a haem/Fe_nS_n protein sulphite reductase, to take the sulphur through bound SO_2 and SO to S^{2-}. There is a parallel reaction for NO_2^- to NH_3 and both reactions require six electron reductions without release of intermediates. The initial steps of the reactions occur in the periplasm, outside cells. Some of the H_2S is then used to produce the many metal (Fe, V, Mo) sulphide units in biology, but we do not know how this is controlled. The synthesis of the Fe_nS_n proteins from initial reactions in mitochondria is described in Section 12.8.1 and the genes for these reactions are now known. Note that some sulphide at least is produced from digestion of

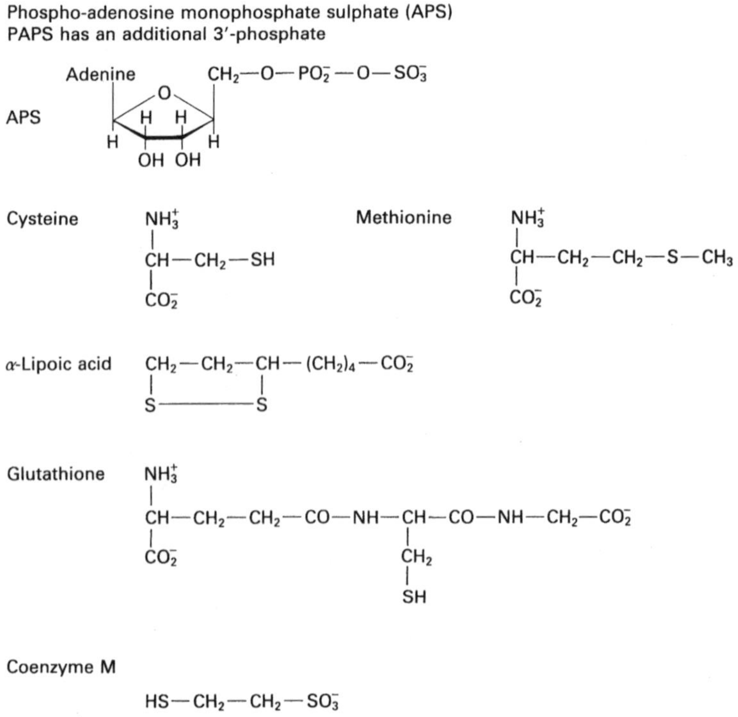

Fig. 19.2 The formula of sulphated adenosine phosphate, APS, which is the form in which active sulphate is transferred, together with formulae of other important forms of sulphur. See also Section 2.17.

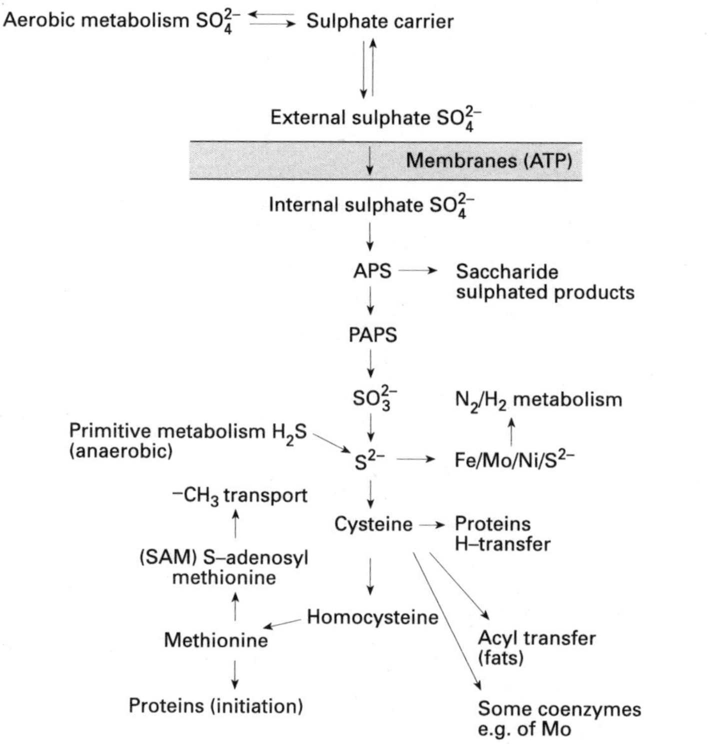

NB Note that S and Se are not directly incorporated in DNA. All the steps except the first two, shown as $\rightleftarrows$, are irreversible. There are also reversible steps of Met and Cys binding to DNA transcription factors.

Fig. 19.3 The uptake and incorporation of sulphate.

sulphur-containing amino-acids in higher organisms. Some of the sulphide is used in energy-capture devices in anaerobes and the reversible reaction from $[-S-S-]_n$ to $2n$–SH is one of the most fascinating in bio-inorganic chemistry. Much sulphur is also incorporated into proteins as thiolate or thioether, that is in cysteine and methionine. Thiolate reactions control protein stability outside cells by the formation of –S–S– bridges, but there are two much more fundamental and primitive roles for thiolate in redox reactions and in enzymes and controls of the cytoplasm, namely in glutathione (Fig. 19.4) and thioredoxin.

We have observed that in membranes there are many Fe_nS_n proteins for the transfer of electrons by oxidation-state changes of the metal ions. We also noted that *in the cytoplasm* there is little transport of electrons by metal-ion-containing enzymes, especially in eukaryotes; radicals such as Fe^{2+}, Cu^{2+}, and Mn^{2+} are too reactive and are therefore used very cautiously. Outside cells the protection from oxidative free radicals is secured by sugars such as ascorbate in higher organisms; inside the cell the exchange of redox equivalents in the cytoplasm at a potential around –0.1 V is managed in large part by glutathione. The reaction generally avoids radicals since it is a two-electron system, but even if $RS^{\cdot}$ radicals are produced they are relatively stable. In fact, all the radicals which biology uses openly are of this stable kind—ascorbate, hydroquinone, and thiols—and they differ from the more aggressive radicals, e.g. of tyrosine, which are deeply screened in proteins. Glutathione is then a protective and homeostatic redox device in the cytoplasm, communicating the redox balance relatively slowly between metabolic pathways at around the redox potential of –0.1 V. It communicates with the low-potential NADPH on

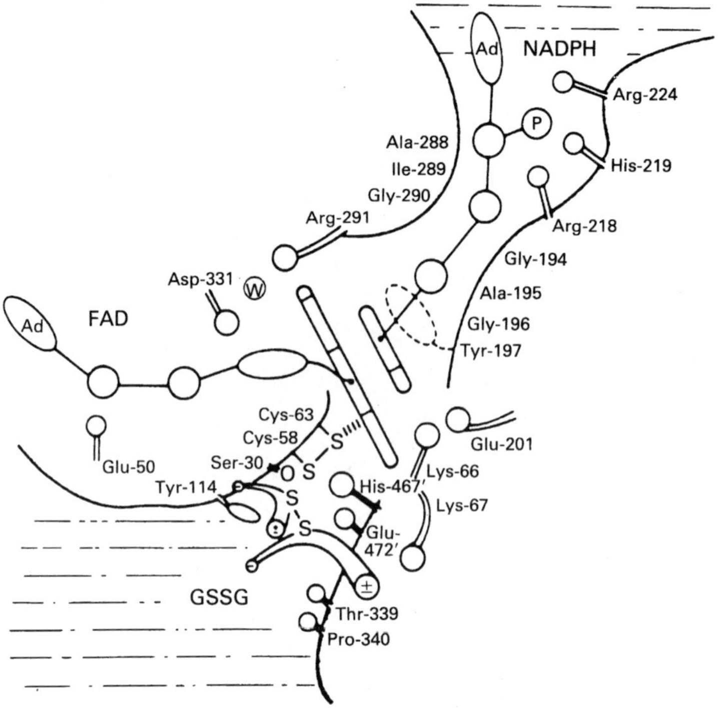

Fig. 19.4 The active site of the enzyme glutathione reductase which transfers reducing equivalents to and from NADPH via flavin. Note the close distance of separation of the active units which transfer hydride (or electrons plus protons) necessitating close approach.

the synthesis side of metabolism only through special enzymes (Fig. 19.4). It does not equilibrate with reducing equivalents which are to be used in degradation and energy generation (NADH) or with those to be used in the cytoplasm via NADPH.

If glutathione is to work in this way then it should be able to communicate reductive synthesis with other reaction pathways; seemingly it does so in part through the enzyme thioredoxin (see Table 19.1 and Aside). This enzyme is directly connected to the synthesis of DNA, since it assists the reduction of ribose to deoxyribose, to the synthesis of proteins, and to regulation of key enzymes in photosynthetic organisms by redox reactions of –S–S– bridges. There is a further connection of the redox reactions of –S–S– bridges via α-lipoic acid (Fig. 19.2) which is a bound coenzyme of great consequence since it acts to take pyruvate to acetate in the pyruvate dehydrogenase complex. This reaction is degradative and acts with NAD, not NADP, indicating again how separated metabolic pools can be compartmentalized by kinetic restrictions (Chapter 3). The structure of this enzyme is not known in detail, but both thioredoxin and glutathione reductase are based on strong β-sheet domains.

The –S–S– bridge is important in more than just redox reactions since it is used as a cross-linking unit in extracellular proteins. The way in which this oxidation occurs is quite complex and the process needs to be easily reversible to avoid folding mistakes. It is likely that this external cross-linking came after the advent of dioxygen in the atmosphere.

A very interesting inorganic reaction involving S–S links is that between FeS and H_2S

$$FeS + H_2S \longrightarrow FeS_2 \text{ (pyrite)} + H_2$$

The stability of FeS_2, a very unusual low-spin inorganic mineral, is such as to drive formation of H_2 in the above reaction. Wächtershauser has proposed that

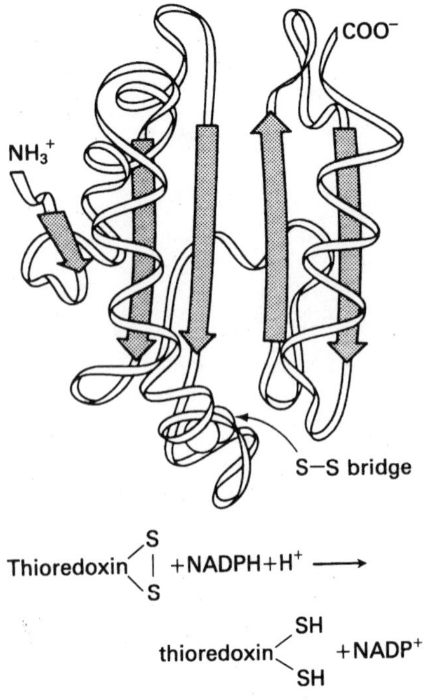

The structure and reaction of thioredoxin. (After Holmgren and C.-I. Brändén.)

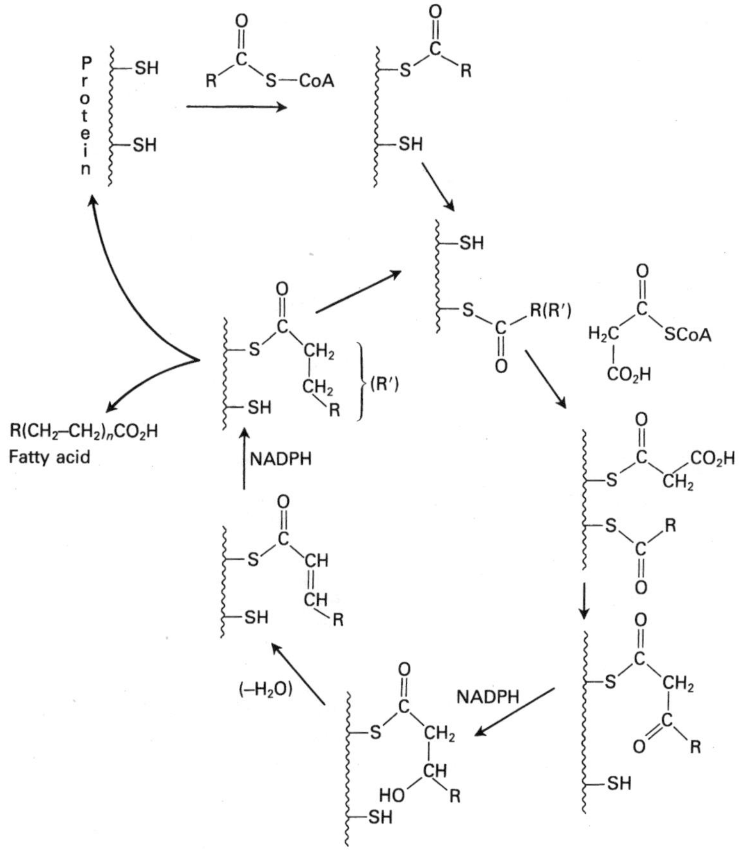

Fig. 19.5 The steps in the synthesis of long-chain fatty acids using transfer of acetyl groups from thiolate centres.

this is a fundamental source of energy for primitive life and he has gone on to show how the reaction can be coupled into metabolism in anaerobic systems. There are alternative reactions of minerals with H_2S which give sulphur and H_2 at higher temperatures.

19.2.2 Acid-base reactions

One of the most interesting features of sulphur chemistry is its reactions as RS^-. In this state, in proteins as cysteinate, it reacts as a base binding H^+ and a range of transition metals including Fe, Zn, Mo, and Cu. The individual chapters on the metals have analysed the active sites so formed. The use of RS^- is extended to both a function as an attacking group in thiol-dependent enzymes, e.g. proteases, and in the transfer of two-carbon units, e.g. acetyl, in coenzymes (Table 19.1 and Fig. 19.5). The reason for the use of sulphur is its ability to act as a good attacking and leaving group so that it gives intermediates of low energy and fast kinetics. The redox reactions of organic sulphides are always linked to an equilibration with protons in cells.

The role of the thioether in methionine is quite closely related to that of thiolate. It acts as a base forming complexes with metals such as copper and iron, e.g. in cytochrome *c*. It can also act as a carrier of one-carbon fragments as in methionyl coenzyme A. Coenzyme M (Fig. 19.2) acts as a simple thiolate.

19.2.3 Group transfer reactions: coupling

Looking back to Chapter 1, where we described some fundamental properties of the chemical elements, we noted that in the periodic table sulphur was the first 'inorganic' element after hydrogen that was involved in all three classes of chemistry: acid/base, redox, and covalent bond formation. Just as phosphorus as phosphate, again in the second row of the periodic table, is ideal for handling the condensation reactions of biology since it has such labile bonds, so sulphur is ideal for the handling of redox transfer reactions involving two-electron steps. In order to understand fully the appropriate nature of biological choices of functions for elements we need to see also the connection in the cytoplasm between redox reactions based on sulphur and condensation reactions. This connection is quite clearly the thioester reactions which transfer carbonium ion fragments,

$$RS - COR \longrightarrow RS^- + {}^+COR.$$

Thus RS^- units are linked into synthesis while they also undergo formation of –S–S– units.

There is a further very important feature of thioester chemistry, RSCOR′. These esters are of 'high energy' in the sense that reaction with water is strongly downhill

$$H_2O + RSCOR' \longrightarrow RSH + R'COOH.$$

In this respect a thioester could be used to drive syntheses which rely on condensation reactions and could therefore have predated pyrophosphate (ATP). Such a pathway may not have required complex template organization since it is possible to give reaction schemes for driving the reaction toward thioester formation using transformations on the surfaces of solids or through organometallic reaction schemes. In primitive life we can picture the series of reactions leading to thioesters as follows. Assuming that both CH_4 and CO were present, then methyl and carbonyl groups can come together on the same metal atom

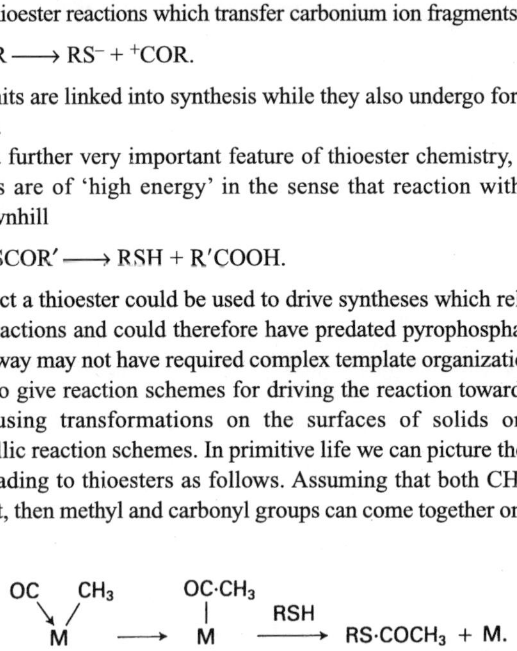

Such a reaction could occur on iron, cobalt or nickel centres, Chapter 16. Today we know that the transfer of, say, acetyl groups from $RSCOCH_3$, i.e. from lipoic acid or coenzyme A, is part of the connection between non-oxidative glycolysis, the oxidative citric acid cycle, and part of the pathway for the synthesis of fats (Fig. 19.5). (Synthesis of fats for the production of membranes was a very early requirement.) The many other thiolate coenzyme-A compounds are involved in further transfers. In archaebacteria there is the very small compound coenzyme M ($HS\text{-}(CH_2)_2 - SO_3^-$), which operates in a similar manner in assisting transfer of one- or two-carbon fragments in metabolism (Fig. 16.3). There is here a hint that this chemistry is very primitive.

The significance of lipoic acid, like that of glutathione, derives from its redox equilibrium between the –S–S– and 2–SH forms, which is such that it connects transfer of carbon fragments to the redox reactions of the cell. However, it is a *bound* coenzyme (compare flavins), while glutathione (compare NADH) is a mobile cofactor. Thus, communication is made from oxido-reduction reactions of NADPH (and, say, glycolysis, NADH) and carbon-fragment transfer using sulphur through the balance between thiolate and disulphide chemistry of mobile glutathione.

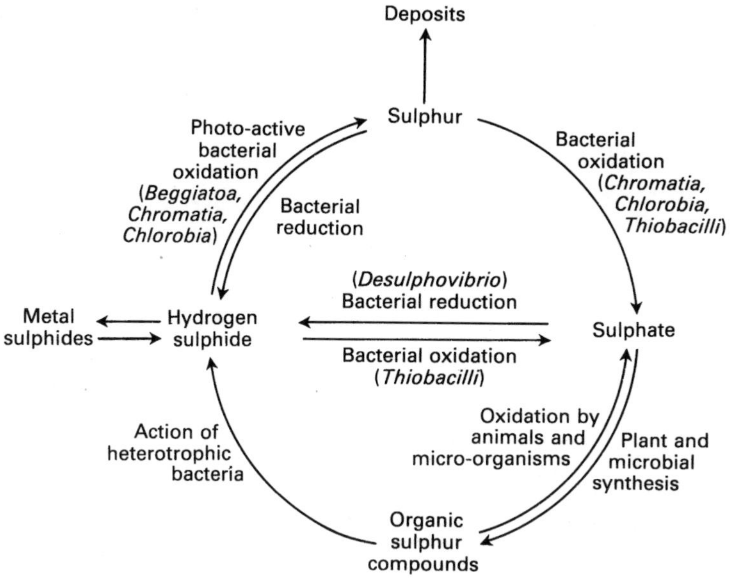

Fig. 19.6 A simple representation of the sulphur cycle.

19.2.4 **The sulphur cycle**

In addition to the reductive reactions of sulphur there is a reversed oxidative series from sulphide to sulphate in organisms such as phototrophic bacteria (Fig. 19.6). The corresponding chemistry does not introduce any new principles differing from those above, but the reduced forms of sulphur can give electrons to the energy-capture electron-transfer chains which terminate in dioxygen.

19.2.5 **Cellular and genetic chemistry of sulphur**

Figure 19.7 outlines the intimate relationship of sulphur metabolism to genetic structure. Both methionine and cysteine levels in cells are controlled by constitutive genetic response (see Section 6.14.3), and in turn their levels control the synthesis of uptake and enzymic conversion of the sulphate to all the sulphur containing compounds of Fig. 19.3. The structure of the extracellular carrier protein for sulphate is shown in Fig. 19.8 and is remarkably similar to that for the uptake of molybdate and to transferrin, the uptake protein for Fe^{3+}, (Fig. 12.5). It is selective, however, based on the difference in size of MoO_4^{2-} and SO_4^{2-}.

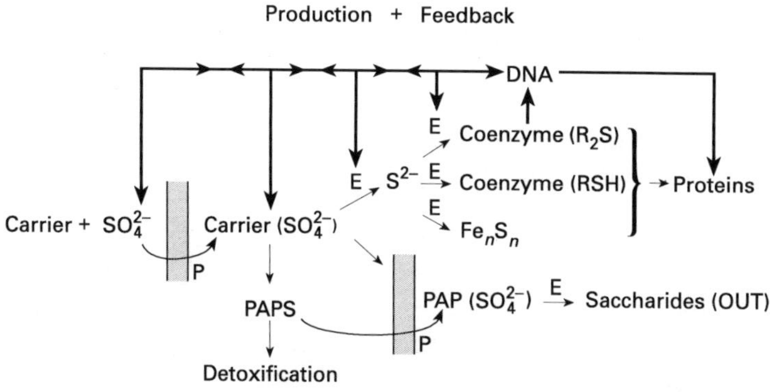

Fig. 19.7 The feedback system controlling sulphur incorporation (see Fig. 19.3). P, pump.

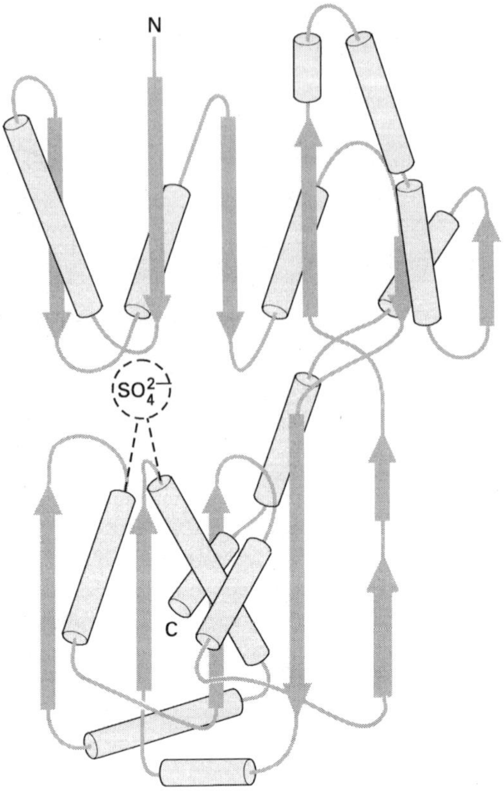

Fig. 19.8 The outline structure of the SO_4^{2-} carrier protein which is closely similar to the carrier for MoO_4^{2-} and $Fe^{3+} \cdot CO_3^{2-}$ (Fig. 12.5).

19.3 Summary of sulphur biochemistry

Closing this short section we stress that there is a huge biochemistry of sulphur compounds. Here we can only refer to some of the most important features which derive from its position in the periodic table. No other available element has such properties in combination.

1. It is a good attacking and leaving group as RS^-. This is of fundamental value in catalysis, e.g. in proteolysis.
2. It can polymerize to –S–S– or S_n units, so it connects acid/base and redox chemistry at $E = -0.1$ V. This is fundamental to the stability of –SH groups, to the reductive synthetic pathways in the cytoplasm of cells, to the stability of extracellular proteins, and in homeostatic control.
3. As a base, RS^- or S^{2-}, it forms complexes with heavy transition metals (V, Fe, Cu, Ni, Zn, Mo) but not with Mn, Mg, and Ca, giving especially metal clusters. Metal sulphides could have supported an organo-metallic chemistry both before and in early life (see 'Further reading').
4. It forms stable organic compounds such as ethers (methionine) and cyclic thioethers, such as biotin (see margin), e.g. in pyruvate carboxylase

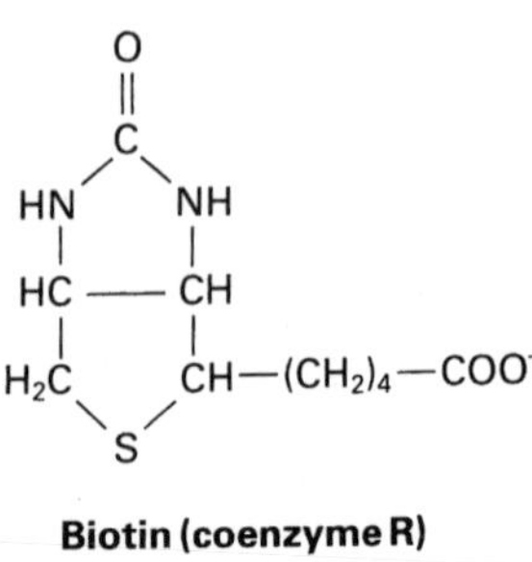

Biotin (coenzyme R)

An element closely related in character is selenium, but it is less available and its redox chemistry is rather different. Selenium has specialist, not general functions in biology.

Table 19.2 Some biological selenium compounds

Compound or enzyme	Source
Selenocysteine	Corn and wheat seed hydrolysates
Selenocystine	Ethanol extracts of red clover and onion
Selenomethionine	Protein hydrolysates of dogs' liver; sheep's wool hydrolysates
Dimethylselenide	Astragalus plants and seeds (Se accumulators)

19.4 The biochemistry of selenium

Selenium is the last of the non-metals to be in a coded representation in an amino-acid in DNA. This gives a taste of the significance of this element in all organisms. It occurs in a number of forms in small compounds (Table 19.2) and has a quite considerable importance in reductive metabolism in bacteria (Table 19.3). The number of proteins in human cells known to contain selenium is shown in Table 19.3 and here selenium is largely associated with oxidative metabolism.

One of the great problems organisms faced with the advent of dioxygen in evolution was the inevitable formation of both superoxide and hydrogen peroxide species. In Chapter 15 we have already described the various Fe, Mn, and Cu superoxide dismutases and their compartmental advantages, but a sim-

Table 19.3 Known selenoproteins and selenoenzymes and their functions*

Selenoprotein (Human)	Function
Glutathione peroxidases, GPx, (GPx1, GPx2, GPx3, GPx4)	Antioxidant enzymes: remove hydrogen peroxide, lipid and phospholipid peroxides (thereby maintaining membrane intergrity, modulating eicosanoid synthesis, reducing inflammation and the likelihood of propagation of further oxidative damage to biomolecules such as lipids, lipoproteins and DNA).
(Sperm) mitochondrial capsule selenoprotein	Form of glutathione peroxidase (PHGPx): shields developing sperm cells from oxidative damage and later polymerises into a structural protein required for stability/motility of mature sperm.
Iodothyronine deiodinases (3)	Production of active thryoid hormone T3, from thyroxine, T4.
Thioredoxin reductase	Reduction of nucleotides in DNA synthesis; maintenance of the intracellular redox state, critical for cell viability and proliferation.
Selenophosphate synthetase, SPS2	Required for the biosynthesis of selenophosphate, the precursor of selenocysteine, and therefore for selenoprotein synthesis.
Selenoprotein P	Found in plasma and associated with endothelial cells. Appears to protect endothelial cells against damage from peroxynitrite.
Selenoprotein W	Needed for muscle function.
Prostate epithelial selenoprotein (15 kDa)	Found in epithelial cells of the ventral prostate. Seems to have a redox function (resembles PHGPx).
DNA-bound spermatid selenoprotein (34 kDa)	Glutathione peroxdidase-like activity. Found in stomach and in nuclei of spermatozoa. May protect developing sperm.
18 kDa selenoprotein	Important selenoprotein found in kidney and large number of other tissues.
Selenoprotein (bacterial)	**Function**
Formate dehydrogenase	$HCOOH \rightarrow H_2 + CO_2$ in *E. coli*
Glycine reductase	Bacterial reaction: Glycine → ?
Xanthine oxidase	Removal of 2H from xanthine
Some hydrogenases	Reversible $H_2 \rightleftarrows 2H^+ + 2e$

* Kindly provided by Dr M.P. Rayman of University of Surrey

ilar problem arose with the removal of peroxides. Catalase (Chapter 13) can handle H_2O_2 itself and many primary peroxides, but it is necessary to have its haem centre deeply removed from aqueous media containing organic molecules since, by single-electron-transfer reactions, H_2O_2, bound to haem in catalase, could act as a dangerous radical generator just as in peroxidases; this must not happen, especially in the cytoplasm. The difficulty with a deep protected pocket is that it will not allow access to secondary or tertiary peroxides. Protection from these peroxides must use two-electron, atom-transfer reactions as far as is possible. Selenium is an ideal reagent for removal of an O atom from any peroxide and the redox potential of the bound oxygen atom is thereby greatly reduced. It is the two-electron covalent chemistry involved in the O atom transfer by selenium which allows it to be exposed to water while it reacts without radical formation, and the safe exposure makes it possible for the selenium-based peroxide reactions to take place not only with primary but also with secondary and tertiary peroxides in water. The mild oxidation catalysis in organic chemistry by SeO_2 is carried out in organisms in the reverse manner by R_2Se in proteins. The cyclic states of the selenium in the enzyme glutathione peroxidase (Fig. 19.9) require a reducing agent and this is again glutathione in the cytoplasm.

Another interesting involvement of selenium with redox chemistry is in the metabolism of the thyroid hormones. These hormones are organic iodine compounds of relatively recent appearance in evolution. The best documented role of selenium is in the enzyme iodothyronine deiodinase type I which removes one iodine atom from the T_4 form of the hormone (thyroxine) to give the more active T_3 form. These are extremely important growth hormones, of course. The involvement of selenium with iodine metabolism is suggestive. They have approximately the same redox potentials, so in new oxidative roles selenium and covalently bound iodine may have been introduced into life during about the same evolutionary period.

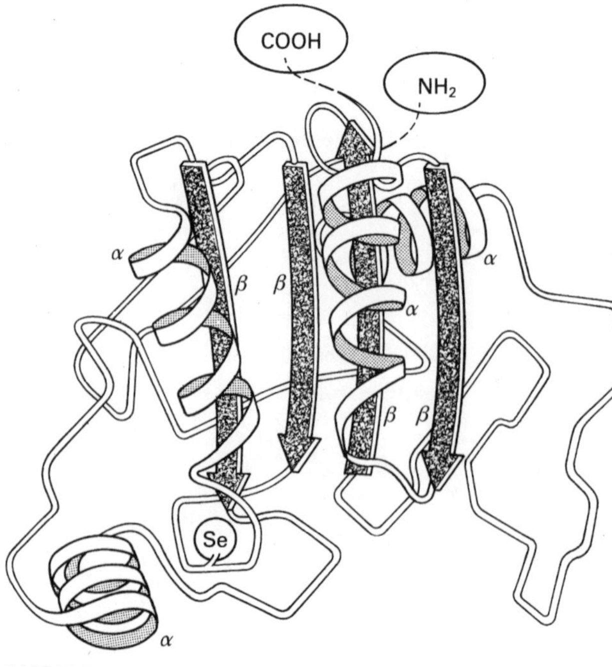

Fig. 19.9 The structure of glutathione peroxidase which has a single atom of Se and is based on a β-sheet. (After Epp, Landenstein, and Wendel; see 'Further reading').

The general tenure of new knowledge concerning selenium and human life is to stress its protective (anti-oxidant) role against diseases, especially cancer. We advise the interested reader to follow up the references at the end of this chapter. All the reactions of selenium and haem-based enzymes with peroxides take place in aqueous media and not in membranes; to protect membranes there is a further scavenging system for peroxide based on the lipid-soluble vitamin E. This is another compound which, like ascorbic acid and glutathione, produces only innocuous free radicals.

There are a few additional proteins which contain selenium. The variety stretches from some hydrogenases to a hydroxylase and there are also some selenium proteins of unknown function in muscle. Table 19.3 lists what is known, but we stress again that much more chemistry may be hidden as yet. The essential nature of selenium could not be better exemplified than by the human deficiency diseases (Kesham and Kashin-Beck syndromes) which cause muscular problems especially in northern central China where the soil is very low in this element. In fact, the soil is now treated with selenate with very satisfactory results. Attention must also be paid to the deficiency of iodine, parallel to that of selenium (see Utiger, R.D. in 'Further reading').

19.5 Evolution and selenium

There is the frequent finding that selenium occurs *in bacteria* in enzymes associated with hydrogen, i.e. in low-potential metabolism, and it is certain that selenium was used in primitive cells, bound to tungsten for example. Selenium in the form of selenomethionine is in the amino-acid DNA code. The advent of dioxygen during evolution switched the available selenide (not much would have been earlier available since many metal selenides are very insoluble) to selenate. The selenate anion could then be taken into cells through the sulphate uptake pathway (Fig. 19.3) thence incorporated in enzymes, and used in the cytoplasm as an O-atom transfer centre. There is a contrast with molybdenum chemistry in that molybdenum became used outside cells as an oxygen-atom transfer reagent, while selenium is used in the cytoplasm. The difference is reflected in the modes of their incorporations. We have described the coenzyme form of molybdenum, but quite unlike this incorporation selenium is taken into an amino acid as selenomethionine. While this fact correlates with use in the cytoplasm it throws up the question of the very early use of selenium in biology. Have modern organisms lost the early importance of selenium and found a new use for it in detoxification from peroxides? A problem with the understanding of selenium biochemistry is that, as stated, the selenium-containing amino acid is coded in DNA so that selenium incorporation is not a post-translational modification. In this feature it is only to be compared with sulphur and again the suggestion arises that its incorporation into biological systems was a very early event in evolution.

19.6 Notes on the use of the halogens

The halide ions, Cl^-, Br^-, and I^-, appear to be activated in vesicles to form organic compounds external to eukaryotic cells, e.g. thyroxine (see margin), or to be used external to prokaryotes in particular as protective poisons. Halogen redox chemistry is a danger to DNA. Organic halogen compounds are often used by man as drugs but there is the danger of side-effects.

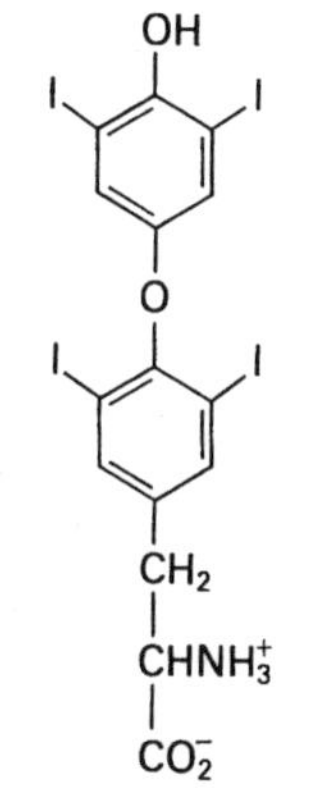

Thyroxine (Tetra-iodo thyronine)

We noted the catalysts for the production of these halogenated organic compounds in Chapters 13 and 17, i.e. haem iron and vanadium peroxidases, respectively, mentioning that these catalysts are probably in vesicles in eukaryotes. Moreover, removal of at least one iodine is due to a selenium enzyme. Such a system could not have developed until dioxygen entered the atmosphere and is probably a relatively recent addition to the armoury of messengers and hormones especially associated with development. At first, however, all such halogen reactions may have been used to produce poisons. A substituent iodine atom has great value in selectivity since there is no comparable unit throughout chemistry in size and shape. It can also be removed relatively easily; bromine and chlorine substitutions are more likely to be just for poisons.

We draw attention to the general contention that evolution develops from defensive strategies. The production of organic halides by cells was probably protective. Subsequently, in a multicellular system the halogenated species could become messengers or hormones, e.g. thyroxine and tri-iodothyronine. A similar point arises in the case of hydroxylation generally. The interesting curious feature concerning the halogens is that they were always available, but organisms rejected Cl^- and ignored the others until late in evolution when they became useful as ionic messengers, Cl^-, or in organic messengers (I), or in poisons (Cl, Br, I). F^- found a use only in insoluble salts, CaF_2 and fluoroapatite $Ca_2F(PO_4)$, which is found in high quantity in krill. Fluoroapatite is the product in teeth resulting from using fluoride-rich drinking water.

19.7 The cellular content of sulphur, selenium, and iodine (the metallome)

When the proteome of a cell compartment is analysed the proteins containing sulphur or selenium can be separately uncovered by mass or radio-active isotope labelling. Such procedures can be used to determine the content of these and any other protein-bound elements, in proteins. However, this information is not adequate for a complete description of a cell compartment since the small molecule content of the elements is not analysed. Just as in the case of iron where we saw that there were different important concentrations of iron and haem iron, so with sulphur and selenium there are important small molecules which have both a relationship to gene expression and to the environment. We may wish to include the levels of these environment related molecules together with simple ions of other elements and compounds of halogens in what we have called the metallome. It is clear, for example, that oxidative stress, connected to the handling of O_2 in an organism, is connected to levels of selenium, sulphur, copper, and iron in the environment—all of which are reflected in both small and large molecules containing these elements. It would be valuable to know how the levels of molecules such as glutathione, cysteine and methionine differ from anaerobic to aerobic organisms. A similar problem arises in the relationship between environmental iodide, iodine in thyroid hormones, and gene expression, and for organisms such as plants concerning the effect of environmental levels of even sources of carbon (CO_2) and nitrogen (NH_4^+ or NO_3^-) on cell contents of simple coenzymes. The connection to coenzymes, proteins and genes is an essential part of the description of the cellular system of C, N, H and O which we have excluded from the metallome.

Further reading

Alexander, N.M. (1984). Iodine. In *Biochemistry of the essential ultratrace elements* (ed. E. Frieden). Plenum Press, New York.

Arthur, J.R., Nicol, F., and Beckett, G.J. (1993). Selenium deficiency, thyroid hormone metabolism and thyroid hormone deiodinases. *American Journal of Clinical Nutrition*, **57** (2 Suppl.), 236S–239S.

Arthur, J.R. and Beckett, G.J. (1999). Thyroid function. *British Medical Bulletin*, **55**, 658–68.

Bates, J.M., St. Germain, D.L., and Galton, V.A. (1999). Expression profiles of the three iodothyronine deiodinases, D_1, D_2 and D_3, in the developing rat. *Endocrinology*, **140**(2), 844–51.

Bernal, J. (1999). Iodine and brain development. *Biofactors*, **10**, 271–6.

Bock, A., Forchhammer, K., Heider, J., Leinfelder, W., Sawers, G., Veprek, B., and Zinomi, F. (1991). Selenocysteine: the 21st amino acid. *Molecular Microbiology*, **5**, 515–20.

Capitani, G., Markovic-Housley, Z., Delval, G., Monu, M., Jansonius, J.N., and Skurmann, P. (2000). Crystal structures of two functionally different thioredoxins. *Journal of Molecular Biology*, **302**, 135–54.

Epp, O., Landenstein, R., and Wendel, A. (1983). Structure of glutathione peroxidase. *European Journal of Biochemistry*, **133**, 51.

Frieden, E. (1991). Iodine and the thyroid hormones. *Trends in Biochemical Sciences*, **16**, 50–3.

Gladyshev, V.N. and Hatfield, D.L. (1999). Selenocysteine-containing proteins in mammals. *Journal of Biomedical Sciences*, **6**, 151–60.

Huber, C., and Wächtershauser, G. (1997). Activated acetic acid by carbon fixation on (Fe, Ni)S under primordial conditions. *Science*, **276**, 245–7.

Kirk, K.L. (1991). *Biochemistry of the elemental halogens and inorganic halides*. Plenum Press, New York.

Kohrle, J. (1999). The trace element selenium and the thyroid gland. *Biochimie*, **81**, 527–33.

Kolbe, M. (2000). Structure of the light-driven chloride pump, bacteriorhodopsin. *Science*, **288**, 1390–6.

Mustacich, D. and Powis, G. (2000) Thioredoxin reductase, *Biochemical Journal*, **346** Pt.1: 1–8.

Patching, S.G. and Gardiner, P.H. (1999). Recent developments in selenium metabolism and chemical speciation: a review. *Journal of Trace Elements in Medicine and Biology*, **13**, 193–214.

Stadtman, T.C. (1990). Selenium biochemistry, *Annual Reviews of Biochemistry*, **59**, 111–27.

Stadtman, T.C. (1996). Selenocysteine. *Annual Reviews of Biochemistry*, **65**, 83–100.

Sun, Y., Gu, Q-P., and Whanger, P.D. (2001). Selenoprotein W expression. *Journal of Inorganic Biochemistry*, **84**, 151–6.

Utiger, R.D. (1998). Kashin–Beck disease—expanding the spectrum of iodine deficiency disorders. *New England Journal of Medicine*, **339**, No. 16, Editorial.

CHAPTER 20

Integrated living systems of elements

20.1 Introduction

As stated at the beginning of Part I our ambition in writing this book is to give in essential outline an account of organisms in terms of general chemistry including all relevant elements. It might appear to be then just another biochemistry book, but our approach has been informed not by the conventional attitude to this chemistry—that, in essence, life can be understood from the analysis of the properties of particular *organic molecules* extracted from living cells—but by the study of the way in which a moderately large group of elements, about twenty (see Aside) including C, N, O and H, come together systematically to form a living organism. There were prerequisites to this study, namely that *it was necessary to know individually and quantitatively the essential elements involved, as well as the properties of them and the isolated molecules in which they occur, and then the networking of each element and its compounds separately, through connections to the genome and the environment.* The first point of departure from conventional biochemistry that we have noted, therefore, is that the essential elements, which form the basis of every living system, are *minimally* the metals K, Na, Mg, Ca, Fe (or possibly Mn), Zn, Mo (or possibly W or V) and the non-metals Cl and Se, in addition, of course, to the very high percentage of 'organic' elements C, H, N, O, S, and P (see Chapter 1). We shall leave to one side for now that many organisms today also contain essential Co, Ni, and Cu. Beside these essential elements it is known that life in one form or another can manipulate and use Sr, Ba, B, Si (Sn), Cd, As, Br, and I and quite possibly many other elements, some of which may be essential in their respective organisms but this has not been demonstrated. We shall also put them to one side at first. In this light we set out in Chapter 2 to see in what selective ways the elements found in organisms could be captured and, apart from noting the special and well-known features of bio-organic compounds described in organic and biochemistry textbooks, we discussed the many known combinations of the 'inorganic' elements and the *quantitative* ways in which they can associate with inorganic or organic molecules in model complexes. The description applied to both surface and free ligands. Now an organism can select the elements by energized uptake or rejection through the cell's limiting boundaries, but once inside cells the elements enter into competition for partners at different biologically selected concentrations. Major further steps in selection possibilities, after the pumping of elements into organisms (and their compartments), were then shown to be either thermodynamically stable combinations in complexes, on surfaces or in minerals, or in kinetically contrived transport systems leading to insertion into molecules and assemblies, both in particular compartments (Fig. 20.1). Cells are able to produce the same wide variety of N, S, O-donor ligands with appropriate stereochemical structures as we observed in models and we could therefore assess their ability to combine selectively with other elements, especially metal ions. We also showed that selectivity was greatly enhanced in a system where the input of elements and these selected ligands was carefully controlled giving the cell managed buffering or homeostasis (Section 2.10). (We shall return to this control in Section 20.3.3). This led us to consider the *expected* (from models) profile of the concentration of free elements and the types of compounds in which they would be found in the simplest cell (Fig. 2.10). This profile we have called the *free metallome*. Of course, the background to these processes was that every organism is a contained energized system behind membrane barriers in a particular environment (Chapter 3) and has further energy capture devices as well as

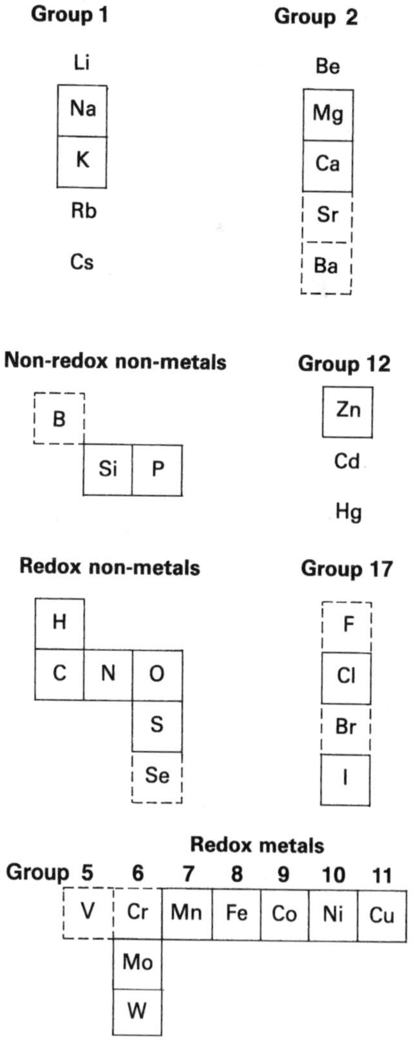

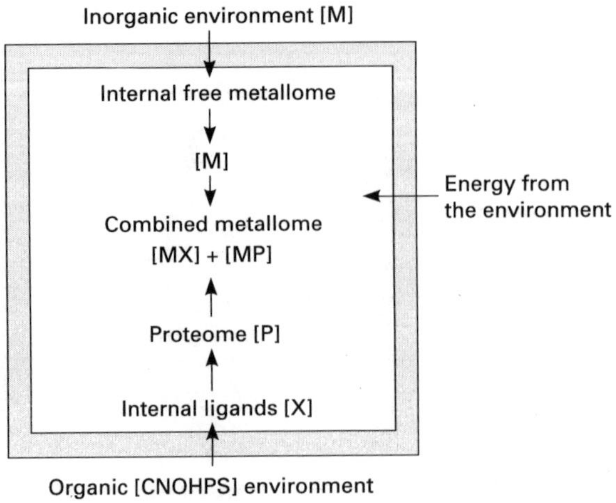

Fig. 20.1 The coming together in a cell of inorganic and organic components of the environment which, with the application of energy, generate (1) the cellular content of small and large organic molecules, including the genome and the proteome, and (2) the essential inorganic elements which comprise the free and the combined metallomes.

pumps to sustain it (Chapter 5). (We did not include pumping in deriving Fig. 2.10. These pumps are under feedback control by the element they pump into the cell.) The energization imposes additional possibilities on the metallome. Now the coming together of all the above elements in a constantly energized cell could not be seen just in this way, that is, as being in stationary local equilibrium systems, since all our subsequent observations of life revealed it to be dynamic in two senses—chemicals were constantly reacting and were therefore turning over and they were also physically transferred from site to site. Hence *there was a required constant flow through the organisms of both chemicals and energy*. Chapter 4 looked at the ways in which individual elements could contribute to this system of activity either by their fluctuating combinations and diffusion in space or by assisting turnover in catalysts or pumps. There were also mechanical activities which were summarized. At this point, the end of Chapter 5, we considered that we had put in place the chemical and physical principles which life's chemistry must obey with respect to the selected essential twenty elements.

The question obviously arises as to how the total system of twenty elements is managed in a sustained way and reproduced which goes far beyond the analysis of the activities of the individual elements. It was shown in Chapters 6 and 7 that the fundamental feature was that there is a centrally coded production of proteins by DNA via RNA and that these proteins are generated in certain quantities under feed-back control from their own levels and those of small chemicals in the cell including the elements. It is this feedback which ensures that proteins are produced in a controlled way and not overproduced and therefore the selectivity of processes is increased. The proteins are the major binding agents for the elements, as ions and small molecules. Hence the cell communication networks and their controls extended from the *genome* to the proteins, the *proteome* (and the products of its activities), to the *metallome*, that is the 'inorganic' element content, especially metal ions, of the cell, and therefore all were linked to the *environment*, which was the source of both elements and energy. Environmental restrictions on the uptake of elements concerns their abundance and availability (see Chapter 1). While we have been able in Chapters 1–7 to give this description of cellular activity in terms of outlined principles and then in Chapters 8–19 in terms of individual elements, we have not shown the full development of integrated cellular chemistry. In effect this is the problem of (temporary) conditions of a

Table 20.1 Approaches to chemical interactions

Molecule approach	**Systems approach* (variables)**
Composition	Composition
Sequence	(Concentration) (C)
Structure (shape)	Temperature (T)
Dynamics (internal)	Pressure (P)
e.g. DNA, proteins, enzymes, elementary ions	Spatial divisions (phases)
	Time (t)
	e.g. *Proteomics, genomics*
	Metallomics

* Stationary systems show emergent properties on change of variables (functions) such as melting points and allotropy, and can be defined by combined functions such as ΔG, ΔH, ΔS, as well as by T, P, and compositoin at equilibrium [see Williams and Fraústo de Silva (1999) in 'Further reading']. Time dependent (irreversible) systems are not easily described in functional terms.

cell, its *homeostasis*, at fixed times in its development, while it changes with growth. The first requirement for a description of integrated homeostasis is to clarify the nature of the task. It is clearly not to be satisfied by describing individual molecules separately but demands that we look at properties which emerge from *systems of many molecules*. That is to say that there are fundamental considerations concerning *systems* as opposed to single *molecule* studies (Table 20.1). As stated we must always remember that our description of cellular activity has to link the internal biological and the external environmental conditions in just one system. Hence we must appreciate the geochemistry of the surface of earth to appreciate the nature of life. We may well conclude that the whole is one huge, always developing, ecosystem, see Williams and Frausto da Silva (1999) in references.

20.2 **The nature of systems**

As stated above, in order to discuss jointly the evolution of the physical and biological parts of earth we need to analyse systems as well as molecules individually. We note that there are essentially two kinds of physico-chemical system—stationary and flowing. A stationary system is unchanging with time, no matter that it is often extremely dynamic in a disordered sense. For example, a contained gas is a stationary system and so is a liquid or a solid. The development of multitudes of solids on the earth from a single gaseous phase can be treated very approximately in terms of the chemical physics of the evolution of equilibrated systems, to which we turn shortly, and where we can follow slow changes as a series of equilibria both between phases, spatial regions, and chemical combinations, while temperature is varied (Fig. 20.2). A flowing (changing) system is quite different in that the physical/chemical variables are not just space, composition, pressure, and temperature as in an equilibrium system, but time has now to be included. Material and energy flow through such systems from and to the environment in a non-cyclical manner. Flow can be completely turbulent or can form a developing series of patterns (an organization) and both require and dissipate energy. Because these flow systems are inevitably linked to the environment for their energy and material, they, plus the environment, form together a whole system and perforce will change with one another. On earth the most obvious case of the evolution of patterned flowing systems is that of life. Whereas we easily appreciate the slow changes in the appearance of open *ordered* systems, for example as temperature is reduced or pressure increased, as one of increasing *static order* resulting in structure, as in the phase diagrams of Fig. 20.2, we have no such easily treated variables in the analysis of the evolution of open patterned flowing systems, although they have produced increasing *organization* rather than order, especially in the varieties of living organisms

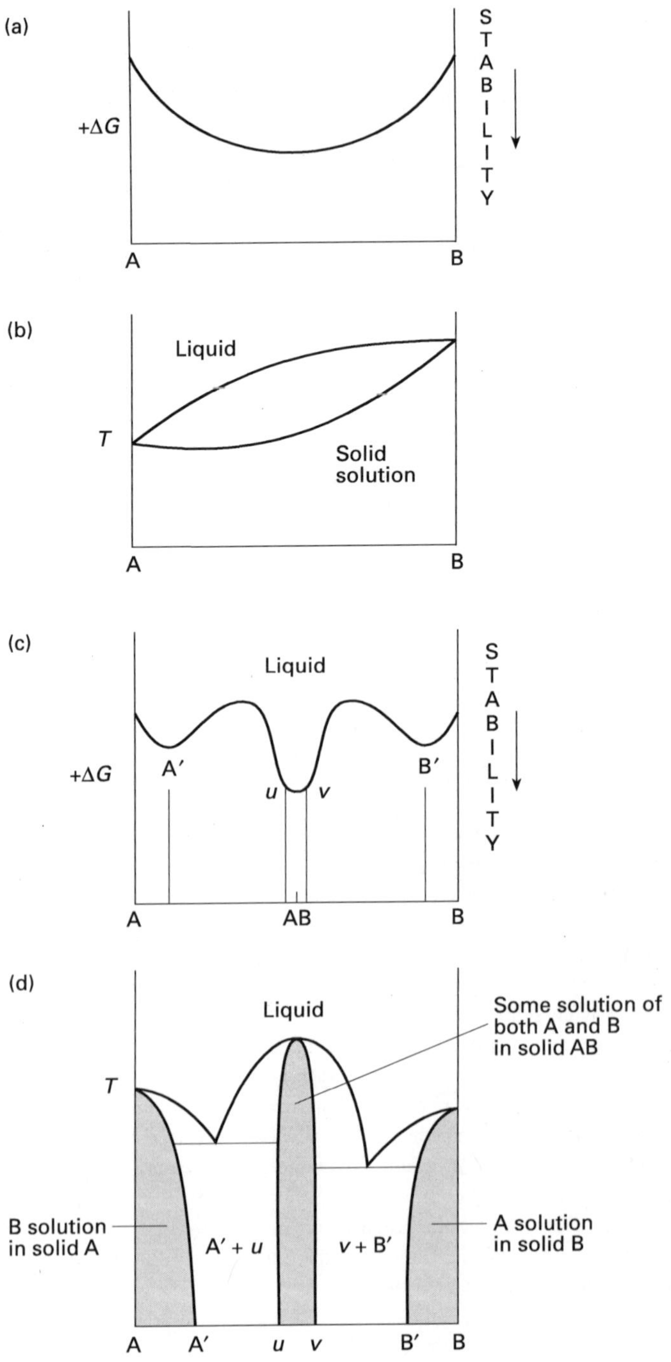

Fig. 20.2 Phase diagrams, composition against a variable such as free energy, ΔG, or temperature, for (a) and (b) miscible liquids and solids where mixing of A and B is complete: and for (c) and (d) liquids which, on forming solid phases, give limited compositional zones a to A to A′, B to B′ and *u* to *v* around AB. Note that composition here breaks down into allowed ranges of non-mixing species.

(Fig. 20.3). There is no doubt that all this organization has structural requirements including boundaries (that is, some ordered units) but its essence lies in controlled energized flow of material through and within these structured limits. Thus, for example, a contained gas cannot by itself develop organization by cooling or increasing its pressure. In order for a gas to flow it requires not just a structured container but a structured inlet and outlet to and from this space and a continuous external application of energy to the material flow which is finally dissipated as heat, largely into the environment. To adjust the gas flow, to alter its organization, it is required to adjust the application of energy and/or the structural constraints. Now, this example concerns just the flow of a single material but life has not only such a physical flow; it also has a patterned flow of chemical transformations in well-known metabolic pathways which are not just due to structural constraints. These are commonly illustrated in biochemical laboratories with the aid of wall charts (see Fig. 20.4). They arise from control over chemical reactions by catalysts (enzymes) and several pathways coexist while acting cooperatively to give the resultant cellular forms of life. The variables of such a physical/chemical *steady state* are given in Table 20.1. However, a *self-organized* flow system needs further internal controls in order to maintain itself or to change systematically in response to external change. The only internal controls possible in such a system of chemical pathways are by feed-back of material, energy and information from and to themselves and from and to other flows about their status. Such a self-generated feedback controlled system remains totally dependent on its environment for both energy and material, of course. The problem here is what is the nature of this feed-back in organisms, (F , in Fig. 20.3), such that it can coordinate a great variety of processes? Before we analyse flowing systems further let us see how we might describe the earth in terms of a succession of (equilibrium) stationary states.

20.2.1 Earth as a series of equilibrated states

The simplest approach to development of open stationary abiotic systems is by resort to equilibrium thermodynamics when change is only permitted at

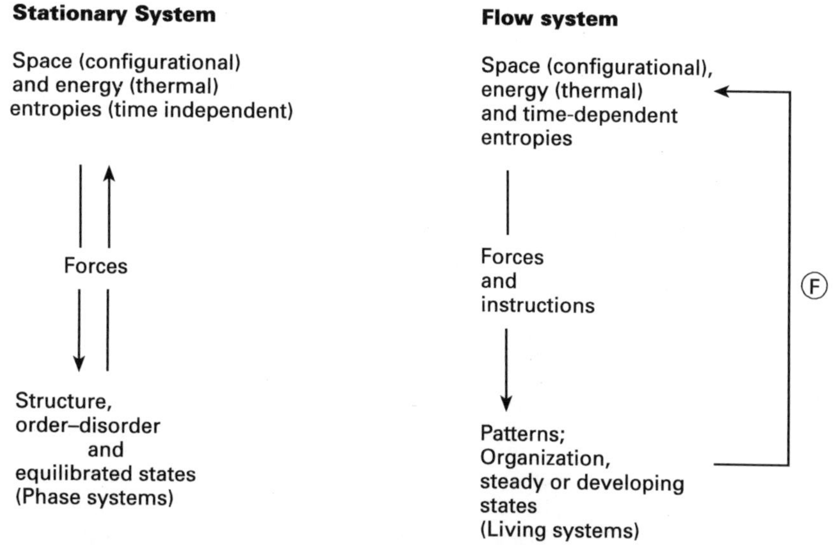

Fig. 20.3 An attempted parallel between the variables of stationary and flow systems and their emergent properties. Ⓕ is feed-back.

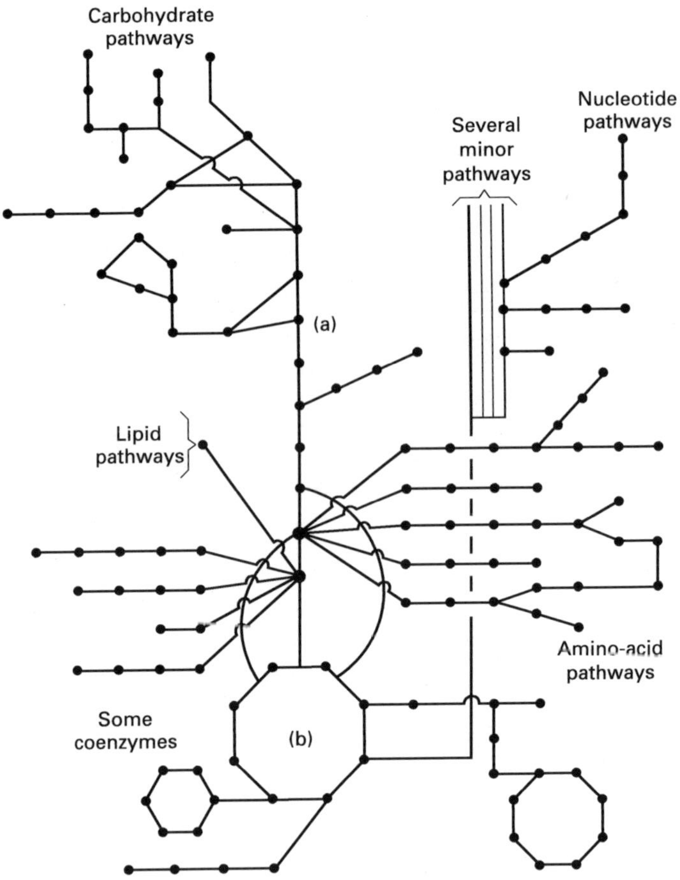

Fig. 20.4 A scheme for the organic pathways in the cytoplasm of all cells illustrating their interconnected nature (after Kauffman, see Further reading). (a) glycolytic (b) Krebs cycle pathways.

infinitesimally slow rates. In this case we can examine the change of state free energy, ΔG, to show us the spontaneous way a system will respond to temperature, pressure, or compositional variation. As stated, the result is illustrated in conventional phase diagrams (Fig. 20.2). In this sense the control of the physical/chemical species which have appeared in the crust of the earth was limited by the abundance of the elements and forces, electrical and gravitational, as the system cooled slowly to a roughly constant temperature and pressure. The formation of the major minerals observed can then be treated to a first approximation as a series of equilibrated steps due to a balance of order and disorder on cooling, [Fig. 20.5(a)]. Unfortunately due to their complexity we can go very little further into molecular detail of the individual chemicals and their phases unless we analyse systems of extreme simplicity—for example the properties of single atomic chemicals such as argon as it cools, though even here final analysis is not fully obtainable. We are therefore left with the description of simple properties of complex condensed materials, such as their composition, the shapes of their crystals, boiling points, and melting points, as *emergent* from the nature of equilibrium thermodynamics of *systems* and without a precise description in terms of single unit, *molecular*, behaviour (see Fig. 20.5 and Table 20.1). The difficulties increase in the description of non-stoichiometric mineral phase diagrams of similar components (Fig. 20.6). Hence, details of evolution of molecular chemical speciation and form on earth even in

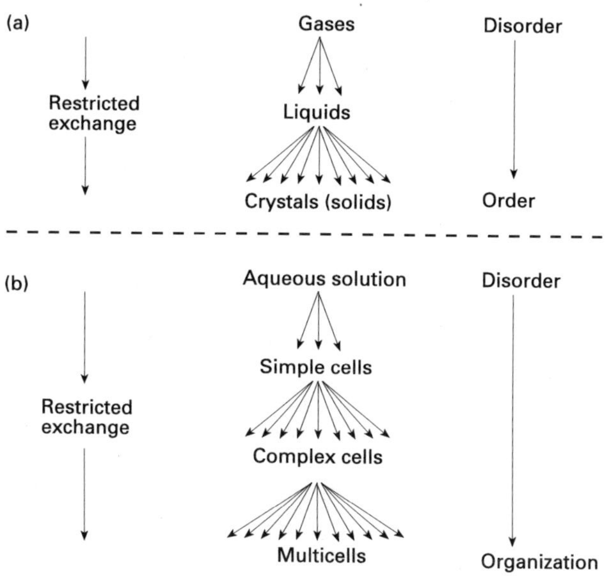

Fig. 20.5 An illustration of the parallel between equilibrium and non-equilibrium flow systems as order and organization develop respectively to give more and more species.

this simple manner defeat us, much though we believe we understand these principles of equilibrium thermodynamics. In passing we see that parameters of systems such as composition, boiling and melting points, and crystal habit are, however, extremely useful for the characterization of chemical species and can be given functional (systems) analysis in these phase diagrams. We shall see later that there is a parallel with the methods of characterization of living organisms by shape, by DNA, or by composition. We shall stress the last of these in what we call survival diagrams (Fig. 1.14 and Fig. 20.20).

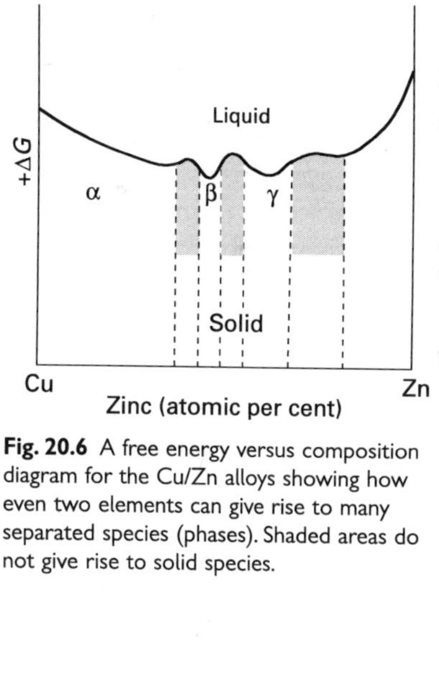

Fig. 20.6 A free energy versus composition diagram for the Cu/Zn alloys showing how even two elements can give rise to many separated species (phases). Shaded areas do not give rise to solid species.

20.2.2 Earth as a series of non-equilibrated domains

Now, this treatment of the formation of earth by equilibrium thermodynamics fails through three additional factors: (1) Cooling was at first so fast that different layers of material were kinetically trapped locally but especially the high-temperature interior was separated from the surface quite far from equilibrium; (2) the earth always radiated heat and lost some material to the cosmos; and (3) the earth has had a constant external source of light energy, the sun, which affected the surface, that is the atmosphere and the crust. As a consequence there were early losses and then continuous compensating changes in the losses and gains of energy so that cooling of the surface of earth ceased some 4.0×10^9 years ago and a remarkable, approximately constant *steady state average surface temperature* of about 300 K was attained (Fig. 20.7). The system as a whole remained with several fluid or gaseous phases on the surface at somewhat different temperatures. As far as materials are concerned there has been continuous internally driven further surface changes from its initial state due to (1) further losses of gases from earth's atmosphere and (2) chemical alternations through the input of substances from one layer of earth to another in an inevitable effort to reach equilibrium. Thus we experience volcanoes and erosion. However, there is a third way (3) in which chemical change has occurred, due to the supply of energy from external sources. We are concerned here with the surface crust and the atmosphere layers of earth since they are exposed to the above-mentioned two sources of energy mentioned above, high temperature heat from inside the earth and light

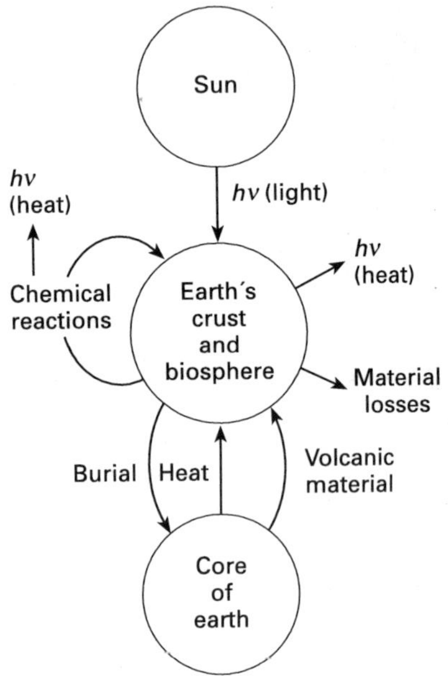

Fig. 20.7 The energy and material transfers to and from the earth's and life's compartments.

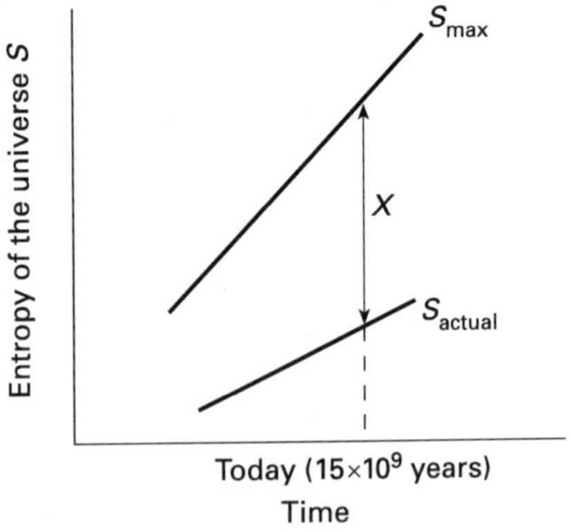

Fig. 20.8 Schematic time-dependence of the entropy of the expanding universe. S_{max} corresponds to the equilibrium value and S_{actual} to the value at a given time. Much of the difference is due to local order and organization. X = capacity of doing work.

from the Sun. While overall changes of the whole system must move slowly toward equilibrium, these radiations have quite the opposite effect locally on the two surface layers since they *excite* quite fast chemical reactions locally, driving chemicals away from equilibrium towards an energized state.

$$\text{INPUT} \rightarrow \begin{array}{c}\text{High energy}\\ \text{source}\\ +\text{ Stable}\\ \text{chemicals}\end{array} \rightarrow \begin{array}{c}\text{Steady}\\ \text{state of}\\ \text{unstable}\\ \text{chemicals}\\ \text{(intermediates)}\end{array} \rightarrow \begin{array}{c}\text{Low energy}\\ \text{source}\\ +\text{ Stable}\\ \text{chemicals}\end{array} \rightarrow \text{OUTPUT} \qquad (20.1)$$

Whether such a steady state can be reached finally is very questionable in such an open system. The existence of some *final Gaia-like condition* is, therefore, not very probable because of losses and redistributions.

This cycling is a simple device for the creation of new chemicals on earth. The unstable chemicals may be generated quickly but have long life-times, decaying slowly, and can be lost (or gained) from (or by) the environment. We note especially the generation of dioxygen from water. However, the final situation is made much more complicated due to the further reactions of the unstable products, such as dioxygen, in the presence of the surface chemicals.

Unstable chemicals are to some degree reactive and will not just cycle as above but over long periods of time they can generate other chemicals with which they can then react and which can also enter into geochemical cycles. Putting to one side the energy input and output, which drives the whole system through the entropy gain of a small number of high energy quanta (light) or of high temperature radiation on going to a larger number of low energy (temperature) quanta (heat) (see Fig. 20.8), we now have the extra coupled material cycles

$$\begin{array}{l} \text{C} \rightarrow \text{C*} \rightarrow \\ \updownarrow \\ \text{B} \rightarrow \text{B*} \rightarrow \text{energy (heat)} \\ \updownarrow \\ \text{Light} + \text{Low energy, } + \text{A chemical} \rightarrow \text{High energy, A* chemical} \rightarrow \text{energy (heat)} \end{array} \qquad (20.2)$$

where A*, B*, and C* revert only slowly to A, B and C and eventually all return to A.

Now, this series of slow changes involving A, B, C, etc. has led inevitably to an ever-increasingly energized surface on earth in an evolutionary sequence while moving toward a possible steady state. The sequence follows from the order in which A* is produced first and B*, C*, etc., follow due to thermodynamic/kinetic control. The progress of changes both geochemical and biological, see below (they went together until man's intervention), has been exceedingly slow, being measured in billions of years (see Fig. 20.9) Notice that the individual elements may be transformed by energy quite independently from other elements, e.g. $O_2 \leftrightarrow O_3$, or the transformations may be coupled, but there is no essential linkage between them all on earth, contrast biological systems below and note also how many separate cycles of chemicals are produced.

20.2.3 Geochemical and biochemical interactions

Complicated as the changes in earth's surface have been they are very slow, so that we can still consider different aeons in the earth's history as stationary or even equilibrated states. For example, the beginnings of the earth's surface and atmosphere can be treated as an anaerobic state which predated stepwise changes to the present-day aerobic state. Even in this anaerobic state many chemicals in unstable form were constantly injected into the crust from the core, as they are today, and then these chemicals energise other chemicals of the crust. In this light we look on Earth's evolution, which must still continue, and then ask how would such stepwise changes be affected by and affect the nature of local small flow systems (such as cells) which have very fast cycling times of material and throughput of energy. The above scheme then becomes [see also Fig. 20.5(b)]:

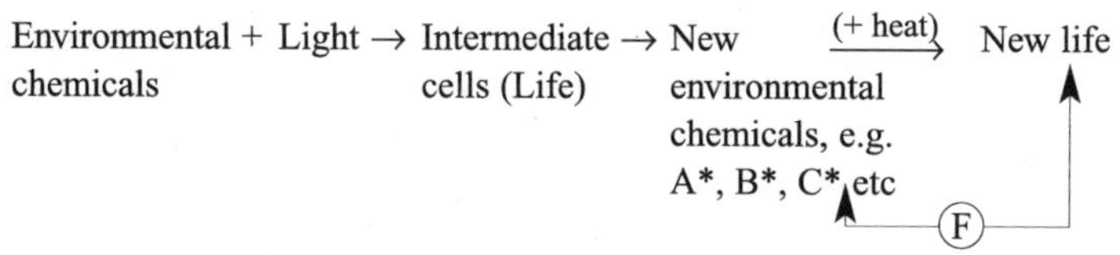

Life is the dominant system of this kind and it became one of the major activities forcing the evolution of the surface of the earth and its atmosphere and making use of environmental changes to give A*, B*, C*, etc. Some of the rejected new environmental chemicals from organisms have long cycling periods which may hardly or may not participate subsequently in life cycles but accumulate as geochemicals, e.g. oil and coal. Because the living cellular systems are efficient in energy capture (and managed to replicate!) they became a major agent of geochemical change (Fig. 20.9). The manner of change of the sea which is in contact with living organisms is shown in Fig. 1.13. We examine first the observed geochemical changes which are easy to date and explain without troubling too much as to whether they did or did not originate from life's or other activities. Notice again that in the description of purely chemical systems on earth we do not consider restrictions on the chemical elements involved except for the limitiations of the periodic table and the relative abundance of the elements in the crust, but when considering biological systems we shall have to look at which *selected* elements entered *combinations* for long enough to achieve *survival* with reproduction.

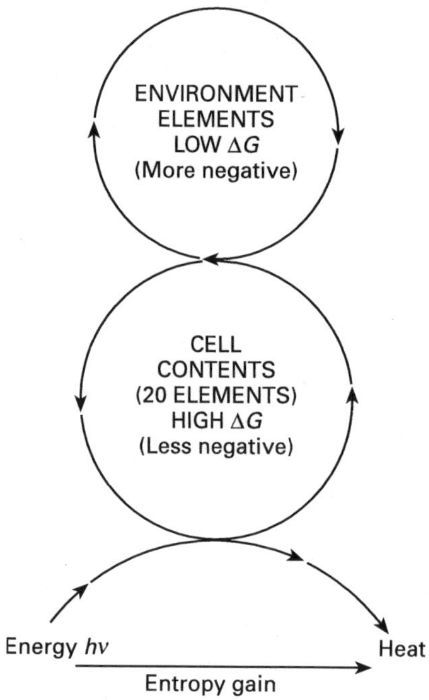

Fig. 20.9 The total entropy drive (see Fig. 20.8), which underlies the cycling of elements in life.

20.3 The detailed steps in the evolution of earth's surface

As stated, the first steps in evolution of earth after the formation of the chemical elements as gases in a giant star were their combination as chemical species and their condensation to form many liquid and solid phases (Fig. 20.2). The surface, largely aqueous, was formed in a reducing atmosphere due to the presence of large amounts of hydrogen (lost progressively) and its compounds, H_2O and H_2S. Based on the abundance of all the elements these three chemicals controlled the *availability* of many of them, especially in the surface liquid (aqueous) and gaseous phases. For example, the insolubility of sulphides in water greatly depressed the presence of zinc, copper, and molybdenum, but elements such as Na, K, Mg, Ca, Cl, P, S, C, H, N, O, Fe, Mn, V, and W were readily available, and Ni and Co somewhat less so (see Fig. 1.6). Much nitrogen remained as N_2 and NH_3 and it is likely that carbon was present largely as CO

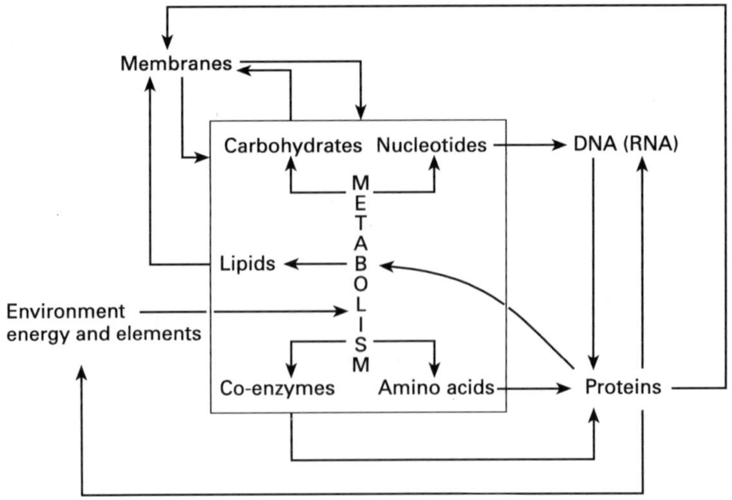

Fig. 20.10 An indication of the necessary interlocking feedback conditions between the pathways of elements in compounds, (see Fig. 20.4).

and CO_2. (Observe however that these gases are constantly being lost to the cosmos to a small degree) Now, as stated, these surface layers of water solutions and gases were exposed to energy from a high temperature internal zone and a very high temperature external zone, the sun. This led the initial inorganic chemical composition to give rise to some organic molecules through using the stable hydrides (H_2S in particular) to reduce CO or CO_2 to unstable compounds such as HCHO. On polymerization and reduction formaldehyde produced lipids which formed micelles and vesicles and therefore generated new compartments in the aqueous layer. Certain organic as well as inorganic products also evolved, and scientists consider that all the essential small ingredients of life could have been produced, (see Fig. 20.10).

Dioxygen (A*) was later and is today the major product of the action of light on water yielding H_2 for other reactions. (We are not concerned with the way in which this came about as yet.) O_2, generated at first in a reducing atmosphere, was a reactive chemical. The reactions follow as shown above in scheme (20.2): O_2 + reducing chemicals $\rightarrow$ B*, and so on. Slowly over 4×10^9 years the air, the sea and the land surface of earth became a huge deposit of energized (oxidized) chemicals in the solid crust and aqueous layers—mostly oxidized since H_2 was lost (or reducing equivalents were stored in living organisms)—yet still lying between a much greater reservoir of reduced chemicals in the sulphide and metal alloy zones of earth's centre, and the atmosphere which also changed slowly to one of just CO_2, N_2, and O_2, that is, no H_2 or NH_3. This surface remained open to radiation from the sun. The subsequent changes followed an inevitable but incomplete time sequence dependent on the redox potential of combinations between all the elements (Fig. 20.11), while available energy sources also changed with time (see Table 20.2). [Note that a final steady state, if it comes about, may well include higher levels of oxidised nitrogen, sulphur, carbon, and halogens than are present today, but we cannot be sure that any such condition will be attained.] Remember too that the sea contained a changing sequence of collected organic chemicals—initially reduced and then increasingly oxidized. A major consideration in such systems as the sea was, and is today, the availability of the elements, and not surprisingly therefore abundance is one major restriction and the changing volatilities and solubilities of combinations are others.

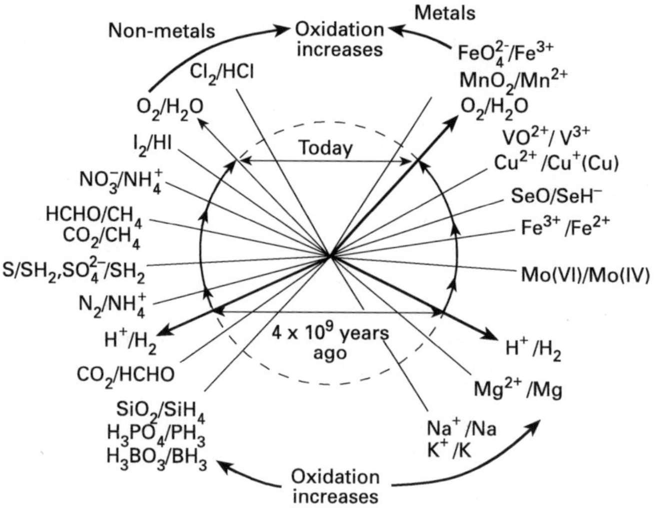

Fig. 20.11 The change from a reducing to an oxidizing atmosphere has moved elements from lower to higher oxidation states in a time-sequence.

Table 20.2 Sources of energy

Period (years ago)	Energy sources
Initial (4.5×10^9)	(a) the sun (b) basic unstable chemicals in the crust (c) chemicals stored at high temperature in the core
After say 1 billion years (3.5×10^9)	(a) as above (b) some oxidized materials some SO_4^{2-}; very little O_2, H_2O_2
After say 2 billion years (2.5×10^9)	(a) as above (b) further oxidized materials; modest O_2, H_2O_2
After say 3 billion years (1.5×10^9)	(a) as above (b) further oxidized materials almost final O_2 pressure
Today	(a) as above (b) man's fuels (c) atomic energy

20.3.1 The evolution of organized flow on earth: patterns

Now we can see the way in which energization and subsequent reactions of chemicals, in the earth's crust, which lies between two higher temperature energy sources, the sun and the molten interior, would generate cycles and irreversible changes of elements. Slowly the surface of earth changed and is still changing as earth generates oxidized chemicals, loses light materials such as CO_2, N_2, H_2O, the sun dies, volcanoes erupt and meteorites strike. ('The end is not yet'). In addition, the earth spins on an axis which causes directional flow of its atmosphere and uneven heating so that there are temperature and pressure gradients vertically and horizontally. The earth itself has irregular land masses also both vertically and horizontally and these cause physical barriers to the liquid and atmosphere flows and generate new temperature gradients. What we observe in the mobile phases is, somewhat surprisingly, not turbulence alone but patterns of flow (currents) in the clouds, the oceans and the rivers on land. These patterns fluctuate but they occur repeatedly. We must note that they require *structured* energy gradients. We note the gradient of light intensity which gives the chemical flow in, for example, the $O_2 \rightleftarrows O_3$ layers. Flow always depends on such spatial irregularities giving boundaries, due to spatial or kinetic barriers.

Finally, we have to remember that on earth many of the changes in chemicals occurred in aqueous solution in the presence of local surfaces which act as catalysts. These surfaces are unevenly heated due to gross surface convolutions or to light intensity. Heating of liquids (or gases) unevenly in the presence of such surfaces inevitably generated both chemical and physical local flow of the liquids (or gases). We must remember also that energization of some minerals or organic chemicals (which were insoluble in water, lipids) in the sea produced vesicles which then generated flow around them. Since differential permeability allows flow in and out of the enclosed space as well as around the surfaces, vesicles give rise to chemical separations. There are indeed millions upon millions of energized flow patterns all around us and they have existed for 4–5 billion years and long before life. Others, such as the vertical distribution of O_3, are of relatively recent origin. They arise inevitably even without the organization we call life but the elements involved must be present as gases, liquids, or solutions in liquids at 300° K. There is then a considerable restriction to the number of possible elements in sizeable flow to about 30, which, as we shall see, is not far from the number living systems have turned to functional use.

Now flow through membranes enclosing aqueous space was bound to differentiate chemicals and then organic ions, produced inside by reaction, would be retained with certain inorganic ions in vesicles by these lipid membranes. It is not surprising that non-living 'cells' developed as trapped ionic inorganic and organic chemicals in water. Nor is it surprising that such a 'cell' could accumulate some ions while rejecting others (see Table 20.4) since this depends on filtering selectivity, selective applications of absorbed energy, and competition for binding sites (see Section 2.3).

Putting together physical (flow) and chemical (flow) it is clear that the nature of flowing systems inevitably evolves on a planet's surface as the energization of chemicals develops and then produces new boundaries, including spherical containers, and new chemicals to flow around and in and out of them. Because of restrictions of diffusion, including binding, we can be certain that enclosed spaces will be chemically different from their environment so that there will be energization in terms of local chemical concentration gradients as well as of chemical composition of ions or compounds. The conclusion is that patterns, organizations, are emergent (inevitable) properties of the dynamics of systems of the kind present on the earth's surface (without coded molecules or reproduction). Moreover, the patterns will become more organized in time.

20.3.2 The persistence of patterns

One outstanding problem remains. Why are some dynamic patterns persistent (metastable) and why do some patterns disappear only to reappear? (This is the problem of 'survival' when we consider organisms.) The reappearing patterns such as those of clouds, waves on the sea and patterns in the sands of sea-shores and deserts depend upon many physical factors, but as we have stated earlier, patterns imply structural constraints on flows and energy input. The only constraints in the atmosphere, where we see cloud formations of different kinds repeated time and time again, are due to feedback from local temperature (energy) gradients in space (structure) upon water molecules which flow and can condense to droplets which then flow in the opposite sense (see Aside). It is the flow downwards of the droplets, that is up the temperature gradient which maintains the vapour pressure at the lower height and the flow upwards of the vapour which maintains the droplets. The upward flow itself generates the cloud and condensation inhibits further cloud development. This is more easily seen in the cyclic flow of water molecules from sea to structured land followed by condensation, rain, to give rivers running back to the sea again. There is heating of the sea and cooling of vapour over land in cycles of evaporation and condensation. As before the upward flow of vapour to form drops by condensation, binding, prevents further upward flow allowing return of the water. The 'structural' constraints, the temperature gradients, met by the molecules are due to the uneven properties of the sea and land. The problems of persistence of patterns of a condensed phase becomes the problem of the persistence of gradients of energy, created by or present in the 'structured' systems. One feature is apparent from this analysis of physical flow patterns of water in clouds and rivers on land—in one part of space the temperature and vapour pressure within the structure must just exceed the values of balance between free molecules and bound molecules of water and in different parts of nearby space these values are slightly different in opposite senses. Thus the water patterns have as a major determinant a thermodynamic equilibrium constant, the dew point vapour pressure, p, of water at the temperature T. We shall find such reversible thermodynamic constraints are common to the formation of all patterns, though the systems are irreversible.

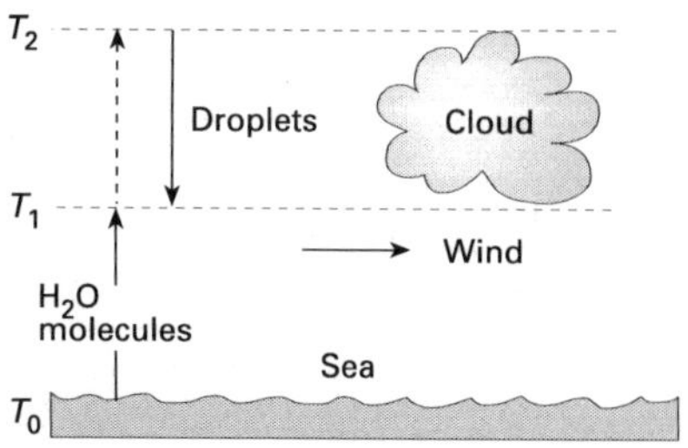

The pattern of a cloud is decided by the temperature and its gradient, $T_0 \gg T_1 > T_2$, the wind speed, and gravity.

The fixed dew point is the temperature at which air has a saturated vapour pressure and is related to a binding constant, K, where $RT \ln K = -RT \ln p$, which is the entropy difference, $T\Delta S$, between the vapour and liquid states. This equation, replacing p by c, concentration, applies equally to the equilibrium between a solid and its solubility in a solvent. The solubility is the equivalent of the dew point vapour pressure. Because the relationship is invariant a solid buffers a solution at a precise value. Precipitation locally occurs on structured surfaces supporting nucleation but only if the solubility of the substance is exceeded. Bone or ferritin, for example, have fixed values of solubility products, K_s, which at a fixed temperature hold the products $[Ca^{2+}][HPO_4^{2-}]$ or $[Fe^{3+}][OH^-]^3$ at fixed values. At fixed pH these constants are very powerful controls over the free flowing component concentrations, i.e. they act as buffers for Ca^{2+} and HPO_4^{2-} and for Fe^{3+} respectively. Note the cooperative interaction of cation and anion (ligand). Thus if a balanced amount of precipitate is present in a flow system the flow must contain ions in concentrations close to those of the solubility products.

20.3.3 Persistence of patterns in an enclosed space

It follows from the above and consideration of patterns of clouds that if more precipitate is to be prevented from forming the entry of the ions into an enclosed space in an open flow system must be stopped by feed-back to the source, but to maintain the precipitate the solubility product must be closely approached. Now the input to an enclosed organism volume, say a cell, is often by an energized pump on a surface. We showed in Section 2.10 that for a binding of any unit material in free solution to a set of independent surface sites, e.g. a pump, we could define an equilibrium constant K_B which bore the relationship to 50 per cent occupancy of the site, $[A]_{50\%}$,

$$K_B = 1/[A]_{50\%}$$

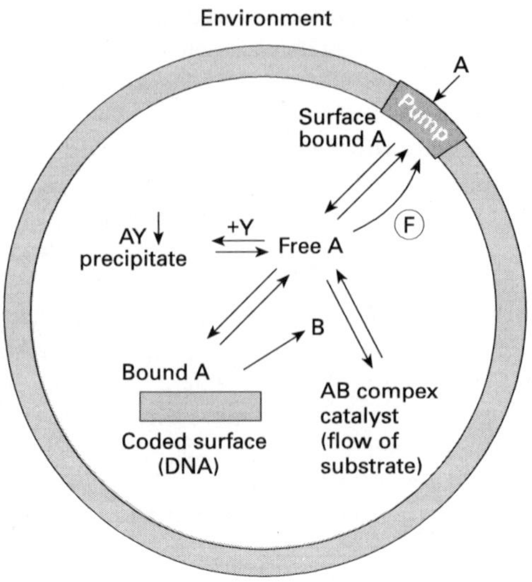

Fig. 20.12 A very simple picture of some possible reversible exchanges (close to equilibrium) inside the cytoplasm or any other compartment. Ⓕ shows a feedback in place of an equilibrium return.

Thus if control buffering is wanted in an internal solution over occupancy then the best control is at 50 per cent bound to a pump protein surface. Thus K_B must be equal to the solubility product K_S if both internal solids and surfaces act together to maintain flow concentrations in an enclosed space, (see Fig. 20.12). Let us call this concentration $[A]_R$.

Now, we showed in Section 2.10 that for a binding between two agents in a solution with an equilibrium constant, K_C [A] [B] = [AB], good buffering of the more dilute of the free partners, say A, is achieved when [AB] = [B] whence [A] for 50 per cent binding is given by $K_C = 1/[A]_{50\%}$. This is the well-known case of pH buffering. For a reaction in the contained volume this value of K_C must be equal to K_S and K_B if a precipitate, a surface and a complex in solution are to come into active balance whether or not they are in a flow system. Suppose additionally that to act as a catalyst A must bind to B (a protein) and that [B] is controlled internally by synthesis again on a surface, now of a coded molecule (to become RNA and DNA where we discuss biological cells). Clearly, to give the appropriate reaction rate due to AB, B must be produced only when A is present, and there is clearly waste if B is overproduced. The production of B (on the surface) must be restricted by feedback of [A] to this synthesis machinery, shutting off B production at the same concentration limit as that which shut off A entry into the container (a cell), i.e. $[A]_R$. Hence A (acting like a transcription factor), must bind to the coded molecule (DNA/RNA) with a constant $K_f = 1/[A]_R$. Returning to the level of production of B for a catalyst, AB, clearly it has to be such that a critical amount of AB is formed so as to maintain catalysis. [AB] cannot exceed a certain value given by K_c[A][B] = [AB]. However if B binds A too strongly, K_C is very large, and it will lower [A] to less than $[A]_R$, so that A will be pumped in and B will be synthesized to too great an extent. Similarly, if A is bound too weakly, K_C is small, then AB is not produced sufficiently at $[A]_R$. The optimum control is clearly when $K_C = 1/[A]_R$, that is when [AB] = [B]. *The best conditions for controlled content of a flow system are the thermodynamic conditions for optimal buffering including formation of phases (precipitates) binding to surfaces and formation of complexes.* 50 per cent occupancy of the binding sites is the most sensitive condition for a feed-back balance between forward and back movements. We then expect that the predicted profile of binding constants and of free metal ions (or another free unit) in an enclosed volume (a cell) will be close to that found in a cell compartment which selects at equilibrium ligands B to keep elements A binding separately. We examined this condition in Chapter 2, (see Fig. 2.10). That this is the case, we shall see when we turn to the analysis of the metal ion profile of the cytoplasm of primitive life and the enclosed volume becomes a real cell (see Fig. 20.14). Note that the concentration of the catalyst, controls the flow of a reaction so that substrate turnover in the enclosed volume is also fixed. *To control a large number of different reactions in the same volume many catalysts and many cooperative feedback networks are required, all associated with selected partners and related to equilibrium constants (see below).*

We see further that in a biological cell reagents, carriers or buffers, pumps, transcription factors and reversible enzymes must all bind approximately equally at a given element concentration $[A]_R$ whether or not free A is an important part of exchange. Hence we may use binding constants to define homeostasis of A, i.e. the metallome. Exactly the same logic applies to an ion, a molecule such as a cofactor or coenzyme *which is used in exchange*, e.g. both as a messenger and in an enzyme, a transcription factor, a pump or whatever. Hence we can characterize a state of a cell by an extended form of free metallome to any exchanging entity in any one compartment.

The proof of these considerations is very clear in the case of the calcium ion, see Chapter 10, where we find that *all* exchanging Ca^{2+} binding proteins in the cytoplasm of cells have $K \approx 10^7\ M^{-1}$ and all exchanging Ca^{2+} binding proteins outside the cytoplasm have $K \approx 10^{3.0}\ M^{-1}$. We return to the general considerations of feedback homeostasis in Section 20.5 but it is necessary to consider also the rates of reactions, Free unit ↔ Bound unit, for which $K = k_{forward}/k_{back}$ and k is a rate constant, since they will restrict the time taken to approach equilibrium, contrast Ca^{2+} and Ni^{2+} (see Section 4.6). Differential rates may allow different temporary switches in states of cells. To operate sensitively and quickly the two rate constants must be fast when K can not be too large, say $>10^7\ M^{-1}$, contrast Ca^{2+} and Zn^{2+} later, but much of biological change is slow.

It is against this background of evolving patterns of flow over millions and millions of years that we must see the possibility of the evolution of life and the effect of this evolution upon the evolution of the environment and, by feed-back, of the changed environment upon the further evolution of life, all in an inevitable sequence. We do not need to be concerned about the precise form the molecules of life took in order to appreciate these principles of survival and of evolution. *The clear suggestion is that complex energised systems formed first without a code.*

20.4 Macromolecules and systems

Before we treat biological systems further we need to look once more at the descriptions of polymeric molecules and approaches to their properties by (a) molecular structure and (b) systems analysis, since polymer molecules are not comparable with simple atomic or molecular units, see Chapter 6.

The term 'molecular biology' might give the impression that knowledge of the structure of a molecule can lead directly to an understanding of its biological properties. That this is clearly untrue is seen as soon as we consider any chemical substance and its extensive properties, as we have stated above. Let us take two examples of considerable significance in organisms, water and

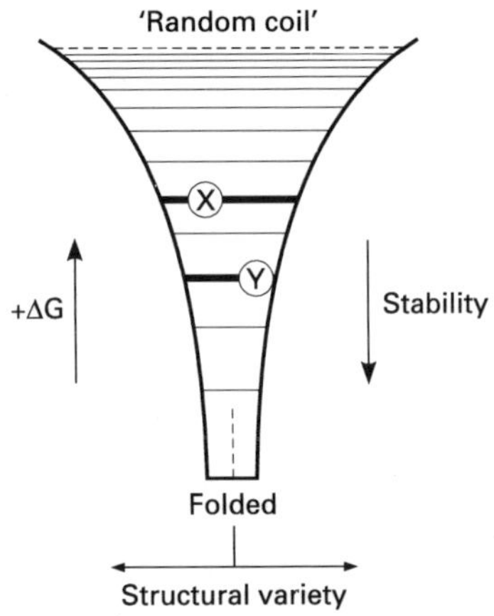

A plot of free energy against strutural variety of states of a protein in water. Heavy line: states of special kinetic stability in fold X and Y, see Section 6.3.1

haemoglobin. Water as a molecule is well defined by composition, H_2O, its molecular structure is well known, and its properties, e.g. bond lengths and dipole moment, are calculable. Any single molecule is very closely independent of ambient physical conditions such as temperature and pressure over considerable ranges and we can claim that we understand it. Now consider water as a bulk substance which is of major importance in the study of life. It has many properties which cannot be calculated from the formula H_2O. These include melting and boiling points, and kinetic properties such as viscosity and conductivity. As stated earlier, these properties are not describable except in terms of *bulk considerations of phases; that is, of a system of a very large number of* H_2O molecules. Since bulk systems are certainly not calculable from the properties of single molecules in any simple way, we get around the problem by discussing *the system, statistical properties, of states of water, not of individual water molecules*. This returns us to the subject matter of reversible, time-independent thermodynamics, which analyses systems in terms of variables (Table 20.1) and observables, often using phase diagrams plotting a variation of free energy, or a variable of it, against composition (Fig. 20.2). Complexity arises here only from the statistical properties of large numbers of very small units. When we consider the *flow* of water in a cell the problems of complexity increase greatly, of course.

Next consider haemoglobin—a protein. It too can be isolated and given a compositional formula as a molecule $Fe_4C_xO_zN_wS_u$ which is exact and in principle it could be given a determined molecular structure at 0°K. Now, this is not the haemoglobin in organisms where it is in flowing solutions. We wish to describe first its non-flowing properties in a solution. Even as a single polymeric molecule in solution at 300 °K it does not have a fixed structure but has an ensemble of related structures due to considerable dynamic configurational variation, (see Figs 6.3 and 6.4; see also Aside), all of which are close in energy. The function of haemoglobin is to carry dioxygen and critically the ensembles of *dynamic structures* differ in the oxygenated (T) and de-oxygenated (R) states. We may now use equilibrium system language, binding constants for the characterization of the behaviour of single polymer molecules (or a bulk solution of molecules) exactly parallel to that used for the study of phases of small molecules—thermodynamic descriptions. Again, haemoglobin as a single molecule has a melting point, and as a 'mini-phase' it can undergo phase transitions. However, on top of this statistical problem, here of a single molecule, the transformations in a cell are not just of precise structural change but of overall free energy change of many molecules, ions, and water in a solution. Moreover, the cells in fact do flow to different environments of pH, etc. The conclusion is that the use of molecular structure, concerning just one molecule, to describe cellular functions is a useful approximation for some of the time-independent properties, but it does not give a full analysis. Note that hemoglobin works best as an O_2-carrier close to the equivalent partial pressure of O_2, such that $K \simeq 1/p_{50\%}$ (see Section 20.3.3), even in flowing systems (blood).

Now we can ask, do large molecules, such as a protein, generate state patterns through their dynamics? These should be seen not in the final *stable* folded states but in the fluctuations of the conformations of the polymers as they flow in and out of a limited number of metastable states (see Aside). The implication is that partial unfolding is not a random rise up a set of excited energy states but that access to certain states is preferred. In a cell these energy flows in and out of conformations occur much like condensation of the many water molecules in clouds. We do not know to what degree these fluctuating states of polymers and multi-molecular units have persistence but we do know that proteins enter a certain limited number of allosteric states.

It is against this background of systems behaviour that we wish to understand chemical change in flowing systems and much though we may look at molecular structures and thermodynamic functions, we need further analysis. Here we attempt to look at the problem through knowledge of the elements involved in living cells.

20.5 The beginnings of life

We shall not pretend that we can give a good explanation of how life began. What is certain is that living organisms contain open flowing *systems* in cells of great chemical complexity and that they require energy and material from the environment for them to grow and reproduce (see Section 20.2.2). At various stages of growth the conditions of the internal cell compartments can be considered as fixed, as described above, in what we call *homeostatic states*. The progression of such states is fast in bacteria but relatively slow in differentiated cells. Our concerns in this chapter are with these cell states but also with the changing sources of energy and the changing availability of the elements with time (that is, in evolution), as outlined in Fig. 20.11 and Table 20.13, against the background of organization in a structured confined volume. We commence discussion with the simplest cells.

20.5.1 The basic cellular flows and element homeostasis: the primitive metallome

The elements we treat first are those we believe to have been essential even for the most primitive cells (Table 20.3). They must all cooperate in a single aqueous cytoplasm in order to achieve survival of the whole (Table 20.4). Our immediate concern therefore is obviously with the content of C, H, N, O, S, and P in small and large molecules. We here use a scheme for the flow of these elements which is generalized from conventional wall-charts in laboratories (see Fig. 20.4). Molecules can pass back and forth along the pathways and many feed back not just onto their own path but onto the paths of others. Only in this way can there be homeostasis of the concentrations of all the intermediates and polymers in different polymer syntheses. These cytoplasmic pathways have changed little during the whole of evolution (Table 20.5 and

Table 20.3 Essential trace elements in primitve life

Element	Function
Na, K, Cl	Osmotic, electrolyte balance
Mg	Phosphate reactions
Fe	Redox reactions, e, H, H_2S, CH_4, CO
Ni, Co*?	H, CO reactions
Mo, (W) ?	O-atom transfer
Mn, Co*?	Cytoplasmic hydrolyses
V?	N_2-fixation (?)
Zn?	Structural role(?)

NB* It is uncertain if Ni, Co, or Mn were required, but they have functions in all early life we know about. We exclude Ca, Cu and Zn(?) since their functions appear to become essential later. Analysis of early fossils, e.g. in stromalites or from primitive plants, for these elements and their isotope composition will help to extend evolutionary perspectives.

Table 20.4 The connected role (with feedback) of inorganic elements in primitive homeostasis

Function	Required elements
Osmotic Controls, Structures	Na^+, K^+, Cl^-
Acid–base catalysis	Mg^{2+}, (Zn^{2+})
Acid–base balance	H^+, HPO_4^{2-}(ATP)
Redox balance	H, Fe, S, Se
Protein synthesis, etc.	(CHNOS), Fe, Mg^{2+}, (Zn^{2+} ?)
Nucleotide synthesis DNA/RNA	(CHNOP), Fe, Mg^{2+}

*NB All element uptake is connected to MgATP, or to ion gradients (see Fig. 20.8).

Fig 20.4). We can analyse a cell to show that in a given state it has such a fixed system of polymers. For proteins this is called its *proteome*. Now, these proteins are made from many smaller molecules. The concentrations of small molecules are of necessity (see above) also in feed-back balance based upon equilibrium binding of small molecules to specific targets, e.g. enzymes and transcription factors, note especially mobile coenzymes. The pathways have two further requirements: energy and incorporated inorganic elements. The energy from the environment (see below) is stored in ion gradients and in ATP (or more generally NTP) which act generally in many pathways together with Mg^{2+}, e.g. as $Mg^{2+}\cdot$ATP (Table 20.6). Hence they coordinate energy flow. As well as Mg^{2+} (Chapter 9) the above pathways require essential Na^+, K^+, Cl^-, at

Table 20.5 Maintained pathways throughout evolution

Pathway	Example
Syntheses and degradation of saccharides	Glycolysis (Mg)
Dicarboxylic acid reaction sequence	CO_2 incorporation in incomplete cycle. Completed later to give energy capture, i.e. Krebs cycle (Fe, Mg).
Amino-acid synthesis	Products of glycolysis and Krebs cycle + NH_3 (Fe).
Protein synthesis	Formyl initiation (Fe) and methionine initiation (Fe, Co).
DNA, RNA, syntheses	Nucleic acid pathways (Mg, Zn, Fe/S, B_{12} or Fe_2O).
Synthesis of fats	β-carbon oxidation/reduction (flavin, Fe).
Nitrogen incorporation	Formation of NH_3 (Mg, Fe, V, Mo) in symbiotic bacteria.
Hydrogen reactions	H_2 as a reductant (Fe, Ni) in anaerobic archaea.
Energy: electron/proton flow in membranes	Energy capture related to ATP (Fe, NADH, flavin, quinones).
Exchange of ions in membranes	Osmotic and electrolyte control (Na, K, Mg, Ca, H).
Light capture	Chlorophyll and haem (Mg, Fe).
Pumps (ATPases)	Most elements as cations or anions
Redox balance	Glutathione, Fe

Table 20.6 Some connections to Mg^{2+}·ATP or proton gradients in anaerobic prokaryotes

Reaction	Connection
Fe/S systems	Production of H^+ gradients prior to ATP formation (?).
Glycolysis	ATP production and feed-back restriction.
Exchange for H^+ or Na^+	Inward movement of substrates and other ions
Pumps ATP-ases (later)	Rejection of Na^+, Ca^{2+}, Cl^-. Inward movement of K^+, HPO_4^{2-}, Fe^{2+}, Mg^{2+}, etc. and substrates.

fixed concentrations for the reasons described in Chapter 8, they also require low Ca^{2+} (Chapter 10) and some free transition metals to act as strong catalysts, e.g. at least Fe^{2+} and possibly a little Zn^{2+} and Mn^{2+} (see Chapters 12, 14, and 11). As we have seen in Section 20.3.3, all the element uptakes or rejections must also be coordinated by feed-back, connected to metabolic pathways, and linked to energy sources (Fig. 20.13 and Table 20.4) in order to maintain a homeostatic flow system. There is then a coordinated inorganic as well as an organic flow system within the primitive cytoplasm and which largely persists to this day. The cytoplasm has both a recognizable proteome and a free metallome, but it also has a homeostatic level of coenzymes and small organic molecules. These are minimal requirements for the survival of a cell but for reproduction and repair it also requires a coded central unit, DNA/RNA, also in feed-back relationship with all concentration terms in the cell (Section 20.6).

Now these requirements are quantitative, as we have seen in Chapters 8 to 19, for all elements in a homeostatic state. The critical conditions are set by the possible modes of *selective binding* (Table 20.7). Hence the metallome profile is a characteristic of a cell in a given state of activity of free metal ions and bound metal ions. The analysis of an active cell such as *E. coli* indicates that the free metallome concentrations cannot be far from 10^{-3}M Mg^{2+}, 10^{-3}M ATP^{4-}, 10^{-1}M K^+, 10^{-2}M Na^+, 10^{-6}M Ca^{2+}, 10^{-2}M Cl^-, 10^{-7}M Fe^{2+}, 10^{-7}M Mn^{2+} and perhaps 10^{-13}M Zn^{2+}. Note that this element content has

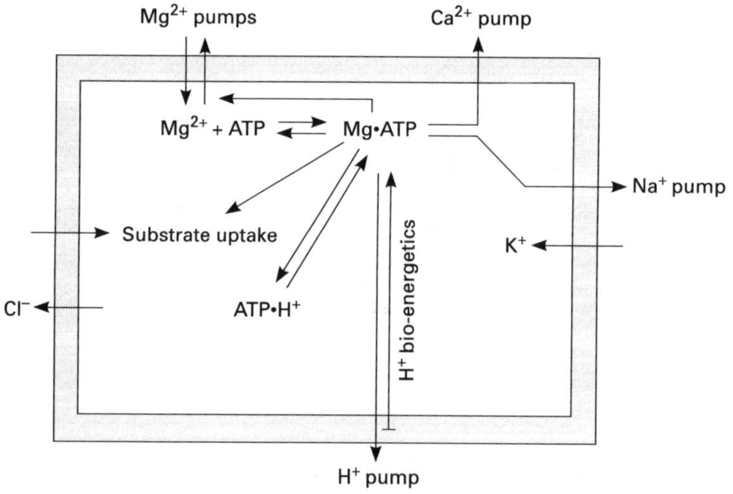

Fig. 20.13 The network of pumps linked to ATP and controlled by selective binding and feed-back.

Table 20.7 Original anaerobic binding competition (sulphide media)

	Mg^{2+}	$Mn^{2+}(10^{-6}M)$	$Fe^{2+}(10^{-7}M)$	Zn^{2+} $(10^{-11}M)$
Ligands	ROH, RO^-	Mainly – CO_2^- → and 1–2 N →	Mainly RS^-(clusters) and some N/O →	Mainly N → or RS^- donors
Geometry	6-coordinate	6-coordinate	4-coordinate clusters (S^{2-}) 6-coordinate (N/O →)	4-coordinate
Binding constant (M^{-1})	$10^{3.5}$	10^7	10^7	10^{13}
Exchange from proteins (sec)	Fast, 10^{-5}	Fast $>10^{-1}$ (Signal to DNA ?)	Fast $>10^{-1}$ Signal to DNA	Slow $>10^{+4}$ No signalling

The exchange rates meant that free iron (or manganese) could act as signalling ions to DNA-bound proteins of the $Fe^{2+}/Fe^{3+}/S^{2-}$ type but that Zn^{2+} could not be useful *in signalling by exchange* within the life-time of bacteria in fast growth (doubling time 30 min).There was probably very little zinc in these primitive cells, but see eukaryotes.

very strong connections with that which we believe to have been present in the sea of the early reducing environment (Fig. 1.13). While pumping affects the individual concentrations, for example Na^+ is pumped out by a factor of ten, Ca^{2+} is pumped out by a factor of 10^4 and K^+ is pumped in by a factor of ten, it does not grossly alter the profile of the free metallome. In fact, the reducing conditions of early life made a cytoplasm close to anaerobic which is only a little different from that we find in later aerobic life (see below). We consider that the cytoplasm of *all* cells, anaerobic and aerobic, has a similar free metallome profile indicating a limitation on life, as we know it, with the necessity for reducing conditions (see Figs 20.14 and 20.24). The feed-back pattern is based on that of the binding constants for the free elements in a reducing environment which, for reasons given earlier, must be the inverse of their free concentrations (see Section 20.3.3). These concentrations, the free cytoplasmic metallome (Fig. 20.14) and indeed those of the proteins, the cytoplasmic proteome, are then characteristic of activity in a cell system. Thus everything in the cytoplasm is reasonably fixed even though, as we have seen in Chapter 6, the turnover of elements and substrates depends today upon the presence of proteins coded in and produced through DNA (RNA).

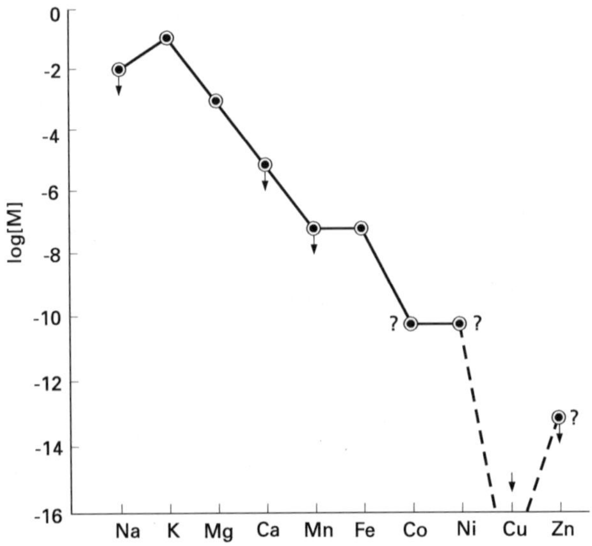

Fig. 20.14 A suggested profile of the free element content of the most primitive organism—its free metallome. Downward arrows show outward pumps.

20.6 Survival, reproduction, and the need for a coded molecule

The survival of reproductive cells today is, and as far as we know always has been, clearly dependent upon a central molecule—the DNA or perhaps RNA. It is conventional to treat DNA as a code for reading the proteins which the cell produces, albeit in a controlled way and with attached machinery (Section 6.14). In fact, it is often regarded as a self-sufficient blueprint for the cell, but this is incorrect in at least three ways. First, the DNA sequence itself has a binding surface for the attachment of various proteins including transcription factors. Thus zones of the DNA have selective binding constants which are not readable from its sequence and are system properties of the cell, whence the DNA is just a part of a total system. Second, the DNA through the proteins (including the proteins of enzymes which cause modification of DNA bases) interacts with small molecules and ions which themselves interact with the environment so that it is the whole system of molecules from DNA to the environment which creates the cellular unity. Third, the DNA is a mobile macromolecule like haemoglobin in having many conformations (Section 20.4) and is a minisystem energetically. Thus while the DNA, the genome, is of central importance to reproduction, its sequence does not lead to an understanding of the survival of an organized system of a given composition. We need an understanding of its place in the cell system. Only certain compositions in a given environment lead to survival (see Fig. 20.20), while DNA has a continuously variable sequence and does not reveal this underlying limitation.

20.7 Evolution: introduction and morphological changes

Now that we have an approach based upon a feed-back-controlled system of the elements to give the organism homeostatic integrated cytoplasmic states, which we take to include its shape, we can look at subsequent evolution of such a system. We shall do so by examining changes in organization as follows: (1) of compartments, (2) of composition of the metallome, free and total, and (3) of communication. The changes in all three are connected to the DNA, which changes, too, but remains overwhelmingly in the cytoplasm. (NB Organelles, which are of symbiotic origin, have their own DNA; vesicles do not.)

20.7.1 The development of internal morphology—evolution of eukaryotes

We have described the most primitive characteristics of prokaryote organisms in Section 20.5. There are three important considerations in the change from prokaryotes to early single-cell eukaryotes (Table 20.8): (1) the introduction of new types of mainly internal proteins, filaments, associated with mechanical devices; (2) the appearance of vesicles; and (3) the incorporation of organelles, units derived from bacteria. These features are all *internal* and fully illustrated in Chapter 3. The change may appear discontinuous, but some bacteria have deeply invaginated membranes and almost certainly localized activity, and some have vacuoles. It is probably only the poor fossil record that prevents us from seeing the change as a succession of very small steps.

We turn next to examine changed external form in eukaryotes, which is again a part of the changed cell homeostasis and indicative of greater division and speciation of chemical use during evolution.

Table 20.8 Levels of self-organization

(a) Prokaryotes (3.5×10^9 years ago to today)
1. Single membrane containment.
2. Differential, inside–outside, distribution of elements.
3. Polymer–polymer synthesis and interactions.
4. Relatively simple messenger-based coordination.
(b) Eukaryotes, single cells (2 to 2.5×10^9 years ago to today)
1. Multiple membrane-limited compartments and later organelles,
2. Differential distribution of elements and large molecules in (b) 1.
3. More complex polymer–polymer interactions including filaments and external mineralization which generate very selective shapes.
4. Complicated messenger systems linking compartments and the environment, especially using calcium
NB Eukaryotes evolved from the archaebacteria plus organelles from eubacteria
(c) Multicellular organisms (about 1.0×10^9 years ago)
The extra organization is now of eukaryotic cells, so we add to (1) to (4) in (b).
5. Multiple cell-limited activities, organs.
6. Differential distribution of all chemical in cells.
7. Increasing control over extracellular body fluids, filaments, and
minerals.
8. Messenger systems linking cells (nerves) via transmitters, hormones, etc., organic in nature.
9. Central coordination (the brain) in more advanced animal organisms ($<0.5 \times 10^9$ years ago).

The reader may wish to add other features to this table.

20.7.2 The development of external morphology: evolution of multicellular organisms up to animals

As stated in Section 3.1, the original prokaryote organisms are single cells with an almost fixed walled shape (see Chapter 3). The next stage involved colonies of them held together by long external filaments and cemented with silica (stromatolites) or calcium carbonate (coral). It is doubtful if these inorganic aggregations need any detailed genetic instruction but they do represent new cooperative phases external to the original cell. Colonies are also found after the nature of cell morphology had changed and for some cells in which there are external filaments these are connected to the changed internal structure, e.g. of the early eukaryotes. Here the introduction of inner filaments connected to outer filaments makes the larger cell and the cell organization more flexible. The organization and impermanent differentiation of cell types occurs next with some greater degree of communication and cell–cell support as in sponges. There is then a steady increase in morphological complexity and organization (see Chapter 3) with a greater degree of permanent differentiation as seen finally today in animal evolution (Table 20.8 and Fig. 20.15). A brief outline of animal eukaryote evolution within multicellular systems is based on the increase in compartments due to layering of cells and then in the appearance of organs. (These developments are also seen as an egg develops into an embryo.) In essence we observe:

One-celled animals: amoeba, vorticella, trypanosoma

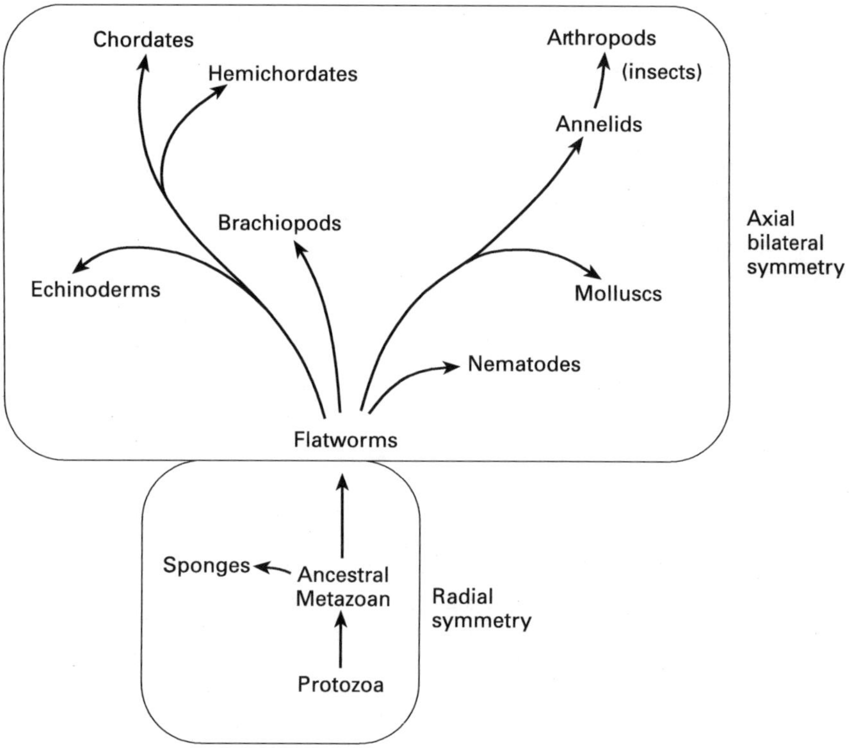

Fig. 20.15 Distribution of symmetry within the animal kingdom [After Barnes R.D. (1987) *Invertebrate Zoology*. Saunders, Philadelphia.]

Two-layer cell structures: some sponges, corals, and jellyfish where differentiation is weak
(The two layers are of internal cells (endodermic) and external cells (ectodermic).)
Three-layer cell structures: flat worms (annelids), round worms, arthropods (spiders, centipedes, crustaceans and insects)
(The new layer is called the mesoderm, from which the nerve system and brain evolved.)
Three-layer cell structures with shells and bones: molluscs (also octopuses) and vertebrates with internal mineralization, e.g. fish, dogs, man.

All these changes can be co-related with changes of DNA but they offer no explanation of evolution of compartments or chemistry opposite survival strength.

A feature of animal life in general is that such morphology allowed mobility but it became more and more dependent upon other organisms in an ecological system. The contrast with the morphology of plants in general is obvious and the development and organization of compartments is quite different (Fig. 20.16). We use this as an example of the fact that change of chemical composition, very largely outside the cytoplasm, goes with change of living form and survival style, since we shall see later that plants generally have a different composition from animals (Fig. 20.26).

The plants illustrate strikingly one further feature of morphological development. These higher organisms evolved in a symbiotic relationship with prokaryotes so that functions were divided out but shared in the cause of survival. Plants require bacteria to supply them with nitrogen in usable form.

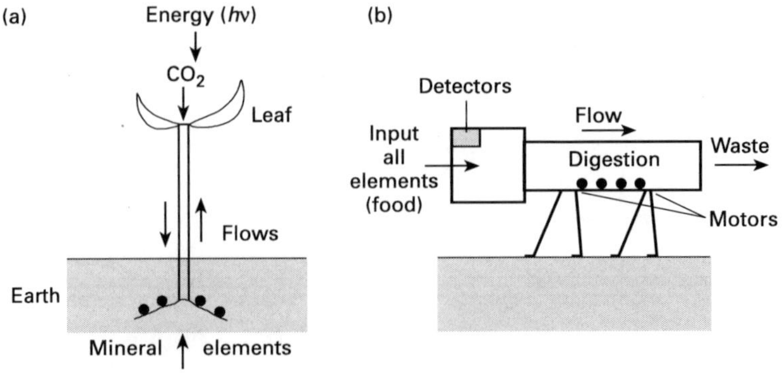

Fig. 20.16 (a) The basic plant structure receives energy from the sun and carbon from the air. All other elements (15–20) come from the earth as inorganic minerals or from bacteria ($N_2 \rightarrow NH_3$) shown as filled cicles. (b) The basic animal structure. Motors are connected to arms, legs, wings, fins, etc. The animal acquires 15–20 elements and energy from food, and requires a multitude of assisting bacteria.

A final step of symbiosis is incorporation which has led to the appearance of mitochondria, chloroplasts, and the separated nucleus within what may well have been an archaebacterium (see Margulis in 'Further reading'). (Higher organisms, e.g. humans, have incorporated lower organisms, even bacteria, and so they have no independent existence and belong to an ecosystem of hardly separate living systems).

Before we turn to the development of compartments and their integration in single organisms we need to see that the network of life extends not just through the way one or two elements are used differently in different organisms. (The given example is the promotion of shared uptake of a simple substance such as nitrogen by bacteria, employing molybdenum or vanadium atoms in very idiosyncratic ways, which is used to supply plants). A quick examination of plants shows how they are also constructed to obtain directly all other essential elements, carbon from CO_2 and the required minerals and water from the soil, Fig. 20.16(a). At the same time they have the apparatus for collecting energy from the sun. Detailed examination of roots and leaves illustrate that these connections to the environment are extremely sophisticated but virtually demand immobility in the soil. Note that their seeds have genetic material but require too all 20 elements since initially the seed has no external source of food or energy. Inheritance is not simple. By contrast the essence of an animal is digestion, Fig. 20.16(b). It is built to scavenge. Thus it obtains all the elements and energy from plants. The more sophisticated the animal the less it does chemically. Hence man needs not just essential minerals but scavenges for amino-acids, fats, sugars and vitamins (coenzymes) which he can not synthesise. Even digestion and health depend on bacteria. An animal is then a sophisticated mobile scavenger with appropriate sensors, incorporating an ecosystem. Not until the arrival of man did the sensor machinery, now including the brain, allow any animal to work in a non-biological manner from the environmental elements and sources of energy. Hence while the whole of life is an extensive ecosystem of compartments man is altering it slowly but surely.

Now the development of compartments in a single organism brings with it advantages and problems. The advantages are the introduction and the separation of new chemical pathways so that selected chemicals and processes can be put in differently energized regions. Therein lies a complication in that now the proteome and metallome of each compartment require their own treatment, see below. The problems are to maintain homeostasis in each compartment, to link each one back to the central cytoplasm and the DNA, and to increase survival time of the individuals since cyclic growth is inevitably

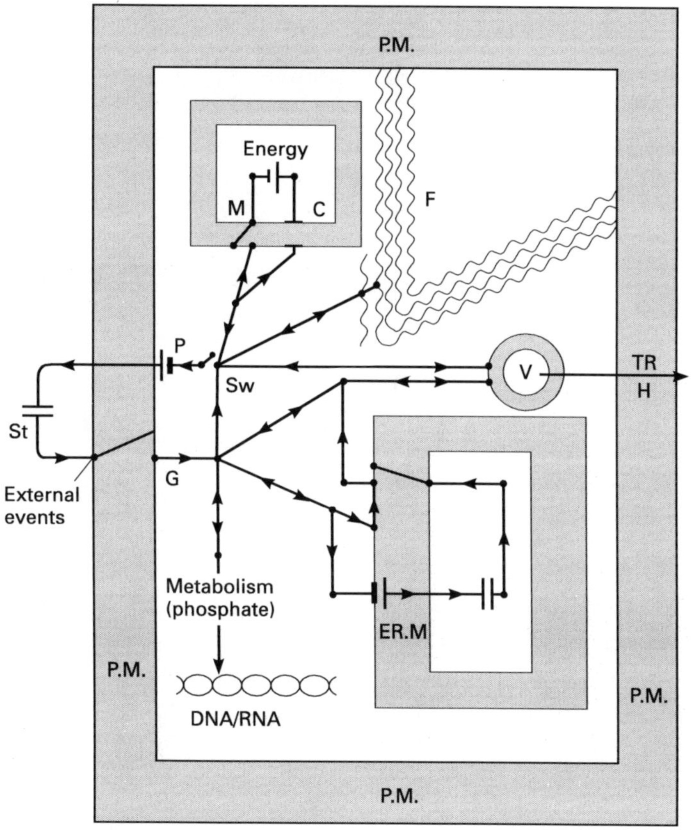

Fig. 20.17 The network of connections required between compartments in a eukaryote cell. P.M., plasma membrane; F, filament; M/C mitochondrion/chloroplast; P, pump; Sw, switch; V, vesicle; ERM, endoplasmic reticulum membrane; P.M., plasma membrane G, entry control; TR/H, vesicle content exit; St, store (condenser); B, battery.

slower. Hence inside the eukaryote cell there have to be also new networks of carriers and communication and new means of protection, on top of those conserved in the cytoplasm, with which they must establish contact to assist survival (see Fig. 20.17).

We must now attempt a description of the way in which all the compartments of one organism can act as a new chemical unity within its environment, which requires the use of messengers (see Section 20.16). The type of network we envisage is shown in Fig. 20.17. It relies upon the use of the same or a very similar set of *elements* as those used by prokaryotes but cleverly diversified in concentration in compartments, both internal and external. We shall describe this in two stages, noting first how the free and total metallome content of the organisms has evolved with compartmentalization and new environmental chemistry and then how all the elements are knitted together in activity through new messengers and feedback controls, both based on the functions of the new elements or of new oxidation states of elements already present in primitive organisms. We base the discussion on the analysis of the profile of the metallomes in widely different species. We stress again that metallomes are systems properties. The analysis is parallel to that of small molecules and their polymers in the description of the proteomes. We have to be careful to analyse the environment simultaneously since as we have explained in Section 20.2.2 it is part of the evolving system (see Fig. 20.7). We must also separate exchanging from non-exchanging parts of the system (see Fig. 20.18).

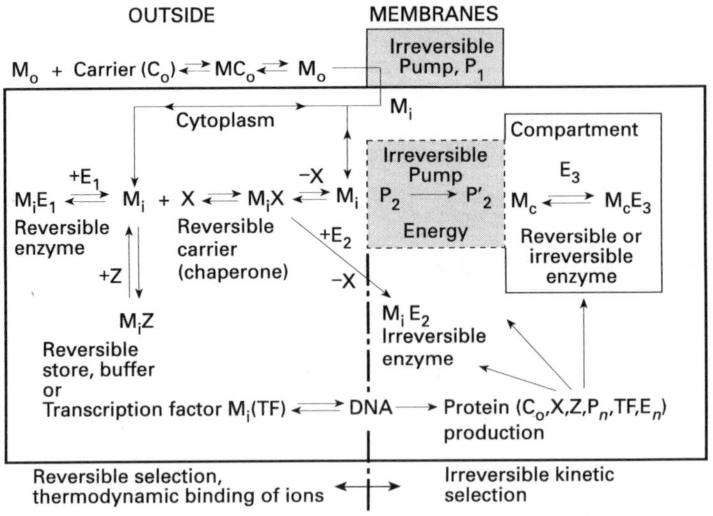

Fig. 20.18 The reversible and the irreversible (or kinetically controlled) steps of a cell. At the top right is the environment (OUTSIDE); the full line is the limit of the cytoplasm; the broken line indicates a pump in a membrane. Irreversible steps are separated at the right of the diagram. M is an element, C, P, X, Z, E, TF, are varieties of proteins for binding Mo (external), Mi (cytoplasmic) or M_c (vesicular compartment), P = pump, C = Carrier, X = Chaperone, E = Enzyme, TF = Transcription Factor, Z is other proteins.

20.8 Evolution and the metallome

Historically the discussion of the evolution of organisms has been based on very different approaches. Classical procedures have depended upon the semi-quantitative description of shape and form of both external and internal features of the organisms and on their behavioural activity. These criteria are

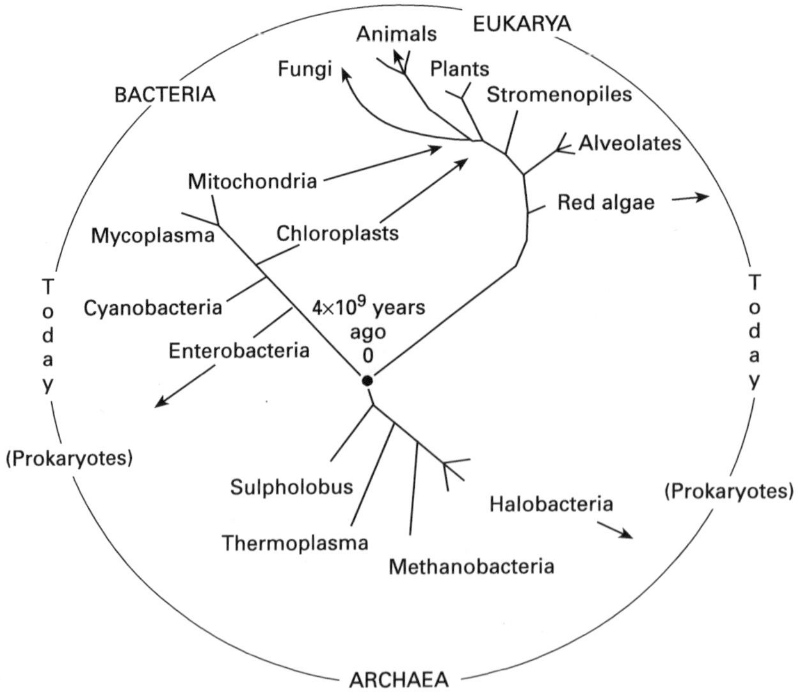

Fig. 20.19 (a) The three domains of life in a radial time sequence based on genetic (RNA) analysis. The distinction between anaerobes and aerobes is not made here. All the domains advanced with time but in very different ways. [See Woese, C. (1990). *New Scientist*, 11 August; Sogin, M. (1993), *Science*, **260**, 340.]

Table 20.9 Involvement of elements in homeostasis during evolution

Primitive anaerobic prokaryotes	**Early (anaerobic) single-cell eukaryotes**	**Later (single-cell) and Multi-cellular eukaryotes (aerobic)**
H, C, N, O, P, S, Se substrates and polymers →	→	→
H^+, Na^+, Mg^{2+}, Cl^-, K^+, Ca^{2+} exchangers →	→	→
Ca^{2+} structural →	→	→
H^+, P, S. Fe signals →	→	→
W enzymes →		
Mn, Fe, Mo, low potential enzymes →	→	high potential enzymes
Ni enzymes (H_2,CO) →		
Ni(urease) →	- - - - - →	plants only
Co(B_{12}) →	- - - - - →	animals only
	Ca^{2+} ATP-ases →	→
(Zn enzymes) → ? →	Zn enzymes →	in vesicles and extracellular
	→	Zn signalling (DNA)
Ca^{2+} rejected →	Ca^{2+} in vesicles and filaments and inner signalling →	Calmodulin, annexin →
Na^+, K^+ osmotic and charge balance →	→	Outer filaments and signalling Na^{+}/K^+ between cells, Na^+/K^+-ATP-ase
		Organic hormones Iodine hormones
	→	Cu enzymes

the basis of the works of Linnaeus, Darwin, and Wallace. In recent times molecular biology has provided an alternative approach based upon sequences of units in molecules—proteins and especially RNA and DNA. In both cases, pictures of evolutionary trees (Fig. 20.19), depend upon relative small differences, e.g. in the sequences between 'related' species but a large number of observations are needed to arrive at a convincing tree. There is at present considerable confusion since the new DNA sequences show much cross-interaction between lines of evolution as drawn from RNA data, (see Aside and see 'Further reading'). Notice that the molecular data are *one-dimensional* sequences and they do not represent *quantitative features* of the organisms. There is no intensity factor though this characteristic of a system was present in the earlier Darwinian shape definition of species. Our approach will be quite different. We shall seek both qualitative and quantitative data concerning the levels of elements in cells (free and bound), which we have called the metallome, in parallel with the *gross developments of evolution* (Table 20.9) before considering small changes. We stress again that this is a concept parallel to that of the proteome; *both are not just molecular but are systems*, since both demand concentration analysis of separate but interacting compartments.

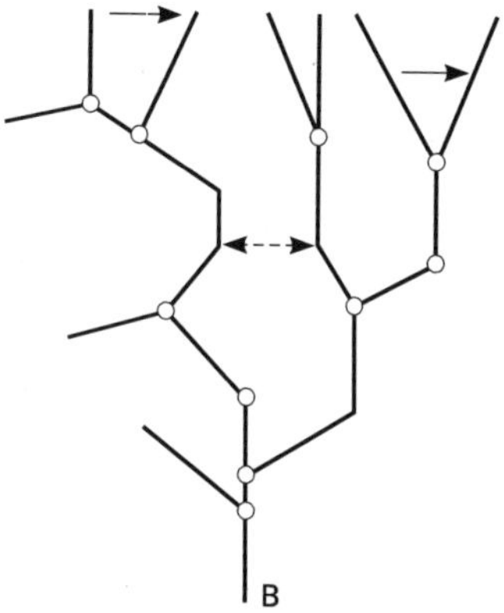

◄- - -► Incorporation (symbiosis)

The base B of this diagram may have been broad, with mixing of cell types

20.8.1 The metallome and survival strength

In sections 1.6 and 1.7. we drew attention to the fact that living systems depend for survival on their ability to accumulate and use certain elements to

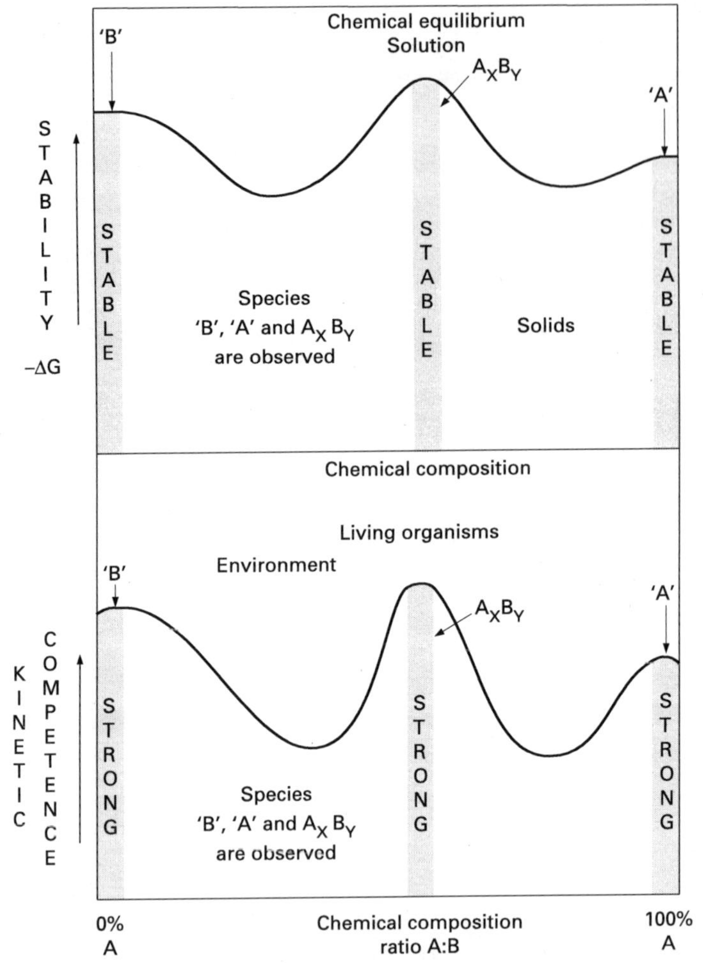

Fig. 20.20 A comparison between a phase equilibrium diagram (see Fig. 20.2) and a survival (kinetic competence) diagram where composition is plotted on the abscissae. One involves solution ⇄ solid-state equilibria while the other involves environment/organism balanced flow.

give a certain composition which is held constant within a given homeostatic state. Different species can then be distinguished by different compositions while intermediate values of composition have low survival values and are not observed. Hence a plot of survival against composition has similarities to a phase diagram, (see Fig. 20.20). We have noted how the composition is related to equilibration and kinetic factors (Section 2.10). Now we wish to discover the compositions which have survival strength, the metallome (free and total), but in which we shall include heavy non-metals which vary amongst species, that is S,P, Se, I, B, but not C, N, H or O, which show little change amongst species. The metallome (see Table 20.10) is then a defining property of a species much as is a genome or a proteome. Now the metallome can be analysed in two ways: (1) the free element composition, or (2) the total element composition, but remember that we have to consider every compartment separately.

The variety of species is very large, but we shall make only the divisions of Table 20.9 and that between anaerobic and aerobic organisms. We shall show that the chemical pattern of development of compartments closely interacts in particular with the effect that the switch to aerobic conditions had upon the chemical usability and availability of the elements in the sea (Fig. 1.13). In

Table 20.10 The metallome approach

1. The location and total concentration of each element selectively can be followed by the use of the electron microscope with X-ray fluorescence.
2. Free metal ions can be discovered using probes.
3. Metal coenzymes and cofactors are readily studied separately.
4. Proteins associated with metals are usually isolated with them and have specific metal-dependent properties. They are included in the proteome.
5. Many proteins for given metal ions have signature sequences which can be followed in the genome.
6. The metals are related to selected activities in enzymes, which can be measured.
7. The metals can be isotopically and/or radiolabelled.
8. The environmental content of metal ions can be related to their internal uptake route in organisms since it is possible to follow transfer using isotopes.
9. The concentration of free elements in different compartments can be related to gradients and energization.

Table 20.11 The networks for individual elements

Element	Network references Sections	Figures
Na^+, K^+, Cl^-	8.10	8.12
Mg^{2+}	9.18	9.17
Ca^{2+}	10.22	10.21, 10.22
Zn^{2+}	11.12, 11.13	11.11
Fe^{2+}	12.17	12.15, 12.18
Haem	13.12, 13.13	—
Mn^{2+}	14.9	—
Cu^+	15.7	15.11
Ni^{2+}/Co^{2+}	16.7	16.7
Mo/W/V	17.5	17.7
P, Si, B	18.2, 18.7	18.2
S, Se, I	19.2	19.7

other words there is a match between geochemical and biochemical evolution which is quantitative, but only in a proportional sense since organisms use energy to control their contents to a much greater degree than does the sea or the crust of the earth. At any period of evolution therefore we shall look at the element concentration in the sea (free ions), and the organisms, free and combined metallomes. Such a relationship is a necessity since organisms and the environment belong to one (eco)- system.

The total metallome must be seen finally as a network of all the elements in all compartments just as the proteome represents the network of all the proteins in an interactive system. The networks for individual elements, even in advanced organisms, have been given in the different chapters, (see Table 20.11). In this chapter we have to show how they combine through communication systems (Section 20.16). We start from changes in the prokaryotes.

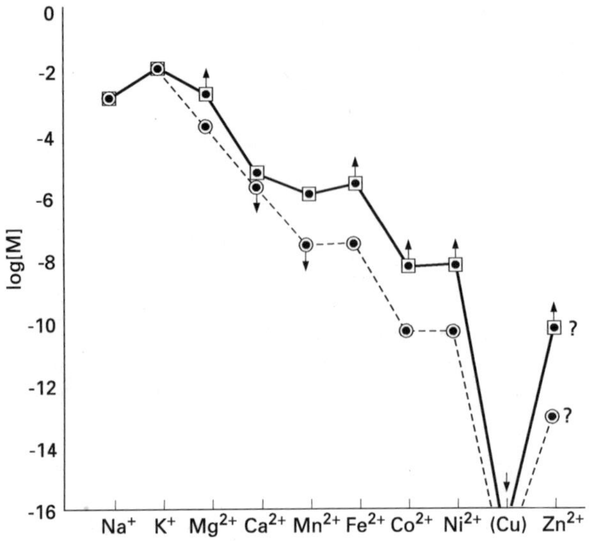

Fig. 20.21 The proposed free metallome profiles of a primitive organism after the synthesis of ring chelates: ○ free metallome; □ combined or total metallome. Arrows (up) indicate the elevation due to the chelates and arrows (down) indicate pumping out.

20.9 Changes of elements in prokaryotes

The outline free metallome as estimated in primitive anaerobic bacteria was given earlier (Fig. 20.14). We believe that the first development of the metallome within the cytoplasm of prokaryotes came with the arrival of new chelating centres for the elements Mg, Fe, Co, and Ni which were in such slow exchange that these new chelates are effectively additional elements—in fact, they are additional coenzymes. (We have already stated that a full description of a cell will have to include the analytical values of the concentration of certain organic small molecules including coenzymes that are related through synthesis to the genome and the proteome but are also connected to the environment, e.g. coenzyme carriers of C, N, H, S, P small units.) These changes represent therefore an increase in the content of Mg, Ni, Co, and Fe in the total metallome but no change in the free metallome (Fig. 20.21). This kinetic 'chemical compartmentalization' allowed the development of new species.

Consider the combination of the organic rings with their metal ions. All the ring systems are made by a common synthetic route from α-aminolevulinic acid giving rise first to uroporphyrin and then to the different rings for the incorporation of Mg, Co, Ni, and Fe. Both Mg and Fe are to be found in more than one ring chelate (Fig. 2.13, and see Fig. 20.22). This system of chelates

Ring chelate	Cobalamin, ←	Heme d_1 Siroheme, (a)	→ F–430	Chlorophyll (b)	Haem (o,b,d_1) (c) →	Haem a (d)
Metal Ion	Co	Fe	Ni	Mg	Fe (Cu)	Fe.Cu
Substrates	RNA → DNA Methylation	SO_3^{2-} NO_2^-	H_2	Light Mn $H_2O \rightarrow O_2$	$O_2 \rightarrow H_2O$ NO (–Cu)	$O_2 \rightarrow H_2O$

Fig. 20.22 A possible picture of ring chelate development is in the sequence (a) (b) (c) (d).

has existed in organisms to this day. The Mg (chlorophyll) and the Fe/haem have been more and more used, while Co (vitamin B_{12}) and Ni (F-430) are now less used or not used at all in some organisms, e.g. higher plants. It would appear that there were two bacterial photosystems (Fig. 5.1) reliant on some of these units, the more primitive being that of green sulphur bacteria which was followed by that of the purple bacteria which became photosystem II in chloroplasts for the evolution of O_2, leading stepwise to the changes of Fig. 20.11.

Notice immediately that all the metals in the rings are linked by the synthesis of the same ring precursors. The exact development is still not clear but it is likely that certain of these synthetic products evolved long before the coming of appreciable amounts of dioxygen. Be this as it may, the iron and magnesium chelates became the centre of extra energy capture, of light, while cobalt and nickel chelates probably took over specialist roles in the metabolism of H_2, CO(Ni) and nucleotides (Co, B_{12}) all of which may have been managed by Fe/S centres originally.

Dioxygen, when it appeared in low concentration, was firstly removed by sulphide to give sulphate, Fig. 20.11, which then could replace sulphur as a source of energy and to give external –S–S– bridges. Note that a particular early haem, sirohaem, is involved here in sulphite reductase. Of course, this and all the above set of new syntheses needed new enzymes and new kinetic controls. The controls over the syntheses rest with iron and phosphates (kinases) which regulate porphyrin synthesis (as does haem itself) and its proteins at the gene level, but note that zinc in α-aminolevulinic dehydratase is also a controlling factor certainly today. The exact relationships of the genes to all the chelates is unsure. The new network is then not fully defined, but all the chelates are undoubtedly connected to one original synthetic path and are in balanced production differently in different organisms with element supply. Here there are clear relationships between metallomes and speciation, related to Mg, Ni and Co contents especially, but in all species there remain necessary connections between the new and the most primitive networks. Note how the separation of Mg as chlorophyll from free Mg^{2+} and of iron porphyrins from free Fe^{2+} allowed the evolution of photosynthetic from non-photosynthetic bacteria separately and eventually plants from animals.

At the beginning it may well have been that Mn^{2+} was also exported from primitive cells with the other larger cation Ca^{2+}. Note that much functioning of Mn^{2+} like that of Ca^{2+} is associated with situations in membranes or outside the cytoplasm (see eukaryotes below) and also that later Mn^{2+} became a much more important element in plants than in animals. There was then little functional use for this element, but like calcium it could bind quite well to the outer surface of a membrane. To make a further speculation once light energy became available using chlorophyll it was possible for photosynthetic layers to use the energy to convert Mn^{2+} to MnO^{2+} and it was this that led to the use of H_2O, not H_2S, to give not just bound H_2 (inside cells) but free O_2 (outside cells) as well (see Section 5.2). We can well believe that Mn^{2+} became an important unit in the genome response once this value of it was discovered in certain plant species.

We see that in evolution the system develops in its combined metallome, in its metabolic chemistry, and in its energy capture cooperatively. The advance in prokaryotes continued with new uses of oxidized elements which we take under eukaryotes with one exception, that of sulphate, since utilization of sulphate, compare chlorophyll and haem, is a novelty relative to the chemistry of sulphide and allowed a new variety of prokaryote to appear.

20.9.1 'Anaerobic' sulphate-using bacteria and protection

We include a short paragraph on sulphate-using bacteria since they may well have arisen before true aerobes of any kind but when some O_2 or H_2O_2 was present to produce sulphate. They illustrate three new features of metallomes—a new source of energy, $SO_4^{2-} \rightarrow S^{2-}$, the need for further novel metabolism and control and the need for protection from O_2 and H_2O_2. Sulphate reduction today uses molybdenum, Fe/S, and haem iron in catalysts. Notice that the availability of sulphate and removal of sulphide took place before the oxidation state changes of any other non-metals but after or closely at the same redox potential as molybdenum (Figs 1.8 and 20.11). Molybdenum replaced tungsten in cells since it has a readily oxidisable sulphide giving MO_4^{2-} from MoS_2 and it became the catalytic centre for the earliest steps in sulphate reduction, see chapter 17. The change in availability of oxygen as H_2O_2 or O_2 and of Mo as opposed to W led to new features requiring the genome to express new proteins. How does this environmental pressure operate at the genome level? We have indicated in our previous books that one possible route arises from the damaging effect of 'poisons' (new elements) in the environment upon proteins and hence the necessary extra production of these proteins. Now new protein production requires DNA to produce RNA and during this synthesis DNA is single-stranded. The single-stranded DNA is open to greater mutation and hence the very proteins to which the poisonous element bound became adjusted by selection pressure to bind and at first to remove this element. However, if possible it is clearly advantageous to utilize the new protein/metal combination outside the cell. For example it looks as if much SO_4^{2-}, VO_4^{3-} MoO_4^{2-} transport is derived from HPO_4^{2-} binding proteins, see Sections 17.6, 17.9.2 and 19.25.

The protective devices of 'anaerobic' sulphate bacteria use Fe in rubreferrodoxin to remove O_2/H_2O_2 coupled to NADH without coupling to energy. Protection from superoxide in them and generally in prokaryotes depends upon either an Fe or a Mn superoxide dismutase. (Later in eukaryotes this enzyme is based on Cu and Zn released from oxidized sulphides.) Control of these protein productions may rest with changes in free Fe^{2+} levels.

Before going further we must mention the new compartment of prokaryotes. The earliest prokaryotes may have had only one aqueous compartment, the cytoplasm, although the cell wall would have been different and would have equilibrated with the element profile of the environment (Fig. 1.5). Later prokaryotes developed with an additional membrane and an outer intra membrane aqueous space—the periplasm (see Fig. 3.11). The periplasm is more or less open to the environment except that large molecules are trapped in it. The cells exported material into this compartment which therefore developed a quite different relationship to the environment from that of the cytoplasm. As the environment changed, for example to become oxidizing, so the periplasm developed with it but the cytoplasm changed little. Much of SO_4^{2-} and MoO_4^{2-} chemistry is not to be found in the cytoplasm. We leave these developments until later since similar developments in compartments introduce the eukaryotes and further novel oxidative chemistry.

20.10 Eukaryotes: the development of new membrane components: new lipids

The oxidation of lipids led to the development of eukaryotic cells. Probably the first major development was the oxidation of squalene derivatives to give cholesterol, which was accomplished by a readily oxidized flavo-enzyme.

Table 20.12 Calcium in evolution

Cellular organization	Calcium function
A. Prokaryotes Single cells.	Externally—wall crosslinks—random mineralization.
	Internally—very slow signal: swimming.
	No message to genetic code except for sporulation under starvation.
B. Early eukaryote single cells. (No organelles).	As for prokaryotes (A) but external events relayed to internal structures. (a) Shape response: contractile devices.
	(b) Metabolic and energy response.
	(c) Increased response from endoplasmic reticullum stores.
	(d) Controlled external mineralization.
	(e) Control over cell death.
C. Late eukaryotes single cells + organelles, e.g. see *Acetabularia*.	As for (B) but (a) Extra response and energy release from mitochondria and chloroplasts.
	(b) Necessary connection to genetic code through phosphate compounds to give growth, see C(a).
	(c) Growth pattern linked to constant circulating calcium current around cell edges: differentiation.
D. Early multicellular animals (nematods). plants are similar but without nerve structures.	As for (C) except (a) link to nerve response as well as chemical response to the environment. (b) Control of circulating fluids inside organism by organs, releasing hormones.
	(c) Connective tissue allows build-up of internal mineralization.
	(d) Protective and immune system organizations.
E. Late multicellular animals with brains.	As for D except (a) the brain is a new plastic organ (unlike other organs) and calcium currents around nerve cell tips of the brain are now linked to local growth to give memory. Apoptosis.

Eukaryote membranes contain cholesterol, prokaryote membranes do not. The oxidation of linear long-chain fatty acids led to arachidonic acid and here ferrous enzymes and cytochrome b_5 are used. [In turn arachidonic acid could become coupled later to novel sensitisation of the calcium message through the in-membrane activation of kinase C (which is calcium and zinc dependent) and then to the prostaglandins (further Fe^{2+} activated oxidation) (see Section 12.11). Notice that the oxidation of fatty acids is somewhat different in animals and fungi compared with plants.] Only after these stages of lipid oxidation could the compartments increase. The very extensive use of the calcium as a messenger then became possible in the eukaryotes (see Chapter 10, Table 20.12, and Fig. 20.23).

Interestingly, the development simultaneously of new membranes with vesicles and internal filaments allowed much larger cells and these cells to accumulate and position vesicles. The mobility of these membranes without

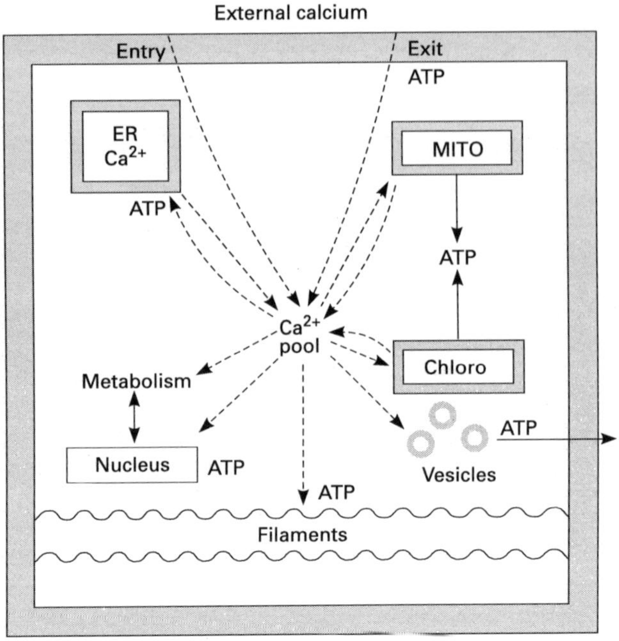

Fig. 20.23 The calcium circuit in an active transient state (see Chapter 10). All the activated proteins have very similar binding constants for calcium. NB ATP is really MgATP.

outer cross-linked walls as in many prokaryotes also allowed endo- and exocytosis of the vesicles stimulated by messages (see the use of calcium again, Section 10.10) and general movement of cells by chemotaxis. The larger cells could respond *locally* due to the positioning of different activities along their outer membranes which now took on diverse shapes. At the same time the total metallome changed (due to the vesicular content, mainly increased calcium) but not the free metallome of the cytoplasm. We have noted that the new uses of calcium are deeply integrated into the whole system through connection to the very primitive phosphate metabolism (see Section 10.22).

20.10.1 The metallome of early 'anaerobic' eukaryotes

Only a very low level of dioxygen may have been sufficient to allow the above modification of cell membranes and vesicles to be formed in cells. These vesicles are almost inevitably inside/out with respect to ion gradients, so that they can contain, by pumping, differentially high levels of Na^+, Cl^-, Ca^{2+}, Mn^{2+}, and SO_4^{2-} but low levels of K^+, Mg^{2+}, Fe^{2+}, and phosphate. They are closer to extracellular fluids than is the cytoplasm. The total free ion content of a eukaryote cell therefore differs from that of prokaryote, as do the combined forms of the elements. As stated before, just as when we consider the proteome it is essential to describe the metallome separately for each type of vesicle. With the incorporation of organelles (captured prokaryotes) mitochondria in most cells and chloroplasts in some cells, further changes in the total bound and free elements occurred. The further division of plants from animals followed and the fungi also developed a new branch of evolution and of chemistry (see Fig. 20.19). Here we wish to stress first one particularly overridingly important early and continuous change in all the total metallomes—the ever-increasing value of calcium in communication, (Chapter 10 and Table 20.12), and in the ability of eukaryotes to manipulate this element in vesicles to give much better protection in the form of external mineral structures.

It is probably not necessary to remind readers first that each time such a new element is pressed into service it can only be used if specialised new proteins are produced, and this means not only new sequences in the genome but new feed-back controls linking genome, proteome, and metallome to the environment in order to control production. For an early eukaryotic cell which was effectively anaerobic few further changes in the proteome were involved other than for the creation of new compartments and their contents, except those proteins connected to calcium and their integration into the whole system. Now these calcium-binding proteins have to be selected to match the thermodynamic limitations on systems as we shall stress in the next section. The whole range of calcium-binding proteins inside the cytoplasm responds to 10^{-7}M free Ca^{2+} since this is the inverse of the binding constant useful to all of them in *the activated state*. As we mentioned in Section 20.3.3, a condition of a homeostatic state is that an element has a fixed binding constant to all the ligands in a given compartment in particular state. The calcium signal knits the primitive cytoplasmic system through the earlier magnesium/phosphate reactions of kinases and even to the iron–zinc phosphatases such as calcineurin. It links in a novel way to mitochondria and chloroplasts, (see Table 10.1). However, it communicates to quite different new proteins in the new internal vesicles all with binding constants of $10^{3.5}$ M^{-1}. The calcium system also sensitizes the whole organism to the environment, $[Ca^{2+}] = 10$ M^{-2} in the sea. This is an essential feature of a full flow system development; the corresponding proteins for all this activity may well have evolved from proteins for protection from Ca^{2+} in prokaryotes due to the extra stresses Ca^{2+} imposes in eukaryotes, see Chapter 10.

20.10.2 The aerobic single-cell eukaryotes and their metallome

The gradual increase in the level of dioxygen gave rise next to changes in availability and oxidation states of the elements in the sea (Figs 1.13 and 20.11). These slow changes interacted with cellular development and have caused the greatest changes in evolution. Very early changes have already been described in the inorganic chemistry of iron, manganese, sulphur, and molybdenum in the organic chemistry of membrane lipids and the use of calcium in signalling. We follow next, as best we can, the further changes of the metallome on top of those of the early vesicular contents of basically anaerobic eukaryotes (see Tables 20.13 and 20.14). We should add that the presence of sulphate and certain sugars now in certain vesicles of the eukaryote cell, the Golgi appara-

Table 20.13 Major element availability changes with the advent of dioxygen*

Non-metals						
	H_2↓					
CO_2↓	N/O↑	O_2↑				
		S/O↑	Cl^-(↑)			
		Se/O↑	Br^-(↑)			
			I^-(↑)			
Metals						
V/O↑	—	—	Fe/O↓	—	Cu↑	Zn↑
-	Mo/O↑					

*Indicates increases in availability. The formulation N/O↑ indicates an increased presence of nitrogen oxides while Cl^-(↑) indicates an increase in covalently bound chlorine. (CH_4 and NH_3 disappeared); (↑) for the halides represents changes in reactivity only; S and Se changed to oxyanions.

Table 20.14 New element biochemistry after the advent of dioxygen

Element	Biochemistry
Copper	Most oxidases outside higher cells; connective tissue finalization; production of some hormones, dioxygen carrier, N/O metabolism
Molybdenum	Two-electron reactions outside cells; NO_3^-, SO_4^{2-}, aldehyde metabolism
Manganese	Higher oxidation state reactions in vesicles, organelles, and outside cells; lignin oxidation (note especially plants); O_2 production
Nickel	Virtually disappears from higher organisms
Vanadium	New haloperoxidases outside cells
Calcium	Calmodulin systems; γ-carboxyglutamate links; general value outside cells and in cell-cell links
Zinc	Zinc fingers connect to hormones produced by oxidative metabolism
Selenium	Detoxification from peroxides, de-iodination?
Halogens	New carbon-halogen chemistry, poisons, hormones
Iron*	Vast range of especially membrane-bound and/or vesicular oxidases; peroxidases for the production of hydroxylated and halogenated secondary metabolites; dioxygen carrier and store

*Note that iron can act as an agent responding to redox potential changes in the cytoplasm.

tus, gave rise to external glycosylated sulphated proteins and polysaccharides through export from the vesicles to become part of the outer filamentous structures of cells.

The next change in the metallome was probably the rise in free zinc to 10^{-11}M, still very low, in the cytoplasm following zinc release from its sulphide to the environment by the oxidation to its sulphate. Zinc became dominantly useful in proteins for export (hydrolytic enzymes) to aid digestion, in protection, and in internal transcription factors. This expansion of the role of zinc is described in Section 11.13. Somewhat later there arose a great increase in the available copper as its sulphide was oxidized, and evolution took advan-

Table 20.15 Zinc and copper enzymes of the external matrix

Enzyme	Metal	Activity
		Hydrolysis of
Elastase	Zn	Elastins
Collagenase	Zn	Collagens
Stromelysin	Zn	Proteoglycans
Gelatinase	Zn	Gelatins
Protease 24.11	Zn	Peptide hormones
		Oxidation
Lysine oxidase	Cu	Cross-links collagen and elastin
Laccase	Cu	Cross-links via phenols (plants)
Phenol oxidase	Cu	Cross-links via phenols (plants)
Amine oxidase	Cu	Removes amine hormones
Glycine oxidase	Cu	Amidated peptide hormones

tage of this new redox catalytic element to produce further new metabolism, now oxidative, outside cells and in vesicles (see Chapter 14). We see this development in both aerobic prokaryotes, usually in the periplasm, and single-cell eukaryotes in vesicles, e.g. in yeast. We also observe increased protection from superoxide in all eukaryotes with a new copper–zinc superoxide dismutase in place of the Mn or Fe dismutases of prokaryotes (see Section 15.5). Simultaneously the concentration of available iron fell as free Fe^{2+} was oxidized to precipitated $Fe(OH)_3$. Our interest first is then in the linked metallome change that is particularly with that of iron, copper, and zinc (see Table 20.13) but it will centre shortly upon their interacting value in the *extra-cellular* metabolism of multicellular organisms, demonstrating the integration of the new aerobic system. We therefore note the interactive Zn and Cu in external developments (Table 20.15) together with the management of iron. They illustrate the many changes in use of other elements which have to be cooperative in the whole cell. The emphasis on Cu and Fe also stresses that they became the dominant elements in cell redox chemistry at high potential. The cytoplasmic link between iron and zinc and then to the oxidative state of the cell can be examined as an example in the Zn, Fe phosphatase, calcineurin, which is activated by calcium. The link between copper and iron can equally be seen in certain uptake systems for iron, FET-3, and also in cytochrome oxidase, in Chapters 12, 13 and 15.

We look now at the problem that oxidation created for the cytoplasm, especially for free Fe^{2+}. Oxidation of it led not to increased but to greatly decreased availability as Fe^{2+} was oxidized to precipitated Fe^{3+}. Iron availability fell by a factor of 10^{10} but the metallome-free Fe^{2+} value in the cytoplasm hardly changed from 10^{-7} M. This remarkable homeostasis shows that the cytoplasmic activity of life has always relied upon a fixed tight control of composition. One element, once incorporated, cannot replace another at all easily, much though some elements in the periodic table may have similar properties. Organisms of all kinds overcame the problem of the source of iron by scavenging, carrying, buffering, and reducing (see Chapter 12). Now the external scavengers and carriers had to have molecular properties opposite this reduced availability. We observe that the binding constants for Fe^{3+} (at pH = 7) outside the cytoplasm must now exceed 10^{+17} M^{-1} (the free Fe^{3+} concentration is 10^{-17} M) but must not be so strong that, on capture of the scavenging complex, in a vesicle for example, Fe^{3+} cannot be released and reduced. (Note that many scavengers are produced by oxidative chemistry.) The binding constants are then systematically adjusted inside the vesicle. The final protein pump from the vesicle to cytoplasm, NRAMP-2, has a binding constant for Fe^{2+} of about 10^7 M^{-1} on the cytoplasmic side (see Section 12.3) to match the free metallome level of this metal. The matching of binding constants to required solution conditions is the essence of the feed-forward and feed-back homeostasis of activity and applies to each element in turn (see Sections 20.3.2 and 20.5). Note that the external transport protein for Fe^{3+}, transferrin, appears to be derived from that for HPO_4^-, SO_4^{2-} and MoO_4^{2-}, Section 19.25.

The difficulty evolution suffered was not only that the fall in available Fe^{2+} [and of elements such as $S(H_2S)$, $Se(H_2Se)$, $N(NH_3)$] and the rise in initially poisonous elements such as Zn, Cu, Cd, and Pb, slow and continuous, but was such that organisms had to be continuously modified in their production of proteins (and DNA) to meet these demands. The eukaryotes (and some prokaryotes) produced first protective buffer and carrier proteins for zinc and copper so that free levels in aerobes were kept very low, (Fig. 20.24). As in the

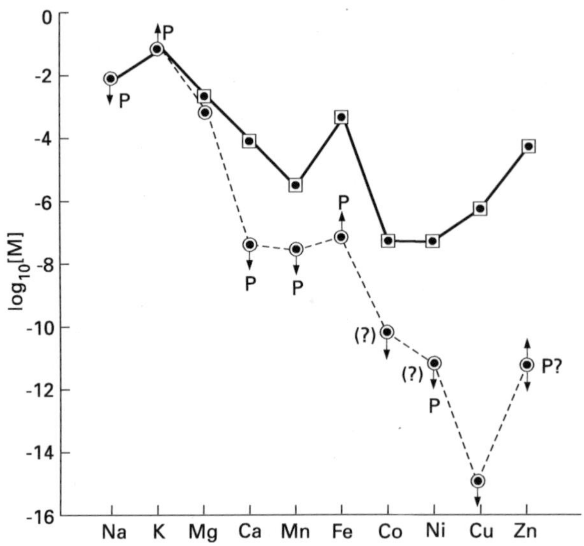

Fig. 20.24 The proposed free metallome profile of the cytoplasm of the aerobic cell ○ and of the combined metallome of the organism □. P represents pump.

case of iron the cytoplasmic conditions changed but slightly while special systems for uptake or rejection and then for use, especially in vesicles or externally, arose which had to be connected back to the central coded DNA in the nucleus (for proteins), to energy and to other metabolic paths. How was this managed in the cytoplasm and in the vesicles? One very important feature of the homeostasis of the vesicles must now be restated. Just as in the cytoplasm, homeostasis in vesicles can only be effective if in each compartment separately free ions equilibrate or bound molecules communicate by exchange. A good example mentioned already is that external and vesicular calcium binding proteins must have a binding constant close to $10^{3.5}$ M in man's body fluid (see Section 10.18). This equilibration is also necessary for the pump-binding inside the vesicles but is connected on the cytoplasmic side to the calcium (10^{-7} M) and then indirectly to the genome so that the calcium proteins for the vesicles as well as the cytoplasm are produced in the correct amounts (see Sections 2.10 and 20.3.3). The basic requirement rests in transcription factor dependence on binding constants, e.g. of phosphatases and kinases, which must match those of buffers, carriers, pumps, and all exchanging sites in the comportment. However, for tightly bound elements such as copper and perhaps zinc, simple ion exchange is not possible and we must consider carriers. We ask how did appropriate proteins arise?

20.11 **Carriers**

A major recent realization is that several metal ions require transport systems in addition to pumps in cells, especially large cells, as well as in extracellular fluids (Table 20.16). These transporters, chaperones, are proteins which combine selective binding with the metal ion to taking it to or from pumps and delivering it to enzymes or transcription factors. Thus the free metal ion may not be present except at very very low concentration and does not exchange directly or diffuse in significant quantity. (NB We can still give binding constants for these associations). The transporters outside cells in multicellular organisms have been recognized for a long time for elements such as P, S, Se, Mo, Ca, Fe, Cu, and Zn. More recently a series of intra-cytoplasmic trans-

Table 20.16 Some carriers and pumps in eukaryotes

Element	Carrier	Pump
H^+	Many	H^+-ATPases
Na^+, K^+, Cl^- Mg^{2+}		ATP-ases
Ca^{2+}	Calbindin	Ca^{2+}-ATPase
Mn^{2+}, Fe^{2+}	SMF-1	NRAMP-1 and -2 (DCT 1)
Fe^{3+} (Mn^{3+})	Transferrin Siderophores	
Haem	Hemopexin	
Zn^{2+}	Albumin	
Ni^{2+}	Histidine	NIC 1
Cu^{2+}	Albumin	
Cu^+	ATX	CCC 2
Mo	MOD A, MOD C	MOD B
O_2	Myoglobin Haemoglobin	

porters have been found (see Table 20.16). Effectively these transport proteins in cells also prevent injury to the DNA and other element sites from metal ions which were poisonous to the earliest organisms, e.g. Ca^{2+} and Cu^+, while allowing distribution. We note also certain pumps for ions (see the chapter on given elements for further details). The combination of pumps and carriers is also an essential part of communication as are transmitters and hormones (see section 20.16). The transporters can be considered to act as conductance channels for the circuitry of individual elements, but observe again that whether or not the element is transported or diffuses freely, there is a thermodynamic binding constant typical of a given element with its given compartmental proteins which can be different in different compartments. This means that there are separate effective cytoplasmic and vesicle, free and total metallomes.

The simple consideration of the thermodynamic binding constants of model complexes and the requirement for *selective* binding by ligands led us to an expectation for the free levels of ions in the cytoplasms (Fig. 2.10) given the range of potential donors. We know from earlier Chapters 8 to 19 that the cytoplasmic proteins generate a similar set of ligand binding sites and we are then not surprised that the cytoplasmic free metallome is closely similar to that expected from the models (see Figs 20.14 and 20.24). In extracellular fluids the range or proteins is very different and the free metallome also differs greatly. We believe that the distinction arises because the cytoplasmic metallome reflects the possibility of a living flow system under primitive reductive environmental conditions while the metallomes outside the cytoplasm are related to later oxidative and more specialized conditions.

Again, not all elements exchange rapidly between bound chelated states and free element pumps or carriers. However, their bound levels must be maintained just as much as their free exchanging levels. Cases in point are the levels of haem, chlorophyll, and vitamin B_{12}. In these cases and within a compartment the bound carriers or pumps must be under separate controlled synthesis as well as connected to the free element concentration. As stated this synthesis is still controlled by the free ions; for example, the synthesis of all the ring chelates for Fe is dependent on FNR, a protein which is a transcrip-

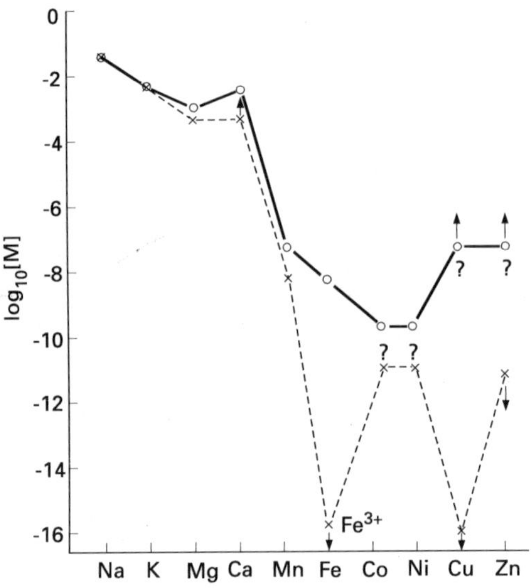

Fig. 20.25 Estimated free and the total metallome profiles of the extracellular fluid of a multicellular animal such as man: × free, ○ total.

tion factor and exchanges with free iron (Section 12.17). However, units such as haem, in fact coenzymes generally, must also bind to carriers, pumps, and transcription factors with particular constants related to the compartment of residence and their cytoplasmic homeostasis must also be maintained.

20.12 The metallome of extracellular and vesicular fluids of multicellular organisms

The next step in organization in evolution is that of multicellular organisms with organs. The first requirement was for the synthesis of connective tissue and for a communication network between cells to maintain organization. We shall follow in this section the parallel changes in the metallome of the compartments of the cell and in particular that of the extracellular fluids (Fig. 20.25). The development here has reached its most delicate point so far in higher animals where the levels of Na^+, K^+, Cl^-, Mg^{2+} and Ca^{2+} are all delicately poised so that diseased states can be readily recognized by a 1 or 2 per cent change in circulating values. However, the extracellular fluids of a huge variety of plants and animals show that in the adult organism all these five ions are managed homeostatically even if not so carefully. Moreover, proteins in the extracellular fluids, if they bind these ions, have binding constants, K, close the inverse of the free ion concentration, i.e. close to 10^3 M^{-1}, opposite millimolar Mg^{2+} and Ca^{2+} in such proteins as fibrillins. The same constants are found for activation of the processes of blood-clotting, immune response, digestion, and so on in these fluids. We showed earlier that a sensitive equilibrium response of a feed-back control required this relationship between free and bound ions (Section 20.3.3). Note that all these ions exchange through pumps and channels with free ions in the cell and can again be managed by direct messages through membranes to the DNA in the cytoplasm.

There is a different but equally important balance in the complement of elements such as Cu, Zn, and Mo as well as of SO_4^{2-} and HPO_4^{2-} and probably several others in the extracellular fluids. We concentrate attention first on the dominant importance of Cu, Zn, Ca and SO_4^{2-} for connective tissue.

Copper, as stated in Chapter 14, and zinc, as discussed in Chapter 11, are central to the cross-linked synthesis (Cu) and its hydrolysis (Zn) of chitin and collagen. Thus the whole of *multicellular growth* is predicated on the uptake and management of Cu and Zn enzymes in the extracellular fluids (see Fig. 20.29 and see also Fig. 20.25 and Table 20.15). The balance between Cu, Zn, and Ca is also shown in Fig. 20.29. The value of zinc is extended within feedback systems to that of the synthesis of the proteins for the management of the extracellular calcium proteins and of the levels of Na, K, and Cl all of which are set by sterol hormones which act at zinc transcription factors (zinc fingers). Of course, it is well known in nutrition studies that zinc, copper, and calcium levels are necessary for the maintenance of connective tissue and of mineralization. We shall return to the link between the hormone sterol levels and those of free metal ions in Section 20.16.

We observe that the multicellular organisms have differentiated cells in different organs and each organ and each differentiated cell has its own metallome, much as it has its own proteome. While noting the difficulty in describing all the compartments, linked in a homeostatic system, we must also observe that there is an ever-increasing variety of calcium, zinc, and copper proteins in extracellular fluids, see tables in Chapters 10, 11, and 15. There is, of course, an increasing variety of all proteins opposite oxidatively managed metabolism and opposite communication. Undoubtedly this variety matches the great variety of species.

The enzymes of zinc and copper are also deeply connected with the synthesis and degradation of small signalling peptides, (see Section 20.16). To maintain whole organism homeotasis it is necessary to have a central organising unit and messengers, which in animals becomes the brain.

20.13 **The metallome of brain extracellular fluid**

The last step in the advancement of organs is the brain, which requires a novel extracellular environment, the cerebrospinal fluid. In a previous book we have

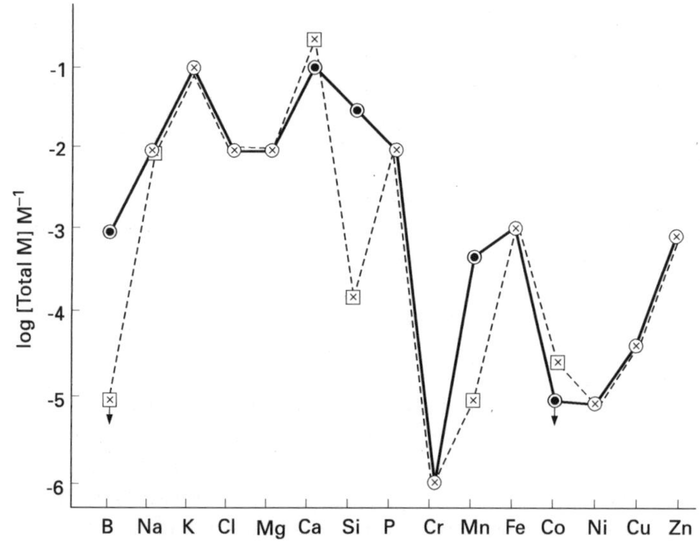

Fig. 20.26 The approximate total element content of advanced plants ⊙ and of animals, ⊠, ⊗ where they are nearly equal.

given some indication of the nature of the brain and have shown how this organ, which, unlike the others, is plastic, developed in evolution, and how even with age of individuals its metallome shows considerable changes. Much future research is needed to allow us to know how the growth patterns of the brain relate to metal ion content. Again, we stress the relationship of the brain's activity to ionic and hormonal messages (see Section 20.16.1).

20.14 Summary of metallome content

It is clear that the changes of the elements in the environment introduced severe stresses upon original (cytoplasmic) cellular life. The living system has responded in two ways. It has increased its protection from the environment and it has utilised environmental change to increase organization in compartments. This has led to the diversification of living species which clearly requires mutual dependence in a vast ecosystem. Strikingly, we can follow these changes and the dependencies by following *quantitatively* the concentrations of metal ions (and non-metal small molecules) in cells. Thus a clear distinction is seen between plants and animals, (Fig. 20.26). This study is unlike that of following single functions in proteins or RNA, or of following sequence changes of DNA. The *quantitative* changes in metal ions indicate the introduction of novelty in the organization (systems) of organisms. There is a parallel change in the values and uses of the non-metals, of course, which we describe next. We shall not elaborate on the change in the element oxygen itself which was instrumental in the major changes during evolution as described above. However, we must remark again that its activity as O_2 or H_2O_2 generated great stress on the primitive reductive system, especially in the cytoplasm. Apart from providing the possibility for change, this stress, still present today from all highly oxidized chemicals as well as O_2, is the source of disease. Dioxygen, hydrogen peroxide, nitric oxide, and similar chemicals are of great value in living organisms but are accompanied by great risk. Use of these chemicals has developed with protection from them and their employment as weapons against invading cells. In these respects O_2 and H_2O_2 are typical of many a chemical introduced since primitive times, e.g. copper.

20.15 Changing non-metals in the proteome and in small organic molecules

A general impression of the changes of the non-metals in the environment, with some changes of prokaryotes, during evolution is given in Fig. 20.27. We shall not be concerned with the major non-metals H, C, N, O, in this book. The proteome contains some proteins which have atoms in addition to H, C, N, O, and here we note P, S, and Se, to which we have referred in Chapters 18 and 19. These elements give specific signatures in the mass spectrometry of the proteome and, if selectively radiolabelled, in its chromatography. It is important to observe that these special features like the binding of a selected metal or a coenzyme in a protein allow part of the proteome to be followed easily during changes in cell growth. For example, this can be especially useful in the case of selenium, the levels of which appear to be critical to the health of humans. However, to define non-metal cell composition we must also describe the levels of small molecules which contain non-metals. These have advanced little in evolution in the cytoplasm, for

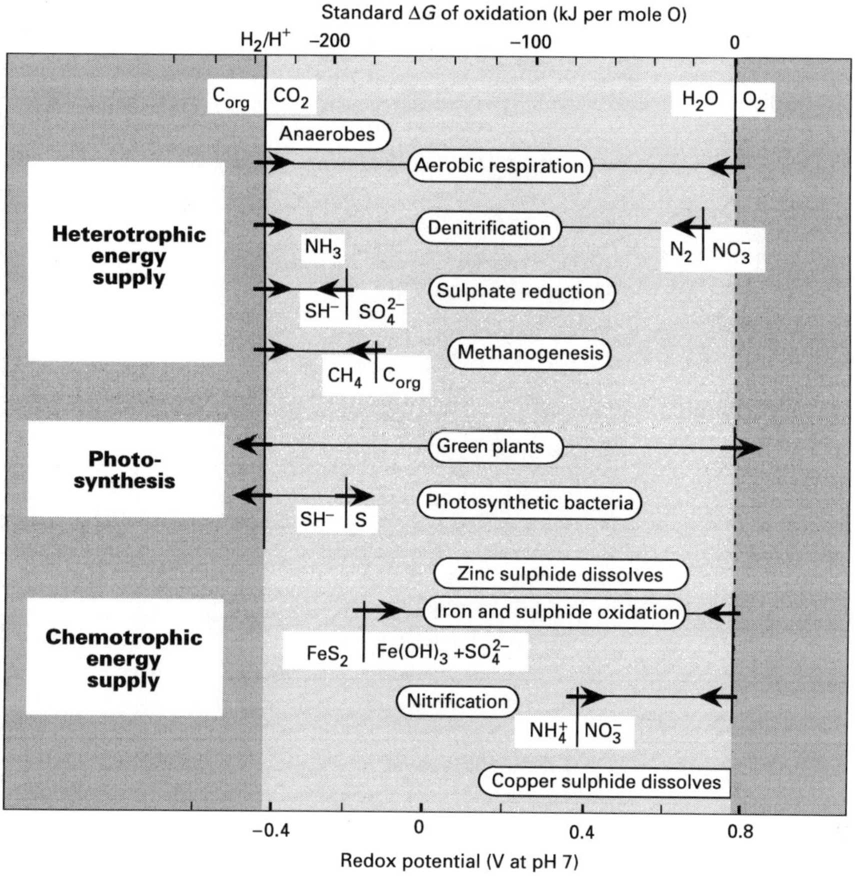

Fig. 20.27 The redox potential ranges of certain organisms and simple reactions allowing concentration changes greater than 10^{20} in element speciation. [From Cox, P.A. (1995). *The elements on earth*. Oxford University Press, Oxford.]

example not in the range of nucleotides or coded amino-acids, but have advanced outside the cytoplasm in the range of chemical modifications of amino-acids, and with the inclusion of saccharides (hexose-based) especially in sulphated forms. The presence of the more oxidized units is of relatively recent origin (Table 20.17). Of course, the number of proteins of a given kind has also greatly increased (Table 20.18) in all compartments, but truly novel proteins are rare in the cytoplasm except in messenger systems. (Of course, all proteins have undergone some amino-acid substitutions and some change in sub-unit structure.) Especially interesting is the use of –S–S– bridges and other organic cross-links due to oxidation outside cells.

Table 20.17 Examples of oxidized proteins and saccharides

Bone proteins	Gla-containing
Blood-clotting proteins	Gla-containing
Connective tissue	Collagen hydroxylation of proline and lysine Chitins and lignins—oxidized phenolics
External polysaccharides	Sulphated saccharides

Table 20.18 Metallo-protein sets in two organisms

Protein set	**Yeast**	**Worm (*C. elegans*)**	**Metal**
Nuclear hormone receptor (Zn)*	0	270	Zn
Binuclear GAL cluster (Zn)*	54	0	Zn
Metalloproteases	0	94	Zn
Na^+ channels*	0	28	Na
Mg^{2+} adhesion*	4	43	Mg
Calmodulin-like proteins*	4	36	Ca
K^+-Channels (voltage gated)*	1	68	K
EGF, Ca^{2+}-binding cysteine-rich repeats*	0	135	Ca
Kinases (tyrosine)	15	63	Mg
Cytochrome P-450	3	73	Fe

*Absent in bacteria. Note that values are also available for man and arabidopsis.

After S.A. Chervitz *et al.*, *Science* (1998) **282**, 2022–7, see also K.N. Deotyarenko and T.A. Kulikova (2001), *Biochemical Society Transactions*, **29**, 139–47.

Although there has been only some change in the coenzymes (different for archaebacteria) and vitamins have replaced self-synthesized compounds, there has been an enormous diversification of oxidized organic compounds to be used as messengers, e.g. adrenaline and other hormones such as steroids, and their synthetic apparatus and receptors in higher animals and plants. We turn to the uses of these messenger molecules in the next section. Notice that most of the changes to messengers took place in vesicles or in extracellular fluids. In this book while we draw attention to the organic messenger systems we note a huge increase in the number and variety of required proteins incorporating individual metals, especially Ca^{2+}, Zn^{2+}, Cu^{2+}, and Na^+/K^+, for the synthesis, degradation, and detection of these messengers which have been documented in earlier chapters (see Table 20.19). Clearly, there have also been changes in structural proteins in the cytoplasm and outside it, (Table 20.20).

A critical change for all the biological non-metals was that of the oxidation states available to organisms as the dioxygen concentration in the atmosphere increased. We have noted the changes of oxygen, nitrogen, and carbon in Fig. 20.28, and especially of sulphur toward sulphate, of selenium toward selenate and of iodide ion toward covalent iodine in Chapter 19. The incorpo-

Table 20.19 Examples of the evolution of metalloproteins associated with messengers

Messenger	**Novel associated proteins**
Na^+/K^+	ATP-ases in animals
Ca^{2+}	Annexins, calmodulins, S-100
Peptides	Peptidases and proteases (Zn^{2+})
Sterols	Zinc fingers
Thyroxine	Zinc fingers
Retinoic acid	Zinc fingers
Amidated peptides	Glycine oxidase (Cu^{2+})
Adrenaline	Phenol oxidases (Fe^{2+}, Cu^{2+})

Table 20.20 Protein evolution: examples

Protein type	Eubacteria and Archaea		Eukaryotes
	Anaerobes	Aerobes	
Tubulins (Mg)	–	–	+
Actinomyosin (Mg, Ca)	–	–	+
EF-hands (Ca)	–	–	+
Oxidases (Cu)	–	+	+
Microfilaments (Mg)	–	–	+
Integrins (Mg)	–	–	+

NB Bacteria have no enzymes for driving mechanical changes in filaments using ATP.

ration of the higher oxidation states allowed remarkable changes in the chemistry in the aerobes relative to the chemistry of the anaerobes, but again these changes had to be largely outside the cytoplasm. We note especially the production of sulphated saccharides in connective tissue; of sulphoxides outside coccoliths; of seleno-peroxidases and of iodinated molecules, thyroxines, in vesicles largely outside cells. Clearly, another way to analyse evolution is to follow the appearance of oxidized non-metals in fossils. This has been done in part for sulphate, using sulphur isotopes. A general impression of changes of non-metals during the course of evolution is given in Fig. 20.28. Clearly all these mainly extra cytoplasmic developments of both non-metals and metals required new extracellular organic communications networks and proteins.

20.15.1 Changes in non-metal metabolism

We have stressed that the basic metabolism of life could not be changed, especially in the cytoplasm. There are fixed sets of bases, amino acids, saccharides and lipids. The minor modifications of these units have been described. However oxidation allowed a great diversity of new small molecule metabolites to be produced, often related to the essential monomers but also many others, related to energy metabolites, e.g. of the Krebs cycle. Thus there are for example sterol hormones, modified peptides, but also alkaloids. Many of the associated enzymes which use metal ions have been described. The major concern in this book is with the function of these new metabolites. Outstanding is the value of them in communication networks as hormones and transmitters, often to be ejected from vesicles. Of course we have to include also the consumption of many organic monomers by oxidation to yield energy.

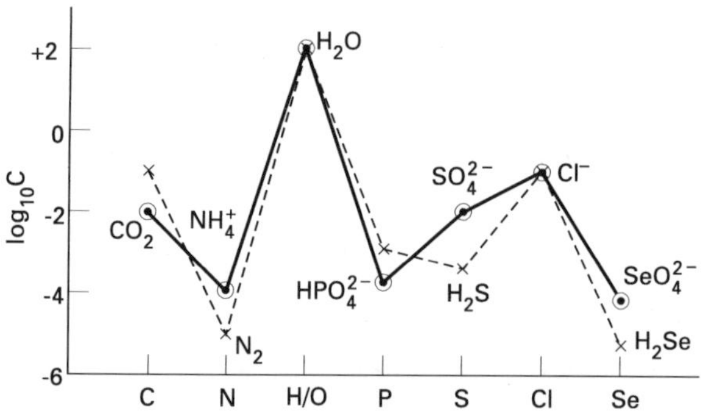

Fig. 20.28 The estimated non-metal element content of the primitive (×) and the modern (⊙) sea.

Table 20.21 Primitive messengers

Messenger	Functional control upon
Mobile coenzymes	Distribution of metabolic fragments H^-, CH^-_3, -$COCH_3$, etc.
Nucleotide triphosphates, Mg^{2+}	Distribution of energy
Fe^{2+}, $2RS^-/(RS)_2$	Distribution of electrons
	Redox state balance
Some simple substrates (feed-back)	Metabolic products, e.g. glutamine, nucleotide bases, amino acids, and upon gene expression
Phosphorylation of proteins, Fe^{2+} (Mn^{2+})	Gene expression

20.16 Extracellular communication networks

In primitive anaerobic prokaryotes, the earliest cells, we have noted that the fixed free metal composition (the metallome) and a given set of small molecules and proteins (the proteome) imply a feed-back system of controls linking all the chemical components of the cytoplasm. The original system in the cytoplasm is based largely on equilibrium binding or fast exchange using mobile ions, Mg^{2+}, Fe^{2+}, H^+, and HPO_4^{2-} in various combinations, and on substrates and coenzymes as messengers amongst pathways, genetic centres and transport routes (Tables 20.1 and 20.4). Hence they are in homeostatic states. All evolutionary changes in cellular activities required incorporation of new pathways but demanded also that these pathways were linked to the primitive system, which had to be almost unchanged.

As we have described in Section 20.5, the first simple change in activity from the most primitive prokaryotes was in the effective number of elements by incorporating four of them in non-exchanging ring chelates. The *internal* messenger control link for them was then in part through the elements themselves, Fe, Co, Ni, and Mg, which were primitive messengers, and in part through the direct use of these new factors in transcription factors. (Note that two of the four have disappeared from advanced plants while in animal cells one has been lost and one has become a vitamin.) Further changes occurred in eukaryotes with advance in number of compartments which needed to send messages back to the cytoplasm and thence to DNA. Here *external* Ca^{2+} oper-

Table 20.22 Oxidized organic messenger chemicals

Messenger	Source
NO	Arginine
Sterols	Squalene (cholesterol)
Adrenaline	Tyrosine
8-OH Tryptophan	Tryptophan
Amidated peptides	Terminal glycine of peptides
C_2H_4	(?)
Plant hormones	Indoles
Retinoic acid	Retinol (vitamin A)
EGF-peptides	Hydroxy aspartate

Note the receptors dependent on $Cu(C_2H_4)$ or Zn (sterols, etc.) and the metabolism of peptides dependent on Zn.

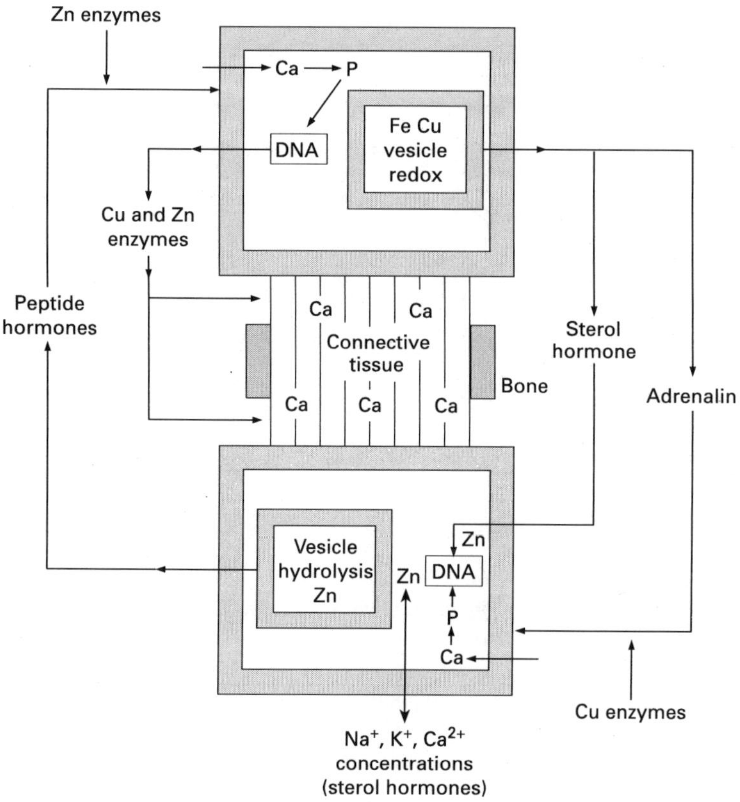

Fig. 20.29 An illustrative diagram to show the complexity of communication networks in a multicellular aerobic organism. Note that 8 metal ions and 4 classes of organic molecule are involved. Mg is active everywhere (not shown).

ated very effectively across membranes in a coordinating network with the primitive H^+, Fe^{2+}, Mg^{2+} and phosphate cytoplasmic systems, probably before dioxygen had increased in the atmosphere to any marked degree. However, the big challenge came with the increased presence of dioxygen and multicellular organisms which demanded a complete new scheme of cell to cell messenger systems in the new aerobic atmosphere. Resort was had to organic oxidative chemistry in the synthesis of messengers as much in vesicles or outside cells as was possible, (Table 20.22). This coincided with (1) the *external* oxidative chemistry for modifying amino-acids, (2) the use of hexoses and sulphate in polysaccharides rather than reducing them in the cytoplasm (or oxidizing them completely in mitochondria), and (3) oxidative cross-linking of the external polymers (Table 20.15). All this oxidative activity required new chemistry of iron and copper. In turn the new essential messengers between the cells had to communicate to the cytoplasm. In some cases the solution was simple in that, acting on membrane receptors, the messengers released calcium to the cytoplasm. Molecules more resistant to reductive destruction, such as sterols, entered the cell and were linked to DNA by a new internal system of a non-oxidative kind, the zinc finger transcription factors. (Remember that zinc was now available through *oxidation* of its sulphide.) Again, zinc could be used to manage a further new set of messengers—peptides, which in turn connected to the interior of cells through calcium (see Fig. 20.29). It may well be that the cytoplasm now included an effective level of free zinc (10^{-11}M) which connected all the zinc exchanges in this compartment including pumps, buffer (metallothioneins) and transcription factors (zinc fingers).

Table 20.23 The evolution of messengers to DNA

Organism	**Messengers**
Primitive anaerobic prokaryotes (Connection to the gene and between pathways of the cytoplasm (a)	Phosphate compounds Some substrates and coenzymes Fe^{2+}, Mg^{2+}, Mn^{2+}?
Advanced anaerobic prokaryotes (As above (a))	As above, plus haem
Anaerobic eukaryotes As above (a) + environment Ca^{2+} (b)	As above plus Ca^{2+} via phosphates.
Aerobic eukaryotes As above (a) + (b)	As above, plus initial responses to dioxygen use; Zn^{2+} and Cu^{+}.
Multi-cellular organisms	As above, plus new organic molecules and receptors (Zn^{2+}, Fe^{2+}, Cu^{+})
Animals with nerves and brains	As above, plus $Na^{+}/K^{+}/Cl^{-}$ conductance.

However the concentration of extracellular or vesicular metalloproteins cannot be determined by direct message feedback to DNA, where ultimately the control over synthesis resides. Since many of these proteins do not exchange metal ions, levels of these ions also cannot be used as a feed-back message. Since all proteins are continuously if slowly destroyed, the only way to manage homeostasis externally is to utilize external conditions related to the activity of the proteins in the cells to produce signals. These signals of activity have to reach the DNA of cells of many organs. The solution to this problem has been found in central glands which produce hormone messengers in external fluids according to the stress the organism experiences. Thus physical conditions such as external light, heat, or even pressure act through monitoring glands which circulate messages to cell receptors. Chemical conditions such as food, recognized invasion by foreign organisms, perceived risks, etc., also produce messengers which cause changes in both short-term and long-term activity down to the level of protein synthesis throughout an organism. Hence the gland/hormonal, endocrine, system monitors the whole external flowing system while interacting locally and differentially within each cell cytoplasm. Here we stress that very much of this hormonal activity is related to organic chemistry which could only exist using metal ions in oxidative conditions especially, that is, the elements copper, zinc and iron. They are involved both in hormonal production and in enzymes responsible for subsequent activity, including growth and morphology, and for the destruction of the hormones. We cannot go further into detail here but the reader can follow the developments in Table 20.23 and in Chapters 10 to 15, or in a more sophisticated account in our book '*Bringing chemistry to life*' (see 'Further reading'.).

20.16.1 **Nerves and brain**

As we have explained in our previous books, the final advances today of multicellular organisms, animals, required fast responses conveying long-range messages from and to muscles and nerves. These messages are described in Chapter 8 and are based upon the novel $Na^{+}/K^{+}/Cl^{-}$ currents, pumps and channels, assisted by Ca^{2+} (perhaps Zn^{2+}) and several organic

Table 20.24 Evolution of simple ionic equilibrium signals

Stage 1	Primitive organisms Prokaryotes	$Mg^{2+}/ATP^{4-}/HP^{2-}$ controls phosphorylations Fe^{2+} controls redox equilibrium $Na^+/K^+/Cl^-$ control osmotic pressure
Stage 2	Single-cell eukaryotes	Ca^{2+} controls activated states and relationship to environment. Mn^{2+} controls development of plant-related organisms.
Stage 3	Multicellular	Zn^{2+} controls hormonal responses relating to growth of organisms and development and connective tissue. $Cu^+(Cu^{2+})$ controls connective tissue and responses. Extended use of Ca^{2+} in excited states. Generation of Na^+ (K^+) signalling and evolution of the nervous system.

transmitters. The final step in this development is the brain. Thus the receipt of information by the brain results in fast ionic responses and can also be followed by slow glandular messages to the whole body (see Section 20.16). Thus the brain contains a *second code*, not linear nor strictly inheritable, in the form of stored electrochemical energy gradients (as in a computer). This code responds immediately to experience, i.e. to the environment, unlike DNA, and can be passed on, and 'inherited', by external communication using additional codes in books and in computers. Use of it has increased survival strength, especially of mankind. We shall not repeat our earlier descriptions here (see Section 8.10, but see also Table 20.24 and our book *Bringing chemistry to life*). It was the changing environment, availability of elements, Table 20.25, which made the initial development of the brain possible just as further exploitation of elements by man using his brain allows the new evolution of the environment not dependent upon DNA changes. Notice particularly how the development of element use is coincidental with novel protein synthesis (here Na^+/K^+ ATP-ases) and a new morphology of cells in the brain allowing new behavioural patterns. We stress that here evolution is not about random production or reproduction of organisms but the systematic development of organisation in and between compartments to increase *the survival of individuals and groups* since the problems of reproduction of species were solved four billion years ago.

Table 20.25 Order of introduction of element changes in evolution

Element change	Time (ago) years
$H_2O/H_2 \rightarrow O_2$ ($O_2^{\cdot-}$, H_2O_2)	3 to 1 × 10^9
$CH_4 \rightarrow CO_2$	3 to 2 × 10^9
$WS_4^{2-} \rightarrow WO_4^{2-}$	3 × 10^9
$MoS_4^{2-} \rightarrow MoO_4^{2-}$	3 × 10^9
$H_2Se \rightarrow SeO_4^{2-}$	3 × 10^9
$NH_3 \rightarrow NO(NO_3^-)$	3 × 10^9
$Fe^{2+} \rightarrow Fe^{3+}\downarrow$, $H_2S \rightarrow SO_4^{2-}$	3 to 2 × 10^9
$ZnS \rightarrow Zn^{2+}$, $FeS \rightarrow Fe_2O_3$	3 to 2 × 10^9
$Cu^+\downarrow \rightarrow Cu^{2+}$	2 to 1 × 10^9

In niches the primitive anaerobic composition of chemicals is virtually undisturbed.

20.17 **Survival of systems: summary of the value of the elements**

In chemistry, major concerns are the stability of a material either in a thermodynamic sense or in a kinetic sense. The systems are stationary. In biological chemistry the relationship of the properties to potential uses of a material is much the same as in chemistry but in place of stability we must place the overriding consideration upon the survival strength of a flowing organization to which each material in it has to contribute (see Figs. 7.9 and 20.30). Elements, for example, are selected in kind and in concentration by cells not just to perform isolated functions but in relationship to the survival strength

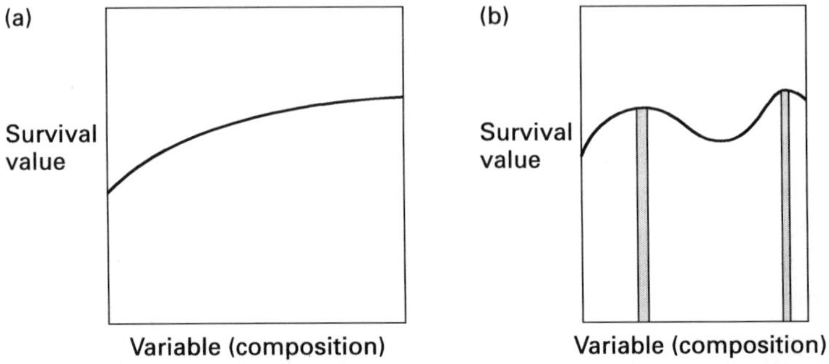

Fig. 20.30 Plots of survival value against a variable such as composition, energy supply or temperature, showing (a) continuous variation, (b) the development of distinct species, shaded regions. Note that the peaks in this graph correspond to higher survival values while in the case of stability ($-\Delta G$) composition plots (Figs 20.2, 20.6 and 20.20) we have plotted in just the opposite way; there is no conflict whatsoever between the two systematic representations—it is all a matter of definition of the reason for existence, thermodynamic stability, or kinetic strength (survival).

of the whole cellular (or higher) organization. Similarly, proteins are selected to interact in a controlled way with these elements. Thus we need to appreciate the factors in the organization which contribute to and those that limit survival. Here we must clearly separate the long-term survival of a species from the average survival time of individuals of that species. Long-term survival of species depends upon the product of individual survival time (lifetime) and reproductive rate. In the earliest organisms and still in their simple descendants today, single-cell prokaryotes, the reproductive rate was and is the dominant part of survival in active conditions so that we must recognize that such a cell devotes virtually all its energy and material to generating daughter cells rapidly. Here in the cytoplasm the chemistry is as close as possible to that initially available (4.0×10^9 year ago) in reducing conditions (Figs 20.14 and 20.28). It may well be also that the earliest cells were able to

Table 20.26 Evolution of dependencies

Bacteria	Single-cell eukaryotes	Multicellular plants	Organisms animals (mankind)
None Note basic one-carbon chemistry -HCO- formyl	Some dependence on folic acid	N_2-fixation CO_2-chemistry	Vitamins Essential fats Essential amino-acids
Environmental elements	Environmental elements	Environmental elements	Essential minerals in correct proportions

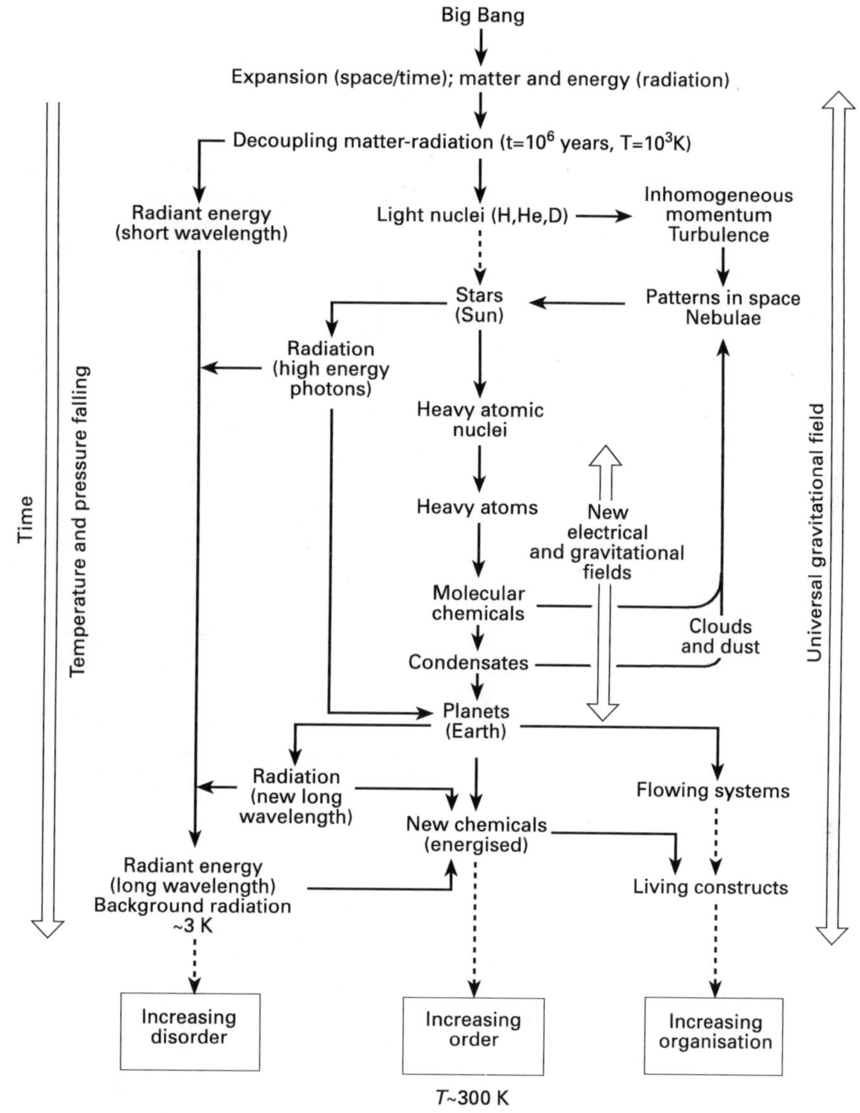

Fig. 20.31 An outline scheme given as an indication of the development of objects and systems around us, and some of the underlying interactions and energies since the beginning of time.

share chemicals, even DNA, so that they formed a continuum of compositional types not species, [see Fig. 20.30(a)] (Here we do not refer to an 'inert' spore, which in principle can remain viable forever, but to an active cell.) Evolution, while keeping such cell types has expanded to specialist cells, in single or multi-cell species, of much longer life-time but with much reduced reproductive rates [Fig. 20.30(b)]. (Within such species, individuals increase in importance.) Such cells have much greater skill in handling their environment and in protecting themselves while they develop. They do this by internal management of compartments while utilizing increasing oxidative chemistry and some new available elements in these compartments (Table 20.25). However, we must not assume that their species survival strength is greater overall. The advances in compartmental chemistry made for extreme complexity in management. Since there is a danger that the consequence would be vulnerability the complexity has found relief by utilizing primitive organisms in symbiosis or relying upon ecosystems (Table 20.26). We stress that this ecosystem is critically dependent upon the environment for material and energy. Thus our picture of evolution is strongly dependent upon changes in the environment. We have to put all of this development and even the ongoing efforts of mankind into a universal framework of the use of chemical elements in ever increasing compartmentalisation (see Fig. 20.31). From the very creation of the elements there has been an inevitable pursuit of order and organization. We have no ability to predict its timing, its whereabouts other than on earth, or its ultimate destination, but we must see it as a consequence of the principles of the physics and chemistry of the limited number of elements of the periodic table. Life and the environment are one ever-changing dynamic system.

Further reading

Anderson, G.M. and Crerar, D.A. (1993). *Thermodynamics and geochemistry: The equilibrium model*. Oxford University Press, New York

Aravind, L., Tatusov, R.L., Wolf, Y.I., Walker, D.R., and Koonin, E.V. (1998). Evidence for massive gene exchange between archaeal and bacterial hyperthermophiles. *Trends in Genetics*, **14**, 442–444.

Attwood, T.K. (2000) The babel of bioinformatics, *Science*, **290**, 471–47.

Baltscheffsky, H. (ed.) (1996). *Origin and evolution of energy conversion*. VCH Publishers, New York

Blackstock W. and Munn, M. (eds) (2000). *Proteomics: a trends guide*, Elsevier, London

Brocks, J.J., Logan, G.A., Buick, R., and Summons, R.E. (1999). Archean molecular fossils and the early rise of eukaryotes. *Science*, **285**, 1033–1036

Cairns-Smith, A.G., Hall, S.J., and Russel, M.J. (1992). Mineral theories of the origin of life and the evolution of the biosphere: an iron-sulphide example. *Origins of life and the evolution of the biosphere*, **22**, 161–80.

Caporale, L.H. (ed.) (1999). Molecular Strategies in Biological Evolution. *Annals of the New York Academy of Sciences* Vol. 870.

Capra, F. (1996). *The web of life—a new synthesis of mind and matter*. HarperCollins, London

Chandler, J.L.R. and Van de Vijver, G. (2000). Closure: Emergent organisations and their dynamics. *Annals of the New York Academy of Sciences*, **901**, 1–349

Cox, P.A. (1995). *The elements on earth—inorganic chemistry in the environment*. Oxford University Press, Oxford

Deamer, D.W. (1997). The first living systems: a bioenergetic perspective. *Microbiology and Molecular Biology Reviews*, **61**, 239–61

Davies, P. (1998). *The fifth miracle*. Allen Lane. The Penguin Press, London

Depew, D.J. and Weber, B.H. (1995) *Darwinism evolving*. MIT Press, Cambridge, Mass.

Demetrius, L. (2000). Thermodynamics and evolution. *Journal of Theoretical Biology*, **206**, 1–16.

Doolittle, W.F. (2000). Uprooting the tree of life. *Scientific American*, **282**, 72–7.

de Duve, C. (1995). *Vital dust*. Basic Books, New York.

Eide, D.J. (1998). The molecular biology of metal ion transport in yeast. *Annual Review of Nutrition*, **18**, 441–69.

Eigen, M. and Schuster, P. (1979). *The hypercycle: a principle of natural self-organization*. Springer-Verlag, New York.

Eigen, M. (1996). *Steps towards life*. Oxford University Press, Oxford.

Gray, M.W., Burger, G., and Lang, B. F. (1999). Mitochondrial evolution. *Science*, **283**, 1476–81

Holm, R.H., Kennepohl, P., and Solomon, E.I. (1996). Structural and functional aspects of metal sites in biology. *Chemical Reviews*, **96**, 2239–314.

Kauffman, S.A. (1993). *The origins of order*, Oxford University Press, Oxford.

Kauffman, S.A. (1995). *At home in the universe*. Oxford University Press, Oxford.

Lazcano, A. and Miller, S.L. (1996). The origin and early evolution of life: prebiotic chemistry, the pre-RNA world and time. *Cell*, **85**, 793.

Maeda, M., Hamano, K., Hirano, Y., Suzuki, M., Takahashi, E., Terada, T., Futai, M., and Sato, R. (1998). Structure of P-type ATP-ase and their gene locations. *Cell structure and function*, **23**, 315–23.

Margulis, L. and Schwartz, K.V. (1998). *Five kingdoms* (3rd edn). W.H. Freeman, New York.

Margulis, L. (1998). *Symbiotic universe*. Basic Books, New York.

Maynard Smith, J. and Szathmáry, E. (1999). *The origins of life*. Oxford University Press, Oxford.

Nelson, N. (1999). Metal ion transporters and homeostasis. *The EMBO Journal*, **18**, 4361–71.

Prantzos, N., Vangioni-Flan, E., and Cass, M. (eds) (1994). *Origin and evolution of the elements*. Cambridge University Press, Cambridge.

Purves, W.K., Orians, G.H., and Craig Heller, H. (1995). *Life—the science of biology* (5th edn), Sinauer Associates, W. H. Freeman. Sunderland, Mass., Chapters 6–8.

Radisky, D.C. and Kaplan, J. (1999). Regulation of transition metal transport in yeast. *Journal of Biological Chemistry*, **274**, 4481–4.

Rossi, C., Bastianoni, S., Donati, A., and Marchettini, N. (eds) (1999). Tempos in science and nature: structures relations and complexity. *New York Academy of Sciences*, Vol. 879.

Rusnak, F. and Reiter, T. (2000). Sensing electrons: phosphatase redox regulation. *Trends in Biochemical Science*, **25**, 527–9.

Russell, M.J., Daniel, R.M., and Hall, A.J. (1993). Emergence of life and iron-sulphide membranes. *Terra Nova*, **5**, 343–7.

Russell, M.J. and Hall, A.J. (1997). The emergence of life from iron monosulphide bubbles at a submarine hydrothermal redox and pH front. *Journal of the Geological Society of London*, **154**, 377–402.

Silver, S. (1998). Genes for all metals. *Journal of Industrial Microbiology and Biotechnology*, **20**, 1–12.

Stearns, S.C. and Hoekstra, R.F. (2000). *Evolution*, Oxford University Press, Oxford.

Theodoridus, G.C. (2000). Selection mechanisms during initial evolution and the role of ionic balance. *Journal of Theoretical Biology*, **207**, 15–20.

Weber, B.H. (1999). Emergence of life and biological selection from the perspective of complex system dynamics in Van de Vijver G. (ed.) *Evolutionary systems*, pp. 59–66. Kluwer Academic Publishers, Amsterdam.

Williams, R.J.P. and Fraústo da Silva, J.J.R. (1996). *The natural selection of the chemical elements*. Oxford University Press, Oxford.

Williams, R.J.P. and Fraústo da Silva, J.J.R. (1999). *Bringing chemistry to life*. Oxford University Press, Oxford.

Williams, R.J.P. and Trausto de Silva, J.J.R. (2000). Distribution of elements in cells. *Coordination Chemistry Reviews*, **200**, 347–8.

Woese, C.R., Kandler, O., and Wheelis, M.L. (1990). Towards a natural system of organisms: proposal for the domains Archaea, Bacteria and Eukaria.

Woese, C.R. (1998). The universal ancestor. *Proceedings of the National Academy of Sciences USA*, **95**, 6854–9.

Wright, S. (1930). Review of the above. *Journal of Heredity*, **21**, 349–56.

Note: The analysis of genomic sequences of a variety of species has revealed some unexpected problems for the simple idea of an evolutionary tree. The bacterial genomes are rather as expected in that regions of the genome are either eventually expressed as proteins or they act as controls over this expression. The gene sequence mutates readily so that adaptation to new environments can be made quite rapidly. In the eukaryotes quite new problems are met in that some genomes, such as that of man, have some 95% 'junk' DNA. Looking for a purpose has to be seen against the background of the fact that all the DNA of the Japanese puffer fish is of clear value as in bacteria. What use can junk DNA have? Again the complexity of man compared with that of the plants arabadopsis, has evolved with only small change in the gene pool. In fact arabadopsis has about 25 000 expressible genes while the human genome has some 35 000. Another strange fact is that some 1% of the genes of human genome appear to be from DNA of bacterial origin leading to the conclusion that perhaps life has not developed in a linear branching tree but can develop by incorporation from distant branches. This is also true of the genes from early eukaryotes. A final peculiarity is that sex, which is gene linked in humans, is dependent on the environment and the temperature in the case of development of the embryo in reptiles. Many of these facts make it almost impossible to conceive that inherited DNA with mutation is the only cause of evolution of individual character. It may be that the system of chemicals which is connected to DNA is to some degree flexible. The environment then affects DNA in many ways and DNA may have ways of changing to meet circumstances, over long periods of time, $>10^3$ years. Do not forget that the metallome is inherited as is a large part of the proteome so that much of the cell's cytoplasmic chemistry belongs to a somewhat flexible system interactive with the DNA, see Rubin G.M. *et al.* (2000). Comparative genomics of the eukaryotes. *Science*, **287**, 2204–12 and the report of The Aribidopsis Genome Initiative (2000) *Nature*, **408**, 796–826. The future lies in the study of the combined *system* of DNA (genomics), its expression (proteomics) and the connection to the environment (largely through the metallomes). How do they feed-back upon one another?

Index